Lecture Notes in Mechanical Engineering

Lecture Notes in Mechanical Engineering (LNME) publishes the latest developments in Mechanical Engineering—quickly, informally and with high quality. Original research reported in proceedings and post-proceedings represents the core of LNME. Volumes published in LNME embrace all aspects, subfields and new challenges of mechanical engineering. Topics in the series include:

- Engineering Design
- Machinery and Machine Elements
- Mechanical Structures and Stress Analysis
- Automotive Engineering
- Engine Technology
- Aerospace Technology and Astronautics
- Nanotechnology and Microengineering
- Control, Robotics, Mechatronics
- MEMS
- Theoretical and Applied Mechanics
- Dynamical Systems, Control
- Fluid Mechanics
- Engineering Thermodynamics, Heat and Mass Transfer
- Manufacturing
- Precision Engineering, Instrumentation, Measurement
- Materials Engineering
- Tribology and Surface Technology

To submit a proposal or request further information, please contact the Springer Editor of your location:

China: Dr. Mengchu Huang at mengchu.huang@springer.com
India: Priya Vyas at priya.vyas@springer.com
Rest of Asia, Australia, New Zealand: Swati Meherishi at swati.meherishi@springer.com
All other countries: Dr. Leontina Di Cecco at Leontina.dicecco@springer.com

To submit a proposal for a monograph, please check our Springer Tracts in Mechanical Engineering at http://www.springer.com/series/11693 or contact Leontina.dicecco@springer.com

Indexed by SCOPUS.

All books published in the series are submitted for consideration in Web of Science.

More information about this series at http://www.springer.com/series/11236

Magd Abdel Wahab
Editor

Proceedings of the 8th International Conference on Fracture, Fatigue and Wear

FFW 2020, August 26–27 2020

Springer

Editor
Magd Abdel Wahab
Faculty of Engineering and Architecture
Ghent University
Ghent, Belgium

ISSN 2195-4356 ISSN 2195-4364 (electronic)
Lecture Notes in Mechanical Engineering
ISBN 978-981-15-9892-0 ISBN 978-981-15-9893-7 (eBook)
https://doi.org/10.1007/978-981-15-9893-7

This Springer imprint is published by the registered company Springer Nature Singapore Pte Ltd.
The registered company address is: 152 Beach Road, #21-01/04 Gateway East, Singapore 189721, Singapore

Organising Committee

Chairman

Prof. Magd Abdel Wahab, Ghent University, Belgium

International Scientific Committee

Prof. S. Abdullah, Universiti Kebangsaan Malaysia, Malaysia
Dr. J. Abenojar, Universidad Carlos III de Madrid, Spain
Dr. R. Das, University of Auckland, New Zealand
Dr. A. H. Ertas, Bursa Technical University, Turkey
Dr. S. Fouvry, Ecole Centrale de Lyon, France
Prof. T. Hattori, Shizuoka Institute of Science and Technology, Japan
Dr. M. Kchaou, University of Sfax, Tunisia
Dr. S. Khatir, Ghent University, Belgium
Dr. C. Le Thanh, Open University Ho Chi Minh City, Vietnam
Dr. K. Masuda, University of Toyama, Japan
Prof. T. Miyazaki, University of Ryukyu, Japan
Prof. H. Nguyen, HUTECH, Vietnam
Prof. K. Oda, Oita University, Japan
Prof. R. V. Prakash, Indian Institute of Technology, India
Prof. T. Qin, China Agricultural University, China
Prof. F. Rezai-Aria, Ecole des Mines d'Albi, France
Prof. T. Rabczuk, Bauhaus University Weimar, Germany
Dr. A. Rudawska, Lublin University of Technology, Poland
Dr. M. S. Prabhudev, Government Polytechnic, Kalgi, India
Dr. Y. Sano, Kyushu Institute of Technology, Japan
Prof. J. Song, Ostwestfalen-Lippe University, Germany
Prof. J. Toribio, University of Salamanca, Spain

Prof. C. Xu, China Agricultural University, China
Dr. T. Yue, Ghent University, Belgium
Prof. C. Zhou, Nanjing University of Aeronautics and Astronautics, China
Dr. X. Zhuang, Leibniz Unversität Hannover, Germany

Preface

This volume contains the proceedings of the 8th International Conference on Fracture Fatigue and Wear FFW 2020. Fracture Fatigue and Wear 2020 is the 8th FFW conference and is held online during the period 26–27 August 2020. Previous conferences were celebrated in Jinan (China, 2010), Kitakyushu (Japan, 2013), Kitakyushu (Japan, 2014), Gent (Belgium, 2015), Kitakyushu (Japan, 2016), Porto (Portugal, 2017) and Gent (Belgium, 2018).

The overall objective of the conference is to bring together international scientists and engineers in academia and industry in fields related to fracture mechanics, fatigue of materials, tribology and wear of materials. The conference covers industrial engineering applications of the above topics including theoretical and analytical methods, numerical simulations and experimental techniques. One of the aims of conference is to promote cooperation between international scientists and engineers from a large number of disciplines, who are involved in research related to fracture, fatigue and wear. The presentations of FFW 2020 are divided into three main sessions, namely (1) fracture, (2) fatigue and (3) wear.

The organising committee is grateful to keynote speaker, Prof. David Nowell, Mechanical Engineering, Imperial College London, for his interesting keynote speech entitled 'Are interfaces a good thing? Fretting fatigue, wear, and frictional damping'.

Special thanks go to members of the Scientific Committee of FFW 2020 for reviewing the articles published in this volume and for judging their scientific merits. Based on the comments of reviewers and the scientific merits of the submitted manuscripts, the articles were accepted for publication in the conference proceedings and for presentation at the conference venue. The accepted papers are of a very high scientific quality and contribute to advancement of knowledge in all research topics relevant to FFW conference.

Finally, the organising committee would like to thank all authors, who have contributed to this volume and to those who have presented their research work at the conference venue in Ghent.

Ghent, Belgium

Prof. Magd Abdel Wahab
Chairman of FFW 2020

Contents

Wear

Fracture

The Folded Wiener Process for Non-negative Non-monotone Degradation Processes

Massimiliano Giorgio and Gianpaolo Pulcini

Abstract The Wiener process is very largely and successfully adopted for modelling degradation phenomena that can exhibit negative increments. However, so far as the observed degradation phenomenon is intrinsically non-negative, as generally occurs, its use may lead to incongruous results, unless (based on the parameter setting) the probability that the considered Wiener process takes negative values is negligible for any time of interest. To overcome such a limitation of the Wiener process, a new Markovian degradation process, called the folded Wiener process, is here proposed that can be used to describe intrinsically non-negative, non-monotone, degradation phenomena. In fact, under the folded Wiener process, the degradation increment over future time intervals is constrained to be no less than the opposite of the current degradation level, so that the total amount of degradation accumulated up to any time t is necessarily non-negative. The main features and properties of the folded Wiener process are then presented and discussed, and closed form expressions for the mean and variance functions of the process are also provided. Finally, to show the feasibility of the proposed folded Wiener process, an applicative example is developed by using real data.

Keywords Non-monotone degradation process · Non-negative degradation · Folded normal distribution

1 Introduction

The Wiener process [1, 2] is a natural choice for modelling degradation phenomena that can exhibit negative increments, and hence it has founded wide application in the analysis of non-monotone degradation data. However, the Wiener process has

M. Giorgio (✉)
Dipartimento Di Ingegneria Industriale, Università Degli Studi Di Napoli Federico II, Naples, Italy
e-mail: massimiliano.giorgio@unina.it

G. Pulcini
Istituto Motori, National Research Council (CNR), Naples, Italy
e-mail: g.pulcini@im.cnr.it

M. Abdel Wahab (ed.), *Proceedings of the 8th International Conference on Fracture, Fatigue and Wear*, Lecture Notes in Mechanical Engineering,
https://doi.org/10.1007/978-981-15-9893-7_1

to be used with particular care when, as mostly occurs, the observed degradation phenomenon is intrinsically non-negative, because under the Wiener process the degradation is allowed to take negative values. In fact, in these cases, the Wiener process can be considered appropriate if and only if, based on the values of its parameters, the probability of observing negative degradation is negligible for any time of interest.

To overcome such a limitation of the Wiener process, we propose a new continuous time and state degradation process, called folded Wiener (FW) process, that can be used to describe intrinsically non-negative, non-monotone degradation phenomena.

In fact, under the FW process:

- the degradation level $W(t)$ at the generic time t is a non-negative random variable, which is distributed as a folded Normal variable, and
- the degradation increment $\Delta W(t, \Delta t) = W(t + \Delta t) - W(t)$, given $W(t) = w_t$ (with $w_t \geq 0$), is constrained to be no less than $-w_t$, so that $W(t + \Delta t)$ is necessarily non-negative.

The FW process is a non-homogeneous Markov process, yet it reduces to a homogeneous one for specific settings of its parameters. Under the FW process, $\Delta W(t, \Delta t)$ depends in stochastic sense on $W(t)$ (that is the process has state dependent increments). The FW process is quite mathematically tractable. In fact, its likelihood function, mean and variance can be expressed in closed form.

The remainder of the paper is organized as follows. Section 2 introduces the FW process and illustrates its main properties. Section 3 discusses the maximum likelihood estimation of model parameters. Finally, Sect. 4 reports the results obtained by applying the FW process to a set of real wear data of the magnetic heads of 10 hard disk drives, which exhibit non-negative and non-monotonic wear paths. Such an application shows the feasibility of the suggested FW degradation process.

2 The Folded Wiener Degradation Process

Let μ_0 and σ_0 be two positive parameters, and $\Lambda(t)$ be a non-negative, increasing function of the operating time t, with $\Lambda(0) = 0$. The degradation process $\{W(t); t \geq 0\}$ is assumed to be a folded Wiener process if (and only if):

1. the degradation level $W(t)$ at any time $t \geq 0$ has a folded Normal distribution [3], with probability density function (pdf):

$$f_{W(t)}(w) = \frac{1}{\sqrt{2\pi}\sigma_0\sqrt{\Lambda(t)}} \times \left(\exp\left[-\frac{1}{2}\left(\frac{w-\mu_0\Lambda(t)}{\sigma_0\sqrt{\Lambda(t)}}\right)^2\right] + \exp\left[-\frac{1}{2}\left(\frac{-w-\mu_0\Lambda(t)}{\sigma_0\sqrt{\Lambda(t)}}\right)^2\right]\right), \; w \geq 0 \quad (1)$$

having location parameter $\mu_0\Lambda(t)$ and scale parameter $\sigma_0\sqrt{\Lambda(t)}$;

2. the joint pdf of $W(t)$ and the degradation increment $\Delta W(t, \Delta t) = W(t + \Delta t) - W(t)$ during the time interval $(t, t + \Delta t)$ is given by:

$$\begin{aligned}
f_{W(t),\Delta W(t,\Delta t)}(w_t, \delta) &= \frac{1}{2\pi\sigma_0^2\sqrt{\Lambda(t)\Lambda(t,\Delta t)}} \\
&\times \left\{\left(\exp\left[-\frac{1}{2}\left(\frac{\delta - \mu_0\Lambda(t,\Delta t)}{\sigma_0\sqrt{\Lambda(t,\Delta t)}}\right)^2\right] + \exp\left[-\frac{1}{2}\left(\frac{-\delta - 2w_t - \mu_0\Lambda(t,\Delta t)}{\sigma_0\sqrt{\Lambda(t,\Delta t)}}\right)^2\right]\right)\right. \\
&\times \exp\left[-\frac{1}{2}\left(\frac{w_t - \mu_0\Lambda(t)}{\sigma_0\sqrt{\Lambda(t)}}\right)^2\right] \\
&+ \left(\exp\left[-\frac{1}{2}\left(\frac{-\delta - \mu_0\Lambda(t,\Delta t)}{\sigma_0\sqrt{\Lambda(t,\Delta t)}}\right)^2\right] + \exp\left[-\frac{1}{2}\left(\frac{\delta + 2w_t - \mu_0\Lambda(t,\Delta t)}{\sigma_0\sqrt{\Lambda(t,\Delta t)}}\right)^2\right]\right) \\
&\left.\times \exp\left[-\frac{1}{2}\left(\frac{-w_t - \mu_0\Lambda(t)}{\sigma_0\sqrt{\Lambda(t)}}\right)^2\right]\right\}, \; w_t \geqslant 0;\, \delta \geqslant -w_t
\end{aligned} \tag{2}$$

where $\Lambda(t, \Delta t) = \Lambda(t + \Delta t) - \Lambda(t)$.

Suitable functional forms for $\Lambda(t)$ are the power-law function $\Lambda(t) = t^\beta$ and the exponential functions $\Lambda(t) = \exp(\beta t) - 1$ and $\Lambda(t) = \exp(\beta t^\gamma) - 1$.

From (1) and (2), the (conditional) pdf of the degradation increment $\Delta W(t, \Delta t)$, given the current degradation level $W(t) = w_t$, is easily obtained:

$$\begin{aligned}
f_{\Delta W(t,\Delta t)|W(t)=w_t}(\delta|w_t) &= \frac{f_{W(t),\Delta W(t,\Delta t)}(w_t, \delta)}{f_{W(t)}(w_t)} \\
&= \frac{1}{\sqrt{2\pi}\sigma_0\sqrt{\Delta\Lambda(t,\Delta t)}} \frac{1}{\exp\left[-\frac{1}{2}\left(\frac{w_t - \mu_0\Lambda(t)}{\sigma_0\sqrt{\Lambda(t)}}\right)^2\right] + \exp\left[-\frac{1}{2}\left(\frac{-w_t - \mu_0\Lambda(t)}{\sigma_0\sqrt{\Lambda(t)}}\right)^2\right]} \\
&\times \left\{\left(\exp\left[-\frac{1}{2}\left(\frac{\delta - \mu_0\Lambda(t,\Delta t)}{\sigma_0\sqrt{\Lambda(t,\Delta t)}}\right)^2\right] + \exp\left[-\frac{1}{2}\left(\frac{-\delta - 2w_t - \mu_0\Lambda(t,\Delta t)}{\sigma_0\sqrt{\Lambda(t,\Delta t)}}\right)^2\right]\right)\right. \\
&\times \exp\left[-\frac{1}{2}\left(\frac{w_t - \mu_0\Lambda(t)}{\sigma_0\sqrt{\Lambda(t)}}\right)^2\right] \\
&+ \left(\exp\left[-\frac{1}{2}\left(\frac{-\delta - \mu_0\Lambda(t,\Delta t)}{\sigma_0\sqrt{\Lambda(t,\Delta t)}}\right)^2\right] + \exp\left[-\frac{1}{2}\left(\frac{\delta + 2w_t - \mu_0\Lambda(t,\Delta t)}{\sigma_0\sqrt{\Lambda(t,\Delta t)}}\right)^2\right]\right), \\
&\left.\times \exp\left[-\frac{1}{2}\left(\frac{-w_t - \mu_0\Lambda(t)}{\sigma_0\sqrt{\Lambda(t)}}\right)^2\right]\right\}, \quad \delta \geq -w_t.
\end{aligned} \tag{3}$$

Such a pdf, being:

$$\begin{aligned}
&\frac{\exp\left[-\frac{1}{2}\left(\frac{w_t - \mu_0\Lambda(t)}{\sigma_0\sqrt{\Lambda(t)}}\right)^2\right]}{\exp\left[-\frac{1}{2}\left(\frac{w_t - \mu_0\Lambda(t)}{\sigma_0\sqrt{\Lambda(t)}}\right)^2\right] + \exp\left[-\frac{1}{2}\left(\frac{-w_t - \mu_0\Lambda(t)}{\sigma_0\sqrt{\Lambda(t)}}\right)^2\right]} \\
&= \frac{1}{1 + \exp\left[-\frac{1}{2}\left(\frac{-w_t - \mu_0\Lambda(t)}{\sigma_0\sqrt{\Lambda(t)}}\right)^2\right] \Big/ \exp\left[-\frac{1}{2}\left(\frac{w_t - \mu_0\Lambda(t)}{\sigma_0\sqrt{\Lambda(t)}}\right)^2\right]} \\
&= \frac{1}{1 + \exp\left(-\frac{2w_t\mu_0}{\sigma_0^2}\right)}
\end{aligned} \tag{4}$$

and:

$$\frac{\exp\left[-\frac{1}{2}\left(\frac{-w_t-\mu_0\Lambda(t)}{\sigma_0\sqrt{\Lambda(t)}}\right)^2\right]}{\exp\left[-\frac{1}{2}\left(\frac{w_t-\mu_0\Lambda(t)}{\sigma_0\sqrt{\Lambda(t)}}\right)^2\right]+\exp\left[-\frac{1}{2}\left(\frac{-w_t-\mu_0\Lambda(t)}{\sigma_0\sqrt{\Lambda(t)}}\right)^2\right]}=\frac{1}{1+\exp\left(\frac{2w_t\mu_0}{\sigma_0^2}\right)} \tag{5}$$

can be also rewritten as:

$$\begin{aligned} f_{\Delta W(t,\Delta t)|W(t)=w_t}(\delta|w_t) &= \frac{1}{\sqrt{2\pi}\sigma_0\sqrt{\Lambda(t,\Delta t)}} \\ &\times \left\{ \frac{\exp\left[-\frac{1}{2}\left(\frac{\delta-\mu_0\Lambda(t,\Delta t)}{\sigma_0\sqrt{\Lambda(t,\Delta t)}}\right)^2\right]+\exp\left[-\frac{1}{2}\left(\frac{-\delta-2w_t-\mu_0\Lambda(t,\Delta t)}{\sigma_0\sqrt{\Lambda(t,\Delta t)}}\right)^2\right]}{1+\exp\left(-\frac{2w_t\mu_0}{\sigma_0^2}\right)} \right. \\ &+ \left. \frac{\exp\left[-\frac{1}{2}\left(\frac{-\delta-\mu_0\Lambda(t,\Delta t)}{\sigma_0\sqrt{\Lambda(t,\Delta t)}}\right)^2\right]+\exp\left[-\frac{1}{2}\left(\frac{\delta+2w_t-\mu_0\Lambda(t,\Delta t)}{\sigma_0\sqrt{\Lambda(t,\Delta t)}}\right)^2\right]}{1+\exp\left(\frac{2w_t\mu_0}{\sigma_0^2}\right)} \right\}, \\ \delta &\geqslant -w_t. \end{aligned} \tag{6}$$

From (1) and (1) [or (6)], we have that, under the proposed FW process: (a) the degradation level at any time t can not be negative, even if the degradation increment $\Delta W(t, \Delta t)$ can be negative, and (b) the degradation growth $\Delta W(t, \Delta t)$ in the future time interval $(t, t + \Delta t)$ depends on the current level w_t.

The proposed FW process possesses the additive property, and hence the marginal pdf of $W(t + \Delta t)$:

$$f_{W(t+\Delta t)}(w) = \int_0^\infty f_{\Delta W(t,\Delta t)|W(t)=w_t}(w - w_t|w_t) f_{W(t)}(w_t)dw_t, \quad w \geq 0 \tag{7}$$

is a folded Normal distribution with parameters $\mu_0\Lambda(t + \Delta t)$ and $\sigma_0\sqrt{\Lambda(t + \Delta t)}$.

From (6), it can be noted that, only when $\Lambda(t)$ is linear with the operating time (so that $\Lambda(t, \Delta t) \propto \Delta t$), the conditional pdf of the increment $\Delta W(t, \Delta t)$ does not depend on the current time t, but only on the width Δt of the time interval. Thus, in general, the proposed FW process is a non-homogeneous Markov process.

Figure 1 shows the behavior of the conditional pdf $f_{\Delta W(t,\Delta t)|W(t)=w_t}(\delta|w_t)$, under the power-law function $\Lambda(t) = t^\beta$, with $\mu_0 = 1.0$, $\sigma_0 = 1.5$, $\beta = 1.5$, $t = 1.0$, and $\Delta t = 2.0$, for different values (i.e., 0, 2, 4, and 1) of the current degradation level w_t.

From (6), we can see that:

1. when $w_t = 0$, the conditional pdf of the degradation increment $\Delta W(t, \Delta t)$ reduces to the folded Normal distribution:

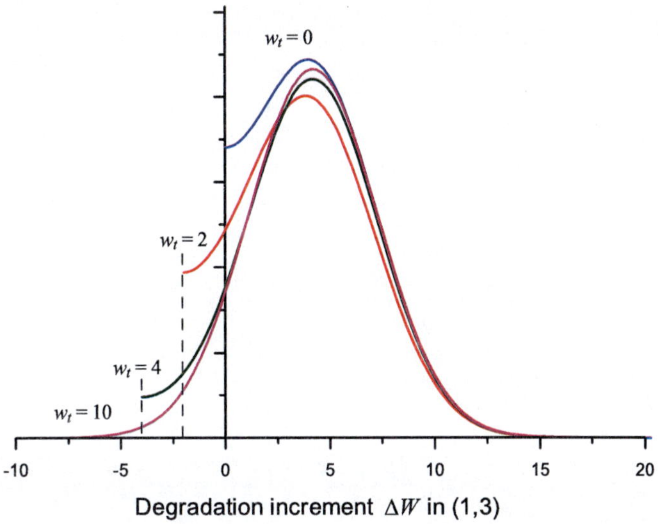

Fig. 1 Conditional pdf of the degradation increment $\Delta W(1, 2)$, for different values (0, 2, 4, and 10) of the current degradation level w_t.

$$f_{\Delta W(t,\Delta t)|W(t)=0}(\delta|0) = \frac{1}{\sqrt{2\pi}\sigma_0\sqrt{\Lambda(t, \Delta t)}}\left(\exp\left[-\frac{1}{2}\left(\frac{\delta - \mu_0 \Lambda(t, \Delta t)}{\sigma_0\sqrt{\Lambda(t, \Delta t)}}\right)^2\right] + \exp\left[-\frac{1}{2}\left(\frac{-\delta - \mu_0 \Lambda(t, \Delta t)}{\sigma_0\sqrt{\Lambda(t, \Delta t)}}\right)^2\right]\right), \quad \delta \geq 0 \qquad (8)$$

with location parameter $\mu_0 \Lambda(t, \Delta t)$ and scale parameter $\sigma_0\sqrt{\Lambda(t, \Delta t)}$;

2. when $w_t \to \infty$, the conditional pdf of $\Delta W(t, \Delta t)$ tends to a Normal distribution:

$$\lim_{w_t \to \infty} f_{\Delta W(t,\Delta t)|W(t)=w_t}(\delta|w_t) = \frac{1}{\sqrt{2\pi}\sigma_0\sqrt{\Lambda(t, \Delta t)}} \times \exp\left[-\frac{1}{2}\left(\frac{\delta - \mu_0 \Lambda(t, \Delta t)}{\sigma_0\sqrt{\Lambda(t, \Delta t)}}\right)^2\right], -\infty < \delta < \infty, \qquad (9)$$

independent of w_t, with mean $\mu_0 \Lambda(t, \Delta t)$ and standard deviation $\sigma_0\sqrt{\Delta(t, \Delta t)}$.

From (8), it is easy to show that the pdf (1) of the degradation level $W(t)$ can be derived from the conditional pdf (3) of the increment $\Delta W(t, \Delta t)$ by setting in (3) $t = 0$, $w_t = 0$, and $\Delta t = t$.

From (1) and (3), the (marginal) pdf of the degradation increment $\Delta W(t, \Delta t)$ is given by:

$$f_{\Delta W(t,\Delta t)}(\delta) = \int_0^\infty f_{\Delta W(t,\Delta t)|W(t)=w_t}(\delta|w_t) f_{W(t)}(w_t)\mathrm{d}w_t, \tag{10}$$

and hence, even when $\Lambda(t)$ is linear with t, the marginal pdf of the increment during the time interval $(t, t + \Delta t)$ depends not only on Δt, but also on t.

From (1), the cumulative distribution function (Cdf) of the degradation level $W(t)$ is:

$$F_{W(t)}(w) = \Phi\left(\frac{w - \mu_0\Lambda(t)}{\sigma_0\sqrt{\Lambda(t)}}\right) - \Phi\left(\frac{-w - \mu_0\Lambda(t)}{\sigma_0\sqrt{\Lambda(t)}}\right), w \geq 0, \tag{11}$$

where $\Phi(z)$ indicates the standard Normal Cdf.

Mean $E\{W(t)\}$ and variance $V\{W(t)\}$ of $W(t)$ (e.g., see [4]) can be expressed as:

$$E\{W(t)\} = \mu_0\Lambda(t)\left[1 - 2\Phi\left(-\frac{\mu_0\sqrt{\Lambda(t)}}{\sigma_0}\right)\right] + \sigma_0\sqrt{2\Lambda(t)/\pi}\exp\left(-\frac{\mu_0^2\Lambda(t)}{2\sigma_0^2}\right) \tag{12}$$

$$V\{W(t)\} = \mu_0^2\Lambda^2(t) + \sigma_0^2\Lambda(t) - E^2\{W(t)\}. \tag{13}$$

From (12) and (13), it results that, for $t \to \infty$, $E\{W(t)\} \to \mu_0\Lambda(t)$ and $V\{W(t)\} \to \sigma_0^2\Lambda(t)$. Figures 2 and 3 show the behavior of the mean and variance functions of the proposed FW process, under the power-law function $\Lambda(t) = t^\beta$, for different values (i.e., 0.5, 0.9, 1.0, 1.2, and 2.0) of the shape parameter β.

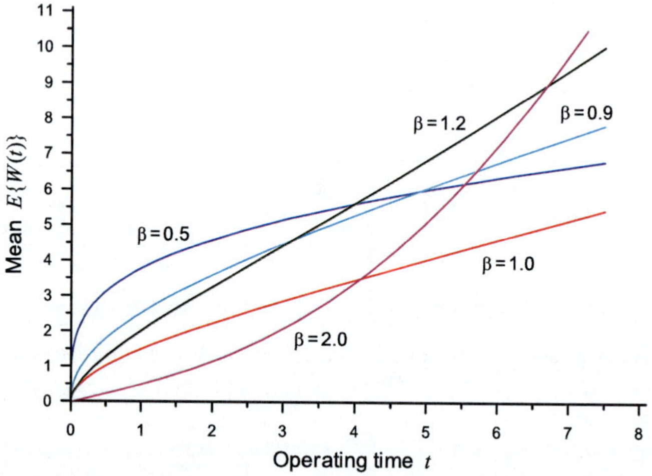

Fig. 2 Behavior of the mean function of the FW process for selected values (0.5, 0.9, 1.0, 1.2, and 2.0) of the power-law parameter β.

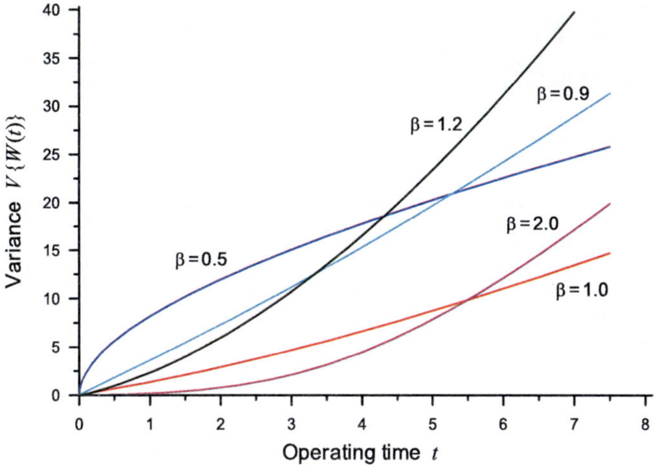

Fig. 3 Behavior of the variance function of the FW process for selected values (0.5, 0.9, 1.0, 1.2, and 2.0) of the power-law parameter β.

From Fig. 2, we can note that the mean curve is: (a) monotonically concave when $\beta < 1$ (that is, when $\beta = 0.5$ and 0.9); (b) initially concave and then asymptotically linear when $\beta = 1.0$; (c) initially concave and then convex when β is a few larger than 1 (that is, $\beta = 1.2$); (d) monotonically convex when $\beta = 2.0$.

Instead, from Fig. 3, we can note that the variance function is: (a) monotonically concave when $\beta = 0.5$; (b) initially convex and then concave when $\beta = 0.9$; (c) initially convex and then asymptotically linear when $\beta = 1.0$; (d) monotonically convex both when $\beta = 1.2$ and when $\beta = 2.0$.

By comparing (1) and (8), we have that the conditional Cdf, mean and variance of the increment $\Delta W(t, \Delta t)$, given $W(t) = 0$, can be obtained from (11) to (13) by substituting $\Lambda(t)$ with $\Lambda(t, \Delta t)$ and $E^2\{W(t)\}$ with $E^2\{\Delta W(t, \Delta t)\}$.

For $w_t > 0$, the conditional Cdf, mean and variance of the increment $\Delta W(t, \Delta t)$ can be easily computed via numerical integration. In Fig. 4 the behavior of the conditional mean $E\{\Delta W(t, \Delta t)|w_t\}$, under the power-law function $\Lambda(t) = t^\beta$, is plotted for $\sigma_0 = 1.5$, $t = 1.0$, $\Delta t = 2.0$, and different values (0.5, 0.7, 1.0, and 1.5) of the power-law parameter β. The parameter μ_0 changes with β in order to ensure a common asymptotic mean equal to $\mu_0\left[(t + \Delta t)^\beta - t^\beta\right] = 2.0$.

From Fig. 4, we have that, regardless of the value of β, the conditional mean first decreases with w_t from the initial value

$$E\{\Delta W(t, \Delta t)|0\} = \mu_0 \Lambda(t, \Delta t)\left[1 - 2\Phi\left(-\frac{\mu_0\sqrt{\Lambda(t, \Delta t)}}{\sigma_0}\right)\right] + \sigma_0\sqrt{2\Lambda(t, \Delta t)/\pi}\,\exp\left(-\frac{\mu_0^2 \Lambda(t, \Delta t)}{2\sigma_0^2}\right)$$

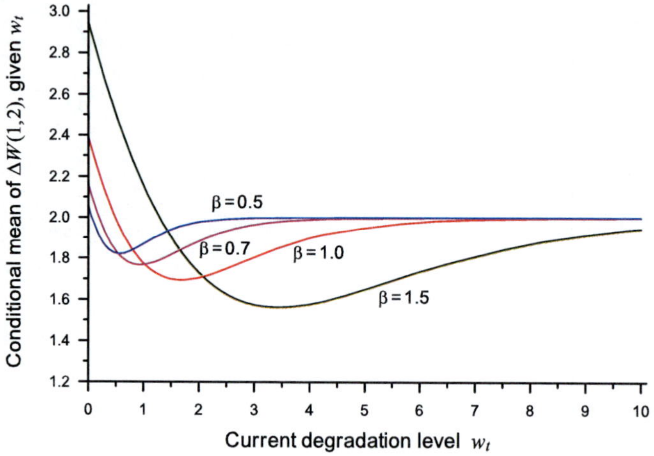

Fig. 4 Behavior of the conditional mean of the degradation increment, for different values (0.5, 0.7, 1.0, and 1.5) of the power-law parameter β, and common asymptotic mean of 2.0

assumed at $w_t = 0$, and then grows monotonically up to the asymptotic value equal to 2.0.

3 Maximum Likelihood Estimation

Given a set $\boldsymbol{w} = \left(w_{1,1}, \ldots, w_{1,n_1}, \ldots, w_{m,1}, \ldots, w_{m,n_m}\right)$ of degradation data, obtained via periodic inspections, from m identical units operating under homogeneous conditions, the maximum likelihood estimates $(\hat{\mu}_0, \hat{\sigma}_0, \hat{\boldsymbol{\xi}})$ of the parameter $(\mu_0, \sigma_0, \boldsymbol{\xi})$ of the FW process, are the values of $\mu_0, \sigma_0,$ and $\boldsymbol{\xi}$ that maximize the log-likelihood function:

$$\ell(\mu_0, \sigma_0, \boldsymbol{\xi}|\boldsymbol{w}) = \sum_{i=1}^{m} \sum_{j=1}^{n_i} \ln\left[f_{\Delta W(t_{i,j-1}, \Delta t_{i,j})|W(t_{i,j-1})=w_{i,j-1}}(\delta_{i,j}|w_{i,j-1})\right] \quad (14)$$

where $\boldsymbol{\xi}$ denotes the parameter vector that indexes the function $\Lambda(t)$, $w_{i,j} = W\left(t_{i,j}\right)$ indicates the measured degradation level of unit i at time $t_{i,j}$ (with $0 < t_{i,j} < t_{i,j+1}$, $i = 1, \ldots, m$ and $j = 1, \ldots, n_i$), $\delta_{i,j} = w_{i,j} - w_{i,j-1}$, $\Delta t_{i,j} = t_{i,j} - t_{i,j-1}$, and $t_{i,0} = w_{i,0} = 0$ for every i.

Unfortunately, the values $\hat{\mu}_0$, $\hat{\sigma}_0$ and $\hat{\xi}$ that maximize the likelihood function (14) cannot be expressed in closed form. Hence, they have to be obtained by using a numerical maximization procedure.

Given $\hat{\mu}_0$, $\hat{\sigma}_0$ and $\hat{\xi}$, the ML estimate of any function of the FW parameters of model, $h(\mu_0, \sigma_0, \xi)$, such as the mean and variance of $W(t)$, can be easily computed by substituting $(\hat{\mu}_0, \hat{\sigma}_0, \hat{\xi})$ to (μ_0, σ_0, ξ).

4 Numerical Application

In order to demonstrate the feasibility of the proposed approach, we have applied the folded Wiener process to the wear data reported in Table 1.

Such data, given in [5], refer to the magnetic heads of m = 10 hard disk drives (HDD) tested under an overdrive of 2 nm, which induces excessive protrusions of the head and accelerates the head wear. Data were collected via periodic inspections performed at $t_1 = 3$, $t_2 = 6$, $t_3 = 9$, $t_4 = 12$, $5t_5 = 15$, $t_6 = 18$, and $t_7 = 21$ (units of time). The inspection times are the same for all the units.

Figure 5 shows the wear paths of the considered units, obtained by linear interpolation of the wear data reported in Table 1.

Such data were initially analyzed by Ye et al. [5], by assuming four different time-scaled Wiener processes, all of them with power-law time-transformation t^β. One of the assumed model was the (simple) time-transformed Wiener process with common parameters, whereas in the other three models some of the parameters are assumed to be random. The simple Wiener process, without random effect, showed to provide the better fit in terms of the Akaike information criterion (AIC) [6] values.

Although these HDD dataset contains negative increments, the wear is an intrinsically non-negative quantity, which can make inappropriate the use of the Wiener process if the (estimated) probability that the wear level $W(t)$ is negative does not assume a negligible value for any time t of interest.

By using the ML estimates of the Wiener process parameters given in [5], we have that the estimated probability $\Pr(W(t) < 0)$ is smaller than 0.05 only for $t > 3$. This

Table 1 The HDD wear data

Unit #	$t_1 = 3$	$t_2 = 36$	$t_3 = 9$	$t_4 = 12$	$t_5 = 15$	$t_6 = 18$	$t_7 = 21$
1	0.000	0.120	0.620	0.745	0.620	0.870	0.745
2	0.075	0.125	0.625	0.625	0.750	0.750	0.875
3	0.250	0.250	0.500	0.625	0.750	0.875	1.000
4	0.250	0.500	0.625	1.000	1.125	1.250	1.375
5	0.250	0.500	0.750	0.875	1.000	1.125	1.125
6	0.250	0.500	0.875	1.000	1.125	1.125	1.000
7	0.250	0.625	0.750	0.750	1.000	1.125	1.375
8	0.250	0.625	0.750	0.875	1.125	1.375	1.750
9	0.500	1.000	1.375	1.625	1.750	2.000	2.250
10	0.750	0.875	1.125	1.250	1.375	1.375	1.375

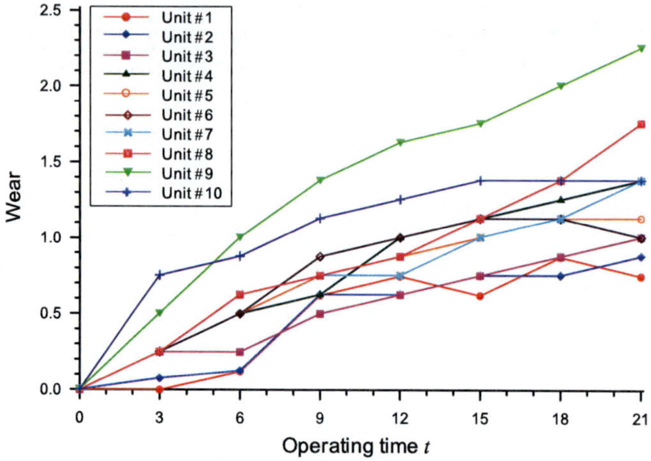

Fig. 5 The (linearly interpolated) wear data of $m = 10$ hard disk drives tested under an overdrive of 2 nm

result gives evidence that the Wiener process is not completely adequate for these data.

Aiming to overcome this limitation, we applied the proposed FW process with power-law function $\Lambda(t) = t^{\beta}$ to the HDD data of Table 1. The ML estimates of the parameters of the FW process are:

$$\hat{\mu}_0 = 0.1476, \hat{\sigma}_0 = 0.1345, \text{ and } \hat{\beta} = 0.7064$$

and the estimated log-likelihood is $\hat{\ell} = 37.52$.

In Fig. 6, the ML estimate of the degradation mean $E\{W(t)\}$ under the FW and the Wiener processes are compared to the empirical mean. The ML estimated curves are almost overlapping, and both are quite close to the empirical estimates of the mean at the inspection times.

Figure 7 shows the ML estimate of the variance $V\{W(t)\}$ under the FW and the Wiener processes, together to the empirical estimates of the variance at the inspection times. Even in this case, the ML estimated curves are quite similar, and both are close to the empirical estimates.

Finally, by comparing the resulting AIC value relative to the FW process, say AIC $= -69.0$, with the AIC value relative to the time-scaled Wiener process given in [5], say AIC $= -67.7$, we can conclude that the proposed FW process fits the observed HDD data better than the Wiener process.

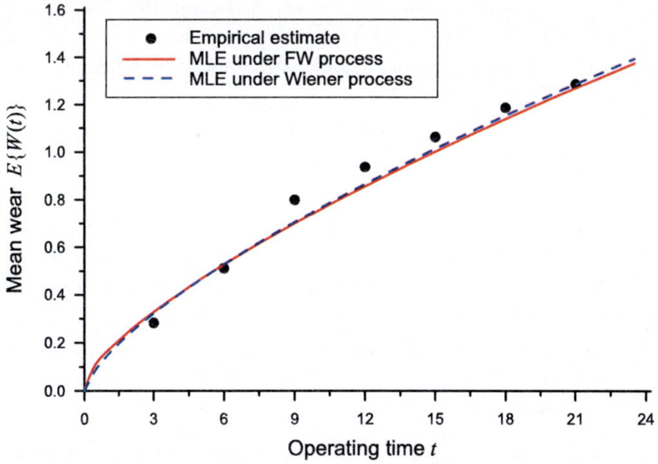

Fig. 6 Empirical (•) estimates of $E\{W(t)\}$ and corresponding ML estimates obtained under the FW process (solid line) and the Wiener process (dashed line)

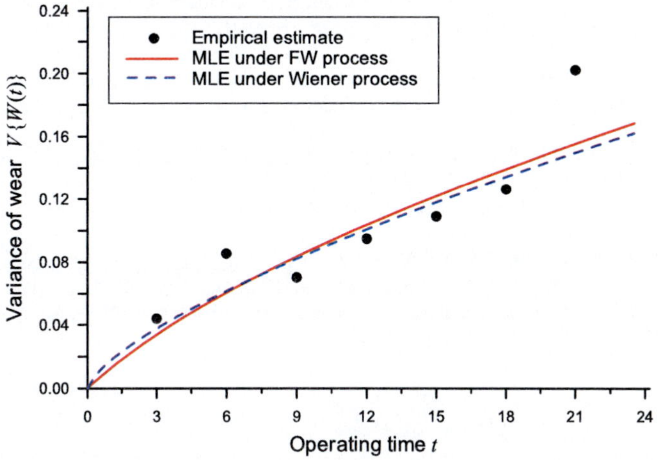

Fig. 7 Empirical estimates (•) of $V\{W(t)\}$ and corresponding ML estimates obtained under the FW process (solid line) and the Wiener process (dashed line)

5 Conclusions

This paper proposes the folded Wiener process, a new non-negative degradation model with possibly negative increments, that has been expressly conceived to overcome the limitation of the Wiener process in describing non-monotonic non-negative degradation phenomena. Differently from the Wiener process, the folded Wiener process has correlated increments. Yet, as the operating time goes to infinity, this correlation tends to vanish and the folded Wiener process tends to a Wiener process.

The folded Wiener process has a good mathematical tractability. In fact, both the conditional distribution of the degradation increment and the distribution of the degradation level at a generic time can be expressed in closed form. In particular, this latter distribution is a folded Normal distribution, a feature that motivated the name of the proposed process.

The main features of the folded Wiener process have been presented and discussed. Closed form expressions for the mean and variance function of the process are also provided. Finally, an applicative example has been developed by using a set of real wear data. Obtained results have shown that, in the considered experimental situation, the folded Wiener process over-performs the Wiener process in terms of fitting ability, and allows to overcome the limitation of the Wiener process in analyzing intrinsically non-negative degradation phenomena.

References

1. Doksum KA, Normand SLT (1995) Gaussian models for degradation processes—part I: methods for the analysis of biomarker data. Lifetime Data Anal 1(2):131–144. https://doi.org/10.1007/bf00985763
2. Whitmore GA, Schenkelberg F (1997) Modelling accelerated degradation data using Wiener diffusion with a time scale transformation. Lifetime Data Anal 3(1):27–45. https://doi.org/10.1023/A:100966410
3. Leone FC, Nelson LS, Nottingham RB (1961) The folded normal distribution. Technometrics 3(4):543–550. https://doi.org/10.1080/00401706.1961.10489974
4. Tsagris M, Beneki C, Hassani H (2014) On the folded normal distribution. Mathematics 2(1):12–18. https://doi.org/10.3390/math2010012
5. Ye ZS, Chen N, Shen Y (2015) A new class of Wiener process models for degradation analysis. Reliab Eng Syst Safety 139:58–67. https://doi.org/10.1016/j.ress.2015.02.005
6. Akaike H (1974) A new look at the statistical model identification. IEEE Trans Autom Control 19(6):716–723. https://doi.org/10.1109/TAC.1974.1100705

Crowd Management for Power Generation: A Critical Analysis on the Existing Materials and Methods. (Structural Modal Analysis)

Abdulaziz O. Alnuman, Muhammad A. Khan, and Andrew Starr

Abstract Energy harvesting by means of different materials and mechanisms is considered an important topic of interest in past decades. Materials such as piezoelectric, electromagnetic and electrostatic in nature are generally used to harvest energy in many of the sensing applications. Mechanisms based on simple deformation, vibration and magnetism are used to harvest energies in power generation applications. The published research shows that the harvested energy from these materials and mechanisms are still far away from practical feasibility and optimisation. However, not a single review article is available which can provide a critical analysis on the existing materials and methods with regards to the mentioned feasibility and optimisation. In this paper, a review attempt has been made to describe the mentioned analysis. All the past research is described in two categories: Energy Harvesting through Materials and Energy Harvesting through Mechanisms. The materials used for energy harvesting contain characteristic to release electric charge under the influence of an external excitation. Several materials such as Multiwalled Carbon Nano Tubes (MWCNT), Polyvinylidene Fluoride (PVDF), Polydimethyl siloxane (PDMS) and ZirconateTitanate (PZT) with slight change in their internal properties and efficiencies are used to harvest charge. In contrast to the materials, several mechanisms are also in use to produce useful energy from available external forces. Their mechanics is principally based on phenomena like structural vibrations, electromagnetic induction, and magnetism. This review concludes that a methodology of energy harvesting which can utilise any random load and converts into maximum useful energy is still not present.

Keywords Power generation · Footsteps · Energy harvesting · Piezoelectric · Generation electrical · Human weight · Sidewalk block · Pedestrian · The sidewalk as the electrical energy · Smart city · Vibration energy harvesting

A. O. Alnuman (✉) · M. A. Khan · A. Starr
Cranfield University, Cranfield, UK
e-mail: A.Alnuman@cranfield.ac.uk

M. Abdel Wahab (ed.), *Proceedings of the 8th International Conference on Fracture, Fatigue and Wear*, Lecture Notes in Mechanical Engineering,
https://doi.org/10.1007/978-981-15-9893-7_2

1 Introduction

The current sources of electricity have far-reaching effects on the environment because of their high-power nature, and also due to the fact that many of the sources are non-renewable [1]. There are limited natural resources and the demand for energy is growing each and every day [2]. There is a lack of devices and systems that can be installed as an alternative and decentralised energy in any city. Hence researchers looking forward for sustainable and better energy efficiency. A large amount of energy is lost in the environment and technical process without measurable impact. However, Bischur and Schwesinger [2] argued that something can be done to these losses to bridge the gap between energy demand and sustainability thereof. Additionally, several issues have emerged on how energy is generated and consumed today, including the cost of generating energy, the ever-rising demand for it, and the ever-declining reserve of the current energy resources. Various researchers and engineers have tried to exploit new sustainable power sources to curb the current limitations. Therefore, currently, there is a considerable and continuing need globally to develop the energy harvesting methods and self-powered electronics which are cost-effective and sustainable.

The process of captivating the lost energy from naturally circulating energy sources, storing and collecting to subsequent use that is energy harvesting or scavenging [3]. In other definition is Energy harvesting is transformation and collecting electrical energy from ambient energy, existing environmental energy [4]. The common sources of energy harvesting are: Mechanical energy such as vibration and mechanical stress and strain; Thermal energy for example waste energy form furnaces, heaters and friction sources; Light energy for instance captured form sunlight, room light via photo sensors, photo diodes and solar panels; Electromagnetic energy as inductors, coils, transformers; Natural energy such as wind, water flow, ocean currents and solar; Human body for instance bio-organisms, walking and sitting; Other energy for example chemical and biological [4]. Whereas energy harvesting has gained more attention in the recent year's studies due to their advantages of energy conservation, sustainability, less maintenance cost, economical way to power an electronic system for a long time and environmentally friendly. However, there are drawbacks such as low power conservation rate, materials can easily get damaged, discomfort during stepping, low economic practicality and the voltage lower than other sources such as fossil fuel. The three components of the energy harvester are the micro-generator, voltage booster and the storage element in Fig. 1

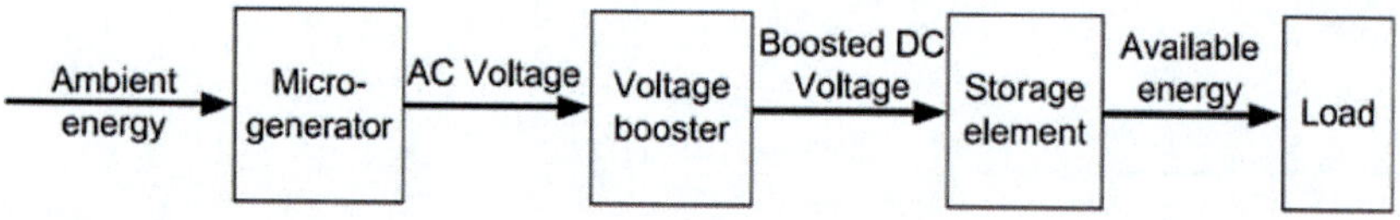

Fig. 1 Block diagram of an energy harvester Clearly [5]

[5]. The micro-generator generates electrical energy by converting it from environmental energy. The Voltage booster pumps the AC voltage and regulates the boosted DC voltage.

Kinetic energy harvesting has main mechanisms that are piezoelectric, electromagnetic and electrostatic or by using magnetostrictive materials [5]. There are numerous methods and various ways have been used by piezoelectric, and find suitable technology for it, through different applications. The instance of it uses includes knee replacements, shoes, car tires, and others [6]. Also, there have various sources of harvestable ambient energy exist as flowing water, wind-exerted stresses, the vibration of structures, and some successful studies which use it from human walking to generate power and among others [6]. The relative motion generated between the conductor and the magnetic flux gradient results in the rise of electromagnetic induction that is utilized by the electromagnetic generators [6]. The energy generated by the relative movements between the capacitor plates that are electrically isolated charged is utilized by the electrostatic generators. The capacitor plates generate the electrostatic force between them, and the work done against this force results in the harvested energy [6].

Harvesting mechanical energy from ground is possible through either electromechanical means (i.e., Faraday's principle) or piezoelectric means (Curie's principle) or electrostatic means (Coulomb's principle). Faraday's principle states that a time-varying magnetic field generates a corresponding electrical field, and vice versa. Curie's principle states that piezoelectric materials generate a charge across their boundaries when a stress is applied to them, and vice versa. Coulomb's principle states that the magnitude of the electrostatic force of attraction or repulsion between tow point charges is directly proportional to the product of the magnitudes of charges and inversely proportional to the square of the distance between them. The force is along the straight line joining them.

This paper presents a critical analysis on the existing materials and methods with the harvested energy from materials and mechanisms and will be dividing into three main groups: energy harvesting by piezoelectric, energy harvesting by electromagnetic, energy harvesting by electrostatic by using magnetostrictive materials. Which includes mechanical energy harvesting tools, Curie that is piezoelectric, electromagnetic (i.e. Faraday). Coulomb which includes electrostatic. It is divided by technological principle and addresses details on mechanisms, methodologies, output.

2 Structure of Transduction

The mechanical systems of the generator combine the environmental displacements with the transduction mechanisms that are required to generate the electrical energy with the help of kinetic energy harvesting from the motion. The increase in the coupling between the transduction mechanism and the kinetic energy source that depends on the environmental motion characteristics is directly proportional to the mechanical system designs [6, 7]. The vibration energy is mainly suitable for the inertial generators having an inertial frame as a reference with the addition of the mechanical component. The relative displacement is produced between the frames due to the transmission of vibrations from the inertial frame to the suspended inertial mass. In order to match with the attributes of the environment frequency, the system designs a possessed resonant frequency. The approach of this resonant system attains the quality factor that expands the vibration amplitude of the environment [6].

The relative displacement or the mechanical strain of the system is utilized by the transduction mechanism in order to generate the electricity. The effects of the mechanical strain typically use active materials such as piezoelectric and [6] system deformations. In the relative displacement, the transduction mechanism is either coupled with the position or the velocity. The electromagnetic transduction velocity mainly lowers the relative position of the electrostatic transduction. In the process of generator modelling, the fact to be considered is that each transduction mechanism has different kinds of damping characteristics. The basic examples to show the complexity increase due to the mechanical systems are the rotary generators with the coupled linear displacements and the magnification of forces and amplitudes by including the hydraulic systems [8].

3 Energy Harvesting by Piezoelectric

Researchers and engineers have made several efforts to harvest energy using piezoelectric technology across a number of applications. For instance, piezoelectric materials have been embedded in shoes, tire automobiles, knee replacement among others. Some studies have investigated a recent field for piezoelectric materials in energy harvesting from a number of sources in the environment such as wind and water-exerted stresses and the vibration of structures [8]. However, other studies have probed the use of piezoelectric energy harvesting technology to generate energy from human movements. Still, another use of piezoelectric materials is realized in powering transducers and sensors to detect by generating energy from vibrations.

Piezoelectric materials have proved to be efficient in scavenging mechanical energy from roadway pavements because of their high-power density and simplicity. The energy produced has been used to power wireless sensors. Several researchers have reported that piezoelectric materials are the best choices for powering wireless sensors and smart road sensors [9].

In the year 1880, J and P Curie discovered the piezoelectric effect. According to them, the exposed crystals associated with the mechanical strain are electrically polarized and this polarization is figured out with a degree depending upon the applied strain. On the contrary, the exposition of the materials to the electric field helps them to attain the deformity. There are various forms of piezoelectric materials such as quartz known as single crystal, sputtered zinc oxide basically known as thin-film, PZT and the lead zirconatetitanate known as piezoceramic, the piezoceramic powder based printable thick films [9, 10] polyvinyllidenefluoride (PVDF) and the polymeric materials [11].

The force directions and the location of electrodes and polarization differentiate the material properties; hence this exhibits the anisotropic characteristics for the piezoelectric materials. Ceramic attaining the anisotropic piezoelectric properties are mainly defined by the notation and the symbol systems [6]. It is mainly related to the direction of applied force or stress and the measurements and the ceramic orientations.

4 The Main Properties and Principles

Piezoelectric materials produce electricity when electric energy when strain or stress is applied on it. These materials can be employed in the generation of electric energy while under stress or equally used as strain sensors as electric excitation. The electric energy store in the piezoelectric materials can be produced by connecting their poles to the external resistance [11].

Piezoelectric materials can be obtained in nature in the form of a single crystal, for example quartz. However, there are also many other synthetic piezoelectric materials, such as ceramics, polymers, composite and quartz analogues. There are two commonly used piezoelectric materials: lead zirconatetitanate (PZT) and polyvinylidene fluoride (PVDF). PZT material falls in the piezo-ceramic family, while PVDF is in the polymer family [10]. PZT based piezoelectric materials have extensively been used for industrial application for energy harvesting due to their large voltage output [10]. The following transducers can be used for energy scavenging: cymbal, biomorph, multilayer, RAINBOW, thin layer unimorph ferroelectric driver and sensor (THUNDER), and macro-fibre composite (MFC). Each of these piezoelectric materials has distinct parameters that greatly influence their performance and efficiency, for instance PZT is an organic delicate ceramic material with a broader piezoelectric coefficient [6]. Conversely, the PVDF composite has an outstanding flexibility and thermal stability. Furthermore, it has ability to resist chemical, and it is also durable; nonetheless, it has smaller piezoelectric coefficient.

Piezoelectric energy harvesting mechanism incorporates two significant elements: the mechanical element that produces electric energy and the electric circuitry that influences the process of energy generation. These two elements integrate and interact to have a significant effect on the efficiency of the material. Moreover, there are several factors that affect the performance of piezoelectric materials in energy scavengings, such as ceramic tolerance, delamination of ceramic, electric permittivity and material constraints.

To effectively harvest energy from the pavements needs piezoelectric devices that meet the stringent requirements for durability, strengths and stiffness [12]. Therefore, in order to protect piezoelectric elements from adverse environmental conditions, vehicular loads so as to encase them in a suitably designed enclosure made from suitable materials such as steel, engineered plastic or other relevant materials. Essentially, there are a number of studies that have been conducted regarding designing piezoelectric materials enclosures, and many of these studies looked into cantilever beam configuration. The researcher reported that cantilever beam configuration cannot effectively handle heavy loads because of their fragile nature. Nonetheless, a few studies revealed that scavenging energy using piezoelectric materials configured with cantilever beam system is feasible and can be recommended to be used to improve efficiency and durability.

The conjunction consisting of the constants series with the notation axes defines the piezoelectric activity level of the material. The constant of piezoelectric strain is denoted by “d” and defined as

$$d = \frac{developed\ strain}{field\ applied}\ \text{m/V}.$$

$$d = \frac{charge\ density\ of\ the\ short\ circuit}{stress\ applied}\ \text{C/N}.$$

The piezoelectric generators attain the d31 material coefficient by applying the transverse strain parallel to the electrodes whereas applying the compressive strain in a perpendicular direction to the electrodes exploits the d33 coefficient of the material. The use of multi-layer stacks or the increase in the thickness of piezoelectric elements improves the power output that is generated due to the compressive mode. Moreover, the compressive loading for the majority of the applications is not considered as a coupling mechanism for harvesting vibration energy. The coupling direction for the bonded substances of piezoelectric elements or piezoelectric films is transverse. The above arrangements result in the mechanical amplification of all the stress applied [6].

The coupling coefficient of the electro-mechanical is denoted by “k” and it is affected by the generated electrical power [13]. The compliance “s” is measured by the produced strain due to the unit of applied strain and material permittivity “ε” is measured by the insulating displacement caused by the unit of an electric field that is considered as the piezoelectric constants [6]. The constant of piezoelectric voltage is denoted by “g” and it is defined as the developed strain for a unit of applied charge density or as the generated electric field for a unit of mechanical stress. These above constants are considered as anisotropic. The reader refers to the IEEE standards in order to attain a complete constants description [6].

The three measures of piezoelectric properties on which it varies are temperature, stress and age. The ageing rate is defined based on the change in piezoceramic properties with the time period and it is mainly dependent on the type of the material and the methods of construction. The property of the material stabilises with the manufactures and the age based on the material changes that are logarithmic with respect to the time.

The applied stress to the ceramics accelerates the process of ageing and it is mostly considered in the applications based on the harvesting of cyclically loaded energy. The PZT-5A is considered as the harder compositions that are less susceptible to the changes in the induced stress than the PZT-5H, which is considered as the soft piezoceramic compositions. The Curie point limits the temperature factors with the

Table 1 Common piezoelectric materials coefficients [14]

Property	PZT-5A	PZT-5H	$BaTiO_3$	PVDF
$d31(10^{-12}\ C\ N^{-1})$	−171	−274	78	23
$d33(10^{-12}\ C\ N^{-1})$	374	593	149	-33
$g_{31}\ (10^{-3}\ V\ m\ N^{-1})$	−11.4	−9.1	5	216
$g_{33}\ (10^{-3}\ V\ m\ N^{-1})$	24.8	19.7	14.1	330
$k31$	0.31	0.39	0.21	0.12
$k33$	0.71	0.75	0.48	0.15
Relative permittivity ($\varepsilon/\varepsilon_o$)	1700	3400	1700	12

piezoceramics. The piezoelectric properties are lost by the piezoelectric materials by effectively attaining the depolarization if they cross the above limits. The Curie temperature is lowered by the stress applications [6].

The Table 1 shoes the common materials attaining the piezoelectric constants, hard lead zirconatetitanate piezoceramics (PZT-5A), barium titanate (BaTiO3), soft lead zirconatetitanate piezoceramics (PZT-5H) and polyvinylidene fluoride (PVDF) [6].

"For over the many years, the conversions of mechanical energy into the electrical energy are done by the piezoelectric ceramics. It was discovered that the crystals such as specific Centro-symmetric and the non-Centro symmetric associated with the mechanical strain are electrically polarised and the strain applied is directly proportional to the degree of polarisation. On the same contrary, these piezoelectric materials attain the deformity as they are exposed to the electrical field. The piezoelectric effect is mostly used to realise the micro-generators and by the ultrasonic sensors and motors. The strained Centro-symmetric crystals do not induce the polarisation by deforming its shape and the displacements of the negative and positive centre of gravity mainly denoted as G- and G+ in Fig. 1a, whereas the non-Centro symmetric crystals induce the polarisation in the above case in order to generate the net electric field [15]. The piezoelectric materials are mainly divided into five different forms such as thin film, polymeric materials, crystals, thick films in terms of printable screen and the piezoceramic. In most of the directions under the electrical field effects, the deformation is obtained due to the anisotropy of piezoelectric materials. The coupling of electromechanical mainly undertakes the three modes of principal. For the plane-parallel bar having a PZT ceramics with the different primary electromechanical coupling is shown in Fig. 2.

The three modes of deformation are:

The 31 mode: the direction of mechanical stress is 1 whereas the direction of voltage is 3. The surface of the electrode collects the charge in the direction of polarisation perpendicularly when the polarisation axis faces the applied force in the perpendicular direction to the axes.

The 33 mode: the direction of the stress and the voltage are same. It mainly indicates the charge collection in the perpendicular direction to the polarisation on

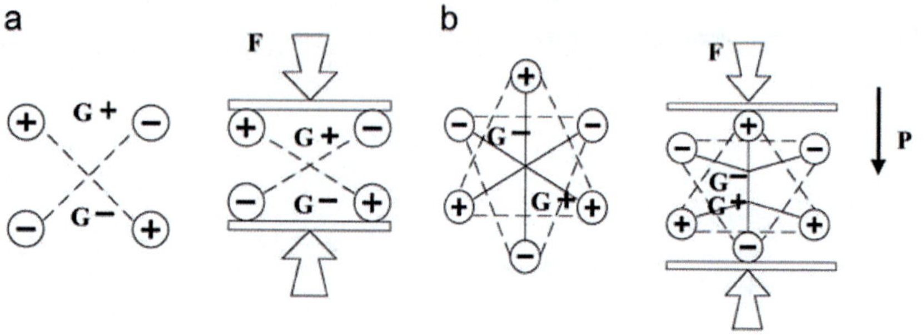

Fig. 2 **a** deformation of Centro-symmetric crystal. **b** deformation of non-Centro-symmetric crystal [16]

the electrode surface when the applied compressive and the tensile forces are acting along the axes of polarisation.

The 15 mode: The induced shear deformation is noticed around the axis 2 when the applied electric field is seen on the axis 1 [17]."

5 Mechanisms and Performance Effectiveness

The schematic representation of the reverse and direct piezoelectric ferroelectric effects.

The established mechanisms are the piezoelectric harvesters that are used for the conversion of vibrations and the motion, thus considering the resonant conditions for the piezoelectric generator and proposing the solutions for the usage of non-resonant and the resonant conditions of piezoelectric structure [17]. The piezoelectric devices attain the proposed conditions in order to improve the conversions of electrical energy from mechanical energy. Including the material with the high conversion properties, the techniques of power management, establishing the condition of novel power and matching the ambient vibration source with the piezoelectric device's resonant frequency are the above-proposed conditions. Compared to all the above conditions, the matching of the resonant frequencies is considered as the most reliable method for the harvesting systems depending upon the piezoelectricity [18].

Xu et al. studied the sophisticated structures as the mode of energy harvester that is used to maximise the off-resonance mode a 20 mW was provided by one of them. The operating frequency of the structure is nearly from 1 to 2 Hz and is made up of the multilayer ceramic stack of PZT [19]. The harvesting of resonant vibration is researched without affecting the ability of such designs to operate at resonance and at off-resonance [18]. The improvements in the conversions of mechanical energy into electrical energy are noticed by adding several conditions to the piezoelectric devices. The above conditions include the combination of ambient vibration sources and the resonant frequency of the piezoelectric devices, the inclusion of the materials that are highly conversable, to establish the novel power conditions and the techniques of the power management. Among all the above conditions, the combination of resonant frequency with the external source of vibration of the piezoelectric devices is considered as the unbiased method to utilise the harvesting systems depending on the piezoelectricity factor [18]. The operation of the resonant frequencies is taken into account for briefing the presentations for the above-mentioned devices. Moreover, we are limiting the results as the design features as it is nearly impossible to describe all the reported devices mentioned in the literature in this review. Even though the resonant devices are important in this presentation we are choosing the non-resonant piezoelectric structures in order to start the presentation.

5.1 The Non-resonant Structures

The human applications are mainly driven by the coupling of direct straining of the piezoelectric elements. The large-amplitude movements at the lower frequencies characterises the human motions. The extensive studied topic includes the human-wearable systems that are powered by the piezoelectric generators [6].

5.1.1 The Piezoelectric Shoe from Media Laboratory in MIT

The device shown in the Fig. 3 [20], was developed by the Paradiso et al. from Massachusetts Institute of Technology (MIT) that generates the energy while walking. Either side of the thick plastic sheets of 2 mm was mounted by the researchers that contained the lamination of the electrodes with the stack of eight layers of Polyvinylidene Fluoride (PVDF). The charge is produced by the 31-mode

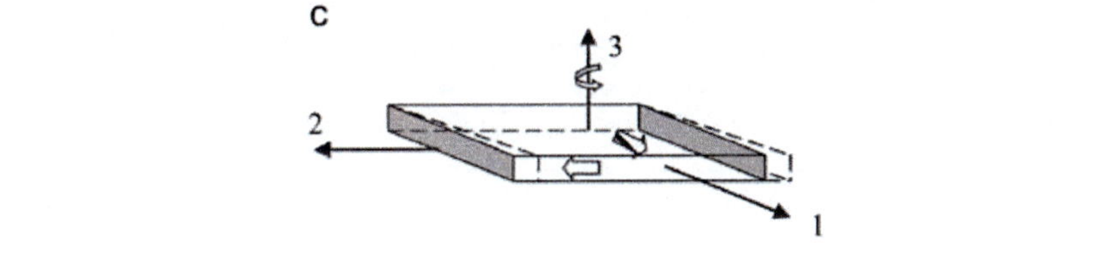

Fig. 3 **a** "31" mode. **b** "33" mode. **c** "15" mode [16]

due to the sole bending that strains both stacks. As a result of the above, the average power equal to 1.3 mW at a frequency equal to the 0.9 Hz was discharged in the load of 250 kΩ. The incorporation between the two piezoelectric transducers is caused by the designs attaining the compressible dimorph. The transducers are mainly the primary section of the stressed assemblies consists of the bond of aluminium, PZT and stainless steel. The materials attaining the differential thermal expansion coefficients bring out the curved structure characteristics with the addition of the compressively stresses PZT layers and hence initiating the deformation to a greater extent as compared to the standard PZT structures. In the 500 kΩ load, the power of about 8.4 mW is produced by the dimorph in order to maintain the pressure at a frequency nearly equal to 0.9 Hz.

5.1.2 Design of Impacts from Renaud et al

The piezoelectric unimorph as shown in the Fig. 4a is used by Renaud [21]. The impacts of energy harvesting and the shock are considered by the piezoelectric element supported by the elastic cantilever between the two electrodes. In the same, the mass is added in the given diagram. In the 31-mode transverse the piezoelectric material produces a state of increased energy in the approximation at the initial level. Figure 4b illustrates the designs governing the physical situations such as impact and the shock. The impact case includes the moving object hitting the bender whereas the shock case includes a frame that contains the piezoelectric benders undertaking the shock that has been dropped on the ground from an appropriate height is illustrated in Fig. 4c.

In the above cases, the behaviour of the benders is noticed right after the impulse inputs. These inputs are in the terms of mechanical energy at the initial quantity that is transferred to the bender and denoted by

Fig. 4 The Piezoelectric shoe [22]

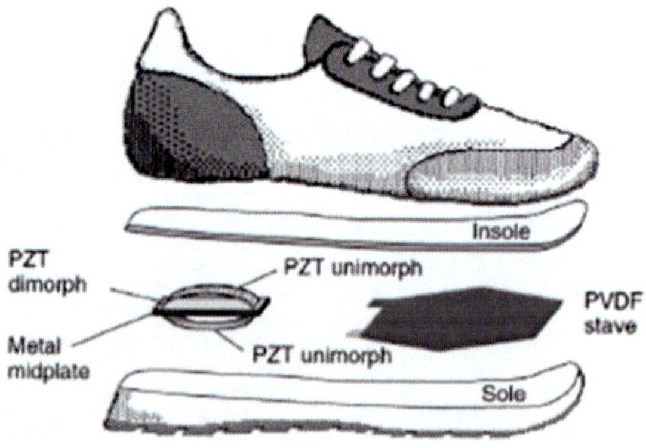

$$E_0 = \frac{1}{2} m v_0^2$$

where

m mass of the system
ν_o the initial velocity given to the system

The Fig. 4d is schematically designed by proposing the principle of an ideal case, where the impulse acting is based on the system's power as the average output, mainly defined as

$$P = E_0 / T$$

where

T the time interval between the two shocks and impacts

The figure shows the consideration of the free sliding mass in a frame that affects the piezoelectric elements. The piezoelectric benders can be seen in the direction of mass movement located at the frame edge. The expected maximum power by the authors was about 40 μW by considering the 1 cm of moving distance, 1 Hz of frequency, 750 mg of sliding mass and the 10 cm of amplitude movement.

5.1.3 Resonant Structures

Figure 5 contains the usage of mass-spring-damper that proves that the resonant frequency is optimal when the structure's output power is possessed with the increased energy with the help of vibration source [6]. The ambient vibration not being static employs the difficulty for such kind of mode. Hence, the energy output is lowered in the smaller amounts by the variation in the excitation frequency that is wrapped around the fundamental frequency of the harvester, thus making the process of energy harvesting inefficient. There are multiple numbers of frequencies in the sources of vibration. The harvester vibration energy includes different kinds of solutions that can be used for tuning the resonant frequencies. These solutions are differentiated under two groups such as power generators with the addition of large bandwidths that brings out the change in frequencies from the source of vibrations and the piezoelectric power generators with the addition of vibration frequencies that are concentrated on the devices having the resonant frequencies.

The geometry of the harvesting vibration energy is considered to be attractive when the piezoelectric material with the addition of cantilever structures are attached at the bottom or top of the surfaces [6]. The mass is added at the tip of the beam in order to lower the resonant frequencies that are provided by the cantilever. Glynne-Jones et al. [15] proposed the principle of the piezoelectric generators and the working that is associated with the model of mass-spring as shown in Fig. 6. The deformation of the piezoelectric layer that is adhering to the beam surface occurs due to the

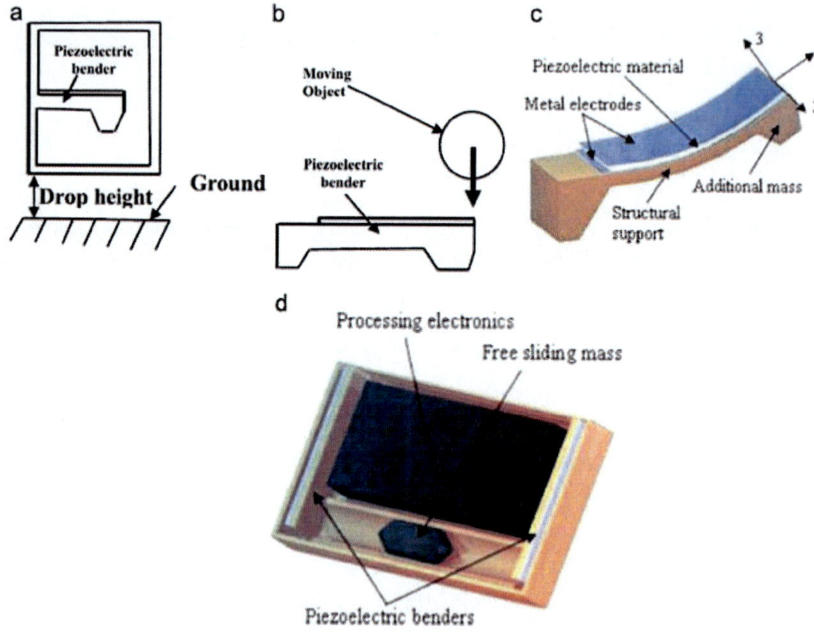

Fig. 5 **a** Piezoelectric structure, **b** illustration of shock and **c** impact excitation impacts. **d** schematic design [16]

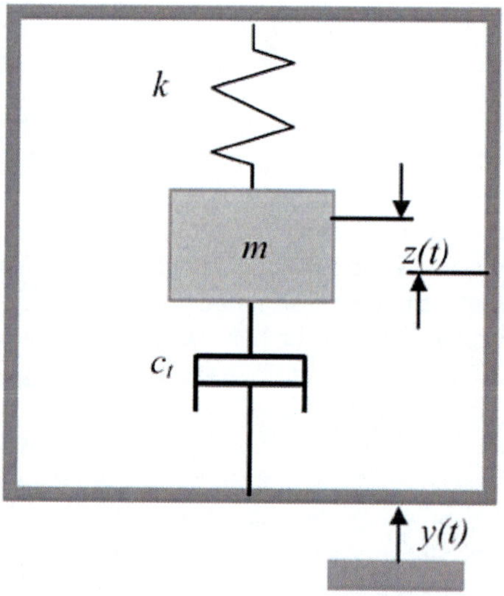

Fig. 6 Model of a inertial and the linear generator [16]

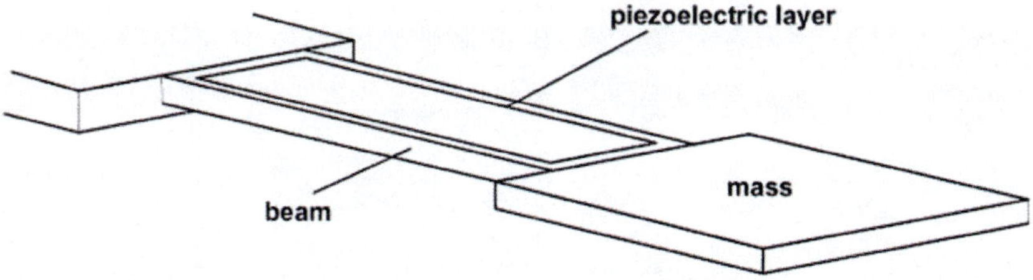

Fig. 7 Schematic diagram of a piezoelectric cantilever beam with tip mass. [16]

oscillations of the mass or beam structures. The above deformation results into the displaced charge for the piezoelectric elements on the electrodes that are located on the surfaces.

The results from the deformation cause the potential difference that is later utilised for the generation of the electrical work. This discussion includes the magnets of the cantilever beams that add the mass at the tip and then exposing it later to the magnetic field [23, 24]. According to the Wickenheiser work, the exposition to the magnetic field allows the beam frequency to tune in order to combine it with the base frequency possessed with the increased energy varying slowly at the extent response that utilises the off-resonance beams [25].

5.1.4 Cantilever Beam with Tip Mass Described by Roundy [13]

Wright and Roundy proposed the designs represented in Fig. 7 that follows the described principle [26]. The PZT elements that are attached to the central brass shim of 0.1 mm thickness is used by the below-mentioned devices. In the Fig. 7, the design 2 produces the 375 μW with the help of 2.5 ms^{-2} of input acceleration at a 120 Hz of resonant frequency by using the 0.28 mm of thickness for the PZT element, the $17 \times 7.7 \times 3.6$ mm^3 of the tungsten-proof mass and the 11 mm of beam length. The demonstration of this generator mainly includes the radio transceiver powering with the addition of the energy storage capacitors.

5.1.5 The Tapered Cantilever Beams

The illustrated design in Fig. 8 is produced by Glynne-Jones et al. [15]. The PZT is used in the formation of the piezoelectric layers. The piezoelectric film attaining the constant strain along with its respective length in the addition of the given displacement ensures the diminished profiles. In order to combine the potential applications, the dimensions are provided to the resonant frequencies. The authors faced various problems between the PZT and the steel in terms of chemical interactions that made them choose the dielectric material in order to eliminate the problems with the help of separated layers. The 333 kΩ optimal resistive load is resulted due to the 3 μW

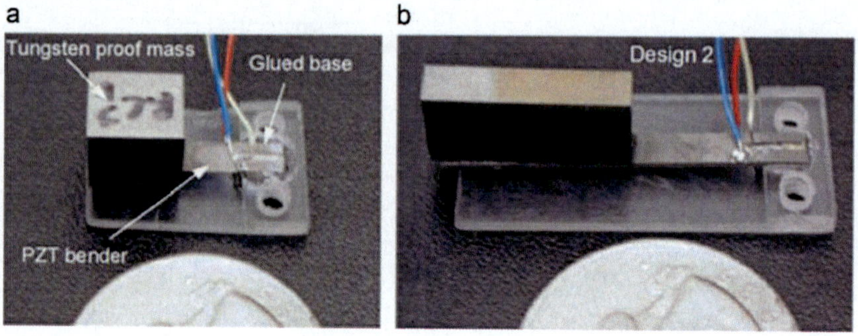

Fig. 8 Designs 1 and 2 [13]

produced power that operates the structure at an 80.1 Hz resonance frequency in order to attain the fundamentals of the bending mode.

5.1.6 Micro Rounded Shape Cantilever Beam

As mentioned by the M'boungui et al., Marzencki proposed the micropower generator with the resonance depending on the one-dimensional models that match with the model of the finite element [17] is illustrated in the Fig. 9 [27]. The composition of the device includes the vibrational mass weighing 1.57 μg of silicon connected by the cantilever substance that is mainly in the shape of the circle. The cantilever beams are the mechanical supports that are made of 5 μm of microcrystalline silicon consisting of two electrodes made of 1 μm piezoelectric layer of aluminium nitride. The device is arranged appropriately in the wafer measuring the 2 mm × 2 mm square that is insulated with silicon. Based on the shape of modes, the mass movement is induced by the ambient vibrations. The collection of the electric charges that are done by the aluminium and silicon electrodes is created by the piezoelectric layers on the first vibratory mode. By stimulating the 89 NW at 0.5 g with the addition of 204 Hz results in the power that helps in powering the wireless sensor node [27].

5.1.7 Double-Layered PZT Cantilever Beam

The solution for the devices having the external vibration variations and are designed in order to work with the resonance conditions was developed by Soon-Jong Jeong [28]. This solution includes the cantilever mass geometry version mainly consisting of the bender devices. The illustration in the Fig. 10 consists of the two layered devices having the various amount of inbuilt thick layers, where the tuning of each layers attains the frequency having the different resonant. The multiple numbers of piezoelectric devices are adapted with the different resonant frequencies in order to combine the variable source of vibration frequency with the resonance modes that

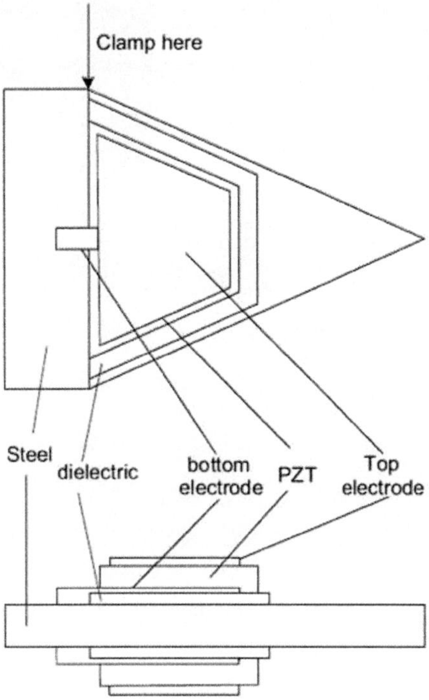

Fig. 9 Schematic diagram of tapered cantilever beam [17]

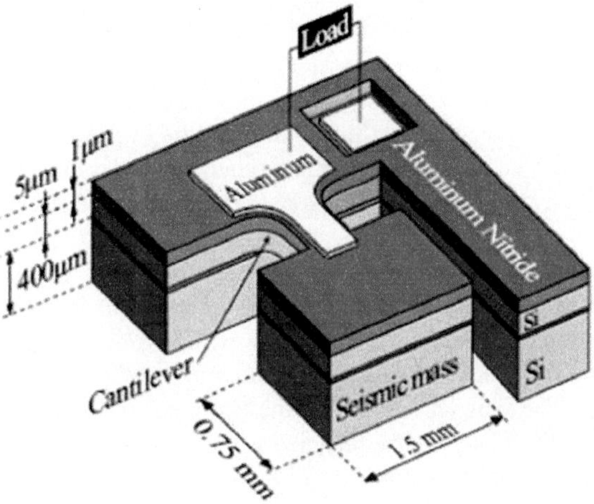

Fig. 10 Schematic diagram of micro rounded cantilever beam [18]

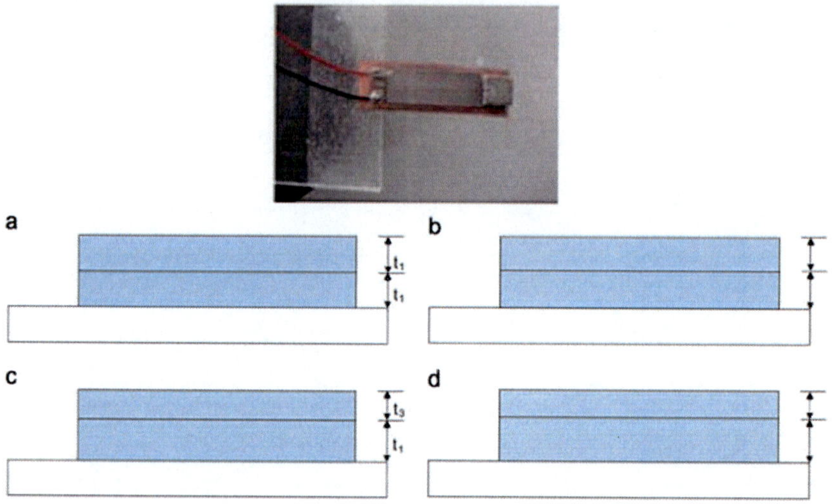

Fig. 11 Schematic diagram of double-layered PZT cantilever beam [19]

covers the range of frequencies. The structure of the beam includes the $40 \times 10 \times t_{i=1,4}$ of two piezoelectric layers where *i* ranges from 100 to 247 μm and the $55 \times 12 \times 0.4$ mm^3 of copper layers, where the payload is equal to 1 g. The illustrated four samples in the Fig. 10 were used to conduct the studies. The pulse type hammer test was used in order to identify the resonant frequencies with the help of the expressions retrieved from the relation of frequency versus vibration voltage.

The internal stress is negligible at the free end when the two piezoelectric layers are attached on the opposite sides based on the boundary conditions. The sample 1 was explained with the thickness of the two layers of the piezoelectric elements that are equal and by comparing them to the others that are attaining the minor pick in the initial stage of bending resonant frequency. The output voltages determined the value of the maximum power by evaluating itself separately as resistive load functions attaining the vibration frequencies that are entered on the identified fundamental frequencies. On the other hand, the resonant frequencies were mannerly helpful to attain the peak powers as they were determined by being away from them. Thus, the power generated gathers the maximum value by combining the loads.

The samples containing the two resonant frequencies are increased due to bandwidth of the peak frequencies as the above result. In the end, the device manifests the 0.5 mW of power and 2 V of peak voltage in the resonance mode that is stored inside the capacitor with the help of 0.1 G of external vibration acceleration at 120 Hz.

5.1.8 Coupling Impact Vibration Energy Harvester

As illustrated in Fig. 11, the vibration energy harvester that is operated in the larger bandwidths and the lower frequencies affected by the coupling and is suitable for the implementation of the microelectromechanical systems (MEMS) was proposed by the Gu et al. [29]. The device consists of the driving beam in addition to the mass available at the tip that resonates in the range of 1 Hz to 30 Hz by exploiting the ambient vibrations. The mass at the tip is considered as the sub-structure and it consists of the two generating beams mounted with the piezoelectric layers for the collection of voltages. The generating beam is separated from the tip mass by a possible air gap between them. The resonance of the self-generating beam is occurred due to the high frequency ranging from more than 100 Hz which is converted from the impacted vibrating mass having the low frequencies. Moreover, the resonant frequency in the current configuration for the generating beam is nearly not more than 30 Hz.

The simulation of the linear dynamics edges of the system and the performance verifications are evaluated by using the model of single DOF spring-mass. The construction of the macro-scale prototype with the 0.51 × 26 × 6.4 mm of two rigid beams of PZT bimorph, the driving beam weighing 8 g and the 1 × 100 × 10 mm^3 of acrylonitrile butadiene styrene beam resulted into the generated beams with the 1 mm of the air gap between them. In comparison for the above, the construction of the harvester prototype was initiated. The two above-mentioned prototypes obtained 20.1 and 18 Hz of resonant frequencies as a result of the acceleration that is nearly equal to 0.4 g. In addition to the appropriate loads in the improved power, impacted devices and the improved power density resulted in the wider bandwidths.

5.1.9 Triple-Morph and Bi-Morph Study from Kim

Kim et al. [30] proposed the cantilever beam's experimental study as illustrated in Fig. 12. This study helped to estimate the structure characteristics based on the parameters such as load resistance, frequency, mass weight, and vibrations. Figure 12 [30] shows the addition of the "*triple-morph*" to the "*bi-morph*" that is manufactured with the help of a screen printable cantilever beam that uses the 25 × 5 × 2 mm^3 of

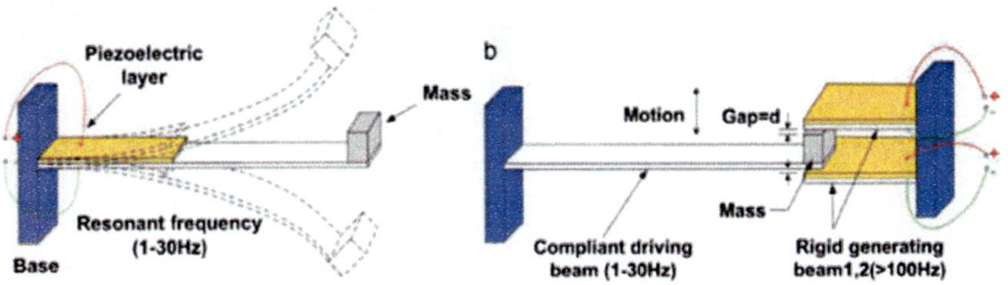

Fig. 12 Schematic diagram of coupling impact vibration energy harvester [20]

aluminium substance between the other shapes. The models for the above structures were proposed by Du et al. [31]. The bi-morph cantilever's output voltage is directly proportional to the mass and the applied load resistance, whereas the resonance frequency is inversely proportional to the mass as compared by the measurements and the results. The "*triple-morph*" cantilever adapts the 0.432 mW of power at 154 Hz and the 6.57 Vrms, whereas the "*bi-morph*" cantilever having a shape of 25 × 5 × 2 mm^3 adapt the 0.23 mW of power with the addition of 0.25 g of mass, 2.75 Vrms of output voltage, 33 kΩ of load resistance and 197 Hz of the resonant frequency (Fig. 13).

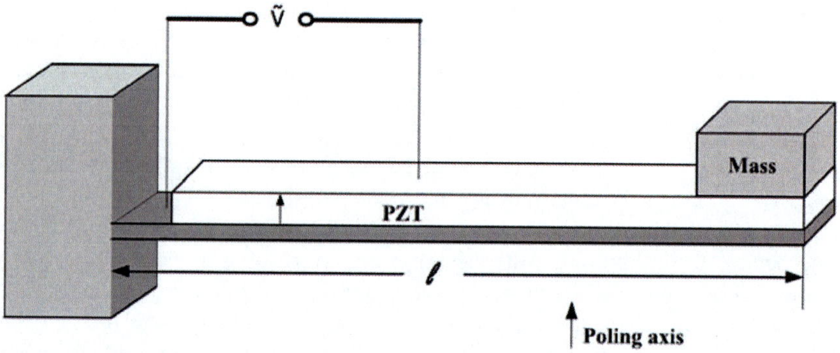

Fig. 13 Schematic diagram of cantilever beam from Kim [21]

Fig. 14 Model of a linear, inertial generator [14]

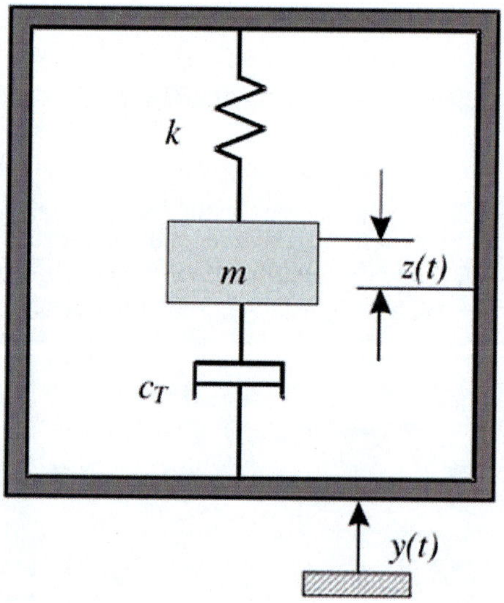

Piezoelectric mechanism is the ability of certain material to produce an electric energy in response to the mechanical strain applied on it. One of the unique characteristics of piezoelectric mechanism is that it is reversible. In other words, that material used to generate electricity when strain is applied on it is the same one that exhibits the converse piezoelectric mechanism. When a mechanical stress is applied on a piezoelectric material, a positive and negative charge centre shift occur in the material, which then results to external electric field [15]. However, when a reverse force is applied, an outer electrical field either compresses of stresses the piezoelectric material.

Several factors affect the produced electric energy, including piezoelectric material, element configuration, position of the source with respect to enclosure, the speed of the moving body, and the depth of the installed piezoelectric material and enclosure. The vertical strain dissipates down the depth, meaning the closure of the piezoelectric material, the more the mechanical power received. It is suggested that when using roadways, the centre of the piezoelectric material be positioned along the wheel path or at a position frequently followed by the moving bodies, which should be 45–60 cm from the end of the traffic in case of the moving vehicle. A study conducted by Xue et al. [23] concluded that if the distance between the moving object of a pavement or a pathway exceeds 4 m, then the generated energy will be zero. Similarly, Sharma et al. [24] conducted a research on the installation of piezoelectric energy scavenger on roadway tunnels and he concluded that the vibration energy from the moving vehicles was not enough to make energy harvesting realistic. The research linked this occurrence to the distance between the wheels of the moving vehicles and the transducers [25]. Moreover, the literature posits that the stiffness of the energy harvester enclosure should be equal to the stiffness of the surrounding roadways or pavement in order to maximise deflections and the pavement control.

The amount of energy generated using piezoelectric energy harvesting mechanism and its efficiency rely on several variables, including the material used, shape, dimensions and number of embedded piezoelectric elements. Thus, electric energy generated can significantly vary from one harvester to the other, following the design. Over the time, there has been development of the piezoelectric energy harvester. For instance, one developed in JR East of Japan where power harvesting using piezoelectric array embedded in Tokyo railway station generated energy that could light up 100 W light bulbs for 80 min. Nonetheless, the energy generated decreased gradually due to degradation of the mechanisms. This indicated that the long-time efficiency of piezoelectric material for energy harvesting largely depends on its maintenance.

6 Energy Harvesting by Electromagnetic

In the year 1831, Faraday discovered electromagnetic induction. It generates the electric current that is placed in a conductor with a presence of the magnetic field. The typical form of the conductor is the coil; it generates the electricity depending upon the magnetic field changes or the relative movement of the coil and the magnet.

The number of turns of the coil, the magnetic field strength and the relative motion velocity is directly proportional to the amount of generated electricity [6].

The production of the electromagnetic induction with the help of a resonating cantilever beam, the coil, and the permanent magnets is considered as one of the prominent ways for energy harvesting. According to the principle, between the coil and magnets, any one of them is organised on the beam and the relative other is considered to be in a fixed position. Moreover, attaching the magnets to the beam attains the inertial mass. The descriptive working principles of the electromagnetic generators are generally illustrated in the Fig. 1 with the schematic diagram. The arrangement of the coil and magnet is effectively represented by the damper denoted by "*c*" as the mechanism of the electromagnetic transduction [6].

Electromagnetic energy harvesting technology captures the mechanical energy of walking and other moving objects on the roadways and generates electricity through them. Every day, roadways are subjected to a number of loads of moving walk. Normally, the mechanical energy from moving walk on pathways dispels as heat and pavement deformation [32]. Electromagnetic technologies can be used to harvest this energy from the pavement and pathways. Characteristically, the electromagnetic devices absorb the mechanical energy from the moving walk and objects and convert them into a movement that actuates an electrical motor that works on Faraday's principle. This electromagnetic technology has also been used to harvest energy from other sources, for instance mechanical energy from ocean waves, railroads, vehicle suspensions and backpacks.

7 The Main Properties and Principles

Almost all electromagnetic technologies energy generators are premised on the basic principle of Faraday's law, which states that the relative movement of an electric conductor with respect to a magnetic field induces an electric current between the ends of the conductors. The Faraday's principle further states that the electromagnetic force (emf) induced in a circuit are proportional to the time rate of change of the magnetic flux linkage of that particular circuit. In other words, the amount of energy produced relies of the velocity of the relative movement, the number of the coils used and the strength of the magnetic field.

Practically, the mechanical energy from sources like human walking, moving vehicles, railroads and/or backpacks applies a force to a mechanical system that results to relative displacement between the coil and the magnet, hence producing electric energy [32].

William et al. [33] from the University of Sheffield, UK described the electromagnetic approach in schematic diagram as illustrated in the Fig. 15.

The generator mainly includes the spring denoted as "*k*" and the vibrational mass denoted as "*m*". The vibration in the generator causes the mass and the generator housing to phase out so that they can attain the net displacement between them. The above displacement has the nature of a sine wave that can generate electrical energy

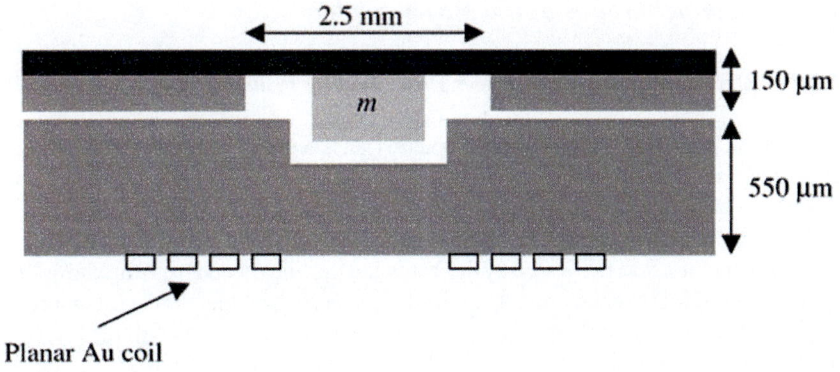

Fig. 15 Cross-section of the electromagnetic generator proposed by Williams et al. [26]

with the help of a suitable electromagnetic transducer. Moreover, the transducers are said to be dashpots as they damp the vibrations during the conversion of electrical and mechanical energy and mainly denoted by "*d*".

8 Mechanisms and Performance Effectiveness

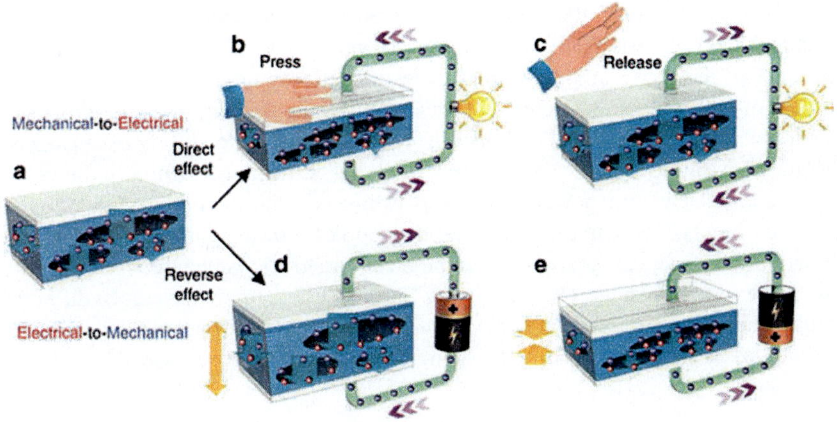

8.1 *Implementations of the Wafer-Scale*

8.1.1 Electromagnetic Transducer Described by Williams et al

Williams et al. described the electromagnetic transducer size as equal to the 5 mm × 5 mm × 1 mm. The assessment of the possibility of the devices is undertaken with the help of the harmonic analysis. The typical device attains the 1 μW of power for 70 Hz of frequency and the power generation of about 100 μW for the frequency of 330 Hz [6]. The generator was fabricated by the Yates and the Shearwood from the University of Sheffield, UK [34] depending upon the design. A circular membrane is comprised that is micromachined bulk on the GaAs substance coated by the polyimide layers of 7 μm. The membrane is attached to the 0.4×10^{-3} kg of the SmCo magnet. The Au coil having the planar properties attained in the 13 turns on the separate patterned wafers. The testing of the device generated the power of 0.3 μW at 0.4 kHz of the excitation frequency. As compared to the existing value, the measured value for the electrical power output was comparatively low. This phenomenon occurred due to the increase in the excitation amplitude with the stiffening of the spring based on the non-linear effects.

8.1.2 Described by Mizuno and Chetwynd [27]

The device consists of the fixed external magnet and the integrated coils. The NdFeB magnet has a size of 30 mm × 10 mm × 6 mm; the cantilever beam has the dimensions of 500 μm × 100 μm × 20 μm whereas the structure attains the resonant frequency of about 58 kHz. The electromagnetic micro-generator as a single element attains the 6 nW of the power output whereas the estimated magnitude was nearly equal to the 1.4 mV of the voltage output. In order to evaluate the purposes, the proposed version was fabricated by the authors. The result was raised to 700 Hz of resonance frequency with the increase in the size by 25 mm × 10 mm × 1 mm in the beam whereas the 0.4 nW of power output was estimated with the help of 0.64 μm of vibration inputs. The authors suggested the problems of the energy harvesting do not attain the practical solutions from the electromagnetic generators based on the low power output results. The conclusion for the authors might have been different if the availability of the increased electrical power would have been evaluated by the addition of the vibration masses for the moving-magnet approaches.

8.1.3 Kulah and Najafi Describe a Silicon-Based Generator that Comprises Two Separate Chips Combined Together [28, 35]

Their system uses two modulated frameworks equipped for up-frequency electromagnetic transition. The authors claim that since bulk of silicone materials has fairly large frequencies of resonance (several kHz) and the resonance frequency is less

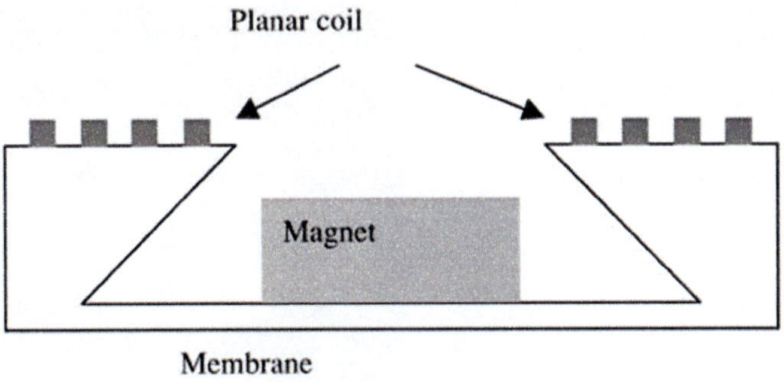

Fig. 16 Inertial generator described by Pérez-Rodríguez et al. [29]

than 100 Hz, a system capable of conversion of up frequencies would then provide an effective solution to the problems of energy recovery. They also developed a silicon motor with a high diaphragm with a vibration frequency of 25 Hz to illustrate that. The NdFeB magnet on the top bladder is being used in vibration via magnetic connection to a low frame. The smaller bladder is nearly 11 kHz resonant. Interconnected spindles are built on the smaller surface. The writers cite 2.5 μW of the maximum theoretical strength, but only 4 NW of a millimeter macro.

8.1.4 Huang et al. Have Addressed the Issue of Human-Powered Electrical Generation [14]

The approach for an electromagnetic harvesting machine comprised of a geometrical coil, a spring component of nickel-iron and a magnet was examined. The equipment has a modulated rate of 100 Hz and is recorded to be capable of producing 0.16 μW for a "finger tap" excitation level. Perez-Rodriguez et al. [36, 37] who used a polyimide movie as the core, NdFeB magnet and a geometrical spindle consisting of a 1.5 μm thick aluminum plate, studied a similar mechanism and Fig. 16 displays a cross-section of the unit. For a 10 μm displacement and 400 Hz frequency, the reported power output amounts to 1.44 μW. Scherrer et al. [38] addressed the potential to manufacture a multi-layer screen printed co-fired (LTCC) co-fired ceramics. The bucket has 96 layers of tape and 576 rotations in all. It lies inside two copper-beryllium cisterns and can be shifted upright when the incoming excitation allows the fluxes of four separately installed magnets to be split. When used at a vibrational frequency of 35 Hz, the estimated average output power shall be 7 mW.

8.1.5 Beeby et al

The silicon-based generator was created by Beeby et al. including micro-machined paddles, four NeFeB magnets and wire-wheel spindles that is depicted in Fig. 16 [39].

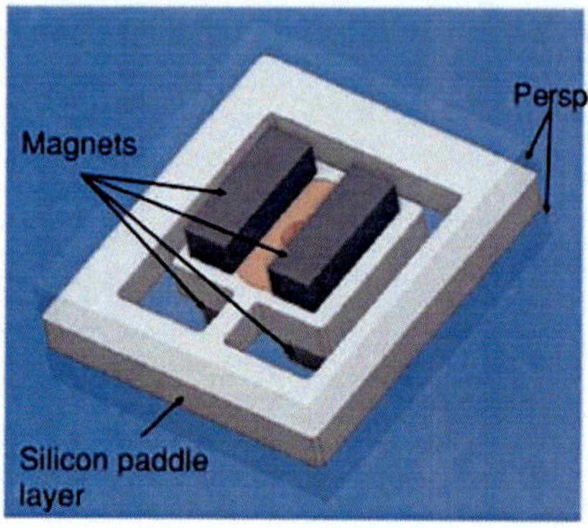

Fig. 17 A silicon electromagnetic generator (after Beeby et al.) [30]

For two Pyrex wafers, that are all anodically joined to either side of a silicone wafer, two of the magnets are contained for engraved recesses. The rotating mechanism is mounted on a silicon-paddle, which is laterally built to vibrate in the wafer axis. The resonant frequency of the unit is 9.5 kHz and 21 nW of electrical energy was seen to be derived from 1.92 m s − 2 rms (Fig. 17).

8.2 *Micro-generator*

See Figs. 18 and 19.

The recorded modeling frameworks are intended to substitute either an optimum voltage source with the mini-generator of a power harvester (Fig. 14a) [40] or a

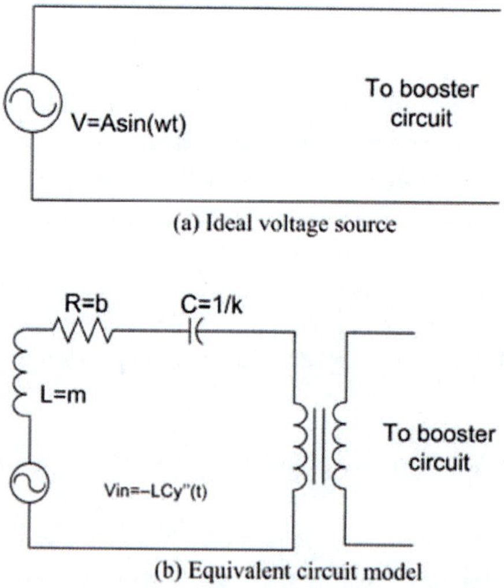

Fig. 18 Micro-generator models [5]

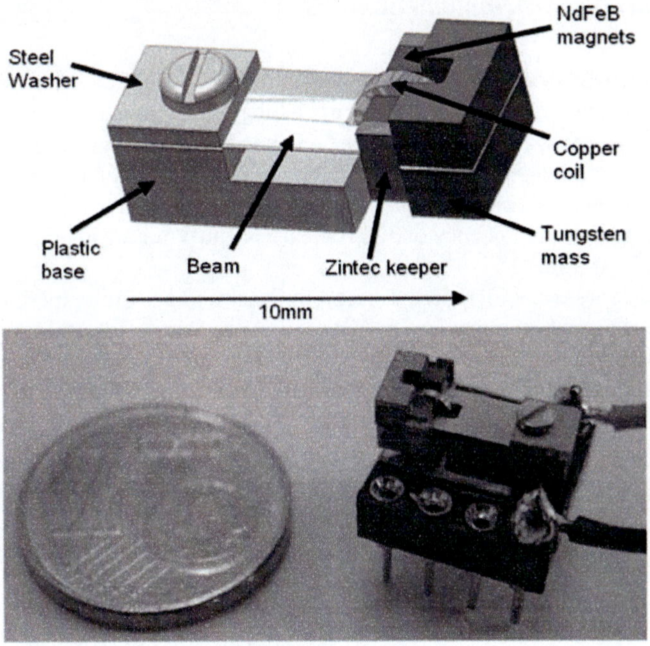

Fig. 19 Cantilever-based electromagnetic micro-generator [31]

circuit pattern comparable (Fig. 14b) [41] when the voltage amplifier is built. Other experiments using a micro-generator, built by Torah et al. [42] focused on vibrations. The structure is based on cantilever structure where the spiral is connected to a the base and four magnets on either side of the spiral form the proven amount (see Fig. 15).

It can be modeled as a large-scale second-order spring damping device [6].

$$mz((t))^{\cdot} + c\_\ p\ z((t))^{\cdot} + k\_\ s\ z(t) + F\_\ em = -\ my((t))$$

In the above equation, z(t) is the relative displacement between the mass and the base, m is the proof mass, ks is the spring stiffness, cp is the parasitic damping factor, Fem is the electromagnetic force and y(t) is the displacement of the base.

According to Faradey's law, the electromagnetic voltage generated in the coil is $v\_em = \Phi(z) \times z((t))^{\cdot}$

where $\Phi(z)$ is the magnetic flux in the coil [5].

8.3 Macro-scale Implementations

El-Hami et al. at the University of Southampton, UK [43] explains a motion electromechanical power generator design, modelling, production and characterisation. The system contains a beam in spring that is connected to one end and that supports a couple of NdFeB magnets. The coil comprises of several turns of enumerated copper cable and is connected inside magnetic poles that are displayed in Fig. 20. The power output at a vibration frequency of 320 Hz reaches 1 mW for a 240 mm^3 volume has been identified (Fig. 21).

Glynne-Jones et al. [46] reported that preliminary findings from experiments made with various magnet combinations by batch tooling and manual assembling. After the analysis of El-Hami et al. [43] two experiments were tested by the Southampton teams focused on a cantilever system, but separate magnets and bobbins were combined; this is shown in Fig. 22. The model was developed to strengthen the magnetic connection between magnets and the coil in order to increase the power output. In general terms,

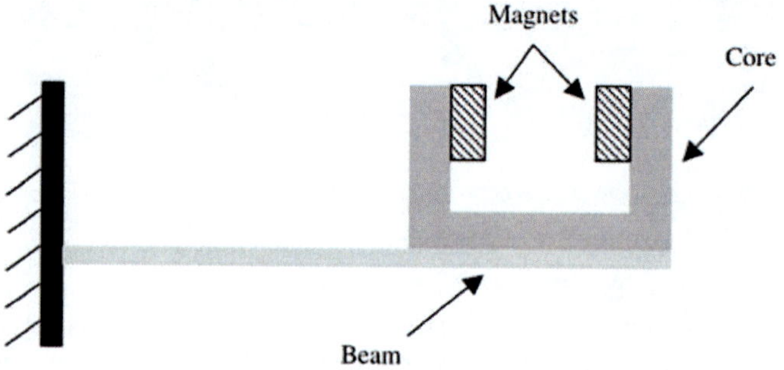

Fig. 20 The electromagnetic generator proposed by El-Hami et al. [44]

Fig. 21 An electromagnetic generator based on fixed coil and the four moving magnets (after Glynne-Jones et al. [45]

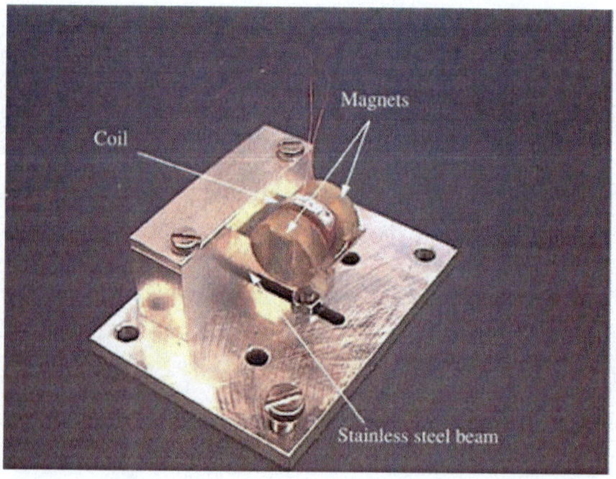

Fig. 22 Schematic of a typical electrostatic energy harvester horizontally vibrating [13]

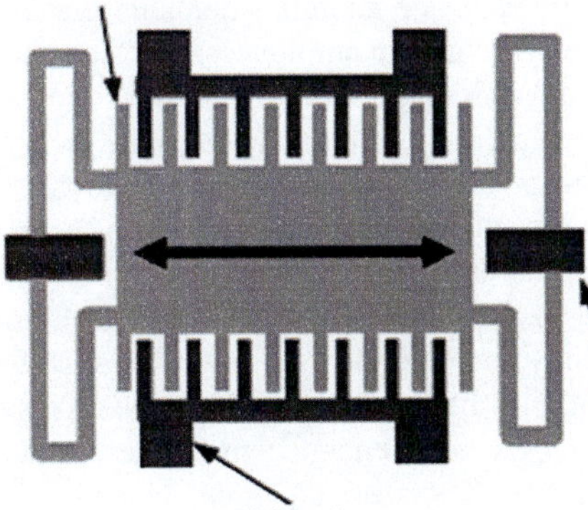

the secondary generator provided a voltage of more than double (1 V opposed to a few hundred millivolts) and therefore more than four times immediate strength for that same input vibration. The researchers analyse the data that demonstrate the reaction of a generator placed on a turbocharger block. Throughout the 1.24 km trip, spontaneous output was estimated at 4 mW, with an average output of 157 μV.

The Massachusetts Institute of Technology's Amirtharajah and Chandrakasan identified an engineered low power Digital Signal Processing (DSP) circuit that involves a traveling electromagnetic coil generator [47]. The turbine resonating frequency was 94 Hz with a maximum output voltage of 180 mV. The authors modelled the generator output and estimated an equivalent of 400 μW from a 2 Cm vibration at a 2 Hz frequency.

Li et al. from Hong Kong Chinese University have constructed an array of magnets and spirals, consisting of an NdFeB and a magnet and a circular pattern magnet tool, which are fixed to the mechanism housings [6]. The instrument is capable of generating 2 V with an oscillation frequency of 64 Hz and comprises a total amount of around 1 cm^3. This leads to excitation amplitude of 10 μW at 100 μm. In the same context, Ching et al. addressed the identical generator developed on a printed circuit board (PCB) [48]. Strength of coil spring leads to a maximum output voltage of 4.4 V and a peak voltage of 830 μW in its oscillation frequency of 110 Hz, and thus powered by a 200 μm torque.

Some researchers have evaluated the performance effectiveness of electromagnetic technology. For instance, Ryu [49, 50] designed a mechanical power harvesting prototype using a speed- bump type of actuator and a rack-and-pinion system. From this experimental study, the highest amount of energy produced was 200 W. In different experimental study Ryu [49, 50] could produce up to 647 W peak power for every moving axle. Ryu continued to conduct an experimental study to test the effectiveness of a new type of electromagnetic energy harvesting mechanism using a cam-arm mechanism to scavenge energy from speeding vehicles. The researchers tested the prototype at different speed and the highest amount of energy generated was 24 W and a voltage of 5 V when a vehicle was passing at 5 mph.

There are three major coupling domains that that harvesting energy through electromagnetic mechanism employees: mechanical, vibration and electric domain. These entire three domains first act on the beam, which results in Fe–Ga beam bending and releases stress in the beam in the form of mechanical domain. Subsequently, the stress induces the Villari effect, which results in magnetic flux. During this phase, the mechanical domain fuses with the magnetic domain, which results in the transformation of mechanical energy into magnetic energy. Any vibration exerted of the magnetic flux leads a Faraday electromagnetic induction mechanism, and subsequently an induction in the electromotive force is produced in the pick-up coils. Meaning that the Fe–Ga energy harvesters scavenges mechanical energy and transforms it into electrical energy in three phases; however, strain and magnetization are the linking effects between the three phases.

9 Energy Harvesting by Electrostatic—Magnetostrictive Material (MsM)

The electrostatic energy harvesting depends on the changing capacitance of the vibration-dependant variables capacitors with mobile electrodes which are separated by a dielectric [51, 52]. The mechanical energy can be converted to electrical energy when a charge is placed on the electrodes. Furthermore, the density of the energy scavenger can be increased by decreasing the spacing of the capacitor, hence facilitating miniaturization. Conversely, the density of the energy harvester can be decreased by decreasing the surface area of the electrodes. Nia et al. [53, 54] provided a comprehensive explanation of the advantages of conversion of both charge and voltage constraint. Theoretically, a slightly more energy could be generated from the voltage constrained system. The electrostatic energy harvesting device attained energy density of 0.23 $\mu W/cm^3$ using the vibration of 2.5 kHz. Roundy et al. [13] established that optimized electrostatic energy harvester to improve the power output density up to 110 $\mu W/cm^3$ at 120 Hz vibration.

Energy harvesting using electrostatic mechanism does not need smart materials and is feasibly embedded with MEMS using relatively mature silicon micromachining methods. Because of the capacitive-based material, electrostatic mechanism produces a slightly high power output of 2– 10 which results in a limited current-supplying capacity. Nonetheless, electrostatic energy harvesting requires the use of external voltage or any other sources of charge for initialization [55]. Moreover, the mechanical constraints stressed on the mobile electrodes are indispensable to prevent electrode shorting.

Recently, magnetostrictive material (MsM) has been considered in the applications of vibration energy harvesting. This energy harvesting mechanism uses Villari effect, in which vibration induces stress on magnetostrictive material which thus results in a change in magnetization of the material. Upon cyclic loading, the magnetization change is converted into electric energy with the help of solenoid placed on the magnetostrictive layer as expressed in the Coulomb's principle.

Researchers have attempted to apply magnetostrictive Terfenol-D in vibration energy harvesting. Particularly, Staley and Flatau [56] operated the Terfenol-D rod in axial mode instead of the commonly used flexural bending mode. These researchers used a bulky dimensional Terfenol-D rod of 1000-turn pick-up and 1500-turn DC actuation coil for producing bias magnetic field. This experimental study produces a maximum energy output of 45 μW at the resonant frequency of 45 Hz and the amplitude of AC output voltage was less than 0.35 V.

10 The Main Properties and Principles

The magnitude between any of the two-point charges is the electrostatic force of repulsion and the attraction between them and it is inversely proportional to the distance multiplied by itself between these point charges and directly proportional to the magnitude value of products of these point charges. Hence, the above statement is defined as the Coulomb's principle where the force acts along the straight line that joins these two-point charges.

The electrostatic generator includes a capacitor that consists of electrically isolated plates that are typically isolated by an insulator, vacuum or the air between them. The plates are charged by the battery of V voltage that creates opposite and equal charges on either of the plates and the source of voltage is disconnected, the Q leads to the charge storage. There is a small separation between the plates in the Micro-Electro-Mechanical Systems (MEMS).

Thus, the capacitance definition of the above-mentioned capacitor is mainly denoted by

$$C = Q/V$$

where C = capacitance in farads.

Q = plate charges in coulombs.

V = plate voltage in volts.

In case of the parallel plate capacitor, is mainly denoted by the

$$C = \varepsilon \frac{A}{d}$$

where ε = permittivity of the material between the plates in F $m^{-1.}$

A = plate area in $m^{2.}$

d = the plate seperation in m.

Considering as the permittivity of free space, the above equation is expressed by the dielectric constant of the insulated materials as $\kappa = \varepsilon/\varepsilon_0$. The equation is

$$C = \kappa\varepsilon_0\frac{A}{d}.$$

Now the parallel plate capacitor attaining the voltage across them is given by,

$$V = \frac{Qd}{\varepsilon_0 A}$$

Therefore, the energy stored in a capacitor is given by,

$$E = 0.5QV = 0.5CV^2 = 0.5Q^2/C$$

Now in the case of perpendicular force being constant between the plates depending upon the plate charges is given by

$$F = 0.5Q2d/\varepsilon A$$

In the similar case, the plate voltage between them is constant and the perpendicular force between the plates is given by

$$F = 0.5\varepsilon AV^2/d^2$$

Thus, the harvested energy is obtained by the work done between the plates against the electrostatic force [6].

11 Mechanisms and Performance Effectiveness

Magnetostrictive energy harvester generates electrical energy from vibration sources in two distinct processes: 1. transforming mechanical energy into magnetic energy through the magneto-mechanical coupling of MsM; 2. converting magnetic energy to electrical energy through the use of electromagnetic coupling on electrical devices and circuits. Duarte and Ferreira [33] described magneto-mechanical coupling using the Stoner-Wohlfarth approximation, in which the MsM is viewed to be a set of non-interacting magnetic domains, as shown in the Fig. 23.

Every magnetic field displays a uniform local magnetization. The bulk magnetization is the total sum of the weight of the local responses, which are dictated by the strain and domain-dependent field orientations. MsM exhibit maximum magneto-mechanical coupling mechanism, which is equal to a 90° field rotation, in the event that the mechanical energy and magnetic energy are at equilibrium [34]. The mechanical strain exhibited in the magnetostrictive kinetic energy harvester is influenced by the source of the kinetic or vibration energy sources. Therefore, a bias in the magnetic domain is used to balance the mechanical energy. However, the bias magnetic field in magnetostrictive vibration energy is often produced by the permanent magnet.

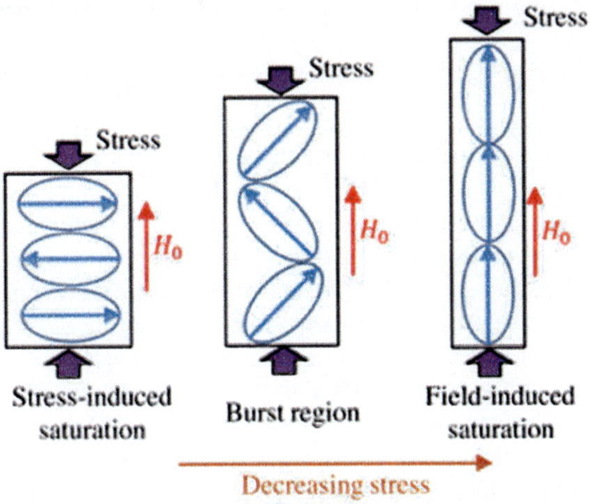

Fig. 23 Magnetic domain rotation in magnetostrictive materials

The classification of electrostatic generators is as follows [57]: In-plane overlap varying, In-plane gap closing and Out-of-plane gap closing. The in-plain mechanism includes overlapping varying, gap closing, patterned varying and full plate varying mechanisms.

Both overlap varying and gap closing forms of 'In-plane' electrostatic generators could generate two capacitances having values of 180° off the phase. The three methods could function one of the two forms that are in charge confined and the voltage confined cycles.

Usually, the voltage confined cycle facilitates additional energy benefits than the charge confined approach. But the energy benefits from the charge confined system could align to the values produced by the voltage confined system by attaching a capacitance parallelly with a capacitor that harvests energy because the energy value would impacted when the capacitance value attached in parallel fashion approaches infinity. The voltages are effectively confined energy harvesting capacitor by the use of a parallel capacitor [58].

According to Roundy [59], the 'In-plane gap closing' mechanism facilitates the availability of the maximum output of the power by having a design that is optimized to accounts the output of a hundred μW cm^{-3}. The 'Out-of-plane gap closing' classification of electrostatic generators mechanism ranks next in production closely succeeded by 'In-plane overlap varying'. However, it has been concluded that the thin dielectric gaps in the generators account for the maximum power generation.

11.1 Introduction to In-Plane Overlap Varying Type and Its Implications

In a study by Meninger et al. [60], designed a form alike the 'In-plane overlap varying' form of electrostatic generator that generates static electricity and power at a continuous current and at a high voltage which was structured in a comb-driven fashion that is a microelectromechanical actuators that are used as a linear actuators and generated 8 μW power output by considering the input frequency of 2.5 kHz. The structure used an integrated circuit process whose voltage value was limited to 8 V. The process of Charging and discharging of an electrostatic generator at some particular points in the cycle of integrated circle process was important in ensuring the use of a technique with sophisticated energy feedback mechanism which triggered the timings according to the energy generated.

Another study by Ma [61] revealed about the charged floating gate generator which generated the power of one μW by considering an optimum load of 58 MΩ for a level of displacement of 5 μm at a frequency of 4.3 kHz which corresponded to an acceleration that is the estimation of the conversion of velocity of a particular object as compared to a given time was of nearly 1825 m s^{-2}. The inertial mass is a parameter measuring the inertial resistance to the acceleration of a body at the time when it responds to all the levels of available forces was estimated to be 2×10^{-7} kg and operating in resonance the damping coefficient β of the device was estimated to be 0.002 kg s^{-1}. An integrated poly silicon floating gate was used in the single wafer device, which was insulated for provision of the required bias voltage. The electron tunnelling was responsible for charging the floating gate and a variable capacitor that generated the power, structured by a mobile gold proof mass and a fixed floating gate. Maximum power of 65 nW could be generated by an array of 20 devices at the frequency of 4.2 kHz attributing to the displacement of 2.2 μm which was equivalent to the level of the acceleration value of 766 m s^{-2}.

11.2 Introduction to In-Plane Gap Closing Type and Its Implications

As explained by Despesse [62], the in-plane gap closing type structure with electrostatic transduction that is normally tested with the variable capacitor, having a high-electrical damping that is used in those systems in which the power is to be damped in the internal parts of the system, function operated over a wide range of low-frequencies of order < 100 Hz. The structural form was selected as an 'in-plane gap closing mechanism' structure having a fixed charge-confined cycle where the electrostatic force was found to be linearly proportional to the inertial mass displacement alike the mechanical spring force which for all displacements of the inertial masses were found to be balanced. This structural form shows the relationship between electrostatic force and mechanical spring force for the inertial mass displacement in the in-plain closing mechanism. Selection of the electrical stiffness that shows the electrical supply's origin impedance close to the value of mechanical stiffness could be associated with high-electrical damping effect. As the stiffer it would be the low impact will be there on the voltages due to changing loads as it is true for the harmonic and power frequencies.

A device with volumetric dimensions of 18 $cm^2 \times 1$ cm and an inertial mass of 0.104 kg, electrically discharge and was machined from the Tungsten to produce a scavenging power of value 1052 μW having the vibration of amplitude, 90 μm at the frequency of 50 Hz which corresponded to the acceleration value of 8.8 ms^{-2} which could be attributed to 60% of scavenging efficiency, with the losses of scavenging efficiency is being accounted by the charging/discharging losses of the electricity discharged and the losses during transduction. A silicon microstructure that is a small shape of a material revealed by an optical microscope of a magnification of more than twenty-five with alike geometric dimensions, 81 $mm^2 \times 0.4$ mm with an inertial mass of 2×10^{-3} kg being excited with the vibration of amplitude, 95 μm at the frequency of 50 Hz and estimated to produce 70 μW of scavenging power.

11.3 Introduction to Out-of-Plane Gap Closing Type and Its Implications

In a study by Tashiro [63], a honey-comb structure as shown in Fig. 24 which was structured up by folding a polyester film strip having a surface dried out of aluminium. The Two sheets of the polyester film with dimensions of five micro meter thickness, thirty millimetre width and five meter in length, were combined with each other with the aid of a double-sided adhesive tape that is use to join the combination applied at 5 mm distances to check the outcomes.

A variable capacitor was formed by combined the sheets using adhesive tape and the resultant accommodated 20 cells on every layer and 100 cells resulted in 50 layers. An inertial mass attached to check the dispersion of one of the acrylic boards to which the capacitor was suspended using 12 springs. The resonator exhibited the spring constant to have the value of 1100 Nm^{-1} with a total mass of 0.78 kg resulting

Fig. 24 Variable capacitor of honeycomb type [64]

in a resonant frequency that is the oscillation of a particular system at its unforced or natural resonance was with value of 6 Hz. Resonance happens when a particular system is capable of storing and transferring energy among various storage models such as potential energy and kinetic energy as in the case of a simple pendulum. Most of the systems only have one resonant frequency along with various harmonic frequencies that gets eventually lower in the amplitude as they step away from the middle or centre.

The capacitor that is an equipment that keep electric energy in the surrounding of an electric field and a capacitor is an electronic equipment with the presence of two terminals was charged in the beginning to 45 V and the stimulated movement shook the generator to produce a motion of wall in left ventricle of a canine heart which lead to the generation of a mean power value of 36 mw or the 15 ma at 2.4 V, having the peak power that shows the highest values as high as 500 mw. The movements exhibited by accelerometers were estimated to be about 1 m s^{-2} at the frequency level of 6 Hz.

The previous experiments performed by the same group [65] demonstrated that after the initial charging to the voltage of 24 V, the power generation of 58 μW from the simulated and measured movements of a goat heart were reported. A generator having honeycomb like structure with an initial capacitance that shows the percentage of conversion in an electric charge of a system as compared to the conversion of the change in the electric potential of 32 nF that will vary to 200 nF was used and had the resonant frequency of 4.76 Hz. Being driven by the monitoring of the harmonic frequency that is the current or voltage at a frequency of the heart, ranging from 1 to 2 Hz, the study of motion showed the frequencies up to 10 Hz to be found. The study revealed that in context to the inertial mass of 1.2 kg and the resonator spring constant was found to have the value of 570 Nm^{-1}. The resonator that is a system or device that shows residence behaviour was eventually tuned to the application of vibrational frequencies by triggering the values of the masses.

In the research by Miyazaki [66], an illustration of an 'out-of-plane' cantilever-based generator having a base capacitor of value 1 nF with a variable capacitance that can be changed repeatedly by electronically or mechanically, ranging between 30 pF and 350 pF. The device with the resonance frequency of 45 Hz and a quality factor of 30 that shows the dimensionless parameter of an under damped resonator or oscillators was very well tested with a displacement value of 1 Micrometre at maximum frequency of 100 Hz. The wall was revealed to possess an acceleration value of 0.08 ms^{-2} recording the power of 120 nW.

11.4 Introduction to Coulomb Force Parametric Generator and Its Implications

Other researches by Miao [106] and Mitcheson [107] detailed about non-resonant electrostatic generators which were fabricated to operate for applications wherein the amplitude associated with the external motion exceeded the maximum value of internal displacement by the generator with the proof mass of 10–4 kg which was fabricated to move only when the maximum acceleration in the cycle was attained. The device operated in four phases which could be stated as the primary stage is prime, secondary is flight, and third is wait and finally discharge. The three-wafer device (glass, silicon and glass) structured by the inclusion of mobile silicon-proof mass affixed to silicon frame by a polyimide suspension, at the bottom plate containing the counter electrode and the discharge studs whose top plate had the discharge contacts.

During the initial prime phase, the proof mass is held against the charging studs by the externally provisioned the initial voltage of nearly 100 V. The type priming voltage induced a force of value little lesser than that of inertial force which was generated by the maximum amount of the acceleration in the application to be addressed for future. The flight phase was accounted when the mass from the charging plate separated towards the drift towards the discharge contacts with a constant charge when the inertial force exceeded the holding force. The work done during the separation enhanced the capacitor voltage till the movable plate made contact with the discharging circuit that occurred at the maximum displacement.

The stored energy that is of 0.5 CV2 increased along with the ratio of the percentage change in initial capacitance to final capacitance thereby, by minimizing the gap between the movable plate and the counter electrode, the initial capacitance could be maximized, which also incorporated the effect of declining the prime voltage necessitated by a holding charge capacitor with the spacing of approximately 6 μm stating the capacitance of force. The device dimensioned to be 20 mm $\times$ 25 mm by 1.5 mm thickness and the capacity of 150 pF. The value of discharge capacitance was revealed to 5.5 pF. The device generated an output voltage of maximum value, 220 V which corresponds to a net power of comparatively of 120 nJ produced as per cycle for accelerations of the order of 50 m s^{-2}. The theory estimates the energy value of 2.6 μJ per cycle, which was equivalent to 80 μW, to be achievable for the device sized at frequency of 30 Hz with an acceleration of 10 m s^{-2}. The tilting motion of the inertial masses and parasitic capacitances were investigated to be the practical hurdles.

Tilting problem could be minimised by sufficient flexibility in the movement of required direction and appropriate stiffness in the direction of undesired movement. In comparison with the comb like structure, it facilitates a higher level of the initial capacitance value and also a greater capacitance ratio between the two end conditions of the capacitor by minimizing the discharge capacitances, by this process the stored energy could be increased to its optimum level. However, for the majority of energy sources is to be discharged in the discharge capacitor that is source of red-hot light bulb that can permit the level of voltage present in the capacitor, it must exceed the parasitic capacitor values. At low parasitic capacitance values and low leakage values, the extraction circuitry thus, has potential to handle high voltage values.

12 Comparison of Mechanisms

This section gives a comprehensive summary comparison of the energy harvesting technologies that have been reviewed in this report. This section will do the comparison by looking at the advantages and disadvantages of every harvesting technology reviewed and also summarises the power density output and maximum power output for these technologies. Most of the energy scavenging technologies reviewed in this study exhibit the potential of harvesting energy from walking. Evidently, some energy harvesting technologies are more feasible and developed than others. On the other hand, utilise any random load and converts into maximum useful energy is still not present.

piezoelectric materials produce electrical energy from vibration or strains caused by passing objects. Normally, this technology generates a relatively lower power voltage compare with other harvesting technologies [67]. According to the past researchers, piezoelectric technology can supply energy to power low—power devices, such as LED lights and electric sensors. The past literature suggests that the amount of energy generated using piezoelectric effect and its efficiency relies on the type of the material used, the shape, dimension and the circuitry used [67]. Nonetheless, other researchers have reported that the efficiency and effectiveness of the piezoelectric technology do not solely depend on the characteristics of the device, but also certain external factors, such as the magnitude and the frequency of the moving object. Additionally, the vigour and the frequency of the passing object which results in higher excitation frequencies, which impact the amount of energy produced. Nevertheless, piezoelectric energy harvester has some drawbacks. For instance, this technology has a low energy output hence they can only power devices which require low power. Additionally, piezoelectric materials need to be placed close to the pavement surface where the strain is maximum [68]. Again, while piezoelectric materials have a stronger and higher compressive strength, they have low tensile strength; hence they can easily be destroyed under eccentric strains or bending strains. Finally, piezoelectric materials require constant maintenance which may not be cost-effective. A study conducted by [44] to investigate the feasibility of piezoelectric energy technology in Tokyo, Japan and reported that the technology was effective in energy harvesting up to sometimes when output began to decline. The researcher discovered that the system was not well maintained, which is why the output declined after some time.

On the other side, electromagnetic energy harvester is capable of producing a large amount of energy independent of the prevailing weather condition. According to the reviewed body of literature, the electromagnetic energy harvester technology with different energy generation mechanism exhibits a high potential to be applied in kinetic energy harvesting and provides cost-effective, renewable sources of energy in crowded areas where power may be limited or unavailable. Nonetheless, The common attributes between the electromagnetic generation and piezo harvesting are that they can be applied in the wearables. For instance, electromagnetic generation through inductive generators can be used in shoes or other places where the motion is highly defined. The piezo harvesters, on the other hand, are excited by the motion of limbs and not necessarily by vibration where the amount of energy generated will be a function of the velocity of joints and bending angles. Further studies review that the above two methods can be heavily be implemented in wearables in order to harvest body motion. The limitations could be the high masses that are likely to make applications uncomfortable.

Of all the three mechanisms, electrostatic generation is the one which is least exploited because of the sophisticated production mechanism involved and complexity of the equipment required in the process. However, they have the potential to be employed in the wearables just as is the case with the piezo mechanism. The cost also plays a very vital role in the power generation method employed.

The difference between the three methods is the equipment and materials necessary. The piezo, for instance, need to use quartz crystal, electrostatic require a capacitor and electromagnetic method heavily relies on magnets for power generation. However, the common denominator is that they are required excitation through movement. More often than not, the challenges most of the involve looking for means to create motion. While running waters has been heavily employed in the electromagnetic harvesting, it remains very unpractical for the piezo and electrostatic generators. Piezo generators are favoured by wearables especially on the feet where both motion and vibrations are generated.

13 Conclusion

Energy sources require expansion in order to increase their reliability and as well as the total output. Among the most feasible ways of expansion is by use of multiple methods of harvesting. As was discussed in the previous section of this study, piezo method, electromagnet and electrostatic harvesting. While piezo harvesting is quite expensive, it easy to harvest a large amount of energy by installing plates in the subways and other areas where people pass through in large numbers.

Use of electromagnetic methods has received wide application because of its easy implementation and long history of its reliability. As a matter of factor, almost 90% of energy is generated using this method. However, its carbon footprint, as well as direct environment degradation caused by this method, calls for alternative sources that will be both reliable and environmentally friendly.

Though energy storage for renewable sources remains a challenge which reduces the dependability and reliability of sources in non-peak seasons, alternative ways of energy generation remain unexplored. Additionally, utilise any random load and converts into maximum useful energy is still not present. Further studies need to be done especially on the piezo energy generation since it has unlimited potential to supply to the grid an affordable way in both the short and long run. While all the above methods remain feasible, they should only be adopted if their cost of production is low and the quality of electricity generated is high.

Further studies on energy generation method should be conducted mainly on the piezoelectricity generation method as well as the electrostatic generation for better tomorrow. Every method of power generation has its own cost, equipment, machinery, quality and production capacity and the country's economy plays an important role for the selection of both suitable and feasible method inters of quality, cost and capacity.

References

1. Patlins A, Hnatov A, Arhun S, Dzyubenko O (2019) Design and research of constructive features of paving slabs for power generation by pedestrians. Transp Res Procedia 40:434–441
2. Bischur E, Schwesinger N (2012) Energy harvestingfrom floor using organic piezoelectric modules. In: 2012 power engineering *and* automation conference, pp 1–4
3. A review on energy harvesting from road. Andriopoulou Symeoni M.Sc. Environmental Engineering & Sustainable Infrastructure, pp 1–39
4. EH Systems No Title
5. Ka TJ, Wang L (2011) Modelling, performance optimisation and automated design of mixed-technology energy harvester systems.
6. Science M (2006) Energy harvesting vibration sources for microsystems applications
7. Gholikhani M, Roshani H, Dessouky S, Papagiannakis AT (2020) A critical review of roadway energy harvesting technologies Appl. Energy 261:114388
8. Maghsoudi Nia E, Wan Abdullah Zawawi NA, Mahinder Singh BS (2019) Design of a pavement using piezoelectric materials. Materwiss Werksttech 50(3):320–328
9. Science M (1977) Thick-film sensors : past, present and future
10. Lovinger AJ (1983) Ci ence (1) 220(4602):1115–1122
11. Baudry H Screen printing piezoelectric devices. pp 71–74
12. Singh S, Gupta VK, Mukherjee S (2018) Piezo electric based energy harvester embedded in shoe for wearable electronics. Mater Phys Mech 37(2):159–167
13. Richards CD, Anderson MJ, Bahr DF, Richards RF (2004) Efficiency of energy conversion for devices containing a piezoelectric component
14. Nakajima T, Okaya K, Ohta K, Furukawa T, Okamura S (2011) Performance of piezoelectric power generation of multilayered poly(vinylidene fluoride) under high mechanical strain. Jpn J Appl Phys50(9)
15. Towards a piezoelectric vibration-powered microgenerator. 148(2)
16. Academy R Performance optimization of piezoelectric
17. M'Boungui G, Adendorff K, Naidoo R, Jimoh AA, Okojie DE (2015) A hybrid piezoelectric micro-power generator for use in low power applications. Renew Sustain Energy Rev 49:1136–1144
18. Vullers RJM, van Schaijk R, Doms I, Van Hoof C, Mertens R (2009) Micropower energy harvesting. Solid State Electron 53(7):684–693
19. Xu TB et al (2013) Energy harvesting using a PZT ceramic multilayer stack. Smart Mater Struct 22(6)
20. Shenck NS, Paradiso JA (2001) Energy scavenging with shoe-mounted piezoelectrics electricity from the forces exerted on a shoe during walking : A, pp 30–42
21. Renaud M, Fiorini P, Van R, Renaud M, Fiorini P, Van Hoof C (2007) Optimization of a piezoelectric unimorph for shock and impact energy harvesting
22. Kiziroglou ME, Yeatman EM (2012) Materials and techniques for energy harvesting Woodhead Publishing Limited
23. Moon FC (1978) Problems in magneto-solid mechanics, vol. 4. Pergamon Press Inc

24. Challa CR, Prasad MG, Shi Y, Fisher FT (2008) A vibration energy harvesting device with bidirectional resonance frequency tunability. Smart Mater Struct 17(1)
25. Fajardo S, García-Galvan FR, Barranco V, Galvan JC, Batlle SF (201) We are intechopen, the world' s leading publisher of open access books built by scientists, for scientists TOP 1%. Intech, vol. i, no. tourism, p 13
26. Roundy S, Wright PK, Rabaey J (2003), A study of low level vibrations as a power source for wireless sensor nodes. 26:1131–1144
27. Marzencki M et al (2005) A MEMS piezoelectric vibration energy harvesting device. PowerMEMS 45–48
28. Jeong SJ, Kim MS, Song JS, Lee HK (2008) Two-layered piezoelectric bender device for micro-power generator. Sens Actuat A Phys. 148(1):158–167
29. Gu L (2011) Low-frequency piezoelectric energy harvesting prototype suitable for the MEMS implementation. Microelectron J 42(2):277–282
30. Kim I, Joo H, Jeong S, Kim M, Song J (2010) Micro power generation of PMN-PZT triple morph cantilever for electric harvesting devices. Phys Status Solidi Curr Top Solid State Phys 7(9):2331–2335
31. Eduard Noel du Toit Modeling and design of MEMS piezoelectric vibration energy harvester
32. Duffy M, Carroll D (2004) Electromagnetic generators for power harvesting, pp 2075–2081
33. Williams CB, Yates RB (1996) s S oRs actuators analysis of a micro-electric generator for microsystems dF-1 z(O ii, vol 52, pp 8–11
34. Optical harmonic upconversion for generation in bidirectional broadba 33(22):1883–1884
35. Kulah H, Najafi K An electromagnetic micro power generator. 237–240
36. Gorlatova M, Sarik J, Grebla G, Cong M, Kymissis I, Zussman G (2014) Movers and shakers : kinetic energy harvesting for the internet of things
37. Fondevilla N, Morante JR, Esteve J (2005) Design of electromagnetic inertial generators for energy scavenging applications
38. Scherrer S, Plumlee DG, Moll AJ (2005) Energy scavenging device in LTCC materials. In: 2005 IEEE work. microelectron. Electron devices, 2005. WMED'05, pp 77–78
39. Beeby SP et al Micromachined silicon generator for harvesting power from vibrations. pp 2–5
40. Yan J, Montero H, Akhnoukh JM, De Vreede A, Burghartz LC (2005) An integration scheme for RF power harvesting. Proc STW Ann (1):64–66
41. Amirtharajah R, Wenck J, Collier J, Siebert J, Zhou B (2006) Circuits for energy harvesting sensor signal processing. Proc Des Autom Conf 639–644
42. Torah R, Beeby SP, Tudor MJ, O'Donnell T, Roy S (2006) Development of a cantilever beam generator employing vibration energy harvestin. Energy 1:4–7
43. El-hami M et al (2001) Design and fabrication of a new vibration-based electromechanical power generator. Sens Actuat A Phys 92(1–3):335–342
44. Yang B, Lee C (2010) Non-resonant electromagnetic wideband energy harvesting mechanism for low frequency vibrations. 961–966
45. Xue Z, Li L, Ichchou MN, Li C (2017) Hysteresis and the nonlinear equivalent piezoelectric coefficient of MFCs for actuation. Chinese J Aeronaut 30(1):88–98
46. Glynne-Jones P, Tudor MJ, Beeby SP, White NM (2004) An electromagnetic, vibration-powered generator for intelligent sensor systems. Sens Actuat A Phys 110(1–3):344–349
47. Generation VP, Amirtharajah R, Chandrakasan AP (1998) Self-powered signal processing using. IEEE J Solid-State Circ 33(5):687–695
48. Ching NNH, Wong HY, Li WJ, Leong PHW, Wen Z (2002) A laser-micromachined multi-modal resonating power transducer for wireless sensing systems. Sens Actuat A Phys 97–98:685–690
49. Ryu J (2017) Energy Fuels 1(10)
50. Sharma K, Kohli P, Pikhan S Footstep power generation using piezo-electric transducers
51. Zhao HD, Ling JM, Fu PC (2013) A review of harvesting green energy from road. Adv Mater Res 723:559–566
52. Ahmad S, Abdul Mujeebu M, Farooqi MA (2019) Energy harvesting from pavements and roadways: a comprehensive review of technologies, materials, and challenges. Int J Energy Res 43(6):1974–2015

53. Nia EM, Zawawi NAWA, Singh BSM (2018) A review of walking energy harvesting using piezoelectric materials. IOP Conf Ser Mater Sci Eng 291(1)
54. Triono AD et al (2018) Utilization of pedestrian movement on the sidewalk as a source of electric power for lighting using piezoelectric censors. IEEE Int Conf Intell Transp Eng ICITE 2018:241–246
55. Williams CB, Yates RB (1996) s S oRs ACTUATORS Analysis of a micro-electric generator for microsystems dF-1 z (O ii, vol 52, pp 8–11
56. Staley ME, Flatau AB (2005) Potential of Terfenol-D and Galfenol, no May 2005, 2020
57. Roundy S, Wright PK, Pister KSJ (2002) Micro-electrostatic vibration-to-electricity converters. ASME Int Mech Eng Congr Expo Proc 487–496
58. 43551823-MIT.pdf
59. Yildiz F, Zhu J, Pecen R, Guo L (2007) Energy scavenging for wireless sensor nodes with a focus on rotation to electricity conversion. ASEE Annu Conf Expo Conf Proc
60. Meninger S, Mur-miranda JO, Amirtharajah R, Chandrakasan AP, Lang JH (2001) Vibration-to-electric energy conversion 9(1):64–76
61. Wong M, Rufer L (2005) Montreux, Switzerland, 01–03 June 2005 Dynamic simulation of an implemented electrostatic power micro-generator, pp 1–3
62. Despesse G et al (2012) Fabrication and characterization of high damping electrostatic micro devices for vibration energy scavenging to cite this version : HAL Id : hal-00748983
63. Tashiro R, Kabei N, Katayama K, Tsuboi F, Tsuchiya K (2002) Development of an electrostatic generator for a cardiac pacemaker that harnesses the ventricular wall motion. J Artif Organs 5(4):239–245
64. Mizuno M, Chetwynd DG (2003) Investigation of a resonance microgenerator
65. Kubo M, Matsuda H, Tanaka M, Kimura Y, Okuda H, Higashino M, Tani T (1986) NII-electronic library service. Chem Pharm Bull 34(1):430–433
66. Miyazaki M et al (2003) Electric-energy generation using variable-capacitive resonator for power-free LSI: efficiency analysis and fundamental experiment. Proc Int Symp Low Power Electron Des 193–198
67. Foot step power generation using piezoelectric material. 4(10):2503–2507
68. Duarte F, Ferreira A (2016) Energy harvesting on road pavements: state of the art. Proc Inst Civ Eng Energy 169(2):79–90

Damage Evaluation of Free-Free Beam Based on Vibration Testing

Duong Huong Nguyen, Viet Long Ho, Thanh Bui-Tien, Guido De Roeck, and Magd Abdel Wahab

Abstract Damage can be detected by vibration responses of a structure. Damage changes the modal properties such as natural frequencies, mode shapes, and damping ratios. Natural frequency is one of the most frequently used damage indicators. In this paper, the natural frequency is used to monitor damage in a free-free beam. The modal properties of the intact free-free beam are identified based on 15 accelerometers setup. A finite element model is used to model the free-free beam. Three models are considered: beam (1D), shell (2D) and solid (3D). The numerical models are updated based on the first seven bending natural frequencies. The free-free beam is damaged by a rectangle cut. The experiment is re-setup and the model properties of the damaged beam are re-identified. The cuttings are modeled in the numerical model. The first seven numerical bending natural frequencies of the damaged beam are compared with the experimental ones. The results showed that the 1D beam element model has the highest differences, while the 2D and 3D models have approximately the same results. Therefore, the 2D representation can be used to model the damaged beam for fast computation.

Keywords Structural health monitoring (SHM) · Damage detection · Vibration-based method · Free-free beam · Finite element method (FEM)

D. H. Nguyen (✉) · V. L. Ho · M. Abdel Wahab
Soete Laboratory, Department of Electrical Energy, Metals, Mechanical Constructions and Systems, Faculty of Engineering and Architecture, Ghent University, Ghent, Belgium
e-mail: huongduong.nguyen@ugent.be

D. H. Nguyen
Department of Bridge and Tunnel Engineering, Faculty of Bridge and Road, National University of Civil Engineering, Hanoi, Vietnam

V. L. Ho · T. Bui-Tien
Department of Bridge and Tunnel Engineering, Faculty of Civil Engineering, University of Transport and Communications Campus in Ho Chi Minh, Ho Chi Minh, Vietnam

G. De Roeck
Department KU Leuven, Department of Civil Engineering, Structural Mechanics, 3001 Louvain, Belgium

M. Abdel Wahab (ed.), *Proceedings of the 8th International Conference on Fracture, Fatigue and Wear*, Lecture Notes in Mechanical Engineering,
https://doi.org/10.1007/978-981-15-9893-7_3

1 Introduction

Structural Health Monitoring (SHM) requires techniques that could detect and locate damage in structures. Vibration-based SHM has based on the principle that damage changes the modal properties of the structures [1]. Some modal properties such as natural frequencies, mode shapes, modal curvatures, damping ratios, were used to detect damage [2–5]. Uniaxial/triaxial accelerometers or velocity meters are traditionally used to collect data from the vibrating structures. Analysis of these data provides the modal properties of the structures. Some recent researches combined the vibration-based method and machine learning algorithm and recognized the effectuation [6, 7]. The Finite Element Method (FEM) was used to get the training data for the machine learning network [8, 9]. The FE model is updated using modal properties of the intact structure. Whether this FE model updating could produce detectably the modal characteristic changes of the damaged structure, as well as intact structure, is a challenging question.

Beam is commonly used in the field of mechanical and structural engineering. Beam is a structural element, which is designed to resist bending, widely used in bridges, buildings, rooftops, etc. In this paper, three FE models in ANSYS: beam (1D), shell (2D) and solid (3D) are considered. The natural frequencies of the intact and damaged beam obtained using FEM are compared with experimental ones (Sects. 2 and 3). The conclusion and discussion about these three FE models are presented in Sect. 4.

2 Steel Beam Experiments

2.1 Experimental Setup

A steel beam, having dimensions of 1 m long × 70 mm wide × 10 mm thick, density of 7820 kg/m^3 and the free-free boundary condition was used for the experiment. On the top of the beam, fifteen accelerometers were attached with sensitivity ranges from 10.13 to 10.50 mV/m/s^2. A hammer was used as an excitation force. A laptop and NI equipment were used to record the vibration data. Figure 1 shows the experimental setup in the laboratory and Table 1 shows the location of the accelerometers. The locations in Table 1 are the distance from the accelerometers to the left end of the beam.

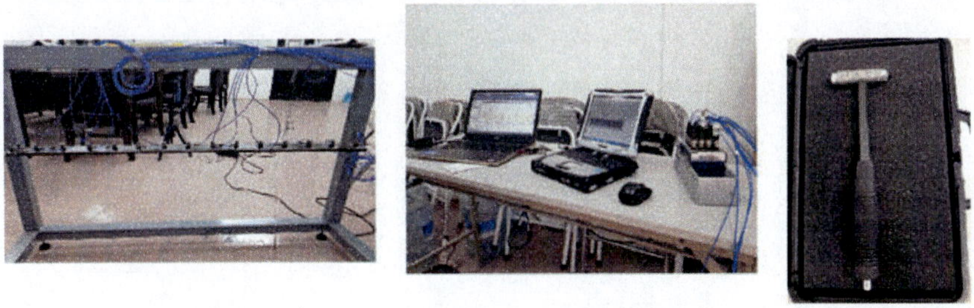

a. Tested beam b. Computers and NI device c. Hammer

Fig. 1 Experimental setup

Table 1 Accelerometers location

Location	1	2	3	4	5	6	7	8
No. (m)	0.01	0.08	0.15	0.22	0.29	0.36	0.43	0.5
Location	9	10	11	12	13	12	15	
No. (m)	0.57	0.64	0.71	0.78	0.85	0.92	0.99	

2.2 *Experimental Results*

Intact Beam

To analysis the collected data, covariance-driven stochastic subspace identification algorithm (SSI-COV) was used thanks to less time-consuming of computation and accuracy. In general, physical modes and spurious modes are distinguished by using a stabilization diagram. A clear stabilization diagram implies that the physical modes are present with consistent modal parameters such as frequency, damping ratio, and mode shapes. For this reason, some preset stabilization criteria are 1% for frequency, 5% for damping ratio and 1% for mode shape vectors. The stable poles in the stabilization diagram should be selected first (Fig. 2). The first seven bending modes were identified based on the stabilization diagram as shown in Fig. 3 and presented in Table 2.

Damaged Beam

The beam was damaged by a cutting on the edge. There are two scenarios. Scenario 1: cut on the left side of the beam, the dimension of the cutting: 5 × 12.5 (mm), and scenario 2: cut on both sides of the beam, the dimension of the cutting: 5 × 12.5 (mm) and 5 × 13.3 (mm). Figure 4 shows the cutting places on the beam. The vibration data was analyzed with the same method as the data from the intact beam. Figure 5 illustrates the stabilization diagram, Fig. 6 plots the mode shapes and Table 3 presents the natural frequencies. From Fig. 6, we can see that the mode shapes of the two damaged scenarios are almost the same, whereas the frequencies show a small difference between the intact beam and the two damaged beams.

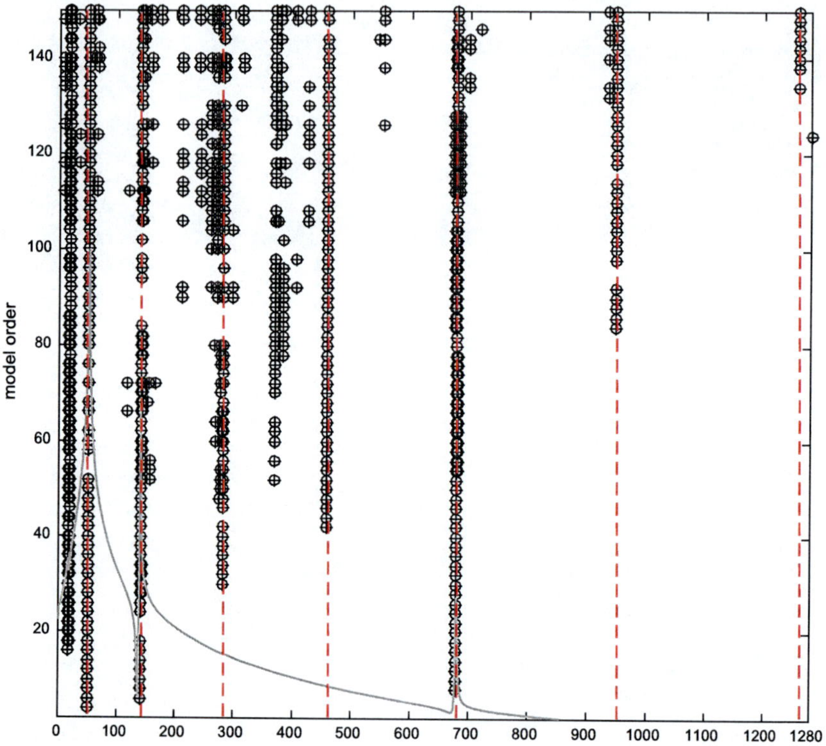

Fig. 2 Stabilization diagram obtained from experiment of intact beam

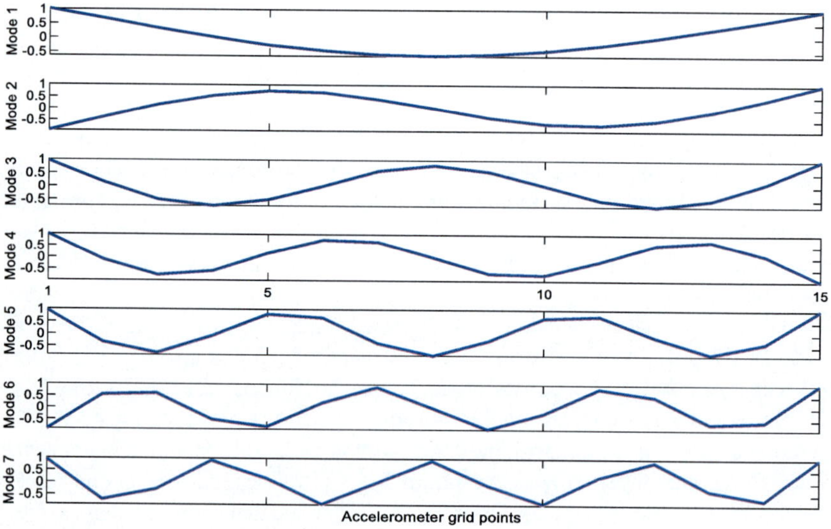

Fig. 3 The first seven bending mode shapes of the intact beam

Table 2 Natural frequencies from measurements for intact beam

Mode	1	2	3	4	5	6	7
$f(Hz)$	50.83	140.40	274.74	456.94	678.90	946.23	1259.75

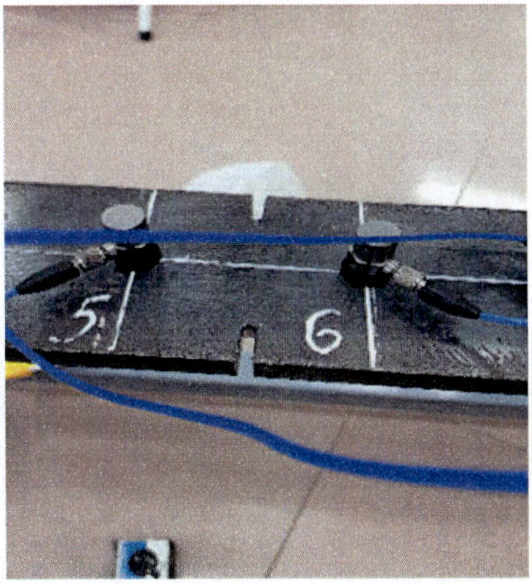

Fig. 4 The cutting places in the damaged beam

3 Numerical Model

The free-free beam structure having dimensions of $1000 \times 70 \times 10$(mm) was modeled in ANSYS. The steel density is = 7820Kg /m^3, and Young's modulus $E = 2.00 \times 10^{11}$N/m^2.

Three types of elements i.e. beam, shell and solid elements, were employed to construct the FE models for modal analysis of the free-free beam. The intact beam and two damaged scenarios were considered.

3.1 Beam Model (1D Model)

The steel beam was modeled by using BEAM189 in ANSYS. It is a quadratic beam element in which each node has six degrees of freedom (DOF), i.e. three translations and three rotations. For the intact beam, each element has a length of 0.07m, and a rectangular cross-section of (70 × 10)mm. For the damaged beam, the damaged elements were modeled by a reduced width of the cross-section. Dimensions of the cross-section are (57.5 × 10)mm and (44.2 × 10) mm for scenario 1 and scenario 2, respectively. The models of the intact and damaged beam are illustrated in Fig. 7.

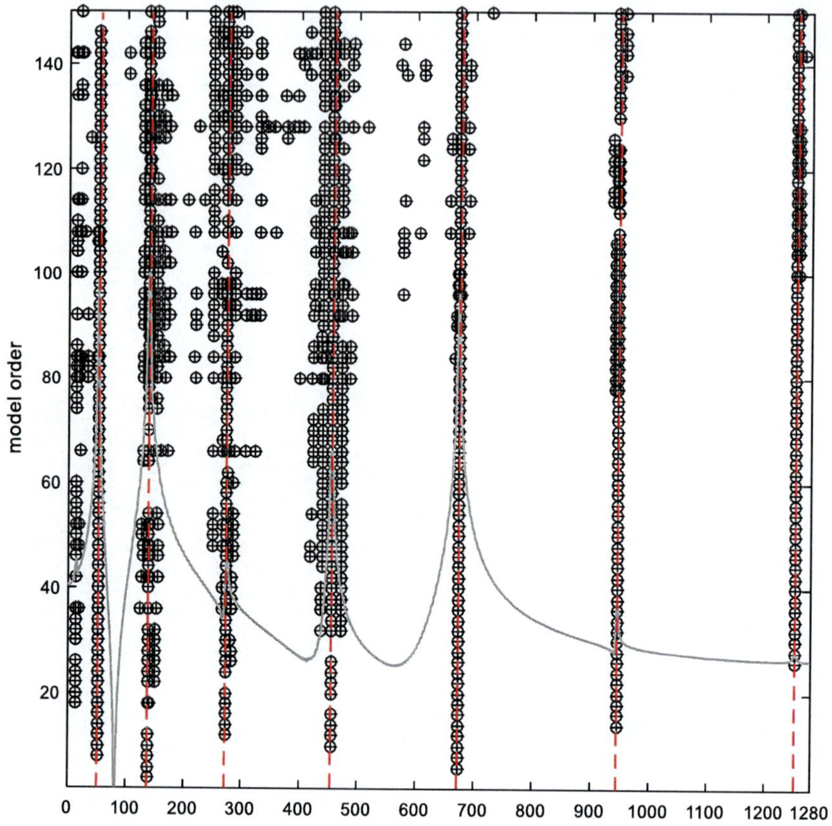

Fig. 5 Stabilization diagram obtained from experiment of damaged beam

3.2 *Shell Model (2D Model)*

A four-node element, SHELL181, was the next element used to identify the vibration of the beam. This element is suitable for analyzing moderately-thick shell structures as the beam in our study. It has six DOFs at each node: three translations in the *x*, *y*, and *z* directions, and three rotations about the *x*, *y*, and *z* axes. Dimensions in the shell model are the same as in the beam model, as mentioned above. The thickness of the surface is defined as 10 mm. The width of the surface is 70 mm. The mesh structure and the FE model of this beam are shown in Fig. 8. The damage is defined by reducing the width of the shell element. The numerical models of the two damaged scenarios are shown in Fig. 8.

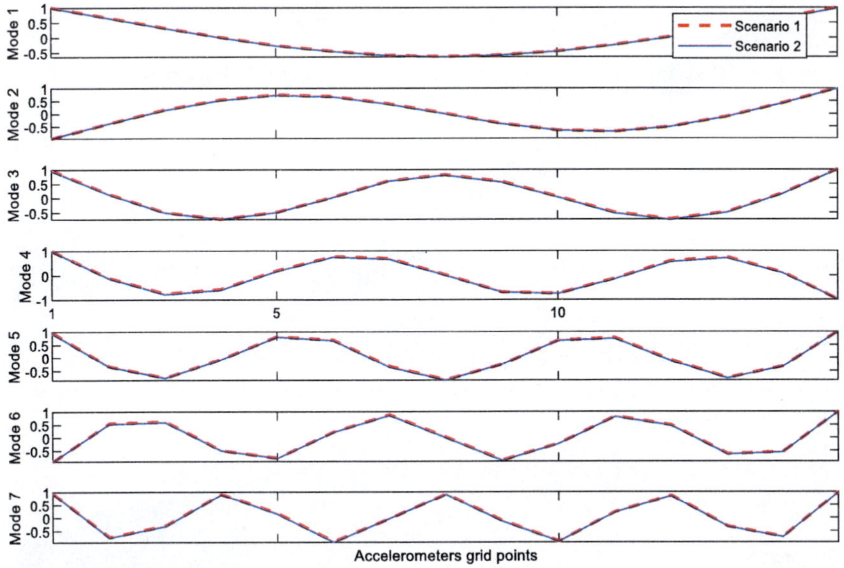

Fig. 6 The first seven bending mode shapes of the damaged beam

Table 3 Frequencies from measurement for damaged beam (Hz)

Mode	1	2	3	4	5	6	7
Scenario 1	50.65	139.69	273.77	456.38	675.99	945.65	1256.19
Scenario 2	50.36	138.64	273.53	454.60	672.58	944.56	1252.65

Fig. 7 The beam element model of the intact and damaged beam

3.3 *Solid Model (3D Model)*

For further modal analysis, a solid model was built with SOLID185. This type of element is often used for 3D modeling of solid structures. The element is defined by eight nodes with three translations at each node. Figure 9 shows the beam model using a solid element with grid meshing.

Fig. 8 The shell element model of intact and damaged beam: **a** Intact beam, **b** damage scenario 1 and **c** damage scenario 2

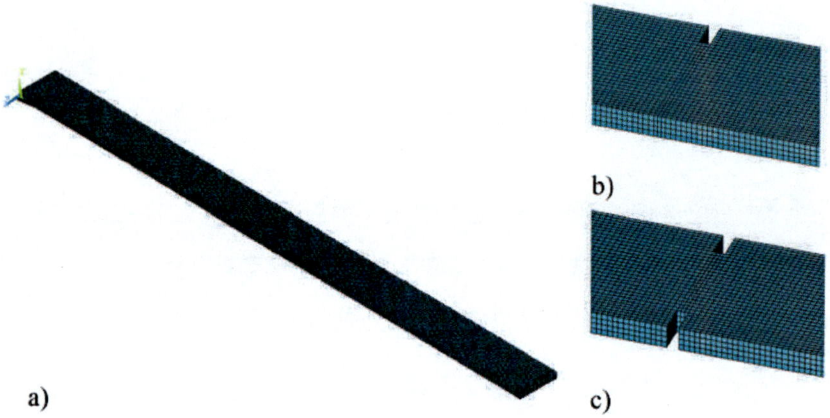

Fig. 9 The solid element model of intact and damaged beam. **a** Intact beam, **b** damage scenario 1 and **c** damage scenario 2

4 Results and Discussion

The results from the experiments showed that, when the damage occurs, the natural frequencies of the beam reduced, whereas the change in mode shapes could not be realized directly. Therefore, only natural frequencies are used for a discussion in this section.

For numerical model, the number of elements in each model affects the natural frequencies. Figure 10 illustrates the frequencies extracted from the three FE models for the first five modes. The natural frequencies converge when the number of elements in the numerical model increases. Moreover, the 1D model uses less number of elements than the 2D model. The convergence curve of frequency from the 1D

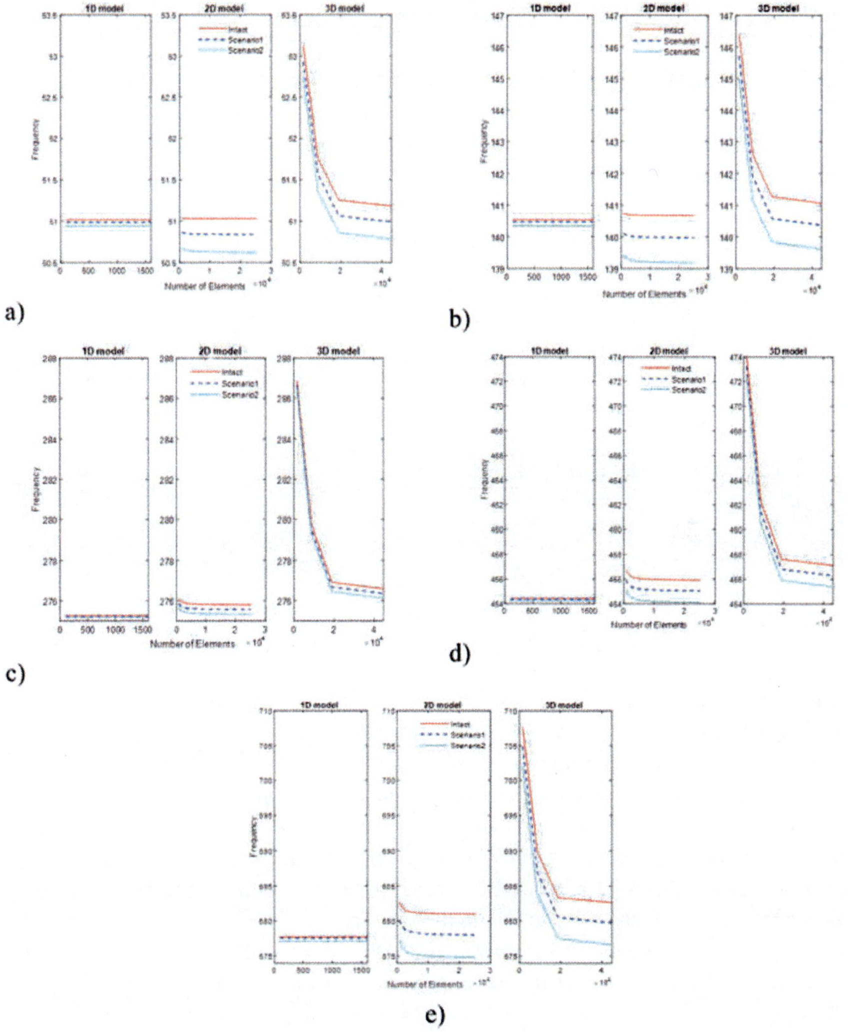

Fig. 10 The frequencies extracted from FEM: **a** Mode 1 **b** Mode 2 **c** Mode 3 **d** Mode 4 and **e** Mode 5

model is approximately a line, whereas the 2D model needs more than 1000 elements and 3D model more than 2000 elements to converge. Moreover, one beam element has two nodes, one shell element is presented by four nodes and one solid element is modeled by 8 nodes. Therefore, the computation time for the 3D model is the longest. The converged natural frequencies extracted from the FE model are presented in Table 4 (intact beam), Table 5 (Scenario 1), and Table 6 (Scenario 2). The differences between FEM and experimental data are then calculated and plotted in Fig. 11.

Table 4 Frequencies [Hz] from FEM for intact beam

Mode	1	2	3	4	5
1D	51.01	140.53	275.23	454.39	677.70
2D	51.02	140.67	275.79	455.90	680.94
3D	51.18	141.06	276.57	457.11	682.60

Table 5 Frequencies [Hz] from FEM for damaged beam, scenario 1

Mode	1	2	3	4	5
1D	50.99	140.48	275.22	454.37	677.59
2D	50.84	139.98	275.58	455.09	678.14
3D	51.00	140.38	276.36	456.29	679.73

Table 6 Frequencies [Hz] from FEM for damaged beam, scenario 2

Mode	1	2	3	4	5
1D	50.93	140.34	275.19	454.26	677.16
2D	50.63	139.21	275.34	454.17	675.00
3D	50.78	139.62	276.13	455.39	676.65

These are less than 1% in all cases; i.e. the FEM can reflect the behavior of the experimental beam.

Observing Fig. 11, although the differences in frequencies of the 1D model for the intact beam are the smallest compared with the other model, it increases steadily for the damaged beam. The more severity of the damage, the higher the differences between numerical and experimental results. For example, at mode 1, the differences are 0.35%, 0.67% and 1.13% for intact, damaged beam scenario 1 and damaged beam scenario 2, respectively. We can obtain the same results for modes 2, 3 and 5. In case the beam having the damaged elements, the 1D model could not reflect the vibration response of the damaged beam, the differences between FEM and experiment are high.

The differences in natural frequencies between the 2D model or 3D model and experiment do not change much for intact, scenario 1 and scenario 2. For example, in mode 1 of 2D model, the differences are 0.35%, 0.375% and 0.53% for intact, scenario1 and scenario 2, respectively. In mode 1 calculated from the 3D model, these differences are 0.68%, 0.69% and 0.83% for intact, scenario1 and scenario 2, respectively. This trend is repeated in all five modes. Except for mode 4, the 2D model and 3D model have the same tendency regarding these the differences. This proved that the damaged element does not affect the results of the 2D model and 3D model. These models can reflect the vibration response of the damaged beam more accurately. Moreover, the differences between the 2D model and the experimental results are less than those from the 3D model. One thing must be noticed that the

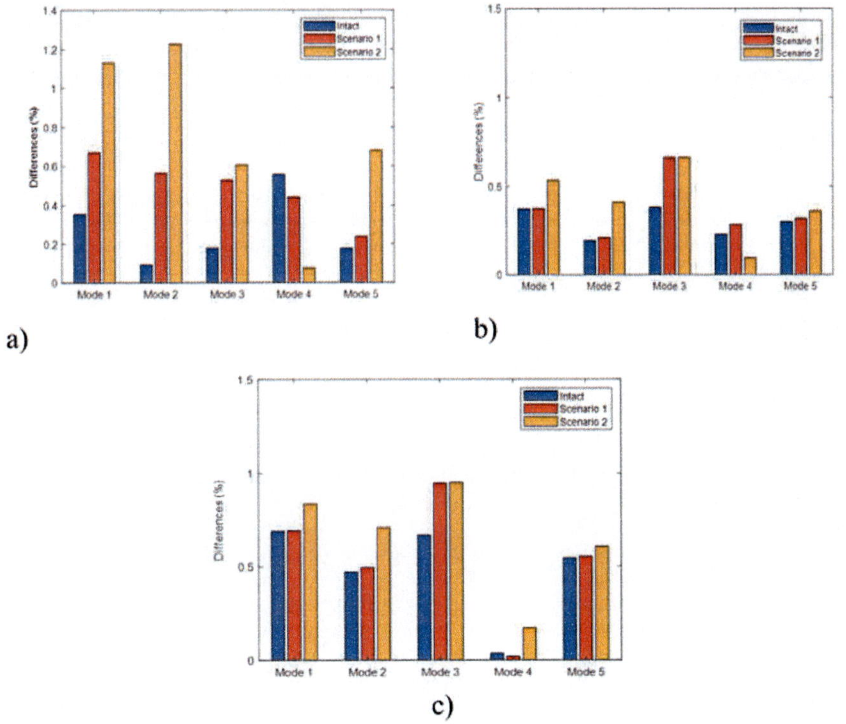

Fig. 11 The frequency differences between numerical and experimental results **a** 1D model **b** 2D model and **c** 3D model

same material properties were used for both models. These material properties need to be updated for each model to get the best results.

Each element working as a plate in the 2D model, and as a solid in the 3D model. Therefore the time used for computing the 3D model is much longer than that for the 2D model. From the discussion above, we conclude that the 2D model should be used for the damaged beam for simplicity and fast computation.

5 Conclusion

In this paper, a free-free beam was modeled using three kinds of ANSYS elements: beam (1D), shell (2D) and solid (3D). Although using the same material properties, each FE model provided different results of natural frequencies. The experiment is conducted in the laboratory. Fifteen accelerometers were set up to measure the first seven bending natural frequencies of the free-free beam. Three states of the beam were considered: intact beam, and two scenarios of damage. The conclusion was made based on the differences between the numerical model and the experiment of each model. For the 1D model, these differences increase steadily when the severity of the damage increase. Whereas the 2D and 3D models have the same trend of changing. Therefore, the 2D model is proposed as the model of the free-free beam because it is an easy model and provided fast computation.

Acknowledgements This research is funded by National University of Civil Engineering (NUCE) under grant number 11-2020/KHXD-TĐ. The authors acknowledge the financial support of VLIR-UOS TEAM Project, VN2018-TEA479A103, 'Damage assessment tools for Structural Health Monitoring of Vietnamese infrastructures' funded by the Flemish Government.

References

1. Carden EP, Fanning P (2004) Vibration based condition monitoring: a review. Struct Health Monit 3(4):355–377
2. Maia N, Silva J, Almas E, Sampaio R (2003) Damage detection in structures: from mode shape to frequency response function methods. Mech Syst Signal Process 17(3):489–498
3. Yan Y, Cheng L, Wu Z, Yam L (2007) Development in vibration-based structural damage detection technique. Mech Syst Signal Process 21(5):2198–2211
4. Thyagarajan S, Schulz M, Pai P, Chung J (1998) Detecting structural damage using frequency response functions. J Sound Vib 210(1):162–170
5. Wahab MA, De Roeck G (1999) Damage detection in bridges using modal curvatures: application to a real damage scenario. J Sound Vib 226(2):217–235
6. Nguyen DH, Bui TT, De Roeck G, Wahab MA (2019) Damage detection in Ca-Non Bridge using transmissibility and artificial neural networks. Struct Eng Mech 71(2):175–183
7. Yu Y, Wang C, Gu X, Li J (2019) A novel deep learning-based method for damage identification of smart building structures. Struct Health Monit 18(1):143–163
8. Tran-Ngoc H, Khatir S, De Roeck G, Bui-Tien T, Nguyen-Ngoc L, Abdel Wahab M (2018) Model updating for Nam O bridge using particle swarm optimization algorithm and genetic algorithm. Sensors 18(12):4131
9. Nguyen HD, Bui, TT, De Roeck G, Wahab MA (2018) Damage detection in simply supported beam using transmissibility and auto-associative neural network. In: International conference on numerical modelling in engineering. Springer

A Comparative Study on Indentation and Flattening Contacts

Qingming Deng, Xiaochun Yin, and Magd Abdel Wahab

Abstract This work presents a comparative analysis of elastic-plastic flattening and indentation contacts by using finite element analysis. The similarity and difference between the two types of single asperity contacts are investigated. Five elastic-ideally plastic materials, whose yield strengths covering a typical range of steel materials used in engineering, are selected as materials of deformation body. The numerical simulation covers a wide range of contact deformation. The results are normalized such that to be valid either for macro contacts or for micro contacts. The contact force and radius are obtained and compared. The similarity of the two contact types is found when the contact deformation is relatively small, even when the contact approaches to the fully plastic deformation. When the interference is large enough, the two contacts behaviors are significantly different. Two prediction formulas for upper limit interference of similarity of contact force and radius are given.

Keywords Similarity · Flattening contact · Indentation contact · Elastic-plastic deformation · Finite element analysis

1 Introduction

Contact problems occur in many engineering applications [1–10]. Inelastic deformation near the contact area complicates the contact behavior. The relationship between contact properties and interference is difficult to predict. For many repeated predictions, numerical simulation is too cumbersome. So many theoretical contact models for fast calculations have been proposed [11–19]. For pure elastic spherical contact, Hertz [20] proposed the prediction formulas of contact force and contact radius.

Q. Deng · X. Yin (✉)
Department of Mechanics and Engineering Science, Nanjing University of Science and Technology, Nanjing 210094, China
e-mail: yinxiaochun2000@aliyun.com

Q. Deng · M. Abdel Wahab
Faculty of Engineering and Architecture, Department of Electrical Energy, Metals Soete Laboratory, Mechanical Constructions and Systems, Ghent University, Ghent, Belgium

M. Abdel Wahab (ed.), *Proceedings of the 8th International Conference on Fracture, Fatigue and Wear*, Lecture Notes in Mechanical Engineering,
https://doi.org/10.1007/978-981-15-9893-7_4

Most current models still use it during the elastic deformation stage. Since contact and collision problems often cause inelastic deformation, later models are trying to supplement the prediction formulas of elastic-plastic and fully plastic phases. Most existing contact laws are derived for spherical contacts [21]. Ghaednia et al. [22] reviewed the contact models of different theoretical basis and application background. And they divided the single asperity contact models into two categories: indentation and flattening model [13]. As shown in Fig. 1, the contact between rigid sphere and an elastic-plastic half-space is the indentation contact. A normal force F acts on top of a rigid sphere with radius R. The parameters a and a' are real contact radius and truncated contact radius, respectively. The ω is interference between rigid sphere and elastic-plastic half-space. And the flattening is the contact of an elastic-plastic hemisphere contact with a rigid flat as shown in Fig. 2.

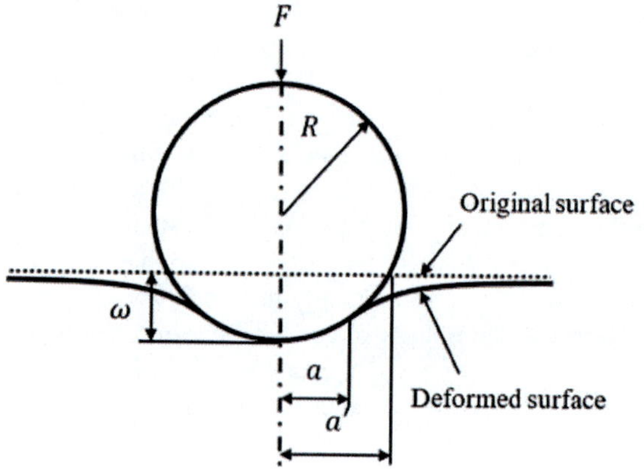

Fig. 1 An indentation model, in which the sphere is rigid and the flat is deformable

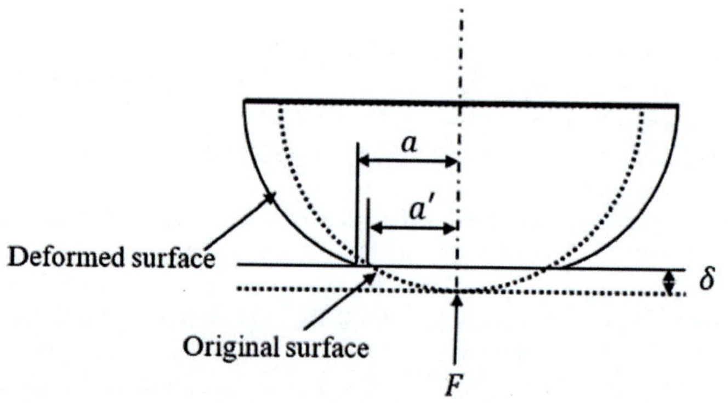

Fig. 2 A flattening model, in which the flat is rigid and the sphere is deformable

The study of indentation contact originated from the hardness test. The hardness H is the resistance of metal to plastic deformation which equals the average pressure when contact surface under fully plastic deformation. When contact involves plastic deformation, the average pressure p_m is an important parameter. To study the ability of materials to resist deformation, different hardness tests were carried out [23, 24]. And Ishlinskii [25] present a an analytical model for the contact of a rigid sphere and a fully plastic half-space by slip lines theory. Johnson [26] defined initial point for yielding and the interference when $p_m = 1.1S_y$ and considers $p_m = 2.8S_y$ in the fully plastic phase. He gave a contact model called simplified spherical expansion model. Then, more analytical models based on Johnson's theory were proposed [18, 19, 27–29]. And the works of Brake [18, 29] and Ma and Liu [19] were verified by comparison with the coefficient of restitution from the collision experiment. Recently, the finite element method has gradually become the main research technique. Compared with experiments, numerical simulations can get more information and can avoid some arbitrary assumptions. Mesarovic and Fleck [30] found that the hardness was not a constant multiple of the yield limit. And more researcher obtained the same result and found that the hardness depended on the material properties of contact [13, 14, 17, 31]. Some new indentation contact models were proposed using data fitting of numerical results [16, 17, 31]. Different deformation stages were given by observing the evolution of the plastic zone below the contact surface.

On the other hand, there are also a lot of researches on flattening contact. Greenwood and Williamson proposed the early model of elastic flattening contact [32]. And a fully plastic flattening model was presented by Abbott and Firestone [33]. Based on their research, more models suitable for flattening contact were proposed [15, 34–37]. Kogut and Etsion (KE model) proposed a new flattening model by data fitting of numerical results [15]. They defined the hardness of the elastic-plastic sphere as a constant multiple of its yield strength ($H = 2.8S_y$). Later, Jackson and Green (JG model) used numerical simulation and found that the hardness of sphere depended on the material properties [14]. So they provided a new prediction formula of the critical interference ω_c causing initial yield inside hemisphere. And loading phase includes two stages: elastic and elastic-plastic. Compared with KE model, JG model is continuous at the boundary between the elastic and elastoplastic stages, and is suitable for a larger deformation range.

By comparing the two indentation models [16, 31] and KE, JG models, Jackson and Kogut found that there were huge differences between the two types of models [38]. However, in some literatures, the two types of models were not distinguished, such as using prediction hardness of flattening model [14] to prove the correct of an indentation model [17]. And the correctness of the indentation model [17] was verified by coincidence with the deformation evolution boundary of a flattening model [15]. In the field of collision research, such misuse and mixing are more serious [39]. Moreover, in the prediction of physical quantities such as impact force, coefficient of restitution, and contact duration, the two types of models did not show a clear difference [21]. So the main purpose of this article is to compare these two types of contacts in detail through finite element simulation and discuss their similarities and differences.

2 Theoretical Background

The Hertzian purely elastic contact theory [40], dimensionless method of JG model [14] are shortly reviewed, here. The contact area and force prediction equations of Hertz theory are:

$$A_E = \pi R' \omega \tag{1}$$

$$F_E = \frac{4}{3} \times E' \sqrt{R'} \times \omega^{\frac{3}{2}} \tag{2}$$

where ω is the interference between two bodies. And in order to consider the feature of both contact bodies, the equivalent elastic modulus, E', and an equivalent radius, R', are expressed as:

$$\frac{1}{E'} = \frac{1 - v_1^2}{E_1} + \frac{1 - v_2^2}{E_2} \tag{3}$$

$$\frac{1}{R'} = \frac{1}{R_1} + \frac{1}{R_2} \tag{4}$$

where E_1, E_2, v_1, v_2, R_1, and R_2 are Young's modulus, Poisson's ratio, and the radius of curvature of sphere and half-space, respectively. After elastic phase, the plastic yield initiates from the elastic-plastic body under the contact surface [14]. And JG model gives the critical interference ω_c resulting in the initial yield as:

$$\omega_c = \left(\frac{\pi \times c \times S_y}{2E'} \right)^2 R \tag{5}$$

where S_y and v are the yield limit and Poisson's ratio of the elastic-plastic body. And parameter c is expressed as:

$$\mathrm{c} = 1.295 e^{0.736 v} \tag{6}$$

And the critical contact area A_c and force F_c at critical interference can be obtained by substituting Eq. (5) into Eqs. (1) and (2):

$$F_c = \frac{4}{3} \left(\frac{R}{E'} \right)^2 \left(\frac{c}{2} \pi S_y \right)^3 \tag{7}$$

and

$$A_c = \pi^3 \left(\frac{c S_y R}{2E'} \right)^2 \tag{8}$$

Equations (5), (7) and (8) are used to normalize the contact physical quantity ($\omega^* = \omega/\omega_c$, $F^* = F/F_c$, and $A^* = A/A_c$) in JG model [14]. This dimensionless method proposed for the flattening case, in order to compare flattening and indentation contact, so this study uses the same dimensionless method for both contact cases.

3 Finite Element Analysis

Figure 1 shows a schematic of a frictionless elastic-plastic half-space indented by a rigid sphere of radius R. And Fig. 2 shows a contact of an elastic-plastic hemisphere with a rigid flat. As other researches in the literature [13–17, 41, 42], for flattening and indentation contact problems, an axisymmetric 2-D model is used to improve the efficiency of computation. The present study utilizes the commercial program ANSYS™. The convergence of half-space is studied. The thickness T and width L are gradually increased, until the contact force and contact area differed by less than 1% between iterations. And finally a half-space with a width of $10R$ and a thickness of $6R$ is selected as shown in Fig. 3. The left side of the half-space is fixed horizontally and can move freely in the vertical direction. The bottom of half-space is completely fixed and a displacement ω is applied on the top of rigid sphere.

And the mesh of flattening contact is shown in Fig. 4. The central axis of the hemisphere is fixed horizontally and is free vertically. Similar to several other studies [70, 73–75, 77, 78], the quadrilateral, eight node element solid 183 is used to mesh the deformable body as shown in Figs. 3 and 4. The mesh size is selected after the study of convergence. The element size of the contact area was decreased until the difference of contact force between the two iterations is less than 1%. The element size of the contact edge should also be small enough so that the error of the contact radius is less than 3.3% as the work of JG model. The resulting FE model consists

Fig. 3 Finite element model of indentation contact

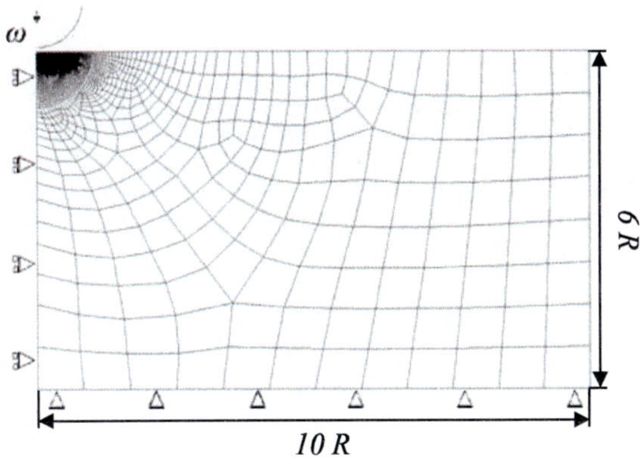

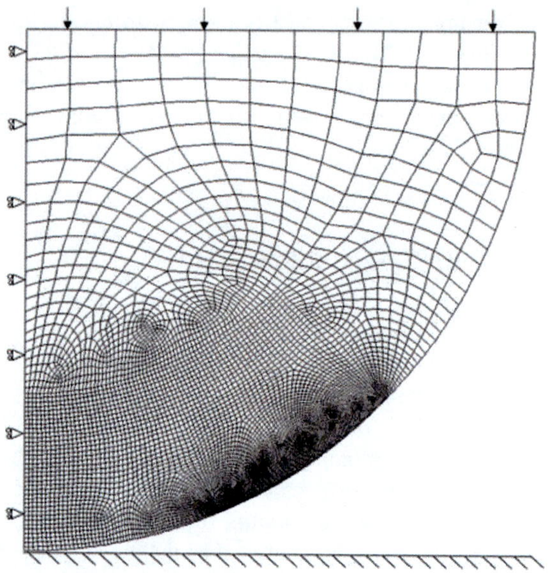

Fig. 4 Finite element model of flattening contact

of at least 17,526 elements, since the number of elements varies with the change of contact area.

The elastic modulus and Poisson's ratio of deformable body are held constant at 200 GPa and 0.32, respectively. And five different material yield strengths S_y are considered, which are 0.210, 0.5608, 0.9115, 1.2653, and 1.619 GPa, respectively. In the following, they are referred to as Mat 1, Mat 2, Mat 3, Mat 4, and Mat 5. For both indentation and flattening contact, the five materials cases are calculated until the contact radius reaches about 0.8 *R*. In this study, more than 1309 simulation cases are finished. The computation takes from 10 min to several hours on a 3.2 GHz PC. And contact force, contact radius and average contact pressure to yield strength ratio $\mathrm{F}/(\mathrm{A}S_y)$ have been analyzed.

4 Numerical Results and Discussion

Two finite element models are shown in Figs. 3 and 4 to analyze the contact problems of different materials and different interferences. This paper verifies the FE model by comparing the simulation results with the Hertz theory prediction in the elastic deformation stage. Then this article compares the two types of contact response. The similarities and differences of the flattening and indentation contacts are discussed.

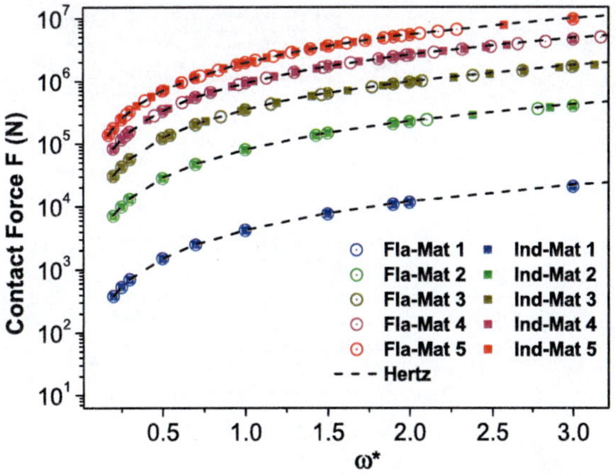

Fig. 5 Comparison of contact force predicted by simulation and Hertz theory in the elastic deformation regime

4.1 FEM Model Validation

Figure 5 shows the comparison of contact force predicted by simulation and Hertz theory in the elastic deformation regime ($\omega^* < 1$). It shows that all simulation results agree with the analytical solution obtained from classical Hertz theory and maximum difference is less than 0.5%. And the simulation results of indentation and flattening contacts are also very close. Up to the dimensionless interference $\omega^* = 3$, this agreement still exist. This verifies the suitability of the finite element mesh shown in Figs. 6 and 7.

4.2 Contact Force

Figure 6a shows the variation of contact force of two kinds of contact with interference. For same deformation body material cases, the indentation contact force is higher than the flattening contact under large deformation. Jackson and Kogut compared four contact models [14–16, 31] and found the contact force of flattening model was larger than indentation model. Contact model predictions are diametrically opposed to numerical simulations and show that the detailed numerical simulation research on this issue is necessary. It can be found that the growth rate of all flattening contact forces decreases around $\omega/R = 0.1$. The relationship of $F^*/(\omega^*)^{3/2}$ and ω^* is shown in Fig. 6b. According to Hertz theory Eq. (2), when the deformation is pure elastic, this value should be equal to 1. The agreement of all cases in Fig. 6b shows that there is similarity between the two kinds of contact under certain interference. This means that two kinds of contact have similar contact force response under a certain interference. The difference between indentation and flattening contact force for same material deformation body cases is shown in Fig. 6c. The difference remains around 0 before the dimensionless interference exceed a certain value. By increasing

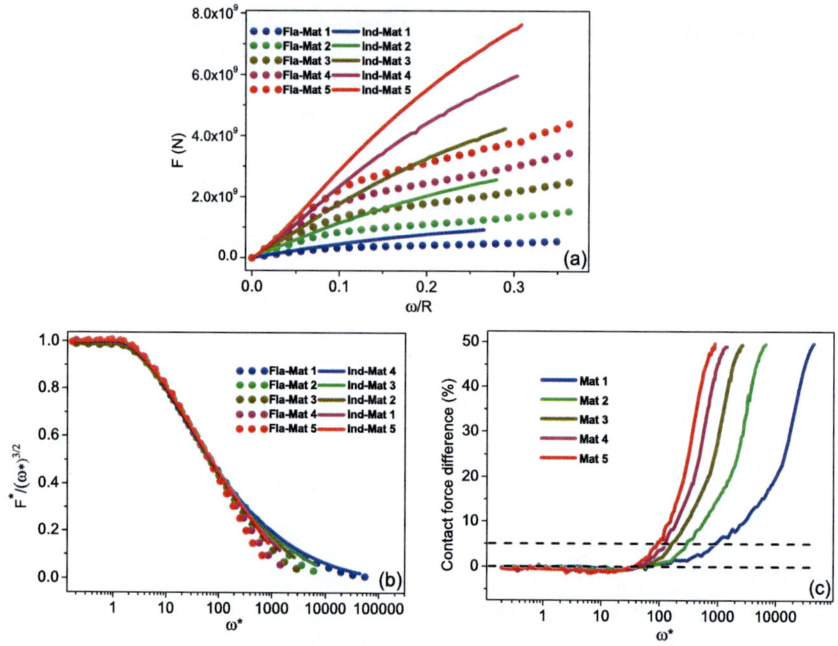

Fig. 6 Comparison of contact force (**a**) Contact force change with ω/R, (**b**) contact force change with ω^*, and (**c**) the contact force difference between indentation and flattening contact

ω^*, the difference will finally exceed 5% and reach 50%. In this paper, the ω^* when the difference exceed 5% is defined as upper limit of similarity ω_{lf}^*. The value of E'/S_y is an important indicator to distinguish material properties [13]. E'/S_y and upper limit of similarity for different material is list in Table 1.

And a prediction formula is given by:

$$\omega_{lf}^* = 31.97 + 0.0855 \times \left(\frac{E^*}{S_y}\right)^{1.334} \tag{9}$$

So a clear upper interference limit is proposed for different material cases here. If the prediction model is reliable, then when the $\omega^* < \omega_{lf}^*$, the indentation and flattening contact models can be used to predict contact force. However, if $\omega^* > \omega_{lf}^*$, specific prediction models can only be used for specific contact problem.

Table 1 Upper limit of contact force similarity for all material

Material	Mat 1	Mat 2	Mat 3	Mat 4	Mat 5
E^*/S_y	1061.03	397.32	244.45	176.1	137.626
ω_{lf}^*	959.8	282.85	160.6	119.82	91.38

4.3 Contact Radius

In addition to the contact force, the contact radius is also a key response predicted by the contact model [14–17] and depends on contact force and radius, the average contact pressure can be got. The maximum average contact pressure is also called hardness.

As shown in Fig. 7, the contact radius of two kinds of contact are compared. The indentation and flattening contact radii of the same deformable material cases show similarity with small interference. However, the contact radius growth rate of all flattening contact cases decreases around $\omega/R = 0.1$. Then the contact radius increases almost linearly with the increase of interference as plotted in Fig. 7a. So in large deformation, the contact radius of indentation case is significantly larger than the flattening contact. The similarity when the interference is relatively small is more obvious in Fig. 7b. For example, for the case where the deformable body is material 5, the contact radius of indentation is still very close to flattening case until the $\omega^* = 15000$. The contact radius difference between two contacts for different material cases are shown in Fig. 7c. And the interference at the inception of fully plastic deformation defined by some classical indentation contact models are $\omega^* = 84$ or $\omega^* = 82.5$ [11, 27]. So, this shows that the contact radii of two very different types of contacts are similar even under the fully plastic contact state.

In practical applications, a formula to quickly assess the similarity of the contact radius of two types of contacts is necessary. The interference, which results in a 5% difference is defined as upper limit ω^*_{lr}.

The upper limit interference for different material is list in Table 2. And a prediction formula for interference which results in a 5% difference is given by:

$$\omega^*_{lr} = 97.754 + 0.0178 \times \left(\frac{E^*}{S_y}\right)^{1.965} \tag{10}$$

Equation (10) is obtained by fitting the numerical results of material 1 to material 5. So, it can predict the interference upper limit of contact radius similarity for the cases with $137.626 < E^*/S_y < 1061.03$.

The contact radius get from all numerical simulation cases include two part: the original horizontal coordinates LocX of the contacting edge node (before deformation) and the horizontal displacement UX of this node (after deformation). And Fig. 8 shows the evolution of LocX and UX with interference. It can be found that the smaller the yield limit, the greater the UX is. This is because of the incompressibility of the plastic zone near the contact surface. And the horizontal displacement of the edge point of two contacts has big difference under large deformation. The UX growth rate of all flattening cases also decreases when ω/R is around 0.1.

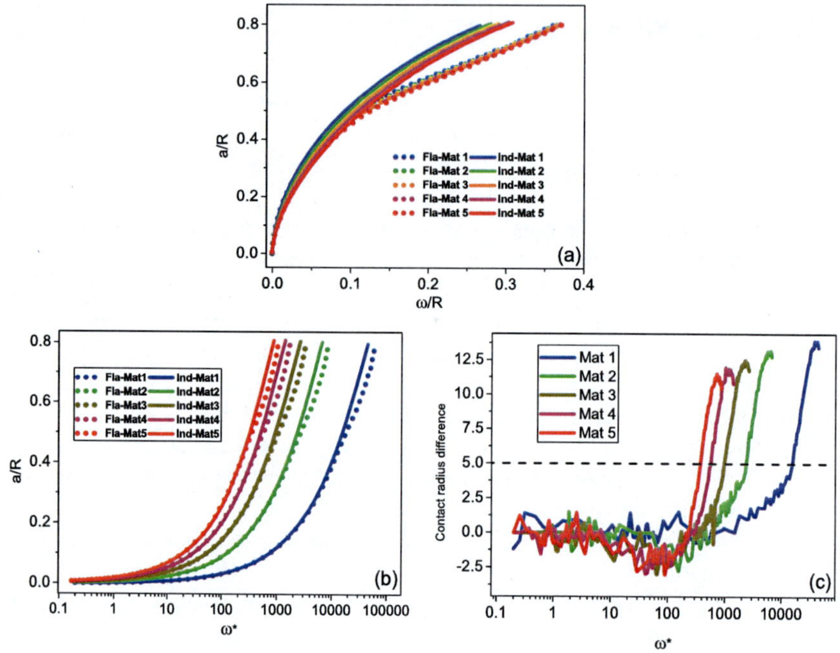

Fig. 7 Comparison of contact radius (**a**) Contact radius change with ω/R, (**b**) contact radius change with ω^*, and (**c**) the contact radius difference between indentation and flattening contact

Table 2 Upper limit of contact radius similarity for all material

Material	Mat 1	Mat 2	Mat 3	Mat 4	Mat 5
E^*/S_y	1061.03	397.32	244.45	176.1	137.626
ω_{lr}^*	15872.45	2387.29	975.71	568.95	377.56

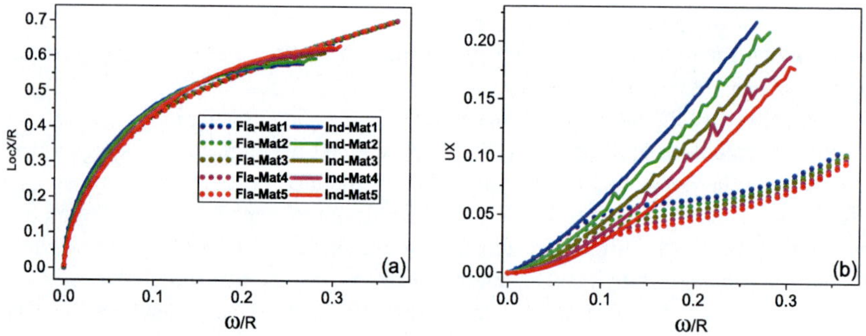

Fig. 8 The evolution of **a** LocX and **b** UX with interference

5 Conclusions

The contact response between indentation and flattening contact is studied by finite element method. The following conclusions are drawn.

1. The contact forces of indentation and flattening contact simulations agree with Hertz solutions in the elastic stage. As the interference increases, the contact force of indentation case exceeds the flattening case. Although the difference always exists after elastic stage, this difference between two contact is less than 5%, even in case of a fully plastic stage. And a prediction formula for upper limit interference of similarity is given.
2. With small interference, the contact radii of the two kinds of contact are very similar. Contact radius of indentation cases exceeds flattening case under large deformation. And even under the fully plastic contact stage, the difference of contact radius is still less than 5%. A prediction formula for upper limit interference of similarity is given.
3. The reason for the widening of the difference between the two types of contacts is that the growth rate of contact force and contact radius and horizontal displacement of the contacting edge node of all flattening cases decreases around $\omega/R = 0.1$.

Acknowledgements The authors would like to acknowledge the financial support of the grants from the China Scholarship Council (201806840127), the National Natural Science Foundation of China (Grant No. 11372138 and No. 11572157).

References

1. Borodich FM, Keer LM (2004) Contact problems and depth-sensing nanoindentation for frictionless and frictional boundary conditions. Int J Solids Struct 41(9–10):2479–2499
2. Mccarthy B et al (2002) A dynamic model, including contact bounce, of an electrostatically actuated microswitch. Microelectromec Syst J 11(3):276–283
3. Pal RK, Awasthi AP, Geubelle PH (2013) Wave propagation in elasto-plastic granular systems. Granular Matter 15(6):747–758
4. Komvopoulos K (2000) Head–disk interface contact mechanics for ultrahigh density magnetic recording. Wear 238(1):1–11
5. Cowles BA (1996) High cycle fatigue in aircraft gas turbines—an industry perspective. Int J Fract 80(2–3):147–163
6. Polycarpou AA, Etsion I (2000) A model for the static sealing performance of compliant metallic gas seals including surface roughness and rarefaction effects. Tribol Trans 43(2):237–244
7. Wang Y et al (2002) A mixed-TEHD analysis and experiment of journal bearings under severe operating conditions. Tribol Int 35(6):395–407
8. Jiang X, Cheng HS, Hua DY (2000) A Theoretical analysis of mixed lubrication by macro micro approach: part i—results in a gear surface contact. Tribol Trans 43(4):689–699
9. Kweh CC et al (1992) Simulation of elastohydrodynamic contacts between rough surfaces. J Tribol 114(3):412–419

10. Tichy JA, Meyer DM (2000) Review of solid mechanics in tribology. Int J Solids Struct 37(1–2):391–400
11. Big-Alabo A, Harrison P, Cartmell M (2015) Contact model for elastoplastic analysis of half-space indentation by a spherical impactor. Comput Struct 151:20–29
12. Dong X et al (2018) Local contact behavior between elastic and elastic–plastic bodies. Int J Solids Struct 150:22–39
13. Ghaednia H et al (2016) A comprehensive study of the elasto-plastic contact of a sphere and a flat. Tribol Int 93:78–90
14. Jackson RL, Green I (2005) A finite element study of elasto-plastic hemispherical contact against a rigid flat. J. Trib. 127(2):343–354
15. Kogut L, Etsion I (2002) Elastic-plastic contact analysis of a sphere and a rigid flat. J Appl Mech 69(5):657–662
16. Kogut L, Komvopoulos K (2004) Analysis of the spherical indentation cycle for elastic–perfectly plastic solids. J Mater Res 19(12):3641–3653
17. Song Z, Komvopoulos K (2013) Elastic–plastic spherical indentation: deformation regimes, evolution of plasticity, and hardening effect. Mech Mater 61:91–100
18. Brake M (2012) An analytical elastic-perfectly plastic contact model. Int J Solids Struct 49(22):3129–3141
19. Ma D, Liu C (2015) Contact law and coefficient of restitution in elastoplastic spheres. J Appl Mech 82(12)
20. Hertz H (1982) Ueber die Berührung fester elastischer Körper.: Journal für die reine und angewandte Mathematik (Crelle's Journal). Journal Für Die Reine Und Angewandte Mathematik
21. Wang H et al (2020) The correlation of theoretical contact models for normal elastic-plastic impacts. Int J Solids Struct 182:15–33
22. Ghaednia H et al (2017) A review of elastic–plastic contact mechanics. Appl Mech Rev 69(6):060804
23. Brinell J (1900) Way of determining the hardness of bodies and some applications of the same. Teknisk Tidskrift 5:69
24. Tabor D (2000) The hardness of metals. Oxford University Press
25. Ishlinsky A (1944) The problem of plasticity with axial symmetry and Brinell's test. J Appl Mathe Mech 201–224
26. Johnson KL (1987) Contact mechanics. Cambridge University Press
27. Stronge WJ (2018) Impact mechanics. Cambridge University Press
28. Yigit A, Christoforou A (1994) On the impact of a spherical indenter and an elastic-plastic transversely isotropic half-space. composites Engineering 4(11):1143–1152
29. Brake M (2015) An analytical elastic plastic contact model with strain hardening and frictional effects for normal and oblique impacts. Int J Solids Struct 62:104–123
30. Mesarovic SD, Fleck NA (2000) Frictionless indentation of dissimilar elastic–plastic spheres. Int J Solids Struct 37(46–47):7071–7091
31. Komvopoulos K, Ye N (2000) Three-dimensional contact analysis of elastic-plastic layered media with fractal surface topographies. J Trib 123(3):632–640
32. Greenwood JA, Williamson JP (1966) Contact of nominally flat surfaces. Mathe Phys Sci 295(1442):300–319
33. Abbott E, Firestone F (1995) Specifying surface quality: a method based on accurate measurement and comparison. Spie Milestone Series MS 107:63
34. Chang W, Etsion I, Bogy DB (1987) An elastic-plastic model for the contact of rough surfaces. J Tribol 109(2):257–263
35. Zhao Y, Maietta DM, Chang L (2000) An asperity microcontact model incorporating the transition from elastic deformation to fully plastic flow. J Trib 122(1):86–93
36. Jacq C et al (2002) Development of a three-dimensional semi-analytical elastic-plastic contact code. J Trib 124(4):653–667
37. Vu-Quoc L, Zhang X, Lesburg L (2000) A normal force-displacement model for contacting spheres accounting for plastic deformation: force-driven formulation. J Appl Mech 67(2):363–371

38. Jackson RL, Kogut L (2005) A comparison of flattening and indentation approaches for contact mechanics modeling of single asperity contacts. J Tribol 128(1):209–212
39. Wang H et al (2017) Experimental and theoretical analysis of the elastic-plastic normal repeated impacts of a sphere on a beam. Int J Solids Struct 109:131–142
40. Hertz H (1882) Uber die beruhrung fester elastische korper und uber die harte (On the contact of rigid elastic solids and on hardness). Verhandlungen des Vereins zur Beforderung des Gewerbefleisses, Leipzig
41. Larsson PL, Olsson E (2015) A numerical study of the mechanical behavior at contact between particles of dissimilar elastic–ideally plastic materials. J Phys Chem Solids 77:92–100
42. Olsson E, Larsson P-L (2016) A unified model for the contact behaviour between equal and dissimilar elastic–plastic spherical bodies. Int J Solids Struct 81:23–32

Impact of Abrasive Blasting Media on the Strength of Steel Sheets Adhesively Bonded Joints

Anna Rudawska, Jakub Szabelski, Magd Abdel Wahab, and Izabela Miturska

Abstract This paper concerns a comparative study assessing the impact of various mechanical surface treatment methods on the strength of adhesively bonded joints. The surface treatment methods used were glass-bead blasting and steel shot blasting. The joints used sheet components of S235JR steel, bonded with one of two epoxy adhesives. Several configurations of adhesive joints were tested, based on specific blasting medium variants. The study included the analysis of substrate surface topography (Hommel-Etamic T8000 RC 120-400) and strength tests (Zwick/Roell Z150). The results show that the highest joint strength was developed by the assemblies with steel shot blasted adherends, which in addition exhibited the highest levels of surface roughness.

Keywords Mechanical surface treatment · Steel · Adhesive joints strength

1 Introduction

The strength of an adhesive joint depends largely on two factors: bonding of the adhesive to the substrate and cohesion, or strength, of the adhesive. Adhesion is a force present at the interfaces of two parts bonded together. Two types of adhesion are classified: mechanical adhesion and specific adhesion [1, 2]. Surface roughness is

A. Rudawska · J. Szabelski (✉) · I. Miturska
Faculty of Mechanical Engineering, Lublin University of Technology, Nadbystrzycka 36, 20-618 Lublin, Poland
e-mail: j.szabelski@pollub.pl

A. Rudawska
e-mail: a.rudawska@pollub.pl

I. Miturska
e-mail: i.miturska@pollub.pl

M. Abdel Wahab
Faculty of Engineering and Architecture, Ghent University, Technologiepark Zwijnaarde 903, 9052 Zwijnaarde, Belgium
e-mail: magd.abdelwahab@ugent.be

M. Abdel Wahab (ed.), *Proceedings of the 8th International Conference on Fracture, Fatigue and Wear*, Lecture Notes in Mechanical Engineering,
https://doi.org/10.1007/978-981-15-9893-7_5

also a major factor (when considering the mechanical adhesion aspect, including the mechanical anchoring of the adhesive within the superficial irregularities [1, 3–5]) and the cleanliness of surface. A surface not clean enough may reduce the strength of the intermolecular bonding. The cleanliness, surface roughness and correct surface activation of the adherends are achieved by suitable treatment [6–10].

Surface treatment is an initial and critical step in adhesive bonding [11–13]. An entire adhesive bonding process comprises several operations, such as cleaning the substrate, forming a suitable surface stereometric structure (texture) of the parts to be bonded and proper activation of their surfaces [14–17]. One surface treatment type is mechanical treatment. Several mechanical treatment variants are used: abrasive machining, abrasive blasting, bead blasting, brushing, and grinding [18–22]. These mechanical treatment processes employ abrasive tools of various grain sizes and abrasive media, such as sand blasting, shot blasting or bead blasting, of various grain hardness levels and sizes. Sand blasting and shot blasting require a suitable medium (grit) delivery pressure to prevent damage to the superficial layer of the substrate and to produce an optimum surface roughness.

Khan et al. [22] investigated the effect of different grit-blasting pressures on the adhesion strength of resin to pure titanium and concluded that a lower grit-blasting pressure might promote adhesion strength under long-term water ageing.

Abrasive blasting involves an abrasive medium, known as grit, which is propelled out of a nozzle under the pressure of compressed air. Abrasive blasting can be wet or dry. Wet abrasive blasting provides a smoother surface finish due to the mix of the grit with a liquid vehicle and compressed air. The medium in dry abrasive blasting features grit without any liquid, which renders a rougher surface on the workpiece. The most common abrasive blasting media include shot, sand, alumina, furnace slag and glass. Sand blasting has been the most common process as sand is relatively inexpensive and easy to source. Another common abrasive blasting process is shot blasting. Here, grit in the form of metal shot is used to produce a surface roughness higher than that obtainable by sand blasting; shot grains are heavier and ‘cold-works’ or ‘upsets’ the superficial layer of the workpiece.

This paper is a comparative study of the strength of adhesively bonded joints based on steel sheets with mechanically treated surfaces; the study also compares selected surface roughness parameters. The steel sheet surface was mechanically pre-treated by sandblasting with glass beads or steel shot before adhesive bonding in either mechanical treatment variant. Each adhesively bonded joint was fabricated with one of two types of epoxy adhesive, in one of a number of various adhesive joint configurations, depending on the type of mechanical treatment.

Table 1 Surface treatment variants

Designation	Surface treatment variants	Medium
A	Degreasing	Ethanol solution
B	Glass bead blasting and degreasing	Glass beads and ethanol solution
C	Shot blasting and degreasing	Broken cast steel shot and ethanol solution

2 Test Method

2.1 Surface Treatment of Adherends

Test specimens made from S235JR steel sheets (according to PN-EN 10025 standard) had the surfaces pre-treated for adhesive bonding by mechanical glass bead blasting or shot blasting, depending on the tested variant. All surfaces were degreased after the mechanical treatment. The surface treatment variants are listed in Table 1.

The first surface treatment variant (A) left the test specimens unmodified. The second surface treatment variant (B) had the surface sand blasted with glass beads within a grain size range of 70–100 μm. The third surface treatment variant (C) had the surface shot-blasted with broken cast steel shot (FH50). Surface treatment variants (B) and (C) were processed in a shot-blasting cab, CABILUX Classic PC-CL 833. The steel sheets were blasted with a BM-SB 200 pneumatic sand-blasting machine complete with a BH-E1 grit blasting lance.

Before each adhesive joint was fabricated after either surface treatment variant, the test surfaces were degreased with Loctite SF 7063, an ethanol solution (manufactured by Loctite, Henkel, Belgium). The degreaser was applied twice by spraying, followed by wiping with no-lint swabs. A third spray application of the degreaser followed, with a flash-off time of approximately 2 min.

2.2 Adhesive Joints

Single-lap adhesive joints were fabricated from the S235JR steel sheet test specimens. Each test specimen had the following dimensions: length, l = 100 mm; width, b = 25 mm; thickness, g = 2 mm. The adhesive lap length was 15 mm (Fig. 1).

The adhesive joints were fabricated with two types of epoxy adhesive: Epidian 57 epoxy resin and Epidian 53 epoxy resin, each using a PAC polyamide hardener. The designations of the epoxy adhesives were: Epidian 53/PAC/100:80 and Epidian 57/PAC/100:80. The properties of the epoxy adhesive components are shown in [23–25].

Each adhesive was prepared by mixing the relevant Epidian epoxy resin with the PAC polyamide hardener in a ratio of 100:80. The components were batched

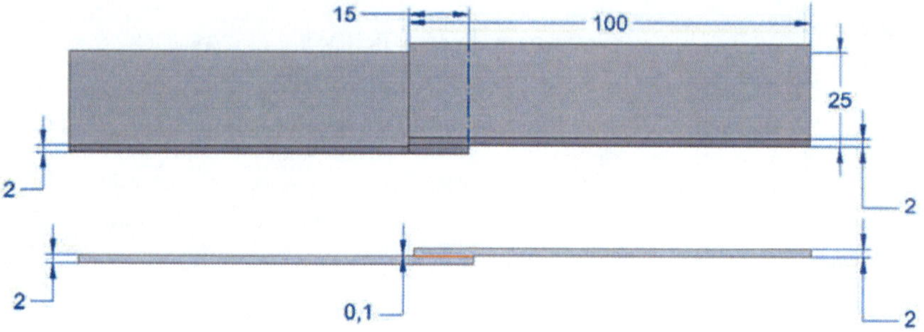

Fig. 1 Single-lap adhesive joint

with an OX-8100 laboratory scale (manufactured by FAWAG S.A., Lublin, Poland) and mixed with an anchor stirrer in a cylindrical mixing vessel made from a polymer plastic material. The mixing time was 2 min at 480 RPM. Within 10 min from mixing, each adhesive was manually applied with a polymer plastic spatula on one of the surfaces to be bonded. The parts to be bonded were brought together to form a joint, and the adhesive bond was left to cure.

The adhesive joints were fabricated according to the configurations in Tables 2 and 3, in reference to the surface treatment variant (ref. Table 1).

The heterogeneous surface treatment adhesive joints were fabricated under the following conditions: temperature 26 ± 2 °C, RH 30 ± 3%. The adhesive joint curing time was 7 days with a pressure applied to the joint of 0.18 MPa. The curing process was completed in a single cold step.

Table 2 Adhesive joint configurations with homogeneous surface treatment types for bonded parts

Adhesive joints configuration	Designation
Degreased/Degreased	A–A
Glass-bead blasted and degreased/Glass-bead blasted and degreased	B–B
Shot-blasted and degreased/Shot-blasted and degreased	C–C

Table 3 Adhesive joint configurations with heterogeneous surface treatment types for bonded parts

Adhesive joints configuration	Designation
Degreased/Glass-bead blasted and degreased	A-B
Degreased/Shot-blasted and degreased	A-C
Glass-bead blasted and degreased/Shot-blasted and degreased	B–C

2.3 Tests and Statistical Analysis

EN 1465 shear strength tests were carried out on various configurations of the steel sheet adhesive joints (ref. Tables 2 and 3) with a Zwick/Roell Z150 strength testing machine. The destructive testing of the adhesive joint test specimens was performed at ambient temperature. The shear strength pre-load was 5 N and the test rate was 5 mm/min. a total of 6 adhesive joints were fabricated for each configuration. The total number of destructively tested adhesive joints was 36 (6 configurations × 6 adhesive joints).

The surface roughness of the adhesively bonded test specimens (3 pieces from each surface treatment variant) was tested with a Hommel-Etamic T8000 RC 120-400 roughness and contour measuring system. These tests combined measurements of surface roughness and 3D topography (each test in 10 iterations). The test parameters measured included: arithmetic mean values of roughness profile deviation, Ra; peak roughness at 10 points, Rz; overall profile height between the peak and valley, Rt; and the average spacing along a specified section, Rsm. All the parameters were determined according to PN-EN ISO 4287, PN-EN ISO 25178 and PN-EN ISO 11562. The Hommel-Etamic T8000 RC 120-400 roughness and contour measuring system used a TKU300 roughness probe for these tests. The test feed rate was 1.3 mm/s with a test resolution of 80 μm. Roughness maps were generated from a total of 241 tests on a mapping surface measuring 4.8 × 4.8 mm.

The shear strength and surface roughness test results were statistically analysed in the Statistica software suite by correlation (to determine the relationship between the variables) and regression (to define the form of relationship). A method of multiple regression was applied to quantify the relationships between multiple dependent variables and one dependent variable. The assumed measure of correlation between one variable and the total of the remaining variables was the multiple correlation factor (R). If R was close, the variables tested were in a linear relationship. Suitable statistical tests were used for the analysis.

3 Results

3.1 Adhesive Joints Shear Strength

Figure 2 shows a comparison of the mean shear strength values from the adhesive joint configurations in which at least one part surface was degreased only: A–A, A-B and A-C.

It was found that the highest shear strength was developed in configuration A-C, where the degreased test specimens (A) were adhesively bonded with Epidian 57/PAC/100:80 to the test specimens treated by broken cast steel shot blasting (C). The mean shear strength for A-C was 20.2 MPa. The lowest shear strength was found in adhesive joint configuration A–A, where both adherends were degreased

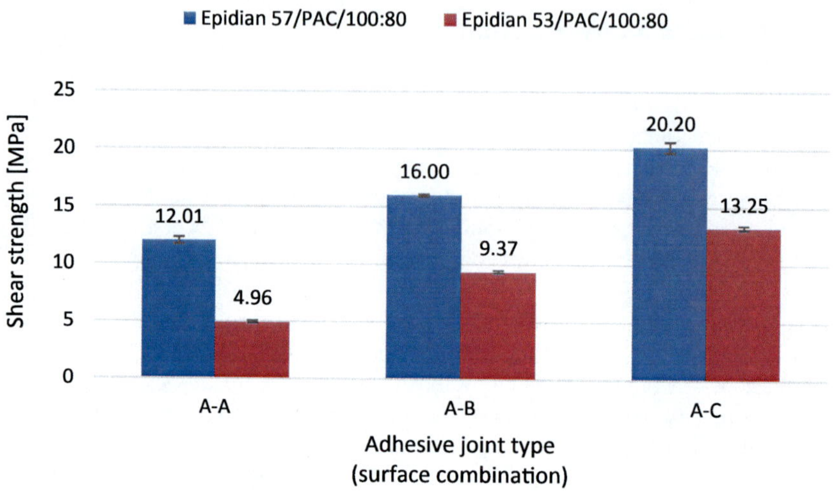

Fig. 2 Adhesive joint shear strength in A–A, A-B and A-C configurations, prepared with two types of epoxy adhesive

only prior to bonding with Epidian 53/PAC/100:80. Those test specimens were not mechanically pre-treated. Here, the mean shear strength for A–A was 4.97 MPa. The difference between the highest and lowest mean shear strength values was 75%.

Note also that of all the tested and analysed adhesive joint configurations shown in Fig. 2, the shear strength values developed with Epidian 57/PAC/100:80 were higher than the shear strength values developed with Epidian 53/PAC/100:80 The differences between the epoxy adhesive types were significant in all adhesive joint configurations: 59% (A–A), 41% (A-B) and 35% (A-C).

Figure 3 shows a comparison between the mean shear strength values of the adhesive joint configurations where both surfaces were mechanically treated, i.e. B–B, B–C and C–C.

An analysis of the shear strength results shown in Fig. 3 revealed that the highest mean shear strength values were found in the adhesive joints where both surfaces were treated by cast steel shot blasting. This conclusion applies to both epoxy adhesives tested, Epidian 57/PAC/100:80 (25.30 MPa) and Epidian 53/PAC/100:80 (23.10 MPa). Adhesive joint configuration B–B produced a higher mean shear strength than adhesive joint configuration B–C with Epidian 57/PAC/100:80 (15.25 MPa). Adhesive joint configuration B–C did not provide significant shear strength differences between Epidian 57/PAC/100:80 (14.65 MPa) and Epidian 57/PAC/100:80 (14.30 MPa). Here, the difference was approximately 2%.

An overall comparison of the shear strength test results from the adhesive joint bonds revealed that the glass-bead blasting treatment was not very effective. Between the adhesive joint configurations treated by glass-bead shot blasting only (B–B) or by glass-bead shot blasting and degreasing (A-B) showed that the shear strength of the bonds varied by a small improvement to aggravation (for Epidian 57) between the

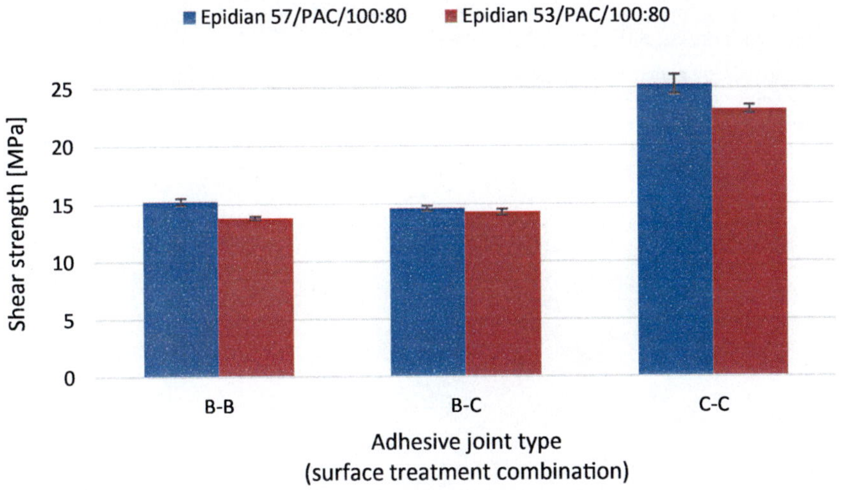

Fig. 3 Adhesive joint shear strength in B–B, B–C and C–C configurations, fabricated with two types of epoxy adhesives

epoxy adhesive types. This led to the conclusion that the glass-bead blasting could not be unquestioningly deemed as providing positive results, at least not in reference to the treatment parameters of the test specimens considered in this paper.

3.2 Surface Roughness

The surface roughness test results (produced as mean values) and the surface topography test results are shown in Table 4. All surface roughness test results are shown in Fig. 4. The following parameters were qualified for post-testing analysis: Ra (mean arithmetic roughness deviation from a mean line), Rz (peak roughness at 10 points of surface profile), and Rt (overall roughness profile height). These parameters were qualified as potential and preliminary predictors for an evaluation of surface treatment. Each parameter was applied to a different feature of surface roughness; this approach facilitated the investigation of each independently from all other qualified parameters.

The results showed in Table 4 and Fig. 4 the surface differences between the specific test specimens from the surface treatment variants. The differences were ascertained by isometric imaging. The highest mean surface roughness was found in the steel sheet surfaces pre-treated by broken cast steel shot blasting (C). The mean Rz values were 44.6 μm, while Rt reached 64.5 μm. The lowest surface roughness was found in the test specimens pre-treated by degreasing only (A). Rz was 5.62 μm and approximately 13% of Rz in surface treatment variant C. Rt was 7.71 μm in A and approximately 12% of Rt in surface treatment variant C. For the glass-bead blasting

Table 4 Test results for qualified surface roughness parameters with isometric test specimen surface imaging for different surface treatment variants

Surface treatment variant	Isometric imaging	Parameters	
A	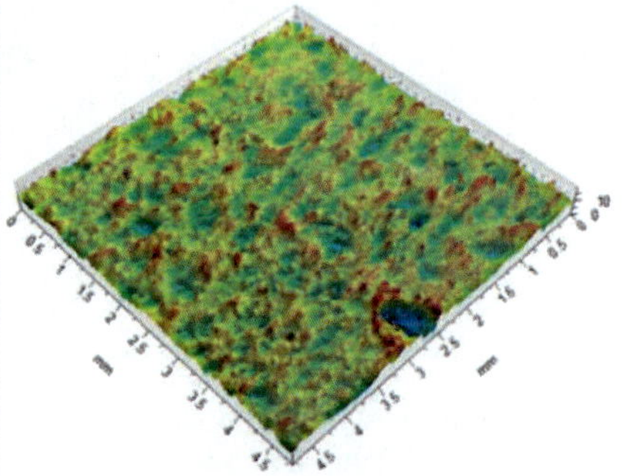	Ra	0.89 μm
		Rz	5.62 μm
		Rt	7.71 μm
		RSm	0.14 mm
B	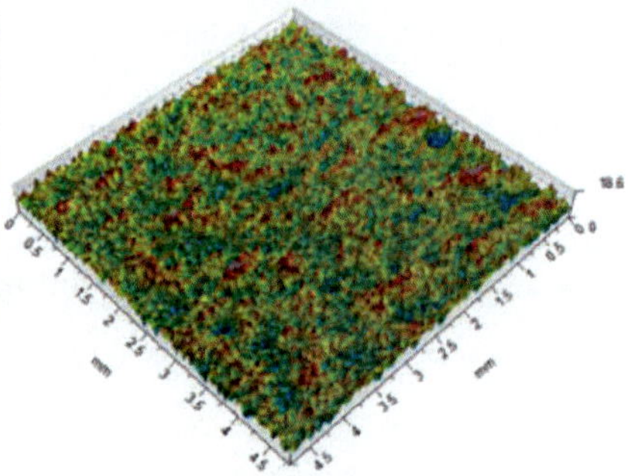	Ra	1.44 μm
		Rz	8.95 μm
		Rt	11.51 μm
		RSm	0.12 mm
C	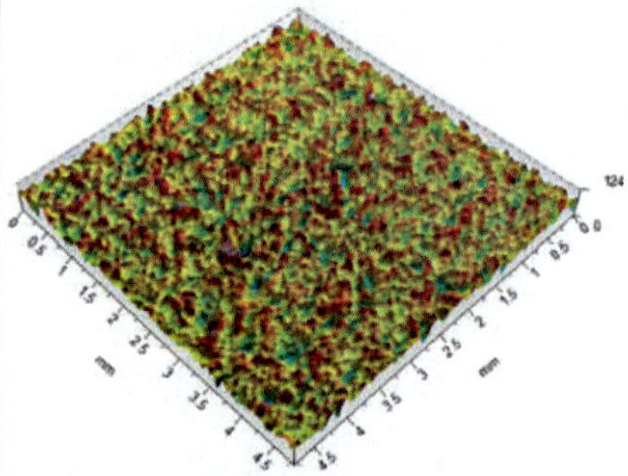	Ra	7.72 μm
		Rz	44.6 μm
		Rt	64.28 μm
		RSm	0.18 mm

treatment, Rz was 8.9 μm, or 37% higher than the Rz in surface treatment variant A and 80% lower than the Rz in surface treatment variant C (with both surfaces processed by glass-bead blasting).

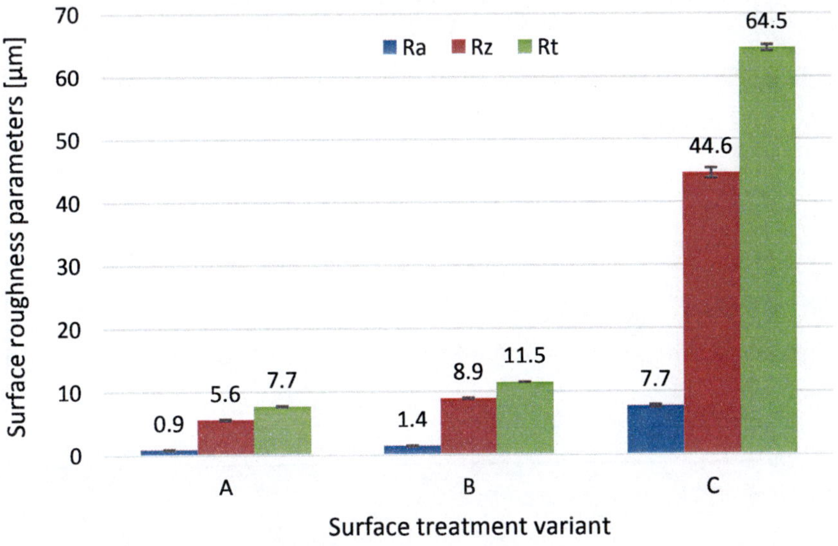

Fig. 4 Comparison of the values of qualified surface roughness parameters

3.3 Discussion of Test Results

The shear strength was analysed for the adhesive joint bonds fabricated with Epidian 57/PAC/100:80. For each adhesive joint configuration studied, this epoxy adhesive composition produced higher mean shear strength values than the adhesive joint configurations joined with Epidian 53/PAC/100:80. The analysis focused on the results from the adhesive joints with a uniform treatment (ref. Table 2) and the results of surface roughness heights produced by specific machine treatments. An attempt was made to identify a correlation between the analysed variables and to create mathematical models which could significantly describe the relationships between the geometrical parameters of surface treatment (i.e. roughness) and the resulting shear strength values of the adhesive joint bonds. The mathematical models could be then applied to estimate the strength parameters for surfaces with similar surface roughness parameters not analysed in this paper.

A statistical analysis was performed on the results from the testing. The TIBCO Statistica 13.3 software suite was used to analyse potential correlations between the three critical surface roughness parameters produced with each of the selected surface treatment methods and the final adhesive joint strength in the test specimens with a uniform surface treatment (A–A, B–B, and C–C). Figures 5, 6 and 7 give the correlation analysis results, while Table 5 presents a cumulative statement of the results.

The Pearson correlation coefficient (PCC) determined for each case analysed was approximately 0.98. This was due to a strong linear relationship between each of the

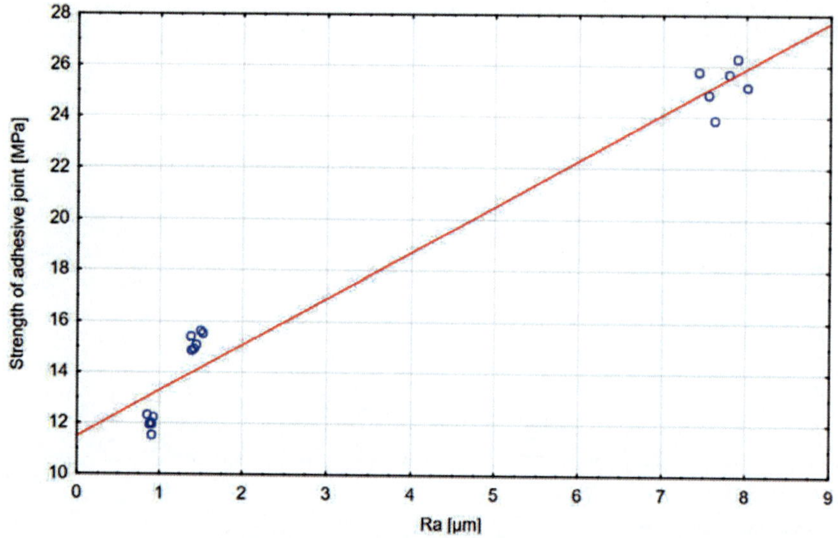

Fig. 5 Adhesive joint shear strength as a function of Ra adherend surface roughness parameter

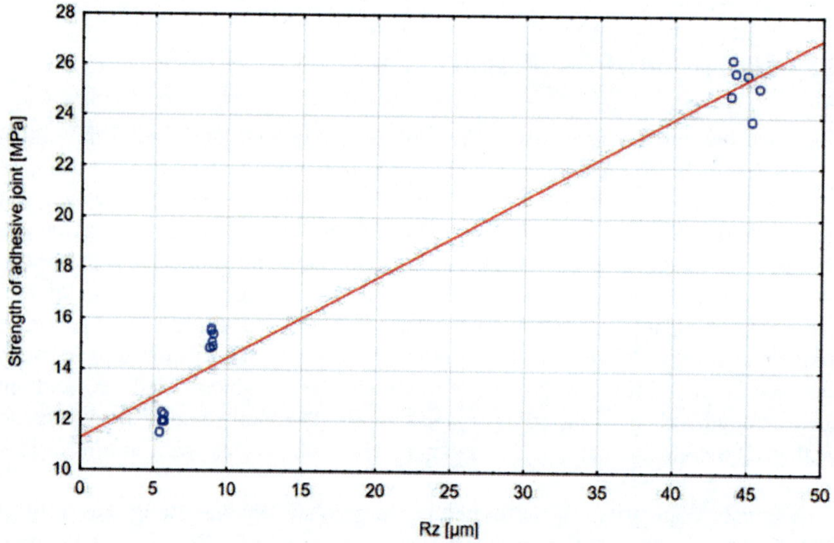

Fig. 6 Adhesive joint shear strength as a function of Rz adherend surface roughness parameter

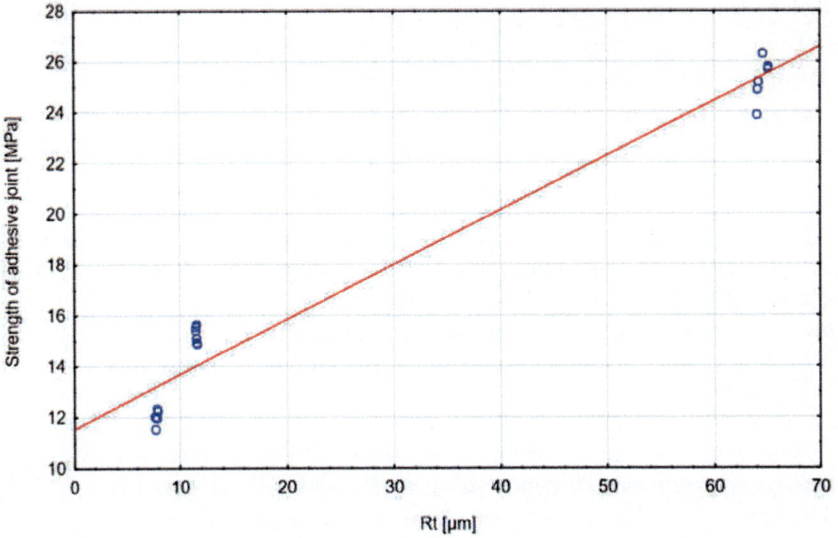

Fig. 7 Adhesive joint shear strength as a function of Rt adherend surface roughness parameter

Table 5 Correlation of surface roughness parameters [µm] to adhesive joint strength [MPa]

X and Y	Mean values	SD	PCC r(X,Y)	R^2	t	p	Regression constant Y to X	Regression coefficient Y to X
Ra [µm]	3.35	3.19						
Adhesive joint strength [MPa]	17.52	5.84	0.983	0.97	21.45	0.00	11.49	1.80
Rz [µm]	19.70	18.17						
Adhesive joint strength [MPa]	17.52	5.84	0.983	0.97	21.26	0.00	11.29	0.32
Rt [µm]	27.90	26.66						
Adhesive joint strength [MPa]	17.52	5.84	0.981	0.96	20.44	0.00	11.52	0.22

Table 6 Redundancy of independent variables

Variable	Tolerance	R^2	Partial	Semip.
Ra [μm]	1.000	0.000	0.983	0.983
Rz [μm]	0.001	0.999	0.047	0.009
Rt [μm]	0.002	0.998	-0.091	-0.017

surface roughness parameters considered (Ra, Rz, Rt) and the adhesive joint shear strength. The coefficient of determination, R^2 was 0.96-0.97 which meant that the adhesive joint strength could be explained by changes in each of the surface roughness parameters considered (Ra, Rz, Rt). For each correlation pair studied, a linear regression coefficient and a regression constant were determined. The regression constants were all very similar and within the range 11.29–11.52, irrespective of the surface roughness parameter analysed.

Each of the three surface roughness parameters was deemed independent of each other and – while they described identical adherends in each case – applicable to different adherend features. An attempt was made to define a collective model with a structure encompassing all three parameters. A reverse analysis of stepwise regression was performed, where a model comprising all potential variables had those variables removed which had the least significant effect on the dependent variable at each step, until the best model was produced [26]. This analysis revealed that the variable parameters of the adherend surface roughness featured redundancy, since the independent variables were a linear combination of other independent variables (Table 6).

Column R2 lists the coefficient of determination between the respective variable and all other independent variables; the semi-partial correlations, which achieved very small values, were partial to a low correlation between those variables and a dependent variable, which was the adhesive joint strength. This led to the conclusion that the best model to explain the mathematical relationship between the adhesive joint strength and a specific group of surface roughness parameters (which were initially investigated as independent variables) would be a linear model which included one of the surface roughness parameters, such as Ra.

The attempts to generate any cumulative statements in the form of 3D plots produced planes; while the surface roughness parameters defining other irregularity features were not deemed to be independent, ref. Figure 8.

A comparison of the results demonstrated that the increasing the surface roughness (as tested) increased the adhesive joint strength. The type of adhesive was important, or more precisely, its properties, with viscosity as the primary one. Jennings [27] underlined that the degree of roughness of a metallic adherend is frequently a design option for an adhesive joint. Rudawska et al. [20] was an investigation into the effects of milling on the adhesive joint shear strength in steel sheets, which also found that the highest adhesive joint shear strength was produced by bonding parts with the highest values of selected surface roughness parameters.

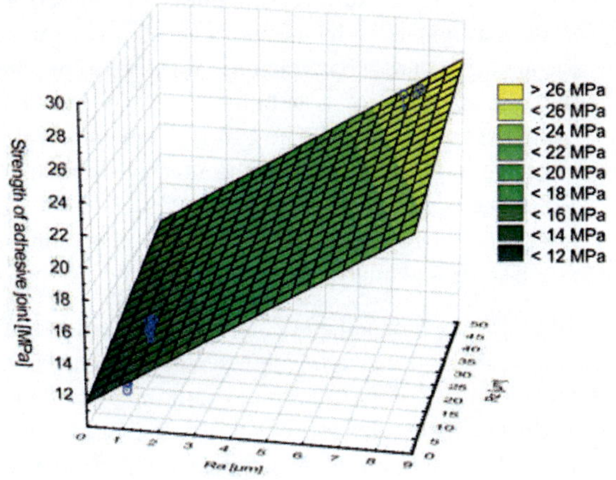

Fig. 8 Adhesive joint shear strength as a function of the adherend surface roughness parameter

Elbing et al. [28] indicated that the adhesive strength of aluminium adherends can be improved significantly with one mechanical treatment method, dry ice blasting. This is a rather complex problem; dry ice blasting causes slight changes only in the topographic surface structure and slightly increases the surface roughness parameters. The primary cause of strength development is the removal of the lubricant: the aluminium coxide and magnesium oxide film.

Staia et al. [4] emphasized that grit blasting produces a sufficient surface roughness to ensure mechanical anchoring between two materials (e.g. the coating and the substrate, which is similar to the effect between the adhesive and its adherend).

4 Summary and Conclusions

The analysis of the experimental test results provided the following conclusions:

- The adhesive joint fabricated with Epidian 53/PAC/100:80 featuring the lowest strength was the bonding of adherends pre-treated by degreasing only (A–A).
- The highest strength was developed by adhesive joints with both adherends (parts of the test specimen) pre-treated by shot blasting with broken cast steel shot (C–C) and bonded with Epidian 57/PAC/100:80.
- In those adhesive joints with one adherend pre-treated by broken cast steel shot blasting and the other adherend glass-bead blasted (B–C) had similar strength values, irrespective of the epoxy adhesive used.
- The highest values of the surface roughness parameters analysed was in the adherends pre-treated by broken cast steel shot blasting (C). The lowest values of the same parameters were provided by adherends pre-treated by degreasing only (A).

- Higher values of surface roughness parameters favoured better filling of the surface irregularities by the adhesives; this generated mechanical anchorage (once the adhesive cured) by which the adhesive joint bonds achieved a higher strength than the adhesive joints with the adherends that featured lower values of surface roughness parameters. The effect of mechanical adhesion on the strength of adhesive joints was higher in the adhesive bonding of rougher adherend surfaces.
- While they defined different surface roughness features, the tested surface roughness parameters (Ra, Rz, and Rt) were treated as independent from one another in the statistical analysis.

To conclude, the method of surface treatment for adhesive bonding is significant; a suitable method can produce a surface roughness of the adherends sufficient to favour anchorage of the adhesive within superficial micro-irregularities. If one of the adherends being bonded is rougher in texture, a stronger adhesive joint can be formed than in adhesive bonding of adherends which both have a low surface roughness. Another contributor is the type of epoxy resin applied for adhesive bonding of the adherends.

References

1. Packham DE (2003) Surface energy, surface topography and adhesion. Int J Adhes Adhes 23:437–448
2. Baldan A (2012) Adhesion phenomena in bonded joints. Int J Adhes Adhes 38:95–116
3. Gent AN, Lin C-W (1990) Model studies of the effect of surface roughness and mechanical interlocking on adhesion. J Adhes 32:113–125
4. Staia MH, Ramos E, Carrasquero A, Roman A, Lesage J, Chicot D, Mesmacque G (2000) Effect of substrate roughness inducted by grit blasting upon adhesion of WC-17% Co thermal sprayed coatings. Thin Solid Films 377–378:657–664
5. Wang B, Hu X, Lu P (2017) Improvement of adhesive bonding of grit-blasted steel substrates by using diluted resin as a primer. Int J Adhes Adhes 48:92–99
6. Mandolfino Ch, Lertora E, Gambaro C (2013) Effect of surface pretreatment on the performance of adhesive-bonded joints. Key Eng Mater 554–558:996–1006
7. Rudawska A (2014) Selected aspects of the effect of mechanical treatment on surface roughness and adhesive joint strength of steel sheets. Int J Adhes Adhes 50:235–243
8. Teng JG, Fernando D, Yu T, Zhao XL (2011) Treatment of steel surfaces for effective adhesive bonding. In: Ye L et al (eds) Advances in FRP composites in civil engineering. Tsinghua University Press, Beijing and Springer-Verlag, Berlin Heidelberg
9. Rudawska A (2012) Surface free energy and 7075 aluminium bonded joint strength following degreasing only and without any prior treatment. J Adhes Sci Technol 26:1233–1247
10. Rudawska A, Nalepa J, Müller M (2017) The effect of degreasing on adhesive joint strength. Adv Sci Technol-Res J 11:75–81
11. Ebnesajjad S, Ebnesajjad C (2013) Surface Treatment of Materials for Adhesive Bonding, 2nd edn. Inc. Norwich, William Andrew, New York
12. Adams RD, Comyn J, Wake WC (1997) Structural adhesive joints in engineering book. Springer, United Kingdom
13. Bouquet F, Cuntz JM, Coddet C (1992) Influence of surface treatment on the durability of stainless steel sheets bonded with epoxy. J Adhes Sci Technol 6:233–242
14. Harris AF, Beevers A (1999) The effect of grit-blasting on the surface properties for adhesion. Int J Adhes Adhes 19:445–452

15. Rudawska A, Danczak I, Müller M, Valasek P (2016) The effect of sandblasting on surface properties for adhesion. Int J Adhes Adhes 70:176–190
16. Sancaktar E, Gomatam R (2001) A study on the effects of surface roughness on the strength of single lap joints. J Adhes Sci Technol 15:97–117
17. Rotella G, Alfano M, Schiefer T, Jansen I (2016) Evaluation of mechanical and laser surface pre-treatments on the strength of adhesive bonded steel joints for automotive industry. J Adhes Sci Technol 30:747–758
18. Ebnesajjad S (2011) In: Handbook of adhesives and surface preparation, Chapter 6. surface prepartaion of metals, Plastic Design Library, Elsevier Inc, pp 83–106
19. Da Silva LFM, Ferreira NMAJ, Richter-Trummer V, Marques EAS (2010) Effect of grooves on the strength of adhesively bonded joints. Int J Adhes Adhes 30:735–743
20. Rudawska A, Reszka M, Warda T, Miturska I, Szabelski J, Skoczylas A, Stancekova D (2016) Milling as method of surface pretreatment of steel for adhesive bonding. J Adhes Sci Technol 30:2619–2636
21. Brack N, Rider AN (2014) The influence of mechanical and chemical treatments on the environmental resistance of epoxy adhesive bonds to titanium. Int J Adhes Adhes 48:20–27
22. Khan AA, Al Kheraif AA, Alhijji SM (2016) Effect of grit-blasting air pressure on adhesion strength of resin to titanium. Int J Adhes Adhes 65:41–46
23. Rudawska A (2020) The influence of curing conditions on the strength of adhesive joints. J Adhes 96:402–422
24. Rudawska A, Wahab MA, Müller M (2020) Effect of ageing process on mechanical properties of adhesive tubular butt joints in aqueous environment. Int J Adhes Adhes 96:1–11
25. Folder of Chemical Ciech Grup (2019) http://ciechresins.com/files/katalogi/EPIDIAN_CATALOG_PL_PAGE_BY_PAGE.pdf. Last Accessed 20 Dec 2019
26. Rabiej M (2012) Statistics with the statistica program (Statystyka z programem statistica). Helion, Poland (in Polish)
27. Jennings CW (1972) Surface roughness and bond strength of adhesives. J Adhes 4:25–38
28. Elbing F, Anagreh N, Dorn L, Uhlmann E (2003) Dry ice blasting as pretreatment of aluminium surfaces to improve the adhesive strength of aluminum bonding joints. Int J Adhes Adhes 23:69–79

Determination of the Effective Stiffness of Half-Open Cross-Section Bars and Orthotropic Steel Deck of a Truss Bridge Using Model Updating

Viet Long Ho, Guido De Roeck, Thanh Bui-Tien, and Magd Abdel Wahab

Abstract Vibration-based model updating is widely accepted as an effective technique to construct a baseline model for structural health monitoring. In this paper, the effects of half-open cross-section and orthotropic deck on the global vibration behaviour of a truss bridge are considered. To do that, an FE model using ANSYS is built with variations of half-open cross-section area of bars and the stiffness of the orthotropic steel deck. The cross-section area of bars are assigned a reduced coefficient, less than and equal to one. Based on a vibration test, a hybrid optimization algorithm, known as Hybrid Particle Swarm Optimization and Gravitational Search Algorithm (PSOGSA), is applied to reduce the differences in frequencies and mode shapes between measurements and FE simulations. Finally, the effective coefficients of the cross-section as well as the stiffness of steel deck are obtained from the updated model.

Keywords Model updating · Hybrid particle swarm optimization and gravitational search algorithm (PSOGSA) · Double-decked bridge · Half-open cross-section

V. L. Ho · M. Abdel Wahab
Soete Laboratory, Faculty of Engineering and Architecture, Ghent University, Ghent, Belgium
e-mail: hoviet.long@ugent.be; longhv_ph@utc.edu.vn

M. Abdel Wahab
e-mail: Magd.AbdelWahab@ugent.be

V. L. Ho · T. Bui-Tien (✉)
Department of Bridge and Tunnel Engineering, Faculty of Civil Engineering, University of Transport and Communications, Hanoi, Vietnam
e-mail: btthanh@utc.edu.vn

G. De Roeck
Department of Civil Engineering, KU Leuven, 3001 Leuven, Belgium
e-mail: guido.deroeck@kuleuven.be

M. Abdel Wahab (ed.), *Proceedings of the 8th International Conference on Fracture, Fatigue and Wear*, Lecture Notes in Mechanical Engineering,
https://doi.org/10.1007/978-981-15-9893-7_6

1 Introduction

One the most important activities for vibration-based Structural Health Monitoring (SHM) is to construct a baseline FE model of considered structures. The FE model then is used as an assessment tool for the structural state. Identification, localization, quantification of damage of structures can be derived from the updated FE model. The indirect model updating based on optimization algorithms shows its dominance on accuracy, reliability, and fast computation.

Some popular optimization algorithms such as Genetic Algorithm (GA), Particle Swarm Optimization (PSO) [1], Simulated Annealing (SA) [2], etc. were applied to reduce errors between numerical and experimental results.

In this approach, modal parameters obtained from vibration measurements, e.g. natural frequencies, mode shapes, and damping ratios, are utilized as a fitness function. After a sufficient number of iterations or fulfillment of a stop requirement, the error between simulations and experiments will be reduced substantially. New values of common updating parameters such as young modulus (E), density (γ), stiffness of bearings, etc. are identified. In turn, these new values are treated as inputs to build a baseline model of the tested structure.

Vibration of structures can be achieved via artificial excitation, ambient excitation [3] or a combination of the two methods [4, 5]. However, for a large structure, it is not easy, even with dedicated devices, to generate an excitation that is large enough to obtain many clear vibration modes. Therefore, ambient loads e.g. traffic loads, wind loads, micro-tremors prevail over artificial loads. Besides, no interruption of traffic is required during measurement. With a significant traffic flow across the bridge, this is an important key-point in considering a suitable excitation method.

In addition to updating parameters, this paper aims to consider the effects of half-open cross-section and orthotropic steel deck on vibration of a truss bridge. Under ambient excitation, dynamic responses of the tested bridge were recorded. The covariance-driven stochastic subspace identification method (SSI-COV[1]) was used to determine modal characteristics. SSI-COV was recommended due to its accuracy as well as straightforward computation procedure [8]. To minimize the gap between experiments and calculations use was made of the hybrid gravitational search algorithm—particle swarm optimization (PSO-GSA). The obtained results of the model updating are improved values of the uncertain physical parameters.

2 Hybrid Particle Swarm optimization- Gravitational Search Algorithm (PSO-GSA)

In 2009, Rashedi et al. developed a new optimization algorithm, namely gravitational search algorithm (GSA) based on the Newton law of gravity and mass interactions

[1]Details of SSI-COV in Ref. [6, 7].

between agents in universe [9]. However, its performance of convergence was not good enough. Therefore, in 2014, Mirjalili et al. studied and proposed a new hybrid, namely PSOGSA [10]. The hybrid optimization algorithm utilized the social thinking of PSO and local search ability of GSA to look for the best fitness. Firstly, the gravitational forces of agent j that applied on agent i were calculated based on random positions x_i, x_j of agents in search space and Euclidian distances between two agents R_{ij} as follows:

$$F_{ij}(t) = G(t) \times \frac{M_{pi}(t) \times M_{ai}(t)}{R_{ij}(t) + \varepsilon} \times \left(x_j(t) - x_i(t)\right) \quad (1)$$

where

$$R_{ij}(t) + \varepsilon \approx norm\left(x_j(t) - x_i(t)\right) \quad (2)$$

where M_{pi}, M_{aj}: passive and active gravitational mass of i and j agents. They can be determined from the best, worst, and current values of the fitness:

$$M_i(t) = \frac{m_i(t)}{\sum_{j=1}^{N} m_j(t)} \quad (3)$$

$$m_i(t) = \frac{fitness_i(t) - worst(t)}{best(t) - worst(t)} \quad (4)$$

Gravitational constant G(t) is identified as following:

$$G(t) = G_0 \times e^{\left(-\alpha.\frac{current-iteration}{\max-iteration}\right)} \quad (5)$$

Then, the total force of N agents on agent i in Eq. (7) was used to identify acceleration of agent i at time t with mass M_i:

$$a_i(t) = \frac{F_i(t)}{M_i(t)} \quad (6)$$

$$F_i(t) = \sum_{j=1, j \neq i}^{N} rand \times F_{ij}(t) \quad (7)$$

The cognitive coefficient c_c and the social coefficient c_s of PSO then were input into Eq. (8) to update velocities of agents with the global best of all particles and inertial coefficient, $gbest$ and ω.

$$v_i(t+1) = \omega \times v_i(t) + c_c \times rand_{(0-1)} \times a_i(t) + c_s \times rand_{(0-1)} \times \left[gbest_i(t) - x_i(t)\right] \quad (8)$$

Finally, after each iteration, positions of agents were determined as follows:

$$x_i(t+1) = x_i(t) + v_i(t+1) \tag{9}$$

Iteration will be stopped when the termination condition is met. The details of PSO and GSA are described in Ref [9]. and [11].

2.1 Measurements Description

Data Acquisition in Situ:

Figure 1 shows the measurement grid required to obtain the vibration of Thang Long bridge, a double-decked truss bridge. Twenty-two setups with eight accelerometers in each setup, were considered in a four-day measurement campaign. Acquisition time for each setup was 20 min under ambient load. Due to the limited number of accelerometers, only half of the points at the upper layer were instrumented (Fig. 1b). In-situ sensor placement is shown in Fig. 2.

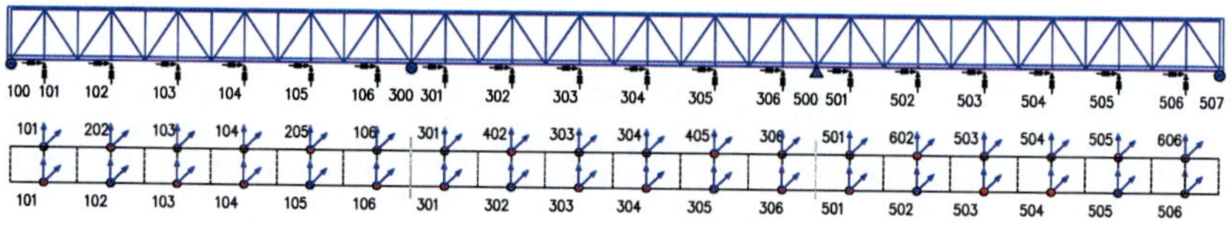

a) At lower layer

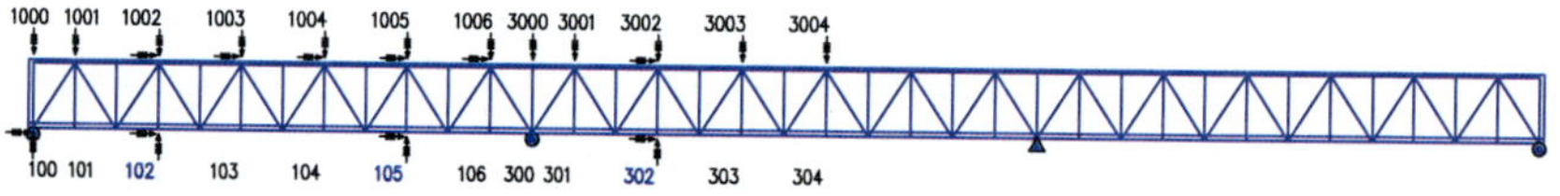

(b) At upper layer

Fig. 1 Sensor placement at both lower and upper layers

Sensor at an upper truss joint

Measuring bearing's movement

Fig. 2 Sensors placement in extra vibration measurements

2.2 System Identification

Data Processing:

The Fast Fourier Transform (FFT) was used to convert the collected signals from time domain to frequency domain as shown in Fig. 3. Then covariance-driven stochastic subspace identification (SSI-COV) algorithm was applied to determine modal properties from dynamic responses of the bridge. The first four vertical bending modes and cross correlation of mode shapes are shown in Figs. 4 and 5.

3 Finite Element Model

Thang Long is a double-decked truss bridge with 5 modules. Each module is constituted of 3 × 12 m continuous spans. The bridge has two layers. At the lower layer there are two railway tracks. The upper road deck is divided in four lanes for trucks and two lanes for pedestrians (see Fig. 6). Because of the similarity of structures, one module is considered in FE model.

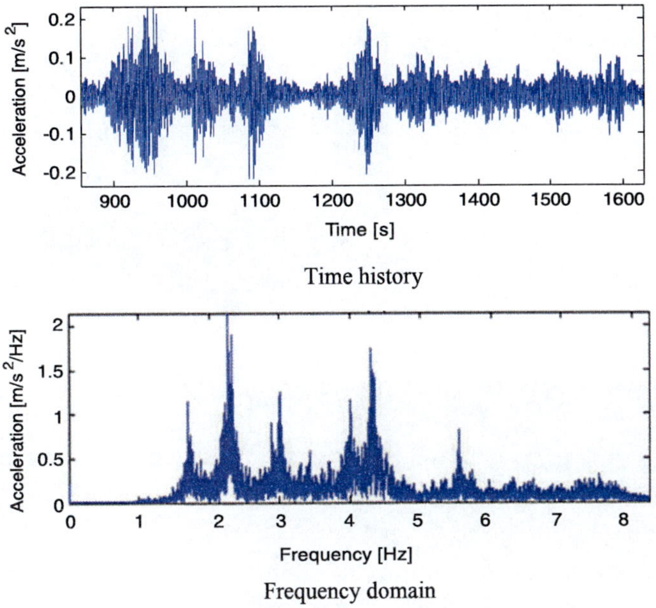

Fig. 3 Dynamic response of a typical vertical sensor

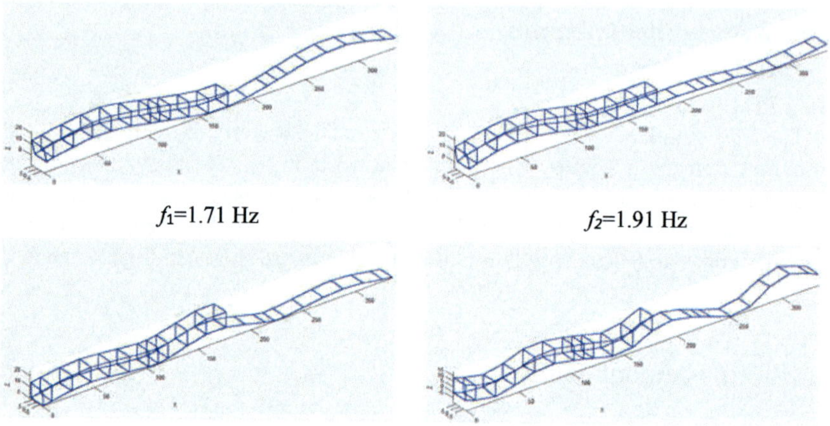

f_1=1.71 Hz

f_2=1.91 Hz

f_3=2.27 Hz

f_4= 3.96 Hz

Fig. 4 Four vertical bending modes

Fig. 5 MAC correlation of the first four identified vertical bending modes. Maximum values of MAC indices are located in the diagonal

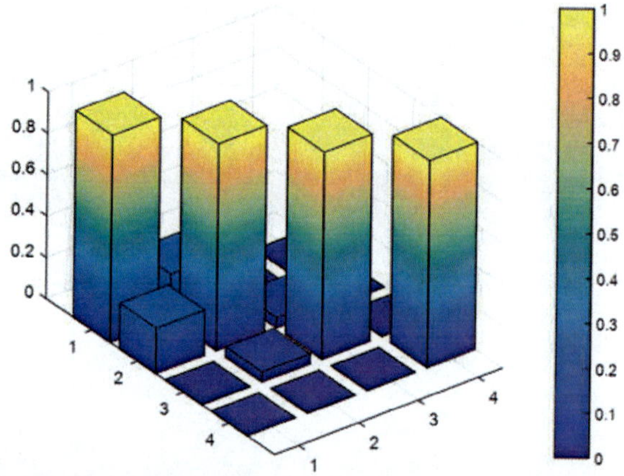

Fig. 6 Overview of Thang Long bridge and typical cross-section (unit *mm*)

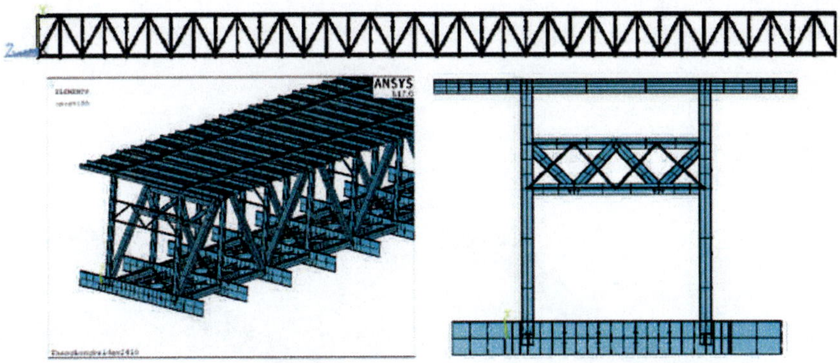

Fig. 7 FE model of one module of Thang Long bridge using ANSYS 17.0

3.1 *Initial Finite Element (FE) Model*

Input Parameters:

Material properties of steel components: Young's modulus $E = 2.1 \times 10^{11}$ N/m^2, Poisson's ratio $\nu = 0.3$, density = 7850 kg/m^3. Cross-sections of truss bars were taken from as-built drawings.

In ANSYS, a FE model was constructed using beam element 189 for truss components and shell element 181 for steel deck. The effects of non-structural components like, handrails, water pipes, rail tracks, wooden sleepers, etc., were simulated as added mass on the element faces. A 3-D model of the truss bridge is shown in Fig. 7.

In this initial FE model, holes (openings) on one side of the bars and extra mass at the truss joints due to steel plates and bolts are not considered. Properties of isotropic materials were used for all truss members. To simplify the model, stiffeners of the steel deck were not included as structural elements in the FE model. The effect of the stiffeners was modelled via an equivalent stiffness of the deck. Initially assumed modulus of steel deck were $E_{deck} = 3.0 \times 10^{11}$ N/m^2, based on composition of the orthotropic deck (Fig. 8).

Boundary Conditions:

Truss joints were assumed to be rigid. Figure 9 shows the arrangement of bridge bearings with two fixed bearings and six movable bearings.

FE Results:

The first four vertical bending modes are shown in Fig. 10. These mode shapes will be used to calibrate the initial FE model with the measurement ones (Fig. 4).

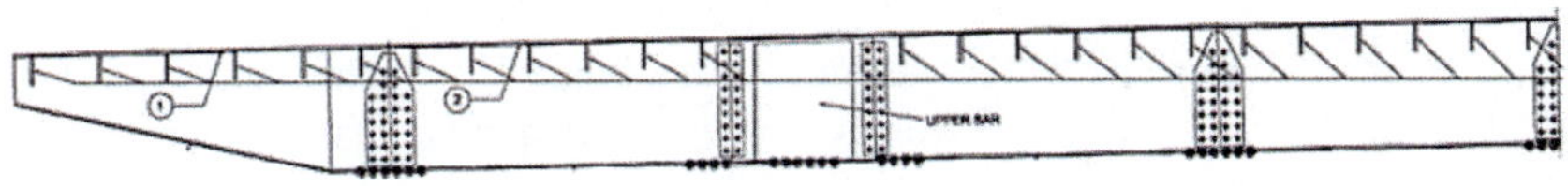

Fig. 8 Typical components of a half of the orthotropic steel deck

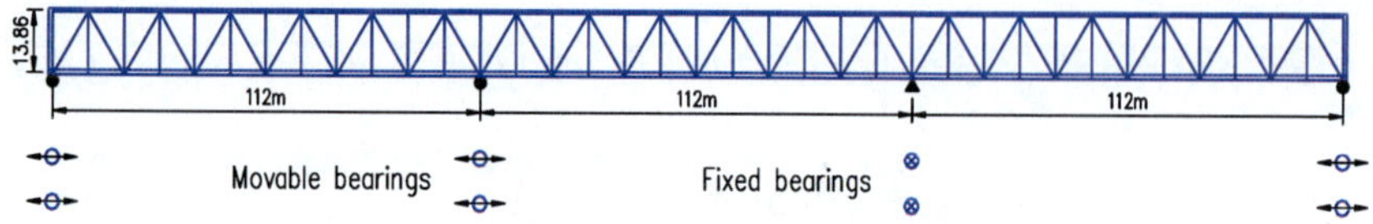

Fig. 9 Bridge bearings layout along the bridge

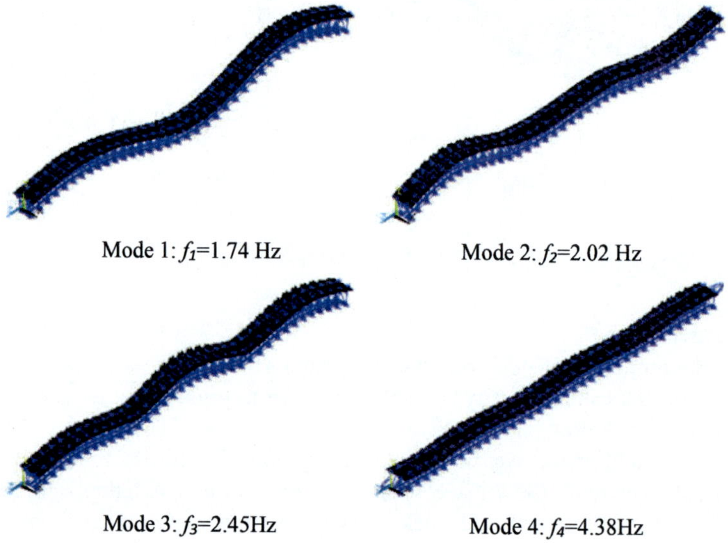

Fig. 10 Four calculated mode shapes

The MAC matrix was used to evaluate the correlation between experimental and simulated mode shapes (see Eq. (10)). Figure 11 shows that the mode shapes of measurement and modelling have a good match with all indices over 0.96. The frequency deviations are shown in Table 1. From the obtained results, errors between experiment and simulation of the last three modes are significant, from 5.76 to 10.61%. Therefore, to reduce the gaps, a model updating procedure is performed in the next section using PSOGSA.

$$MAC_{ij} = \frac{\left|(\varphi_i^{ex})^T \times \varphi_j^{sim}\right|^2}{\left\{\left(\varphi_i^{ex}\right)^T \times \varphi_i^{ex}\right\} \times \left\{\left(\varphi_j^{sim}\right)^T \times \varphi_j^{sim}\right\}} \tag{10}$$

where, MAC_{ij} is calculated between measured mode i and calculated mode j (j could be equal to i) φ_i^{ex}, φ_j^{sim}: experimental and simulated mode shape vectors, mode i and mode j.

Model Updating:

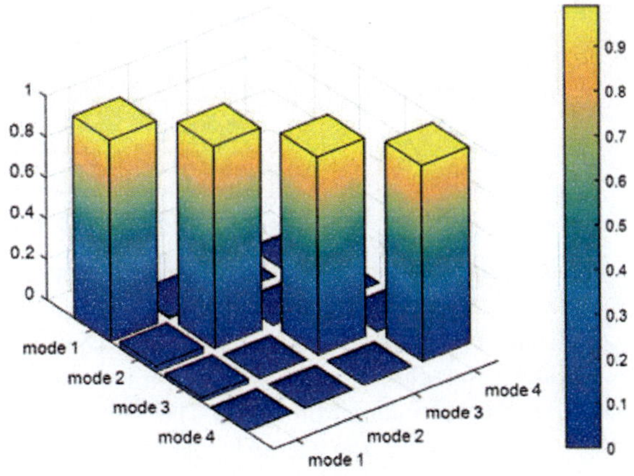

Fig. 11 MAC correlation between measured and calculated mode shapes (0.99, 0.99, 0.97, 0.96 are the diagonal MAC values)

Table 1 Discrepancies between calculated and measured frequencies of the first four vertical bending modes

Mode	Calculated frequency f_{cal} (Hz)	Measured frequency f_{meas} (Hz)	Deviation (*) (%)
1	1.74	1.71	1.75
2	2.02	1.91	5.76
3	2.45	2.27	7.93
4	4.38	3.96	10.61

Note: (*) difference = $[f_{cal} - f_{meas}] \times 100\%/f_{meas}$

From the obtained results above, it is necessary to update some parameters to achieve a smaller error between measurement and calculation. Four updating parameters were proposed: Young's modulus E and density γ of steel, effective cross-section of the half-open bars and stiffness of the orthotropic deck.

As mentioned above, uncertain stiffness of the orthotropic deck was modeled by changing the Young's modulus in intervals $E_{deck} = [2.6 \times 10^{11}–3.4 \times 10^{11}]$ N/m^2.

The effect of openings in the bars was taken into account via a coefficient α less than or equal to 1. The value of 1 implies a full cross-section. Approximate reduction values were determined based on cross-section area of bars. Firstly, a reduced volume of one bar (Fig. 12) was calculated by subtracting the oval holes. Then, this volume was divided by the length of the bar to obtain an equivalent area per meter. Finally, this

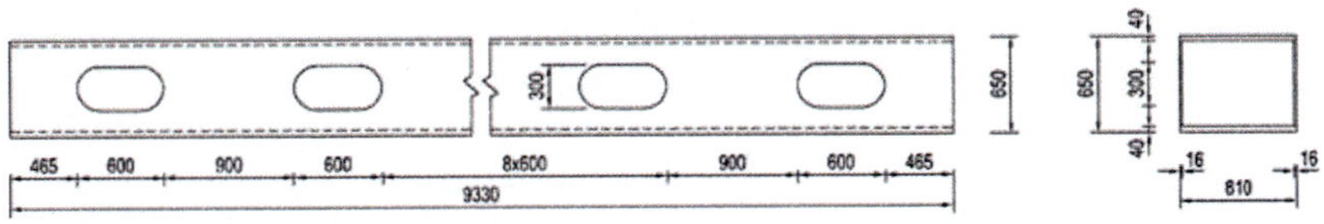

Fig. 12 An opening lower bar with oval hollows on one side

Table 2 Variation of reduction coefficient for half-open bars

Name	Value	Range of change	Note
α_1	0.93	0.89—0.94	Diagonal bars 1 to 4, lower bars
α_2	0.95	0.91–0.96	Diagonal bars 5, 6
α_3	0.98	0.94–0.99	Upper bars

obtained area was compared to a full area for determining the reduction coefficient α. Due to variations in manufacturing as well as real length of truss bars, these coefficients are assumed to vary in the range shown in Table 2.

Young's modulus $E = 2.1 \times 10^{11}$ and weight density $\gamma = 7850$ kg/m^3 were considered in an interval of $\pm$ 5%. The extra mass at the truss joints due to steel plates and bolts were considered by using mass21 element in ANSYS.

Some preset parameters for the PSO-GSA algorithm are $c_c = 0.5$, $c_s = 1.5$, $\alpha = 20$, $G_0 = 1$. Number of agents (particles) was 30 and number of iterations was 35. A fitness function of frequencies and MAC values was proposed as:

$$\text{fitness} = \sum_{i=1}^{m}\left|1 - \frac{f_{sim}^{i} x f_{meas}^{i}}{\left(f_{meas}^{i}\right)^{2}}\right| + \sum_{i=1}^{m}(1 - MAC_{ii}) \tag{11}$$

where f_{sim}^{i}, f_{meas}^{i}: simulated and measured frequencies, *m*: number of considered modes.

Convergence curve of PSOGSA is plotted in Fig. 13. The updated values of parameters are shown in Table 3.

After model updating, discrepancies between calculated and measured frequency of the first three modes are under 5%. Frequency deviation of the fourth mode is also reduced in the updated FE model compared to the initial one, down to 5.56%. MAC

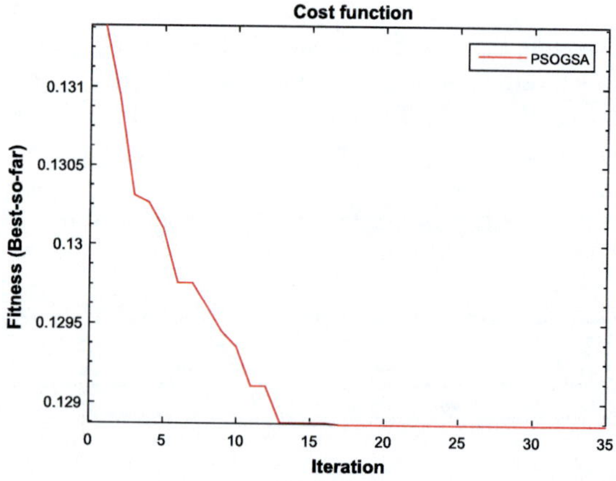

Fig. 13 Relationship between the fitness function and iteration number of PSOGSA

Table 3 Parameters after updating model

No	Updating parameters	Before updating	After updating
1	Young's modulus, E	2.1×10^{11}	2.048×10^{11}
2	Density, γ	7850	8202
3	Young's moduli of deck, E_{deck}	3.0×10^{11}	3.103×10^{11}
4	Sectional reduction factors		
	α_1	0.98	0.9514
	α_2	0.95	0.9327
	α_3	0.93	0.8928

indices also were slightly improved for this mode from 0.96 to 0.97. New values of frequencies and MAC indices are shown in Table 4 and Fig. 14.

Table 4 Updated values of frequencies and MAC indices (*) = $[f_{cal} - f_{meas}] \times 100\%/f_{meas}$

Mode	Frequency				MAC	
	Measured Hz	Initial Hz	Updated Hz	Deviation (*) %	Initial	Updated
1	1.71	1.74	1.65	-3.51	0.99	0.99
2	1.91	2.02	1.91	0	0.99	0.99
3	2.27	2.45	2.30	1.32	0.97	0.97
4	3.96	4.38	4.18	5.56	0.96	0.97

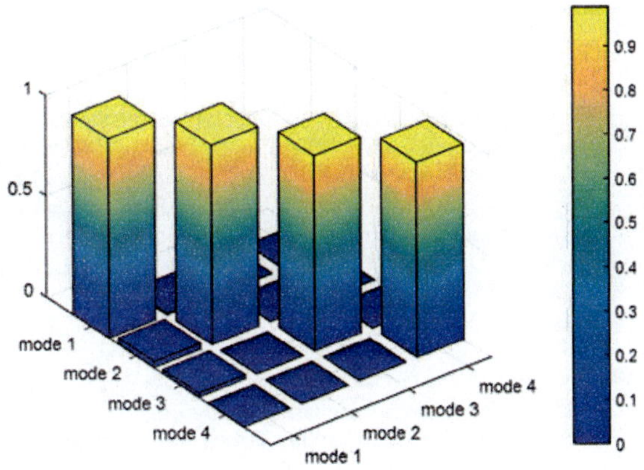

Fig. 14 MAC values after updating model (0.99, 0.99, 0.97, 0.97)

4 Conclusions

The hybrid optimization algorithm PSOGSA showed a promising ability for model updating. From the obtained results, with a moderate number of agents (particles) and iterations, the fitness function converged quickly, after 12 to 15 iterations.

On the other hand, it can be seen that effects of half-open cross-section and orthotropic steel deck should to be taken into account in building a FE model of the tested truss bridge. These are intermediate results in a bridge repairing project. The baseline FE model will be used to assess structural health after repairing bridge deck.

Acknowledgements The authors acknowledge the financial support of VLIR-UOS TEAM Project, VN2018TEA479A103, 'Damage assessment tools for Structural Health Monitoring of Vietnamese infrastructures', funded by the Flemish Government.

Authors would like to express appreciation to Mr. Minh Tran-Quang and colleagues from UCT company, Vietnam for supporting the measurement.

References

1. Tran-Ngoc H, Khatir S, De Roeck G, Bui-Tien T, Nguyen-Ngoc L, Abdel WM (2018) Model updating for nam o bridge using particle swarm optimization algorithm and genetic algorithm. Sensors 18(12):4131
2. Astroza R, Nguyen LT, Nestorović T (2016) Finite element model updating using simulated annealing hybridized with unscented Kalman filter. Comput Struct 177:176–191
3. Ároch R, Sokol M, Venglár M (2016) Structural health monitoring of major danube bridges in bratislava. Procedia Eng 156:24–31
4. Wahab MA, Roeck GD (1998) Dynamic testing of prestressed concrete bridges and numerical verification. J Bridge Eng 3(4):159–169
5. Reynders E, Roeck GD (2008) Reference-based combined deterministic–stochastic subspace identification for experimental and operational modal analysis. Mech Syst Signal Process 22(3):617–637
6. Peeters B, De Roeck G (1999) Reference-based stochastic subspace identification for output-only modal analysis. Mech Syst Signal Process 13(6):855–878
7. Ho VL, Tran NH, De Roeck G, Bui TT, Abdel Wahab M (2018) System identification based on vibration testing of a steel I-beam. In: Proceedings of the 1st international conference on numerical modelling in engineering, pp 254–268
8. Peeters B, De Roeck G (2001) Stochastic system identification for operational modal analysis: a review. J Dyn Syst Meas Contr 123(4):659–667
9. Rashedi E, Nezamabadi-pour H, Saryazdi S (2009) GSA: a gravitational search algorithm. Inf Sci 179(13):2232–2248
10. Mirjalili S, Wang G-G, Coelho L dos S (2014) Binary optimization using hybrid particle swarm optimization and gravitational search algorithm. Neural Comput and Applic, 25(6):1423–1435
11. Kennedy' J, Eberhart R (1995) Particle swarm optimization. In: Proceedings of ICNN'95—international conference on neural networks, Perth, WA, Australia, vol 4. pp 1942–1948

A Study On Optimal Design of Longitudinal Shape For Improving Small-Overlap Performance

Nguyen Phu Thuong Luu and Ly Hung Anh

Abstract This paper presents a study on the optimal design of longitudinal shape for improving the small-overlap performance, based on a computer-based crash simulation model. The small-overlap frontal impact (SOFI) event was simulated using explicit finite element method. The models were developed for the simulation according to the Insurance Institute for Highway Safety IIHS real test conditions with the Flat 150 mm radius rigid barrier and 25% overlap. Several different cross sections of longitudinal members were subjected to dynamic compression load, which occurs in small overlap frontal impact. The different shapes of longitudinal structure were compared initially to obtain the cross section that fulfills the small overlap performance criteria. The evaluated performance parameters included the absorbed crash energy, crush force efficiency, ease of manufacture and cost. Once the cross section was selected, the design was further enhanced for better crashworthiness performances by investigating the effect of material characterization, increasing the wall thickness and by introducing a trigger mechanism. Real experiments were also performed for .... The results of this study showed that the multi edges profile with 2 mm wall thickness and steel material was a good candidate for energy absorption in SOFI condition.

Keywords Small-overlap · Crashworthiness · Frontal impact · Vehicle structure · Longitudinal · IIHS crash test

N. P. T. Luu (✉)
Automotive Engineering Department, Ho Chi Minh City University of Technology (HUTECH), Ho Chi Minh City, Vietnam
e-mail: npt.luu@hutech.edu.vn

L. H. Anh
Department of Aerospace Engineering, Faculty of Transportation Engineering, Ho Chi Minh City University of Technology (HCMUT), Ho Chi Minh City, Vietnam

Vietnam National University, Ho Chi Minh City, Vietnam

M. Abdel Wahab (ed.), *Proceedings of the 8th International Conference on Fracture, Fatigue and Wear*, Lecture Notes in Mechanical Engineering,
https://doi.org/10.1007/978-981-15-9893-7_7

1 Introduction

Small-overlap frontal impact (SOFI) test was released by the Insurance Institute for Highway Safety (IIHS) in 2012 to challenge auto makers for safer production as one of the most serious crashes [1]. In this test, the vehicle collided with a rigid barrier at 64 km/h at 25% overlap. This is similar to a vehicle crashing into a pole or flat object, but only on the part of the headlight from the left side of the driver [2]. The barriers for crash tests were shown in Sect. 2.2 [3].

In SOFI test, the crash forces are applied on the outer of the vehicle's longitudinal frame rails [4, 5], which caused serious compartment intrusion. Sherwood et al. recommended that frontal structure designs had to improve their ability to protect passengers in case small-overlap frontal crashes occur [6]. Furthermore, forces are transferred to the front suspension, the left wheel, and the base of the A-pillar. These zones are damaged severely because of more energy absorb in the frontal structure crashworthiness as in [7–9]. Due to passenger safety, researchers proposed the self-protection and partner-protection for new vehicles [10].

In recent related studies on the small-overlap crashes, offsetting oblique and small overlap impact tests were proposed by the national highway traffic safety administration (NHTSA). The IIHS also studies on moderate and small overlap tests, but in their test the overlap reduces up to 25%. The test methods from IIHS and NHTSA had differences regarding the consequence in vehicle deformation, vehicle kinematics, and occupant kinematics [11]. Results of Nguyen (2018) showed that the rails absorbed over 50% of full vehicle energy when the crash occurred in the IIHS test [12]. Based on study on-site of crash accident, Park et al. [13] has concluded that the offset-frontal crash test can be applied to represent the full-frontal crash. In test of crash accident, frontal structure of vehicle strikes partly to barrier so the car is pushed back side away from the barrier due to reaction. In the results of IIHS, it was show that the SOFI could lead to critical dummy injury. Hence, researchers released a study of occupant protection simulation for IIHS small overlap frontal crash test to predict the dummy injury [14]. In 2014, Lee et al. researched female dummy in new car assessment program (NCAP) full frontal crash test to improve the severity of the star rating [15].

A study on the extent of injury in many automotive front crash tests has been proposed to improve safety performance [16]. In addition to passenger vehicles, light truck and bus structures [17] are also studied to improve rigidity of structure. Similar research has been conducted by numerical simulation in [18] to optimize crash capability for light truck. The methods are found here is filling structural foam in the A-columns and side panels, adding a roof crossbar and reinforcing the rear wall of the cabin. The deceleration pulse is recommended to amend the design of an automobile for crashworthiness [17, 19].

Currently, there are many researchers have studied on vehicle structures to improve the IIHS rating. In 2012, Halloway et al. proposed classifiers to indicate the number of longitudinal members [20]. In 2013, Jakobsson et al. illustrated the significance and developments of car design for frontal crashes with partial overlap [21]. Based on the frontal offset crash protocol in IIHS test, the body structure crashworthiness was investigated to improve occupant safety [22]. In small overlap crash test, the longitudinal beams were almost misses so that [author name] investigated their effects on vehicle crashworthiness [23]. They suggested new design and improvement for vehicle longitudinal beams. In 2007, Oh et al. suggested the size optimization of an engine room member for crashworthiness using response surface method (RSM) [24]. The crashworthiness design using topology optimization method was shown in [25–28]. In order to increase the safety of the structure and minimize the risk of pedestrian leg injury in crash accidents, an optimization of the sturdy design of the fore structure is proposed as shown in [29]. The energy absorption can be controlled to design the collision potential based on the use of compliance mechanisms as described in [30]. The optimal thickness design and reinforcement designs, which are proposed based on the IIHS small-overlap research program [31], are applied to the vehicle model to be used in SOFI simulations. The aim of this study is to improve the overall ratings for frontal structure, performance in the event of a SOFI based on the crash test simulations. Following the SOFI test simulation results, the sensitive components will be identified to strengthen the vehicle structure. In this paper, the best case among of them is chosen to improve the vehicle structure in the SOFI test. The reinforcements and frontal structures are enhanced by using a thickness optimization method which is proposed to improve the structural resistance of vehicle to an SOFI and to maintain the constant mass of the vehicle model. The optimized designs with the best crashworthiness intrusion results are proposed for the comparison with the original model. In this study, a vehicle computer model [32] is applied to real IIHS crash test condition [33–38] using LS-DYNA [39] software for the crash test simulations.

Following the introduction, Sect. 2 presents the finite element modeling for small overlap frontal impact, which is developed for the SOFI test simulation. The SOFI simulation is conducted based on experimental model in Sect. 3. The most appropriated structure is evaluated from results of numerical simulation. After that, the design of longitudinal structure is optimized. Section 4 presents the results and discussion of the optimize cross section of longitudinal member.

2 Finite Element Modeling for Sofi Analysis

2.1 Finite Element Modeling of Longitudinal Column

In this paper, the longitudinal parameters obtained from a minivan FE model which is available in NCAC, 2005. Crash simulations is based on the IIHS small-overlap crash test (IIHS 2012). Different profiles of longitudinal member following SOFI condition are described in Table 1. In this paper, the length of the tube is considered L = 1500 mm, the perimeter of the cross section is P = 520 mm and the thickness of the longitudinal is 2.1 mm. The impact load is applied by the rigid wall and

Table 1 Geometry and dimensions of longitudinal with different profiles

Profile	Specimen ID	Perimeter [mm]	Length [mm]	Cross section	Dimension [mm]	Thickness [mm]	3D
Rectangle	T_rect	520	1500		150 × 110	2.1	
Square	T_squ	520	1500		130 × 130	2.1	
Circular	T_cir	520	1500		Radius = 83	2.1	
Ellipse	T_ell	520	1500		95 × 70	2.1	
Hexagonal	T_hex	520	1500		Each side 86.7	2.1	
Octagonal	T_oct	520	1500		Each side 65	2.1	
12 edges	T_edg	520	1500		Each edge 43.3	2.1	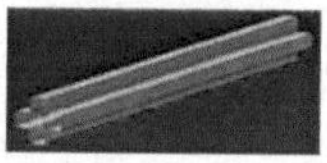

impact velocity v as the same as IIHS test condition. An additional mass of 550 kg is attached to the rigid base wall to generate enough kinetic energy similarly to vehicular crashing. The top of longitudinal member is attacked to the Flat 150 fixed barrier. Such geometry is obtained from typical dimensions of the front rail of a passenger car and the material of longitudinal was given in (Nguyen et al. 2015 and Nguyen 2017 and 2018) as show in Fig. 1.

ABAQUS was used to analyze the crashing behavior and energy-absorbing capacity of the longitudinal member which was modeled using the shell elements. To validate the longitudinal model, the comparison of longitudinal in full vehicle model with longitudinal member only in SOFI simulation was made as shown in Fig. 1, the behavior of longitudinal model the same as the longitudinal in SOFI event. The energy absorption performance of longitudinal in SOFI of full vehicle and longitudinal member only with the same time history as shown in Fig. 2.

2.2 Longitudinal Column in SOFI Crash Test Simulation

In this study, the full vehicle longitudinal model in SOFI simulation was analyzed as shown in Fig. 1 (Nguyen et al. 2015 and Nguyen 2018). The longitudinal member

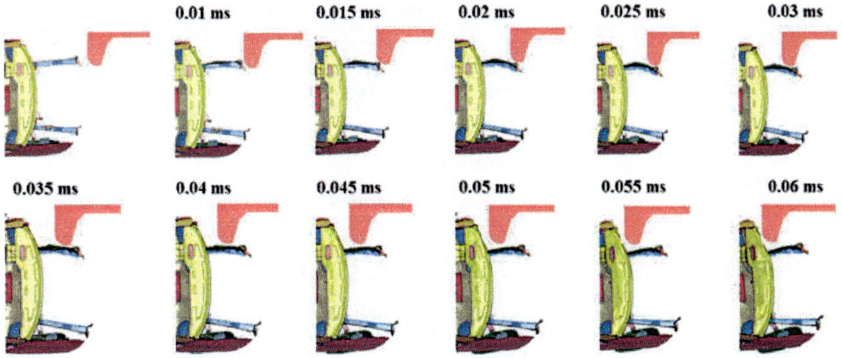

Fig. 1 Longitudinal modeling in SOFI (Nguyen et al. 2015)

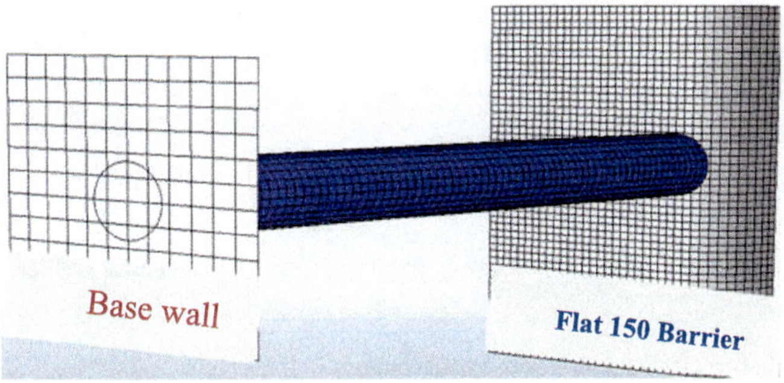

Fig. 2 Finite element analysis setup for longitudinal in SOFI performance

components for the simulations were made using the CATIA software. ABAQUS was then used to derive their FE models. As shown in Fig. 3, these three models were used to develop new FE simulation models as the same as SOFI event.

IIHS uses a rigid barrier called as Flat 150 in small overlap frontal impact tests. FEM is well-known to be utilized for solving problem related to real engineering structure; however, it should take much time for optimum design. Thus, barrier is assumed to be modeled as rigid which does not absorb impact energy, the effect of collision is more severe in consequence. The equation of energy conservation for collision of a vehicle to a Flat 150 barrier can be written as

$$\frac{1}{2}\mathrm{m}v_0^2 = E_{vint} + E_{vke} + E_{Flat\,150} \quad (1)$$

where v_0 is the initial velocity of the wall,
m is mass of vehicle,
E_{vint} is the energy absorbed by the longitudinal structure,
E_{vke} is the final kinetic energy of the longitudinal structure after impact,

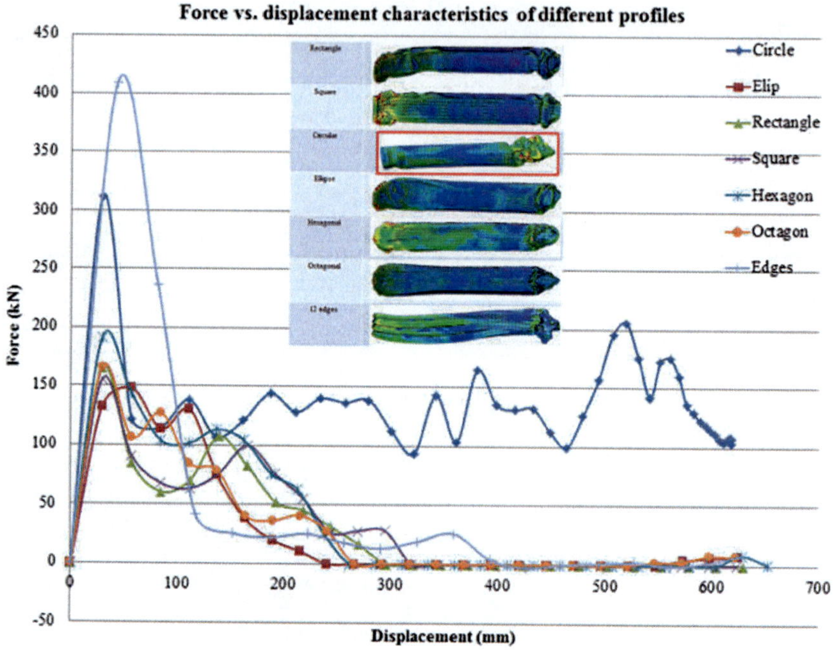

Fig. 3 The force versus displacement characteristics of different longitudinal profiles

$E_{Flat150}$ is the energy absorbed by the barrier.
$E_{Flat150} = 0$ in this study due to rigid barrier.

If the vehicle with a different mass strikes to a rigid barrier with similar velocity, v_0, such that the same amount of energy, E_{vint}, is absorbed by the vehicle, the equation becomes

$$\frac{1}{2}m'v_0^2 = E_{vint} + E'_{vke} \tag{2}$$

Here, m_0 is the equivalent vehicle mass and E_{0vke} is the final kinetic energy of the vehicle after impact. It now follows from Eqs. (3) and (4) that

$$m' = \mathrm{m} - \frac{2}{v_0^2}\left(E_{Flat150} + E_{vke} - E'_{vke}\right) \tag{3}$$

Assuming E_{vke} E_{0vke}, the above relation becomes

$$m' \approx m - \frac{2}{v_o^2}(E_{Flat\,150}) \tag{4}$$

According to Deb et al. [40], the ratio of the energy absorbed by the barrier in a conventional offset test to the vehicle mass remains almost constant for various car. The total mass of the vehicle model is 2000 kg (NCAC 2005). The mass was assumed to be 25% of a compact vehicle model. As an assumption, each tubular structure could absorb an equivalent kinetic energy of 500 kg mass. Since two tubes in reality can absorb a maximum energy of less than 50%, the effective mass, m_0, is 500 kg. In the simulations, the impact speed value was taken from the Insurance Institute for Highway Safety test, where the longitudinal model hits a rigid wall with an initial velocity of 64 km/h (17.8 m/s) and 25% offset.

2.3 Initial and Boundary Conditions for Longitudinal Modeling in SOFI Performance

The boundary and initial conditions defined in the finite element model are expected to resemble the actual crash test. The striker was modeled as a rigid body with only one translational displacement allowed while all other translational and rotational degree of freedom (DOF) were fixed. An initial velocity of 64 km/h (17.8 m/s) was assigned to all parts of the model except the base wall. The rigid barrier, to which the longitudinal member hits, is fixed in all DOFs. The rear end of the longitudinal member, the intermediary base wall and the point mass are constrained to move only in the direction of the initial velocity. This helps prevent relative movement of the blocks in the transverse directions, considering that the transverse displacements are almost zero during the initial phases of the crash while rotation of the vehicle occurs only at later stages of the crash (Fig. 1). Because the geometry of the loading is symmetric with respect to the horizontal as well as vertical mid-planes, only one-quarter of the assembly is analyzed. Corresponding symmetry conditions are applied on the corresponding interfaces.

2.4 Meshing

Considering that the thickness of longitudinal is small in comparison to the other dimensions, shell elements are used to model the end closes of the longitudinal member. The element type used in the model is S4R, a 4-node quadrilateral shell element with reduced integration and a large-strain formulation, which explain for both finite membrane strains and arbitrary large rotations. These elements allow transverse shear deformation. Thick shell elements are based on thick-walled shell theory while thin-walled shell elements are based on Kirchhoff theory. Simpson's integration rule is used with five integration points through the thickness. Finite element FE models of empty tubes were developed using the non-linear FE code ABAQUS-explicit. Based on a mesh convergence study, the element size of 10 mm was chosen for the shells elements. This mesh size offers an accurate capture of the deformation process with a sufficient mesh density. The 'general contact algorithm', which is less intense in terms of computational time, was used to simulate contact interaction among the components and avoid interpenetration. The Coulomb friction coefficient of 0.2 was set for all contact surfaces. The longitudinal FE model for SOFI test is shown in Fig. 2.

2.5 Design Objective

To evaluate the crashworthiness characteristics of different structures, several different criteria are commonly used (Nguyen 2015 and 2018), e.g. energy absorption, specific energy absorption (E_{Sp}), peak crushing force (F_{max}), and mean crushing force (F_{mean}) etc. The energy absorption or internal energy E_{int}, denotes the absorbed energy via plastic deformation of structures as

$$E_{int} = \int_{0}^{d} F(x)dx \tag{5}$$

where d is deformation distance, F(x) is the instantaneous crushing force. The specific energy absorption (E_{Sp}) is often considered a more commonly used criterion to measure energy absorption capacity per unit mass of material defined as

$$E_{Sp} = \frac{E_{int}}{m} \tag{6}$$

where m is total mass of structure. A higher specific energy absorption indicates a higher capacity of energy absorption of a structure.

In crashworthiness design, this paper studied on the longitudinal of the energy-absorption capacity which is measured by specific energy absorption (SEA) the same as Eq. (6) but written with different way especial for the crash energy absorbed per unit mass of the longitudinal. Therefore, the first objective of this optimum design was to maximize SEA for the longitudinal. As shown in Eq. (7).

$$SEA = \frac{\text{Energy Absorbed (EA)}}{\text{Longitudinal Weight}} \tag{7}$$

Typically, crushing forces fluctuate around mean values before they soar up when deformation capacity is exhausted at effective stroke length, which is considered to be equal to 150 mm for all cases. Therefore, mean crushing force F_{mean} was defined by

$$F_{mean} = \frac{1}{d} \int_{0}^{d} F(x)dx \tag{8}$$

The second objective was crush force efficiency (CFE). While F_{max} and F_{mean} are closely related to the occupant's survival rate when impact occurs, F_{max} could be easily varied due to initial imperfections. CFE as is an indicator for characterizing load consistency, which was defined as

$$CFE = \frac{F_{mean}}{F_{max}} = \frac{1}{F_{max}d} \int_{0}^{d} F(x)dx \tag{9}$$

where F_{max} refers to the highest force on the force–displacement curve. The higher the CFE value, the better the load consistency value. For ideal crash absorber, CFE is equal to %100. CFE is the ratio of the average crushing force (F_{mean}) to peak crushing force (F_{max}), which measures the load consistency and uniformity. The higher the ratio, the better the load consistency/uniformity. The same as specific energy absorption (Esp), CFE can be influenced by the loading angle and impacting velocity.

3 Results and Discussion

3.1 Longitudinal Crash Test Simulations

The contact area in a small-overlap crash is very small as shown in Fig. 1; thus, the longitudinal structure is seriously damaged due to the crash. This behavior is shown in Fig. 3 for differential profiles of longitudinal structure. Based on this result, the absorption energy of the circle profile is found higher than the others. The comparison results of the energy absorbed of longitudinal with various profiles as shown in Figs. 4, 5 and 6.

The intrusion of the frontal door was selected as the parameter for measuring the damage severity. Figure 4 presents the intrusion of longitudinal with different profiles which were bent and sliced on the side of longitudinal and divided into two areas: bending and slides. The detailed analysis of the simulation results are presented below. It can be seen from Fig. 1 that the left longitudinal rail was not damaged. This means that the majority of the loading was outside of the longitudinal structures such as the rocker panel and A-pillar. In the SOFI with 25% overlap, the main longitudinal rail was missed to cause damage to the frontal door, so the left wheel was directly loaded and pushed rearward into the toe pan (Nguyen et al. 2015 and Nguyen 2017).

According to the design objective in Sect. 2.5: maximum energy absorb and crush force efficiency (CFE) in Figs. 3 and 6. The circle profile was choose for optimal the design by adding the triggers. The results in Fig. 7 demonstrate that the circle profile with triggers in red square dot line and the circle profile without triggers (original profile) in blue diamond dot line; shown that more energy absorbed by the new design.

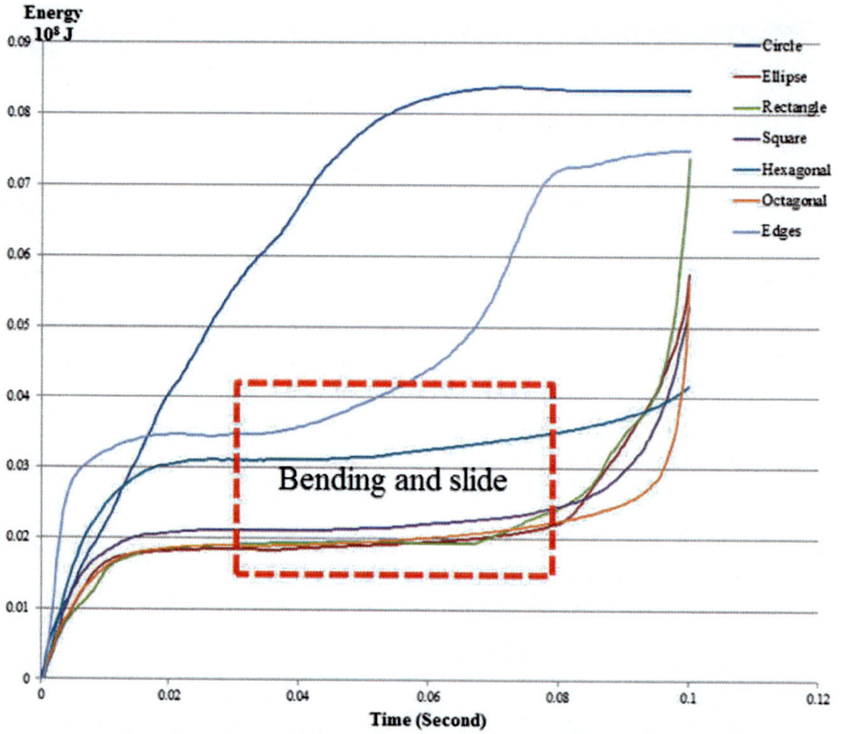

Fig. 4 The energy absorbed of vehicle model with different longitudinal profiles

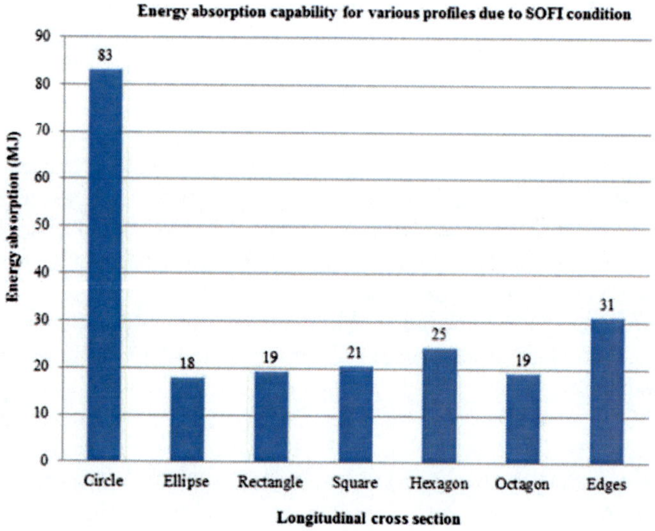

Fig. 5 The energy absorption capability for various profiles due to SOFI condition

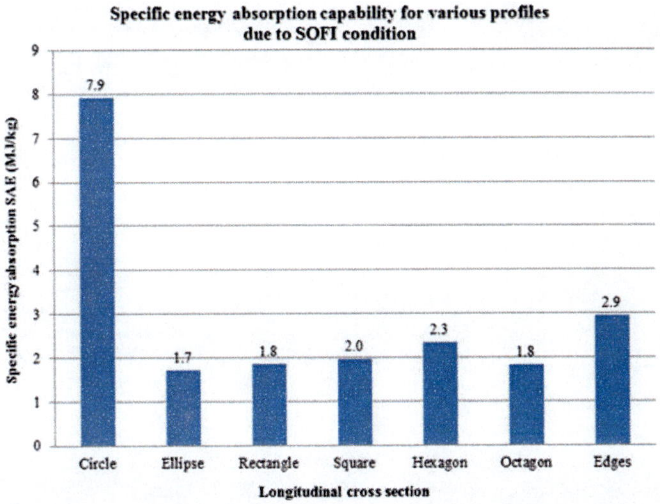

Fig. 6 The specific energy absorption (SEA) capability for various profiles due to SOFI condition

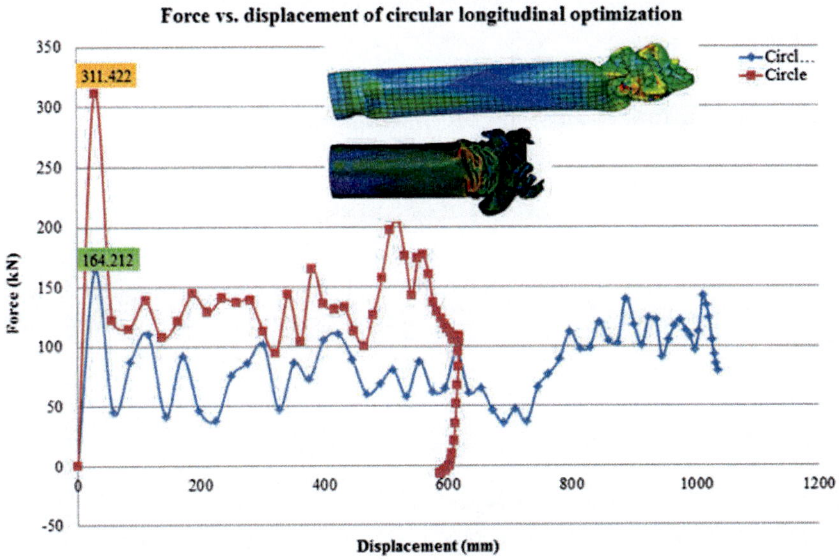

Fig. 7 The force versus displacement characteristics of circular longitudinal profiles with and without triggers

Fig. 8 Longitudinal profiles sample for mechanical experiments

3.2 *Longitudinal Crash Test Experiments*

Due to the experiment machines ability, the real test was made with only the longitudinal with the profiles as shown in Fig. 8. In this paper, the longitudinal with circle profile was skipping according to the manufacture prototype. Based on the test machine capacity, the dimension of the longitudinal was chosen with the length of 200 mm, the thickness of 2 mm, with the cross-section area of 1.067 mm^2, the mass is 1.765 kg.

3.2.1 Facilities and Setting Experiment

In this paper, the mechanical testing of the material machine was used which had compression at the center and compression stroke about 15 mm as shown in Fig. 9. The test results are the relationship between deformation and force as the same as the simulation which was shown in Fig. 3. From the results we compare and select the logical section.

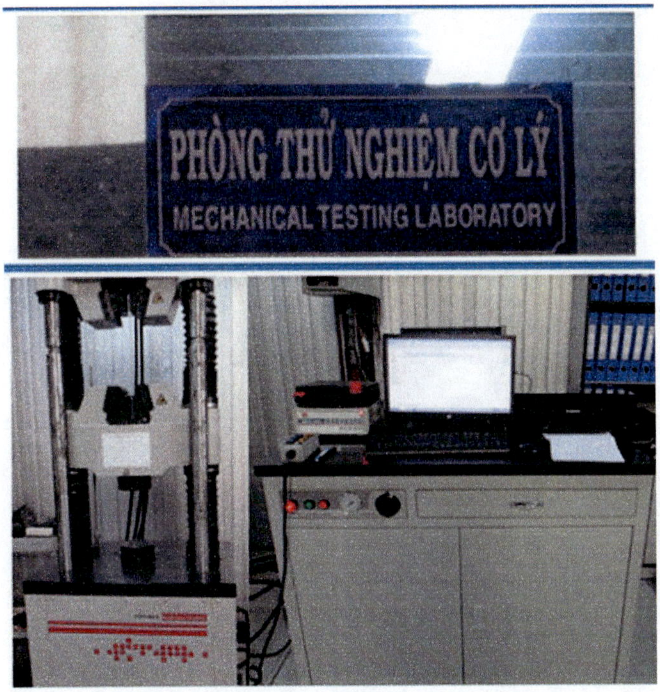

Fig. 9 The Mechanical testing laboratory and machine setting for experiment

3.2.2 Mechanical Testing of Longitudinal Structure with Different Profiles

In this section, different profiles of longitudinal as shown in Fig. 8 were applied for real test. Due to the length of the paper only some profiles were shown in this part such as cross and original profile in Figs. 10 and 12. The Relationship between force and displacement of rectangular and original profiles were shown in Figs. 11 and 13.

Fig. 10 The longitudinal structure with cross profile before and after the mechanical test

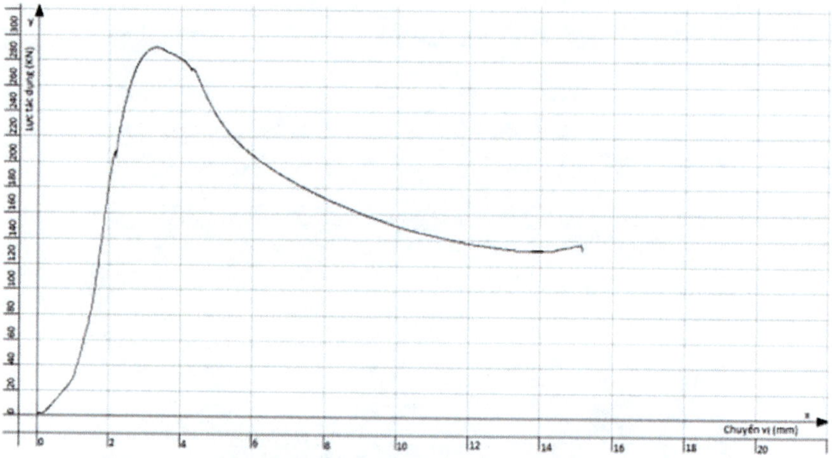

Fig. 11 The relationship between force and displacement of rectangular sample

Fig. 12 The longitudinal structure with original profile before and after the mechanical test

And The results of intrusion of longitudinal with different profiles after one stroke mechanical test were shown in Fig. 14.

3.2.3 Real Test Results Analysis

The results in the Fig. 15 were shown that with the same displacement (about 15 mm), the force is gradually increased from the rectangular cross section (line 1) to the cross section of the square (line 5) to the cross section of the ladder (line 6) to the cross section the hexagonal section (line 3) to the octagonal cross section (line 4) to the section cross section (line 2).

In this section, different profiles of longitudinal as shown in Fig. 8 were experiments. Due to the length of the paper only some profiles were shown in this part such as cross and original profile in Figs. 16 and 17. The Relationship between force

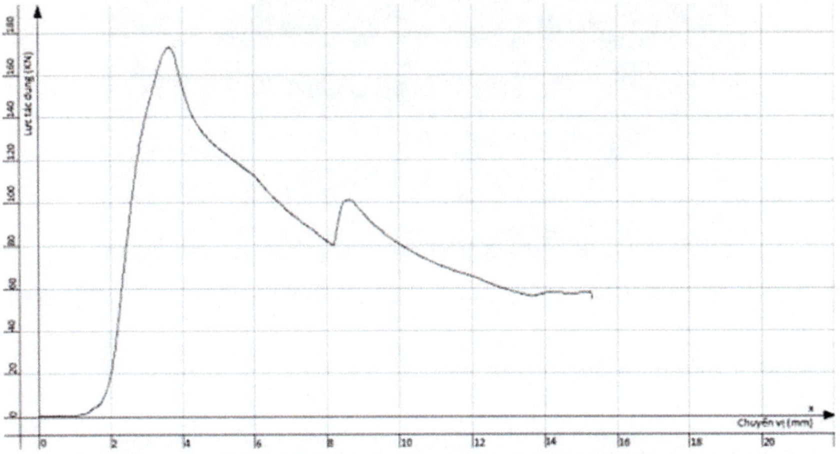

Fig. 13 The relationship graph between force and displacement of original longitudinal profile

Fig. 14 The results of intrusion of longitudinal structure with different profiles after one stroke mechanical test

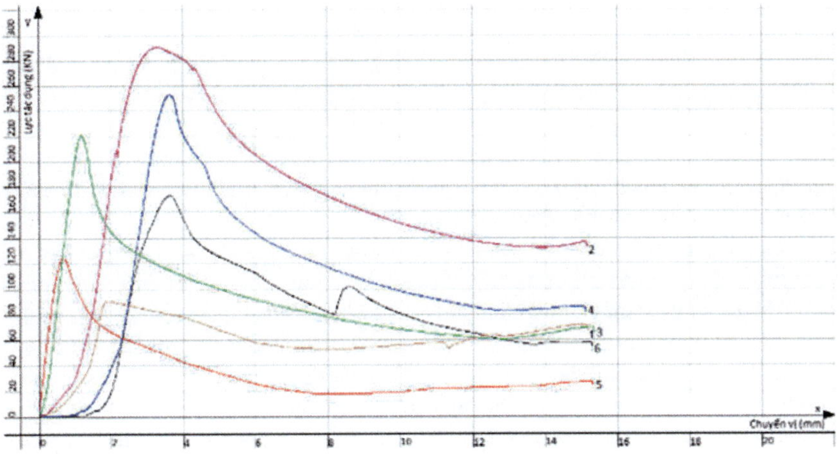

Fig. 15 The general relationship between the force and displacement of the samples

Fig. 16 The longitudinal structure with cross profile before and after the mechanical test with full compression stroke

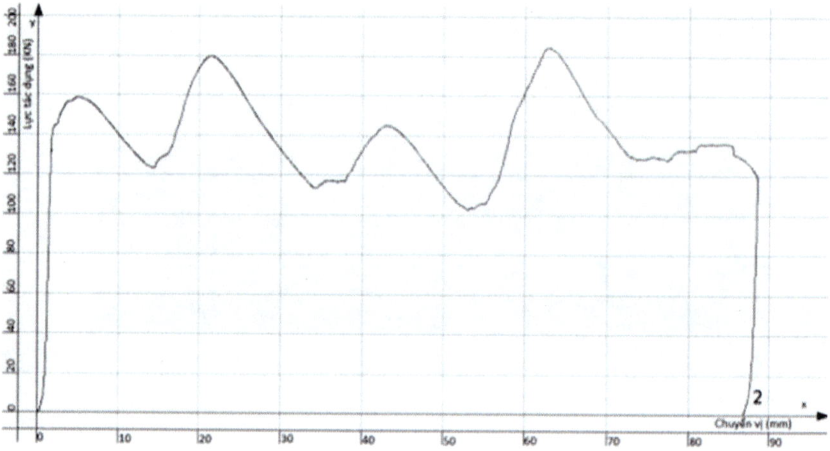

Fig. 17 The relationship between force and displacement of the cross-sectional longitudinal structure

and displacement of cross and original longitudinal profiles were shown in Figs. 17 and 19. And The results of intrusion of longitudinal with different profiles after full stroke mechanical test were shown in Fig. 20.

According to the results we can comparison of local mechanical test results. In order to compare the results of the test, the experimental computer of different profile should in the concurrent graphs. Comparison of the deformation (compression stroke of about 100 mm) and the maximum force was performed, selecting the specimen with the largest force and the overlapping folds overlap.

According to the Fig. 21, with the same displacement (about 100 mm), the force is gradually increased from the cross-section of the cross-section (line 5) to the sectional profile (line 1) to the sectional section of the octagonal section (line 4) to the cross section the trapezoidal profile (line 6) to the section of the hexagonal section (line 3) to the section cross section (line 2). As the results were shown in Figs. 15, 16,

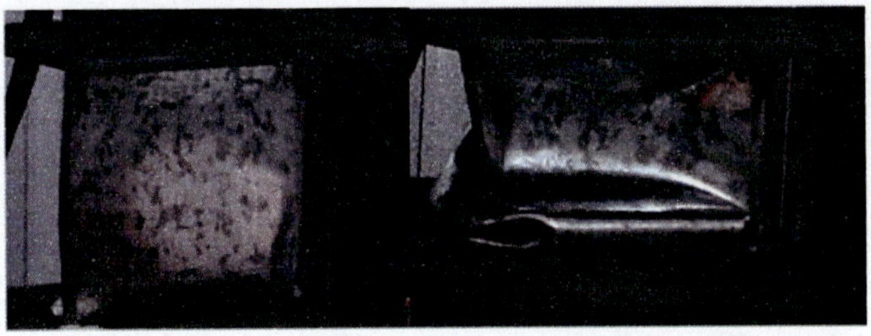

Fig. 18 The longitudinal structure with original profile before and after the mechanical test with full compression stroke

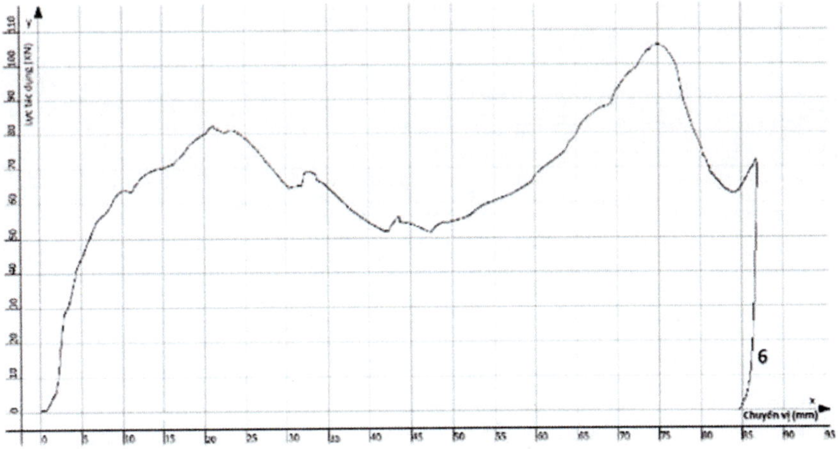

Fig. 19 The relationship between force and displacement of the original longitudinal structure

Fig. 20 The results of intrusion of longitudinal structure with different profiles after full stroke mechanical test

17, 18, 19, 20, the cross sectional area of the longitudinal was influence to the force peak, displacement, and local stress.

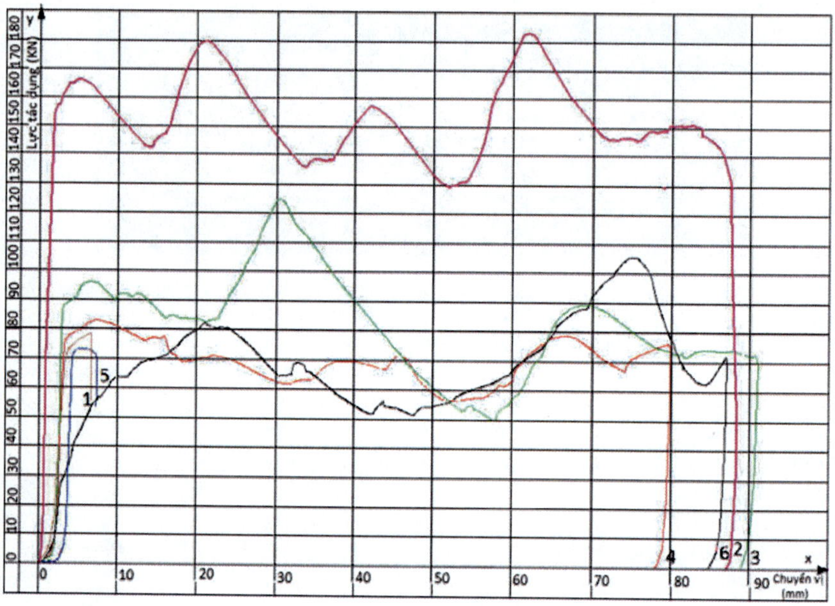

Fig. 21 The general relationship between the force acting on and displacement of the local mechanical sample

4 Conclusion

In the present study, the shape of a longitudinal structure is optimized for the SOFI condition to maximize its crashworthiness performance. Different longitudinal profiles were investigated by numerical simulations and experiments. The parameters defining the shapes of the cross section of the longitudinal profiles as well as the thickness could be chosen as the optimization variables. The longitudinal is modeled as a deformable body in full detail. The formation of the objective function, the number of variables, and the range of values that can be assigned to the variables have high influences on the resulting optimum shape. The results from simulations and experiments showed that the intrusion and energy absorbed could be improved by changing the profiles of the longitudinal structure.

Last but not least, a hole is induced as the trigger mechanism in the longitudinal structure. Its function is to keep peak force in lower value and to improve the crash force. It is concluded that the longitudinal structure made by steel with 12 edges cross section (similar to multiple cells) and 2 mm wall thickness has proven to be a good potential energy absorber candidate for crashworthiness application.

Acknowledgements This work was supported in part by the research program 2020 of Ho Chi Minh City University of Technology (HUTECH) in VIET NAM.

References

1. Sherwood C (2012) An update on the IIHS small-overlap research program. Insurance institute for highway safety IIHS report, pp 1–35
2. Sherwood C (2013) An update on the IIHS small-overlap ratings program. IIHS report, Insurance institute for highway safety, pp 1–13
3. Brewer J, Patel S, Summers S, Prasad A, Mohan P (2011) Accuracy of AHOF400 with a moment-measuring load cell barrier. In: International technical conference on the enhanced safety of vehicles, Washington, DC, USA
4. Choi WM, Kim JS, Jung HS, Kwon TS (2011) Effect of punch angle on energy absorbing characteristics of tube-type crash elements. Int J Automot Technol 12(3):383–389
5. Kikuchi T, Naokao T, Watanabe T, Saeki H, Okabe T (2012) An investigation of injury factors concerning drivers in vehicles involved in small-overlap frontal crashes. SAE Paper No 2012-01-0599
6. Sherwood C, Nolan JM, Zuby DS (2009) IIHS. Characteristics of small overlap crashes. No. 09–0423
7. Bois PD, Chou CC, Fileta BB, Khalil TB, King AL, Mahmood HF, Mertz HJ, Wismans J (2004) Vehicle crashworthiness and occupant protection. Michigan, American iron and steel institute Southfield, pp 11–176
8. Brumbelow ML, Zuby DS (2012) Impact and injury patterns in frontal crashes of vehicles with good ratings for frontal crash protection. In: Proceedings of the 21 st international technical conference on the enhanced safety of vehicle, IIHS, Paper No 09–0257
9. Nolan J (2012) Results of small-overlap frontal crash tests. Berlin, Germany, IIHS, pp 1–18
10. Chauvel C, Faverjon G, Bertholon N, Cuny S, Delannoy P (2011) Self-protection and partner-protection for new vehicles (UNECE R94 Amendment). In: International technical conference on the enhanced safety of vehicles, Washington, DC, USA
11. Joyce L, Nate JD, Jeff D, Martin L, Ryuuji O (2014) A comparison of the NHTSA research offset oblique and small overlap impact tests and the IIHS moderate and small overlap tests. SAE Paper No 2014-01-0537
12. Nguyen PTL (2018) Vehicle frontal impact to pole barrier simulation using computer finite element model. In: IEEE proceeding on Green technology and sustainable development. pp 273–277. https://doi.org/10.1109/gtsd.2018.8595702
13. Park BT, Partyka SC, Morgan RM, Hackney JR, Lee J, Summers L (2000) Comparison of vehicle structural intergrity and occupant injury potential in full-frontal and offset-frontal crash tests. SAE Paper No 2000-01-0879
14. Han JN, Kim JS, Park JW, Son CK, Kim YS (2012) A study of occupant protection simulation for IIHS small overlap frontal crash test. Trans Korean Soc Automot Eng 1833–1836
15. Lee KW, Lim JM (2014) Comparison on rating methods for female dummy in NCAP frontal impact test. Int J Automot Technol 15(6):919–925
16. Kim TW, Jeong HY (2010) Stochastic analysis of the variation in injury numbers of automobile frontal crash tests. Int J Automot Technol 11(4):481–488
17. Luu NPT (2019) Analysis of bus structural performance during full frontal impact. In: IEEE proceeding on System science and engineering, pp 635–638. https://doi.org/10.1109/icsse.2019.8823416
18. Chen DY, Wang LM, Wang CZ, Yuan LK, Zhang TY, Zhang ZZ (2015) Finite element based improvement of a light truck design to optimize crashworthiness. Int J Automot Technol 16(1):39–49
19. Gepner B, Bojanowski C, Kwasniewski L, Wekezer J (2014) Effectiveness of ECE R66 and FMVSS 220 standards in rollover crashworthiness assessment of paratransit buses. Int J Automot Technol 15(4):581–591
20. Halloway DE, Pintar F, Saunders J, Anelli AB (2012) Classifiers to augment the CDC system to distinguish the role of structure in a frontal impact taxonomy. SAE Paper No 2012-01-0575

21. Jakobsson L, Mcinally G, Axelson A, Lindman M, Kling A, Broberg T, Fermer M, Wagstrom L (2013) Severe frontal collisions with partial overlap-two decades of car safety development. SAE Paper no 2013-01-0759
22. Jeong HM, Han BK (2011) A study on the body structure crashworthiness for small overlap crash. Trans Korean Soc Automot Eng 2032–2038
23. Zhida S, Xin Q, Haishu C (2012) BIW safety performance research based on vehicle frontal crash. FISTA 9:13–26
24. Oh S, Ye BW, Sin HC (2007) Size optimization of an engine room member for crashworthiness using response surface method. Int J Automot Technol 8(1):93–102
25. Tovar A, Patel NM, Niebur GL, Sen M, Renaud JE (2006) Topology optimization using a hybrid cellular automaton method with local control rules. ASME J Mech Des 128(6):1205–1216
26. Patel NM, Kang B-S, Renaud JE, Tovar A (2009) Crashworthiness design using topology optimization. ASME J Mech Des 131(6):061013
27. Patel NM, Penninger CL, Renaud JE (2009) Topology synthesis of extrusion-based nonlinear transient designs. ASME J Mech Des 131(6):061003
28. Deepak SR, Dinesh M, Sahu DK, Ananthasuresh GK (2009) A comparative study of the formulations and benchmark problems for the topology optimization of compliant mechanisms. ASME J Mech Rob 1(1):011003
29. Lee YH, Joo YJ, Park JS, Kim YS, Yim HJ (2014) Robust design of optimization of frontal structures for minimizing injury risks of flex pedestrian legform impactor. Int J Automot Technol 15(5):757–764
30. Punit B, James PS, Andres R (2013) Design of crashworthy structures with controlled energy absorption in the hybrid cellular automaton framework. ASME J Mech Des 135(1):091002
31. Sherwood C, Mueller B, Nolan JM, Zuby DS, Lund AK (2013) IIHS Development of a frontal small-overlap crashworthiness evaluation test. Traffic Injury Prev https://doi.org/10.1080/15389588.790539
32. National Crash Analysis Center (NCAC) (2005) FE model of dodge grand caravan, The George Washington University. pp 1–13
33. Insurance institute for highway safety (2012) Front small-overlap crashworthiness evaluation crash test protocol. IIHS report (version I, II), pp 1–25
34. Insurance institute for highway safety (2012) Small-overlap program protocol and rating guidelines. pp 1–17
35. Nguyen PTL, Lee JY, Yim HJ, Lee SB, Heo SJ (2015) Analysis of vehicle structural performance during small-overlap frontal impact. Int J Automot Technol 16(5):799–805
36. Nguyen PTL, Lee JY, Yim HJ, Kim HK, Lee SB, Heo SJ (2015) Optimal design of vehicles structure for improving small overlap rating. Int J Automot Technol 16(6):959–965
37. Nguyen PTL (2015) A study on optimal design of vehicle structure for small overlap frontal impact. Phd Thesis of Automotive Engineering Kookmin University, Seoul, Korea
38. Luu NPT (2017) An optimisation approach to choose thickness of three members to improve IIHS small-overlap structural rating. Int J Crashworthiness 22(5):518–526
39. Hallquist JO (2007) 'LS-DYNA keyword user manual. Livermore Software Technology Corporation
40. Deb A, Naravane A, Chirwa EC (2006) An offset rigid barrier-based test: equivalence to the insurance institute for highway safety frontal offset impact safety test. Int J Crashworthiness 11(4):281–290. https://doi.org/10.1533/ijcr.2005.0407

Investigation of Flow Forming Process and Heat Treatment Effects on 2024 Aluminium Tubes

Acar Can Kocabıçak, Aptullah Karakaş, Güneş Aydın, and Senai Yalçınkaya

Abstract The aim of this study is to investigate mechanical and microstructural properties of 2024 aluminium tubes (2024 AA) which are produced by flow forming technique and make comparisons between the flow formed 2024 AA, full annealed (2024-O) and solution heat treated then artificially aged (2024-T6) temper aluminium tubes. For this purpose, 70% reduction rate was applied on the 2024-O tempered samples by the flow forming process. The 2024-T6 samples demonstrated higher mechanical properties than 70% thickness reduced 2024 AA due to formation of Al2Cu precipitation. However, it was found that the mechanical properties of the flow formed 2024 AA was almost two times higher than that of the 2024-O sample. Furthermore, grain orientations of these three samples were investigated by the optical microscope. Although any grain orientations were not seen in 2024-O and 2024-T6 aluminum samples, severely grain orientations were observed after the flow forming process.

Keywords 2024 aluminum alloy · Flow forming · Cold working · Tensile test · Hardness · Microstructure · Reduction rate

1 Introduction

Flow forming is a cold working technique that is used to produce high precision tubular or hollow parts and many different forms of axisymmetric products. The main principle of the flow forming technique is to significantly reduce the wall thickness of the starting material called "preform" with forming rollers while providing an axial elongation. While wall thickness of the material reduces, the material flows plastically along a mandrel during the flow forming process.

A. C. Kocabıçak (✉) · A. Karakaş · G. Aydın
Repkon Machine and Tool Industry and Trade Inc., Incistanbul 34980, Turkey
e-mail: aptullah.karakas@repkon.com.tr

S. Yalçınkaya
Department of Mechanical Engineering, Technology Faculty of Marmara University, Istanbul 34722, Turkey

M. Abdel Wahab (ed.), *Proceedings of the 8th International Conference on Fracture, Fatigue and Wear*, Lecture Notes in Mechanical Engineering,
https://doi.org/10.1007/978-981-15-9893-7_8

The preform can be produced by many different production methods such as deep drawing, spinning, hot or cold extrusion, forging and machining [1]. The flow forming is a plastic deformation process but without a generating chip that makes a possible to produce the parts with minimum waste of materials compare to conventional machining process. Generally, the production times of the flow forming process are faster than those of various conventional machining processes [2]. Furthermore, it increases the yield and ultimate tensile strength of the parts since the flow forming enhances the mechanical properties of the workpieces. It makes possible to produce lighter and stronger parts compared to conventional machining methods. Moreover, the flow forming process makes possible to manufacture long cylindrical parts with different wall thicknesses and complex contours.

It is used in aerospace, defense fields in order to produce drive shaft for jet engines and helicopters, control the piston rods, aero hydraulic cylinder, accumulators, rocket and missile casing, pressure vessels, cartridge case, and hollow tubes. There are many parameters such as thickness reduction ratio, rotational speed of the mandrel, axial feed, preform dimensions, workpiece material properties and heat-treatment condition of the preform that affect the quality of final parts [3]. Besides, roller geometry, roller setup configuration and friction conditions also influence the flow forming process [4]. Illustration of the flow formed and preform part and the general flow forming process parameters are shown in Fig. 1a and b and Table 1 respectively.

The preform should be prepared in suitable dimensions and surface should be smooth before the flow forming operation. As first step of the process, the preform is clamped between the mandrel and the tailstock, after that the preform is rotated with a driven motor. The preform is pushed by the mandrel under the forming rollers.

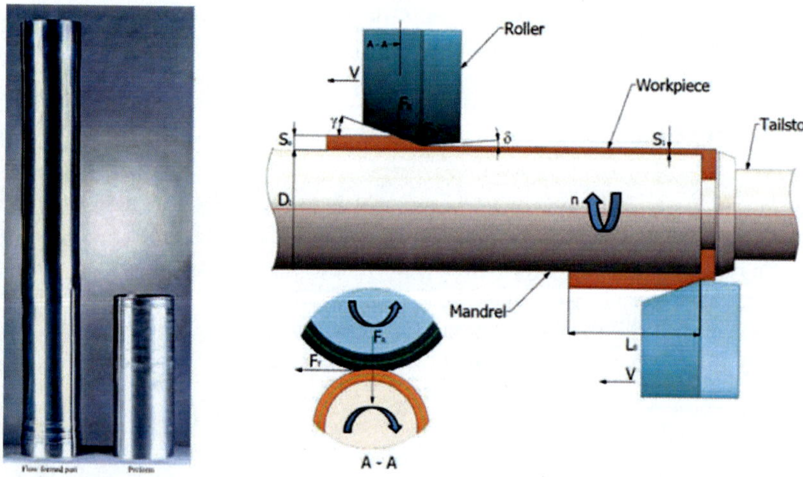

(a) (b)

Fig. 1 Illustration of flow formed part, preform (**a**) and flow forming principle (**b**)

Table 1 Descriptions of flow forming process parameters

S_O Starting wall thickness (mm)	γ Leading angle (degree,°)	F_A Axial force (N or Kg)
S_1 Finished wall thickness (mm)	δ Trailing angle(degree,°)	F_R Radial force (N)
L_O Starting length (mm)	R Nose radius(mm)	F_T Tangential force (N)
D_I Inside diameter (mm)	V Feed rate (mm/minute)	n Spindle speed (rpm)

The forming rollers are driven in a radial direction by electro mechanic or hydraulic motor. The rollers are in contact with the preform (workpiece) during the flow forming process. The roller force should be in contact with the outer surface of the mandrel in order to make a straight forming. This contact point allows the reduction of the wall thickness and the material flows through the axial direction thanks to the axial forces from the headstock of the machine. Meanwhile, the wall thickness of the preform reduces and the length of the workpiece increases as well [5].

Apart from these significant effects of the flow forming process parameters, the maximum reduction ratio depends on the mechanical and material properties of the preform. Both material properties and heat treatment condition of the preform are playing an important role to obtain the optimum flow properties. The flow forming process reduces the wall thickness of the preform and reduction percentage (R) can be calculated using the dimension of starting wall thickness (So) and final wall thickness (S1).

The reduction percentage formulation is defined by the following formula,

$$R = [(So - S1)/So] \times 100 \tag{1}$$

All ductile metals and alloys include stainless steels and aluminum alloys, as well as solution annealed condition materials can be used with the flow forming process. 2024 aluminum alloys (AA) were used for this experimental study. 2024 AA are often used in the aircraft industry due to their high strength to weight ratio and good fatigue resistance. The 2024 AA includes copper alloys, which is the primary alloys used mainly in airframe structural applications. As the 2024 AA have high damage and fatigue tolerances, they are mainly used for the lower wing skins and fuselage structure of commercial aircraft. Although 2024-T3 and T6 have a moderate yield strength, it has very good resistance to fatigue crack growth and good fracture toughness [6].

In this study, a comparison was made between the strength increased 2024 AA which was cold formed by the flow forming method and the precipitation hardened Al2024-T6 AA. The preform that was annealed at 410 °C and hold at this temperature for 2 h in order to produce the 2024-O temper according to AMS2772 standard. The cold working and heat treatment effects were analyzed on the 2024 aluminum alloy. The precipitation hardening and strain hardening processes were applied on 2024-O to investigate strength increasing mechanisms of aluminum alloys. The tensile tests, hardness and microstructural investigation were performed to compare strength increase properties of the flow formed 2024 and 2024-T6 AA.

2 Materials and Experimental Work

Full annealed 2024 aluminum is designated as 2024-O, solution heat treated then artificially aged is designated as 2024-T6 and a flow formed tube were used in this experimental study. A comparison was made between mechanical and microstructural properties of these alloys. The chemical composition of the workpiece material is examined in terms of weight percentages according to ASTM E34-11e1 specification and the results are shown in Table 2 [7].

The aging and solution treatments are performed according to AMS2772 specification, which designates that the solution heat treatment applied at 470 °C for 1 h and aging treatment at 191 °C for 12 h, as shown in Table 3 [8]. After the solution heat treatment, the tube is quenched in water, which includes 5 wt% Polyalkylene Glycol (PAG) at room temperature. 2024-O annealed tubular preform was used before the flow forming process.

The part was loaded on the mandrel and three forming rollers with 120° interval centered were used in present experimental study. During the flow forming process, 70% reduction ratio was applied by the rollers on annealing condition 2024-O the preform. The outside diameter of the used preform is about 130 mm, the thickness is a 10 mm and the length is approximately 300 mm. The final outside diameter is about 116 mm, the final thickness is a 3 mm and the final length is around 1000 mm with 70% flow forming reduction ratio. For the ongoing process, feed rate was selected about 25 mm/ minute and spindle speed of 250 rpm were applied on the 2024-O condition preform. The hardness test was carried out using micro indentation method (HV0,1), which is provided by Emco Test Dura Scan-20 equipment. Furthermore, the tensile tests were performed according to ASTM E8M standard with Instron 3382 device which applies 100kN force [9]. As for the microstructure investigation, samples were taken from surfaces of the tubes. All samples were cold mounted, grinded, polished and electro-etched by Struers A2 solution in order to prepare optical microscope test.

Table 2 The chemical compositions of 2024 aluminum alloy

Element type	Al	Cu	Fe	Si	Mn	Mg
Weight (%)	92.50	4.4	0.5	0.5	0.6	1.5

Table 3 The heat treatment parameters of 2024-T6 aluminum alloy

Element	Solution temp. (°C)	Holding duration (min)	Aging temp. (°C)	Aging duration (hour)
2024-T6	490	90	191	12

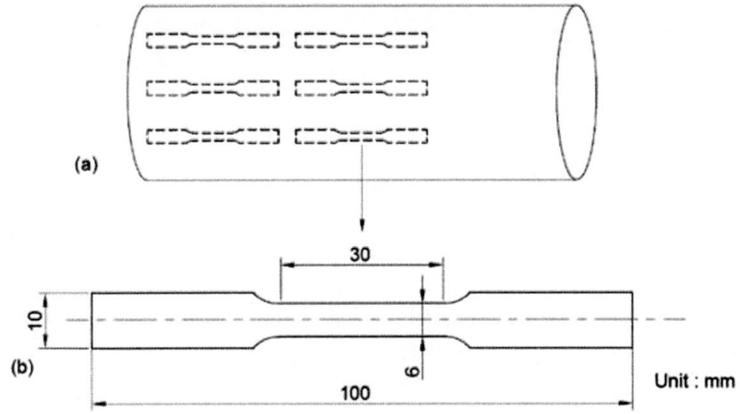

Fig. 2 **a** Illustration of specimen position and **b** specimen dimensions

Olympus BX53M Optical Microscope were used to examine microstructure behavior of the annealed (2024-O), flow formed and heat treated (2024-T6) materials. Tensile testing specimens were taken from three tubes, namely; 2024-O, flow formed 2024 and 2024-T6 alloy. The specimens were extract from the circular tube walls by water jet cutting along the longitudinal direction of the parts and the illustration of the positions and standard dimensions of the tensile specimens are presented in Fig. 2a, b, respectively. For mechanical testing for each tube, five samples were repeated.

As shown in Fig. 2, the specimen length is about 100 mm and width is 10 mm for the tensile test. The gauge length is about 30 mm and width is 6 mm. The thickness of the workpieces is 10 mm before the flow forming process and it reduces to 3 mm after the process thus 70% reduction ratio was applied. The tensile curves with stress and elongation were predicted by using the tensile test device. Yield stress, ultimate tensile stress and elongation are figured out using the tensile curves. Yield stress represents the stress point that a material starts to deform plastically. The yield point is the area where nonlinear deformation begins [6–11]. The tensile strength is defined as a stress and it is measured as force per unit area. Elongation is a measurement of the deformation that presents the displacement between grains in the part regarding the reference length.

3 Results and Discussion

3.1 Mechanical Characterization

The rollers in contact with the tube during the flow forming process as shown in Fig. 3, the surface of formed part is hardened because of the cold working effect. An

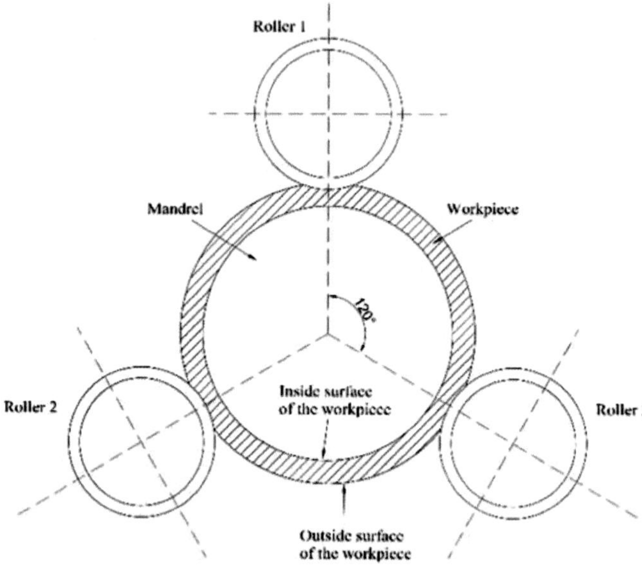

Fig. 3 Roller contact points on the workpiece

inhomogeneous grain orientations occurs during the flow due to the local deformation between rollers and the workpiece surface [11]. During the flow forming, the contact area of the roller is called "outside surface" and the contact surface area of the mandrel is called "inside surface". The outside surface of the flow formed tube is harder than the inside surface of the workpiece.

Load is applied from the rollers to the outside surface and this load is unfolding along the inside surface depending on the concentration gradient. This load should reach up to the mandrel, otherwise two different zones would occur on the microstructure and grains during the forming process. In order to achieve homogeny of microstructure, there must not be two different formed and unformed zones in microstructure [12]. For this reason, the radial force should be stronger for the same material at larger wall thicknesses.

The Hardness values does not change according to the sections in O and T6 temper conditions because the recrystallization occurs after the heat treatment in the microstructure and eliminates grain and phase orientations. 5 longitudinal tensile samples were extracted from each tubes and tensile tests were performed at a speed of 1 mm/ s. Average values of the tensile testing results is presented in Fig. 4. It can be observed that maximum tensile stress was obtained about 450 MPa from T6 sample because of the forming mechanism, which is followed by FF aluminum around 378 MPa but the minimum stress occurred about 210 MPa for O type sample. As for the case of strain rate, the opposite behavior was provided from this graph. For example, the minimum and maximum strain rate was obtained about 6.4% and 14% for the T6 and FF samples, respectively.

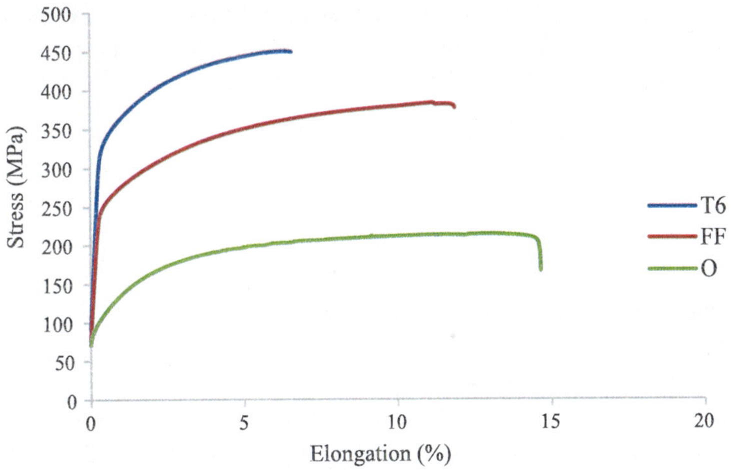

Fig. 4 Tensile properties of 2024 tubes

Two of the used strength increasing mechanisms of metallic materials are precipitation hardening and strain hardening processes. These strengths increasing mechanisms were performed on 2024 AA. Firstly, the precipitation hardening process was performed by aging process in order to carry out to 2024-T6 AA condition. During the aging process of the 2024 AA, "Al_2Cu" precipitates are occurred formed in the grains and these precipitates provide strength increase by dislocation cutting and bowing as described in Orrowan mechanism. The yield and tensile strength increases after aging treatment thanks to the Orowan mechanism [13]. Besides, flow forming process performed on annealed (2024-O) AA and increase the strain properties. The number of dislocations increase after the cold forming process for the metal alloys. After a certain level, dislocations lock each other's and create a dislocation tangle. This dislocation tangle blocks the other dislocation movements thus it causes the strength increase of the material. As seen in Table 4, the yield strength, tensile strength and hardness increase after the flow forming process because of dislocation generation and dislocation pile-up during plastic deformation. The thickness of the flow formed material was reduced by 70% hence it is stronger than the 2024-O condition material.

According to the tensile test results, the mechanical properties obtained by the precipitation hardening process for the 2024 AA were found to be is higher than the

Table 4 Tensile properties of 2024 aluminum alloy

Temper	Yield strength (MPa)	Tensile strength (MPa)	Strain (%)	Hardness (HB)
T6	330	450	6.5	149
FF	250	380	11.9	103
O	114	220	14.67	60

mechanical properties obtained by the hardening in the flow forming method. This result shows that the Al2024 alloy cannot achieve mechanical values obtained by artificial aging even by cold forming at high reductions up to 70%. The reason of this result might be aluminum alloys have high stacking fault energy. Metals with high stacking faults occur cross-slip during deformation therefore dislocations cannot lock each other. The dislocation movement cannot be restricted and serious increases in mechanical values cannot seem during the cold working process.

3.2 Microstructure Characterization

Representative specimens were taken from longitudinal sections of three materials such as preform (2024-O) AA, flow formed 2024 AA and solution heat treated (2024-T6) AA in order to analyze microstructural characterization. The 500X and 1000X magnifications were used in order to take microstructure of the tested sample by optical microscope. Figure 5a shows the microstructure view of "O" condition

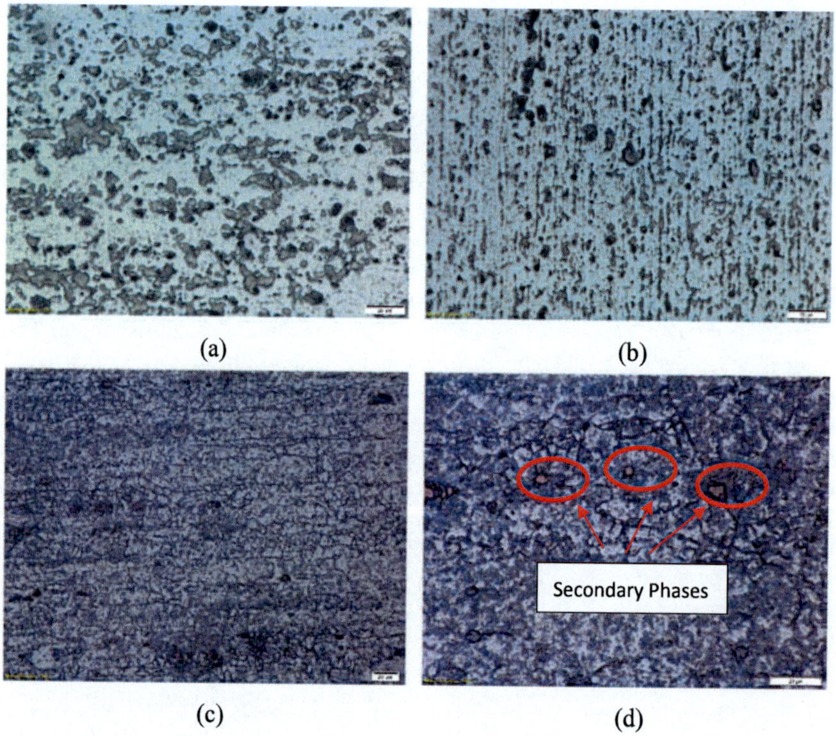

Fig. 5 Grain orientation examinations of microstructures of 2024 AA. 2024-O (**a**), Flow formed (**b**) and 2024-T6 (**c** and **d**) specimens

before the flow forming operation under an optical microscope. Figure 5b shows the microstructure of the material after the reduction of 70% applied thanks to the flow forming method. Representative material microstructures were taken from the 2024-T6 tube, which are shown in Fig. 5c, d respectively. Orientations were not determined on 2024-O condition material, as seen in Fig. 5a.

After the reduction of 70% through the flow forming opration, orientations were observed in the microstructure as indicated in Fig. 5b. The flow formed material shows a fine "fiber-like" microstructure that contains sever elongated aluminium matris. As a result of an important plastic deformation, secondary phases can be orientated during the flow forming processing. This elongated microstructure directly affects the mechanical and strain-hardening properties because dislocations are piled-up on the surface. These piled up dislocations and orientations occurred during the cold working process as during the flow forming method [14]. The microstructures of the solution heat treated tube 2024-T6 alloy are indicated in Fig. 5c and d, respectively. According to the microstructure of the 2024-T6 tube, elongated and oriented phases disapeared due to the recrystallization process during the solution heat treatment, as seen in Fig. 5c and d, respectively. In addition, there are some secondary phases detected in the microstructure that could not be dissolved in the crystal structure of the aluminium matris during the solution heat treatment as shown in Fig. 5d. If these secondary phases had been able to be avoided in the crystal structure during the quenching process, there would have consist in precipitations during the aging treatment hence the mechanical properties could be higher in this situtation [13].

4 Conclusions

Three types of specimens such as 2024-O, flow formed tube thickness reduced by 70% and 2024-T6 aluminum alloys were investigated in this experimental study. The mechanical, hardness and microstructural properties were compared with these three aluminum alloys. It is well understood that the hardness, tensile and yield properties of the aging treatment 2024 AA were higher than flow formed material because Al_2Cu precipitates during the aging process. The artificial aged 2024 AA indicated a higher mechanical property than 70% thickness reduced 2024 AA properties. However, the mechanical properties of the flow formed 2024 AA was almost twice higher than that of the 2024-O condition and the severely grain orientations occurred during the flow forming process. All of these samples exhibited that 2024-O tube could be plastically deformed by applying more than 70% thickness reduction. However, the 2024-T6 applied AA alloy could be more dissolved in order to reduce the secondary phase rate during the solution heat treatment.

Acknowledgements The authors would like to thank Repkon Machine and Tool Industry and Trade Inc. which financially supported this study. This study was prepared by the Repkon Inc. engineers and was performed in Repkon Metalworking Technology Center in Istanbul/Turkey.

References

1. Chang S-C, Huang C-A, Yu S-Y, Chang Y, Han W-C, Shieh T-S, Chung H-C, Yao H-T, Shyu G-D, Hou H-Y, Wang C-C, Wang W-S (1998) Tube spinnability of AA 2024 and 7075 aluminum alloys
2. Bhatt RJ, Raval HK (2016) in Procedia CIRP (Elsevier B.V.) 146–151
3. Joseph Davidson M, Balasubramanian K, Tagore GRN, J Mater Process Technol 203:321
4. Bhatt RJ, Raval HK (2015) Bonfring. Int J Ind Eng Manag Sci 5:46
5. Sivanandini M, Dhami S, Pabla B (2012) Int J Sci Eng Res 3:1
6. Campbell FC (2006) In: Manufacturing technology for aerospace structural materials
7. ASTM E34-11e1 Standard Test Methods for Chemical Analysis of Aluminum and Aluminum-Base Alloys (2017)
8. AMS 2772 Heat Treatment of Aluminum Alloy Raw Materials (2018)
9. ASTM, ASTM E8/ E8M Standard Test Methods for Tension Testing of Metallic Materials (2013)
10. Lin CW, Chen KJ, Hung FY, Lui TS, Chen HP (2019) J Mater Res Technol 8:3137
11. Chang UY, Han W-C, Hung Y-T (1997) In: Bains T, MacKenzie DS (eds) Proceedings of the first international non-ferrous processing and technology conference proceedings, Ohio
12. Külekçi İ, Biro CD, Aydın G (2015) In: 13th International cold formed congress 195
13. Mackenzie DS (2003) In: Handbook of aluminum
14. Gungor MN, Ucok I, Kramer LS, Dong H, Martin NR, Tack WT (2005) Mater Sci Eng A 410–411:369

Application of Improved Artificial Neural Network to Stiffness Reduction Analysis of Truss Joints in a Railway Bridge

H. Tran-Ngoc, L. Nguyen-Ngoc, H. Ho-Khac, A. Le-Thuc, Guido De Roeck, Thanh Bui-Tien, and Magd Abdel Wahab

Abstract Railway bridges are susceptible to the problems of fracture and fatigue because they endure millions of stress cycles under moving train load during their service life. This may lead to stiffness reduction in truss joints of the truss bridges and reduce the operational effectiveness of the bridge. In this paper, a hybrid algorithm based on Artificial Neural Network (ANN) coupled with an evolutionary algorithm, i.e. Genetic Algorithm (GA) is proposed to analyze the stiffness reduction of truss joints in a railway bridge. GA is employed to determine training parameters and overcome the local minimum problems of ANN. Natural frequencies are selected as input data, whereas output data is damage characteristics (locations and levels) of truss joints. The results demonstrate that ANN-GA determines vibration behavior and damages of the considered structure accurately, comprising single stiffness reduction of truss joints as well as multiple stiffness reduction scenarios.

H. Tran-Ngoc (✉) · M. Abdel Wahab
Department of Electrical Energy Metals, Mechanical Constructions and Systems, Faculty of Engineering and Architecture, Ghent University, 9000 Ghent, Belgium
e-mail: hoa.tran@ugent.be

M. Abdel Wahab
e-mail: magd.abdelwahab@ugent.be

H. Tran-Ngoc · L. Nguyen-Ngoc · A. Le-Thuc · T. Bui-Tien (✉)
Department of Bridge and Tunnel Engineering, Faculty of Civil Engineering, University of Transport and Communications, Hanoi, Vietnam
e-mail: btthanh@utc.edu

L. Nguyen-Ngoc
e-mail: nngoclan@utc.edu.vn

A. Le-Thuc
e-mail: btthanh@utc.edu.vn

H. Ho-Khac
Department of Transportation Nghe An, Le Hong Phong Street, Vinh City, Nghe An, Vietnam
e-mail: hokhachanh@gmail.com

G. De Roeck
Department of Civil Engineering, KU Leuven, B-3001 Louvain, Belgium
e-mail: guido.deroeck@kuleuven.be

M. Abdel Wahab (ed.), *Proceedings of the 8th International Conference on Fracture, Fatigue and Wear*, Lecture Notes in Mechanical Engineering,
https://doi.org/10.1007/978-981-15-9893-7_9

Keywords Railway bridge · Artificial neural network · Genetic algorithm truss joints · Stiffness reduction

1 Introduction

Nowadays, because of the rapid development of the economy, transport infrastructure is facing overload situations. Especially, railway bridges are enduring millions of stress cycles under moving train load during their service life [1]. This leads to an increase in damage problems of the bridges. When damages occur in a truss bridge, it tends to cause a decrease in stiffness and affect directly the serviceability of the bridge. Therefore, Structural Health Monitoring (SHM) is vital to detect damages happening in structures [2–5].

Different researches have looked into the stiffness reduction analysis of truss joints to solve damage detection problems. Furukawa and Otsuka [6] used a micro shaking device to identify the stiffness reduction of a truss bridge. The identification technique involved the use of Fourier response function (FRF), which could be obtained by harmonic excitation force generated by a micro shaking device. The research concluded that damage elements on a truss bridge were correctly located by the mentioned method. Al-Qayyim et al. [7] applied the Two-Point Condensation (TPC) technique to detect damages in truss structures. The research concluded that although the mentioned technique showed good results in damage detection, analytical and experimental representations of the structure were not always similar due to the error in the selection of degrees of freedom. Wang et al. [8] proposed a method using multi-layer GA to detect damages in truss bridges under different damage scenarios. The research showed feasible results in damage detection of truss bridges under the effect of measurement noise.

The evolution of machine learning (ML) has changed our lives significantly. Being a subset of ML, ANN has become more and more popular recently, especially in SHM. Researchers have explored the use of ANN for SHM and the results have been promising so far [9, 10]. Choi and Kwon [11] developed a system using neural network to detect damages in a truss bridge, which was tested and showed effective results in the numerical model of the bridge. ANN was applied in [12] and [13] for damage detection systems with a high level of accuracy and efficiency. Da Silva [14] used deep machine learning algorithms such as autoencoders and stacked autoencoders on existing data of real bridges to detect and localize structural damages. The results were then compared to those of traditional methods to prove the superiority in the accuracy of the proposed method.

Nevertheless, since ANNs in particular, and ML in general, apply backpropagation (BP) techniques based on gradient descent (GD), they tend to have major disadvantages relating to local minimum issues. This problem especially occurs when the network produces several best local solutions. To overcome this issue, many researchers employed evolutionary algorithms combined with ANN to look for the global best. Tran-Ngoc et al. [15] used Cuckoo Search (CS) algorithm to improve training parameters of ANN. The objective function included natural frequencies and mode shapes. Yaghini et al. [16] used a hybrid algorithm combining ANN and an evolutionary algorithm to optimize complex production units. Toghyani et al. [17] employed both ANN based on Levenberg–Marquardt algorithm and a combination of ANN and Particle Swarm Optimization (PSO) to determine the energy used for machines. Being inspired by those ideas, in this paper, we apply a combination of ANN and GA (ANN-GA) to determine the stiffness reduction of truss joints in a truss bridge. Training parameters of the network are calculated utilizing global search techniques of GA instead of utilizing common BP techniques of ANN. The discrepancies between calculated outputs and real ones are chosen as an objective function. To validate the efficiency of ANN-GA, Nam O bridge (a railway bridge in Vietnam) is employed. Measurement data of the bridge is collected to obtain the vibration behavior of the structure. Finite Element (FE) model of the bridge is created to obtain the training data. Model updating is conducted to obtain a baseline model that will be used for damage identification (stiffness reduction analysis of truss joints).

The paper is arranged as follows. After the introduction, an improved artificial neural network is presented in Sect. 2. Section 3 introduces the stiffness reduction analysis of truss joints in Nam O bridge. Before damage detection of the bridge is considered, model updating is employed to reduce the discrepancies between the calculated and measured results. Finally, some main conclusions are highlighted.

2 Improved ANN (ANN-GA)

ANN is a subset of models utilized in ML that provides trained network the capacity to automatically learn from experience and improve its results without being explicitly programmed. The structure of ANN consists of:

- One input layer: The input data is inserted into the input layer before transferring to other layers.
- Hidden layers: A transformation to the data is applied.
- One output layer: Data from the hidden layer is received and output data is provided.

An example of the architecture of ANN can be seen in Fig. 1.

A network always has one input layer, one output layer, and hidden layers. These layers comprise different neurons (nodes), in which properties are represented by weights (ω) and bias (b). Weight is the strength of the connection between neurons and bias is an additional parameter used to adjust the output. In the forward process, the input layers transfer neurons to the output layers utilizing training parameters to change the values of neurons. If the discrepancies between calculated results and target ones are still large and do not satisfy the objective function, the backward process is conducted applying BP techniques. This process is iterated until the objective function is satisfied. Training parameters of the network are changed via

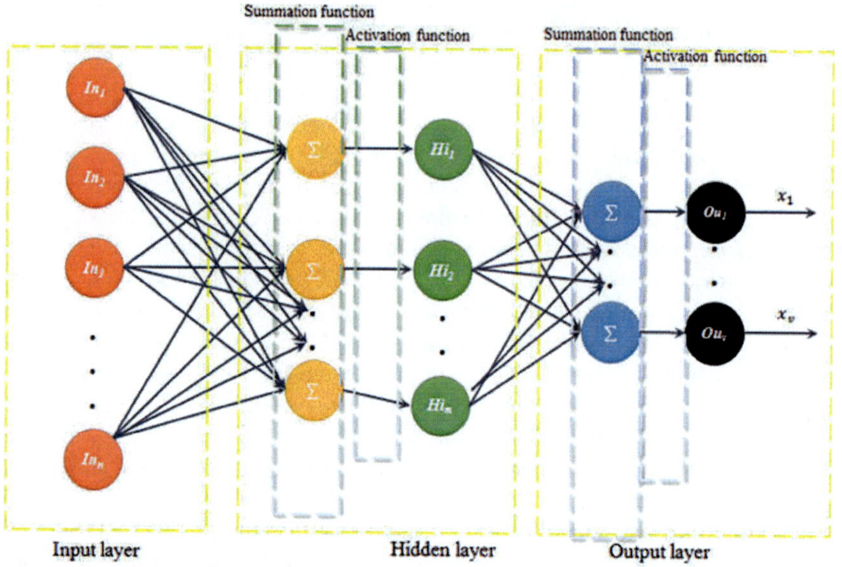

Fig. 1 Example of the architecture of ANN

iterations to reduce the discrepancies between the real and target results. Nevertheless, due to the application of BP techniques utilizing GD techniques, ANN may trap into local minima. This problem especially occurs when the network produces too much the best local solutions. To overcome this issue, in this paper, we propose combining GA with ANN to tackle optimization problems.

3 Stiffness Reduction Analysis of Truss Joints in Nam O Bridge

3.1 *Description of the Bridge*

To investigate the effectiveness of ANN-GA, a railway bridge is used. Nam O bridge (Fig. 2) is located at Danang city (the middle of Vietnam) opened to the traffic in 1985. This bridge includes 4 spans with the same length of 75 m. The rail track is attached directly to the stringer beams. Roller and pin bearings are used to support the span. The general layout of the bridge is shown in Fig. 2.

The truss members consist of lower chords, upper chords, lower wind bracings, upper wind bracings, diagonal chords, vertical chords, stringers and struts (Fig. 3). Because of a long time for exploitation and the effect of heavy loads from trains, the bridge is degraded with stiffness reduction in truss members, especially, at truss joints.

Fig. 2 Nam O bridge

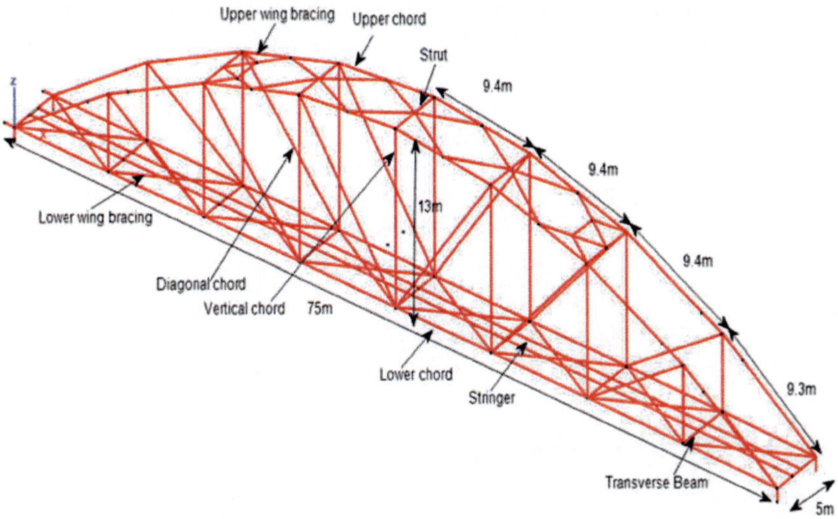

Fig. 3 Main structural elements [18]

3.2 *Measurements*

To make the references easier, we named the bridge's abutments located in the two sides of the river $A0$ and $A1$, while the three piers are considered as $P1$, $P2$ and $P3$, respectively, as shown in Fig. 4. In the context of this paper, only the span connecting abutment $A0$ and pier $P1$ is selected for data measurements.

Fig. 4 Nam O truss bridge with annotated abutments and piers [19]

We considered two main sources of excitation of the bridge: bridge free vibration after train passing and ambient vibration due to the wind. Piezoelectric sensors were utilized to record vibration data. For safety reasons (the bridge without a handrail), only some reachable nodes on the top were installed sensors. Those nodes were then joined by nodes measured at the bottom chords to form the measurement grid. Figure 5 shows the measurement grid including 8 setups (4 reference points and 6 roving points).

In the next step, measured data was processed by MACEC software [20] before it can be used for modal analysis. We only considered the first lower modes for modal analysis since they affect largely the dynamic properties of the bridge and they are also easier to be extracted accurately. The natural frequencies of the first five modes are obtained by the application of the operational modal analysis method. The results obtained will be used in the next part, which is model updating.

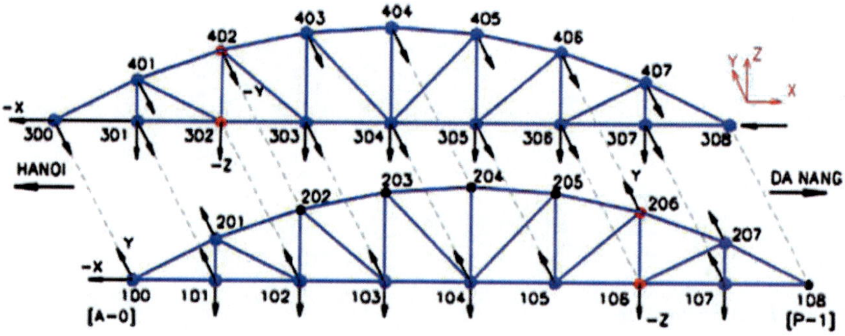

Fig. 5 Measurement grid of the truss bridge. Blue: roving points; Red: reference points

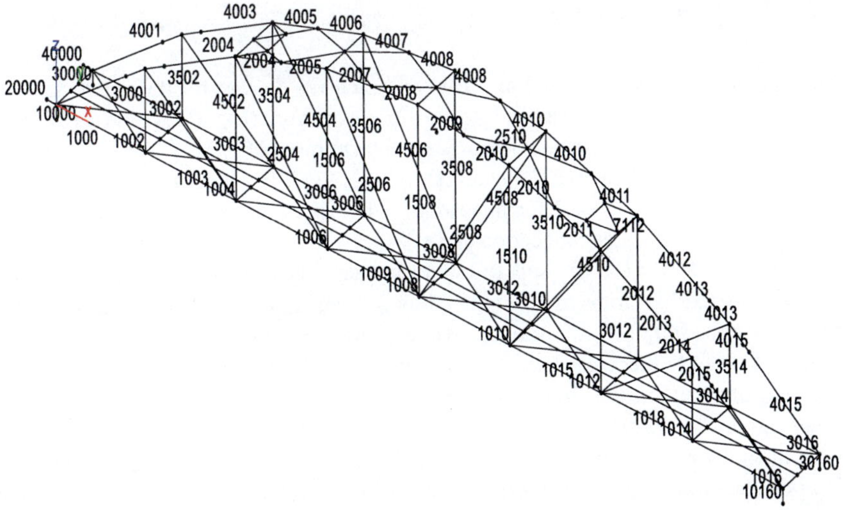

Fig. 6 FEM of bridge [18]

3.3 Baseline FEM

A FE model of the bridge is constructed as shown in Fig. 6. This model consists of 45 nodes and 146 elements. Those elements used in the model have 6 degrees of freedom (DOFs) including translational and rotational displacements in *x, y, z*. The pier (*A0*) uses roller bearing, whereas pier (*P1*) applies a pin one. The roller bearing permits rotational displacement and the pin one allows translational and rotational displacements. The non-structural components comprise handrails, the rail track, maintenance paths, etc. are considered as concentrated loads.

3.4 Model Updating

Before applying ANN-GA to identify structural damages, the numerical model is updated to minimize the differences between the numerical and experimental results. The purpose of this step is to determine the uncertain parameters exactly. Uncertain parameters of Nam O bridge include the stiffness of rotational springs at truss joints, the stiffness of bearings and Young's modulus of truss members. The objective function includes the first five modes. After model updating, the discrepancies between FEM and measured results reduce significantly, i.e. the highest difference between the two results is less than 10%. For more details about the model updating of the bridge, the readers may refer to Ref. [18].

3.5 Damage Detection

The updated FE model is used to create an extensive dataset for the network with various damage scenarios. The architecture of the network (Fig. 7) includes one input layer, one hidden layer and one output layer. The number of hidden layers is selected based on the iteration to obtain the best results [21–23]. The natural frequencies (the first fifteen modes) are used for input data, whereas the output data is the damage characteristics, i.e. damage location and severity.

The trained network is used to determine damages in the considered bridge. Damages in the FE model are introduced by decreasing the stiffness of one or more specified truss joints of the bridge. Two stiffness reduction scenarios at one truss joint and several other truss joints are considered. In the case of single reduction, the stiffness of truss joints is reduced from 0 to 50% with an interval of 1%, whereas others are intact. There are 2500 data samples created from the basic model used to train the network. In case of reduction in several truss joints, the stiffness reduction is assigned randomly for 3 truss joints at the same time. The stiffness is reduced from 0 to 50% with an interval of 1%. Totally, there are around 10 500 damage cases generated from the model.

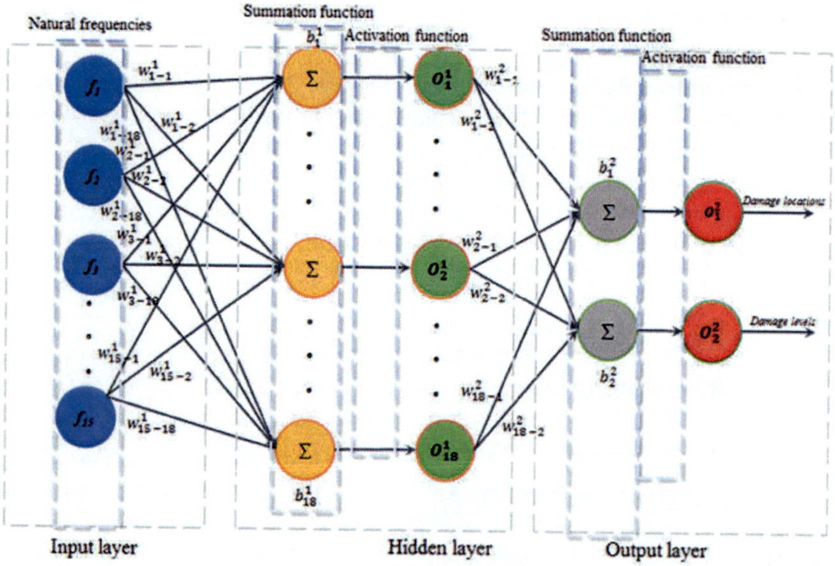

Fig. 7 The architecture of the network used for damage identification in Nam O bridge

3.5.1 The Reduction in Stiffness at One Truss Joint

Figure 8 shows that the regression line is close to the 45-degree line and the R values are higher than 0.98. Thus, the network is trained well and can be applied to identify damages in the considered bridge. Comparison between indices of all algorithms is listed in Table 1.

Table 1 shows that *R*-value calculated by ANN-GA (0.985) is higher than that calculated by ANN (0.982). The difference between calculated and real values (error) of ANN-GA is lowest, at 1.008, whereas those of ANN and GA are is 1.201 and 4.501, respectively. In terms of computational time, ANN and ANN-GA spend 105 s and 119 s, respectively to look for the best solution, whereas GA expends 55423 s for this process.

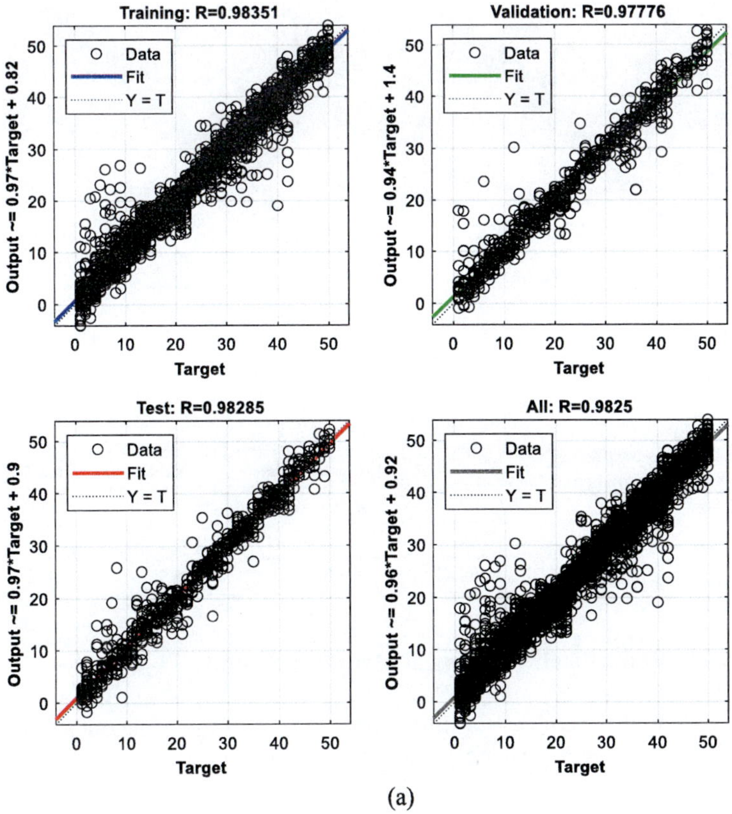

Fig. 8 Training process **a** ANN and **b** ANN-GA

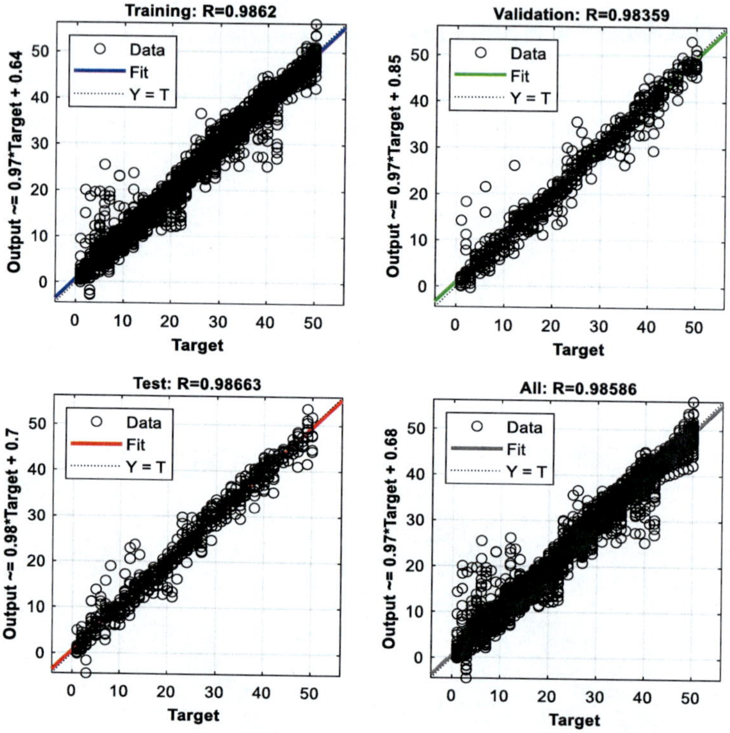

(b)

Fig. 8 (continued)

Table 1 Indices used to consider the performance of algorithms

Algorithm	MSE-Values	*R*-Values	Time (sec)
GA	4.501		55423
ANN	1.201	0.982	105
ANN-GA	1.008	0.985	119

3.5.2 The Reduction in Stiffness at Many Truss Joints

Figure 9 shows that the regression line is close to the 45-degree line and *R*-values are higher 0.97. This means that a good agreement between calculated and real results are achieved. Comparison indices of all algorithms are listed in Table 2.

Table 2 shows that *R*-value calculated by ANN-GA (0.9758) is higher than that calculated by ANN (0.9765). The difference between calculated and real values (error) of ANN-GA is lowest, at 3.0503, whereas those of ANN and GA are is 3.2134 and 4.831, respectively. In terms of computational time, ANN and ANN-GA

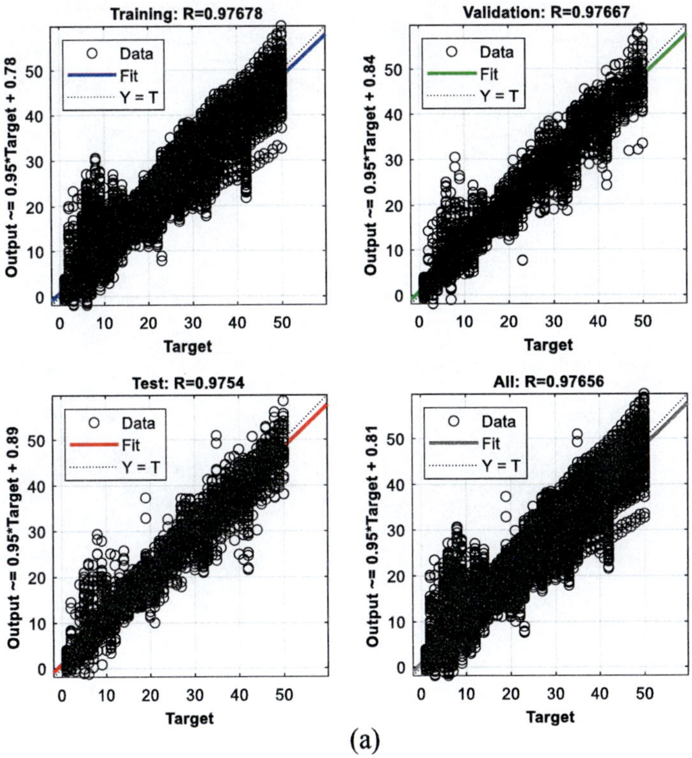

Fig. 9 Training process **a** ANN and **b** ANN-GA

spend 1342 and 1465 s to look for the best solution, whereas GA expends 56324 s for this process.

4 Conclusion

A hybrid algorithm (ANN-GA) is proposed in this research to identify the stiffness reduction of truss joints. ANN is a subset of models of machine learning used BP to look for the best solution. However, because BP applies GD techniques, the network may trap into local minimal if the network generates several local best. GA with capacity for looking globally is used to calculated training parameters, which plays a vital role in remedying the shortcomings of ANN (local minimal). To compare with ANN-GA, ANN alone and GA alone are also applied. Some main conclusions can be summarised as follow:

- All algorithms consisting of GA, ANN, and GA-ANN can determine exactly the damage location of the considered bridge.

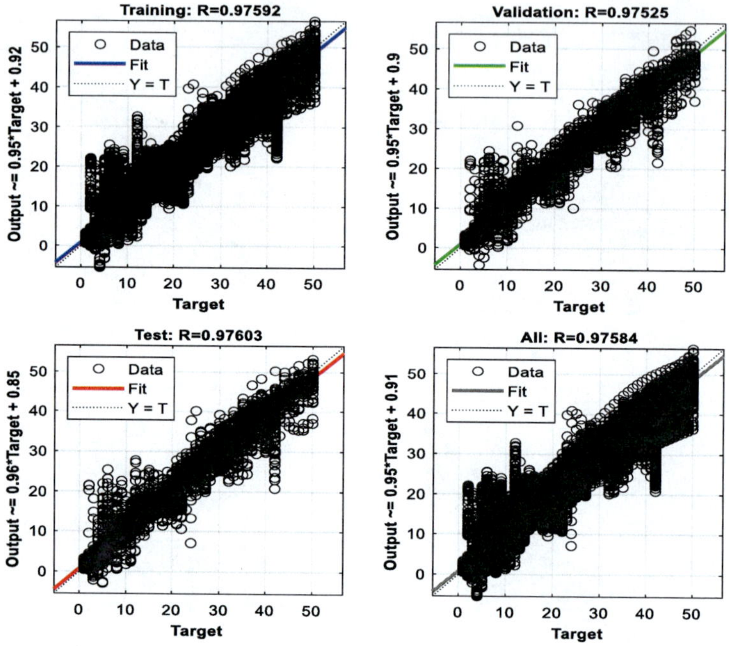

(b)

Fig. 9 (continued)

Table 2 Indices used to consider the performance of algorithms

Algorithm	MSE-Values	*R*-Values	Time (sec)
GA	4.831		56324
ANN	3.2134	0.9765	1342
ANN-GA	3.0503	0.9758	1465

- GA spends too much time for looking for the best solution because it is based on stochastic techniques and too many parameters have to be adjusted in the search process.
- ANN makes some errors in the identification of damage level because it traps into local minima.
- GA-ANN outperforms GA and ANN in terms of both computational time and accuracy because it applies both advantages of ANN and GA. ANN has advantages in reducing computational time, whereas GA can deal with the problem of local minimal of machine learning.

Acknowledgements The authors acknowledge the financial support of VLIR-UOS TEAM Project, VN2018TEA479A103, 'Damage assessment tools for Structural Health Monitoring of Vietnamese infrastructures' funded by the Flemish Government. Moreover, the first author needs to acknowledge the financial supports from Ministry of Education and Training (MOET) under the project research "B2020 – GHA - 02" and Bijzonder Onderzoeksfonds (BOF) of Ghent University.

With the support of THE BELGIAN DEVELOPMENT COOPERATION

References

1. Tran-Ngoc H, He L, Reynders E, Khatir S, Le-Xuan T, De Roeck G, Bui-Tien T, Wahab MA (2020) An efficient approach to model updating for a multispan railway bridge using orthogonal diagonalization combined with improved particle swarm optimization. J Sound Vibr 115315
2. Ho VL, Tran NH, De Roeck G, Bui TT, Wahab MA (2018). System identification based on vibration testing of a steel I-beam. In: International conference on numerical modelling in engineering, Springer, Singapore, pp 254–268
3. Hoa TN, Thanh BT (2019) Damage detection in a steel beam structure using particle swarm optimization and experimentally measured results
4. Tran-Ngoc H, Khatir S, De Roeck G, Bui-Tien T, Wahab MA (2020) Damage assessment in beam-like structures using cuckoo search algorithm and experimentally measured data. In: Proceedings of the 13th international conference on damage assessment of structures, Springer, Singapore, pp 380–385
5. Hoa TN, Khatir S, De Roeck G, Long NN, Thanh BT, Wahab MA (2020) An efficient approach for model updating of a large-scale cable-stayed bridge using ambient vibration measurements combined with a hybrid metaheuristic search algorithm. Smart Struct Syst 25(4):487–499
6. Furukawa A, Otsuka H (2008) Stiffness reduction identification of steel truss bridge using micro shaking device. In: 14th World conference of earthquake engineering, China
7. Alqayyim Ahmed, Caglayan Barlas (2019) Damage detection of a truss bridge via vibration data. Bridge Struct 14:109–118. https://doi.org/10.3233/BRS-180140
8. Wang FL, Chan TH, Thambiratnam DP, Tan AC, Cowled CJ (2012) Correlation-based damage detection for complicated truss bridges using multi-layer genetic algorithm. Adv Struct Eng 15(5):693–706
9. Khatir S, Khatir T, Boutchicha D, Le Thanh C, Tran-Ngoc H, Bui TQ, Capozucca R, Abdel-Wahab M (2020) An efficient hybrid TLBO-PSO-ANN for fast damage identification in steel beam structures using IGA. Smart Struct Syst 25(5):605–617
10. Khatir S, Boutchicha D, Le Thanh C, Tran-Ngoc H, Nguyen TN, Abdel-Wahab M (2020) Improved ANN technique combined with Jaya algorithm for crack identification in plates using XIGA and experimental analysis. Theor Appl Fract Mech 102554
11. Choi MY, Kwon I-B (2000) Damage detection system of a real steel truss bridge by neural networks. 3988:295–306. https://doi.org/10.1117/12.383151
12. Nguyen DH, Bui TT, De Roeck G, Wahab MA (2019) Damage detection in Ca-Non Bridge using transmissibility and artificial neural networks. Struct Eng Mech 71:175–183
13. Weinstein J, Sanayei M, Brenner B (2018) Bridge damage identification using artificial neural networks. J Bridge Eng 23. https://doi.org/10.1061/(ASCE)BE.1943-5592.0001302
14. Da Silva MFM (2017) Machine learning algorithms for damage detection in structures under changing normal conditions. PhD Dissertation
15. Tran-Ngoc H, Khatir S, De Roeck G, Bui-Tien T, Wahab MA (2019) An efficient artificial neural network for damage detection in bridges and beam-like structures by improving training parameters using cuckoo search algorithm. Eng Struct 199:109637
16. Yaghini M, Khoshraftar MM, Fallahi M (2013) A hybrid algorithm for artificial neural network training. Eng Appl Artif Intell 26(1):293–301
17. Toghyani S, Ahmadi MH, Kasaeian A, Mohammadi AH (2016) Artificial neural network, ANN-PSO and ANN-ICA for modelling the stirling engine. Int J Ambient Energy 37:456–468

18. Tran-Ngoc H, Khatir S, De Roeck G, Bui-Tien T, Nguyen-Ngoc L, Abdel Wahab M (2018) Model updating for Nam O bridge using particle swarm optimization algorithm and genetic algorithm. Sensors 18(12):4131
19. Tran-Ngoc H, Khatir S, De Roeck G, Bui-Tien T, Nguyen-Ngoc L, Wahab MA (2019) Stiffness identification of truss joints of the Nam O bridge based on vibration measurements and model updating. In: International conference on arch bridges 2019 Oct 2, Springer, Cham, pp 264–272
20. Reynders E, Schevenels M, Roeck GD (2011) MACEC 3.2: a matlab toolbox for experimental and operational modal analysis. user's manual. Report BWM-2011-XX. 1.2011
21. Dang VH, Hoa TN, Tung VN, Bui-Tien T, Guido DR, Nguyen HX Data-driven structural health monitoring using feature fusion and hybrid deep learning. IEEE Trans Autom Sci Eng 1–17
22. Tran-Ngoc H, Khatir S, Le-Xuan T, De Roeck G, Bui-Tien T, Abdel Wahab M (2020) A novel machine-learning based on the global search techniques using vectorized data for damage detection in structures. Int J Eng Sci 157:103376
23. Nguyen DH, Tran-Ngoc H, Bui-Tien T, De Roeck G, Wahab MA (2020) Damage detection in truss bridges using transmissibility and machine learning algorithm: Application to Nam O bridge. Smart Struct Syst 26(1):35–47

Flow Forming Process for Annealed AISI 5140 Alloy Steel Tubes

Aptullah Karakaş, Acar Can Kocabıçak, Senai Yalçınkaya, and Yusuf Şahin

Abstract Flow forming process performed on annealed AISI 5140 steel material which called "preform" by 50% reduction ratio in this experimental study. The mechanical properties such as tensile strength, yield strength and hardness and microstructures were investigated in order to see the effect of flow forming process on the tested materials. The changes of these properties were evaluated in terms of microstructural observation and hardness tensile measurement. The hardness distribution and microstructural characterization were investigated throughout the wall thickness of the workpiece. The experimental results indicated that the hardness differences occurred because of the roller force effect on the outer surface was more than the force on the inner surface of the workpiece and faster coolant rate. In addition, severely grain orientations were observed, hardness, yield and tensile strength of the workpiece enhanced significantly because of dislocation generation and pile-up during plastic deformation. Although the yield and tensile strength increased, elongation properties reduced after the flow forming operation. Besides, the homogeneous grains orientations were observed throughout the wall thickness of the workpiece due to application the flow forming.

Keywords Flow forming · CNC flow forming machine · Cold working · 5140 steel · Tensile test · Hardness distribution · Microstructure · Reduction ratio

A. Karakaş · A. C. Kocabıçak (✉)
Repkon Machine and Tool Industry and Trade Inc, Istanbul 34980, Turkey
e-mail: acar.kocabicak@repkon.com.tr

S. Yalçınkaya
Department of Mechanical Engineering, Technology Faculty of Marmara University, Istanbul 34722, Turkey

Y. Şahin
Faculty of Engineering and Architecture, Nişantaşı University, Nişantaşı, Istanbul, Turkey

M. Abdel Wahab (ed.), *Proceedings of the 8th International Conference on Fracture, Fatigue and Wear*, Lecture Notes in Mechanical Engineering,
https://doi.org/10.1007/978-981-15-9893-7_10

1 Introduction

Flow forming is an incremental metal-forming process that makes possible to produce high precision thin-walled rotational symmetric parts. Flow forming is a plastic deformation process that increases significantly mechanical properties of the finished part. The wall thickness of the starting material which name is preform reduces by radial forcing thanks to the forming rollers. The preform is clamped between mandrel and tailstock before the flow forming operation. Both the mandrel and preform are rotated by main spindle while the forming rollers contact the workpiece. While the wall thickness of the material reduces, axial elongation is provided by headstock thrust force. This technique enhances the strength of material severely, mainly due to strain hardening [1]. It provides high dimensional accuracy and roughness in order to produces high precision complex contours metal parts. The flow forming increases the mechanical properties of the workpiece considerably and this technique makes possible to produce lighter and stronger parts compared to conventional machining operations [2]. This process is mainly used in aerospace & defense fields in order to produce drive shaft for jet engines & helicopters, control & piston rods, rocket & missile cases and high precision hollow tubes [3].

Thickness reduction ratio, rotation speed of the mandrel, axial feed, preform dimensions, workpiece material properties, heat-treatment condition of the preform, rollers geometry, roller setup configuration and friction conditions influence the flow forming technique [4]. General flow forming process parameters are shown in Fig. 1 and descriptions were given Table. 1 respectively.

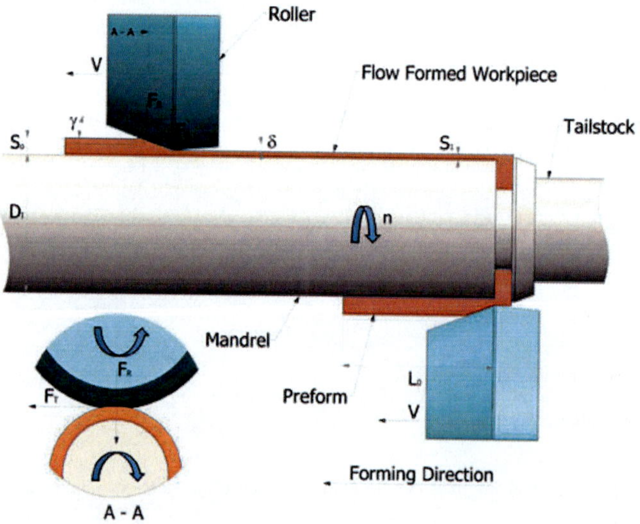

Fig. 1 Principle of flow forming

Table 1 Descriptions of flow forming process parameters

S_O Starting wall thickness (mm)	γ Leading angle (degree,°)	F_A Axial force (N or Kg)
S_1 Finished wall thickness (mm)	δ Trailing angle(degree,°)	F_R Radial force (N)
L_O Starting length (mm)	R Nose radius(mm)	F_T Tangential force (N)
D_I Inside diameter (mm)	V Feed rate (mm/minute)	n Spindle Speed (rpm)

The forming rollers force on the material from the outer surface of the workpiece towards the inside surface and the mandrel. The intensity of this force affects the hardness homogeneity and reduction ratio of the workpiece. The roller force acts on the outer surface is more than on the inner surface of the workpiece. This roller force and the number of forming passes cause different grain orientations which take place throughout the wall thickness of the workpiece [5]. These differences cause to have different mechanical and hardness properties from the outer surface to the inner surface.

The reduction ratio and heat treatment condition of the preform is playing an important role to produce high mechanical and hardness properties material after the flow forming process [6]. The reduction ratio (R) formulation is defined by the following formula,

$$R = [(So - S1) / So] \times 100 \quad (1)$$

In this study, the gained mechanical properties of AISI 5140 material after the 50% reduction ratio flow forming process were investigated Moreover, the hardness and microstructure distribution across the wall thickness of the workpiece were analyzed.

2 Materials and Experimental Work

Ductile metals and alloys include stainless steels and aluminum alloys, as well as annealed or solution annealed condition materials can be used in the flow forming process. A common method for moderating the residual stresses before the cold formed parts is to carry out annealing heat treatment. Annealing condition preform was carried out in this study because this condition improves the plastic forming capability of the metal materials [7]. The used material is annealed AISI 5140 steel for this experimental investigation. The major chemical composition of the workpiece material according to ASTM A29/A29M measured is shown in Table 2 [8]. The material has high load, low impact and good wearing resistance. AISI 5140 steel is to produce crankshaft, knuckle, gear wheel, front axle, steering shaft etc. parts in vehicle and engine components [9].

Table 2 The chemical compositions of 5140 alloy according to ASTM A29/A29M standard (in weight %)

Element type	C	Cr	Si	Mn	P	S
ASTM A29/A29M	0.38–0.43	0.70–0.90	0.15–0.35	0.70–0.90	≤0.035	≤0.040
Measured	0.40	0.80	0.20	0.71	0.024	0.009

The preform and flow formed parts used in this research are shown in Fig. 2(a) and (b) respectively. Before the flow forming process, the preform was loaded on the mandrel. Three forming rollers with 120° interval centered were used in this experimental study. 50% flow forming reduction ratio was applied on the process. The wall thickness of the preform is reduced to the final thickness by elongating it without change in the inside diameter of the tube. During the process, 60 mm/minute feed rate and 300 rpm mandrel speed was used on the workpiece. High-speed rpm causes the rollers rotates faster throughout the mandrel and increase the friction between roller and workpiece. Cooling water (boron oil) was used in order to minimize the friction. Besides, the radial force of the rollers influence the friction factor on the workpiece [10]. Roller forces were powered by a motor servo control system in order to make high dimensional accuracy process.

In order to investigate flow forming effects on the hardness of the materials, the hardness test was performed with Emco Test Dura Scan-20 equipment and a micro

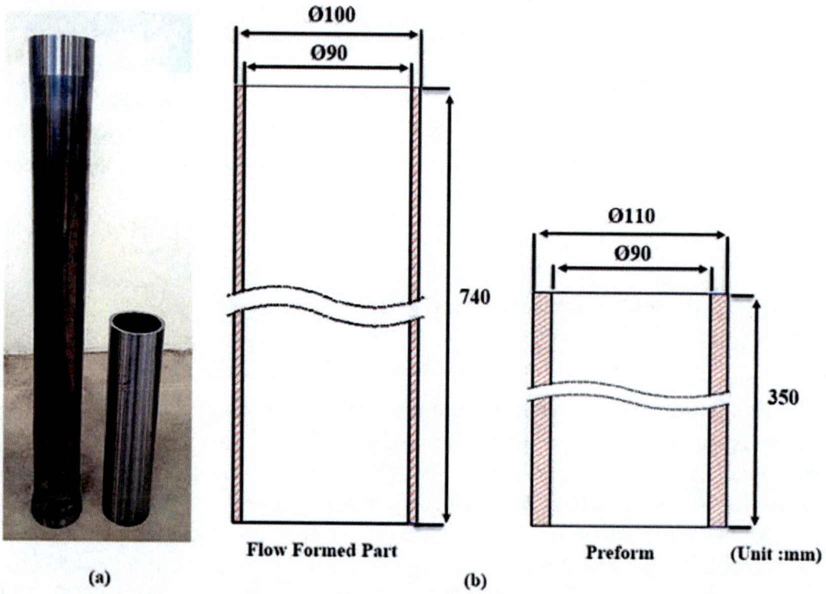

Fig. 2 The preform and flow formed parts (**a**) and illustration of dimension and size of parts (**b**)

indentation method (HV0,1) was used. The hardness tests were applied thorough the wall thickness of the flow formed material to analyze the hardness distribution.

The tensile curves with stress and elongation were predicted by using the Instron 3382 tensile test device. The Instron 3382 test unit applies maximum 100kN force and the test was performed according to ASTM E8M standard [11]. Yield stress (YS), ultimate tensile stress (UTS) and elongation properties are figured out using the tensile curves. YS represents the stress point that a material starts the plastic deformation and nonlinear deformation begins. The tensile stress (TS) is described as a stress and it is measured as force per unit area. UTS is designated as the maximum engineering stress in a tensile test. UTS indicates the end of uniform elongation and beginning of the delimited necking elongation section which is a measurement of the deformation [12].

An Olympus CX43 Optical Microscope was used in order to examine microstructural behavior of the annealed AISI 5140 steel and flow formed AISI 5140 steel materials. Tensile testing specimens were extracted from annealed condition AISI 5140 tube and flow formed AISI 5140 tube. The specimens were taken from the circular tube walls by water jet cutting operation along the longitudinal direction of the parts. This cutting sample directions are illustrated in Fig. 3(a) and the dimensions of the tensile specimens are shown in Fig. 3(b), respectively.

As illustrated in Fig. 3 the used specimen length is 100 mm and width is 10 mm for the tensile test. The length of gauge is 30 mm and width is 6 mm. Before the flow forming process 10 mm wall thickness was used for the preform and the thickness was reduced to 5 mm by single-pass. To summarize, 50% reduction ratio was fabricated on the preform by a computer numerical control (CNC) Repkon flow forming machine.

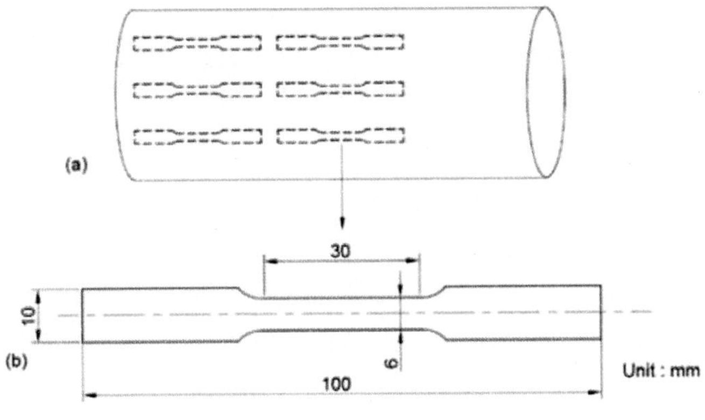

Fig. 3 (**a**) The specimen's cutting position for tensile testing, (**b**) specimen dimensions for the testing

2.1 Process Description

Figure 4 illustrates the rollers contact points with the workpiece and their location to workpiece during the flow forming process. The hardness of the workpiece and mechanical properties increase after the flow forming process. Initially, the rollers begin the contact with the workpiece from outside surface and the forces reach to the mandrel. The applied radial forces should reach up to the mandrel in order to make forming process properly, otherwise different zones would occur on the microstructure and grains during the forming process. In order to achieve homogeneity of microstructure, two different formed and unformed zones should be avoided through the wall thickness [13]. Otherwise, an inhomogeneous grain orientations might occur during the flow forming due to the local deformation between rollers and the workpiece surface. For this reason, the radial force should be stronger for the same material at larger wall thicknesses. The contact area between the roller and workpiece is called "outside surface" and the contact surface area between the workpiece and mandrel is called "inside surface".

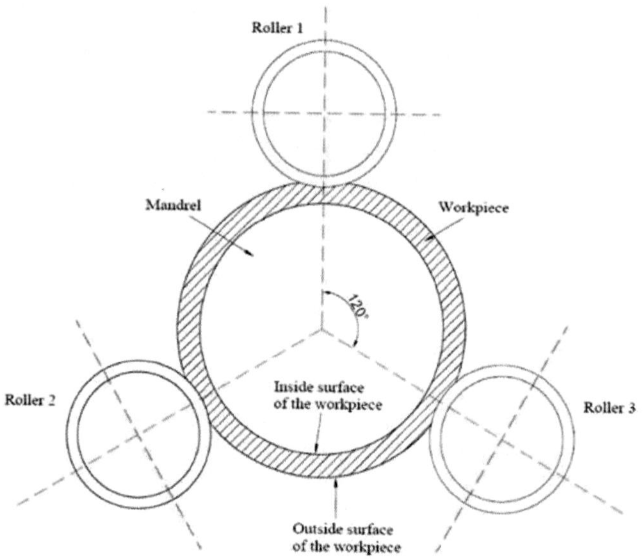

Fig. 4 Roller contact points on the workpiece

3 Results and Discussion

3.1 *Mechanical Characterization*

10 longitudinal tensile samples were extracted from each tubes and tensile tests were performed with a speed of 1 mm/sec. average values of the tensile testing results is illustrated in Fig. 5. It was observed that the obtained YS is about 350 Mpa, maximum TS is about 745 Mpa of the annealed AISI specimen. YS properties of the flow formed specimen reached about 900 MPa, TS is about 1190 MPa because of the application of cold forming mechanism. As for the case of strain ratio, the opposite behavior was provided from this graph. As shown Fig. 5 respectively, strain (elongation) was obtained about 18% for the annealed AISI 5140 and strain reached about 9% for the flow formed specimen.

Mechanical properties of metal alloys such as YS, TS and hardness increase after the cold working process because of strain-hardening mechanism (dislocation generation and dislocation pile-up during plastic deformation). As can be seen in Table 3, the YS and TS increased significantly after the flow forming process. The

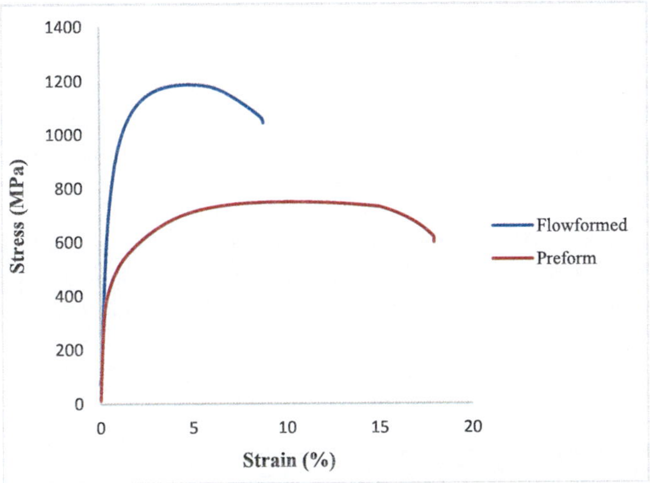

Fig. 5 Tensile properties of AISI 5140 steel tubes

Table 3 Tensile properties of AISI 5140 steel tubes

Temper	Yield strength (YS) (MPa)	Tensile strength (TS) (MPa)	Strain rate (%)	Average hardness (HV)
Annealed (preform)	350	745	18	250
Flow formed	900	1190	9	320

thickness of the flow formed material was reduced by 50% for this reason, it is stronger than the preform material condition. As presented in Table 3, the hardness increased due to the plastic deformation. The main reason of this result might be the AISI 5140 alloy have low stacking fault energy hence, it has high work hardening capability compare to aluminum alloys. The dislocations cannot make cross sliding movement in the low stacking fault energy materials during the plastic deformation thus they accumulate heavily one area. Due to the low stacking fault energy feature in AISI 5140 Steel, the hardening capabilities of the alloy is high. It was observed that the YS properties of the AISI 5140 steel increase approximately 3 times and TS properties increased about 60% after the flow forming process.

3.2 *Hardness Characterization*

The hardness test was carried out by Vickers method. The hardness tests were repeated 10 times and performed at 500 μm (micron) intervals in order to investigate hardness distribution along the wall thickness of the flow formed tube. The specimen is taken from the flow formed tube in the cross section of the deformed part as and the hardness test positions were presented in Fig. 6, respectively. The measured points on section A-A like 1, 2, 3, …, 10 and cross-section of the flow formed tube are indicated in Fig. 6(a), (b), respectively. In Fig. 6(a), 1, 2, 3 points show the outside surface while 8, 9 and 10 points exhibit the inside points. The outside surface of the flow formed tube is found to be harder than the inside surface of the workpiece because of the force transmission along the wall thickness. Figure 7 exhibits the hardness results from different points along the wall thickness of the steel samples. It was observed that the hardness values reduced gradually when approached the inside points, but its hardness increased with approaching outside points. In other words, maximum hardness was obtained from the location of 1 point while minimum hardness was observed from the location of 10 point. This might be due to lower/higher cooling

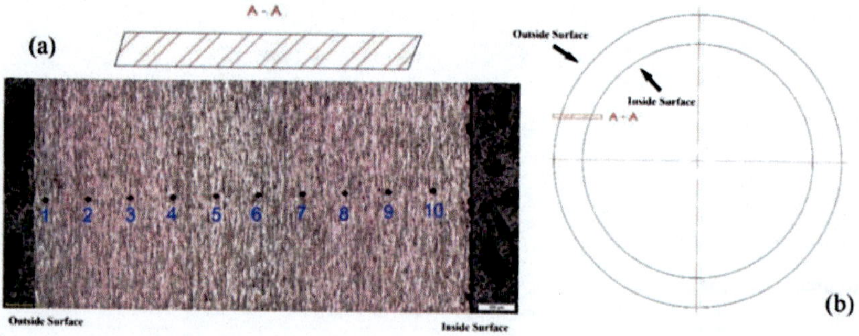

Fig. 6 **a** Hardness test positions and measured points (1, 2, 3, …, 10) on section A-A, **b** Cross-section of the flow formed tube

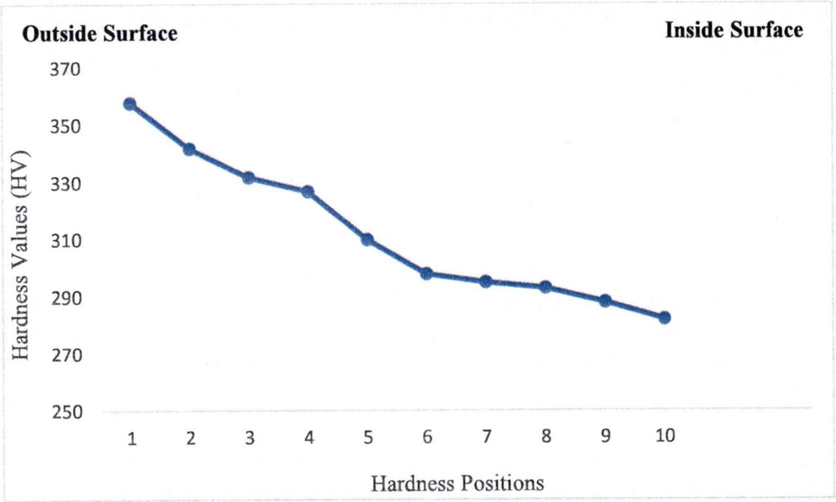

Fig. 7 Hardness test positions versus measured hardness results

rate and this also controls the size of crystal grain/growth. In other words, the region of outside surfaces like 1, 2 exposed to the highest force and the region of inside surface like 8, 9, 10 exposed to the least radial force.

Beside of the hardness distribution along the wall thickness, dislocations occurred more because of the applied force on the material. As the amount of dislocations and dislocation locked increase, hardness value is increased accordingly. As a result of this, the grain orientations were observed obviously in Fig. 8, respectively. The grain orientations occurred perpendicularly to the applied radial force direction and these orientations continued to the inner surface in the same way. The hardness result is demonstrated that higher deformation takes place in the outer surface than in the internal surface [14].

3.3 Microstructure Characterization

Tensile specimens were extracted from longitudinal sections of the annealed AISI 5140 (annealed) and flow formed 5140 tube to examine the microstructure characterization by optical microscope. Figure 8(a), (b) shows the microstructure of annealed condition of the AISI 5140 before the flow forming operation under 200X and 500X magnifications, respectively. It was observed that ferritic (white) and pearlitic (brown) structure and carbide/ferrite were evident at different scale. Its ferritic size ranged from 10 to 50 μm while pearlitic was about 20–80 μm. Shapes of the phases are varied throughout the cross-section. In addition, Fig. 8(c, d) shows the microstructure of the material after the reduction of 50% applied thanks to the flow forming method at 200X and 500X magnification, respectively. The microstructures of the steels were taken from AISI 5140 tube. More uniform and smoother microstructures were observed because the process led to all phase/grains are oriented in the longitudinal direction. The length of ferritic and pearlitic phases were reached to 20–120 μm and 20–60 μm, respectively. The ferritic size was larger than those of

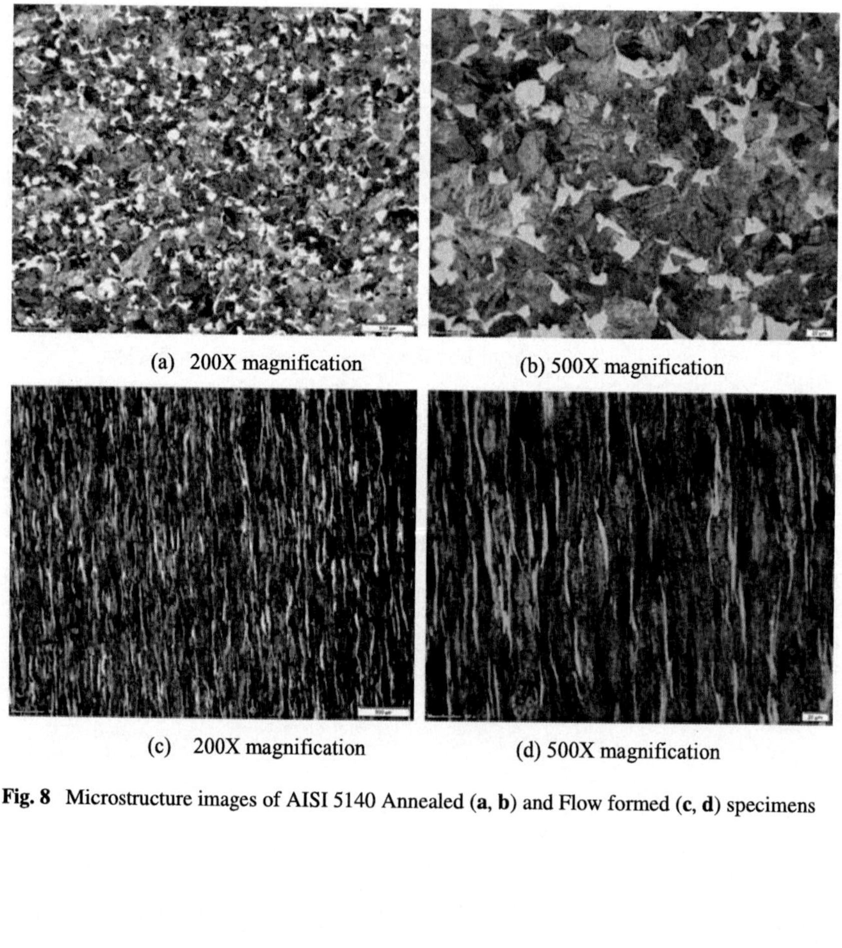

(a) 200X magnification (b) 500X magnification

(c) 200X magnification (d) 500X magnification

Fig. 8 Microstructure images of AISI 5140 Annealed (**a**, **b**) and Flow formed (**c**, **d**) specimens the other phases due to forming a soft one. However, the grain orientations were not determined on annealed condition AISI 5140 steel material, as can be seen in Fig. 8(a) and (b).

After the reduction of 50% through the flow forming opration, orientations were observed in the microstructures as indicated in Fig. 8(c) and (d), respectievely. In other words, the flow formed material shows a fine "fiber-like" microstructure that contains sever elongated steel matris. As a result of an important plastic deformation, secondary phases can be orientated during the flow forming processing. This elongated microstructure directly affects mechanical and strain-hardening properties because dislocations are piled-up on the surface. These piled up dislocations and orientations occurred during the cold working process as during the flow forming method [15]. This phenomenon indicates that the hardness properties are depending on the deformation history of the materials after the flow forming [16].

4 Conclusion

Two types of specimens such as annealed AISI 5140 and flow formed AISI 5140 alloys were investigated in this experimental study. The mechanical properties such as TS, YS, hardness and microstructure were compared with each other. The microstructure properties of the specimens were defined by using optical microscope. The results revealed that the outside surface of the flow formed tube hardness is higher than the inside surface of the tube because of the force transmission. Although the hardness value of the flow formed tube gradually decrease from outer surface to inner surface, the homogeneous grains orientations were obtained along the wall thickness of the flow formed part. As a result, the flow forming process improved significantly the tensile and yield properties of the starting material. The mechanical properties of the flow formed AISI 5140 is higher than the annealed AISI 5140 material. However the hardness distribution might be variable along the wall thickness of the flow formed part. This study demonstrated that annealed AISI 5140 material could be plastically deformed with flow forming process by applying 50% thickness reduction rate.

Acknowledgements The authors would like to thank Repkon Machine and Tool Industry and Trade Inc. which financially supported this study. This study was prepared by the Repkon Inc. engineers and was performed in Repkon Metalworking Technology Center.

References

1. Haghshenas M, Klassen RJ (2015) Mechanical characterization of flow formed FCC alloys. Mater Sci Eng A 641:249. https://doi.org/10.1016/j.msea.2015.06.046
2. Tsivoulas D, Quinta da Fonseca J, Tuffs M, Preuss M (2015) Effects of flow forming parameters on the development of residual stresses in Cr–Mo–V steel tubes. Mater Sci Eng A 624:193. https://doi.org/10.1016/j.msea.2014.11.068

3. Sivanandini M, Dhami S, Pabla B (2012) Flow forming of tubes-a review. Int J Sci Eng Res 3(5):1. ISSN 2229-5518
4. Marini D, Cunningham D, Xirouchakis P, Corney JR (2016) Flow forming: a review of research methodologies, prediction models and their applications. Int J Mech Eng Technol 7:285
5. Wang XX, Zhan M, Fu MW, Gao PF, Guo J, Ma F (2018) Micromechanical behaviour of TA15 alloy cylindrical parts processed by multi-pass flow forming. J Mater Process Technol 261:86. https://doi.org/10.1016/j.msea.2018.09.062
6. Molladavoudi HR, Djavanroodi F (2011) Experimental study of thickness reduction effects on mechanical properties and spinning accuracy of aluminum 7075-O, during flow forming. Int J Adv Manuf Technol 52:949–957. https://doi.org/10.1007/s00170-010-2782-4
7. Srinivasulu M, Komaraiah M, Rao CSKP (2012) Experimental studies on the characteristics of AA6082 flow formed tubes. J Mech Eng Res 4(6):192–198
8. ASTM A29/A29M (2016) Standard specification for general requirements for steel bars. Carbon and Alloy, Hot-Wrought
9. Nie D, Chen X, Wu Q (2020) Mixed-mode fracture behaviors of AISI 1045 and AISI 5140 steels in compressor crankshafts—Part II: Mode II-III. Eng Fail Anal 110:104406. https://doi.org/10.1016/j.engfailanal.2020.104407
10. Bhatt RJ, Raval HK (2016) Influence of operating variables during flow forming process. Elsevier BV
11. ASTM E8/E8M (2013) Standard test methods for tension testing of metallic materials
12. Lin CW, Chen KJ, Hung FY, Lui TS, Chen HP (2019) Impact of solid-solution treatment on microstructural characteristics and formability of rotary-swaged 2024 alloy tubes. J Mater Res Technol 8:3137. https://doi.org/10.1016/j.jmrt.2018.12.029
13. Külekçi İ, Biro CD, and G. Aydın, 13th Int. Cold Form. Congr. 195 (2015).
14. Varela S, Santos M, Arroyo A, Pérez I, Puigjaner JF, Puigjaner B (2017) AIP Conf Proc 1896
15. Gungor MN, Ucok I, Kramer LS, Dong H, Martin NR, Tack WT (2005) Microstructure and mechanical properties of highly deformed Ti–6Al–4V. Mater Sci Eng A 410–411:369. https://doi.org/10.1016/j.msea.2005.08.141
16. Wang XX, Zhan M, Gao PF, Zhang HR (2018) Micromechanical behaviour of TA15 alloy cylindrical parts processed by multi-pass flow forming. Elsevier Ltd

Hydrogen Degradation Effects on Crack Propagation in High-Strength Steels: A Fully Coupled Approach

B. Sobhaniaragh, S. H. Afzalimir, and C. Ruggieri

Abstract This study presents a fully coupled cohesive zone model (CZM)-based computational framework to simulate crack propagation in high strength steels. The model comprises initially zero thickness, cohesive-interface elements with the constitutive response described by a hydrogen-degraded Park-Paulino-Roesler (PPR) model. The linear dependence on hydrogen concentration according to a phenomenological decohesion model is chosen for the critical cohesive traction. Moreover, the value of the cohesive energy of the traction separation law (TSL) is adopted from the experimental data of hydrogen-charged specimen. The computational framework accounting for both hydrogen enhanced localized plasticity (HELP) and hydrogen enhanced decohesion (HEDE) mechanisms is employed to simulate crack growth in a C(T) specimen made of AISI 4130 high-strength steel. The parameters included in the PPR model are satisfactorily calibrated with experimental data for the uncharged and hydrogen-charged specimens. It has been concluded that the lattice hydrogen has the dominating factor in the hydrogen degradation compared with trapped hydrogen.

Keywords Hydrogen embrittlement · Ductile crack growth · Cohesive-interface elements · Park-Paulino-Roesler (PPR) model · Hydrogen enhanced localized plasticity

1 Introduction

Hydrogen embrittlement (HE) is one of the most complicated phenomena of degradation in metallic materials, signifying a hydrogen-induced transition from a high-toughness ductile fracture (micro-void coalescence) to a brittle-type of fracture (either by a transgranular or intergranular mechanism). Indeed, HE manifests as loss in structural integrity of a vast range of metals and alloys, which may result in unex-

B. Sobhaniaragh (✉) · S. H. Afzalimir · C. Ruggieri
Fracture Mechanics and Structural Integrity Research Group (NAMEF), Department of Naval Architecture and Ocean Engineering, University of Sao Paulo, Sao Paulo, Brazil
e-mail: behnam@usp.br

M. Abdel Wahab (ed.), *Proceedings of the 8th International Conference on Fracture, Fatigue and Wear*, Lecture Notes in Mechanical Engineering,
https://doi.org/10.1007/978-981-15-9893-7_11

pected and premature catastrophic failures. To date, the definite mechanisms underlying HE are still under debate; at least two crucial mechanisms have been proposed [1]. The hydrogen enhanced decohesion (HEDE) mechanism assumes interstitial atomic hydrogen reduces the cohesive strength of the lattice and, as a result, enhancing decohesion. According to the hydrogen enhanced localized plasticity (HELP) model, the presence of solute hydrogen enhances the mobility of dislocations leading to enhancing localized plastic deformation at the crack tip. Either model is promoted by a myriad of atomistic observations [2] and experimental data [3] and is contemplated as the underlying factor in particular fracture scenarios.

The numerical simulation of hydrogen-induced fracture, as a multi-physics approach, which couples hydrogen transport, plastic deformation, fracture and their interactions, has gained a substantial attention. Ahn et al. [4] conjectured that the material softening takes place owing to the presence of hydrogen, which affects the void process ahead of the crack tip. They employed a unit cell containing a single void and used the results to adjust the traction separation law (TSL). Using similar methodology, the effects of hydrogen-induced material softening and lattice dilatation on void growth have been investigated by Liang et al. [5]. Olden et al. [6] investigated a three-step FE procedure including elastic-plastic stress analysis, hydrogen diffusion analysis and hydrogen-informed cohesive zone model (CZM) for HE simulation. To include the hydrogen influence in the HEDE manner into the analysis, the hydrogen degradation law proposed by Serebrinsky et al. [7] was established. A two-step analysis has been introduced by Díaz et al. [8] for the HE problem; first hydrogen distribution was obtained with diffusion equations modified by the stress-strain state and afterwards, a CZM was included to model damage initiation. Moreover, the fully coupled and weekly coupled FE-CZM adopting the thermal-stress analysis and the analogy between hydrogen diffusion and heat transfer have been presented by Gobbi et al. [9]. Furthermore, Jemblie [10] simulated hydrogen-induced fracture initiation in a hot rolled bonded clad steel pipe using the CZM with a polynomial TSL. The same authors [11] carried out a comprehensive review on the CZM for numerically assessing HE of steel structures.

The primary aim of this work is to present a hydrogen-induced cracking methodology incorporating the fully coupled problems of elastoplastic deformation and hydrogen transport within cohesive-interface elements framework to predict tearing in high strength steels. The model has comprised initially zero thickness, cohesive-interface elements with constitutive response described by a hydrogen-degraded Park-Paulino-Roesler (PPR) model [12]. Moreover, the value of the cohesive energy of the PPR-based TSL is adopted from the experimental data of hydrogen-charged specimen. The computational framework accounting for both HEDE and HELP mechanisms is employed to simulate ductile crack extension in a C(T) specimen made of AISI 4130 high strength steel. The parameters included in the PPR model are satisfactorily calibrated with experimental data for the uncharged and hydrogen-charged specimens. Then, the influence of hydrogen concentration (normal interstitial lattice sites (NILS) and trapped) on the hydrogen coverage profile ahead of the crack tip is investigated in detail.

2 Transport of Hydrogen

In this work, we assume that hydrogen atoms are generally considered to reside either at NILS or being trapped at microstructural heterogeneities such as dislocations, grain boundaries, precipitates, inclusions, voids, and interfaces. Traps normally lower the amount of mobile hydrogen, hence reducing the apparent diffusivity and raising the local solubility of the system. The hydrogen atoms in lattice and trap sites are assumed to be in equilibrium with the mean stress and equivalent plastic strain. Based on Oriani's equilibrium theory [13], the occupancy of the trapping sites, θ_T, with a corresponding trap binding energy is related to the NILS occupancy, θ_L, through

$$\frac{\theta_T}{1-\theta_T}=\frac{\theta_L}{1-\theta_L}K_T \tag{1}$$

where $K_T=\exp(\mathrm{W_B}/RT)$ is the equilibrium constant, W_B denotes the trap binding energy, R designates the universal gas constant, and T is the absolute temperature. The hydrogen concentration in NILS, C_L, measured in hydrogen atoms per unit volume, is expressed by $C_L=\beta\theta_L N_L$ where β designates the number of NILS per solvent atom, $N_L=N_A/V_M$ denotes the density of solvent atoms per unit lattice volume, N_A is Avogadro's number, and V_M, measured in unit of volume per mole, denotes the molar volume of the host metal. The concentration of hydrogen in the trapping sites is written as $C_T=\alpha\theta_T N_T$, where α denotes the number of sites per trap and N_T is the corresponding trap density which varies point-wise dependent on the local plastic strain level. In this work, the trap density model proposed by Sofronis et al. [14] is considered as follows

$$N_T=\sqrt{2}\rho/\tilde{a} \tag{2}$$

where $\tilde{a}$ denotes the lattice parameter and ρ is the dislocation density, which varies with the equivalent plastic strain such that [15]

$$\rho=\begin{cases}\rho_0+\gamma\varepsilon^p, & \varepsilon^p\le 0.5\\ \rho_0+0.5\gamma, & \varepsilon^p>0.5\end{cases} \tag{3}$$

where ρ_0 is the initial dislocation density for annealed material with $\varepsilon^p=0$, and $\gamma=2\times10^6$ line length/m^3 for body-centered cubic (bcc) iron.

Based on Fermi-Dirac statistics [16], NILS hydrogen concentration, C_L, is calculated with regard to initial NILS hydrogen concentration in the stress-free solid, ${C_L}^{ini}$, and the local hydrostatic stress, σ_{kk}, through

$$\frac{\theta_L}{1-\theta_L}=\frac{{\theta_L}^{ini}}{1-{\theta_L}^{ini}}K_L \tag{4}$$

where $\theta_L{}^{ini} = C_L{}^{ini}/\beta N_L$ is the initial NILS occupancy, and $K_L = \exp(\sigma_{kk} V_H/3\text{RT})$. Using Eqs. 1 and 4, one can obtain the total hydrogen concentration, $c = (C_L + C_T)/N_L$, measured in hydrogen atoms per solvent atom, as a function of initial NILS hydrogen concentration, $C_L{}^{ini}$, the effective plastic strain, ε^p, and the local hydrostatic stress as

$$c = \beta\left(\theta_L(\sigma_{kk}) + \theta_{TL}(\sigma_{kk}, \varepsilon^p)\right) \tag{5}$$

where

$$\theta_L(\sigma_{kk}) = \frac{\theta_L{}^{ini} K_L(\sigma_{kk})}{\left((1 - \theta_L{}^{ini}) + \theta_L{}^{ini} K_L(\sigma_{kk})\right)} \tag{6}$$

and

$$\theta_{TL}(\sigma_{kk}, \varepsilon^p) = \frac{\alpha}{\beta}\frac{N_T(\varepsilon^p)}{N_L}\frac{K_T\theta_L(\sigma_{kk})}{1 - \theta_L(\sigma_{kk}) + K_T\theta_L(\sigma_{kk})} \tag{7}$$

It should be noted that values of both the hydrostatic stress and the effective plastic strain depend on the associated hydrogen concentration, which makes the derivation of hydrogen populations, plastic strains, and stresses fully coupled.

The underlying mechanism for the diffusion of hydrogen in steel is lattice diffusion as a result of interstitial jumps from one interstitial site to a adjacent one. The governing equation for transient hydrogen diffusion taking both hydrostatic drift and trapping into account is expressed by

$$\frac{DV_H}{3RT}C_L\sigma_{kk,i} + \alpha\theta_T\frac{dN_T}{d\varepsilon^p}\frac{d\varepsilon^p}{dt} + \frac{D}{D_{eff}}\frac{dC_L}{dt} - DC_{L,ii} = 0 \tag{8}$$

in which ',' denotes a spatial derivative, d/dt is a time derivative, σ_{ij} is the Cauchy stress, D denotes the hydrogen diffusion coefficient through NILS, and D_{eff} is an effective diffusion coefficient defined by

$$\frac{D_{eff}}{D} = \left(1 + \frac{\partial C_T}{\partial C_L}\right)^{-1} \tag{9}$$

By making use of Eqs. 1 and 2, Eq. 9 is expressed through

$$\frac{D_{eff}}{D} = \left(1 + \frac{K_T\alpha\beta N_T N_L}{(\beta N_L + (K_T - 1)\,C_L)^2}\right)^{-1} \tag{10}$$

It is worth noting that given the fast trap filling kinetics according to Oriani's model, the effective diffusion coefficient, D_{eff}, is less than NILS diffusion coefficient, D, while the traps are not saturated or as new traps are created due to plastic straining. All hydrogen related parameters are listed in Table 1.

Table 1 Hydrogen related parameters for the model

Properties	Symbol	Value
Number of NILS per host atom	β	6
Partial molar volume of hydrogen	V_H	$2 \times 10^{-6}\,\mathrm{m^3/mol}$
Variation of Gibbs free energy	ΔG_b^0	$3 \times 10^4\,\mathrm{J/mol}$
Avogadro constant	N_A	6.023×10^{23} atoms per mole
Gas constant	R	8.31 J/molK
Dislocation density for annealed material ($\varepsilon^p = 0$)	ρ_0	$10^{10}\,\mathrm{m^{-2}}$
Coefficient for calculating dislocation density	γ	$2 \times 10^{16}\,\mathrm{m^{-2}}$
Trap binding energy	W_B	21 kJ/mol
Number of trapping sites per dislocation	α	1
Molar volume of the host lattice	V_m	$7.116 \times 10^{-6}\,\mathrm{m^3/mol}$
Lattice parameter	$\tilde{a}$	$2.86 \times 10^{10}\,\mathrm{m}$

3 Elastoplastic Constitutive Law

From a continuum mechanics standpoint, the softening influence of atomic hydrogen may be explained by a local flow stress, σ_Y, which declines with rising the concentration of atomic hydrogen. Following Sofronis et al. [14], this correlation between the hydrogen effect and the local flow stress may be expressed by a phenomenological constitutive relation as

$$\sigma_Y\left(\varepsilon^p, c\right) = \sigma^{ini}(c)\left(1 + \frac{\varepsilon^p}{\varepsilon^{ini}}\right)^N \tag{11}$$

in which c is the total hydrogen concentration calculated in hydrogen atoms per solvent atom, $\sigma^{ini}(c)$ denotes the initial yield stress in the presence of hydrogen, $\varepsilon^{ini} = \sigma^{ini}/E$ is the initial yield strain, and N the hardening exponent, which is assumed to remain unaltered by the presence of hydrogen atoms. A possible suggestion for $\sigma^{ini}(c)$ is

$$\sigma^{ini}(c) = ((\chi - 1)\,c + 1)\,\sigma^{ini}(0) \tag{12}$$

where $\chi \leq 1$ is a softening parameter. Note that in a continuum sense, Eqs. 11 and 12 imply hydrogen-induced softening reflecting the experimental observations of hydrogen impact on enhanced dislocation mobility at the micro-scale. Based on the proposed model for including HELP and von Mises yielding with associated flow rule [17], the constitutive law taking hydrogen-induced lattice dilatation and material softening into account can be expressed as

$$\overset{\nabla}{\sigma}_{ij} = L_{ijkl} D_{kl} \tag{13}$$

where D_{kl} is the (symmetric) deformation rate tensor.

4 Cohesive Zone Modelling

While solid elements is considered for the bulk material, which is defined by a finite-strain, incremental plasticity (Mises) constitutive model, the cohesive-interface elements are governed by the PPR model. This work employs the intrinsic implementation, in which the cohesive-interface elements are assigned to the bulk element mesh to capture the crack propagation process. The PPR potential, Ψ, a function of normal and tangential separations, Δ_n and Δ_t, is expressed as [12]

$$\begin{aligned}\Psi(\Delta_n, \Delta_t) = \min(\phi_n, \phi_t) + \left(\Gamma_n\left(1 - \frac{\Delta_n}{\delta_n}\right)^{\bar{\alpha}}\left(\frac{m}{\bar{\alpha}} + \frac{\Delta_n}{\delta_n}\right)^m + \langle\phi_n - \phi_t\rangle\right)\\ \times\left(\Gamma_t\left(1 - \frac{|\Delta_t|}{\delta_t}\right)^{\bar{\beta}}\left(\frac{n}{\bar{\beta}} + \frac{|\Delta_t|}{\delta_t}\right)^n + \langle\phi_t - \phi_n\rangle\right)\end{aligned} \tag{14}$$

where $\langle\ \rangle$ denotes the Macauley bracket, $\bar{\alpha}$ and $\bar{\beta}$ are the shape parameters. Γ_n and Γ_t are energy constants, ϕ_n and ϕ_t denote fracture energies in Mode I and Mode II, which are equal to the areas under the traction-separation curves by considering the normal and tangential separation cases, respectively. λ_n and λ_t are the initial stiffness indicators, and $\bar{\alpha}$ and $\bar{\beta}$ are shape parameter. $T_n(\Delta_n, \Delta_t)$ and $T_t(\Delta_n, \Delta_t)$ are traction vectors in normal and tangential direction.

What makes the cohesive technique particularly appealing in the numerical modelling of hydrogen-induced degradation is the capability of implementing degraded TSL, which delineates hydrogen degradation of the material. Using a renormalization procedure [18] in order to scale the atomic-level cohesive properties up to the continuum scale, Serebrinsky et al. [7] implemented the HEDE approach in a CZM of fracture, explaining the impact of hydrogen segregation by a quantum-mechanical treatment. Established on the formulation presented by Jiang and Carter [19] on the ideal cleavage energy of Fe and Al in the presence of different concentration values of hydrogen, the coupling between hydrogen coverage, φ_H, and the critical cohesive traction (cohesive strength), T_c, has been suggested by

$$T_c(\varphi_H) = T_c(0)\left(1 - 1.0467\varphi_H + 0.1687{\varphi_H}^2\right) \tag{15}$$

where $T_c(0)$ denotes the critical cohesive traction with no hydrogen influence. To correlate the total bulk hydrogen concentration C with the hydrogen coverage, the Langmuir-McLean isotherm [20] is applied as

$$\varphi_H = \frac{C}{C + \exp(-\Delta G_b^0/\mathrm{RT})}) \tag{16}$$

where G_b^0 is the Gibbs energy difference between surface and bulk material. In contrast to the approach used by Raykar et al. [21] and Gobbi et al. [22], where a hydrogen damaging effect both on the cohesive strength and on the critical separation was considered, in the present work, a linear dependence on hydrogen concentration according to Eq. 15 is chosen only for the critical cohesive traction. In addition, the value of the cohesive energy of the PPR-based TSL is adopted from the experimental data of hydrogen-charged specimen.

5 Computational Model

A schematic representation of the FE model of a C(T) specimen is demonstrated in Fig. 1. As can be seen, crack growth takes place through the 8-node cohesive elements positioned along the symmetry plane, where cohesive-interface elements reside only over the initial uncracked ligament, thereby confining propagation to the Mode I fracture. The cohesive-interface elements along the crack growth path

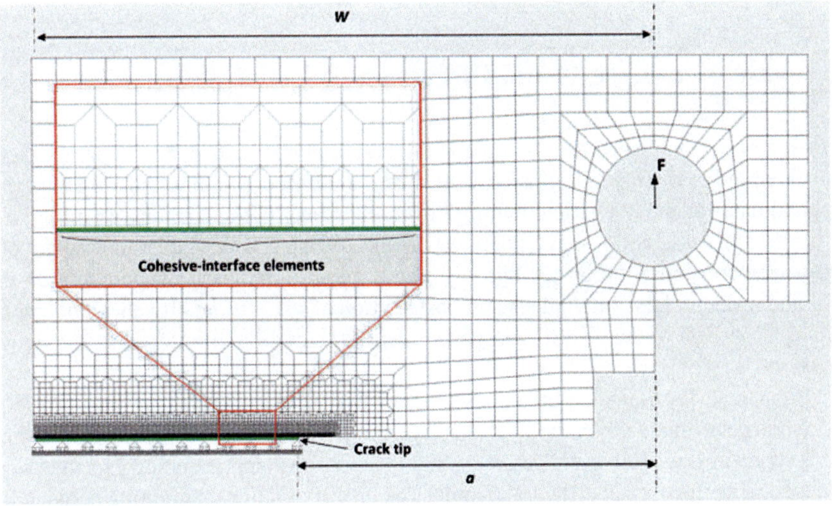

Fig. 1 Finite element model including zero thickness cohesive-interface elements incorporating ductile tearing for the C(T) specimen with $a/W = 0.6$

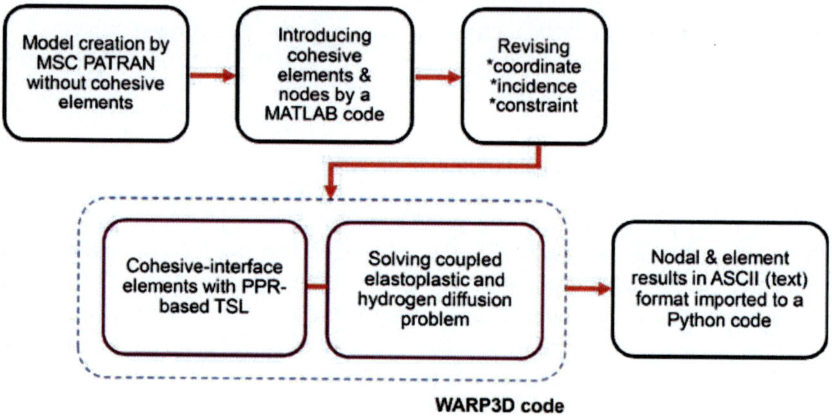

Fig. 2 Flowchart of numerical modelling procedure proposed for prediction of HE

have the mesh size of $L_e = 0.1563\,\mu\text{m}$ to ensure adequate resolution of results. The FE procedure is performed in this work using a research code, WARP3D [23]. A graphic representation of the entire numerical modelling procedure is depicted in the flowchart in Fig. 2.

In this work, the nonlinear FE analysis incorporating ductile crack growth deals with a plane-strain model of a C(T) fracture specimen with $a/W = 6$, where the crack length (*a*) and the width (*W=25* mm) are measured from load line. The AISI 4130 high-strength steel considered in the present analysis has a yield stress of 715 MPa, the modulus of elasticity of 220 GPa, and a Poisson's ratio of 0.3.

6 Numerical Results and Discussion

Having been established the computational model, in this section, firstly parameters involved in the hydrogen-degraded PPR model are properly calibrated with experimental data for the uncharged and hydrogen-charged C(T) specimens. In addition, the role of the softening shape parameter is investigated. Afterwards, the impact of hydrogen concentration on the hydrogen coverage profile ahead of the crack tip is studied in detail. Last but not least, the distribution of stress triaxiality ahead of crack tip as a function of the NILS hydrogen contents and hydrogen-induced softening is assessed.

Sung et al. [24] selected the cohesive energy as the J-integral at the crack onset, which is determined when the load-displacement curve initiates to deviate from the experimental one. However, detecting the location of crack initiation is somehow questionable. It is because the trapezoidal TSL used has a linear softening behaviour with limited number of parameters to fit properly with the experimental results. In contrast, in the present study, the proposed PPR model, which has four physical frac-

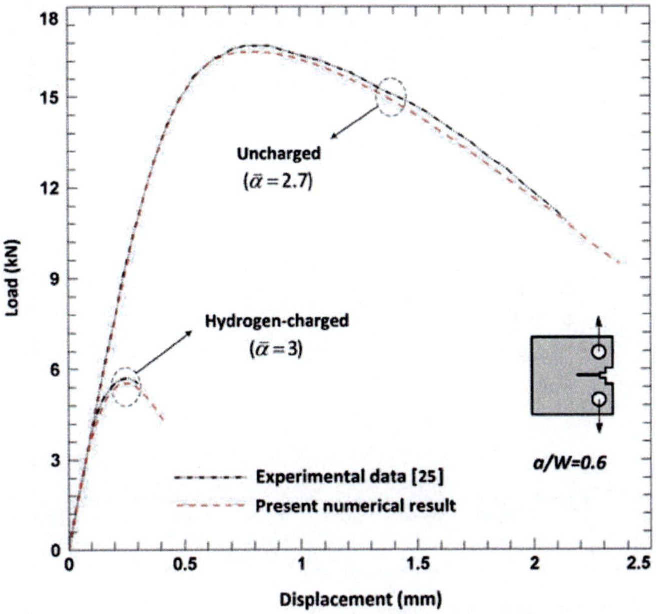

Fig. 3 Comparison between experimental and numerical results with calibrated parameters for the uncharged and hydrogen-charged C(T) specimens

ture parameters including cohesive energy, cohesive strength, shape, and initial slope, enables flexibility in the softening shape in TSL. Accordingly, the numerical results are fittingly matched with experimental results of Colombo et al. [25]. As shown in Fig. 3, a best fit to the load-displacement curve is achieved for a cohesive strength of 2250 MPa and 615 MPa for the uncharged and hydrogen-charged C(T) specimens, respectively. The corresponding cohesive energy 215.5 kJ/m^2 and 22 kJ/m^2 for the uncharged and hydrogen-charged tests, respectively, is determined based solely on the experimental test, similar to the procedure proposed by Roychowdhury et al. [26]. To adequately calibrate the PPR model with experimental data in Fig. 3, the shape parameter is determined as $\bar{\alpha} = 2.7$ and $\bar{\alpha} = 3$ for the uncharged and hydrogen-charged specimens, respectively. Also, $\lambda_n = 0.01$ is adopted for both conditions.

To clarify the role of the softening shape parameter in hydrogen-degraded PPR model, firstly, the effect of $\bar{\alpha}$ on the traction-separation curve for Mode I is studied in Fig. 4. Note that the degradation process in the PPR model is achieved by two means; by reducing the critical cohesive strength and by altering the softening shape parameter, $\bar{\alpha}$, which alters the softening shape of the TSL. It can be inferred from Fig. 4 that increasing the shape parameter, the softening part of the TSL transforms from a convex brittle ($\bar{\alpha} > 2$) into a concave ductile ($\bar{\alpha} < 2$) shape, plateau-like post-peak behavior. It is interesting to note that selecting $\bar{\alpha} = 2$ leads the PPR model to the triangular TSL.

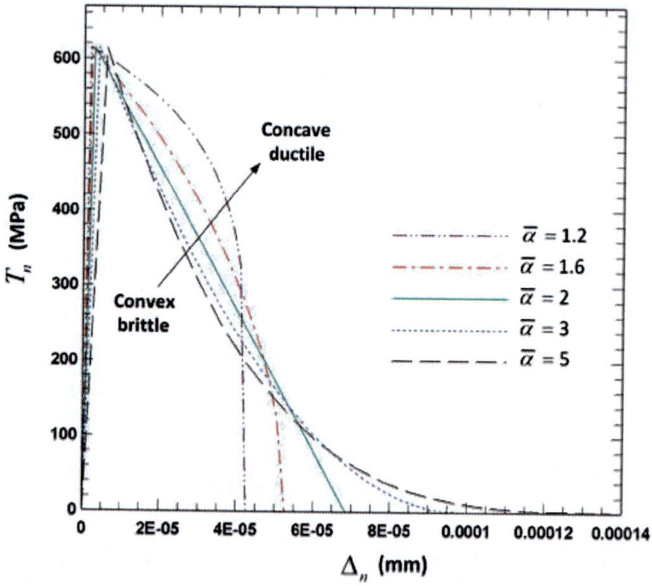

Fig. 4 Impact of softening shape parameters on the hydrogen-degraded traction-separation curve ($\lambda_n = 0.01$)

In Fig. 5 the effect of hydrogen concentration on the hydrogen coverage profile ahead of the crack tip is demonstrated. In this figure, the hydrogen coverage is determined in terms of either the lattice hydrogen concentration, $c = c_L$, or total hydrogen concentration, $c = c_L + c_T$. As can be seen, trapped hydrogen has a negligible influence on the trend of the progress of the hydrogen coverage profile ahead of the crack tip for various values of the initial NILS concentration. It can be concluded from Fig. 5 that the lattice hydrogen has the dominating factor in the hydrogen degradation compared with trapped hydrogen. Similar finding has been elaborated by Ayas et al. [27]. Notwithstanding, Novak et al. [28] supposed that low-binding energy dislocation traps have the leading contribution underpinning hydrogen-induced fracture. Furthermore, Jemblie et al. [10] found that the level of trap binding energy in excess of 23 kJ/mol plays a crucial role in making the dislocation trap sites being dominant factor in hydrogen degradation. As can be observed in Fig. 5, the peak of hydrogen coverage profile increases and tends to the crack tip with increase in initial lattice hydrogen concentration. For ${c_L}^{ini} = 0.1$ ppm, a maximum coverage of 39.68% is achieved at the hydrogen coverage peak.

In the following discussion, we turn our attention to the investigation of the distribution of stress triaxiality ahead of crack tip with the NILS hydrogen contents and hydrogen-induced softening. The quantity σ_{kk}/σ_e defines a convenient assessment of stress triaxiality associated with the crack propagation rate of micro-scale voids, where σ_{kk} denotes the hydrostatic stress and σ_e is the von Mises stress. Figure 6

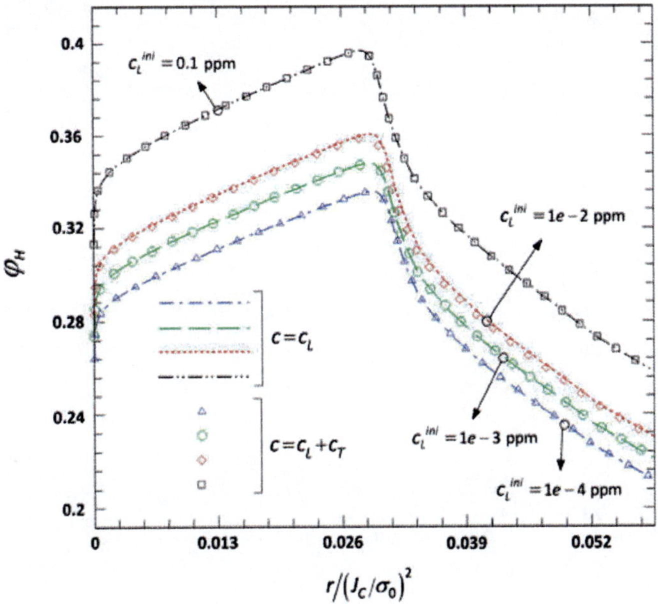

Fig. 5 Effect of initial lattice hydrogen concentration on the hydrogen coverage profile ahead of the crack tip just prior to the crack initiation ($\chi = 1$)

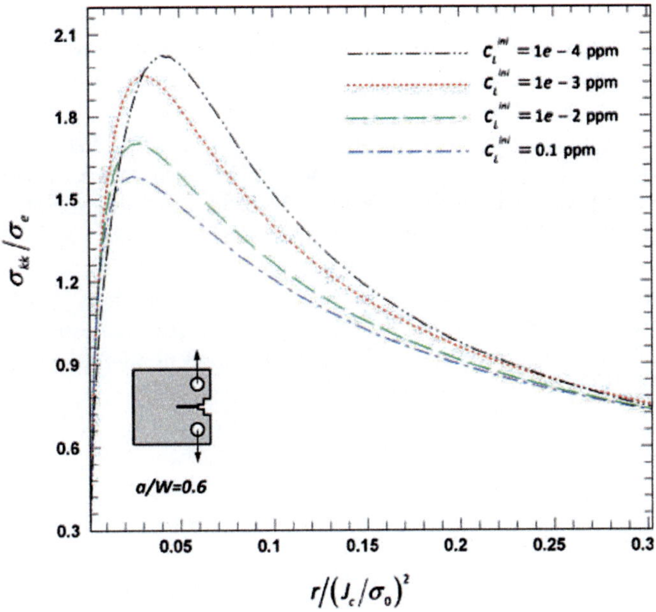

Fig. 6 Variation of stress triaxiality ahead of the crack tip with initial lattice hydrogen concentration just prior to the onset of crack growth ($\chi = 1$)

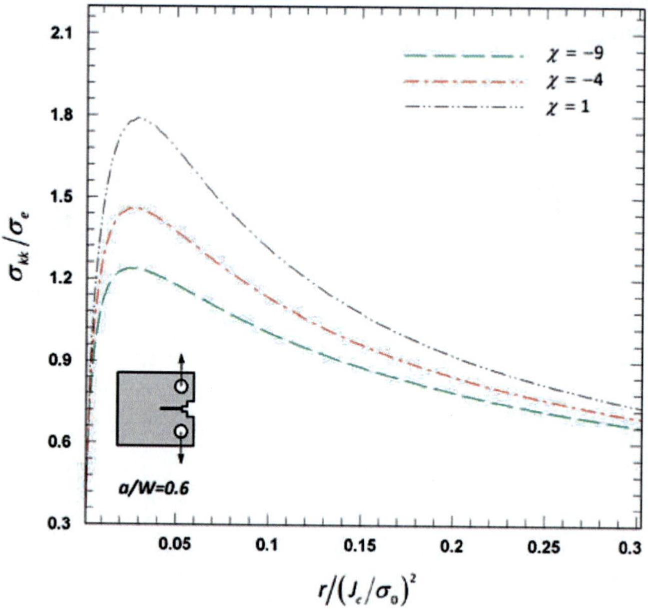

Fig. 7 Evolution of stress triaxiality ahead of the crack tip with hydrogen-induced softening just prior to the crack initiation ($c_L{}^{ini} = 0.0016\,\text{ppm}$)

depicts the variation of stress triaxiality ahead of the crack tip with initial lattice hydrogen concentration, taken only before the crack initiates. As can be seen from this figure, the apex of the triaxiality noticeably declines by increasing the initial NILS concentration, followed by shifting towards the crack tip. However, it is worth noting that the impact of NILS concentration of hydrogen on the stress triaxiality becomes less appreciable by receding from the crack tip. Furthermore, the evolution of stress triaxiality ahead of the crack tip with hydrogen-induced softening just prior to the crack initiation is displayed in Fig. 7. It is inferred from this plot that decreasing softening parameter from 1 to −9 and −4 value, the apex of the stress triaxiality decreases by 18.3% and 30.9%, respectively. It is interesting to note that comparing Figs. 5 and 7 reveals that high hydrogen concentrations are located in the same places as high stress triaxiality.

7 Conclusion

This work addressed a hydrogen-induced cracking methodology incorporating the fully coupled problems of elastoplastic deformation and hydrogen transport within cohesive-interface elements framework to predict ductile tearing in high-strength steel. In this work, on one hand, the flow stress declined with escalating hydrogen concentration and, as a result, taking into account the hydrogen impact on enhancing plastic deformation in the lattice through the HELP model. On the other hand, to implement hydrogen impact into the cohesive-interface elements in the HEDE manner, a phenomenological decohesion model was adopted, leading to combining the both mechanisms (HELP + HEDE). Furthermore, a nonlinear traction-separation relationship based on the PPR model determined the constitutive response of the zero thickness cohesive-interface elements while the finite-strain, incremental plasticity model was accounted for the bulk material. This computational framework was established to simulate ductile crack extension in a C(T) specimen made of AISI 4130 high-strength steel. The main results can be summarised as follows:

- The degradation process in the PPR model has been achieved by two means; by reducing the critical cohesive strength and by altering the softening shape parameter, which alters the softening shape of the TSL. The results show that increasing the shape parameter, the softening part of the TSL transforms from a convex brittle into a concave ductile shape, plateau-like post-peak behavior.
- It has been concluded that the lattice hydrogen has the dominating factor in the hydrogen degradation compared with trapped hydrogen.
- Results reveal that the apex of the stress triaxiality noticeably declines by increasing the initial NILS concentration, followed by shifting towards the crack tip. Notwithstanding, the impact of NILS concentration of hydrogen on the stress triaxiality becomes less appreciable by receding from the crack tip.

Acknowledgements This work has been supported by Fundação de Amparo à Pesquisa do Estado de São Paulo (FAPESP) through research grant 2018/23217-9.

References

1. Barrera O, Bombac D, Chen Y, Daff T, Galindo-Nava E, Gong P, Haley D, Horton R, Katzarov I, Kermode J et al (2018) Understanding and mitigating hydrogen embrittlement of steels: a review of experimental, modelling and design progress from atomistic to continuum. J Mater Sci 53(9):6251–6290
2. Hoagland R, Heinisch H (1992) An atomic simulation of the influence of hydrogen on the fracture behavior of nickel. J Mater Sci 7(8):2080–2088
3. Barnoush A, Vehoff H (2006) Electrochemical nanoindentation: a new approach to probe hydrogen/deformation interaction. Scripta Materialia 55(2):195–198
4. Ahn D, Sofronis P, Dodds R (2007) Modeling of hydrogen-assisted ductile crack propagation in metals and alloys. Int J Fract 145(2):135–157
5. Liang Y, Ahn D, Sofronis P, Dodds R Jr, Bammann D (2008) Effect of hydrogen trapping on void growth and coalescence in metals and alloys. Mech Mater 40(3):115–132
6. Olden V, Alvaro A, Akselsen OM (2012) Hydrogen diffusion and hydrogen influenced critical stress intensity in an api x70 pipeline steel welded joint-experiments and fe simulations. Int J Hydrogen Energy 37(15):11474–11486
7. Serebrinsky S, Carter E, Ortiz M (2004) A quantum-mechanically informed continuum model of hydrogen embrittlement. J Mech Phys Solids 52(10):2403–2430
8. Díaz A, Alegre J, Cuesta I (2017) Numerical simulation of hydrogen embrittlement and local triaxiality effects in notched specimens. Theoret Appl Fract Mech 90:294–302

9. Gobbi G, Colombo C, Miccoli S, Vergani L (2019) A fully coupled implementation of hydrogen embrittlement in FE analysis. Adv Eng Softw 135:102673
10. Jemblie L, Olden V, Mainçon P, Akselsen OM (2017) Cohesive zone modelling of hydrogen induced cracking on the interface of clad steel pipes. Int J Hydrogen Energy 42(47):28622–28634
11. Jemblie L, Olden V, Akselsen OM (2017) A review of cohesive zone modelling as an approach for numerically assessing hydrogen embrittlement of steel structures. Phil Trans R Soc Math Phys Eng Sci 375(2098):20160411
12. Park K, Paulino GH, Roesler JR (2009) A unified potential-based cohesive model of mixed-mode fracture. J Mech Phys Solids 57(6):891–908
13. Oriani RA (1970) The diffusion and trapping of hydrogen in steel. Acta Metallurgica 18(1):147–157
14. Sofronis P, Liang Y, Aravas N (2001) Hydrogen induced shear localization of the plastic flow in metals and alloys. Eur J Mech A/Solids 20(6):857–872
15. Gilman JJ Micromechanics of flow in solids
16. Hirth J, Carnahan B (1978) Hydrogen adsorption at dislocations and cracks in fe. Acta Metallurgica 26(12):1795–1803
17. Sofronis P, McMeeking RM (1989) Numerical analysis of hydrogen transport near a blunting crack tip. J Mech Phys Solids 37(3):317–350
18. Nguyen O, Ortiz M (2002) Coarse-graining and renormalization of atomistic binding relations and universal macroscopic cohesive behavior. J Mech Phys Solids 50(8):1727–1741
19. Jiang D, Carter EA (2004) First principles assessment of ideal fracture energies of materials with mobile impurities: implications for hydrogen embrittlement of metals. Acta Materialia 52(16):4801–4807
20. Hondros E, Seah M (1977) The theory of grain boundary segregation in terms of surface adsorption analogues. Metall Trans A 8(9):1363–1371
21. Raykar NR, Maiti SK, Raman RS, Aryan S (2013) Study of hydrogen concentration dependent growth of external annular crack in round tensile specimen using cohesive zone model. Eng Fract Mech 106:49–66
22. Vergani L, Gobbi G, Colombo C (2014) A numerical model to study the hydrogen embrittlement effect on low-alloy steels. In: Key engineering materials, vol 577, Trans Tech Publ, pp 513–516
23. Healy B, Gullerud A, Koppenhoefer K, Roy A, RoyChowdhury S, Walters M, Bichon B, Ne Cochran K, Carlyle N, Sobotka J et al (Warp3d release 17.5.9): 3-d nonlinear finite element analysis of solids for fracture and fatigue processes, University of Illinois, Urbana-Champaign. http://code.google.com/p/warp3d
24. Sung S-J, Pan J, Korinko PS, Morgan M, McWillliams A (2019) Simulations of fracture tests of uncharged and hydrogen-charged additively manufactured 304 stainless steel specimens using cohesive zone modeling. Eng Fract Mech 209:125–146
25. Colombo C, Fumagalli G, Bolzoni F, Gobbi G, Vergani L (2016) Fatigue behavior of hydrogen pre-charged low alloy cr-mo steel. Int J Fatigue 83:2–9
26. Roychowdhury S, Roy YDA, Dodds RH Jr (2002) Ductile tearing in thin aluminum panels: experiments and analyses using large-displacement, 3-d surface cohesive elements. Eng Fract Mech 69(8):983–1002
27. Ayas C, Deshpande V, Fleck N (2014) A fracture criterion for the notch strength of high strength steels in the presence of hydrogen. J Mech Phys Solids 63:80–93
28. Novak P, Yuan R, Somerday B, Sofronis P, Ritchie R (2010) A statistical, physical-based, micro-mechanical model of hydrogen-induced intergranular fracture in steel. J Mech Phys Solids 58(2):206–226

Sensitivity of Damping for Diagnostics of Damage in Structure

A. Bovsunovsky and E. Soroka

Abstract Available researches on the sensitivity of damping capacity of specimens and structures as applied for the diagnostics of damage are quite conflicting. Certain of researchers revealed sufficiently high sensitivity of damping capacity for the reliable diagnostics of damage, but others declared that the change of damping due to damage is negligibly small. In the presented study this contradiction was attributed to the fact that the change of damping capacity of damaged structures is the function of great number of factors. The influence of these factors on the change of damping of damaged structure was investigated with the developed fracture mechanics-based procedure. In such a way, the sensitivity of damping for the diagnostics of crack in a beam-like structure in the case of bending and axial vibrations was investigated. In particular, it was revealed that the sensibility of damping to crack is dependent on structure's stiffness and on damping capacity of structure in the undamaged state. In addition, the intensity of stress in the damaged area should be sufficiently high to induce the dissipation of energy of vibration. As a result, a simple formula was developed to estimate the sensitivity of damping capacity as applied for the damage detection.

Keywords Damping · Simply supported beam · Edge crack · Diagnostics of damage

A. Bovsunovsky (✉)
Department of Cogeneration Installations of Thermal and Nuclear Power Plants, National Technical University of Ukraine "Igor Sikorsky Kyiv Polytechnic Institute", 37 Prosp. Peremohy, Kyiv 03056, Ukraine
e-mail: apbovsunovsky@gmail.com

E. Soroka
Department of Strength of Structures Made of Brittle Materials, G.S.Pisarenko Institute for Problems of Strength, Timiriasevskaia Str. 2, Kyiv 01014, Ukraine

M. Abdel Wahab (ed.), *Proceedings of the 8th International Conference on Fracture, Fatigue and Wear*, Lecture Notes in Mechanical Engineering,
https://doi.org/10.1007/978-981-15-9893-7_12

1 Introduction

The variation of damping capacity may be causally connected with structural damage [1]. The problem here is that the effect of crack on damping capacity of structures depends on number of factors, such as damping capacity of structure in the undamaged state, the slenderness ratio, the level of stress in the cracked area, etc. In this connection, the experimental evaluation of sensitivity of damping capacity for the diagnostics of damage is extremely time-consuming.

The practice of using damping capacity for the diagnostics of crack is highly controversial. In some researches, a significant increase of damping was observed caused by the origination and growth of crack in specimens. Thus, a 63- and 125-time increase of damping caused by crack with relative depth $a/h = 0.5$ and 0.7 (a is the crack depth; h is the height of cross-section) was observed [2]. A 70-fold increase of damping was observed when crack in a cantilever steel specimen reached 50% of the cross-section [3]. A 30-fold increase of damping capacity for the steel specimen while relative crack depth reached up $a/h = 0.7$ was reviled in [4]. In tests of steel simply supported beam with a crack located in the middle, the increase of damping up to 12 times caused by the crack of relative depth $a/h = 0.75$ was reported in [5]. According to the research [6] the crack of relative depth $a/h = 0.5$ causes the increase by 40% of damping capacity of beam-like structures from dialuminium.

At the same time, some other researchers concluded that the change of damping capacity of structures is insufficient for the trustworthy diagnostics of damage. Thus, the tests of a three-span bridge demonstrated that the potentially dangerous damage caused the change of its dissipation ability up to 10% [7]. Quite similar results were observed in [8]. Vibration tests of airplane structures demonstrated the increase of damping capacity up to 23% as a result of significant fatigue crack growth [9].

The above contradiction may be attributed to the fact, that the change of damping of damaged structure is affected by number of factors. Indeed, as it was shown in [10], to such factors belongs size and location of crack. Besides, the influence of crack on the intensity of damping change is affected by the initial damping capacity of intact structure [4]. The level of stress in damaged area is one more factor that affects the intensity of damping change [6]. As well, this intensity is different for different types of vibration (axial, bending or torsional) and is dependent on slenderness ratio [11].

Thus, the intensity of damping change depends on at least five factors. Under this condition the estimation of sensibility of damping as applied for damage detection is not a simple problem. One of possible way to solve this problem was to develop fracture mechanics-based procedure, which used experimentally-analytical relation between the dissipated in a crack energy and the stress intensity factor (SIF) range [12].

So, the aim of the study was to determine conditions under which damping capacity is sensitive enough for the trustworthy structural damage detection.

2 Analytical Model for the Determination of Damping for Cracked Structure

Simply supported beam was used as an example of structure under investigation (Fig. 1). If the crack is closing (such crack periodically opens and closes every half cycle of vibration), the logarithmic decrement of vibration (LDV) can be calculated according to the formula [13]

$$\delta_c(\sigma) = \delta(\sigma) + \Delta U_c(\sigma_c)/2U(\sigma) \quad (1)$$

where δ is the LDV of the beam in the intact state; U is the maximal strain energy; ΔU_c is the dissipated in crack energy; σ is the maximal stress amplitude lengthways the beam.

The specific dissipated by crack energy depends on the SIF range as follows [12]

$$\Delta \overline{U}_c = 8.634675 \times 10^{-5} \Delta K_I + 3.87315 \times 10^{-4} \Delta K_I^2 - 1.29826 \times 10^{-5} \Delta K_I^3, \quad (2)$$

where $\Delta \overline{U}_c$ is the dissipated in a crack energy per unit of crack front length; ΔK_I is the SIF range. In case of the symmetrical loading, the SIF range is determined by the maximal SIF, that is $\Delta K_I = K_I$.

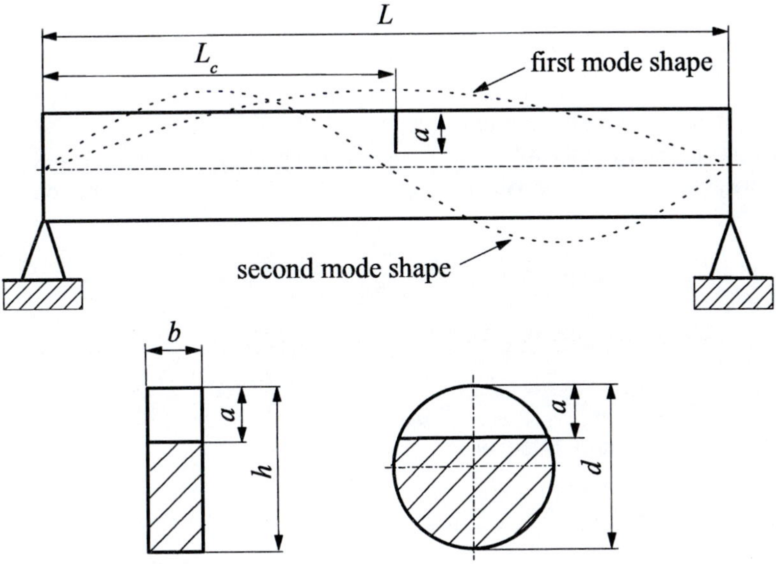

Fig. 1 Scheme of the beam with crack

For the edge crack in the prismatic beam at bending vibration the maximal SIF is determined by the formula [14]

$$K_I = \sigma_c \sqrt{\pi a}(1.122 - 1.40\alpha + 7.33\alpha^2 - 13.08\alpha^3 + 14.0\alpha^4), \tag{3}$$

where σ_c is the stress in the cracked area; $\alpha = a/h$.

At axial vibrations of the prismatic beam [14]

$$K_I = \sigma_c \sqrt{\pi a}(1.12 - 0.231\alpha + 10.55\alpha^2 - 21.72\alpha^3 + 30.39\alpha^4). \tag{4}$$

At bending vibrations of the cylindrical beam [15]

$$K_I = \sigma_c \sqrt{\pi a}(1.1105 - 2.6475\gamma + 5.6875\gamma^2), \tag{5}$$

where $\gamma = a/d$; d is the beam's diameter.

Stress amplitude in the cracked area depends on the mode shape in the following way

$$\sigma_c = \sigma \sin(k_i L_c), \tag{6}$$

where L_c is the crack location. The multiplier $\sin(k_i L_c)$ considers the effect of mode shape on the stress in the cracked area; $k_i = i\pi/L$ (L is the beam's length).

The dissipated in a crack energy was determined by the following equation

$$\Delta U_c = 2c\Delta\overline{U}_c, \tag{7}$$

where $2c$ is the length of crack front. For the prismatic beam $2c = b$ (b is the beam's width), and for the cylindrical beam $2c = 2\sqrt{a(d-a)}$.

The strain energy of the prismatic beam at bending and axial vibrations and of the cylindrical beam at bending vibrations can be calculated, respectively, by formulas

$$U(\sigma) = \frac{bhL}{12E}\sigma^2; \cdots U(\sigma) = \frac{bhL}{4E}\sigma^2; \ldots U(\sigma) = \frac{\pi d^2 L}{64E}\sigma^2 \tag{8}$$

where E is the modulus of elasticity.

3 Results of Calculations

To investigate the effect of slenderness ratio on the LDV, the length of beam was taken to be in the range $L = 0.2 \ldots 2$ m. The LDV of the intact beam was taken to be $\delta = 0.002$; 0.02 and 0.2. The above range of LDV was chosen because it involves practically all structural materials.

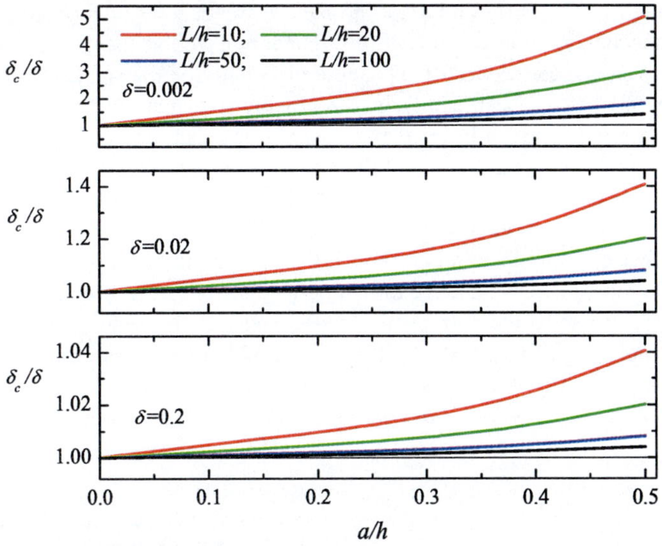

Fig. 2 Crack depth dependence of the LDV change for the prismatic beam ($L_c/L = 0.5$, first bending mode shape)

The dimensions of the beam were $h = 0.02$ m; $b = 0.02$ m; $d = 0.02$ m. Mechanical properties of the beam were $E = 200$ GPa; $G = 80$ GPa (shear modulus); $\rho = 7800$ kg/m^3 (density); $\nu = 0.26$ (Poisson's ratio).

In calculations of the LDV for the cracked beam the stress amplitude was assumed to be $\sigma = 10$ MPa. This level of stress made it possible to prevent crack propagation during testing [12].

The influence of crack depth on the LDV for the prismatic beam with various slenderness ratio in the case of low ($\delta = 0.002$), average ($\delta = 0.02$) and high ($\delta = 0.2$) level of the LDV for the intact beam is demonstrated in Fig. 2. It is evident, that the increase of slenderness ratio suppresses the intensity of LDV changes caused by the crack growth.

The reliable damage detection based on damping is possible if its change caused by damage exceeds the relative error of experiment. Based on the results of investigations demonstrated in [13], this error was estimated to be 20%. So the relative change of the LDV $\delta_c/\delta = 1.2$ served to determine minimal crack, which is possible to detect with this procedure (they are denoted as α_{det}, γ_{det}).

The change of LDV can respond to the origination of quite negligible crack. Thus, in the case of low damping and rigid beam, that is $\delta = 0.002$ and $L/h = 10$, the minimal detectible crack is $\alpha_{det} = 0.040$. At average damping ($\delta = 0.02$), one can detect only relatively big crack, that is $\alpha_{det} = 0.35$. And in the case of high level of the LDV ($\delta = 0.2$) the detection of crack practically is not possible.

To compare the sensitivity of damping and natural frequencies to damage, the natural frequencies of cracked beam was calculated [16]. The results of calculations are presented in Fig. 3 (f_c and f are the natural frequencies of cracked and intact beam, respectively). For the trustworthy damage detection the change of natural frequencies should exceeds 5% [17]. Under this condition the smallest crack which can be detected is $\alpha_{det} = 0.23$. Thus, the sensitivity of natural frequencies to the presence of crack is much lower than the sensitivity damping under the condition $\delta = 0.002$, but is comparable under the condition $\delta = 0.02$. At this, the decrease of slenderness ratio leads to the increase of the sensibility of damping capacity and natural frequency as applied for the detection of cracks.

The effect of crack location on the LDV of prismatic beam is demonstrated in Fig. 4. The most intensive change of damping capacity can be seen in cases, when crack locates close to the sufficiently loaded areas of the beam. These areas are evident from Fig. 1. In such a way, the change of LDV is directly proportional to the intensity of stress in the cracked areas. Consequently, it is hardly possible to reveal crack in the vicinity of the supports or lightly loaded areas of the beam using damping.

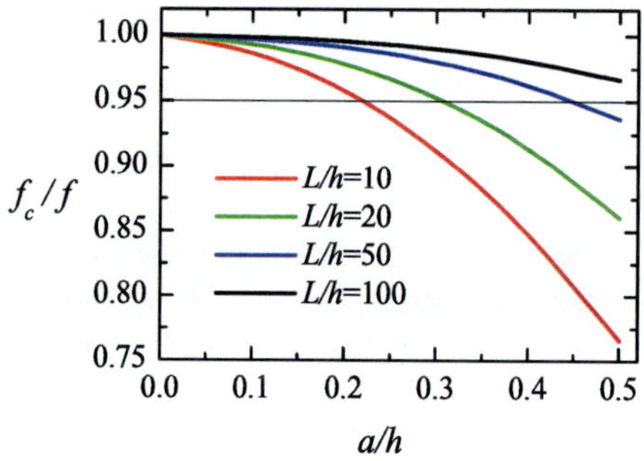

Fig. 3 Crack depth dependence of first natural frequency for the prismatic beam ($L_c/L = 0.5$, first bending mode shape)

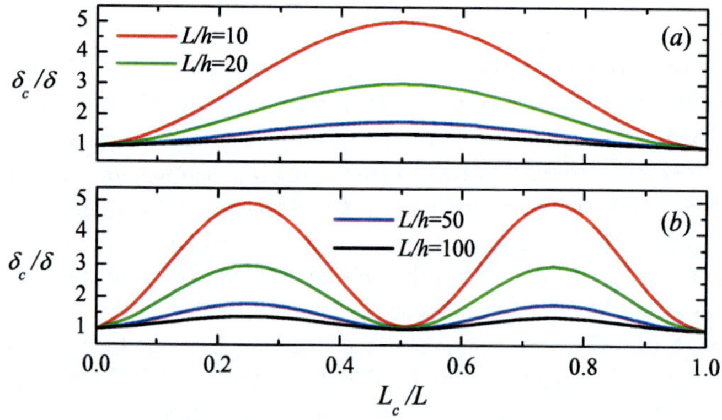

Fig. 4 Crack location dependence of the LDV change for the prismatic beam ($\delta = 0.002$; $a/h = 0.5$): **a** first bending mode shape; **b** second bending mode shape

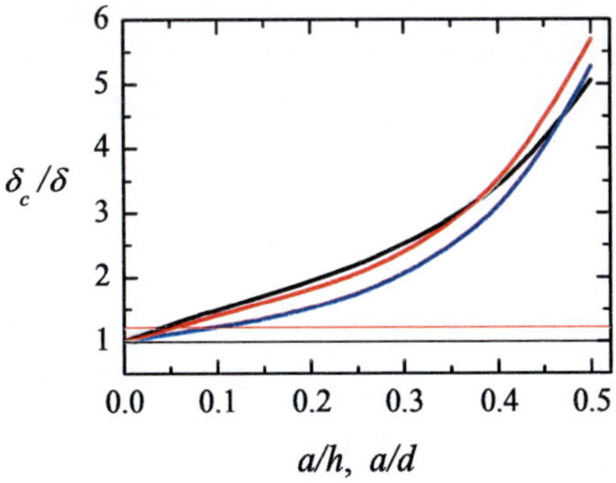

Fig. 5 Crack depth dependence of the LDV change of first bending (black and red lines) and axial (blue line) mode of vibration for the prismatic (black and blue lines) and cylindrical (red line) beam (case $L/h = 10$; $L/d = 10$; $L_c/L = 0.5$; $\delta = 0.002$)

The immediate comparison of the LDV at different types of vibrations demonstrates that damping of bending vibrations is most sensitive to small cracks in the prismatic beam (Fig. 5). Diagnostics of relatively big cracks ($\gamma > 0.4$) when using bending vibrations is more sensitive for cylindrical beam. Damping of axial vibrations is the least sensitive to damage.

Table 1 demonstrate the limit of sensitivity of damage detection on condition of 20% change of the LDV and 5% change of first natural frequency (case $L_c/L = 0.5$).

In contrast to the LDV, the change of frequency is independent on the level of damping for the structure in the intact state. This is the evident benefit of natural frequencies change as applied for the detection of damage. At this, the change of LDV is more sensitive to damage than frequency at relatively low initial damping for the intact structure ($\delta = 0.002$). The sensibility of frequency and LDV change to the crack is practically the same in the case of average initial damping for the intact structure ($\delta = 0.02$). And in the case of high level of damping for the intact structure ($\delta = 0.2$) the change of frequency is more sensitive than the change of damping characteristic.

4 Discussion of Results

The analysis of data in Table 1 made it possible to conclude that the change of LDV is sufficiently sensitive as applied for the detection of crack on condition that the intact structure possesses quite low damping capacity ($\delta = 0.002 \ldots 0.02$). As far as most structural materials possess damping capacity in the above range, structural crack detection using damping capacity can be considered as perspective.

The decrease of slenderness ratio causes the increase of beam's strain energy. Therefore, the caused by crack change of damping is the bigger the bigger the stiffness of structure. In such a way, the change of damping is sensitive to damage in structures with small compliance. At this, the cracked area should be sufficiently stressed to insure periodical closure and opening of crack in the process of vibration or to make

Table 1 Limit of crack detection sensitivity

Type of beam	Type of vibration	δ	L/h, L/d	α_{det}, $\gamma_{det}(\delta_c/\delta = 1.2)$	α_{det}, $\gamma_{det}(f_c/f = 0.95)$
Prismatic	Bending	0.002	10	0.040	0.228
			20	0.082	0.315
			50	0.210	0.457
			100	0.354	>0.5
		0.02	10	0.351	0.228
			20	0.497	0.315
			50	>0.5	0.457
			100	>0.5	>0.5
		0.2	10	>0.5	0.228
			20	> 0.5	0.315
			50	>0.5	0.457
			100	>0.5	>0.5
Prismatic	Axial	0.002	10	0.098	>0.5
			20	0.172	>0.5
			50	0.297	>0.5
			100	0.397	>0.5
		0.02	10	0.397	>0.5
			20	0.492	>0.5
			50 … 100	>0.5	>0.5
		0.2	10…100	>0.5	>0.5
Cylindrical	Bending	0.002	10	0.054	0.231
			20	0.104	0.441
			50	0.246	>0.5
			100	0.367	>0.5
		0.02	10	0.367	0.231
			20	0.75	0.441
			50 … 100	>0.5	>0.5
		0.2	10	>0.5	0.231
			20	>0.5	0.441
			50 … 100	> 0.5	> 0.5

crack open permanently. Crack which remains closed during the process of vibrations does not cause the change of damping and, therefore, can not be revealed by this method. Permanently open crack can be revealed by the change of damping much easily than the closing one [13].

Based on the analysis of above results, the simple condition of sufficient sensitivity of damping for the diagnostics of crack was developed, namely $\Delta U_c/2U \geq \kappa\delta$. The coefficient κ is equal to the relative error of experimental determination of damping characteristics.

To assess the applicability of damping for the diagnostics of damage, the following factors should be considered: crack depth, crack location, type of crack (mode I, mode II or mode III), type of vibration (axial, bending or torsional), damping capacity of intact structure, and the slenderness ratio.

To solve this problem, the experimentally analytical procedure was developed which takes into consideration the above factors. Based on this procedure, there was demonstrated that the sensibility of damping is higher the bigger depth of crack and the higher loading in damaged area. Damping characteristic of axial vibration is influenced by crack less intensively than that of bending one. Thus, in the case of low damping intact structure, the crack of depth $a/h = 0.1$ intensifies damping of axial and bending vibrations up to 39% and 49%, respectively. Such a change of damping is sufficient for the reliable diagnostics of crack. At the same time, in the case of high damping intact structure, the increase of damping caused by crack of depth $a/h = 0.1$ does not exceed 6%. This is far from enough for the trustworthy detection of crack. Besides, the structure's compliance affects the intensity of damping capacity change due to crack. Thus for low damping structure the crack of depth $a/h = 0.1$ causes the increase of LDV for the compliant beam ($L/h = 100$) by less than 1% and for the more rigid beam ($L/h = 10$)—by 49%.

As it was noted in Sect. 1, the main reason of energy dissipated in crack of mode I is plastic zone in the vicinity of crack tip. Consequently, dissipated in crack energy is more intensive in deeper crack. In addition, cracked area should be highly stressed. Thus, even big crack originated in the area of structure which at vibrations deforms slightly can not cause the change of damping ample for the trustworthy diagnostics of crack.

Practically all damping characteristics including the LDV are relative values: the dissipated energy is related to the strain energy of mechanical system. The change of structure's dimensions (and, therefore, the change of slenderness ratio) causes the change of structure's strain energy and thus affects relative change of damping due to crack.

Most frequently vibration methods of damage detection are founded on the use of natural frequencies of structures. But under certain conditions the change of damping can be much more sensitive feature of damage than the change of natural frequencies. At the same time, the determination of damping for real structures is not an easy task. However, at present time a lot of modal analysis techniques are available to assess modal damping along with the natural frequencies of structures [18].

The continuation of this research suggests the verification of the above technique for the detection of cracks in disks and blades of turbines for various purposes.

5 Conclusions

The fracture mechanics based procedure is developed to calculate the change of damping of bending and axial vibrations for the cracked beam-like structure. The application of above procedure made it possible to formulate the condition of sufficient sensibility of damping characteristic for the diagnostics of crack-like damage.

The change of damping capacity is higher the lower the initial damping and compliance of structure. Damping can be effective feature of damage only if the loading of damaged area is sufficient to make crack closing one or permanently open during vibration of structure.

The change of damping is sensitive enough for the reliable damage detection under the following condition: the ratio of dissipated in crack energy to the doubled maximal strain energy of structure should be more or equal to the product of initial damping ability of structure by the relative error of damping capacity determination.

References

1. Česnik M, Slavič J, Boltežar M (2012) Uninterrupted and accelerated vibrational fatigue testing with simultaneous monitoring of the natural frequency and damping. J Sound Vibr 331:5370–5382. https://doi.org/10.1016/j.jsv.2012.06.022
2. Imregun M, Sanliturk KY (1990) Natural frequency and damping changes produced by fatigue cracks. In: Proceedings of of the 15th international seminar on modal analysis. Leuven, Belgium, pp 791–805
3. Rytter A, Brincker R, Kirkegaard PH (1992) An experimental study of the modal parameters of a cantilever. Fracture & Dynamics. Paper No. 37, Department of Building Technology and Structural Engineering, University of Aalborg, Denmark, 76 p
4. Bovsunovsky AP (1999) Comparative Analysis of sensitivity of vibration damage indicators by the results of laboratory tests. In: Proceedings of the 17th IMAC. Kissimmee, Florida, USA, 2, pp 1909–1915
5. Zhang W, Testa RB (1999) Closure effects on fatigue crack detection. ASCE J Eng Mech 125:1125–1132. https://doi.org/10.1061/(ASCE)0733-9399(1999)125:10(1125)
6. Sidorov OT (1983) Determination of the zones of distribution of cracks in flexible samples. Strength Mater 15:858–860. https://doi.org/10.1007/BF01524780
7. Baldwin JW, Salane HJ, Duffield RC (1978) Fatigue test of a three span composite highway bridge. Final Report 73–1, Missouri Highway Reasearch Programme, Colambia, Mo
8. Sun X, Hardy HR (1992) An investigation on applicability of modal analysis as non-destructive evaluation method in geotechnical engineering. In: Proceedings of the 10th IMAC. San Diego, California, 1, pp 9–19
9. Sidorov OT (1982) Influence of fatigue damage to elements of prefabricated structure on its damping. Russian J Nondestr Testing 6:55–59
10. Bovsunovsky AP (2017) Efficiency analysis of vibration based crack diagnostics in rotating shafts. Eng Fract Mech 173:118–129. https://doi.org/10.1016/j.engfracmech.2017.01.014
11. Daily JS, Klingbeil NW (2010) Plastic dissipation energy at a biomaterial crack tip under cyclic loading. Int J Fatigue 32:1710–1723. https://doi.org/10.1016/j.ijfatigue.2010.03.010
12. Bovsunovsky AP (2004) The mechanisms of energy dissipation in the non-propagating fatigue cracks in metallic materials. Eng Fract Mech 71:2271–2281. https://doi.org/10.1016/j.engfracmech.2004.02.003

13. Bovsunovsky AP (2019) Efficiency of crack detection based on damping characteristics. Eng Fract Mech 214:464–473. https://doi.org/10.1016/j.engfracmech.2019.04.005
14. Murakami CY (ed) (1987) Stress intensity factors handbook, vol 3, vol. I. The Society of Materials Science, Pergamon Press Japan
15. Kiuchi A, Aoki M, Kobayashi M, Ikeda K (1982) Fracture strength of surface notched round bar. Iron Steel Inst Japan 68:1830–1838. https://doi.org/10.2355/tetsutohagane1955.68.13_1830
16. Bovsunovskii AP (1999) Numerical study of vibrations of a nonlinear mechanical system simulating a cracked body. Strength Mater 31:571–581. https://doi.org/10.1007/BF02510893
17. Chen G-W, Beskhyroun S, Omenzetter P (2016) Experimental investigation into amplitude-dependent modal properties of an eleven-span motorway bridge. Eng Struct 107:80–100. https://doi.org/10.1016/j.engstruct.2015.11.002
18. Ewins DJ (2009) Modal testing: theory, practice and application, 2nd edn. Wiley, New York

Effect of Cross-Over Fraction on Objective Function Value in Genetic Algorithm Optimization of Machining Parameters in Diamond Turning of Aluminum Alloy (RSA-431) for Optical Applications

Funsho Oyekunle and Khaled Abou-El-Hossein

Abstract The ever-escalating industrial production competition, coupled with the use of highly automated industrial machinery has seen the emergence of artificial intelligence-based process optimization models that are perceived as being reliable. In this present study, the effect of cross-over fraction parameter on objective function value in genetic algorithm optimization of machining parameters is assessed. Single point diamond turning is carried out on RSA 431. Primary cutting parameters namely; cutting speed, feed rate and depth of cut are optimized in this study for minimum surface roughness. A genetic model fitness function was developed using multiple regression modelling in Excel environment to establish the quality of each population member. Validation of the resultant model is achieved through determination of its prediction accuracy. The results indicate the presence of a linear relationship between the cross-over fraction parameters. The linear model has a prediction accuracy of 95.05% indicating its effectiveness in predicting objective function value.

Keywords Genetic algorithm · Cross-over fraction · Objective function value · Single point diamond turning (SPDT)

1 Introduction

The profit margin in high-precision manufacturing processes is dependent upon product surface quality. This is due to the mutuality of surface roughness, production cost and product turnover. Additionally, attributes such as functionality, mechanical strength, service life and aesthetics are largely dependent on the quality of product surface finish [1]. Achieving precise surface roughness is one of the central challenges in machining processes. It is therefore critical that manufacturing industries device

F. Oyekunle · K. Abou-El-Hossein (✉)
Precision Engineering Laboratory, Department of Mechatronics, Nelson Mandela University, Port Elizabeth 6031, South Africa
e-mail: khaled@mandela.ac.za

M. Abdel Wahab (ed.), *Proceedings of the 8th International Conference on Fracture, Fatigue and Wear*, Lecture Notes in Mechanical Engineering,
https://doi.org/10.1007/978-981-15-9893-7_13

reliable and accurate process monitoring methods and modelling techniques to ensure the prescribed surface roughness is reached. The use of reliable modelling techniques has proved to be beneficial for the improvement of product quality as well as the reduction of production time and cost [2]. Diverse factors interact and contribute towards the surface quality of single point diamond turned products. Cutting speed, depth of cut and feed rate are some of the conventional cutting parameters used in surface roughness control. In the last years, the use of Non-Destructive Testing (NDT) techniques in surface roughness prediction is gaining much ground [3, 4].

In the last few years, the use of soft computing methods in surface roughness prediction has increased tremendously. These are evolutionary computing techniques that imitate human minds in solving problems. The application of computer aided approaches reduces process time limits, complexity and error rate [5]. Some of the prominent soft computing methods in use are Artificial Neural Network (ANN), Adaptive Neuro-Fuzzy Inference System (ANFIS), Genetic Algorithms (GA) and Particle Swarm Optimization (PSO) [6, 7]. These methods have been successfully employed by various researchers to develop models that could accurately predict and optimize machining performance such as surface roughness, tool wear, cutting force, etc.[8, 9]. However, it is generally recognized that the performance of any soft computing method depends largely on numerous modelling conditions.

Krishna et al. [10] employed both ANN based on multilayer perceptron and ANFIS to develop models for cutting parameters, that is speed, feed and depth of cut, to accurately predict surface roughness of hard turning machining of aluminium alloy, AA 6351. According to the prediction accuracy, both models were concluded to have a good agreement with the experimental values. However, ANFIS provided a higher prediction accuracy of 95.43%. ANFIS is an implementation of fuzzy neural network framework suitable for non-linear mapping operations by using hybrid learning procedure. Kumar et al. [11] have developed a methodology using ANFIS for predicting surface roughness and GA for optimization of parametric settings in thermally drilled holes. In their study, GA was used to determine fitness function using stochastic uniform and executed while altering the GA parameters. The assessment of generations and the best fitness values performed by GA shows a minimum error compared to experimental values. Genetic algorithm (GA) and Particle Swarm Optimization (PSO) are employed by Ganesan et al. [12] in optimizing machine parameters for minimum operation time in multipass turning. Both techniques produced good result however using PSO machining time can be further minimized. Reddy and Rao [13] studied the effect of tool geometry and cutting conditions - cutting speed and feed rate on the machining performance in end-milling of medium carbon steels. A GA model was developed to optimize the surface roughness model using binary coding system to represent the various variables. The results proved that GA can be successfully used for optimizing surface quality.

This paper is a presentation of genetic algorithm-based optimization of machining parameters in single point diamond turning of RSA 431. The uniqueness of GA in comparison to other optimization algorithms include its ability to make global search such as in rule sets, neural network architectures, etc. over a short computation time.GA has found a wide application in situations where the only information is a measurement of performance. Lastly, GA can perform learning in multi-agent systems and support multi-objective optimization [14, 15].

2 Principles of Genetic Algorithm

The idea of Genetic Algorithm (GA) proposed by John Holland in 1975 [16], is simply inspired by natural biological evolution processes and applied in soft computing for the prediction of behaviours of processes. In GA, high-quality solutions are usually generated to optimize and search problems by relying on bio-inspired principles as it does in nature. This optimization technique is suitable for solving multi-objective problems, hence, its application in modelling and optimization of machining processes could be well established. The working principle of genetic algorithm is based on key operations such as population, selection and reproduction [9, 17]. Figure 1. shows the flow chart of the GA process where the principles of GA are illustrated.

A subset of solutions in the current generation is known as population. To evolve a solution in GA, a population of strings (chromosome) must be initiated. These strings which are coded in binary search n space are represented as a set of parameters or features (genes) that defines an individual solution. To each string, a fitness value is given according to the desired targets. By implication, the higher the fitness is the higher the quality of the solution is. Fitness of an individual can be calculated using Eq. (1) while formula for the probability of selecting an ith chromosome in a population is given Eq. (2) as [19]:

$$Fitness_i = \frac{1}{1 + f\_obj_i} \tag{1}$$

$$P_i = \frac{Fitness_i}{TotalFitness} \tag{2}$$

Selection is a process of choosing the "parents individual" from a population which combines and creates new individuals for the next generation. Selection is significant to the convergence rate of an GA model. Selection is generally fitness-based, therefore good parent selection motivates individuals to a better and fitter solution. Some of the methods of selecting the best chromosomes in GA include roulette wheel, Boltzman, tournament, rank, and steady-state selection [18].

Reproduction in GA (just as in biological terms) is a process through which individual strings (chromosome) are transferred according to their objective values (fitness). Reproduction determines how the next generation of GA is being created. Crossover and mutation are the two background operations involved in reproduction.

Crossover is an important phase in genetic algorithm that involves the creation of new individuals from the genetic code of two parents by combining (mating) their various pairs of chromosomes in the population. During crossover, strand of

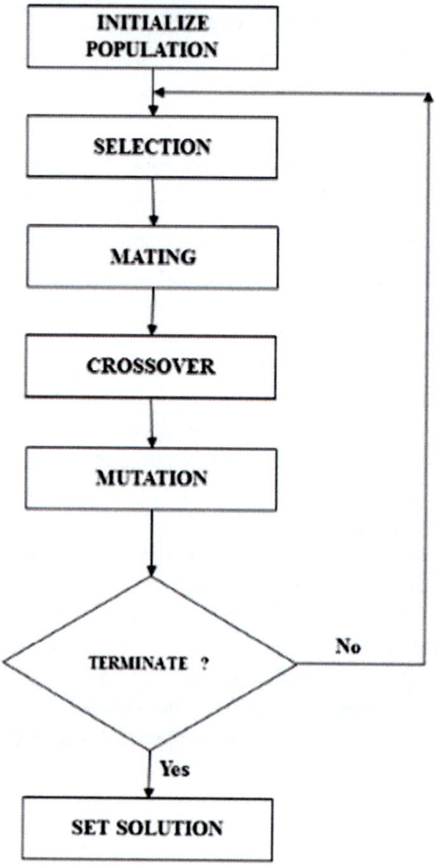

Fig. 1 GA process [18]

chromosomes for each parent are swapped to produce better quality than the previous generation. Widely used crossover operator include: one-point crossover, two-point crossover, uniform crossover and blend crossover [19]. An example of crossover concept and one-point crossover is shown in Fig. 2a and b [18].

Fig. 2 An example of **a** crossover operation, and **b** one-point crossover

Fig. 3 An example of mutation operation [20]

Original offspring 1	1101111000011110
Original offspring 2	1101100100110110
Mutated offspring 1	1100111000011110
Mutated offspring 2	1101101100110110

This is simply a random change in strings of chromosomes (base on the fitness of the selected candidate) to create a new solution [20]. Otherwise, it can also be described as a process of exploiting the power of crossover. With the introduction of mutation, diversity is introduced and maintained in the genetic population. Figure 3 is an illustration of the mutation operation [19].

Both crossover and mutation process of reproduction is important in GA, even though they play different roles. Crossover is a convergent operation which pulls the population towards a local minimum or maximum space while mutation is a divergent operation that breaks out one or more members of a population from a local minimum or maximum. From this point of view, crossover probability (Pc) should be kept high and mutation probability (Pm) low in genetic algorithm since the objective of the algorithms is to bring the population to a convergence. General values for the probability of crossover and probability of mutation are 0.8 and 0.01 respectively, indicating their importance [21].

3 Experimental Setup and Procedure

SPDT was carried out for machining Rapidly Solidified Aluminium alloy—RSA 431 with monocrystalline diamond tool used as a cutting tool. The workpiece is a 60 mm diameter homogenous crystalline material used for a wide range of optical application in industries such as such medical, aerospace, automotive, electronics, etc.[22]. The levels of primary cutting parameters used in this study are presented in Table 1. The design of experiment was performed using Box-Behnken technique of response surface methodology RSM). 15 experimental runs were carried out following the proposed experimental design method. The Box-Behnken design of experiments has the following advantages: it permits building of sequential designs; it detects lack of fit of the model and it permits the use of blocks [23]. The surface roughness of diamond-turned workpiece was measured using Taylor Hobson PGI Dimension

Table 1 Levels of process parameters

Parameter	Level		
	1	2	3
Cutting Speed (rpm)—*v*	750	1375	2000
Feed (mm/min)—*f*	5	15	25
Depth of Cut (μm)—*d*	5	15	25

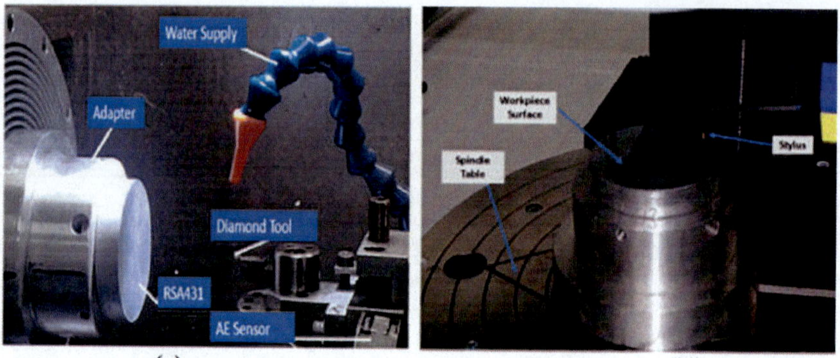

(a) (b)

Fig. 4 **a** Ultra high precision diamond turning setup **b** profilometer for surface roughness

XL surface Profilometer. Figure 4a and b shows the single point diamond turning experiment setup and surface roughness measurement.

4 Results and Discussion

Table 2 is a presentation of the surface roughness received on RSA443 after ultra-high precision diamond turning. The experimental run order was determined by the Box-Behnken design. A genetic algorithm objective function, that links the machining parameters to surface roughness, was developed using multiple linear regression modelling in an Excel environment and is represented by Eq. (3) below.

$$Ra = 17.13 + (0.075) * d + (0.5625) * f + (-0.0086) * v \tag{3}$$

Equation (3) is valid for the use of cutting parameters within the ranges of cutting conditions shown in Table 1.

In this study, the genetic algorithm's critical parameters are shown in Table 3.

Keeping the critical parameters constant while changing the cross-over fraction, yields results presented in Table 4. Crossover fraction is one of the important options in GA that helps the reproduction process to control how the next generation is created. It is the fraction of individuals in the next generation that is the product of the crossover, except the elite children [24]. This implies that a crossover fraction of 1 would conclude that all individuals are crossover offspring and a fraction of 0 means all offspring are mutants. Linear regression analysis in Excel is used to ascertain the relationship between cross-over fraction and mean objective function values. The relationship is represented by Eq. (4). The resultant equation is validated by using it to predict the mean objective function value and finally determining its prediction accuracy.

Table 2 Surface roughness experimental results

Exp. Run	Depth of Cut, d (μm)	Feed, f (mm/min)	Speed, v (rpm)	Surface Roughness, R_a (nm)
1	25	25	1375	27
2	25	5	1375	12
3	25	15	750	25
4	15	25	750	50
5	25	15	2000	16
6	5	15	2000	13
7	15	15	1375	13
8	5	5	1375	10
9	15	5	2000	9
10	15	25	2000	15
11	15	15	1375	13
12	5	15	750	19
13	15	5	750	12
14	15	15	1375	13
15	5	25	1375	17

Table 3 Genetic algorithm critical parameters

Parameter	Value
Population size	40
Mutation rate	0.8
Crossover rate	0.9
Bit number	16
Number of iterations	1500

$$Y = 2.7984 + 1.1110X_1 \tag{4}$$

where Y = mean objective function value, X_1 = cross-over fraction.

The high prediction accuracy of 95.09% shows the suitability of Eq. (4) to predict mean objective function value for a given cross-over fraction. From the plot of cross-over fraction values against objective function values in Fig. 5, it is evident that the objective function value is constant from a cross-over fraction of 0.1 up to 0.5, after which there is a general increase between 0.6 and 0.9 inclusive. Cross-over fraction of 0.7 yields the best result from Table 4 due to the less absolute error. Although a crossover fraction of 0.3 is noticed to show lower error percentage and would have been preferred, however, the mean objective function is lower in comparison with 0.7 as shown in Fig. 5.

Table 4 GA-based cross-over fraction and mean objective function values

Cross-Over Fraction	Mean Objective Function Value	Predicted Objective Function Value	Absolute Error (%)
0.1	3.1179	2.9095	6.68
0.2	3.1182	3.0206	3.13
0.3	3.1178	3.1317	0.44
0.4	3.1184	3.2428	3.99
0.5	3.1175	3.3539	7.58
0.6	3.3218	3.4650	4.31
0.7	3.5564	3.5761	0.55
0.8	3.4557	3.6872	6.69
0.9	4.2611	3.7983	10.86
Mean Absolute Error (%)			**4.91**
Prediction Accuracy (%)			**95.09**

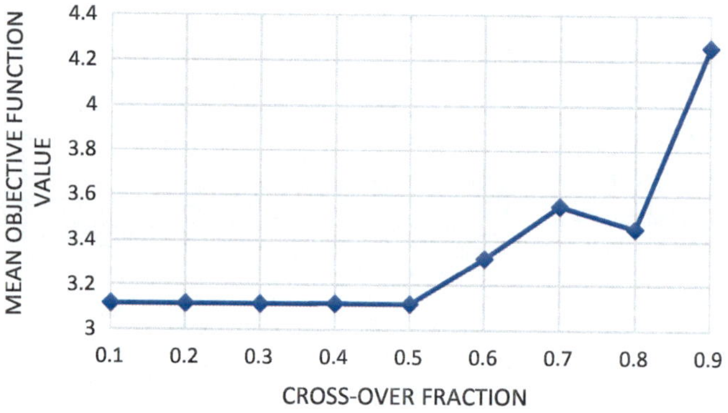

Fig. 5 Objective function value against cross-over fraction values

5 Conclusion and Recommendations

In this study, the Box-Behnken approach has been adopted in designing the experiment. The overall aim of this study was to ascertain whether the cross-over fraction value has any effect on GA modelling objective function value. Regression analysis results indicate a linear relationship between cross-over fraction and objective function value. The resultant linear equation has a prediction accuracy of 95.09%. This derived linear model is valuable by saving time through forecasting of minimum achievable surface roughness value. Further research works can be carried out to investigate the contribution of other genetic algorithm parameters in the optimization of cutting parameters during single point diamond turning of RSA-431.

Acknowledgements The authors would like to acknowledge the financial support provided by the National Research Foundation (NRF) through the National Equipment Program. The financial support provided by Research Capacity Development (RCD); Nelson Mandela University is also acknowledged. The experience and knowledge gained during this study are valuable and will always be treasured.

References

1. Jayswal DSC, Taufik M (2011) Prediction of surface roughness and optimization of cutting parameters of stainless steel turning based on RSM. Int J Eng Sci Technol 3(7):5809
2. Berglind L, Ozturk E (2019) Reviews advances in the application of information and communication technology (ICT) in machine tools and manufacturing. In: Armendia M, Ghassempouri M, Ozturk E, Peysson F (eds) Twin-Control 57–93
3. Otieno T, Abou-El-Hossein K (2018) Cutting forces and acoustic emission in the diamond turning of rapidly-solidified aluminium. Insight-Non-Destr Test Condition Monit 60(1):11–18
4. Bhuiyan M, Choudhury I (2014) Review of sensor applications in tool condition monitoring in machining. Compr Mater Process 13:539–569
5. Beemaraj BK, Vijayan V, Senthilkumar G (2018) Performance analysis of surface roughness modelling using soft computing approaches. Appl Mathe Inf Sci 12:1209–1217. https://doi.org/10.18576/amis/120616
6. Feng CX, Yu ZG, Kusiak A (2006) Selction and validation of predictive regression and neural network models based on designed experiments. IIE Trans 38:13–23
7. Perwej DY, Akhtar N, Dixit (2017) 4th International conference on computing for sustainable global development. India
8. Radha KB (2020) Review of surfae roughness prediction in machining Process by using various parameters. Int J Recent Trends Eng Res 6(1)
9. Chandrasekaran M, Muralidhar M, Krishna C, Dixit U (2009) Application of soft computing techniques in machining performance prediction and optimization: a literature review. Int J Adv Manuf Technol 46:445–464
10. Krishna SH, Satyanarayana K, Raju KB (2011) Surface roughness prediction model using Ann & Anfis. Int J Adv Eng Res Stud 1:102–113
11. Kumar R, Hynes NRJ (2020) Prediction and optimization of surface roughness in thermal drilling using integrated ANFIS and GA approach. Int J Eng Sci Technol 23(1):30–41
12. Ganesan H, Mohankumar G, Ganesan K, Ramesh KK (2011) Optimization of machining parameters in turning process using genetic algorithm and particle swarm optimization with experimental verfification. Int J Eng Sci Technol 3
13. Reddy NSK, Rao PV (2005) Selection of optimum geometry and cutting conditions using surface roughness prediction model for end milling. Int J Adv Manuf Technol 26(11–12):1202–1210
14. Shapiro J (1999) Advanced course on artificial intelligence. Springer
15. Yang X-S (2014) Nature-inspired optimization algorithms. Elsevier, Oxford
16. Yang X-S (2013) Optimization and metaheuristic algorithms in engineering. In: Yang X-S, Gandomi AH, Talatahari S, Alavi AH (eds) Metaheuristics in water, geotechnical and transport engineering. Elsevier, Oxford, pp 1–23
17. Chakraborty S, Chattopadhyay PP, Ghosh SK, Datta S (2017) Incorporation of prior knowledge in neural network model for continuous cooling of steel using genetic algorithm. Appl Soft Comput 58:297–306
18. Hassanat A, Almohammadi K, Alkafaween E, Abunawas E, Hammouri A, Prasath V (2019) Choosing mutation and crossover ratios for genetic algorithms—a review with a newdynamic approach. Information 10(12)

19. Nayyar A, Le D-N, Nguyen NG (2018) Advances in swarm intelligence for optimizing problems in computer science. CRC Press
20. Meyer-Baese A, Schmid V (2014) Chapter 5—genetic algorithms. In: Meyer-Baese A, Schmid V (eds) Pattern recognition and signal analysis in medical imaging, 2nd edn. Academic Press, Oxford, pp 135–149
21. Jones DF, Mirrazavi SK, Tamiz M (2002) Multi-objective meta-heuristics: an overview of the current state-of-the-art. Eur J Oper Res 137(1):1–9
22. RSP Technology, https://www.rsp-technology.com/
23. Ferreira SLC, Bruns RE, Ferreira HS, Matos GD, David JM, Brandão G, Da Silva EGP, Portugal LA, Dos Reis PS, Souza AS, Dos Santos WNL (2007) Box-Behnken design: an alternative for the optimization of analytical methods. Anal Chim Acta 597:179–186
24. Symbolic Math Toolbox (2019). [Online]. Available: https://www.mathworks.com/help/symbolic/

A Heat Transfer Finite Element Model for Wire-Arc-Additive-Manufacturing Process

Y. Ling, J. Ni, Magd Abdel Wahab, J. Antonissen, and J. Vande Voorde

Abstract This paper presents a part of work to develop a suitable numerical tool to define printing strategy for Wire-Arc-Additive-Manufacturing (WAAM) process. Based on previous research work, A three dimensional metallo-thermo-mechanical model on welding process are transplanted through Abaqus CAE and user subroutines for the simulation of WAAM process. A test plate manufactured by WAAM processing is used as the validation of the simulation. In this paper, heat transfer modelling for WAAM process is presented, which includes two key issues, namely a modified Goldak model for heat source and interval layer forced cooling parameters. The modified Goldak model is capable to simulate energy input of filler material correctly. The inter layer idle time and forced cooling flux are very crucial to the geometry, and the quality of the fabricated part by WAAM process. The ambition of this project is to use desired properties as input parameters of the simulation tool and obtain output parameters to define the optimized strategy for WAAM process.

Keywords WAAM · FEM simulation · Strategy · Welding · Additive manufacturing

---

Y. Ling (✉) · J. Ni · M. Abdel Wahab
Laboratory Soete, Faculty of Engineering and Architecture, Gent University, 9052 Zwijnaarde, Belgium
e-mail: Yong.Ling@UGent.be

J. Antonissen
Guaranteed B.V., 9060 Zelzate, Belgium

J. Vande Voorde
OCAS.N.V, 9060 Zelzate, Belgium

M. Abdel Wahab (ed.), *Proceedings of the 8th International Conference on Fracture, Fatigue and Wear*, Lecture Notes in Mechanical Engineering,
https://doi.org/10.1007/978-981-15-9893-7_14

1 Introduction

Additive manufacturing is one of the intelligent manufacturing topics in recent years. Wire-Arc-Additive-Manufacturing (WAAM) is the most efficient process in deposition rate (2–10 kg/h) [1] and one of its big advantages is its capability to make large metal structure parts especially in offshore engineering [2]. Currently, the international state of the art in wire arc additive manufacturing (WAAM) of large scale steel components mainly focuses on making good shape and less defects in the components. Many uncertainties remain in correlations of the final functional properties to the printing material and strategy. The research goal therefore is using the desired properties for the outputs of printing parameters in order to control the quality of the components. The plan envisages to develop both the missing modelling tools as well as the experimental screening capabilities to facilitate a swift industrial uptake of WAAM for safe and cost efficient production of large scale industrial components which bear a safety-critical function [3]. Figure 1 shows the frame diagram of the objectives.

Typically, WAAM process is very similar to automatic gas metal arc welding process(GMAW) [8]. It was called as shape welding(SW), rapid prototype welding(RP), as well as 3D welding. Modelling welding process is related to describe multi physical phenomena mathematically which includes arc burning and high temperature gas flow, heat energy transfer, melting metal drop forming and weld molten pool moving, and so on. Of course it is not possible to model whole complex process. However many research works have been done and some knowledge currently have been adopted to WAAM process. Simulation tool becomes efficient and economical approach to play a key role in making the strategy for WAAM process. The numerical tool to be developed here is a kind of AI (artificial intelligent) operator. Ideally inputting desired phases fractions or mechanical properties of the part to the tool one can obtain the optimized output parameters such as current, voltage, moving speed and filler melting velocity for the printing processing. Rodrigues et al. published the current status and perspectives on WAAM.

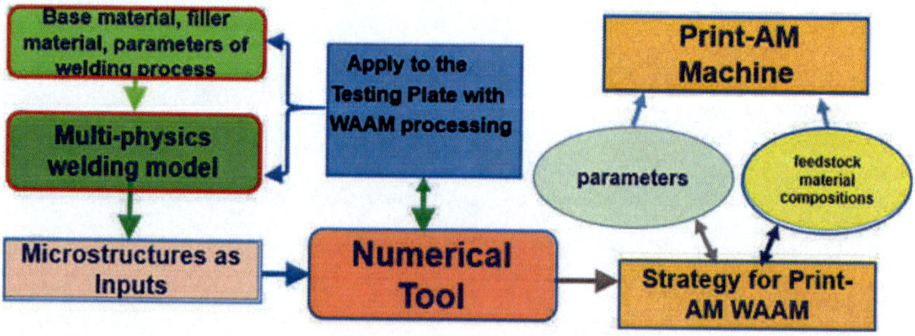

Fig. 1 Frame diagram of the objectives

Many researchers worked on multi-physical modelling on Additive manufacturing and many publications were reported [4–7].

Practically, WAAM is multilayer and multi pass welding process. In this work the 3D test plate was simulated by using Abaqus CAE 2019 and user subroutines with the thermo-metello-mechanical model developed by Ni for welding process simulation [9]. To examine the effects of models on WAAM process heat source Goldak model and modified Goldak heat source model are applied. Heat transfer simulations with real printing parameters were devoted, which include arc current, voltage, moving velocity and wire feeding speed. To simulate the results of the experiments, algorithms in Abaqus for convection and film sink for interval layer cooling were adopted. Obviously the quality and functionality of the AM component is affected directly by the selected parameters of printing processing. Due to the complexity of WAAM process, it is necessary to define the right strategy to obtain a desired functional component despite of the existence of many challenges.

2 Heat Source Input

2.1 Heat Source Models

Good Modelling of heat source input is a significant issue, which relates to WAAM parameters directly. As reviewed in the PhD dissertation of Ni, many research works have been done [10]. The first distributed heat source model was proposed by Pavelic [11], in which a Gaussian distribution of heat flux was assumed to be deposited on the surface of the sample. This was appropriate for modelling heat source of low power density. *Krutz* and *Segerlind* [12] proposed a modified form of this model. If the heat source moves at x–z plane and along x direction, the heat flux q_{in} is expressed as [13]:

$$qin = \frac{3Q_{\text{in}}}{\pi c^2} \exp\left(-\frac{3[x + v(t - t_0)]^2}{c^2} - \frac{3z^2}{r_c^2}\right) \quad (1)$$

where r_c is the characteristic radius of the heat flux distribution. t_0 is the time at which the heat source original action point and t is the current time, and v is the moving speed.

Goldak et al. [14] developed a non-axisymmetric three dimensional (3D) heat source model that could simulate complex weld pool. As shown in Fig. 2, the most general form was created by distinguishing ellipsoid semi-axes in front and rear quadrants. This model is known as double ellipsoid model and widely used in the simulation of welding process. Equation (2) is the power density distribution function:

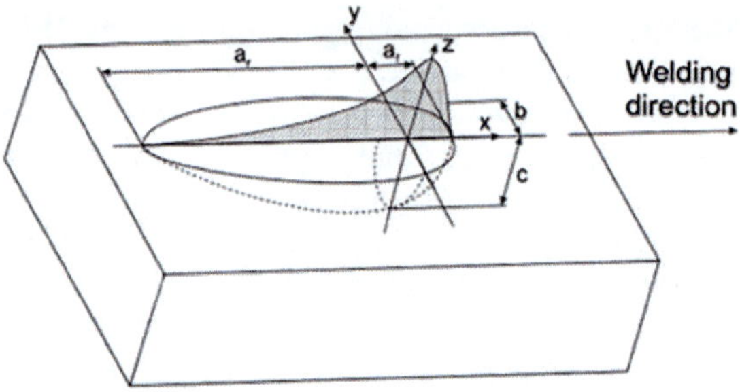

Fig. 2 Goldak double ellipsoid model

$$\dot{q}_v = \frac{6\sqrt{3}\dot{Q} f_{f \cdot r}}{a_{f,r} bc\pi\sqrt{\pi}} \exp\left(-\frac{3x^2}{a_{f \cdot \gamma}^2} - \frac{3y^2}{b^2} - \frac{3z^2}{c^2}\right) \tag{2}$$

This heat source model defined two different power distribution functions for positive and negative x semi axes. It modeled asymmetric heat power distribution over the molten pool. $a_{f,r,}$ b and c are the semi axes of two ellipsoids centered in the origin of the frame of reference, as described in Fig. 2. The parameter a has different values depending on x sign (a_f if positive and a_r if negative) [15]. Usually the values of ellipsoids semi axes are set according to molten pool dimensions [16]. $f_{f.r}$ are the distribution factors, which are f_f the frontward and f_r backward ellipsoids, and the following condition is fulfilled [14]:

$$f_f + f_r = 2 \tag{3}$$

where $\dot{Q}$ is the heat input rate computed as the product of welding current I, welding voltage V and arc efficiency η.

$$\dot{Q} = \eta \text{IV} \tag{4}$$

The integrating form of the two power density functions in spatial coordinates gives the total energy rate generated by the heat source as [14].

$$\int_{-\infty}^{+\infty}\int\int \dot{q}_v(x, y, z)dxdydz = 2\dot{Q} \tag{5}$$

In this model, the real power transmitted to workpiece depends upon the relative position between the heat source frame of reference and the base material. If the origin is positioned on the top surface of the substrate the heat input is transmitted to both of the base and filler material. According to Eq. (5) the heat power transmitted is higher than the actual process heat input $\dot{Q}$. It will result in overheat of molten pool temperature. On the other hand, if the origin positioned on the top of filler material

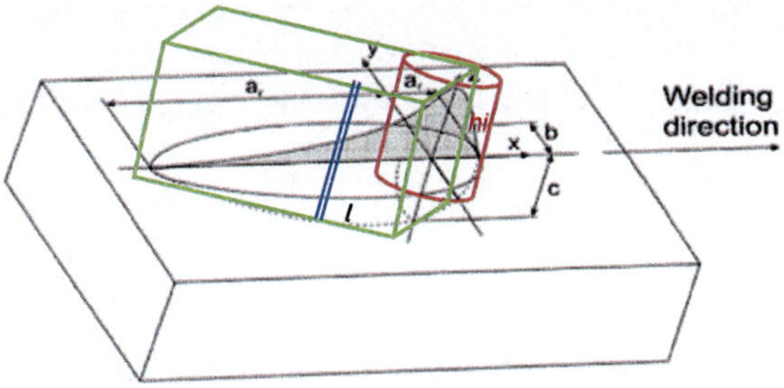

Fig. 3 The modified Goldak model

the heat power transmitted would be good but it will result in a poor weld penetration on the base material [16]. Since most of heat is delivered to the filler material.

Montevecchi et al. proposed modified Goldak model [17]. This model divides heat source power into two parts. The power for melting filler metal is about 50% of total arc power [18]. The main improvement is capable to take the parameter of filler melting into simulations, for instance, the wire feeding speed. As shown in Fig. 3 the part of the power density is delivered to the base material by Goldak Gaussian distribution, while another part of power density is delivered to filler material with a volume of a cylinder or a rectangle block. This allows to obtain the steep temperature gradients in the molten pool, accurately positioning two heat sources on the top of filler material and the base material respectively. In reality the correct amount of heat to the filler material elements is transmitted.

The power density function is presented in Eq. (6)

$$\begin{cases} \dot{q}_b = \frac{6\sqrt{3} f_f \dot{Q}_{bf,r}}{a_{f,r} bc\pi\sqrt{\pi}} \exp\left(-\frac{3[x+v(\tau-t)]^2}{a_{f\cdot r}^2} - \frac{3y^2}{b^2} - \frac{3z^2}{c^2}\right) \\ \dot{q}_w = \frac{\dot{Q}_w}{Vel} \end{cases} \tag{6}$$

where Q_b and Q_w are the total powers transmitted respectively to the base and filler material. According to literature, the amount of total power consumed in melting filler material is roughly 50% [18], Hence Q_b and Q_w can be written as:

$$\dot{Q}_w = \dot{Q}_b = \frac{nIV}{2} \tag{7}$$

In Fig. 3 the cylinder bloke represents the volume of wire filler elements to be melted. The power density value $\dot{q}_w$ is defined as the ratio of the wire melting power to the volume of the elements currently heated by the heat source, as shown in Eq. (8). The volume depends on the heat source size along feed direction, where l is feeding length. To transmit the correct amount of energy to filler FE elements the value of such parameter must be defined, i.e.

$$\dot{q}_w = \frac{\dot{Q}_w}{v} = \frac{\dot{Q}_w}{v_f w_L h_i} \tag{8}$$

where v_f represents the moving speed of the arc torch. This value should be equal to the energy per unit volume actually transmitted to the workpiece in the FE model. In practical thermal FE analysis the energy per unit volume transmitted to filler can be computed as

$$\dot{q}_w = \frac{\dot{Q}_w}{v_{e_l}}\tau = \frac{\dot{Q}_w}{w_l h_l d_{e_l}} \frac{l}{v_f} \tag{9}$$

where v_{el} is the volume of molten filler elements, τ is the time interval from the centroids of the FE elements corresponding to a single segment to the next centroids. Therefore, the torch head moving speed and filler feeding velocity are related to heat power density rate. The heat source dimension in feed direction must be equal to the correspondent size of filler FE elements [17].

2.2 *Validation of Heat Source Model*

Both Goldak double ellipsoid and modified Goldak heat source models are implemented in Abaqus CAE 2019 user subroutine DFLUX. Comparison of the simulation results are presented in Fig. 4.

An additive manufacturing plate model for validation of heat source is created in Abaqus CAE 2019, using Python script to form sets and steps for thermal analysis. Element activation and interaction algorithm (model change) are used for material deposition. The detail parameters and steps will be discussed in section of validation of the experiment. Figure 4a shows the temperature distribution computed using modified Goldak model. Figure 4b is the result by using Goldak double ellipsoid

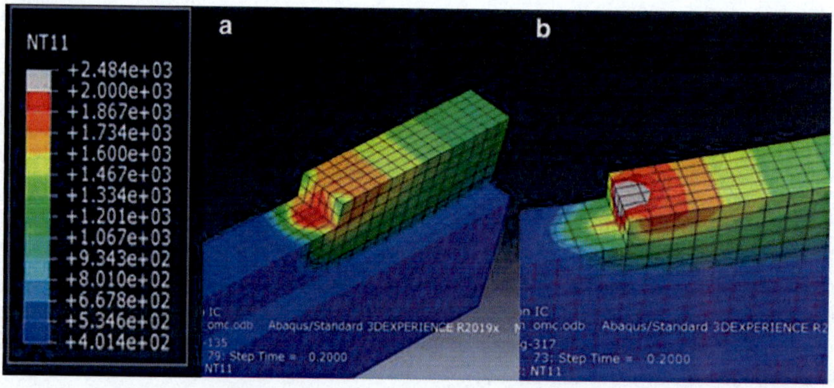

Fig. 4 Temperature distributions: **a** modified model **b** Goldak model

Table 1 Parameters for heat source model [mm]

a_f	a_r	b	C	f_f	f_r	ι
1.8	5.0	3.0	5.0	1.0	1.0	1.8

heat source model. Table 1 lists the 3D parameters for heat source model validation corresponding to Figs. 2 and 3. The heat power density Eqs. (6) and (9) were implemented in Abaqus using user routine DFLUX with all key parameters.

The test plate is 6 mm in the width, base material is low carbon alloy steel S700. The melting temperature.

From simulation results the temperature distribution of molten pool are apparently different. Goldak model Fig. 4b has higher temperature in the upper region (grey color) than the one of modified model Fig. 4a. As discussed in the previous Sect. 2.2 the modified Goldak model distribute heat power to filler and base. In the simulation two troch heads are positioned on the top of the filler and above the base actually in the filler elements. Therefore overheat zone appears on the result by Goldak model (Fig. 4b) Correct temperature distribution is very important for the calculations of thermal stress, deformation, cracks, and microstructure as well as mechanical properties of the component printed by WAAM.

3 Validation of the Heat Transfer Model

3.1 Experimental Test

An experimental test plate made by WAAM processing is simulated. Using Abaqus CAE and the user subroutines, Python scripts thermal-metalogic-mechanical models are fully implemented. Validations of thermal analyses are presented as follows [19].

1. The configuration used in tests:
   - Substrate: 15 mm thickness mild carbon steel
   - Wire material: G4Si1/AWS 5.18 ER70S-6 Ø1.0 mm Böhler Union K56
   - Shielding gas: Air Liquide Arcal Force (92/8% Ar-CO_2) flow rate 20 l/min
2. See Table 2.
3. Printing strategy
   Printing of each layer, waiting with an incremental time (+2 s every layer) before and after forced cooling. Figure 5 is the procedure of the printing. The interval cooling time each layer is about two minutes.
   The inter-layer temperature was measured by pointing the thermal cursors (always in the same pass) just before the passage of the weld pool. The temperature-time curves measured are shown in Fig. 6.
4. The printed test wall

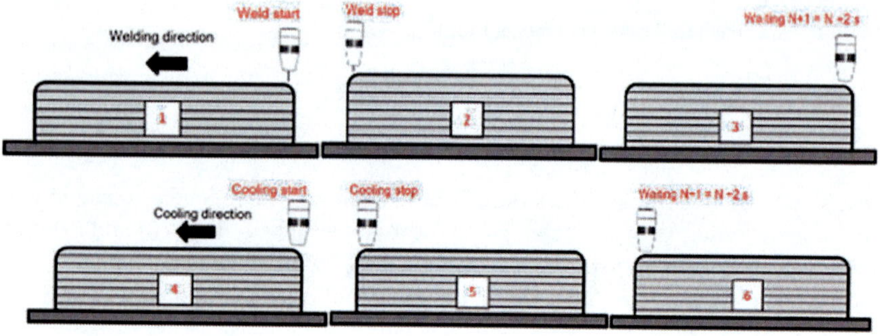

Fig. 5 The procedure of printing process

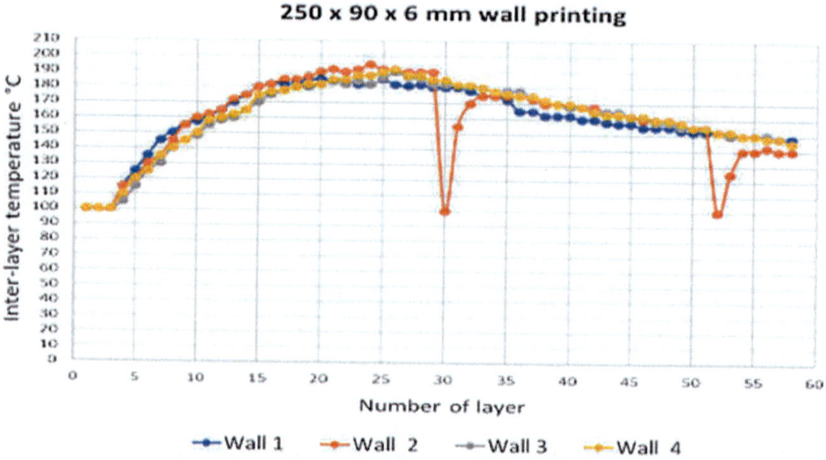

Wall	Arc time (min)	Cooling time (min)	Total time (min)
1	24.5	131	155.5
2	24.5	146	170.5
3	24.5	131	155.5
4	24.5	131	155.5

Fig. 6 Inter layer temperature curves measured in the printing process

Figure 7 shows a test wall printed by wire arc additive manufacturing process with strategy presented above Table 2 [20].

Fig. 7 The test wall printed with parameters above

Table 2 Printing parameters for multi layer walls 120 × 25 × 6 mm

	Root	Start	Middle	End
Job number	1	2	3	4
Length (mm)	120	5	110	5
WFS (m/min)	6.0	5.0	6.0	7.0
U (V)	16.5	15.4	16.5	16.8
I (A)	148	127	148	156
CTWD (mm)	15	15	15	15
TS (cm/min)	60	72	60	48
AL Correction	0	+5	0	-5
DYN Correction	0	0	0	+5
Change Job Ramp (s)	0	0.2	0.2	0.1
Correction S	0	5.0	0	0
t-S	0.2	0.2	0.1	0.2
I-S	50	20	50	20
SL1	0.2	0.1	0.1	0.1
SL2	0.2	0.1	0.1	0.3
t-E	0.2	0.5	0.1	0.5
I-E	30	60	50	25
Correction E	0	0	0	-5

(continued)

Table 2 (continued)

Acronym	Parameters	Range
CTWD	Contact tip to work distance (mm)	5; 20
WFS	Wire Feed Speed (m/min)	0; 20
TS	Robot Travel Speed (cm/min)	0; 100
S	Correction arc length for arc ignition	-10; 10
t-S	Duration of arc ignition (s)	0; 10
I-S	Percentage of current used for arc ignition (%)	0; 200
SL1	Duration of the ramp between arc ignition and welding state (s)	0; 10
SL2	Duration of the ramp between welding state and arc extinction(s)	0; 10
I-E	Percentage of current used for arc extinction (%)	0; 200
t-E	Duration of arc extinction (s)	0; 10
E	Correction arc length for arc extinction	-10; 10
AL Correction	Correction arc length during welding	-10; 10
DYN Correction	Correction of arc dynamic	-10; 10

3.2 *Thermal Simulation Modeling*

By using Abaqus CAE 2019 this teat wall (250 × 90 × 6 mm) is built as a model for simulation. Heat transfer analysis for WAAM process in Abaqus CAE (no plug in modulus) requires some different aspects as follows:

1. Build the part model and choose regional meshing in order to control element numbering and order. This manner will assist to make sets of elements and steps through Python scripts as shown in Fig. 8. Historically adding material is performed by using activation and deactivation of sets of elements. However the

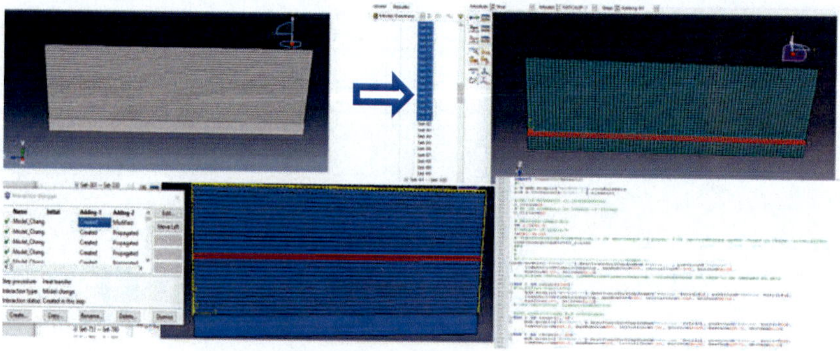

Fig. 8 Main steps required for WAAM simulation

traditional algorithm is not so efficient and the new technique has been in the development in Aabaqus CAE Additive plug in modeler [21].

2. Heat source moving and heat up under layer is achieved by user subroutine DFLUX. As mentioned in the previous sections a modified Goldak ellipsoid model was implemented in FORTRAN code. There are some limitations for WAAM processing, especially for interval parameter variations, multiple torch head moving with different materials, and interval layer forced cooling and so on.
3. Using three algorithms in Abaqus CAE to simulate interval cooling which are film condition, natural convection, cooling flux by DFLUX subroutine.

3.3 Results and Discussion

Heat transfer analyses on the test wall were performed with main parameters in the experiments. Validation is performed by comparing the interval layer temperature measurements of experiments to the results of simulations. Three different approaches for interval cooling have been used in Abaqus/CAE2019. The data of FE models are as follows.

- total number of nodes (type: DC3D8): 9477.
- total number of elements: 6240
- sets of elements for adding materials: 2359, each set includes 2 elements.
- time steps: 2400
- interactions with model change to activate set of elements.

Using natural convection and cooling time for interlayer cooling.

Results of heat transfer simulation during WAAM process are shown in Fig. 9. Four contour plots of temperature distribution when printing, respectively, are shown: (a) layer.5, (b) layer 7, (c) layer 23 and (d) layer 27. Wall temperature increased layer by layer (about 150 °C) when cooling time set as 120 s and internal natural convection

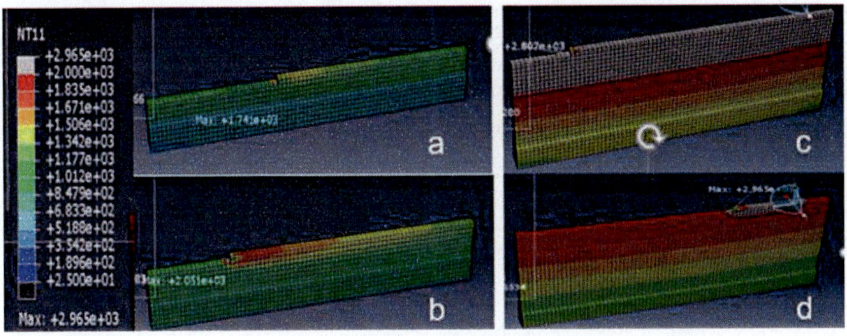

Fig. 9 Temperature distribution during the printing process

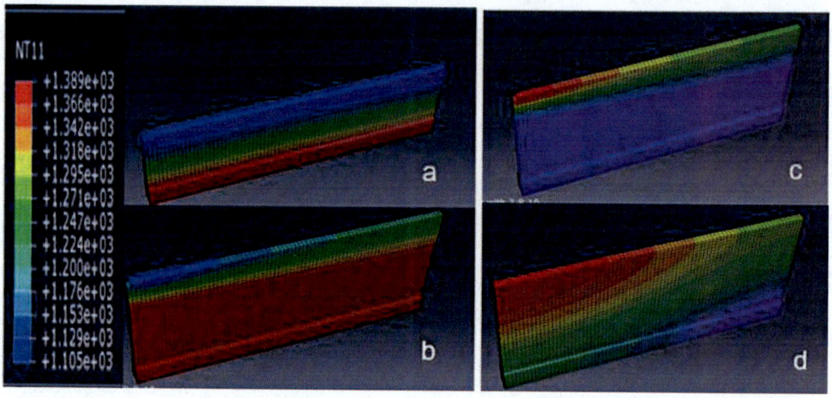

Fig. 10 Forced cooling (**a**), **b** chilling layer, **c** adding layer, **d** final cooling

for cooling. The wall temperature is over molten at layer 17 as well, which means wall printing collapse practically. Therefore interval layer cooling is one of crucial and effective parameters for WAAM process.

Due to lack of parameters for forced cooling we tried to use subroutine DFLUX to set up chilling flux and the jet head movement. For quickly testing, the approach of element activation and material properties were kept. The flux for chilling is just given a negative constant value q_c. Figure 10 presents the results of interval forced cooling. Figure 10a and c show that temperature drops down rapidly under the jet flux. If Abaqus AM plug in with user subroutine is able to simulate multi-torch movement and different material properties the approach will be very useful for variant WAAM process.

Eventually surface film condition in Abaqus interactions modeler was used for interval forced cooling (Fig. 11). By adjusting embedded film coefficient f_c, sink temperature and cooling time the temperature curves measured in the experiment could be fitted to although the real jet head movement could not be simulated.

Table 3 summarized main parameters in heat transfer analyses. In Fig. 12 the temperature curve L3 has general agreement with the experimental data in Fig. 6.

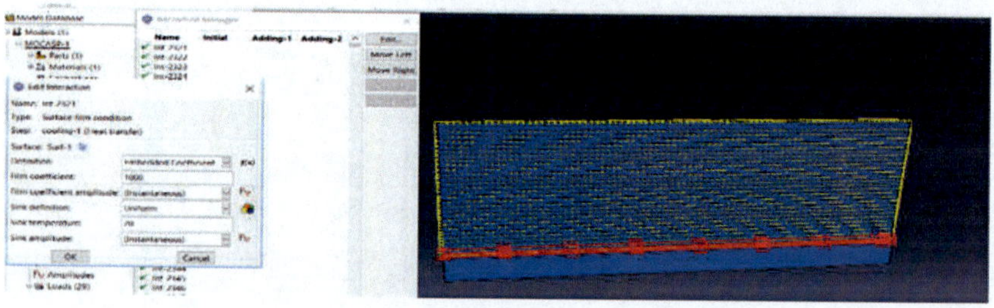

Fig. 11 Surface film condition settings for interval cooling

Table 3 Parameters in the analysis

U (v)	I (A)	V_L (cm/min)	V_f (m/min)	L (mm)	t_{inc} (S)	f_c
16	150	60	6.0	240	120	100

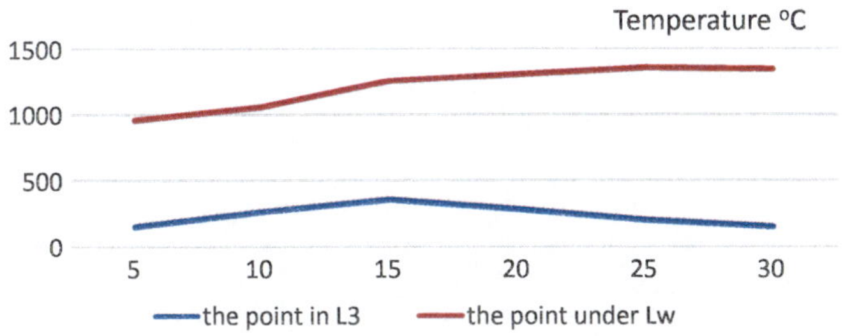

Fig. 12 Temperature curves measured in simulation

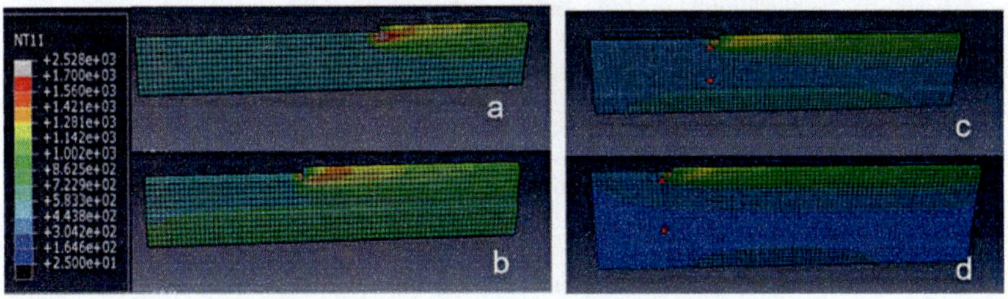

Fig. 13 The temperature distribution in the printing process

The curve L_w is the temperature measured under layer of the current weld pass. Figure 13a–d represent the temperature distribution in the printing process. Therefore thermal FE modelling for WAAM process is successfully implemented via user subroutine. However for simulation on real parts the computation capacity and time consuming are still in challenges.

4 Conclusions

In this work the modified Goldak heat source model is applied to the wire arc additive processing simulation by using Abaqus CAE 2019 and user subroutine. The advantage of modified Goldak model allows to take into account the feeding speed of filler material and resulting in correct energy input for heat transfer FE analyses.

Element activation algorithm is successfully used in WAAM simulation through Abaqus Python scripts and user subroutines. Hopefully the operation and time consuming will be improved by using Abaqus CAE Plugin modrler later.

Key parameters play key roles in the printing process such as heat input energy and interval cooling (or cooling flux and time).

The future work will do further improvement on the efficiency and accuracy for simulation, meanwhile microstructure, properties prediction models and residual stress analysis will be applied to the experimental model plate.

Acknowledgements This research work was carried out in the framework of the research project PRINT-AM ICON, 'Predictable Integrity of Large Scale Additive Manufactured Components through Intelligent Printing Strategies', funded by Flemish government.

References

1. Ding D, Pan Z, Cuiuri D, Li H (2015) Wire-feed additive manufacturing of metal components: technologies, developments and future interests. Int J Adv Manuf Technol 81:465–481
2. Ding D, Pan Z, Cuiuri D, Li H (2015) A multi-bead overlapping model for robotic wire and arc additive manufacturing (WAAM). Robot Comput Integr Manuf 31:101–110
3. Application Form for SIM ICON Project Version of this document. January (2016)
4. Rodrigues TA, Duarte V et al (2019) Current status and perspectives on wire and arc additive manufacturing (WAAM). Materials 12(7):1121
5. Martina F, Ding J, Williams S, Caballero A, Pardal G, Quintino L (2018) Tandem metal inert gas process for high productivity wire arc additive manufacturing in stainless steel. Addit Manuf 2018(25):545–550. https://doi.org/10.1016/j.addma.11.02
6. Almeida PMS, Williams S (2010) Innovative process model of Ti–6AL–4V additive layer manufacturing using cold metal transfer (CMT). In: Proceedings of the 21st international solid freeform fabrication symposium. Austin, TX, USA, 1–9
7. Xie Y, Zhang H, Zhou F (2016) Improvement in geometrical accuracy and mechanical property for arc-based additive manufacturing using metamorphic rolling mechanism. J Manuf Sci Eng 138:111002. https://doi.org/10.1115/1.4032079
8. ASTM F2792-12a (2012) Standard terminology for additive manufacturing technologies (Withdrawn 2015). ASTM International: West Conshohocken, PA, USA
9. Ni J, Voorde J Vande, Antonissen J, Wahab MA (2018) Dependency of phase transformation on the prior austenite grain size and its influence on welding residual stress of s700 steel. Weld World 1–14
10. Ni J (2019) Improved residual stress prediction of welded steel components. Ph.D. thesis, Soete Laboratory, Ghent University
11. Pavelic V (1969) Experimental and computed temperature histories in gas tungsten arc welding of thin plates. Weld J Res Suppl 48:296–305
12. Krutz WG, Segerlind JL Finite element analysis of welded structures. Weld Res Suppl 211–216
13. Goldak JA, Akhlaghi M (2016) Computational welding mechanics. Springer Science & Business Media
14. Chakravarti GJA, Bibby M (1984) A new finite element model for welding heat sources. Metall Trans B 15(2):299–305
15. Hao M, Sun Y (2013) A FEM model for simulating temperature field in coaxial laser cladding of TI6AL4V alloy using an inverse modeling approach. Int J Heat Mass Transf 64:352–360
16. Goldak J, Chakravarti A, Bibby M (1984) A new finite element model for welding heat sources. Metall Trans B 15:299–305
17. Montevecchi F, Venturini G, Scippa A, Campatelli G (2016) Finite element modelling of wire-arc-additive-manufacturing process. Procedia CIRP 2016(55):109–114. https://doi.org/10.1016/j.procir08.024

18. Dupont JN, Marder AR (1995) Thermal efficiency of arc welding processes. Weld Res Suppl 406–416
19. Goldak J, Bibby M, Moore J, House R, Patel B (1986) Computer modeling of heat flow in welds. Metal Trans B 17:587–600
20. Antonissen J, Vande Voorde J et al (2020) Internal Experimental test report
21. Abaqus (2017) Introduction to user subroutines for AM

Failure Analysis of a Fuel Control Tube from an Aircraft Engine

María García-Martínez, María Pilar Valles González, Ana Pastor Muro, and Alejandro González Meije

Abstract Multiple cases of fuel control pressure tube fractures from aircraft engines have been reported. The aim of the present study is to determine the cause of this tube fracture. The studied set was composed by the mentioned tube, a welded connecting pipe, where the fracture has been produced, and a union nut. The fracture has been produced in one most critical zones of the tube, in a region near to the supporting body of the union nut to the connector. Visual examination and microscopic observation through a stereo microscope of the tube fracture surface has been carried out to determine the macrofractographic features. The results revealed a plastic macrodeformation of the tube, a damaged surface and signs of a possible corrosion process. The material has been chemical, mechanical and microstructural characterized. The results confirmed that specifications were fulfilled. Fracture surface was also inspected by scanning electron microscopy to determine the microfractographic features in order to find out the failure mechanism involved in the fracture. Fatigue striations, which are typical from a progressive fracture by a fatigue mechanism have been observed. The origin of the fracture has been placed in defects located in the outer wall of the tube, leading to a final overload fracture.

Keywords Fuel control tube · Aircraft failure analysis · Fractography · Fracture · Fatigue · Striations

1 Introduction

The X6CrNiTi18-10 type steel is an austenitic stainless steel which has good properties as corrosion resistance, conformability and weldability. Furthermore, this steel, when is stabilized with titanium, improves its behavior against intergranular corrosion and high temperature oxidation. Due to all these reasons, it has multiple applications in the heat treatment industry, using them as burners, ovens, tubes, trays, etc., but also in the petrochemical and food processing industry, as well as in tubes,

M. García-Martínez (✉) · M. P. Valles González · A. Pastor Muro · A. González Meije
Metallic Material Area, National Institute of Aerospace Technology (INTA), Madrid, Spain
e-mail: garciamam@inta.es

M. Abdel Wahab (ed.), *Proceedings of the 8th International Conference on Fracture, Fatigue and Wear*, Lecture Notes in Mechanical Engineering,
https://doi.org/10.1007/978-981-15-9893-7_15

vessels and pipelines of nuclear reactors. It is non-magnetic in an annealed state and can only be hardened by cold working [1]. Besides, this type of steel is also used for the manufacture of fuel or coolants pipes in aircraft systems. The fracture of these pipes can occur for many reasons, but when it takes place in a pipe carrying fuel, the situation can be especially dangerous as it can lead to a catastrophic failure [2]. Numerous cases are cited in the literature with the main causes of failure being fatigue, corrosion, wear, creep and overload mechanisms. But the most common failure mechanism in engine parts is usually fatigue induced by a thermomechanical cyclic load, which can be accelerated by a corrosive atmosphere [3, 4]. Fatigue can also be caused by an incorrect manufacture, design or assembly in service of the tubes. The failure in this article has been detected during the aircraft's runway to take-off, so no irreparable damage has occurred. Due to the many similar cases detected, some precautions have already been established in the engine maintenance manual that must be considered during its handling.

2 Materials and Methods

The chemical analysis of the tube material, shown in Table 1, was carried out by an X-ray fluorescence equipment (XRF), Panalytical model PW2404, and by fusion and combustion techniques LECO equipment. This material composition corresponds to a stainless steel type X6CrNiTi18-10 (UNE EN 10,886–1:2006).

A hardness test was carried out on a cross section of the tube in an area far from the fracture, using a durometer Future FM7, resulting in hardness values shown in Table 2.

The metallographic samples used for microstructural characterization were obtained mounting the areas of interest in conductive resin and, after polishing and etching with Superpicral etching agent (10 g picric acid, 5 ml HCl in 100 ml of ethanol), the microstructure was observed in an optical microscope Leica MEF4M.

To observe the fracture surface, a stereo microscope Leica Wild M10 and a field emission scanning electron microscope Jeol 6500F equipped with an X-ray dispersive energies microanalysis system Oxford were used.

Table 1 Chemical composition (% weight)

Fe	C	Si	Mn	P	S	Cr	Ni	Mo	Ti	Cu	N
Base	0.03	0.57	1.63	0.024	0.002	17.1	10.2	0.42	0.48	0.30	0.0126

Table 2 Hardness (HV300 g)

Sample	Hardness (HV_{300g})
Cross section	212 ± 6

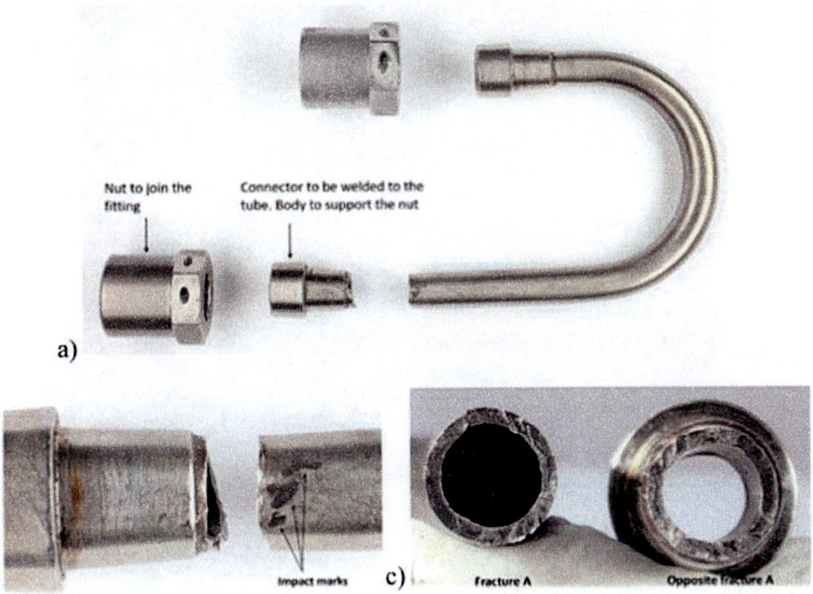

Fig. 1 **a** General view of the fractured parts as received. **b** Side view of the fracture area. **c** The fracture surface (A) and its opposite fracture surface

3 Results and Discussion

3.1 *Visual Observation and Macrofractographic Study*

The fracture has been produced in a cross-section plane forming a 30° angle with the normal plane to generatrix. It is located in the vicinity of the area where the connector terminal is joined to the tube. It presents some plastic deformation in the diameter and most of the surfaces of fracture show a bright and soft texture (see Fig. 1).

The fracture surface (A) shown in Fig. 1c was inspected and six zones on this surface were selected for further detailed inspections, as shown in Fig. 2.

3.2 *Microstructural Characterization*

Two samples from a longitudinal and a cross-section of the tube were mounted from an area far from the fracture.

The tube material microstructure (see Fig. 3) corresponds to equiaxial austenite grains with carbides and carbonitrides of titanium, and molybdenum carbides, which were analyzed by EDX resulting the spectra that are shown in Fig. 4.

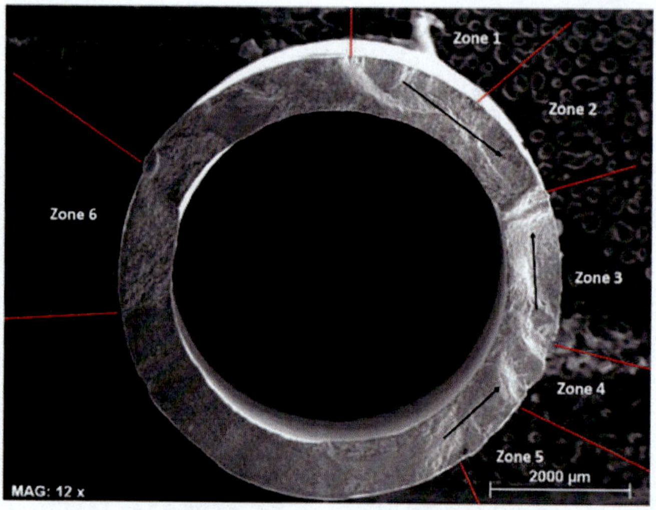

Fig. 2 Front view of fracture surface A. The black rows indicate the fracture propagation

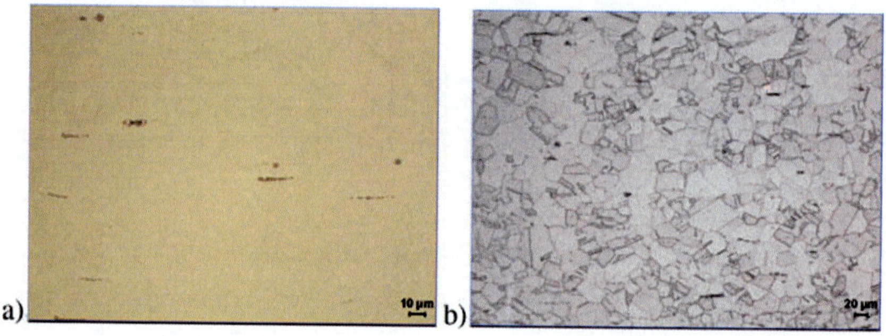

Fig. 3 **a** Carbides and carbonitrides chains aligned in the longitudinal direction of the tube, **b** Austenitic microstructure

3.3 Microfractographic Study

The fracture surface A was inspected by first following the black observation line shown in Fig. 2, between zones 1 and 2. The origin of the fracture was located in zone 1 on account of propagation lines and fatigue striations observed (Fig. 5a–d) that were also presented in zone 2 (Fig. 5e). Some flattened microvoids were identified, indicating an overload fracture, Fig. 5f. Afterwards, another observation line was followed between zones 3 and 5, verifying the existence of another crack whose origin was placed in zone 5, although it has not been possible to identify it due to the fact that in this zone there were multiple crushes and it was severe oxidized. Fatigue striations in zone 4 and some crushed microvoids in zone 3 were observed

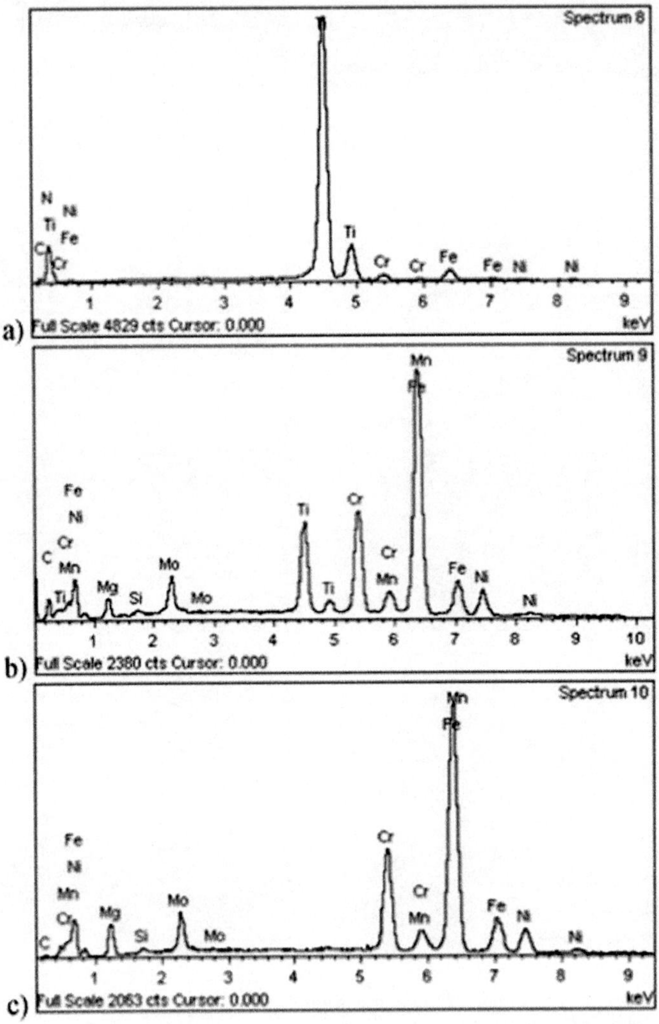

Fig. 4 EDX spectra: **a** Titanium carbonitrides, **b** Titanium and molybdenum carbides, **c** Molybdenum carbides

(Fig. 6). Due to the multiple crushing in zone 6, no microfractographic features could be identified. On the external lateral surface in the vicinity of zone 1, a surface with multiple defects and marks was observed (Fig. 7).

In this study, at least two origins have been located. In both cases, fatigue striations were observed, which are typical of progressive fracture, leading to a final static overload fracture.

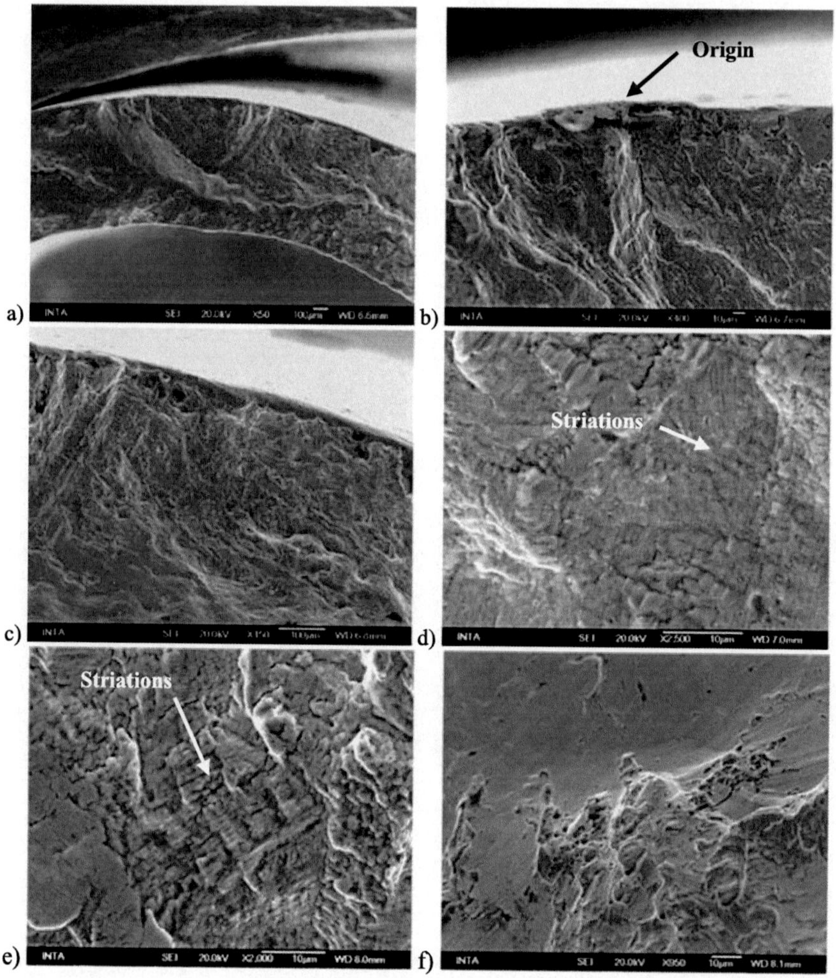

Fig. 5 **a** Zone 1. General view of the crack origin in fracture A, **b** Fracture origin, **c** Fracture propagation zone, **d** Fatigue striations, **e** Zone 2. Fatigue striations, **f** Crushed microvoids

The fracture mechanism has been fatigue, which may be originated in the multiple defects found on the external surface of the pipe and where the cracks may have progressed by the vibrations to which the pipe was subjected.

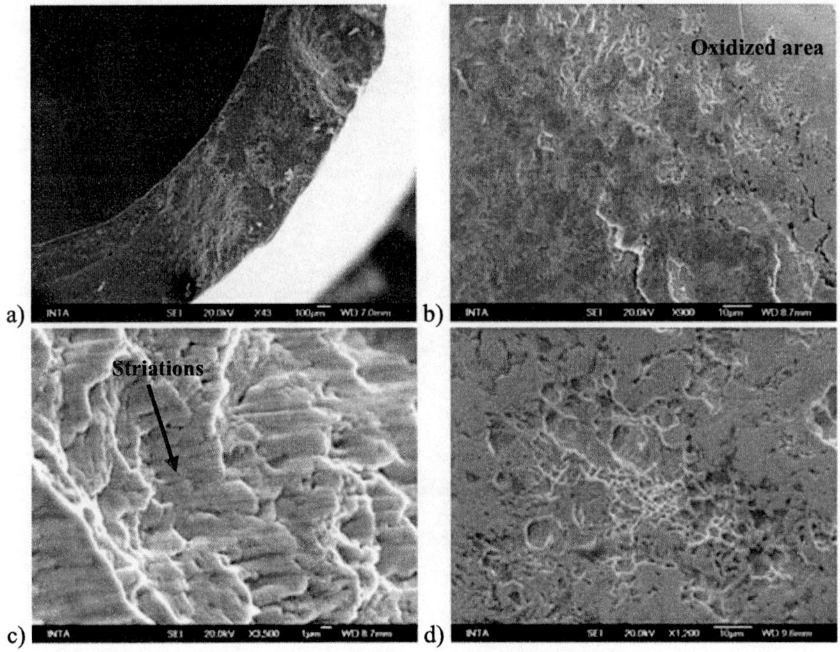

Fig. 6 **a** Zone 5. General view of the crack origin in fracture A, **b** Oxidized zone, **c** Zone 4. Fatigue striations, **d** Zone 3. Crushed microvoids

Fig. 7 Zone 1. Multiple defects on the surface

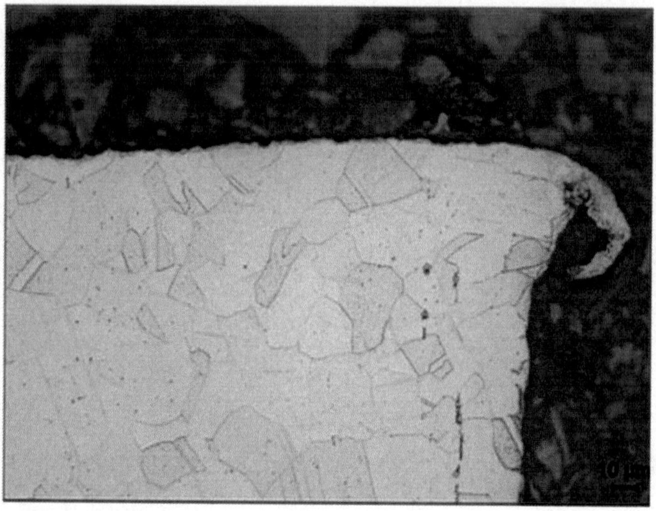

Fig. 8 Opposite fracture surface. Zone 1. Transgranular profile

3.4 Fracture Profile

In order to analyze the grains of the fracture surface in the zone 1, a longitudinal section of the opposite fracture surface was mounted and etched to reveal the grain crystallinity. Figure 8 shows that the fracture presented transgranular profile.

4 Conclusions

The pipe, manufactured from X6CrNiTi18-10 stainless steel according to the standard UNE EN 10,886–1, presented an austenitic microstructure with a hardness value of 212 HV_{300g}, which corresponds to an annealing heat treatment state.

Both macro and microfractographic features revealed that the pipe fracture has been produced by fatigue mechanism, probably as a result of the vibrations to which it has been subjected to. At least, two fracture origins have been located, one in the outer wall of the pipe where an incorrect surface finished has been observed, leading to a final static overload fracture. Furthermore, the stresses introduced during the assembly of the pipe may also have contributed to the fracture.

Although the type of the steel and its state of treatment were suited for the type of application in which it has been used, an incorrect torque or non-installation of clamping flanges to absorb the engine vibrations, may have led to incubations of cracks in the surface defects and to propagate thorough a fatigue mechanism until a final fracture was reached.

Acknowledgements Authors want to thank all Metallic Materials Area staff and especially, to the Microanalytical and Microstructural Characterization Laboratory. This research did not receive any specific grant from funding agencies in the public, commercial or not-for-profit sectors.

References

1. Schwind M, Källqvist J, Nilsson J, Agren J, Andrén H-O (2000) σ-Phase precipitation in stabilized austenitic stainless steels. Acta Mater 48:2473
2. Adib AML, Baptista CARP, Barboza MJR, Haga C, Marques CCF (2007) Aircraft engine bleed system tubes: material and failure mode analysis. Eng Fail Anal 14:1605–1617
3. Clark G (2005) Failures in military aircraft. Eng Fail Anal 12:755–771
4. Bokwon L, Karkkainen RL (2016) Failure investigation of fuel tubes of a jet engine augmenter assembly. Eng Fail Anal 62:39–48

Detection of Damage in RC Beams Strengthened with NSM CFRP Rectangular Rod by Finite Element Modeling

Erica Magagnini and Roberto Capozucca

Abstract The Near-Surface Mounted (NSM) reinforcement method has been shown to be one of the most important technique for the rehabilitation of damaged concrete structures or ones that are not adequate to support expected loads. However, despite decades of scientific research centered on the optimization of construction details, on models for bond behaviour and on design methods for flexural and shear strengthening, the effects of damage on FRP reinforcements have yet to be adequately investigated. In this work, the influence of different typology of damage such as loss of adhesion in the reinforcement-concrete interface, concrete's cracking, and, finally, damages in FRP due to notches have been analysed by Finite Element Modeling (FEM). An incremental non linear analysis is developed using 3D Finite Element (FE) in the ANSYS code, considering both the static and dynamic behaviour of RC beams strengthened with one CFRP rectangular rod inserted into a groove according to NSM technique in relation to the different types of damage. Both progressive local failure and nonlinear stress–strain relationship of concrete are taken into account. For the CFRP, a model that takes into account the debonding phenomena at the interface between concrete-adhesive has been adopted. Finally, the developed theoretical model is calibrated and validated by comparison with results from experimentation and interesting aspects are remarked and discussed.

Keywords NSM technique · CFRP rod · Notch/bond damage · FE analysis

1 Introduction

Maintaining adequate level of security in reinforcement concrete (RC) structures during their cycle of life is as important theme in the field of civil engineering. In recent years, the retrofitting of existing RC beams and frames using composite materials is became an increasingly common method with the aim to increase the element's capacity under bending and shear. There are several strategies that can be

E. Magagnini (✉) · R. Capozucca
Struct. Section DICEA, Università Politecnica Delle Marche, Ancona, Italy
e-mail: e.magagnini@univpm.it

M. Abdel Wahab (ed.), *Proceedings of the 8th International Conference on Fracture, Fatigue and Wear*, Lecture Notes in Mechanical Engineering,
https://doi.org/10.1007/978-981-15-9893-7_16

followed in the strengthening with Fiber Reinforced Polymer (FRP). Among these, the near surface mounted (NSM) technique seems to be one of the most suitable. According to this method the fiber reinforced polymer (FRP) with configurations of rods or lamina is inserted into grooves made on concrete covers [1–4]. In literature, many researches underlined the capacity of this method to solving a number of negative aspects that externally bonded reinforcement (EBR) FRP [5, 6] often shown such as damage deriving from impacts, high temperature, fire and debonding failure [5–7].

The efficiency of the strengthening system with NSM FRP rods is strongly depends on the development of cracking of concrete under service loading as well as to the width and depth of cracks [8–11]. At the same time, the bond between concrete-adhesive-FRP is negative influenced by the presence of damage that may reduce the capacity of strengthened beams [12]. Moreover, the structural performance of composite materials can be affected by a wide range of damages that involves, for example, cracks in fibers and matrix and, in the interfaces of fibers and matrix [13]. Wrong execution or notches caused by impact can determine localized section reduction shapes. Interesting analytical strategies [14–16] and experimental investigations [12, 17–19] have been carried out by many authors for predicting the stress state and fracture behavior of damaged composite laminates. Unfortunately, there are only few existing studies in literature that analysed the effects of damage affected composite materials when they are used as strengthening in RC beams [12, 19].

In recent years, Finite Element (FE) modeling of RC elements with FRP strengthening has been extensively performed by many authors in addition to experimental works. Many numerical/theoretical investigations have been proposed based on the experimental outcomes, focusing the attention on Externally Bonded (EB) technique. The most of these models is based on the assumption of perfect adhesion between concrete-epoxy and FRP-epoxy interfaces [20, 21]; a few authors has proposed numerical modeling approaches that involved the use of interface elements with zero-thickness able to take into account damage due to the debonding mechanism [22, 23]. On the other hand, FE analysis of RC beams with NSM FRP is rarely performed [24, 25]. All the works present in literature do not consider the effect of damage affected the FRP strengthened elements.

In this paper an assessment of three beams with NSM Carbon-FRP lamina as strengthening has been investigated by FE analysis taking into account the presence different type of damages, such as concrete cracking, local bond loss and notches of rod. A non linear three dimensional FE model has been developed using ANSYS code, considering both the static and dynamic response. Both progressive local failure and nonlinear stress–strain relationship of concrete are taken into account. In addition, the debonding phenomenon has been considered through the proper use of interface elements.

Finally, the developed analytical model is calibrated and validated by comparison with experimental results and interesting aspects are remarked and discussed.

2 Experimental Tests and Results

The experimental campaign involves static and dynamic tests performed on three RC beam models, labeled as RB0, RB1 and RB2. The specimens had a length of 1100 mm with a rectangular cross-section of 80 mm · 120 mm. All prototypes were reinforced with 4 steel bars and vertical stirrups having a diameter of 6 mm. A CFRP lamina 1.4 mm · 15 mm · 000 mm made by unidirectional fibers was inserted into a groove on tensile side and glued using epoxy adhesive ($E_m \sim 6 \cdot 10^3$ N/mm^2). The geometrical configuration of experimental models is shown in Fig. 1.

Prelimary tests on materials have been carried out and the main results are summarized below. The concrete had an average cylinder strength equal to $f_{c,av} \sim 40$ N/mm^2 ($E_c \sim 35.3 \cdot 10^3$ N/mm^2). For the steel bars the average yielding stress was equal to $f_{y,av} \sim 500$ N/mm^2 ($E_s \sim 2.1 \cdot 10^5$ N/mm^2). Tensile tests were carried out on two CFRP specimens following the ASTM-D 3039 Standard [26], in order to define the mechanical properties of the composite material. In particular, it was obtained tensile strength $f_{CFRP} = 2500$ N/mm^2 and Young's modulus $E_{CFRP} = 3.53 \cdot 10^5$ N/mm^2. Different configuration of damage was considered. In beam RB0, the strengthening was inserted into the groove without damage. For beam RB1, damage was due to a gap of epoxy resin on two symmetric sections of length equal to 30 mm, as shown in Fig. 2a. In this case, the CFRP rectangular rod was without damage. Finally, in beam RB2 the CFRP lamina was damaged by double notching at the midspan section. Detail of damage is shown in Fig. 2b.

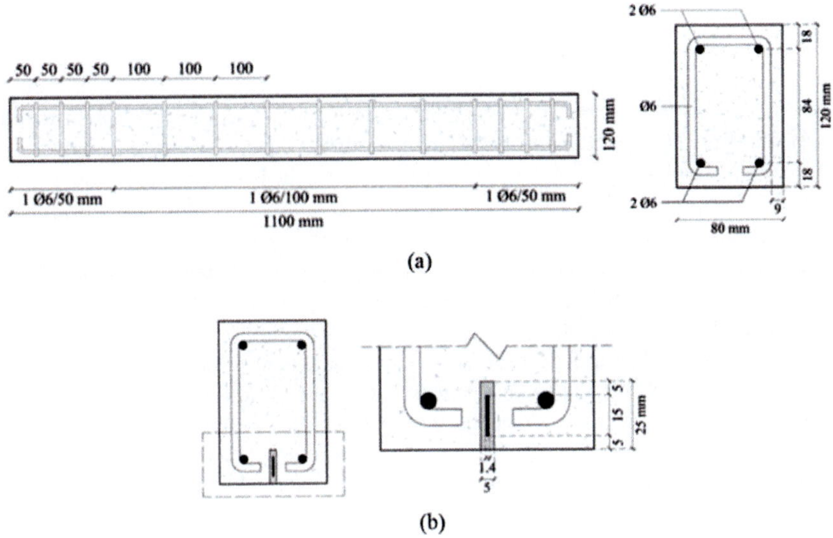

Fig. 1 **a** Geometrical characteristics of RC beam model; **b** detail of CFRP lamina inserted into groove

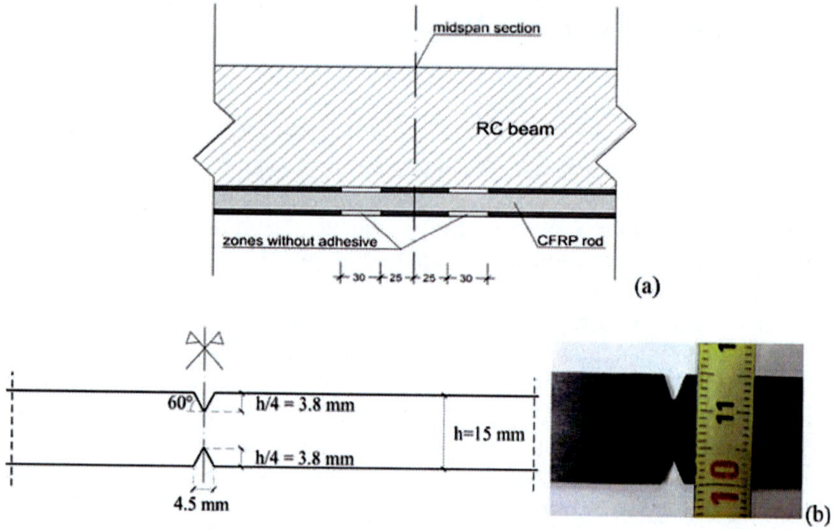

Fig. 2 Configuration of damage for beams **a** RB1 and **b** RB2

2.1 *Experimental Response of Beams*

Experimental static bending tests were performed on three beams, RB0, RB1 and RB2. All the models were tested under cyclic increasing load P applied at two points 300 mm from the middle of the beam until failure. The set up of experimental bending tests involves: a vertical jack to transmit the load P with a load cell to measure P; strain gauges applied on the CFRP lamina to record the evolution of deformation during tests: in RB0 and RB1 one in the midspan of the rod and in RB1 two at 53 mm from the midspan; three horizontal linear transducers (LVDTs) used to measure deformation of concrete—one on the compressed edge and two on the lateral surface—and lastly, two LVDTs to record displacement at the midspan section and close to the support.

Six loading cycles with vertical loads equal to P = 4 kN, 8 kN, 12 kN, 14 kN, 16 kN and 27 kN was adopted and, successively, the beams were tested until failure. For each load step a configuration of the damage degree labeled as D1, D2, D3, D4, D5, D6 and Du was defined.

Failure was reached in RB0 for a load values equal to P $\approx$ 35.44kN with crash of compressive concrete at the beam's midspan, without exhibiting any loss of adherence. In beam RB1, failure occurred at P $\approx$ 30.04kN due to the delamination of the strengthening and the subsequent collapse of the concrete. At first, it was observed a bond slip inside the lamina-adhesive interface and, subsequently, at the adhesive-concrete interface, without any detachment of the concrete cover (Fig. 3). For beam RB2, failure occurs at P $\approx$ 37.02kN with crushing of concrete. It wasn't observed any significant damage on the epoxy adhesive except for the same cracks that interested all of the concrete. At the end of the bending test, the CFRP lamina was pulled out

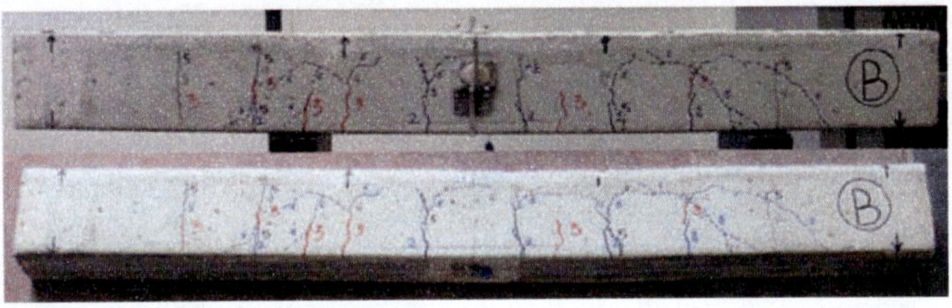

Fig. 3 View of beam RB1 at failure

and it was observed the presence of two slits extending in the longitudinal direction that the notches have triggered causing a reduction of the section of the same reinforcement.

For all the beams, the moment, M, versus curvature, χ, diagrams were experimentally obtained and they are contained in Fig. 4. It is possible to state that the presence of damage, both in the case of loss of bond and localized notches, the presence of damage leads to a reduction of the stiffness of the damaged beams. This effect is more evident in beam RB2. Furthermore, it can be observed a loss of resistance in the case of beam RB1 with two areas without adhesive, as compared to RB0 without damage. In the case of beam RB2, the presence of notches on CFRP at midspan doesn't impairs the resistance capacity of beam, that shows the greatest curvature at failure.

In the second phase of the experimental campaign, free vibration tests were carried out on beams RB0, RB1 and RB2 at each step of loading Di; the entire set up of dynamic tests is shown in Fig. 5. Flexible springs were used to hang the beam models with the aim of simulate the free–free boundary conditions. The beams were triggered using an impact hammer (Brüel & kjær Impact Hammer Type 8202) at a fixed point

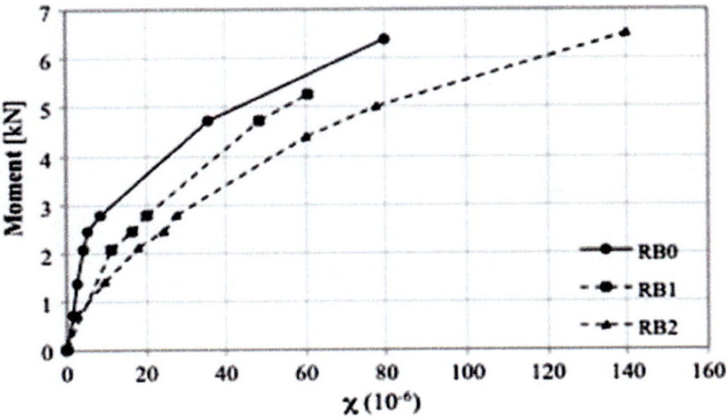

Fig. 4 Bending moment, M, versus curvature, χ, diagrams experimentally obtained for beams RB0, RB1 and RB2

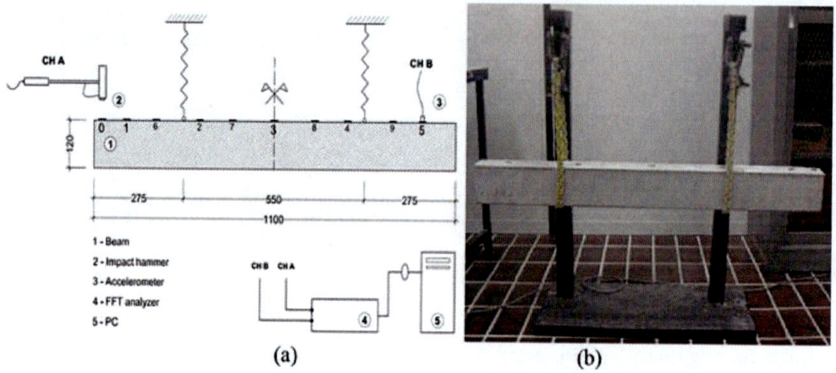

Fig. 5 **a** Set up of dynamic vibrational tests; **b** view of suspension system

on the edge of the beam, the dynamic response was acquired with a piezoelectric accelerometer placed at different points i with i = 1, …, 9.

For each position of the accelerometer a set of 10 hits was performed and the average value was evaluated. Fast Fourier Transform (FFT) technique using Pulse software was used to extract the frequency values from transformed signals in frequency domain. The suitability of the dynamic measurements was checked by evaluation of the coherence function.

From the analysis of the experimental results it is possible to state that, for all tested beam models, the frequency values tend to decrease progressively parallel to the increase of damage by concrete's cracking. In Fig. 6 it can be noted the frequency variations in percent, for the first four vibration modes, at different damage degree Di in relation to the undamaged condition D0 for beams RB1 and RB2. For beams RB0, RB1 and RB2 a frequency variation level of the order of 20% is reached only after the damage degree D5. In the case of damaged beam RB1 after the application of the first loading step it is possible to notice a little higher frequency variation in

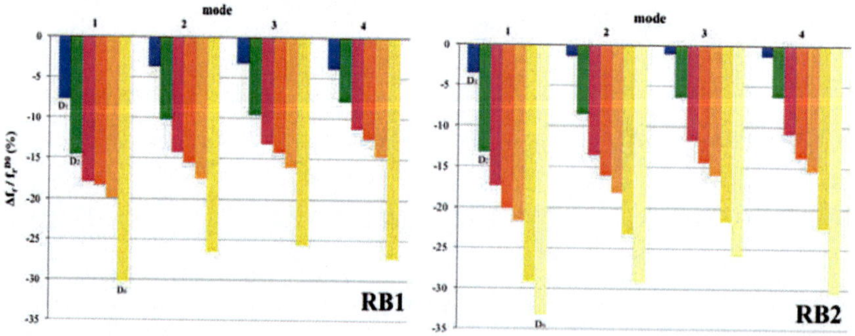

Fig. 6 Frequency variations at different damage degree Di with i = 1, …, 5–6 in relation to D0, for the first four vibration modes

relation to the other beam models. Comparing the diagrams Fig. 6, it is possible to notice similar values of the frequency variations between both the cases investigated. It can be stated that the damage deriving from the absence of adhesion locally or notches in the CFRP lamina don't lead to a consistent reduction of the frequency values experimentally recorded. This is due to the nature of dynamic analysis and measurement of frequency values that are global analysis, in this way, possible defects and local damage are taken into account only in a global sense.

3 Analysis by FE Modeling

Finally, the behaviour of RC beam models with NSM CFRP strengthening was assessed by a Finite Element Modeling carried out using 3D ANSYS code with the aim to further study the effect of damage on the structural response. A comparison between numerical and experimental results was done in order to check the suitability of the developed model.

3.1 FE Model

With the aim of simulate beams, three different models were defined. The presence of all materials including concrete, longitudinal steel reinforcements, CFRP rectangular rod, and adhesive was taken into account.

For what concern the modeling of concrete, the element Solid 65 was adopted. This solid element is characterized by eight nodes with three degrees of freedom at each node. The most important aspect of this element is the capability of plastic deformation, cracking in three orthogonal directions and crushing (Fig. 7). For what concern the steel reinforcement, it is possible to adopt three different techniques: the discrete, the embedded, and the smeared models. In the first case, the steel reinforcement is modeled by beam elements connected to the concrete mesh nodes. In

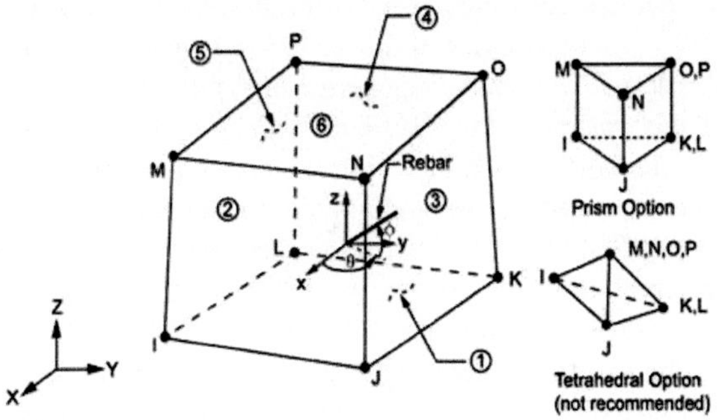

Fig. 7 Definition of Solid 65 element

the embedded model the concrete mesh restrictions are overcome with the increase in the number of nodes and degree of freedom. Finally, in the smeared model it is assumes that reinforcement is spread uniformly throughout the concrete elements in a defined region. This approach is suitable in the case of large-scale models where the overall response of the structure is not significantly affected by reinforcement. In this study, it was chosen to adopt the discrete model to represent steel reinforcement. In particular, the 3D spar element, labeled as Link 180, was used. The behaviour of both the adhesive and the CFRP strengthening was modeled using the Solid 185 element. 8 nodes with three degrees of freedom at each node, plasticity, stress stiffening, large deflection, and large strain capabilities characterize this. With the aim to consider the debonding phenomena at the interface between concrete and epoxy resin, a surface-to-surface contact elements were applied. Targe 170 was adopted for the target surface on concrete and Conta173 was placed on the adhesive surface to represent the contact.

Finally, in addition to the elements described above in order to reproduce the same boundary conditions in vibration, element Combine 14 was chosen and the beams' suspension springs were modeled. Two nodes, a spring constant, assumed equal to k $\sim$ 0.01N/m, and damping coefficients, defined the main properties of this element.

For what concerns the definition of the materials, in order to adequately model the concrete, it was chosen to adopt a model that takes into account the linear and non-linear material properties together with the concrete model proposed in ANSYS [27]. In this way, the concrete's failure is defined by the Von Mises failure criterion together with the Willam and Warnke model. For the concrete material, it was adopted a Poisson's ratio of 0.15. The multi-linear isotropic material was defined by an uniaxial stress–strain curve in compression using the equations given by [28]:

$$F = \frac{E_c\varepsilon}{1+\left(\frac{\varepsilon}{\varepsilon_0}\right)^2};\quad \varepsilon_0 = \frac{2f_c'}{E_c};\quad \mathrm{E_c} = \frac{\mathrm{f}}{\varepsilon} \tag{1}$$

where f is the stress evaluated for any strain ε and ε_0 is the strain reached for the ultimate compressive strength f_c'. The value of f_c', taken from experimental compression test was assumed equal to 40 MPa. The concrete's non-linear stress–strain curve in compression is given in Fig. 8a. As already introduced, a smeared crack model has been introduced through the use of the William-Warnke failure surface [29]. The peculiarity of this criterion is the possibility to consider the capability of cracking under tension and crushing under compression. In Fig. 9 the failure surface that characterized the William-Warnke criterion is shown.

The William-Warnke failure surface is defined by five input strength parameter: f_c, uniaxial compressive strength, f_{ct}, uniaxial tensile strength, f_{cb}, biaxial compressive strength, strength and, two additional parameters f_1 and f_2, respectively, the ultimate compression strengths for states of biaxial and uniaxial compression superimposed on hydrostatic stress state. However, when the hydrostatic stress is less than $\sqrt{3}$ f_c, the failure surface can be specified with a minimum of two constants, f_t and f_c, because the other three constants default to Willam and Warnke:

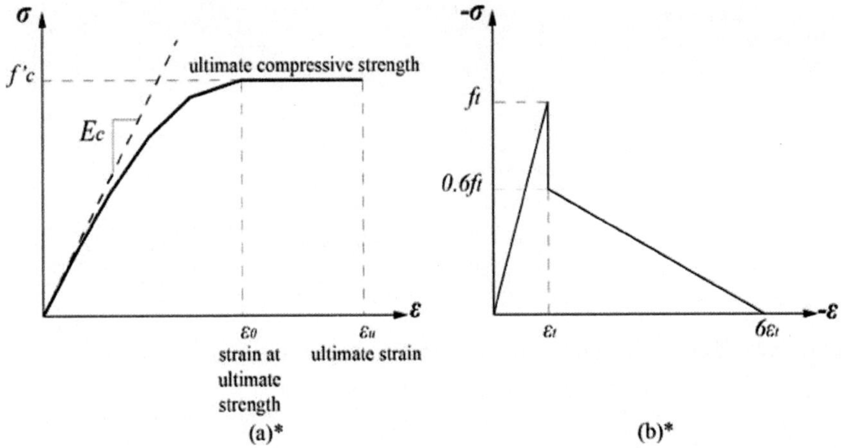

Fig. 8 Strain–stress relation for concrete under **a** compression and **b** tension. *The diagrams are in different scale.

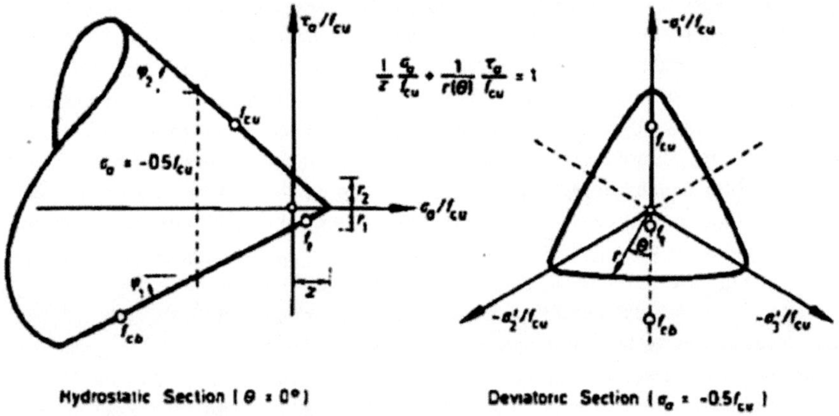

Fig. 9 William-Warnke failure surface [29]

$$f_{cb} = 1.2f_c; \quad f_1 = 1.45f_c; \quad f_2 = 1.725f_c \tag{2}$$

It is possible to define other two coefficients, labeled as β_t and β_c, that are, respectively, the shear transfer coefficients for an open crack and for a closed crack.

Finally, it is possible to define another parameter, T_c, that considers the entity of the tensile strength's reduction after its maximum value. This coefficient permits to take into account the contribution of the tensile concrete to the stiffness of the entire beam. In this way a bilinear stress–strain relation was assumed to reproduce the tension-stiffening phenomenon (Fig. 8b). By the properly definition of all parameters, both

Table 1 Mechanical parameters adopted for the FE modeling of concrete

Young's modulus	E_c	$35.3 \cdot 10^3$ N/mm^2
Poisson's coefficient	ν	0.15
Uniaxial compressive strength	f_c	40 N/mm^2
Uniaxial tensile strength	f_{ct}	3.53 N/mm^2
Open shear transfer coefficient	β_t	0.3
Closed shear transfer coefficient	β_c	0.9
Reduction of tension strength coefficient	T_c	0.6

cracking and crushing failure modes are taken into account through a smeared model and cut-off conditions complete the crisis surface.

The mechanical parameters adopted for the FE modeling of concrete are summarized in Table 1.

For what concern the steel reinforcement, the nonlinear stress-strain curve was inserted through the definition of a multi-linear material model. The CFRP was modeled as isotropic, linear and elastic with a Young's modulus of 350 GPa and a Poisson's ratio of 0.35. Finally, the behaviour of the epoxy resin was modeled by a nonlinear strain–stress curve with a Poisson's ratio of 0.29.

In order to reduce the computational time with a maintaining of results' accuracy, it was chosen to make thicker the mesh only in the bottom, near the groove and near the loading points. The numerical code recommends assuming a rectangular mesh to obtain suitable results from the Solid 65 element. So, the mesh was setup in order to obtain square or rectangular elements. Figure 10 contains an overview of the tridimensional model of the strengthened beams with the detail both of the absence of adhesive in two areas in beam RB1 (Fig. 10b) and of the damage due to notches in RB2 (Fig. 10c).

The FE model was analyzed statically and dynamically with the same steps as the experimental tests. A non-linear static analysis was carried out, by considering a progressive increase of loading. In order to obtain a results' accuracy and to avoid convergence problems, it was chosen to apply the load gradually, by breaking down the total load in many load increments called load-steps. In this way, the load was applied in a gradual way with the assignment for each load-step of a maximum and minimum numbers of the sub-steps. The criteria of convergence were set to be based on force and displacements. Initially the numerical code assigned the convergence tolerance limits by default; then it was decided to increase them with the aim to obtain the solutions' convergence.

The dynamic response of the strengthened beams with damage was investigated by a modal analysis performed for every FE beam model considering the presence of the suspension system made by elastic springs. The aim was to analyzed the dynamic response of the system through the evaluation of the frequency values and the modular forms as beams' damage degree varies.

The presence of the concrete's cracking was considered by the adoption of a variable Young's modulus evaluated for each damage degree. The value of the Young's

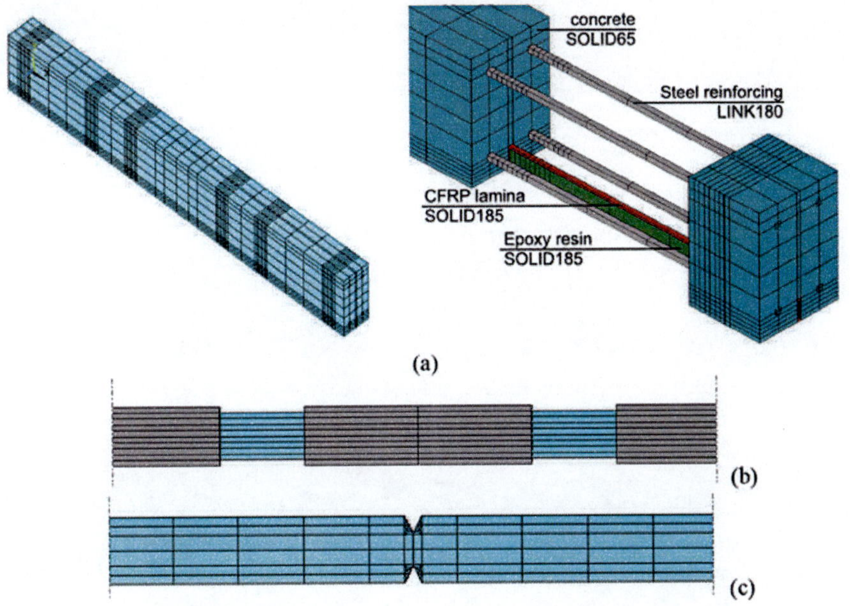

Fig. 10 **a** Typical mesh for FE analysis of RC beam with NSM CFRP rectangular rod; modeling of damage in **b** RB1 and **c** RB2

modulus of damaged concrete was determined using the expression of the frequency value obtained following the Euler–Bernoulli theory for a beam with both end free and assuming known the frequency from the dynamic experimentation.

3.2 Results and Discussion

The suitability of the numerical results was checked through a comparison with experimental data.

The first analysis and the relative comparisons were carried out on the load versus displacement curves, considering the displacements evaluated at the midspan section of beams. From the analysis of the comparison diagrams given in Fig. 11, it can be noted that the ultimate load and displacement given by the analytical model is in agreement with that from the experiments. This confirms the correct calibration, therefore the adequacy of the developed models for all type of beams. The final failure was indicated in the ANSYS program when the solution failed to converge. For beam RB1, the ultimate load varied slightly between the model and the real beam. The failure load predicted by the analysis is 36.05 kN which is good agreement with the experimental data of 36.44 kN. Finally, also the ultimate displacement is well

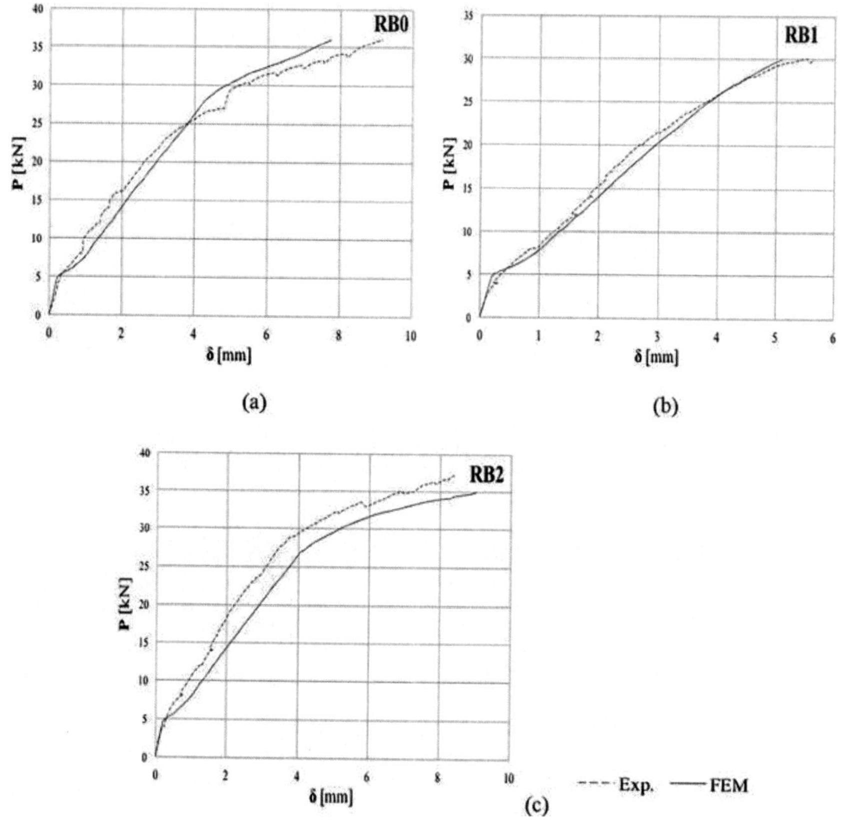

Fig. 11 Comparison between experimental and theoretical FEM diagrams load, P, versus displacements, δ

captured by the numerical model. The same comments can be done for the other beams.

The evolution of cracking has been analysed. The numerical code carries out a change of the stress–strain relations by introducing a plane of weakness in a direction normal to the crack face in order to take into account the presence of a crack at an integration point. The cracking is considered as a "smeared band" of cracks, rather than discrete cracks, by an adjustment of material properties. In particular, the crack propagation has been tracked graphically by the calculation of the principal stresses at the integration point. When the ultimate tensile strength of concrete is reached by a principal tensile stress, cracking signs appear by a circle in the plane of the crack. For what concern crushing, it appears when all the principal stresses are of compression and exceed the limit value. For beam RB2, the cracking load predicted by the FE analysis is equal to F = 6.65 kN which is in good agreement with the experimental results equal to F = 6.865 kN. The cracking development obtained by numerical FE

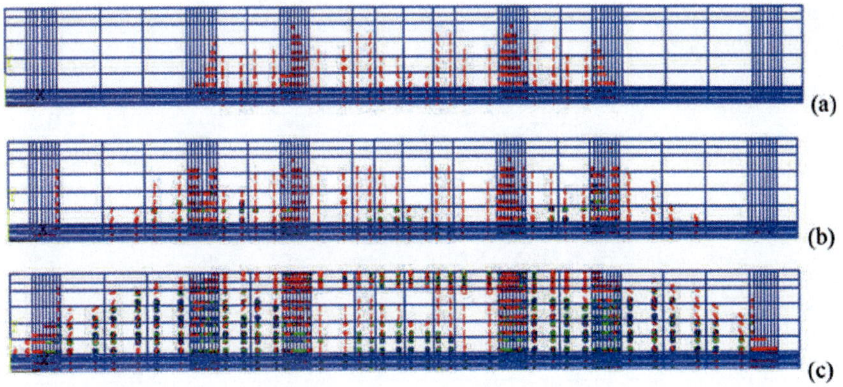

Fig. 12 Cracking pattern development in theoretical FE model **a** D_2, **b** D_4 and **c** D_6

model is given in Fig. 12 and it is possible to notice that it is consistent with what observed during the experimentation Fig. 3.

Finally, the first four natural frequencies of the beams were derived by modal analyses for each FE beam model. Figure 13 shows the first four bending vibration modes extracted for beam RB1. The natural frequency values were determined for each damage degree using a variable Young's modulus E of the concrete to consider damage degree due to cracking of the concrete beam as load cycles increase, as already described. The results obtained by FE numerical modeling were compared to the experimentally derived data in terms of frequency percent $\Delta f_r = 100\frac{f_{r,exp} - f_{r,FEM}}{f_{r,exp}}$ variations for the first modes r = 1, ..., 3 (Tab. 2). By analyzing the frequency variations for each damage steps it can be observed that the theoretical modeling gives results that is reasonable compared with the experimental evidences for all beam models with a percent deviation vary between 0.43 and 5.26% for the first mode. The variations between experimental and FE models are similar for all beams also for the other vibration modes, with maximum percent variation equal to 9.19%.

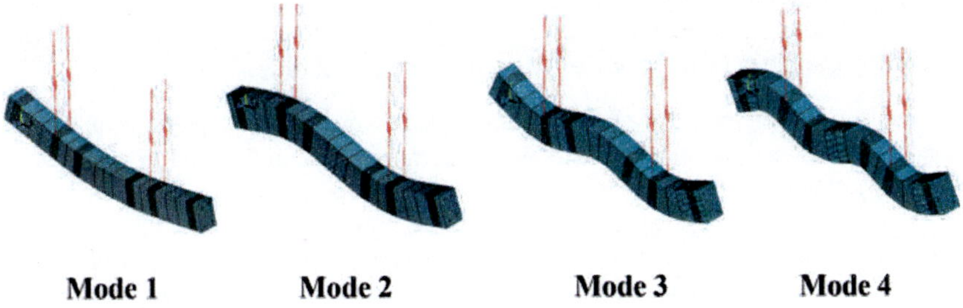

Fig. 13 View of first four vibration modes r = 1, ..., 4 by FE analysis

Table 2 Variation of frequency values between experimental and FE data for first three mode

	$\Delta f/f_{exp\%)}$								
	Mode 1			Mode 2			Mode 3		
	RB0	RB1	RB2	RB0	RB1	RB2	RB0	RB1	RB2
D_0	3.10	−1.58	3.11	3.31	0.60	2.50	1.81	2.71	1.76
D_1	5.26	−1.44	2.55	4.75	0.61	3.91	2.78	2.39	3.62
D_2	3.50	−0.43	4.26	6.83	2.49	6.32	8.26	3.47	6.77
D_3	2.63	−0.55	3.33	5.80	1.96	4.73	7.79	3.51	5.36
D_4	2.26	1.52	2.70	4.68	3.11	4.44	6.65	5.01	5.08
D_5	2.12	1.09	2.48	4.02	2.49	2.80	5.95	4.57	2.12
D_{5bis}	–	–	−4.16	–	–	−1.78	–	–	−2.97
D_6	−1.77	−0.77	−0.45	5.87	3.37	1.38	9.19	5.46	5.02

4 Conclusions

In this paper a research about the assessment of three beams in scale strengthened with NSM Carbon-FRP lamina in presence of damages due to cracking, local bond loss and notches of rod has been carried out by FE analysis. A non linear three dimensional FE model has been developed, considering both the static and dynamic response. Both progressive local failure and nonlinear capacity of concrete are taken into account. The developed analytical model is calibrated and validated by comparison with results of experimentation and the main conclusions may be summarized as follows:

- The modeling approach that consider a smeared crack model for concrete together with linear and non-linear properties, captures the experimental behaviour well both for elastic and plastic range. The experimental failure loads and load–displacement curves are satisfactorily reproduced with the adopted models;
- The evolution of the cracking pattern on concrete is well captured by the FE analysis given results in term of frequency values that are comparable with experimental data;
- Modal analysis carried out by numerical FE modeling considering the cracking of concrete gives results comparable with experimentation.

Results of this research suggests that advanced computational analyses can significantly give a contribution to the detection of damage of strengthened RC beams with NSM CFRP lamina.

Acknowledgements The Authors would thank everyone, technicians and students, who collaborated in the performing of the experimental campaign. Founds provided by Polytechnic University of Marche, Italy allowed the experimental research to take placed.

References

1. El-Hacha R, Rizkalla S (2004) Near-surface mounted fiber reinforced polymer reinforcements for flexural strengthening of concrete structures. ACI Struct J 101(5):717–726
2. De Lorenzis L, Teng JG (2007) Near-Surface mounted FRP reinforcement: an emerging technique for strengthening structures. Compos Part B Eng 38:119–143
3. Szabò ZK, Balazsw GL (2007) Near surface mounted FRP reinforcement for strengthening of concrete structures. Civil Eng 51(1):33–38
4. De Lorenzis L, Nanni A (2001) Shear strengthening of reinforced concrete beams with NSM fiber-reinforced polymer rods. ACI Struct J 98(1):60–68
5. Smith ST, Teng JP (2002) FRP strengthened RC beams. I: review of debonding strength models. J Eng Struct 24(2):385–395
6. Smith ST, Teng JP (2002) FRP strengthened RC beams. II: Assessment of debonding strength models. J Eng Struct 24(2):397–417
7. Ceroni F (2010) Experimental performances of RC beams strengthened with FRP materials. Constr Build Mater 24(9):1547–1559
8. Hassan T, Rizkalla S (2004) Bond mechanism of NSM FRP for flexural strengthening of concrete structures. ACI Struct J 101(6):830–839
9. Capozucca R (2013) Analysis of bond-slip effects in RC beams strengthened with NSM CFRP rods. Compos Struct 102:110–123
10. Capozucca R (2009) Static and dynamic response of damaged RC beams strengthened with NSM CFRP rods. Compos Struct 91(3):237–248
11. Capozucca R (2014a) On the strengthening of RC beams with near surface mounted GFRP rods. Comp. Struct. 117:143–155
12. Capozucca R, Bossoletti S (2014) Static and free vibration analysis of RC beams with NSM CFRP rectangular rods. Compos Part B Eng 67:95–110
13. Berthelot JM (1999) Composite materials: mechanical behaviour and structural analysis. Springer-Verlag, New York
14. Hassan NM, Batra RC (2008) Modelling damage in polymeric composites. J Compos B Eng 39:66–82
15. Maa R, Cheng J (2002) A CDM-based failure model for predicting strength of notched composite laminates. J Compos B Eng 33:479–489
16. Batra RC, Lear MH (2004) Simulation of brittle and ductile fractures in an impact loaded prenotched plate. Int J Fract 126:179–203
17. Capozucca R (2014b) Vibration of CFRP cantilever beam with damage. Comp Struct 116:211–222
18. Capozucca R, Bonci B (2014) Notched CFRP laminates under vibration. Comp Struct 122:367–375
19. Capozucca R, Bossoletti S, Montecchiani S (2015) Assessment of RC beams with NSM CFRP rectangular rods damaged by notches. Compos Struct 128:322–341
20. Rafi MM, Nadjai A, Ali F (2007) Analytical modeling of concrete beams reinforced with carbon FRP bars. J Compos Mater 41(22):2675–2690
21. Camata G, Spacone E, Zarnic R (2007) Experimental and nonlinear finite element studies of RC beams strengthened with FRP plates. Compos Part B: Eng 38(2):277–288
22. Buyle-Bodin F, David E, Ragneau E (2002) Finite element modelling of flexural behaviour of externally bonded CFRP reinforced concrete structures. Eng Struct 24(11):1423–1429
23. Kishi N, Zhang G, Mikami H (2005) Numerical cracking and debonding analysis of RC beams reinforced with FRP sheet. J Compos Constr 9(6):507–514
24. Soliman SM, El-Salakawy E, Benmokrane B (2010) Flexural behaviour of concrete beams strengthened with near surface mounted fibre reinforced polymer bars. Can J Civil Eng 37(10):1371–1382
25. Omran HY, El-Hacha R (2012) Nonlinear 3D finite element modeling of RC beams strengthened with prestressed NSM-CFRP strips. Costr Build Mater 31:74–85

26. ASTM D 3039/D 3039 M-08 (2008) Standard test method for tensile properties of polymer matrix composite materials. American Standard of Testing and Materials
27. ANSYS Inc (2019) ANSYS manual. Southpoint, USA
28. Desayi P, Krishnan S (1964) Equation for stress–strain curve of the concrete. ACI J 61(3):345–350
29. Willam KJ, Warnke EP (1974) Constitutive model for triaxial behaviour of concrete. Seminar on concrete structures subjected to triaxial stresses, international association of bridge and structural engineering conference. Italy, Bergamo, p 174

Particle Swarm Optimisation-Based Support Vector Regression Model to Estimate the Powder Factor of Explosives in Groundwater Tunnel Driving

E. de Miguel-García, K. Martín-Chinea, and J. F. Gómez-González

Abstract In many parts of the world, especially in arid or semi-arid areas, they find drinking water in the subsoil. One way to reach it, it is through small horizontal tunnels (Qanat) built on the mountain using explosives. Civil engineers must design projects where they estimate the budget necessary to undertake the work, taking into account the amount of explosives needed, number of blasts, duration of the civil work and powder factor among other data. However, there is not artificial intelligence-based models that help to forecast the amount of explosive needed to drill a tunnel. In this work, a hybrid regression model based on support vector machine (SVM) and particle swarm optimization (PSO) trained with real data (types of lithologies, geomechanical characteristics of the rocks and the amount of explosives used by engineers based on their previous experiences) obtained from a volcanic groundwater tunnel driving in the island of Tenerife (Spain), is proposed to predict the advance, the amount of explosives, the number of blasts and the powder factor in new tunnels or expansion of existing ones. The results show that a new, simpler regression model has been obtained that reproduces the experimental data and it will reduce the effort of the engineers in the study of a new tunnel driving work.

Keywords SVM · PSO · Tunnel · Blast · Powder factor · Lithology

E. de Miguel-García (✉)
Department of Agrarian, Nautical, Civil and Maritime Engineering, University of La Laguna, 38071 La Laguna, Tenerife, Spain
e-mail: emiguelg@ull.edu.es

K. Martín-Chinea · J. F. Gómez-González
Department of Industrial Engineering, University of La Laguna, 38071 La Laguna, Tenerife, Spain
e-mail: kmartinc@ull.edu.es

J. F. Gómez-González
e-mail: jfcgomez@ull.edu.es

M. Abdel Wahab (ed.), *Proceedings of the 8th International Conference on Fracture, Fatigue and Wear*, Lecture Notes in Mechanical Engineering,
https://doi.org/10.1007/978-981-15-9893-7_17

1 Introduction

A large part of the fresh water used is underground water and horizontal tunnels or wells are used to reach it [1]. These tunnels have their origin in the Qanat of the Middle East that goes back to the 8th century BC [1]. Nowadays, it is a common practice to use explosives to drive groundwater tunnels for what is necessary a civil engineering study that will define the drilling and blasting designs will be used, schedule the civil work and forecast the budget necessary.

The forecast of amount of explosive used in the blast and the powder factor (ratio between the kilograms of explosive used and the volume of extracted material) is important to determinate the duration and the budget of the civil work. Drilling and blasting designs (D&B) for small-section tunnels have been defined to parallel hole-cuts by other authors [2–4] where the geomechanical characteristics of the rocks are taken into account.

In this field, there have not been really new advances related with the develop of computational tools that help the engineer to forecast better the amount of explosive needed to drill a tunnel, according to the expected lithology along the tunnel and the D&B used. In general, the no-linear relations between the different variables (geo-mechanical properties of the lithology, powder factor, D&B, etc.) and the limited number of experimental measurements make difficult to find a simple mathematical relationship [5].

The aim of the present work is to combine of the Support Vector Regression (SVR) [6–9] with Particle Swarm Optimization (PSO) [10, 11] to find a non-linear regression among the rock parameters, the number of blasts and powder factors to drive a tunnel. SVR is a machine learning based on the theory of statistical learning used in different fields of science and engineering to have predictions in complex problems [7, 12–16]. To optimize the parameters of the SVR the algorithm PSO is applied. PSO is a metaheuristic algorithm that was developed based on social systems of the swarm theory inspired by the movement of flocks of birds [12]. Therefore, this hybrid PSO-SVR model helps the civil engineer to predict the duration of the driving and to estimate the human and material resources necessary to execute it. In this study, this methodology is applied to a groundwater tunnel in the island of Tenerife, Spain.

2 Materials and Methods

2.1 Characterization and Measurement of Lithological Data

Data used in the present work were recorded on a real civil work in an underground water tunnel in the island of Tenerife, located in the Canary Islands (Spain), in the North East Atlantic Ocean off the coast of Africa. The civil work aims to 85.75 m plus a tunnel already built of 4000 m in length by blasting using a gelatin-based explosive

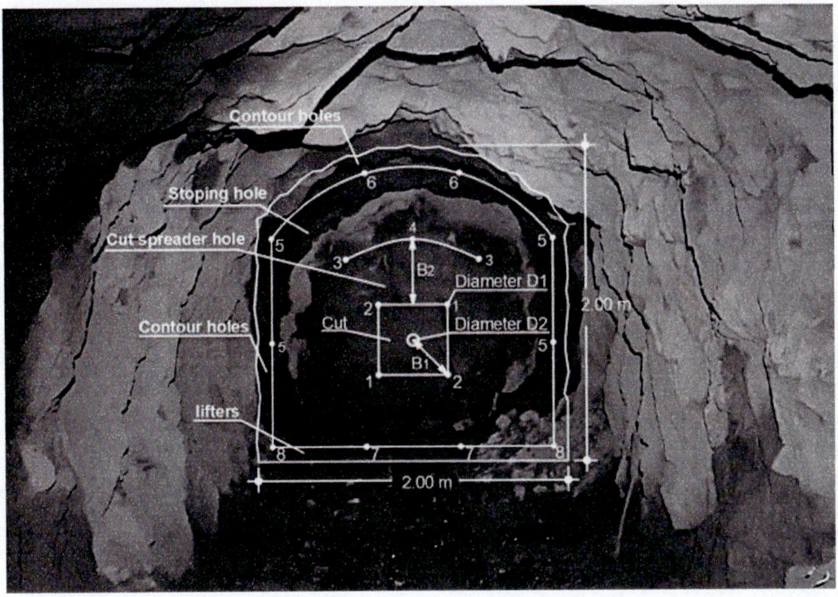

Fig. 1 Geometric drill pattern. **a** Example of a blast pattern with 17 boreholes used in the tunnel of the present work based on the blasting manual, Geological and Mining Institute of Spain. This photography is the front of a tunnel after a blast

for civil and mining purposes (RIODINTM, MAXAM Europe, S.A. Madrid, Spain). This explosive is selected because the impedance of the rock (the impedance of the rock is the product of its density by the propagation velocity of the wave [17]) is similar to the impedance of the explosive to have an optimal fragmentation [18].

The drilling and blasting designs followed the specifications described by Langefors and Kihlstrom [2], commonly applied to these hydraulic works in tunnels on the island of Tenerife. The charge calculation for tunnel blasting, when the cross section is small, is defined by the 'cut', 'cut spreader hole', 'contour hole' and the 'lifters' for a cut of four sections with parallel holes (Fig. 1). The drill plan and the charging and firing pattern are based on the spacing between boreholes as well as the linear load per borehole, taking into account the characteristics of the lithology. In the case described here, the tunnel technicians used three drilling and blasting designs for the tunnel with 14, 16 or 17 blast-holes depending on the type of rock [19].

The mean average advance is limited in this design by the deviation of the loaded boreholes. Given this deviation is below 2%, the estimated advance is 95% of the length of the boreholes [20]. Typically, the water tunnels in the island of Tenerife are perforations in straight line with a gradient of 2% and a section of about 3.94 m^2 with a tunnel height of 2.00 m, a side wall height of 1.80 m, a rise of the arch 0.20 m and a width of 2.00 m. The tunnels cross a wide variety of volcanic rocks as it was described in [5].

The amount of the explosive used, the advance and an undisturbed rock with suitable dimensions (larger than 40 cm) were recorded from each blast. The rocks were tested in a licensed laboratory (Laboratories and Quality of Construction of the Ministry of Public Works and Transport, Vice-Ministry of Infrastructure and Transport of the Government of the Canary Islands, Spain) to determinate their geological and geotechnical properties.

Samples (n = 48) were identified with eight different lithotypes (Fig. 2a): (a) aphanitic massive basalt, (b) altered and highly altered aphanitic massive basalt, (c) vacuolar aphanitic basalt (vacuole <0.05 mm), (d) vacuolar aphanitic basalt (vacuole <0.3 mm), (e) phonolite, (f) red colour ignimbrite, (g) highly altered red colour ignimbrite (vacuolar basaltic fragments), (h) agglomerate basaltic materials [21]. In order to characterize the eight lithotypes found, the following averaged parameters were calculated: point load strength index Is (MPa), density (g/cm^3) and open porosity P (%). To compare the values, data were normalized by the averaged maximum found in the eight lithologies (Fig. 2a). In addition, the Fig. 2b shows the averaged quantity of explosive and pull (advance) and the powder factor for each lithotypes.

2.2 Support Vector Regression

The support vector machine (SVM) [7, 22] for regression (SVR) is used to forecast the advance (pull) when it is known the geomechanical properties of the lithology and the kilograms of explosive needed to excavate a tunnel. SVR tries to found the hyperplane that is near as many data as possible minimizing the sum of the distances from the data points to the hyperplane based on statistical learning theory [16]. When the error ε between the predicted and the real values is assumed based on the experience, the used method is named the ε-SVR. In this article, the υ-support vector regression (υ-SVR) [23–25] a modification of ε-SVR [8] is used, where the parameter $\upsilon \in (0,1]$ controls the number of support vectors and the training errors [6, 25] minimising the error ε automatically. The SVR is implemented using the library LIBSVM [26].

The optimization problem of υ-SVR is defined as follows.

$$\min_{w,b,\xi,\xi^*,\varepsilon} \left\{ \frac{1}{2} \parallel \mathbf{w} \parallel^2 + C\left(\upsilon\varepsilon + \frac{1}{l}\sum_{i=1}^{l}(\xi_i + \xi_i^*) \right) \right\} \quad (1)$$

$$(\mathbf{w}^T\phi(\boldsymbol{x}_i) + b) - y_i \le \varepsilon + \xi_i$$

$$y_i - (\mathbf{w}^T\phi(\boldsymbol{x}_i) + b) \le \varepsilon + \xi_i^*$$

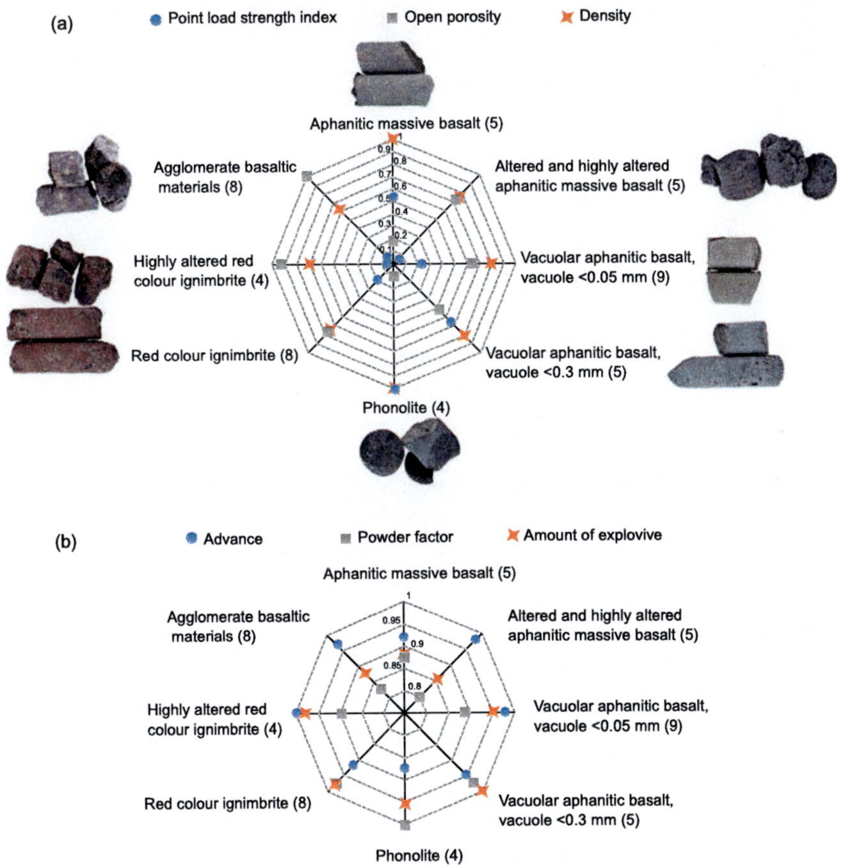

Fig. 2 Characteristic parameters for each lithotype and blasting information. Values are normalized by maximum averaged value for each parameter: point load strength index 5.84 MPa, density 2.63 g/cm^3, open porosity 37.70%, amount of explosive 6.63 kg, powder factor 1.86 kg/m^3 and advance 0.99 m. The number of samples are in parenthesis for each lithotype

$$\xi_i, \xi_i^* \geq 0, i = 1, ..., l, \varepsilon \geq 0$$

where, $\mathbf{x}_i \in R^n$ are the inputs, $y_i \in R^1$ is the target output, w are the support vectors, ϕ is nonlinear map function, C is the regularization parameter, ξ_i and ξ_i^* are slack variables that represent the distance from the value to the corresponding boundary values of ε-tube, and l is the number of training vectors.

Therefore, the dual problem is

$$\min_{\boldsymbol{\alpha},\boldsymbol{\alpha}^*} \left\{ \frac{1}{2}(\boldsymbol{\alpha} - \boldsymbol{\alpha}^*)^T \mathbf{K}(\boldsymbol{\alpha} - \boldsymbol{\alpha}^*) + \mathbf{y}^T(\boldsymbol{\alpha} - \boldsymbol{\alpha}^*) \right\} \tag{2}$$

$$\mathbf{e}^T(\boldsymbol{\alpha} - \boldsymbol{\alpha}^*) = 0, \ \ \mathbf{e}^T(\boldsymbol{\alpha} + \boldsymbol{\alpha}^*) \leq C\upsilon$$

$$0 \leq \alpha_i, \ \alpha_i^* \leq \frac{C}{l}, \ \ i = 1, ..., l$$

where α_i and α_i^* are the dual variables and K is the Kernel function ($K = K(x_i, x) = \phi(x_i)^T \phi(x_i)$). Then, the optimal solution is the function

$$f(x) = \sum_{i=1}^{l} (\alpha_i - \alpha_i^*)\mathbf{K}(x_i, x) + b \tag{3}$$

In the present work, the Kernel selected has been the radial basis function.

$$K_{i,j} = K(\boldsymbol{x_i}, \boldsymbol{x_j}) = exp(-\gamma|\boldsymbol{x_i} - \boldsymbol{x_j}|^2) \tag{4}$$

The values of C, υ and γ are determinate by the Particle Swarm Optimization (PSO) to have regression model that reduces the median absolute percentage error (MdAPE).

2.3 Particle Swarm Optimization for the SVR Model

The values of C, υ and γ for the υ-SVR model must be found in such a way that the MdAPE of the SVR model is minimal. This non-linear optimization problem is resolved with computational intelligence-based techniques. In this work, a meta-heuristic algorithm, particle swarm optimization (PSO) [11, 27] is used, based on the social behaviour of the movement of birds flock or swarm of bees [10, 11, 27–30].

Each particle i of the swarm has a position x_i defined in the 3-dimensional space (C, υ and γ), whose values are a possible solution to minimize the MdAPE of the SVR model. In an iterative way, each particle moves randomly towards the direction where it has found its best personal solution until that moment and in the direction where the best global position x_g has been found by any of the particles of the swarm. Therefore, all particles explore the region where the best solution has been found. This process follows until the stopping criterion is met.

In our case study, the parameters used to run the PSO are: the number of particles is 100, the ranges of each parameter are C ϵ [0.01, 1000], υ ϵ [0.001, 1] and γ ϵ [0.01, 10] and the maximum number of iterations is fixed to 50. Figure 3 shows a flow chart of the PSO-SVR set to obtain the prediction model.

The velocity v_i^t and position x_i^t of each particle i in each iteration t are updated with the following equations,

$$v_i^t = \omega \cdot v_i^{t-1} + \theta_l \cdot r_l \cdot (x_{il} - x_i^t) + \theta_g \cdot r_g \cdot (x_g - x_i^{t-1}) \tag{5}$$

$$x_i^t = x_i^{t-1} + v_i^t$$

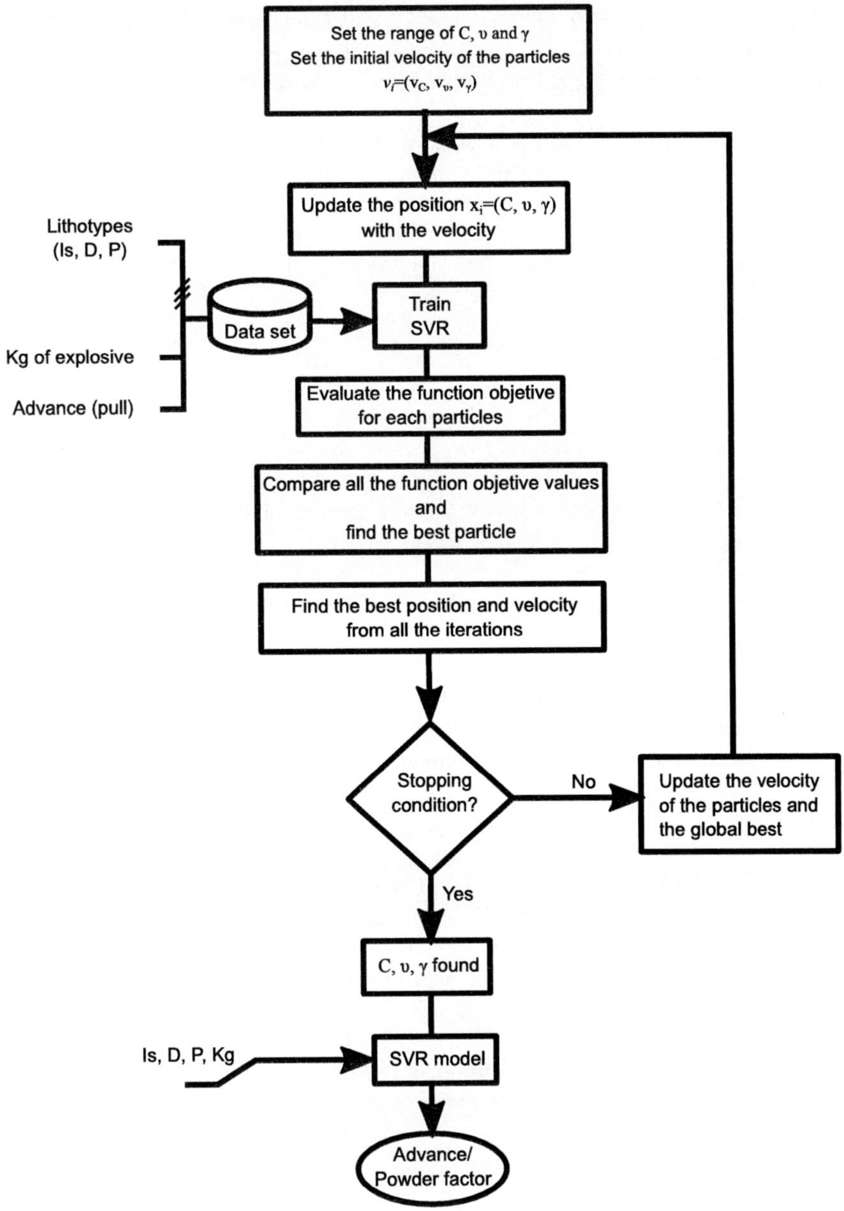

Fig. 3 Flowchart depicting the algorithm for obtaining the PSO-SVR model. The Flowchart shows the steps used in the Particle Swarm Optimization (PSO) algorithm to determine the C, υ and γ values that make the median absolute percentage error (MdAPE) of the SVR model to be minimal (stopping condition)

where r_l and r_g are uniform random numbers in the range between 0 and 1, ω is the inertial factor $\omega = 1/|2 - \varphi - \sqrt{\varphi^2 - 4\varphi}|$ with $\varphi = \varphi_1 + \varphi_2 \geq 4$ [27, 29]. θ_l is the correction factor to the local best $\theta_l = \omega\varphi_1$ and θ_g is for the global best $\theta_g = \omega\varphi_2$ with $\varphi_1 = \varphi_2 = 2.05$.

The stop condition is that the MdAPE of the global best does not change during Δk iterations. Therefore, "Stop" in iteration $t = k$ if

$$\sum_{t=k-\Delta k}^{k} |MdAPE_g^t - MdAPE_g^{t-1}| = 0 \quad (6)$$

with

$$MdAPE_g^t = \min_i \left\{ median \left(\left| \frac{y_{i,j}^t - y_{model,i,j}^t}{y_{i,j}^t} \right|, \forall j = 1, \ldots, n_i \right)_i, \forall i = 1, \ldots, n_i \right\}$$

where n_j and n_i are the number of test samples and the number of particles, respectively. In our case study, $n_j = 8$, $n_i = 100$ and $\Delta k = 5$.

2.4 Statistical Analysis

Table 1 shows the average values of point load strength index (Is), open porosity (P), hydrostatic balance density (D), amount of explosive (Ex) and the advance (Ad) for each type of rock. Therefore, if $X \epsilon \{Is, P, D, Ex, Ad\}$, the average value of each lithotype is $E_j(X_{j,i})$ with $i = 1, \ldots, n_j$, where n_j is the number of samples for the lithotype j. The maximum values are $X_{max} = \max\{E_j(X_{j,i})\}$. These maximum values, $\{X_{max}, X \epsilon \{Is, P, D, Ex, Ad\}\}$, will be used as base values to normalize the training data set, $x_{j,i}$, for the PSO-SVR algorithm. The normalized average values from each lithotype, $\{\overline{x}_j, j\epsilon \{8\ lithotypes\}\}$, are used for forecasting the advance, the powder factor and the number of blasts in computer-simulated tunnels where $x_{j,i} = X_{j,i}/X_{max}$ and $\overline{x}_j = E_j(X_{j,i})/X_{\max}$.

Figure 4 shows that the probability densities of data do not follow a normal distribution. Therefore, Weibull probability density functions, $f(X|a, b)$, were applied with the scale and shape parameters that are showed in Table 2, where

$$f(X|a, b) = \frac{b}{a}\left(\frac{X}{a}\right)^{b-1} exp\left\{-\left(\frac{X}{a}\right)^b\right\} \quad (7)$$

$$X \geq 0,\ a > 0\ (scale),\ b > 0\ (shape)$$

Table 1 Characteristic mean parameters for each lithotype. Is, Point load strength index; P, Open porosity; D, Hydrostatic balance density; Ex, Explosive, Ad, Advance

Lithotype (samples)	Is (MPa)	P (%)	D(g/ cm^3)	Ex (kg)	Ad (m)
Aphanitic massive basalt (5)	3.17	6.70	2.63	5.84	0.91
Altered and highly altered aphanitic massive basalt (5)	0.36	27.39	1.99	5.67	0.97
Vacuolar aphanitic basalt, vacuole <0.05 mm (9)	1.35	24.20	2.09	6.32	0.97
Vacuolar aphanitic basalt, vacuole <0.3 mm (6)	3.82	19.57	2.12	6.63	0.93
Phonolite (4)	5.84	3.78	2.60	6.32	0.86
Red colour ignimbrite (9)	1.06	28.81	1.96	6.49	0.91
Highly altered red colourignimbrite (4)	0.29	35.05	1.82	6.50	0.99
Agglomerate basaltic materials (8)	0.41	37.70	1.65	5.81	0.96
Maximum value	5.84	37.70	2.63	6.63	0.99

2.5 Simulation

A simulation application with the new PSO-SVR model and a user-friendly interface (Fig. 5a) (https://data.mendeley.com/datasets/276fsp2zv2/draft?a=b17cf3bb-70f6-43d8-9772-724ea3c0119b) has been implemented in GNU Octave [31]. The simulation methodology steps are an adaptation of the method defined by [5] and are:

1. Load files with the training data set and the test data set to the PSO-SVR algorithm (steps 1, 2 and 3 in Fig. 5a). The test data set is obtaining with the average values calculated for each lithology (Table 1). This test data is used as lithological information to simulate a tunnel (see next steps).
2. Run the PSO-SVR algorithm to compute the regression model (step 4). Figure 5b shows an example of the movement evolution of some swarm particles towards solving the optimization problem and the Fig. 5c shows as the MdAPE of the global best solution changes with the number of iterations. When the error does not change in last five iterations means that the solution has been found and the PSO-SVR model is created. Figure 5d shows expected advance for the training and test data set for the best PSO-SVR model.
3. Run simulations with the PSO-SVR model. Firstly, a computer-simulated tunnel is defined with information on the distribution of the lithological layers and their thickness. This information is loaded from a file (step 5a in Fig. 5a) and then, after model training, the simulation is ready to run (step 6). The values Is, P, D and Kg of explosive for each lithology are given in the test data set loaded in the step 2. In the area between two different lithotypes, the characteristics Is, P, D and Kg of explosive are taken according to the proportion in which they are found in the simulated drill. Finally, the simulation runs and the final results

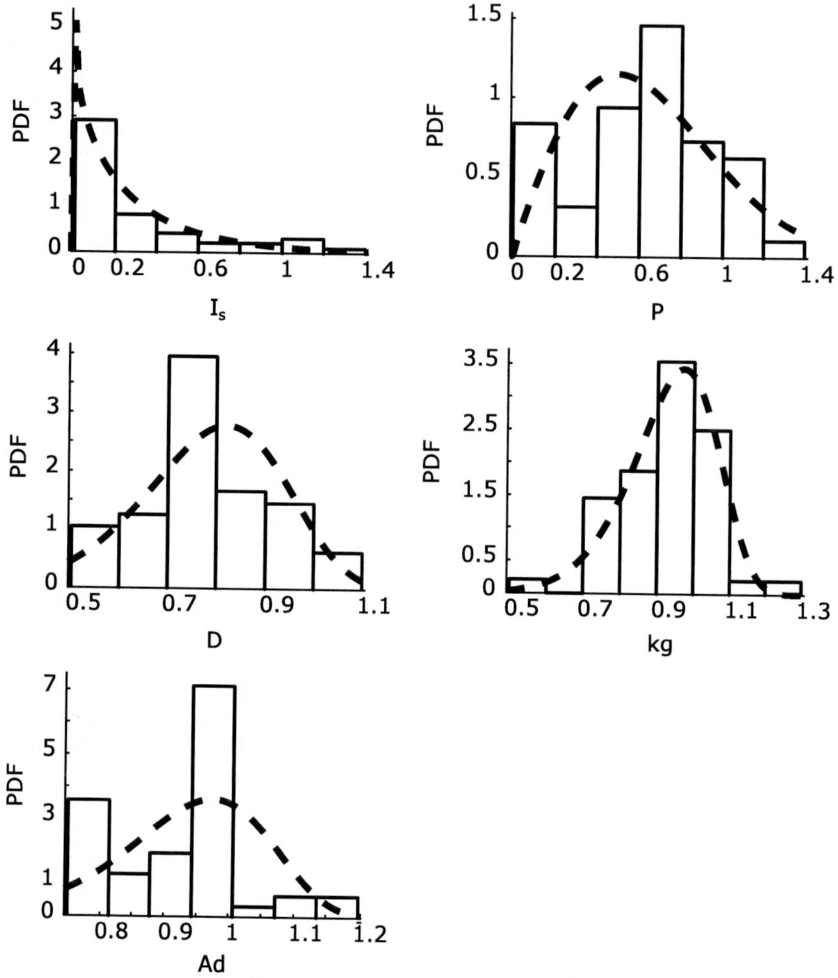

Fig. 4 The probability density functions (PDF) for the normalized data set used to train the PSO-SVR. The PDFs are adjusted with Weibull functions, which parameters are in Table 2

Table 2 Characteristic mean parameters for each lithotype

Parameters	a (scale)	b (shape)
Point load strength index (Is)	0.2924	0.9259
Open porosity (P)	0.7250	1.9181
Hydrostatic balance density (D)	0.8438	6.2883
Explosive	0.9847	9.1331
Advance	0.9857	9.6412

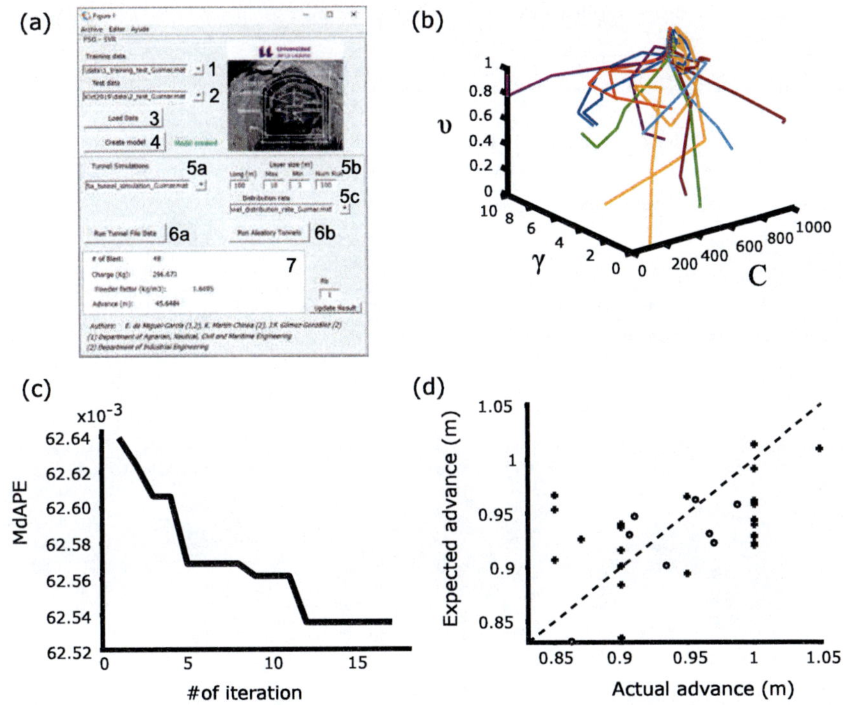

Fig. 5 PSO-SVR algorithm training and testing. **a** User interface of Octave application developed to compute the PSO-SVR model and to run the simulations. The numbers one to seven indicate the steps to follow to create the model, to run the model and to display results. See the text to have more information. **b** Movement of some particles of the PSO algorithm in the space (C, γ, υ) toward the solution. **c** The median absolute percentage error (MdAPE) of the SVR model decreases with the number of iteration of the PSO algorithm. **d** Expected advances versus actual advances are plotted for the training data (crosses) and the test data set (dots) using the PSO-SVR model

(number of blast, the amount of explosive (kg), the advance (m) and the powder factor (kg/m^3)) are displayed on a panel (step 7).

If the lithological distribution of a tunnel is unknown, a probable solution will be calculated using a set of computer-simulated tunnels. In this case, the needed parameters to build the computer-simulated tunnels are the length of the tunnel, the maximum and the minimum layer sizes and number of tunnels to do the statistical study (step 5b). The characteristics Is, P, D and Kg of explosive for each lithology is taken from the test data set (step 2) and the percentage of the distribution expected to be found of each lithology defined in the test data, has to be loaded from a file (step 5c). In our study case, the percentage of each lithology is given in Table 3. After providing these parameters, the calculation can be executed (step 6b) and the results will be displayed on the panel (step 7). These results are computed as the average of the individual results of all tunnels. Figure 6a shows three examples of computer-

Table 3 Real lithotype distribution of the Güimar tunnel for the simulation using a computer-designed tunnel

Lithotype	Total length (m)	Probability (%)
Aphanitic massive basalt	4.55	10
Altered and highly altered aphanitic massive basalt	4.85	11
Vacuolar aphanitic basalt vacuole <0.05 mm	8.70	19
Vacuolar aphanitic basalt, vacuole <0.3 mm	4.67	10
Phonolite	3.45	8
Red colour ignimbrite	7.25	16
Highly altered red colour ignimbrite	3.95	9
Agglomerate basaltic materials	7.65	17

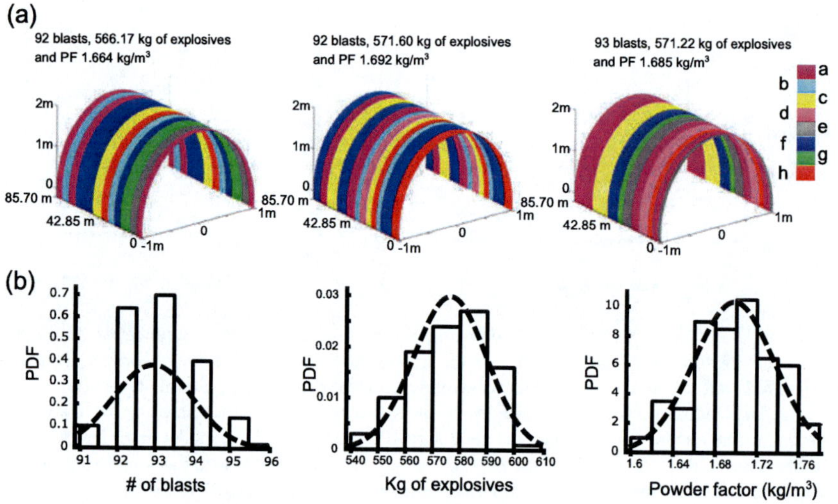

Fig. 6 Computer-simulated tunnels. **a** Three examples of computer-simulated tunnels with different lithologies used to computed the most probable value for the number of blasts, the quantity of explosives and the powder factor for a tunnel of 85.70 m. The colour bar for lithotypes: (a) aphanitic massive basalt, (b) altered and highly altered aphanitic massive basalt, (c) vacuolar aphanitic basalt, vacuole <0.05 mm, (d) vacuolar aphanitic basalt, vacuole <0.3 mm, (e), phonolite, (f) red colour ignimbrite, (g) highly altered red colour ignimbrite and (h) agglomerate basaltic materials. **b** The probability density functions (PDF) and the Gaussian adjustments of the number of blasts, the quantity of explosives and the powder factor calculated using 100 tunnels

simulated tunnels and the probability density function for the amount of explosive (kg), number of blasts and the powder factor (kg/m^3) calculated using 100 tunnels.

Simulations and data analysis were performed using the application of Excel spreadsheets (Microsoft® Office), GNU Octave [31] and MATLAB® (Mathworks Inc., Natick, MS, USA). The SVM model is implemented with the library LIBSVM [26].

3 Results and Discussion

3.1 Relationship Between Lithotypes and Powder Factor

A representative image of the eight lithotypes extracted from the stretch of 85.70 m, at a distance of 4000 m from the entrance of the tunnel, are showed in Fig. 2a. The lithotypes with their geomechanical features (point load strength index Is (MPa), open porosity P (%) and hydrostatic balance density D (g/cm^3)) are shown in Fig. 2a where the values correspond to the mean values normalized by maximum averaged value.

Figure 2b shows the normalized mean explosive charge used, the normalized mean advance and the normalized mean powder factor (relationship between how much explosive (kg) and volume of rock broken (m^3)) for each lithotype. The normalizations were done by the maximum averaged values (to see Table 1).

Data show that the relationship between the powder factor and the geomechanical features of each lithotype is not straightforward. In fact, Fig. 2 shows that highly altered red colour ignimbrite has a lower Is, a lower D and a higher P than phonolite, but their powder factors are similar. Another example is the case of altered and highly altered aphanitic massive basalt and red colour ignimbrite that have similar P and D, however, they have a different powder factor. Therefore, non-linear relationships between statistical values between the characteristics of the lithotypes and the powder factor should be sought as it was done by [5].

In this work, it is proposed that methods used in Machine Learning and Data Science, such as SVR and PSO, could be successful in finding a regression model that provides the powder factor if the geomechanical characteristics of the rock and the usual explosive charge applied by technicians and/or engineers are known.

3.2 Forecasting of the Advance and the Powder Factor with a PSO-SVR Model

In our case study, to 85.70 m in a real underground water volcanic tunnel, not only blasts were used, but also excavations by an overshot mucker and pneumatic breakers, since they did not need explosives. Therefore, a percentage of length perforated by explosives, Rb, is defined. In our case, this value is Rb = 52.59%, 45.07 m pulled with explosives of the 85.70 m driven.

Experimentally, 48 blasts 297.25 kg of explosive with a powder factor of 1.67 kg/m^3 were needed to 45.07 m using a drill of 1.2 m. Theoretically, the commonly used formula defined by Langefors and Kihlstrom (Langefors and Kihlstrom, 1973) applied to the design of small tunnel drilling, forecasts 40 blasts assuming that the advance is 95% of the drilling length (1.2 m). Where the traditional formulation assumes a constant of the rock 'c' of 0.4 and estimates that the advance is 1.14 m per pull. Hence, the difference between the actual and the predicted values could be due to the specific geomechanical characteristics of the rock that are not taken into account in each blast. When a tunnel designed by computer with the same type, order and length of rock slices presented in the Table 3 and taken account the characteristics of Is, P, D and Ex given in Table 1, is applied to the PSO-SVR model the results are: 49 blasts, 302.58 kg of explosive with a powder factor of 1.69 kg/m^3. Therefore, the number of blasts predicted by the PSO-SVR model is significantly close to that obtained in the real civil work. In addition, the amount of explosives and the powder factor predicted by the model are very close to the real ones.

If the study is carried out for a single type of rock (Tables 4 and 5), the estimated actual advances and powder factors for different lithologies can be compared with those predicted by the PSO-SVR model and the formulation of Langefors and Kihlstrom (L-K). In all the cases, the results show that the PSO-SVR model gives the best estimations with a relative deviation of the estimated that goes between 0.57% and 7.91%, however the L-K goes between 12.29% and 23.47%.

The probability of distribution of the different lithologies in a tunnel (see Table 3 case of Güimar tunnel) can be estimated with the rock samples taken. Hence, the amount of explosives and the powder factor needed to drive a tunnel can be predicted using the lithological distribution, the PSO-SVR model and the percentage of driving expected by explosives. Because the order in which the different layers are distributed and their thicknesses are unknown, multiple computer-simulated tunnels are made to obtain statistically significant values. Where each tunnel is constructed with randomly arranged layers following the probabilistic distribution of the experimentally obtained lithology (Table 3) with randomly chosen thicknesses (for example, between 3 and 10 m in our study case). Table 6 shows the results obtained using 100, 1000 and 10,000 simulations for 85.70 m tunnel. These results are similar to the experimental

Table 4 Comparison of the actual estimated advances for different lithologies with the predicted advances by PSO-SVR model and the formulation of Langefors and Kihlstrom. The average values for the advance are shown for each blast as function of the lithology and amount of explosive per blast used in the study case to drive the tunnel. Ex, Explosive; Ad, Advance; PSO-SVR, Hybrid model based on Particle Swarm Optimization and Support Vector Regression; L&K, forecast based on the formulation of Langefors and Kihlstrom

Lithology	Actual		PSO-SVR	L&K
	Ex (kg)	Ad (m)	Ad (m)	Ad (m)
Aphanitic massive basalt	5.84	0.91	0.95	1.14
Altered and highly altered aphanitic massive basalt	5.67	0.97	0.92	1.14
Vacuolar aphanitic basalt, vacuole <0.05 mm	6.32	0.97	0.93	1.14
Vacuolar aphanitic basalt, vacuole <0.3 mm	6.63	0.93	0.90	1.14
Phonolite	6.33	0.86	0.83	1.14
Red colour ignimbrite	6.49	0.91	0.93	1.14
Highly altered red colour ignimbrite	6.50	0.99	0.96	1.14
Agglomerate basaltic materials	5.81	0.96	0.96	1.14

Table 5 Comparison of the actual estimated powder factors for different lithologies with the predicted powder factors by PSO-SVR model and the formulation of Langefors and Kihlstrom. The average values for the powder factor are shown for each blast as function of the lithology and amount of explosive per blast used in the study case to drive the tunnel. PF, powder factor; PFDev, relative deviation of the estimated PF versus the experimental PF; PSO-SVR, Hybrid model based on Particle Swarm Optimization and Support Vector Regression; L&K, forecast based on the formulation of Langefors and Kihlstrom

Lithology	Actual	PSO-SVR		L&K	
	PF (kg/m^3)	PF (kg/m^3)	PFDev (%)	PF (kg/m^3)	PFDev (%)
Aphanitic massive basalt	1.63	1.57	3.96	1.30	20.23
Altered and highly altered aphanitic massive basalt	1.48	1.56	5.43	1.26	14.71
Vacuolar aphanitic basalt, vacuole <0.05 mm	1.66	1.72	3.87	1.41	15.21
Vacuolar aphanitic basalt, vacuole <0.3 mm	1.73	1.87	7.91	1.48	14.68
Phonolite	1.84	1.93	5.03	1.41	23.47
Red colour ignimbrite	1.74	1.77	1.77	1.44	16.99
Highly altered red colour ignimbrite	1.65	1.72	4.35	1.45	12.29
Agglomerate basaltic materials	1.54	1.53	0.57	1.29	16.06

Table 6 Computation of the number of blasts and the powder factor using a computer-simulated tunnel with a lithological distribution based on that found in a real tunnel and the PSO-SVR model estimated with raw data set. Long, length of the tunnel to drive; Ex, Explosive; PF, powder factor; Rb is the percentage of tunnel to drive using explosive. For the simulation the type of lithology, arrangement and lengths of the rock slice are chosen randomly following the probabilistic distribution found in the Güimar tunnel. The slice length is between 3 and 10 m

# of runs	Long (m)	Rb (%)	# of blasts	Ex (kg)	PF (kg/m^3)
100	85.70	52.59	49	301.5	1.69
1000	85.70	52.59	49	303.17	1.70
10000	85.70	52.59	49	303.2	1.70
Real data	85.70	52.59	48	297.25	1.65

ones and there are no differences between the numbers of simulations used. Figure 6b shows the results obtaining for a simulation with 100 tunnels 85.70 m where the probability density function (PDF) of the number of blasts, kilograms of explosives and the powder factor have a Gaussian probability density functions. Therefore, from this Gaussian PDF is able to obtain the more probable prediction of number for blasts (93) and total of kilograms of explosives (576.4 kg) needed to drive the tunnel.

Table 7 Computation of the number of blasts and the powder factor using a computer-simulated tunnel with a lithological distribution based on that found in a real tunnel and the PSO-SVR model estimated with average data set (Table 1). Long, length of the tunnel to drive; Ex, Explosive; PF, powder factor; Rb is the percentage of tunnel to drive using explosive. For the simulation the type of lithology, arrangement and lengths of the rock slice are chosen randomly following the probabilistic distribution found in the Güimar tunnel. The slice length is between 3 and 10 m

# of runs	Long (m)	Rb (%)	# of blasts	Ex (kg)	PF (kg/m^3)
100	85.70	52.59	48	295.19	1.65
1000	85.70	52.59	48	296.04	1.66
10000	85.70	52.59	48	296.09	1.66
Real data	85.70	52.59	48	297.25	1.65

3.3 PSO-SVR Model Trained with a Small Average Data Set

In a new civil work where a survey cannot be carried out to obtain information on the lithological distribution, a PSO-SVR model could be obtained if the following information is available: the average values of the geomechanical characteristics of the existing lithologies in the area, the probability to find each type of lithology (obtained by geological studies of the area, probes or nearby tunnels) and the amount of explosives that is generally used in each explosion depending on the type of rock. In this case, although, the number of data is small, it is possible to obtain a PSO-SVR regression model to make predictions. To verify that a PSO-SVR model can be obtained, we suppose that the training data set is the average values Is, P, D, Ex and Probability given in Tables 1 and 3. Table 7 shows predictions for computer-designed tunnels build randomly using the PSO-SVR model and the computed results using the average data set reproduce significantly the experimental ones.

4 Conclusion

Engineers must design civil work projects for the construction of new tunnels or increase the length of existing ones, as in our case study. In these civil engineering projects, the duration of civil works, the number of blasts, the amount of explosives required, etc. should be evaluated to better estimate the necessary project budget.

Until now, the formulation of Langefors and Kihlstrom had been one of the best known methods to estimate the advance and the powder factor. However, as the results showed, their predictions were far from the actual data that had been obtained in our case study. The need for a better estimate of the 'c' parameter greatly influences this prediction. In the present study, it has been shown that a SVR-based model can give a good prediction of the advance and the powder factor without the need to develop a complex formulation. To set the parameters of the SVR, a PSO algorithm was applied. Therefore, this study proposes a new tool (a hybrid PSO-SVR model) that will help engineers develop their projects giving predictions closer to reality about the amount of explosives, the number of blasts needed and the duration of the civil works.

Acknowledgements The authors would like to thank the owners of the La Paloma water tunnel and the Service of Laboratories and Quality of Construction of the Ministry of Public Works and Transport, Vice-Ministry of Infrastructure and Transport of the Government of the Canary Islands, Spain for their collaboration in this research work.
Conflict of Interest Statement. None declared.

References

1. Nasiri F, Mafakheri MS (2015) Qanat water supply systems: a revisit of sustainability perspectives. Environ Syst Res 4(13):1–5
2. Langefors U, Kihlstrom B (1973) The modern technique of rock blasting, 2nd edn. Wiley, New York. https://trove.nla.gov.au/work/10826003
3. Persson PA, Holmberg R, Lee J (1993) Rock and explosives engineering. CRC Press, New York
4. Holmberg R (1982) Charge calculations for tunneling. In: Hustrulid WA (ed) Underground mining methods handbook, SME, pp 1580–1589. U.S. Department of Energy Office of Scientific and Technical Information
5. de Miguel-García E, Gómez-González J (2019) A new methodology to estimate the powder factor of explosives considering the different lithologies of volcanic lands: a case study from the island of Tenerife, Spain. Tunnell Underground Space Technol 91:103023. https://www.sciencedirect.com/science/article/pii/S088677981830659X?via%3Dihub
6. Smola AJ, Schölkopf B (2004) A tutorial on support vector regression. In: Statistics and computing. Kluwer Academic Publishers, pp 199–222
7. Vapnik V, Golowich SE (1996) Support vector method for function approximation, regression estimation, and signal processing. Advances in neural information processing systems, pp 281–287 (1996)
8. Vapnik VN(1995) The nature of statistical learning theory. Springer New York. http://link.springer.com/10.1007/978-1-4757-2440-0
9. Boser BE, Guyon IM, Vapnik VN (1992) A training algorithm for optimal margin classifiers. In: Proceedings of the annual conference on computational learning theory, pp 144–152. ACM Press, Pittsburgh, PA. http://www.svms.org/training/BOGV92.pdf
10. Gaing ZL (2004) A particle swarm optimization approach for optimum design of PID controller in AVR system. IEEE Trans Energy Convers
11. Eberhart R, Kennedy J (1995) A new optimizer using particle swarm theory. In: Sixth international symposium on micro machine and human science. IEEE, New York, pp 39–43
12. Li Q, Fu Y, Zhou X, Xu Y (2010) A hybrid support vector regression for time series prediction. In: 3rd international conference on knowledge discovery and data mining, WKDD 2010, pp 506–509 (2010)
13. Liu S, Tai H, Ding Q, Li D, Xu L, Wei Y (2013) A hybrid approach of support vector regression with genetic algorithm optimization for aquaculture water quality prediction. Math Comput Modell 58:458–465
14. Kaneda Y, Mineno H (2016) Sliding window-based support vector regression for predicting micrometeorological data. Expert Syst Appl
15. Shamshirband S, Petkovic D, Javidnia H, Gani A (2015) Sensor data fusion by support vector regression methodology - A comparative study. IEEE Sens J 15:850–854
16. Trafalis T, Ince H (2000) Support vector machine for regression and applications to financial forecasting. IEEE, New York, pp 348–353

17. Zukas JA, Walters WP (1998) Explosive effects and applications. Springer, New York
18. International Society of Explosives Engineers (1998) The Blasters' Handbook. ISEE, Cleveland, USA
19. Lopez Jimeno C, Lopez Jimeno E, Ayala Carcedo FJ, Visser de Ramiro Y (2017) Drilling and blasting of rocks. Taylor & Francis, London. http://www.worldcat.org/title/drilling-and-blasting-of-rocks/oclc/1020789548
20. Jimeno CL, Bermúdez PG (2003) Avances tecnológicos en el campo de la perforación y voladura de rocas. Ingeopress, p 7
21. Füster JM, Páez A, Sagredo J (1969) Significance of basic and ultramafic rock inclusions in the basalts of Canary Islands. Bulletin Volcanologique 33(3):665–693
22. Cortes C, Vapnik V (1995) Support-vector networks. Mach Learn 20:273–297
23. Chang CC, Lin CJ (2002) Training v-support vector regression: theory and algorithms. Neural Comput 14(8):1959–1977
24. Tan M, Song X, Yang X, Wu Q (2015) Support-vector-regression machine technology for total organic carbon content prediction from wireline logs in organic shale: a comparative study. J Nat Gas Sci Eng 26:792–802
25. Schölkopf B, Bartlett P, Smola A, Williamson R, First GMD, Chaussee R (1999) Shrinking the tube: a new support vector regression algorithm. In: Advances in neural information processing systems, pp 330–336
26. Chih-Chung C, Chih-Jen L (2011) LIBSVM : a library for support vector machines. ACM Trans Intell Syst Technol 2:1–27. http://www.csie.ntu.edu.tw/~cjlin/libsvm
27. Clerc M, Kennedy J (2002) The particle swarm-explosion, stability, and convergence in a multidimensional complex space. IEEE Trans Evol Comput 6(1):58–73
28. van den Bergh F, Engelbrecht AP (2004) A cooperative approach to participle swarm optimization. IEEE Trans Evol Comput 8(3):225–239
29. del Valle Y, Venayagamoorthy GK, Mohagheghi S, Hernandez JC, Harley RG (2008) Particle swarm optimization: basic concepts, variants and applications in power systems
30. Varthanan PA, Murugan N, Kumar GM (2012) A simulation based heuristic discrete particle swarm algorithm for generating integrated production-distribution plan. Appl Soft Comput J 12:3034–3050
31. Eaton JW, Bateman D, Hauberg S, Rik W (2019) GNU Octave version 5.1.0 manual: a high-level interactive language for numerical computations. https://www.gnu.org/software/octave/doc/v5.1.0/

Study of Pre-strain Effect on Bendability in Three-Point Bending Test

M. M. Shahzamanian, Amir Partovi, and P. D. Wu

Abstract The bendability of sheet metals can be estimated by the three-point bending test. Several factors and parameters such as pre-strain, can affect the bendability of sheet metals. This effect is investigated using the Gurson–Tvergaard–Needleman model through finite element method. Bending strain at the initiation of a fracture is calculated to assess bendability. Pre-strain is observed to reduce maximum force in the bending test and cause early fracturing.

Keywords Three-point bending test · Bendability · Pre-strain · Finite element method (FEM)

1 Introduction

The three-point bending test is a simple method that provides out-of-plane bending to investigate the tensile and compressive properties of various materials [1]. According to [2], failure occurs at the outer surface of a tensile test specimen under bending and the true strain at the outer surface equals that at the fracture point. The surface topography, particle cracking, and voiding of a material are developed during bending. The surface roughening effect on the outer surface of a sheet being bent can be studied using the finite element method (FEM), which was developed by Shi et al. [3]. Grains on the top surface layers of a sheet have a significant effect in controlling the roughness of the outer surface. A void coalescence model was used in [4] to evaluate the bendability of sheet metals. The Gurson–Tvergaard–Needleman (GTN) model, in which void growth and coalescence are considered, was used in [5, 6] to investigate ductile fractures in sheet metals in a bending test. Three-point bending test experimentation and finite element (FE) simulation of 6063 aluminum alloy sheets were performed under different heat treatment conditions in [7], and the results indicated that FEM could accurately describe the bending response of these sheets. In [8], the GTN model was used to simulate the effect of cladding on bendability through

M. M. Shahzamanian (✉) · A. Partovi · P. D. Wu
Department of Mechanical Engineering, McMaster University, Hamilton, ON L8S 4L7, Canada
e-mail: mmshahzamanian@gmail.com; Shahzamm@mcmaster.ca

M. Abdel Wahab (ed.), *Proceedings of the 8th International Conference on Fracture, Fatigue and Wear*, Lecture Notes in Mechanical Engineering,
https://doi.org/10.1007/978-981-15-9893-7_18

FEM. An analytical approach for a sandwich beam with aluminum face sheets and a metal foam core under three-point bending was presented in [9]. The analytical load–deflection results obtained using experiments and FEM were comparable.

Several parameters and factors, such as pre-strain, affect bendability in three-point bending tests. Sarkar et al. [10] studied the influence of pre-strain on the tensile and bending properties of AA5754 aluminum alloys and found that the true stress–strain curve increases while the stress gradient decreases with increasing pre-strain percentage. However, this process causes fractures to occur early in tensile tests. Moreover, bendability decreases with increasing pre-strain percentage. The pre-strain effect on aluminum sheets under bending was studied in [11], and results showed that bendability decreases with increasing pre-strain percentage. The objective of the present study is to carry out a numerical investigation on the effect of pre-strain on sheet metal fractures under bending. All simulations presented in this study are conducted using ABAQUS/Explicit [12] with the GTN model. The effect of pre-strain on bendability is also explained in detail. Excellent agreement between the results obtained numerically and those obtained experimentally is found in this study.

2 Constitutive Model

This study uses the GTN model [13–16], which considers the effects of void growth, nucleation, and coalescence to describe damage in metals. The GTN model is a modification of the Gurson model [17], which does not consider nucleation and coalescence. The relationship between void growth and plastic strain rate ($\mathbf{D}^P$) is as follows:

$$(\dot{f})_{growth} = (1 - f)\mathbf{I} : \mathbf{D}^P \tag{1}$$

The relationship between strain-controlled void nucleation and effective plastic strain rate ($\dot{\bar{\varepsilon}}^P$) is shown in Eq. (2).

$$\dot{f} = \bar{A}\,\dot{\bar{\varepsilon}}^P \tag{2}$$

Chu and Needleman suggested that the process of nucleation follows a normal distribution [18]. This normal distribution may be described according to the plastic strain as follows:

$$\bar{A} = \frac{f_N}{S_N\sqrt{2\pi}} exp\left[-\frac{1}{2}\left(\frac{\bar{\varepsilon}^P - \varepsilon_N}{S_N}\right)^2\right] \tag{3}$$

where ε_N is the average void nucleating strain, f_N is the volume fraction of void nucleating particles, and S_N is the standard deviation of the void nucleating strain.

The total void volume fraction may be calculated as follows:

$$\dot{f} = (\dot{f})_{growth} + (\dot{f})_{nucleation} \tag{4}$$

The effect of coalescence in the GTN model is considered through the following relationship:

$$f^* = \begin{cases} f \quad for \quad f \le f_c \\ f_c + \frac{f_u^* - f_c}{f_f - f_c}(f - f_c) \quad for \quad f > f_c \end{cases} \tag{5}$$

Coalescence and final failure occur when f reaches f_c and f_f, respectively. Furthermore, $f_u^* = \frac{1}{q_1}$. Void growth and nucleation in ABAQUS occur only when an element is under tension. The approximate yield function used in the GTN model to predict when plasticity will occur is:

$$\Phi(\sigma, \tilde{\sigma}, f) = \frac{\sigma_e^2}{\tilde{\sigma}^2} + 2f^* q_1 cosh\left(\frac{3q_2\sigma_H}{2\tilde{\sigma}}\right) - \left[1.0 + (q_1 f^*)^2\right] = 0 \tag{6}$$

where $\boldsymbol{\sigma}$ is the macroscopic Cauchy stress tensor; σ_e, σ_H, and $\tilde{\sigma}$ are the equivalent stress, hydrostatic stress, and matrix stress, respectively; and q_1 and q_2 are fitting parameters. The matrix stress–strain curve, which is undamaged, is as follows:

$$\bar{\varepsilon} = \begin{cases} \frac{\tilde{\sigma}}{E}, \quad for \quad \tilde{\sigma} \le \sigma_y \\ \frac{\sigma_y}{E}\left(\frac{\tilde{\sigma}}{\sigma_y}\right)^n, \quad for \quad \tilde{\sigma} > \sigma_y \end{cases} \tag{7}$$

where $\bar{\varepsilon}$, E, n, and σ_y are the equivalent strain, Young's modulus, hardening parameter, and yield stress, respectively.

3 Problem Formulation and Method of Solution

A two-dimensional (2D) schematic configuration of a sheet metal with length L_o and thickness t_o is shown in Fig. 1. The sheet considered is adequately wide with no deformations in the width direction to enable the accurate determination of the plane strain state. The punch and mandrel are rigid bodies with radii R_p and R_m, respectively. The mandrel, with a length span of L_m, is stationary, and the punch inserts force to the middle section of the sheet. The effect of pre-strain on bendability is examined via ABAQUS in three steps. First, a tensile load is applied to the X direction of the sheet. Second, the sheet is unloaded. Finally, the punch advances and deforms the sheet.

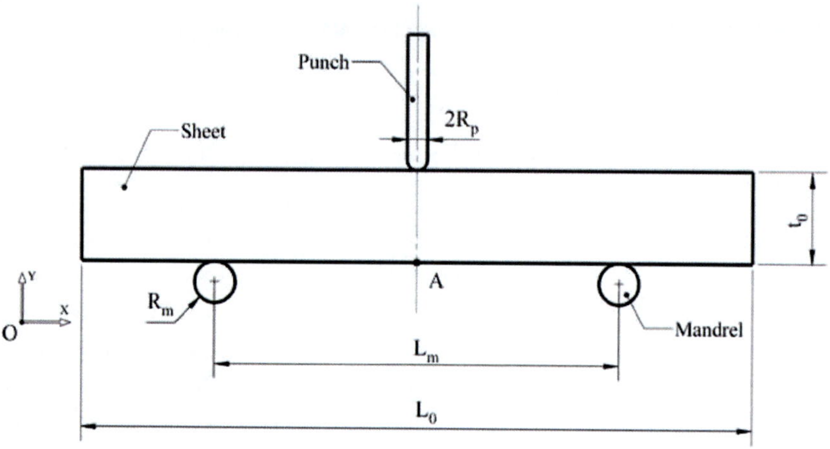

Fig. 1 2D schematics of three-point bending test

4 Results and Discussion

The matrix material stress–strain curve constants are: $\sigma_y/E = 0.0033$, $\nu = 0.3$, and $n = 10$. The initial void volume fraction is considered to be zero, and the fitting parameters of the GTN model in Eq. (6) are $q_1 = 1.5$ and $q_2 = 1.0$. The nucleation void volume fraction is $f_N = 0.04$, the mean strain for nucleation is $\varepsilon_N = 0.3$, the corresponding standard deviation is $S_N = 0.1$, and f_c and f_f are 0.15 and 0.25, respectively. These mechanical property values for the GTN model are taken from Tvergaard and Needleman [13]. Note that the results of this study are not dependent on these parameters because the main aim is to understand the effect of pre-strain on bendability through a numerical analysis.

ABAQUS/Explicit includes several important elements for the current analysis. The CPE4R element, which is a fully reduced plane strain quadrilateral integration element, is considered. R2D2, which is a rigid body element, is also used for the punch and mandrels. Table 1 presents the geometric sizes of the simulated bending test. The failure of the GTN model is not provided by ABAQUS/Standard. Thus, ABAQUS/Explicit is used to provide the necessary failure, and the mass scaling method is utilized to eliminate the dynamic effect of the metal sheet. Figure 2 shows the FE configuration of the three-point bending test of a mesh in which 60 elements in the Y direction and 110 elements in the X direction are assigned (60 × 110). Fracture occurs in the middle section of the specimen. Therefore, the element distribution in

Table 1 Geometry sizes in simulated bending test

Sheet geometry parameters (mm)	L_o	t_o	Plane strain thickness	R_m	R_p	L_m
	20	2.5	10	0.25	0.2	13

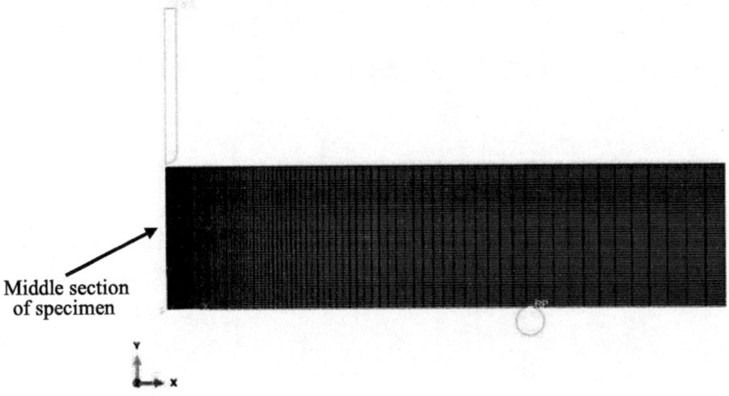

Fig. 2 FE configuration of three-point bending test

the refined area is biased to the middle section of the specimen. Only half of the sheet is simulated because of the geometric symmetry of the specimen. Figure 3 shows the force–displacement curve obtained under the effect of mesh sensitivity. Two different meshes of 60×110 elements and 100×150 elements are investigated.

The effect of pre-strain on the force–displacement curve is shown in Fig. 4. The maximum force decreases with increasing pre-strain, and fracture occurs early. The thickness of the sheet decreases when it experiences a tensile load as the pre-strain, which causes a decrease in maximum force in the bending test. Although the sheet hardens because of the pre-strain, the effect of thickness reduction on the decrease

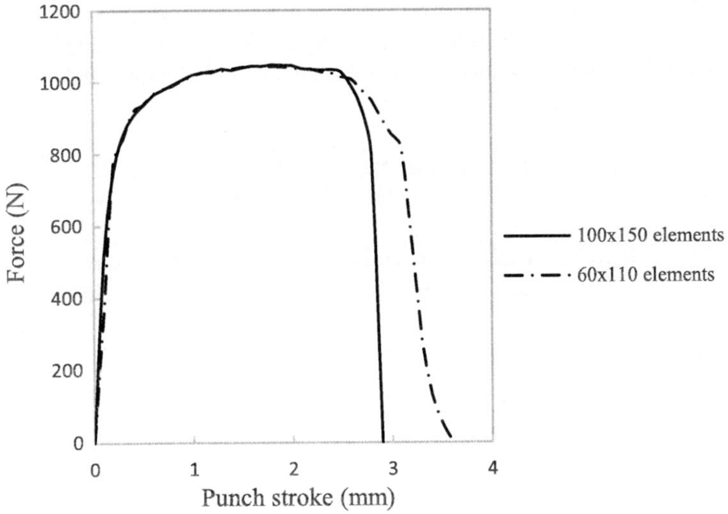

Fig. 3 Mesh sensitivity effect on force–displacement curve

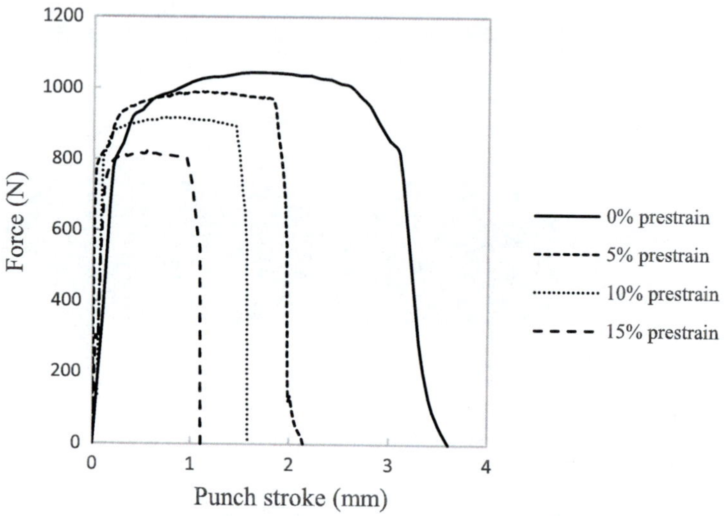

Fig. 4 Pre-strain effect on force–displacement curve

in maximum bending force is fairly obvious. The effect of pre-strain on the bending force–displacement curve of AA6111 and AA 5754 aluminum sheets was shown in [11]. The maximum bending force in these sheets first increases and then decreases with increasing pre-strain. Pre-strain causes thickness reduction and hardening of the sheets during pre-strain loading. In bending tests, thickness reduction decreases force while hardening increases force. Thus, thickness reduction and hardening compete to decrease or increase the bending force. Reference [11] shows that the bending force in AA6111 and AA5754 sheets first decreases and then increases with increasing pre-strain. This finding can be justified by the competition between thickness reduction and hardening to decrease or increase the bending force. Hardening efficiently occurs under low pre-strain values, while thickness reduction develops under high pre-strain values. However, the effect of hardening from pre-strain loading on the bending force depends on the stress–strain curve. The present study finds that the effect of thickness reduction on the bending force dominates that of hardening for all pre-strain values tested.

The effect of pre-strain on the void volume fraction at point A (see Fig. 1) during the bending test under various pre-strain values is shown in Fig. 5. The void volume fraction generated during pre-straining is considered the initial void volume fraction in the bending test. As expected, the initial void volume fraction increases in the bending test with increasing pre-strain, which causes early fracture. Although the initial void volume fractions are very low, they have a substantial effect on the bending fracture strain. The effect of pre-strain on the bending fracture strain is shown in Fig. 6; as expected, bending fracture strain decreases with increasing pre-strain.

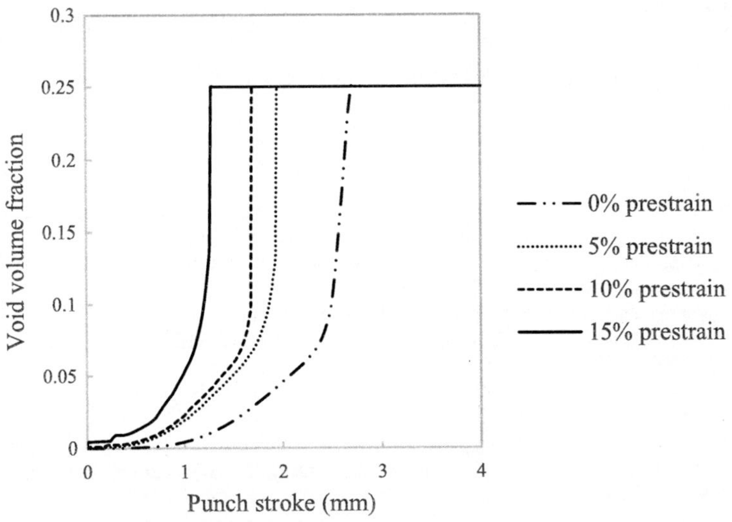

Fig. 5 Pre-strain effect on void volume fraction at point A

Fig. 6 Effect of pre-strain on bending fracture strain

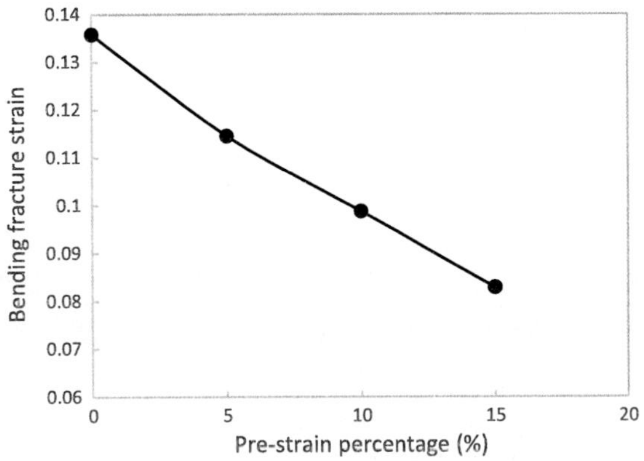

5 Conclusion

The simulation of a sheet metal under bending is performed using the commercial FE software ABAQUS, and elastic–plastic response. Also, ductility and fracture are analyzed when parameters such as those in Tvergaard are considered. Moreover, the effect of pre-strain on bendability is analyzed. Pre-strain is found to reduce maximum force in the bending test and cause early fracturing.

References

1. Kim J, Ryou H, Kim D, Kim D, Lee W, Hong SK, Chung K (2008) Constitutive law for AZ31B Mg alloy sheets and finite element simulation for three-point bending. Int J Mech Sci 50(10–11):1510–1518
2. Datsko J, Yang CT (1960) Correlation of bendability of materials with their tensile properties. J Eng Ind 82(4):309–313
3. Shi Y, Zhao PZ, Jin H, Wu PD, Lloyd DJ (2016) Analysis of surface roughening in AA6111 automotive sheet under pure bending. Metall Mater Trans A 47(2):949–960
4. Ragab A, Saleh CA (2005) Evaluation of bendability of sheet metals using void coalescence models. J Mater Sci Eng A 395(1–2):102–109
5. Soyarslan C, Gharbi MM, Tekkaya AA (2012) A combined experimental–numerical investigation of ductile fracture in bending of a class of ferritic–martensitic steel. Int J Solids Struct 49(13):1608–1626
6. Cha WG, Kim N (2014) Quantification of micro-cracks on the bending surface of roll formed products using the GTN model. Met Mater Int 20(5):841–850
7. Liu Z, Li L, Jie YI, Li S, Wang Z, Guan W (2017) Influence of heat treatment conditions on bending characteristics of 6063 aluminum alloy sheets. Trans Nonferrous Metals Soc China 27(7):1498–1506
8. Shahzamanian MM, Lloyd DJ, Wu PD (2020) Enhanced bendability in sheet metals produced by cladding a ductile layer. Mater Today Commun J 23:100952
9. Tagarielli V, Fleck N, Deshpande V (2004) The collapse response of sandwich beams with aluminium face sheets and a metal foam core. Adv Eng Mater 6(6):440–443
10. Sarkar SJ, Kutty TRG, Conlon KT, Wilkinson DS, Embury JD, Lloyd DJ (2001) Tensile and bending properties of AA5754 aluminum alloys. Mater Sci Eng A 316(1–2):52–59
11. Lloyd DJ, Evans D, Pelow C, Nolan P, Jain M (2002) Bending in aluminium alloys AA 6111 and AA 5754 using the cantilever bend test. Mater Sci Technol 18(6):621–628
12. ABAQUS Inc., A.M.V. (2014)
13. Tvergaard V, Needleman A (1984) Analysis of the cup-cone fracture in a round tensile bar. J Acta Metall 32(1):157–169
14. Tvergaard V (1981) Influence of voids on shear band instabilities under plane strain conditions. Int J Fract 17(4):389–407
15. Tvergaard V (1982) On localization in ductile materials containing spherical voids. Int J Fract 18(4):237–252
16. Shahzamanian MM (2018) Anisotropic Gurson-Tvergaard-Needleman plasticity and damage model for finite element analysis of elastic-plastic problems. Int J Numer Methods Eng 115(13):1527–1551
17. Gurson AL (1977) Continuum theory of ductile rupture by void nucleation and growth: part I—yield criteria and flow rules for porous ductile media. J Eng Mater Technol 99(1):2–15
18. Chu C, Needleman A (1980) Void nucleation effects in biaxially stretched sheets. J Eng Mater Technol 102(3):249–256

Fracture Mechanics Performance of Through-Thickness Crack of Polymeric 3D Printed Components

Waleed Ahmed, Essam Zaneldin, and Souzan Kabbani

Abstract This experimental study aims to explore the failure behavior of a pre- and post-cracked polymeric 3D printed components subjected to tensile mode. A set of through-thickness pre-cracked specimens of different cracks patterns and geometry was designed and implemented in the 3D printed parts. The specimens are then subjected to a tensile test mode. Besides, analogous intact samples were produced by 3D printing technology where the through-thickness post-cracks were created using laser cutting process of a geometry with cracks similar to those of the pre-cracked specimens. It has been observed that the pre-cracked samples initially introduced, and 3D printed cracked specimens have more resistance to fracture mechanics failure due to crack-bridging caused by the 3D printing filament profile around the crack profile. On the other hand, the samples with post-cracks made by laser cutting demonstrated a significant drop in the fracture failure resistance due to the interruption of the 3D printed filaments of the intact specimens. In conclusion, this study revealed that pre-cracked 3D printed components did not show the actual failure and fracture mechanics behavior. This is because the cracks could be introduced in the components after the additive manufacturing process during the service life and that would damage the 3D printed filament path of the components and, hence, will cause high-stress concentration that leads to unpredicted and fast failure.

Keywords FDM · Fracture mechanics · Crack · 3D printing

W. Ahmed (✉)
ERU, College of Engineering, UAE University, Al Ain, UAE
e-mail: w.ahmed@uaeu.ac.ae

E. Zaneldin
Civil and Environmental Engineering Department College of Engineering, UAE University, Al Ain, UAE
e-mail: essamz@uaeu.ac.ae

S. Kabbani
Finance and Banking Department, College of Economics, Aleppo University, Aleppo, Syria

M. Abdel Wahab (ed.), *Proceedings of the 8th International Conference on Fracture, Fatigue and Wear*, Lecture Notes in Mechanical Engineering,
https://doi.org/10.1007/978-981-15-9893-7_19

1 Introduction

Additive manufacturing (AM) technology has witnessed rapid progress over the past decades in terms of the development and applications due to a wide range of applications and benefits in different fields including the manufacturing and construction industries and education [1]. Nowadays, there are a wide range of different materials that can be used for additive manufacturing (3D printing) such as polymeric, metallic, non-metallic, biomass, ceramics and many more [2, 3]. There are also different AM technologies used for 3D printing [4]. One of these technologies, named "Fused Deposition Modeling (FDM)", produces functional parts by the sequential deposition of fused materials. This particular AM technology is considered one of the most cost-effective and widely used technology in additive manufacturing since the process is simple and less costly when compared with other technologies such as "Selective Laser Sintering (SLS)" and "Stereolithography (SLA)". The FDM process is mainly used for cost-effective printing plastic prototypes since it shows excellent capabilities in producing more durable fiber-reinforced composite components that have a unique potential in aerospace, automotive, and biomedical applications. AM technology has attracted many researchers and scholars to study its effect on the mechanical properties [5] and failure behavior of produced components. Such studies were conducted mainly to understand the failure mechanism and to eliminate the consequences caused by the manufacturing process due to internal flaws and stress concentration. The main objective of these studies was to increase the durability and safety levels of components [6] especially using fracture mechanics since it is concerned with the analysis of the material degradation and the failure caused by internal flaws and cracks [7]. This can determine the level of stresses with known cracks geometry and configurations that can propagate through a material to trigger failure, whether with big scale or micro level structures [8–10] analytically and by using a finite element method [11]. Another study conducted by [12] investigated the fracture behavior of U-shape notched made by 3D-printing technology under different failure modes. In this study, experiments were conducted using polymeric specimens made using 3D-printing fused deposition modeling (FDM). The fracture behavior of 3D printed components under flexural loading were investigated by [13] for different curvatures used to develop a fracture criterion to describe the functioning of materials by considering tensile and compressive stresses. The deformation and fracture behavior of the 3D printed model was presented to investigate a disordered lattice material subjected to uniaxial tensile stress. For predicting the response of the crack pattern and load–displacement for the printed lattice materials, a lattice model was developed by [14]. In this model, two-directional 3D printed components were considered to explore the influence of the layered structures of 3D printed materials [14]. In another study, it has been noticed that filament-scale geometric characteristics are more significant than filament bonding especially designed 3D-printed compact tension specimens that are tested regularly along the direction of extruded filaments. Higher toughness and strength are also noticed in the filament direction [15]. A new criterion based on mechanical stress was introduced by [16] where the filaments' directions follow the principal directions of the stress in the tested samples. Several cracked samples were 3D-printed considering the new criterion and without considering it to investigate the fracture behavior of cracked specimens [16]. The study of [17] revealed that the fracture resistance and interlayer adhesion of fused deposition modeling 3D printed materials were characterized using double cantilever beam specimens made of ABS. These specimens were designed with a pre-crack at the layers' interface under opening mode to detect the critical load at the crack initiation [17]. The effort of [18] investigated the height of the deposition on the fracture behavior of a 3D printed PLA samples under uniaxial tensile loading and the results of this study revealed that the energy needed for interlayer fracture is dependent on the height of the deposition. On the other hand, the work of [19] investigated, experimentally and numerically, the fracture modes of 3D printed nylon samples made by two different 3D printing technology. This study was conducted to figure out the load-carrying capacity of U-notched plates fabricated by the 3D printing techniques using equivalent material concept with J-integral failure criterion to investigate the failure of the notched samples [19]. The work of [20] investigated the mechanical response of polymer sheets produced by additive manufacturing technology and the essential work of fracture. The study was conducted to explore the role of fiber reinforcement on the fracture resistance of additively manufactured polymer sheets using double edge notched tensile tests on different polymeric material [20]. This study investigates the impact of the manufacturing process of cracked samples using 3D printing FDM process on the mechanical properties of printed samples. Besides, Linear Elastic Fracture Mechanic (LEFM) was used to analyze the failure behavior of the 3D printed samples. The study reveals a critical conclusion related to the failure of the 3D printed parts.

2 Sample Preparation and Material Properties

In this investigation, specimens have been produced using a 3D printer model Ultimaker S5 to figure out the impact of through-thickness crack on the mechanical performance of the 3D printed samples. The samples were made of standard PLA filament material approved by the manufacturer Ultimaker B.V., the mechanical properties are shown in Table 1.

Table 1 PLA properties [20, 21]

Properties	Experimental value
Tensile modulus	2.34 GPa
Tensile stress at	Yield 49.5 MPa
Tensile stress at break	45.6 MPa
Elongation at break	5.2%

The tensile test sample configurations adopted in the study are based on ASTM 638 and were considered for the intact and the cracked samples. The sample dimensions are shown in Fig. 1.

The through-thickness crack configuration adopted in the study is 1 mm in length, which has an elliptical shape with an aspect ratio of 1:0.5, as illustrated in Fig. 2.

The printing parameters were considered carefully to maintain the consistency in the printing process of all samples and to assure the reliability and quality of the printed samples. Printing parameters adopted in the production of the intact and defected samples are listed in Table 2.

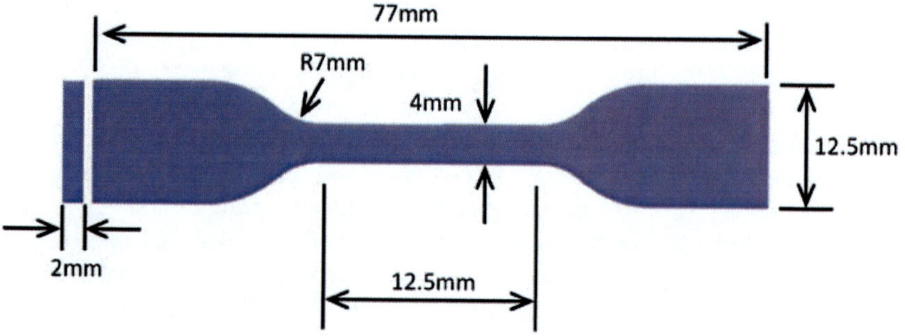

Fig. 1 Sample dimensions according to ASTM 638

Fig. 2 Through-thickness cracked sample

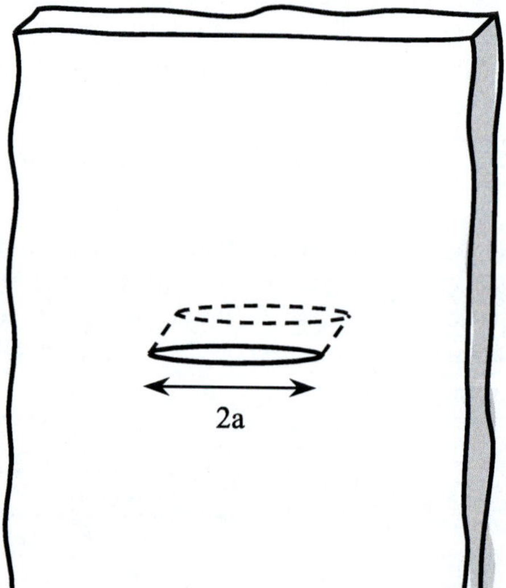

Table 2 Ultimaker S5 printing parameters

3D printing parameters	Implemented characteristics
Layer height	0.2 mm
Wall thickness	1 mm
Infill density	25%
Infill pattern	Lines
Top layer thickness	1 mm
Bottom layer thickness	1 mm
Print speed	70 mm/min
Printing temperature	200 °C
Build plate temperature	60 °C

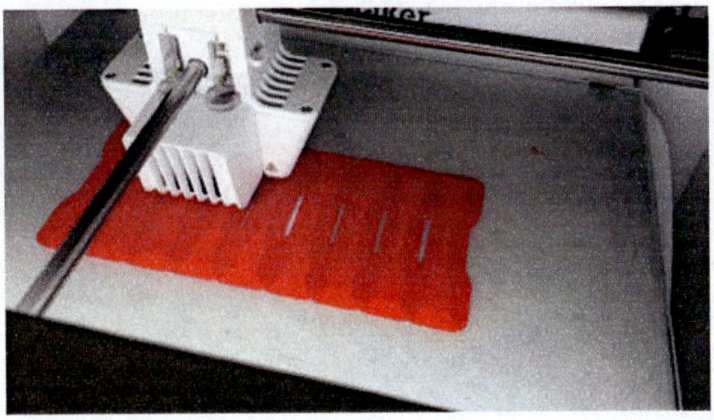

Fig. 3 Printing process to produce tensile samples using Ultimaker S5

Three sets of 3D printed samples are considered in this study in order to investigate the impact of the crack initiation procedure on the failure and the performance of the 3D printed samples. An open-source slicing software for 3D printing named "CURA 4.6.1" is used. The software works by slicing the CAD model file into layers and generating a specific G code that can be used by the printer for printing the physical part, as shown in Fig. 3.

The samples were tested using a universal tensile testing machine of 10 kN load cell at a testing speed of 1 mm/min and a room temperature of 23 °C. In this study, to distinguish the samples in the manufacturing process, these samples are identified as follows:

Intact: The samples are produced without a crack.

Printed hole: A central through-thickness crack of aspect ratio 1: 0.5 was produced by 3D printing for all samples.

Laser hole: Laser cutting machine was utilized to produce the through-thickness cracks using intact samples.

3 Results and Discussion

The tensile tests were performed for the three printed sets and the results are illustrated in Fig. 4. It was observed that the intact specimens failed at a max load of 418.2 N with a standard deviation of 8.3 N, whereas a drop of 4.7% of the failure load was noticed for the printed cracked samples. This explains the positive impact of 3D printing contours produced through the slicing process (shown in Fig. 5), where they allow smooth stress stream flow around the crack profile. Besides, due to the side, longitudinal extruded filaments that occupy 50% of the constructional section of the sample, assist the cracked sample to resist the failure and raise the failure load more than the laser cracked samples. Moreover, the samples with cracks produced by laser showed significant reduction of around 26.8% in the failure load. The laser cutting process demonstrated a negative effect on the performance of the cracked samples, the reasons behind this is that laser cuts the 3D printed extruded filament profile and this disrupts the stress stream flow and causes increases in the stresses at the crack tip and, therefore, produces failure at a lower load.

On the other hand, a failure strain of 10.3% has been estimated for the intact samples with a standard deviation of 0.16, as illustrated in Fig. 6. For the 3D printed holes and laser holes, it is observed that the trend is similar to that of the failure load with a reduction of 39% in the ductility of the 3D printed holes while the reduction in the ductility was 52% for laser cracked samples. This explains the deterioration of the stiffness of the laser cracked samples more than the 3D printed cracked specimens due to the impact of the laser cutting process on the extruded filament by the 3D printing process.

It is anticipated that the resistance to impact will have the same behavior as the statin, and this is apparent in Fig. 7. A significant decrease of 51% in toughness of the printed cracked sample as compared with the intact sample, whereas a 76.4% reduction in toughness was noticed for the laser cracked sample. The drop in the toughness came because of the drop in the strain as well as the load at failure of

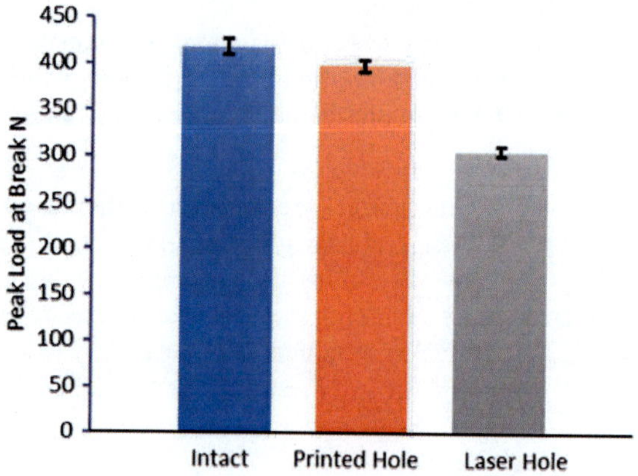

Fig. 4 Maximum load measured at the failure point of the 3D printed samples

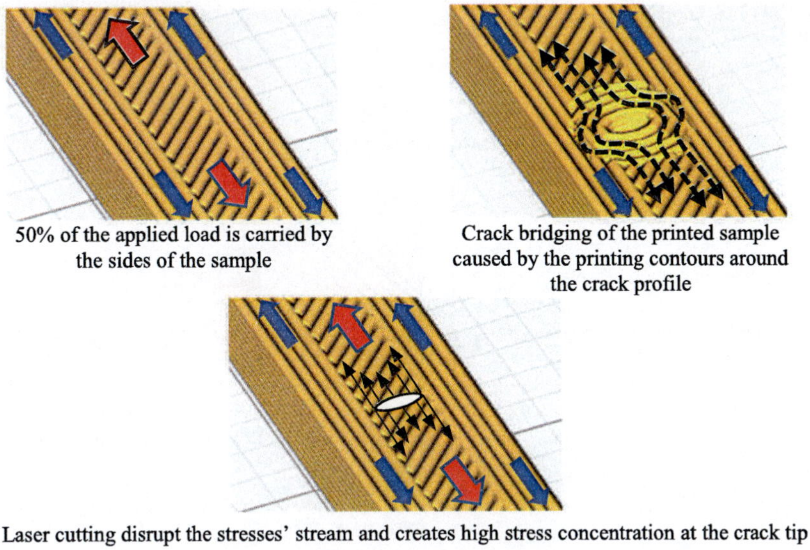

Fig. 5 Demonstration of the stresses stream for the three samples

Fig. 6 Strain at the failure condition (ductility) of the tested samples

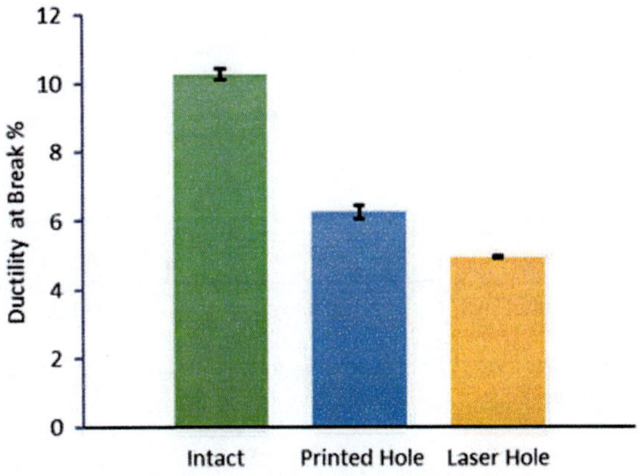

the cracked samples. It is also worthwhile mentioning that the effect of the type of specimen is dramatically higher on the "toughness" and "ductility" as compared to its effect on the "maximum load", as shown in Fig. 8.

In order to stand on the main reasons behind the low load at the failure point of the cracked samples, two analysis were conducted for this purpose in terms of stress concentration and Linear Elastic Fracture Mechanics (LEFM). The maximum stress was estimated for the three sets using the following formula [23]:

$$\sigma_{max} = \sigma_{nom} \cdot k_t \tag{1}$$

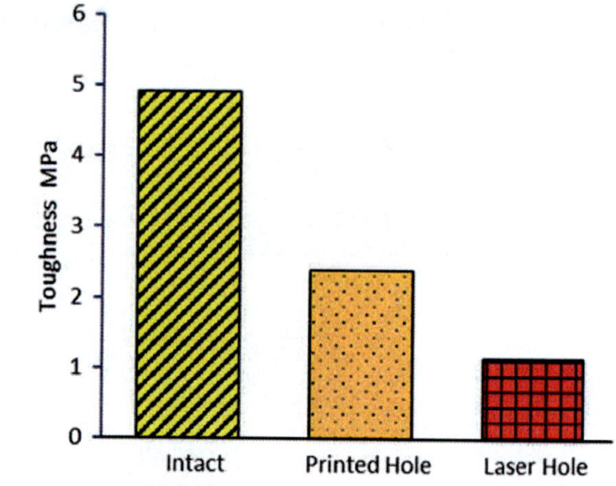

Fig. 7 Toughness properties of the tested 3D printed samples

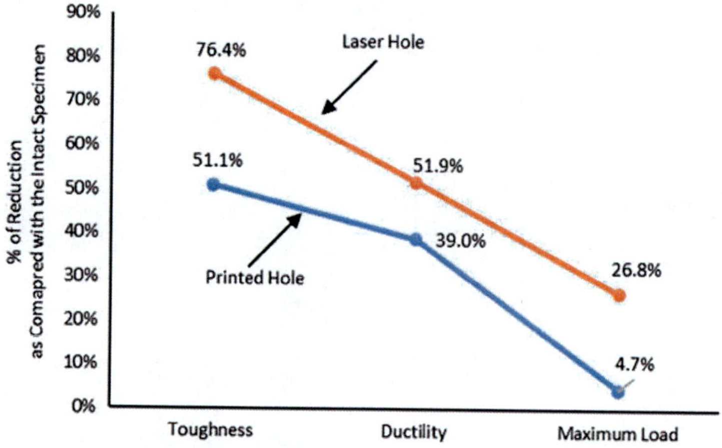

Fig. 8 Percentage of reduction in maximum load, ductility, and toughness as compared with the intact specimen

where,

σ_{max} is the max stress (MPa).
σ_{nom} is the nominal stress (MPa).
k_t is the stress concentration factor.

For the elliptical crack configuration of the cracked samples that has an aspect ratio of 1: 0.5, the estimated stress concentration factor is 3.95. This means that for both the printed cracked and the laser cracked samples, the stresses are around to be four times the nominal stresses, but it was observed that the laser cracked samples failed at a load lower than the printed cracked samples. Figure 9 demonstrates the normalized stresses as well as the maximum stresses for the three printed samples.

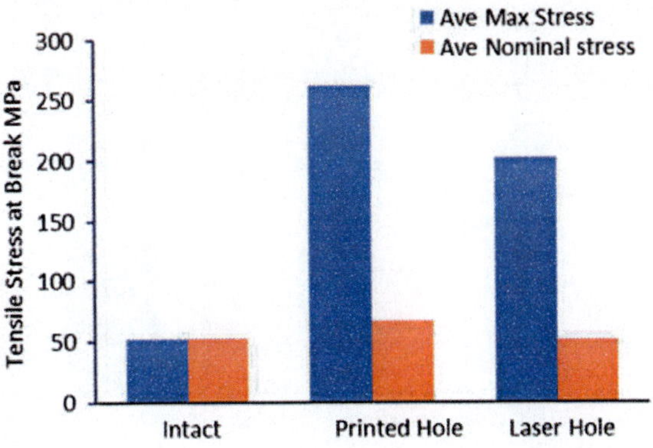

Fig. 9 Maximum tensile stress at the breakpoint

It is apparent that, by considering the nominal stress, the printed cracked samples have maximum stresses more than the laser cracked samples. Due to that, the failure load for the laser cracked samples are lower than the printed cracked samples, as discussed earlier and demonstrated in Fig. 4.

In fact, the stress concentration justifications have a contradiction with the reality, that is, laser cracked samples are supposed to have higher stresses or at least the same stresses as those of printed cracked samples. Therefore, Linear Elastic Fracture Mechanics has been adopted to investigate the failure reasons. Mode I Stress Intensity Factor (SIF) is calculated for the three sets of the printed samples using the following formula [22]:

$$K_I = \sigma \sqrt{\pi a} Y \quad (2)$$

where,

K_I Mode I Stress Intensity Factor MPa.m$^{1/2}$.
σ Applied stress (MPa).
a Crack length (m).
Y Specimen geometry dependent function of the crack length.

It is apparent from Fig. 10 that the normalized stress intensity factor of intact samples using different manufacturing methods is always more significant than the cracked samples whether the samples are produced using 3D printing or a laser cutting machine. This means that it is going to resist crack extension more than the flawed parts. Besides, the normalized stress intensity factor for the 3D printed cracked samples (i.e., 3D hole) is more than the cracked samples made by the laser cutting machine (i.e., laser hole), which indicates that the latter has less tendency to crack extension, and this explains why the cracked samples made by the laser cutting machine. Moreover, it has been observed that, for the same crack configuration adopted in this study, the 3D printing process made a crack profile cordoned by extruded filament contours that created bridging to the stress stream around the crack profile. This adds more resistance to the crack extension, whereas for the cracked

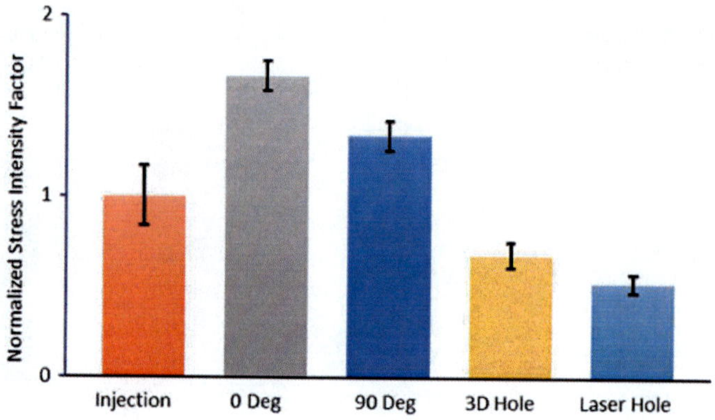

Fig. 10 Normalized SIF for PLA using different production methods

samples made by laser cutting process, the extruded filaments already cut by the laser process promotes the crack extension. Therefore, the latter suffered from lower failure load as well as smaller stress intensity factor, which explains the reasons behind the failure.

4 Conclusion

In this experimental analysis, an investigation has been carried out to study the impact of cracks on the behavior of 3D printed samples using different manufacturing methods to produce these cracks. The 3D printing process was used to produce the cracked sample while a laser cutting machine was used to produce the second sample with the same crack profile and configuration, through-thickness crack. It was observed that the laser cutting process has a more negative impact on the failure and the degradation of the mechanical properties of the samples as compared with 3D printing. The negative impact was noticed on the failure load, ductility, and toughness. The 3D printed cracked samples showed better features than the laser cutting process due to the extruded filaments contours that cordon the crack profile caused by the slicing process performing a crack-bridging for the stress streams. Moreover, it was noticed that when using LEFM, the stress intensity factor of the laser cutting cracks has lower values than the printed cracks. This explains the reason behind the more moderate load failure for the cracks made by the laser cutting process. The authors are currently working on another research to investigate the impact of using laser cutting process on the failure of the 3D printed parts as well as on the printing parameters and, when ready, the findings of this study will be presented in a future research paper.

References

1. Issah MA, Waleed KA, Hayder ZA, Hamad A (2019) 3D printing applications in mechanical engineering education. Brill|Sense, Leiden, The Netherlands, pp 90–131
2. Ahmed WK, Al-Douri Y (2020) 17—three-dimensional printing of ceramic powder technology. In: Al-Douri Y (ed) Metal oxide powder technologies. Elsevier, pp 351–383
3. Wang XC, Wei J, Yi XB, Zhang J, Shang K, Wang Q (2014) 3D printing technology and the adaptability of printing material. Appl Mech Mater 633/634(Part 1):569–573
4. Hao B, Lin G (2020) 3D printing technology and its application in industrial manufacturing. IOP Conf Ser Mater Sci Eng 782:022065
5. Khawaja HA, Alabdouli H, Alqaydi H, Mansour A, Ahmed W, Jassmi HA (2020) Investigating the mechanical properties of 3D printed components. In: 2020 Advances in science and engineering technology international conferences (ASET), pp 1–7
6. AlDarmaki A, AlShamsi A, Ahmed WK, Jassmi HA (2019) Designing and developing innovative structural engineering failure experiment using additive manufacturing technology. In: 2019 advances in science and engineering technology international conferences (ASET), pp 1–6
7. Broek D (1978) Elementary engineering fracture mechanics. Sijthoff & Noordhoff, Alphen aan den Rijn
8. Ahmed W (2020) Wind turbine composite blade: fracture mechanics assessment. J Theor Appl Mech 60(3):238–258
9. Ahmed WK, Al-Rifaie WN (2015) The impact of cracked microparticles on the mechanical and the fracture behavior of particulate composite. J Nano Electron Phys 7(3):03008-1–6
10. Ahmed WK, Teng H (2014) Characterisation of fractured particulate reinforced composite. Int J Microstruct Mater Prop 9(2):160–175
11. Wahab MA (2010) Finite elements in fracture mechanics: for general power type singularity problems. LAP LAMBERT, Saarbrücken, Germany
12. Khosravani MR, Zolfagharian A (2020) Fracture and load-carrying capacity of 3D-printed cracked components. Extreme Mech Lett 37
13. Ishii K, Todoroki A, Mizutani Y, Suzuki Y, Koga Y, Matsuzaki R, Ueda M et al (2019) Bending fracture rule for 3D-printed curved continuous-fiber composite. Adv Compos Mater 28(4):383–395
14. Xu Y, Zhang H, Šavija B, Chaves Figueiredo S, Schlangen E (2019) Deformation and fracture of 3D printed disordered lattice materials: experiments and modeling. Mater Design 162
15. Allum J, Gleadall A, Silberschmidt VV (2020) Fracture of 3D-printed polymers: crucial role of filament-scale geometric features. Eng Fract Mech 224
16. Gardan J, Makke A, Recho N (2018) Improving the fracture toughness of 3D printed thermoplastic polymers by fused deposition modeling. Int J Fract 210(1–2):1–15
17. Aliheidari N, Tripuraneni R, Ameli A, Nadimpalli S (2017) Fracture resistance measurement of fused deposition modeling 3D printed polymers. Polym Test 60:94–101
18. Noori H (2019) Interlayer fracture energy of 3D-printed PLA material. Int J Adv Manuf Technol 101(5–8):1959–1965
19. Zolfagharian A, Khosravani MR, Kaynak A (2020) Fracture resistance analysis of 3D-printed polymers. Polymers 12(2)
20. Cuesta II, Martinez-Pañeda E, Díaz A, Alegre JM (2019) The essential work of fracture parameters for 3D printed polymer sheets. Mater Des 181:107968
21. Gendviliene I, Simoliunas E, Rekstyte S, Malinauskas M, Zaleckas L, Jegelevicius D, Bukelskiene V et al (2020) Assessment of the morphology and dimensional accuracy of 3D printed PLA and PLA/HAP scaffolds. J Mech Behav Biomed Mater 104
22. Ultimaker PLA data sheet, https://ultimaker.com/materials/pla. Accessed on 21 July 2020
23. Beer FP, Johnston ER, DeWolf JT (2006) Mechanics of materials. McGraw-Hill Higher Education, Boston
24. Broek D (1997) The practical use of fracture mechanics. Kluwer Academic Publishers, Dordrecht

D-Beam Theory for Functionally Graded Double Cantilever Beam Analysis

Calogero Orlando

Abstract A formulation that takes advantage of both Layer-Wise and Equivalent Single Layer approaches for modeling Functionally Graded beams is presented. Such alternative formulation, referred to as D-Beam, is here applied to model the Double Cantilever Beam specimen in order to estimate the relevant fracture opening mode of metallic graded beams made up by means of Additive Manufacturing technique. Numerical results are presented.

Keywords Double cantilever beam · Functionally graded · Layer-wise · Equivalent single layer

1 Introduction

Functionally graded (FG) materials are considered of relevant interest in aerospace engineering applications because of their high stiffness-to-weight and strength-to-weight ratios. They belong to the class of advanced composite materials and their properties are usually varied gradually along the thickness direction when beams and plate-like structures are taken into account. Such gradual variation leads to the advantage, with respect to layered composites, of having reduced interfacial stresses [1]. Generally, two different materials are mixed accordingly to a certain rule in such a way that the desired property is let vary gradually along a particular direction. To build such materials novel approaches, that are alternative with respect to the time-consuming traditional methods, are represented by the Additive Manufacturing (AM) techniques [2]. AM can be simply defined as the process of joining, point by point and layer by layer, materials in such a way to make an object. Nowdays, AM is not only used for rapid prototyping but it is considered to realize actual structural components saving time, effort and materials with respect to traditional approaches [3]. Among the AM techniques, the direct laser deposition has been used by Liu

C. Orlando (✉)
Universitá di Enna Kore, Enna 94100, Italy
e-mail: calogero.orlando@unikore.it

M. Abdel Wahab (ed.), *Proceedings of the 8th International Conference on Fracture, Fatigue and Wear*, Lecture Notes in Mechanical Engineering,
https://doi.org/10.1007/978-981-15-9893-7_20

et al. [4] to realize metallic FG material ranging from 100% of Ti6Al4V to 100% of AlSi10Mg to meet the light weight and high strength specification required by industry. Despite the stiffness and hardness properties obtained, cracks were found at layers transition and this evidences the need for fracture mechanics analysis of metallic FG structure made up by AM technologies. The attention is here focused on the modeling and fracture mechanic analysis of FG beam structure. The 3D modeling of FG beams can be considered the proper modeling approach for the accurateness of the solution; however a great modeling or computational effort are generally requested. This motivates the use of 1D beam models but a proper level of modeling accuracy must be ensured. In particular, beam models can belong to two main group, namely the Equivalent Single Layer (ESL) and the Layer-Wise (LW) models. The first are preferable to reduce the computational burden while the latter to increase the accuracy of the solution. A third class is represented by the Zig-Zag beam models where a layerwise discontinuous function is a-priori selected to enrich the kinematical model of ESL beam [5]. Such approaches can be used to model the Double Cantilever Beam (DCB) specimen allowing to compute the specimen compliance and from this the Energy Release Rate (ERR) [6]. In this work, an alternative approach for laminated composite beams is used [7, 8]. A layer-wise kinematical model that fulfill the point-wise equilibrium equations, the layer interface continuity conditions and the traction-free conditions at the top and bottom faces of the beam is obtained. Then, generalized variables that represent the kinematics of the beam as a whole are introduced and, by manipulating the interface continuity conditions, the problem degrees of freedom are reduced obtaining an ESL model that preserves the accuracy of the assumed LW kinematics. The proposed approach is used in combination with the discrete layer technique to solve the FG DCB problem.

2 D-Beam Model

The D-Beam model holds for layered linear elastic beams. Without lack of generality, let us assume that the beam has n layers, has rectangular cross-section of height h and width B and its length is L. The beam is loaded at its ends by generalized stress resultants. To describe the beam, a global reference system $\{x, z\}$, centered at the left end of the beam, is used. The x-axis lies along the beam midline. Local reference systems $\{x, z_l\}$ are also used to describe the kinematics of each layer l. The thickness of each layer is h_l while the distance between the beam and the layer midlines is $\bar{h}_l$. The LW kinematics is written as

$$\begin{aligned} u_l(x, z_l) &= u_{0l}(x) + \theta_l(x) z_l + B_l(x) z_l^2 + C_l(x) z_l^3 \\ w_l(x, z_l) &= w_{0l}(x) + A_l(x) z_l + D_l(x) z_l^2 \end{aligned} \quad (1)$$

being u_l, w_l and θ_l the axial and transverse displacement and the section rotation of each layer. In Eq. 1, A_l, B_l, C_l and D_l are LW functions of x.

In particular, by using Eq. 1 along with linear strain-displacement and linear elastic constitutive relations it is obtained a set of LW ordinary differential equations that are solved by imposing interface continuity conditions (in terms of continuity of displacement and equilibrium between adjacent layers) and by taking into consideration the stress free condition of the top and bottom beam surfaces. Such solutions allow to write the LW functions A_l, B_l, C_l and D_l in terms of the layer midline kinematics u_{0l}, w_{0l} and θ_{0l}. A set of useful relationships are also obtained that allow to manipulate the interface displacement continuity conditions in such a way that the LW kinematic variables of each layer u_{0l}, w_{0l} and θ_{0l} are written in terms of the midline kinematics of the specific layer m that contains the beam midline at $z_m = \tilde{z}_m$. By introducing the axial U and transverse W displacements of the multilayered beam midline and the beam cross section rotation Θ as generalized variables, namely

$$u_m(x, \tilde{z}_m) = U, \quad w_m(x, \tilde{z}_m) = W, \quad \Theta = \left.\frac{\partial u_l}{\partial z_l}\right|_{\tilde{z}_m} \tag{2}$$

the LW kinematical model Eq. 1 is then rewritten as

$$\begin{aligned} u_l(x,z) &= U(x) + f_l^{u\Theta}(z)\Theta(x) + f_l^{uU''}(z)U''(x) + f_l^{u\Theta''}\Theta''(x) \\ w_l(x,z) &= W(x) + f_l^{wU'}(z)U'(x) + f_l^{w\Theta'}(z)\Theta'(x) \end{aligned} \tag{3}$$

where $f_l^{\star}(z)$ are known function of z and of the layers' material constants and primes denote derivative with respect to x. For more detail about the formulation the interested Reader is referred to [7].

Looking at Eq. 3 it appears that the number of unknowns in the present formulation is independent on the number of layers. In addition, as stated before, the continuity and equilibrium conditions at layers' interfaces are met as well as the point-wise balance equations are fulfilled. It follows that the present D-Beam model has the accuracy of the LW models and the computational effectiveness of the ESL ones.

At this point, by using Eq. 3 along with the strain-displacement and the layer constitutive equations, it is possible to write the beam constitutive equations as

$$\begin{Bmatrix} N \\ M \\ T \end{Bmatrix} = \begin{bmatrix} k_m & z_G k_m & 0 \\ z_G k_m & k_f & 0 \\ 0 & 0 & \chi^2 k_s \end{bmatrix} \begin{Bmatrix} U' \\ \Theta' \\ \gamma \end{Bmatrix} \tag{4}$$

where N and T are the axial and shearing stress resultant forces, respectively, and M is the bending moment per unit length. In Eq. 4, k_m and k_f are the membrane and the flexural stiffness per unit width, respectively, z_G is the position of the beam section elastic centroid with respect to the beam midline, k_s is the shear stiffness per unit width, $\gamma = \Theta + W'$ is the generalized shear strain of the beam and χ^2 is the shear correction factor. The interested reader is referred to [7] for the definitions of all the aforementioned quantities that have been here omitted for the sake of conciseness.

3 Double Cantilever Beam

The proposed D-beam model is used to compute the energy release rate (ERR) of a double cantilever beam (DCB) specimen. It is assumed that the DCB presents a delamination of length a, each arm of the DCB has thickness h_{arm} while the beam width is B. The uncracked part of the DCB is assumed to be perfectly rigid, thus the arms cross-section at the crack tip are modeled as clamped ends. Each DCB's arm is loaded with a tip shearing force F and this results in a mouth opening Δ. The Energy Release Rate (ERR) is the amount of the total potential energy Π that is released as consequence of an infinitesimal advancement of the crack. Taking into account that the material is linear elastic, it is defined as

$$ERR = -\frac{1}{B}\frac{d\Pi}{da} = \frac{F^2}{2B}\frac{dC}{da} \tag{5}$$

being $C = \Delta/F$ the compliance of the DCB. In particular, taken into consideration the beam constitutive relationships Eq. 4 along with the cantilever beam loading conditions at $x = 0$, namely $\{N(0) = 0,\ M(0) = 0,\ T(0) = -F\}$, and its constraints at $x = a$, i.e. $\{U(a) = 0,\ W(a) = 0,\ \Theta(a) = 0\}$, it is possible to compute the mean-line curve deflection of each arm of the DCB as

$$W(x) = -T(0)\frac{a^3}{6(k_f - z_G^2 k_m)}\left[(\frac{x}{a} - 1)^2(\frac{x}{a} + 2) + \frac{6(k_f - z_G^2 k_m)}{\chi^2 k_s a^2}(\frac{x}{a} - 1)\right] \tag{6}$$

Assuming now that material is graded symmetrically with respect to the DCB mean-line, the specimen compliance is easily written as

$$C = \frac{2W(0)}{F} = \frac{2}{B}\left[\frac{a^3}{3(k_f - k_m z_G^2)} + \frac{a}{\chi^2 k_s}\right] \tag{7}$$

It follows that the ERR can be written as

$$ERR = \frac{F^2}{B^2}\left[\frac{a^2}{(k_f - k_m z_G^2)} + \frac{1}{\chi^2 k_s}\right] \tag{8}$$

For homogeneous DCB Eq. 8 simplifies as

$$ERR = \frac{12F^2a^2}{EB^2h_{arm}^3}(1 + f) \tag{9}$$

being E the young modulus along the length direction. The factor f equals zero under the Euler-Bernoulli beam assumption, it is worth $f = Eh_{arm}^2/12\chi^2Ga^2$ under the first order shear deformation Timoshenko's theory and it is worth $f = Eh_{arm}^2/8Ga^2$

using the formulation proposed in this work. In particular, the ERR relationships obtained by the present formulation is equal to one given by Tada et al. [6] taking into account the correction due to shear deformation for a isotropic beam having Poisson's ratio $1/3$.

4 Applications

Each arm of the DCB has length $a = 10\,\text{mm}$, width $B = 2\,\text{mm}$ and thickness $h_{arm} = 0.25\,\text{mm}$. It undergoes a shearing resultant force at the free end $T_0 = 1\,\text{N/m}$. The DCB's arm are assumed to be realized by functionally graded material with Ti6Al4V and AlSi10Mg as constituent phases. Material properties of Ti6Al4V $E = 110\,\text{GPa}$, $G = 40\,\text{GPa}$ and $\nu_z = 0.31$ while material properties of AlSi10Mg are $E = 68\,\text{GPa}$, $G = 26.5\,\text{GPa}$ and $\nu_z = 0.33$. The grading law for the upper arm is $V_B = \frac{(\bar{h}_l + z_l + h_{arm}/2)^\eta}{h_{arm}^\eta}$ and $V_A = 1 - V_B$ and the Young's and Shear moduli and Poisson's ratio are computed by the rule of mixtures. The DCB lower arm is symmetric to the upper one with respect to the DCB mean-line where the crack is assumed to lie. The exponent η in the grading law is the grading index and it is let vary as $0 \leq \eta \leq 2$.

To set the number of discrete layer that allows the appropriate modeling of the FG beam, a convergence analysis on the tip deflection, cross section rotation and ERR is carried for different grading index η. Results are collected in Table 1 along with computation time showing that 100 layers are enough to gain converging results but 1000 layers can be used considered the very low computational requirements of the method.

The through the thickness distribution of the FG material properties are shown in Fig. 1a, b for two different grading index, the number of layers is set to 1000 when the Ti6Al4V is at the DCB mean-line.

Figure 1c reports the value of the ERR computed by using the proposed formulation as function of the grading index η for two different phase arrangements. In particular, the continuous line of Fig. 1c shows the trend of the ERR versus the grading index when the Ti6Al4V is assumed to lay at the DCB mean-line while the AlSi10Mg is at top and bottom surface of the DCB's arms, namely the Ti6Al4V is the phase A. It appears that as the grading index increases, i.e. when the presence of the more stiffer phase is greater across the DCB mean-line where is the crack,

Table 1 Discrete layer approximation convergence for $\eta = 0.5$

Number of layers	10	50	100	500	1000
$W \times 10^{-6}$ (m)	3.115	3.107	3.107	3.107	3.107
$\Theta \times 10^{-4}$ (rad)	4.669	4.658	4.657	4.657	4.657
ERR (Pa m)	233.5	232.9	232.9	232.9	232.9
CPU-time (s)	0.039	0.041	0.058	0.112	0.369

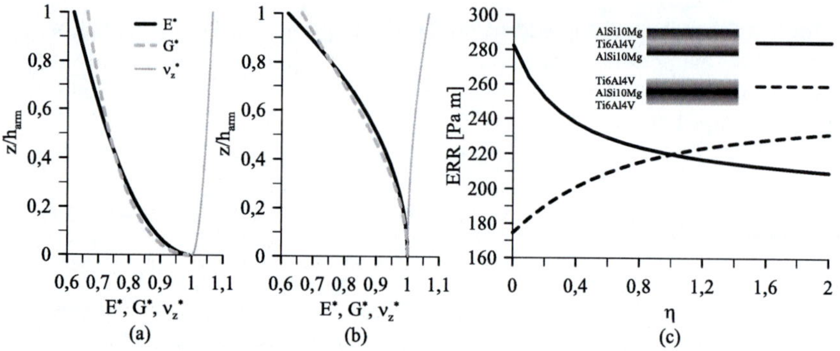

Fig. 1 Through the thickness distribution of material constants ($E^* = E/E_{Ti6Al4V}$, $G^* = G/G_{Ti6Al4V}, \nu_z^* = \nu_z/\nu_{Ti6Al4V}$) for **a** $\eta = 0.5$ and **b** $\eta = 2$ in the upper arm of the DCB, and **c** EER as function of η

the ERR decreases. The opposite trend is obtained for the DCB realized with the AlSI10Mg on the mean-line and Ti6Al4V on the top and bottom surfaces as shown by the dashed line in Fig. 1c.

5 Conclusions

An alternative beam formulation, called D-Beam, has been presented in this work for the analysis of DCB FG specimen. The D-Beam approach uses a LW kinematics that fulfill the point-wise equilibrium equations as well as the interface continuity conditions. The LW relations are then lumped into an ESL model by introducing generalized kinematic variables. The FG material properties have been modeled by means of discrete layer technique and the proposed D-Beam approach has been employed to study the dependency of the strain ERR as function of the grading law that can be obtained through AM process.

Acknowledgements I would like to express my deep gratitude to Professor Giuseppe Daví for his valuable support and discussion about the paper.

References

1. Nikbakht S, Kamarian S, Shakeri M (2019) A review on optimization of composite structures part II: functionally graded materials. Comp Struct 214:83–102. https://doi.org/10.1016/j.compstruct.2019.01.105
2. Yan L, Chen Y, Liou F (2020) Additive manufacturing of functionally graded metallic materials using laser metal deposition. Add Manufact 31:100901. https://doi.org/10.1016/j.addma.2019.100901
3. Gibson I, Rosen DW, Stucker B (2014) Additive manufacturing technologies, vol 17. Springer, New York. https://doi.org/10.1007/978-1-4419-1120-9

4. Liu Y, Liu C, Liu W, Ma Y, Zhang C, Cai Q, Liu B (2018) Microstructure and properties of Ti/Al lightweight graded material by direct laser deposition. Mater Sci Tech 34(8):945–951. https://doi.org/10.1080/02670836.2017.1412042
5. Carrera E (2003) Historical review of zig-zag theories for multilayered plates and shells. Appl Mech Rev 56(3):287–308. https://doi.org/10.1115/1.1557614
6. Tada H, Paris PC, Irwin GR (2000) The stress analysis of cracks handbook, vol 130. ASME press, New York. https://doi.org/10.1115/1.801535
7. Davì G, Milazzo A, Orlando C (2014) An analytical solution for multilayered beams subjected to ends loads. Comp Struct 116:772–781. https://doi.org/10.1016/j.compstruct.2014.06.003
8. Alaimo A, Davì G, Milazzo A, Orlando C (2017) Analytical solution for composite layered beam subjected to uniformly distributed load. Mech Adv Mat Struct 24(16):1315–1324. https://doi.org/10.1080/15376494.2016.1227512

Numerical Treatment of Fractional Differential Models

Angelamaria Cardone, Dajana Conte, and Beatrice Paternoster

Abstract Several fracture and fatigue problems are modelled by fractional differential equations. This paper deals with the numerical solution by a class of one and two step spline collocation methods. These methods have higher order of convergence with respect to most numerical methods for fractional differential equations. The methods are illustrated, and their convergence properties are discussed. Several numerical experiments confirm theoretical results and compare one and two step spline collocation methods.

Keywords Fractional differential equations · Collocation · Two-step collocation · Convergence

1 Introduction

Several models of fractures, damages and fatigue of materials involve fractional differential operators. Some interesting models have been recently proposed in [8, 26, 45, 52, 55]. Furthermore, fractional differential equations model many problems regarding anomalous diffusion and visco-elastic materials behavior. These and other important applications may be found in [5, 6, 31, 44, 48, 53].

A. Cardone (✉) · D. Conte
Associate Professor, Dipartimento di Matematica, Università di Salerno, Via Giovanni Paolo II n. 132, 84084 Fisciano, Salerno, Italy
e-mail: ancardone@unisa.it

D. Conte
e-mail: dajconte@unisa.it

B. Paternoster
Full Professor, Dipartimento di Matematica, Università di Salerno, Via Giovanni Paolo II n. 132, 84084 Fisciano, Salerno, Italy
e-mail: beapat@unisa.it

M. Abdel Wahab (ed.), *Proceedings of the 8th International Conference on Fracture, Fatigue and Wear*, Lecture Notes in Mechanical Engineering,
https://doi.org/10.1007/978-981-15-9893-7_21

In the present paper we treat the numerical solution of fractional differential problems of this form

$$\begin{cases} D^\alpha y(t) = f(t, y(t)), \; 0 \le t \le b \\ y^{(i)}(0) = \gamma_i, \qquad\qquad i = 0, \dots, n-1 \end{cases} \tag{1.1}$$

where $n-1 < \alpha < n$, $n \in \mathbb{N}$, $\gamma_i \in \mathbb{R}$, $f : [0, b] \times \mathbb{R} \to \mathbb{R}$ is a given continuous function. The fractional derivative is of Caputo type, i.e.:

$$D^\alpha y(t) := \frac{1}{\Gamma(n-\alpha)} \int_0^t \frac{y^{(n)}(s)}{(t-s)^{\alpha+1-n}} ds.$$

The analytical solution of fractional differential equations (FDEs) is known only in very simple cases. Therefore a numerical approach is necessary. Due to the increasing interest in fractional differential models, many numerical methods have been introduced, as fractional linear methods [47], product integration methods [34, 38, 39], collocation methods [2, 5, 9, 12, 42, 46, 50, 51, 54, 57], Adomian decomposition methods [28, 37, 43], spectral methods [5, 15, 56, 57]. One of the main drawbacks of most methods is the low order of convergence, due to the lack of smoothness of the solution near the origin. However, this difficulty may be overcome by spline collocation methods, which can reach high order of convergence when applied on suitable graded meshes. Moreover, collocation methods have also other desirable properties, as dense output and strong stability properties, as proven in the case of ODEs and of Volterra integral equations [4, 7, 10, 11, 19, 23, 30, 49]. Here we focus on a class of one and two step spline collocation methods, recently proposed and analyzed in [9, 12, 50, 51].

In the following section, we illustrate the formulation of one and two step spline collocation methods. In Sect. 3 we recall main convergence results and give some hints on the choice of method parameters. In Sect. 4 we provide some numerical experiments, which confirm the theoretical order, and compare accuracy and efficiency of both classes of spline collocation methods.

2 Spline Collocation Methods

This section describes: one-step spline collocation methods, introduced by Pedas and Tamme in [51] and further analyzed in [9]; two-step spline collocation methods proposed by Cardone, Conte and Paternoster in [12]. Both classes of methods will be referred to as '*spline collocation methods*'.

The fractional differential problem (1.1) is equivalent to

$$z = f(t, J^\alpha z + Q). \tag{2.1}$$

where $z = D^\alpha y$, $J^0 z := z$,

$$(J^\alpha z)(t) = \frac{1}{\Gamma(\alpha)} \int_0^t (t-s)^{\alpha-1} z(s)\, ds, \quad t > 0, \tag{2.2}$$

$$Q(t) = \sum_{i=0}^{\lceil\alpha\rceil-1} \frac{\gamma_i}{i!} t^i, \tag{2.3}$$

$\lceil\alpha\rceil$ is the least integer not less than α. Then the solution of (1.1) is

$$y = J^\alpha z + Q. \tag{2.4}$$

Put $I_N = \{0 = t_0 < \cdots < t_N = b\}$, with $h_j = t_j - t_{j-1}$ and $\sigma_j = [t_{j-1}, t_j]$, for $j = 1, \ldots, N$. Let us define the piecewise polynomial space

$$S_k^{(-1)}(I_N) = \left\{ v : v|_{\sigma_j} \in \pi_k,\ j = 1, \ldots, N \right\}, \tag{2.5}$$

where π_k is the space of algebraic polynomials of degree less or equal to k.

Spline collocation methods provide an approximate solution $v \in S_k^{(-1)}(I_N)$ of Eq. (2.1). Then, the numerical solution of problem (1.1) is computed as

$$y_N = J^\alpha v + Q. \tag{2.6}$$

The collocation solution $v(t)$ satisfy the FDE (1.1) at the collocation points

$$t_{jk} = t_{j-1} + \eta_k h_j, \quad k = 1, \ldots, m,\ j = 1, \ldots, N, \tag{2.7}$$

where

$$0 \le \eta_1 < \eta_2 < \cdots < \eta_m \le 1, \tag{2.8}$$

are fixed collocation parameters. In the case of two-step collocation methods $(\eta_1, \eta_m) \ne (0, 1)$.

2.1 *One-Step Spline Collocation Methods*

In the one-step collocation methods, the approximate the solution v of (2.1) belongs to $S_{m-1}^{(-1)}(I_N)$, and can be expressed as

$$v(t) = \sum_{j=1}^{N} \sum_{k=1}^{m} z_{jk} L_{jk}(t), \quad t \in [0, b]. \tag{2.9}$$

For each $j = 1, \ldots, N$,

$$L_{jk}(t) = \begin{cases} k\text{-th Lagrange fund. pol. wrt } \{t_{jk}\}_{k=1}^m, & t \in [t_{j-1}, t_j], \\ 0 & \text{elsewhere}, \end{cases} \quad k = 1, \ldots, m,$$

while the unknowns $z_{jk} = v(t_{jk})$, $k = 1, \ldots, m$, are the solution of the nonlinear system

$$z_{jk} = f\left(t_{jk}, (J^\alpha v)(t_{jk}) + Q(t_{jk})\right), \quad k = 1, \ldots, m. \tag{2.10}$$

Nonlinear system (2.10) is equivalent to

$$z_{jk} = f\left(t_{jk}, \sum_{\mu=1}^{m} (J^\alpha L_{j\mu})(t_{jk}) z_{j\mu} + \sum_{l=1}^{j-1} \sum_{\mu=1}^{m} (J^\alpha L_{l\mu})(t_{jk}) z_{l\mu} + Q(t_{jk})\right), \tag{2.11}$$

$k = 1, \ldots, m$. Furthermore, the fractional integrals appearing in (2.11), can be expressed as follows

$$(J^\alpha L_{j\mu})(t_{jk}) = h_j^\alpha \sum_{\nu=0}^{m-1} a_\nu^{(\mu)} c_k^{\nu+\alpha} \frac{\Gamma(1+\nu)}{\Gamma(1+\nu+\alpha)},$$

$$(J^\alpha L_{l\mu})(t_{jk}) = \frac{1}{\Gamma(\alpha)} h_l^\alpha \sum_{\nu=0}^{m-1} a_\nu^{(\mu)} \left(\frac{t_{j-1} + c_k h_j - t_{l-1}}{h_l}\right)^{\nu+\alpha} \mathrm{B}\left(\frac{h_l}{t_{j-1} + c_k h_j - t_{l-1}}; 1+\nu, \alpha\right), \quad l = 1, \ldots, j-1,$$

$k = 1, \ldots, m$. In the previous formula $\mathrm{B}(z; a, b)$ is the incomplete beta function:

$$\mathrm{B}(z; a, b) := \int_0^z \sigma^{a-1} (1-\sigma)^{b-1} \, d\sigma,$$

As regards the definition and the computation of coefficients $a_\nu^{(\mu)}$, compare [9, Sec. 3].

2.2 *Two-Step Spline Collocation Methods*

Two-step collocation methods approximate the solution of (2.1) as the piecewise polynomial $v \in S_{2m-1}^{(-1)}(I_N)$ with

$$v(t) = v_1(t) + \sum_{j=2}^{N} \left(\sum_{k=1}^{m} z_{jk} L_{j,m+k}^{(2)}(t) + \sum_{k=1}^{m} z_{j-1,k} L_{jk}^{(2)}(t)\right), \quad t \in [0, b]. \tag{2.12}$$

For each $j = 2, \ldots, N$,

$$L_{j\mu}^{(2)}(t) = \begin{cases} \mu\text{-th Lagrange fund. pol. wrt } \{t_{j-1,k}, t_{jk}\}_{k=1}^{m}, & t \in [t_{j-1}, t_j], \\ 0 & \text{elsewhere}, \end{cases}$$

$\mu = 1, \ldots, 2m$; $\{z_{jk} = v(t_{jk}) | k = 1, \ldots, m\}$ satisfy the nonlinear system

$$z_{jk} = f\left(t_{jk}, (J^{\alpha}v)(t_{jk}) + Q(t_{jk})\right), \quad k = 1, \ldots, m. \tag{2.13}$$

Finally,

$$z_{1k} = v_1(t_{1k}), \quad k = 1, \ldots, m,$$

where function v_1 is chosen by the starting procedure and satisfies $v_1(t) = 0$, $t \in (t_1, b)$. For more details compare [12].

The nonlinear system (2.13) can be written as

$$\begin{aligned} z_{jk} = \\ f\Bigg(& t_{jk}, (J^{\alpha}v_1)(t_{jk}) + \sum_{\mu=1}^{m} z_{j\mu}(J^{\alpha}L_{j,m+\mu}^{(2)})(t_{jk}) + \sum_{\lambda=2}^{j-1}\sum_{\mu=1}^{m} z_{\lambda\mu}(J^{\alpha}L_{\lambda,m+\mu}^{(2)})(t_{jk}) \\ & + \sum_{\mu=1}^{m} z_{j-1,\mu}(J^{\alpha}L_{j\mu}^{(2)})(t_{jk}) + \sum_{\lambda=2}^{j-1}\sum_{\mu=1}^{m} z_{\lambda-1,\mu}(J^{\alpha}L_{\lambda\mu}^{(2)})(t_{jk}) + Q(t_{jk})\Bigg), \end{aligned} \tag{2.14}$$

$k = 1, \ldots, m$. The fractional integrals appearing in the above system can be computed as follows [12]

$$(J^{\alpha}L_{j\mu}^{(2)})(t_{jk}) = h_j^{\alpha}\sum_{l=0}^{2m-1} a_l^{(j\mu)}\eta_k^{l+\alpha}\frac{\Gamma(1+l)}{\Gamma(1+l+\alpha)}, \quad \mu = 1, \ldots, 2m.$$

$$\begin{aligned}(J^{\alpha}L_{\lambda\mu}^{(2)})(t_{jk}) = \frac{h_\lambda^{\alpha}}{\Gamma(\alpha)}\sum_{l=0}^{2m-1} a_l^{(\lambda\mu)}\left(\frac{t_{j-1}+\eta_k h_j - t_{\lambda-1}}{h_\lambda}\right)^{l+\alpha} & \\ \mathrm{B}\left(\frac{h_\lambda}{t_{j-1}+\eta_k h_j - t_{\lambda-1}}; 1+l, \alpha\right), & \quad \lambda = 2, \ldots, j-1.\end{aligned}$$

Coefficients $a_l^{(j\mu)}$ have been defined in [12, (3.18)].

3 Convergence Analysis

It is the purpose of this section to provide some simple information to set the optimal method parameters, to achieve the maximum order of convergence of spline collocation methods. Thus, we briefly summarize the convergence analysis carried out in [12, 51] and point out most significant results in Tables 1 and 2.

To this aim, we recall an important theorem on the smoothness of the solution. Further results on existence, uniqueness and regularity of the solution are contained in [32, 33, 44, 51, 53].

First let define the functional space $C^{q,\nu}(0, b]$, with $q \in \mathbb{N}$ and $\nu \in (-\infty, 1)$. A real-valued function y, defined in $[0, b]$, belongs to $C^{q,\nu}(0, b]$ if it belongs to $C^q(0, b]$ and

$$|y^{(i)}(t)| \leq c \begin{cases} 1, & \text{if } i < 1-\nu, \\ 1 + |\log t|, & \text{if } i = 1-\nu, \\ t^{1-\nu-i}, & \text{if } i > 1-\nu, \end{cases} \quad t \in (0, b], \quad i = 1, \ldots, q.$$

Theorem 1 [51] *Assume that: the real-valued function* $f \in C^q([0, b] \times \mathbb{R})$*; there exists* $\nu \in [1-\alpha, 1)$ *such that:*

$$\left| \frac{\partial^{i+j}}{\partial t^i \partial y^j} f(t, y) \right| \leq \phi(|y|) \begin{cases} 1, & \text{if } i < 1-\nu \\ 1 + |\log t|, & \text{if } i = 1-\nu \\ t^{1-\nu-i}, & \text{if } i > 1-\nu \end{cases}, \quad (t, y) \in (0, b] \times \mathbb{R},$$

$\forall i, j \in \mathbb{N}$ *with* $i + j \leq q$*. For* $\alpha \in (0, 1)$*, assume also that*

$$\left| \frac{\partial^{i+j}}{\partial t^i \partial y^j} [f(t, y_1) - f(t, y_2)] \right| \leq \phi(\max\{|y_1|, |y_2|\}) |y_1 - y_2| \begin{cases} 1 & \text{if } i = 0 \\ t^{1-\nu-i} & \text{if } i > 0 \end{cases},$$

$(t, y_i) \in (0, b] \times \mathbb{R}$, $i = 1, 2$*; where the real valued function* ϕ *is monotonically increasing in* $[0, \infty)$*. Assume that there exists a solution* $y \in C[0, b]$ *of the problem* (1.1)*, with* $D^\alpha y \in C[0, b]$*. Then* $y \in C^{q,\nu}(0, b]$ *and* $D^\alpha y \in C^{q,\nu}(0, b]$*.*

The crucial ingredient of spline collocation methods for FDEs, to obtain high order of convergence, is the graded mesh, i.e. the mesh of points satisfies

$$t_j = b\left(\frac{j}{N}\right)^r, \quad j = 0, \ldots, N, \tag{3.1}$$

where $r \in \mathbb{R}$, $r \geq 1$ is called grading exponent. The optimal choice of r, with respect to the order of convergence, depends both on the number of collocation parameters m and on the smoothness of the analytical solution. For a rigorous exposition of the convergence analysis the reader may compare [12, 51].

We define

$$E_N(m,\nu,r) = \begin{cases} N^{-r(1-\nu)}, & \text{if } 1 \le r < \frac{m}{1-\nu}, \\ N^{-m}(1+\log N), & \text{if } r = \frac{m}{1-\nu} = 1, \\ N^{-m}, & \text{if } r = \frac{m}{1-\nu} > 1 \text{ or } r > \frac{m}{1-\nu}. \end{cases} \tag{3.2}$$

Let assume that Theorem 1 holds with $q = m$ and $\nu \in [1-\alpha, 1)$. If additional suitable hypotheses are fulfilled, it can be proven that for the one step spline collocation method (2.9) and (2.10):

$$\|y_N - y\|_\infty \le C E_N(m,\nu,r),$$

where the constant C does not depend on N. (cfr. [51, Th. 4.1]).

Moreover, if hypotheses of Theorem 1 are fulfilled with $q = m + \min\{m, \lceil\alpha\rceil\}$ and $\nu \in [1-\alpha, 1)$; if collocation parameters $\eta_1, \ldots, \eta_m$ are the nodes of the Gauss-Lagrange rule in [0, 1] and if other suitable assumptions are satisfied, then the error the one step spline collocation method (2.12) and (2.13) satisfies

$$\|y_N - y\|_\infty \le C\Phi_N(m,\nu,r),$$

where $\Phi_N(m,\nu,r) \le const N^{\min\{m+\alpha,2m\}}$. All details are provided in [51, Th. 4.2].

Now let consider the two step spline collocation method (2.12) and (2.13). If Theorem 1 holds with $q = 2m$ and $\nu \in [1-\alpha, 1)$, and if other suitable hypotheses are fulfilled, then

$$\|y_N - y\|_\infty \le C E_N(2m,\nu,r),$$

where the constant C does not depend on N (cfr. [12, Th. 4.5]).

As a useful guidance to the reader, we outlined in Table 2 the maximum attainable order of spline collocation methods, if the grading exponent r is set as suggested by Table 1.

Table 1 Optimal choices for the grading exponent

Optimal choices for the grading exponent r
• one-step method with arbitrary nodes: $r = \dfrac{m}{1-\nu} > 1$ or $r > \dfrac{m}{1-\nu}$
• one-step method with Gaussian nodes:
– if $\alpha \in (0,1)$, $r = \frac{m+\alpha}{1+\alpha-\nu} > 1$ or $r > \frac{m+\alpha}{1+\alpha-\nu}$;
– if $\alpha \in (1,m)$, $r > \frac{m+\alpha}{2-\nu}$,
• two-step method: $r = \dfrac{2m}{1-\nu} > 1$ or $r > \dfrac{2m}{1-\nu}$

Table 2 Maximum attainable order of convergence of spline collocation methods

Collocation parameters $0 \leq \eta_1 < \cdots \leq \eta_m \leq 1$
• arbitrary points in [0, 1]
$-p = m$ for one-step collocation method
$-p = 2m$ for two-step collocation method[a]
• Gauss-Legendre nodes shifted in [0, 1]
$-p \leq \min(m + \alpha, 2m)$ for one-step collocation method
$-p = 2m$ for two-step collocation method

[a]applicable with $(\eta_1, \eta_m) \neq (0, 1)$

4 Numerical Experiments

We now compare the performances of one and two step collocation methods, by showing the numerical results obtained on the following test problems:
Problem 1 [51]

$$D^{1/2}y(t) = y^2(t) + \frac{1}{\Gamma(1.5)}t^{0.5} - t^2, \ t \in [0, 1] \quad y(0) = 0,$$

Exact solution: $y(t) = t$. The hypotheses of Theorem 1 are fulfilled by $\nu = 0.5$ and any $q \in \mathbb{N}$
Problem 2 [51]

$$D^{1.2}y(t) = y^4(t) + \frac{\Gamma(2.8)}{\Gamma(1.6)}t^{0.6} - (t^{1.8} - t + 1)^4, \quad t \in [0, 1],$$
$$y(0) = 0, y'(0) = -1,$$

Exact solution $y(t) = t^{1.8} - t + 1$. The hypotheses of Theorem 1 are fulfilled by $\nu = 0.4$ and any $q \in \mathbb{N}$.
Problem 3 [35]

$$D^{\alpha}y(t) = \frac{40320}{\Gamma(9-\alpha)}t^{8-\alpha} - 3\frac{\Gamma(5+\frac{\alpha}{2})}{\Gamma(5-\frac{\alpha}{2})}t^{4-\frac{\alpha}{2}} + \left(\frac{3}{2}t^{\frac{\alpha}{2}} - t^4\right)^3 + \frac{9}{4}\Gamma(\alpha+1) - (y(t))^{\frac{3}{2}}, \quad t \in [0, 1],$$
$$y(0) = 0,$$

$\alpha = 1/2$. Exact solution $y = t^8 - 3t^{4+\alpha/2} + \frac{9}{4}t^{\alpha}$. The hypotheses of Theorem 1 are fulfilled by $\nu = 0.5$ and any $q \in \mathbb{N}$.

In the following tables we list

- the error e_N, computed as

$$e_N = \max_{j=1,\dots,N} \max_{k=0,\dots,10} |y_N(\tau_{jk}) - y(\tau_{jk})|,$$

with $\tau_{jk} = t_{j-1} + k(t_j - t_{j-1})/10,\ k = 0, \dots, 10, j = 1, \dots, N$;
- the number of correct significant digits

$$\text{cd} = \log_{10}(e_N);$$

- the experimental order

$$p_{\text{eff}} = \frac{\log_{10}\left(e_{\frac{N}{2}}\right) - \log_{10}(e_N)}{\log_{10} 2};$$

- 'fval', that is the number of evaluations of the function f.

In Tables 3, 4, 5, 6, 7 and 8 we show the results obtained by spline collocation methods with $m = 3$. In Tables 3, 4 and 5 the collocation parameters are equal to Gauss-Legendre nodes shifted in [0, 1], i.e.

- $\eta = \left[\frac{5-\sqrt{15}}{10}, \frac{1}{2}, \frac{5+\sqrt{15}}{10}\right]$,

while in Tables 6–8 the collocation parameters are equidistant points in [0, 1], in particular

- for $m = 3$, $\eta = \left[\frac{1}{4}, \frac{1}{2}, \frac{3}{4}\right]$,
- for $m = 6$, $\eta = \left[\frac{1}{7}, \frac{2}{7}, \frac{3}{7}, \frac{4}{7}, \frac{5}{7}, \frac{6}{7}\right]$.

Both for the one-step and for the two-step methods, we adopt a graded mesh with exponent r such that the attainable order is maximum, i.e. the order is m for the one-step methods with equidistant nodes, $m + \alpha$ for one-step methods with Gauss-Legendre nodes and $2m$ for the two-step methods (compare Table 2). Therefore, by convergence results summarized in Sect. 3 and in particular in Table (1),

- for the one-step methods with equidistant collocation parameters we set $r = \dfrac{m}{1-\nu}$;
- for the one-step methods with Gauss-Legendre collocation parameters we set $r = \dfrac{m+\alpha}{1+\alpha-\nu}$;
- for the two-step methods we set $r = \dfrac{2m}{1-\nu}$.

In all the experiments the theoretical order of convergence is numerically confirmed.

From the results in Tables 3, 4 and 5 we observe as, on Problems 1 and 2, two-step methods with the same number of stages $m = 3$ (and so a comparable number of function evaluations) are able to achieve a smaller error with respect to one-step methods, despite the one step method is superconvergent for this choice of collocation parameters.

Table 3 Results obtained on Problem 1 with Gauss-Legendre collocation parameters. The expected order for two-step method is $p = 6$; the expected order for one-step method is $p = 3.5$

N	two-step $m = 3$				one-step $m = 3$			
	e_N	cd	p_{eff}	fval	e_N	cd	p_{eff}	fval
8	0.05	1.27		348	4.71e−05	4.33		396
16	2.09e−04	3.68	8.01	636	3.83e−06	5.42	3.62	744
32	1.35e−06	5.87	7.27	1200	3.14e−07	6.50	3.61	1416
64	1.25e−08	7.90	6.76	2292	2.66e−08	7.57	3.56	2712
128	1.53e−10	9.81	6.35	4464	2.44e−09	8.61	3.45	5124
256	1.93e−12	11.71	6.32	8568	2.20e−10	9.66	3.47	9456

Table 4 Results obtained on Problem 2 with Gauss-Legendre collocation parameters. The expected order for two-step method is $p = 6$; the expected order for one-step method is $p = 4.2$

N	two-step $m = 3$				one-step $m = 3$			
	e_N	cd	p_{eff}	fval	e_N	cd	p_{eff}	fval
8	0.04	1.35		276	6.61e−06	5.18		384
16	7.05e−03	2.15	2.66	480	3.56e−07	6.45	4.22	708
32	3.61e−04	3.44	4.29	780	1.87e−08	7.73	4.25	1356
64	8.87e−06	5.05	5.35	1524	9.74e−10	9.01	4.27	2460
128	1.50e−07	6.82	5.88	2952	5.10e−11	10.29	4.25	4596
256	2.09e−09	8.68	6.16	5796	2.67e−12	11.57	4.26	9204

Table 5 Results obtained on Problem 3 with Gauss-Legendre nodes. The expected order for two-step method is $p = 6$; the expected order for one-step method is $p = 3.5$

N	two-step $m = 3$				one-step $m = 3$			
	e_N	cd	p_{eff}	fval	e_N	cd	p_{eff}	fval
8	2.83e−03	2.55		312	0.02	1.80		336
16	9.06e−06	5.04	8.29	624	2.56e−03	2.59	2.64	636
32	3.97e−08	7.40	7.83	1188	3.03e−04	3.52	3.08	1176
64	2.62e−10	9.58	7.25	2316	3.08e−05	4.51	3.30	2256
128	4.84e−12	11.31	5.76	4500	2.90e−06	5.54	3.41	4380
256	8.49e−14	13.07	5.83	8676	2.64e−07	6.58	3.46	8568

In case of equidistant collocation parameters, in Tables 6, 7 and 8, the two step method with $m = 3$ and $p = 6$ is compared with one-step method having the same number of stages $m = 3$ (with $p = 3$) and the one-step method having the same order $p = 6$ (with $m = 6$). The error of two-step method exhibits a significantly smaller error with respect to the one step method with $m = 3$ on all the three problems; while it shows a similar error with a relevantly smaller number of function evaluations, when compared to one-step method with $m = 6$.

Table 6 Results obtained on Problem 1 with equidistant nodes

N	two-step $m = 3$				one-step $m = 3$				one-step $m = 6$			
	e_N	cd	p_{eff}	fval	e_N	cd	p_{eff}	fval	e_N	cd	p_{eff}	fval
8	0.13	0.87		312	1.85e−03	2.73		372	6.84e−04	3.16		1176
16	1.44e−03	2.84	6.54	600	1.46e−04	3.83	3.66	708	1.66e−05	4.78	5.37	2184
32	1.78e−05	4.75	6.34	1176	1.20e−05	4.92	3.61	1320	2.85e−07	6.54	5.86	4200
64	2.40e−07	6.62	6.21	2268	9.74e−07	6.01	3.62	2544	4.53e−09	8.34	5.97	8148
128	3.43e−09	8.46	6.13	4404	8.29e−08	7.08	3.56	4872	7.06e−11	10.15	6.00	15,624
256	5.09e−11	10.29	6.08	8532	7.07e−09	8.15	3.55	9312	8.85e−13	12.05	6.321	30,408

Table 7 Results obtained on Problem 2 with equidistant nodes

N	two-step $m = 3$				one-step $m = 3$				one-step $m = 6$			
	e_N	cd	p_{eff}	fval	e_N	cd	p_{eff}	fval	e_N	cd	p_{eff}	fval
8	0.02	1.79		312	5.05e−04	3.30		348	1.37e−04	3.86		1404
16	1.73e−04	3.76	6.53	600	3.31e−05	4.48	3.93	660	2.36e−06	5.63	5.86	2268
32	2.12e−06	5.67	6.35	1176	2.16e−06	5.67	3.94	1284	3.75e−08	7.43	5.97	4284
64	2.91e−08	7.54	6.19	2316	1.38e−07	6.86	3.96	2472	5.90e−10	9.23	5.99	8190
128	4.25e−10	9.37	6.10	4452	8.76e−09	8.06	3.98	4632	9.16e−12	11.04	6.01	15,918
256	6.43e−12	11.19	6.05	8652	5.51e−10	9.26	3.99	9120	7.92e−14	13.10	6.85	30,744

Table 8 Results obtained on Problem 3 with equidistant nodes

N	two-step $m = 3$				one-step $m = 3$				one-step $m = 6$			
	e_N	cd	p_{eff}	fval	e_N	cd	p_{eff}	fval	e_N	cd	p_{eff}	fval
8	0.11	0.94		240	0.12	0.93		276	7.79e−03	2.11		882
16	0.02	1.71	2.56	480	0.03	1.50	1.91	528	7.59e−04	3.12	3.36	1554
32	1.06e−03	2.97	4.18	780	4.86e−03	2.31	2.68	1020	2.83e−05	4.55	4.75	2988
64	2.78e−05	4.56	5.26	1524	5.74e−04	3.24	3.08	1932	6.03e−07	6.22	5.55	5544
128	5.00e−07	6.30	5.80	2964	5.90e−05	4.23	3.28	3744	9.75e−09	8.01	5.95	10,710
256	7.53e−09	8.12	6.05	5796	5.66e−06	5.25	3.38	7320	1.39e−10	9.86	6.13	20,580

5 Concluding Remarks

In this paper we considered a class of one and two step collocation methods for FDEs, introduced in [12, 51]. We described the formulation of the methods in some detail and briefly resumed main convergence results. Then we showed their performances on some significant test problems from the specialized literature.

Further developments may be carried out in the field of numerical solution of FDEs. One possibility consists in exploring suitable extensions of other numerical methods for ODEs, e.g. peer methods and general linear methods [3, 17, 18, 20, 22, 25, 41]. Another area of future investigation may regard adapted methods for FDEs with oscillatory behavior [1, 27, 36]; in such cases suitable numerical methods may be introduced by exploiting exponential fitting techniques (compare [13, 14, 16, 21, 24, 29, 40]).

Acknowledgements The authors are members of the INdAM Research group GNCS. This work is supported by GNCS-INDAM project and by PRIN2017-MIUR project.

References

1. Belmekki M, Nieto JJ, Rodríguez-López R (2009) Existence of periodic solution for a nonlinear fractional differential equation. Bound Value Probl pp. Art. ID 324,561, 18. DOI https://doi.org/10.1155/2009/324561
2. Blank L (1996) Numerical treatment of differential equations of fractional order. Tech. rep., University of Manchester, Department of Mathematics. Numerical Analysis Report
3. Braś M, Cardone A, Jackiewicz Z, Pierzchała P (2018) Error propagation for implicit-explicit general linear methods. Appl Numer Math 131:207–231. https://doi.org/10.1016/j.apnum.2018.05.004
4. Brunner H (2004) Collocation methods for Volterra integral and related functional differential equations, Cambridge Monographs on Applied and Computational Mathematics, vol 15. Cambridge University Press, Cambridge. https://doi.org/10.1017/CBO9780511543234
5. Burrage K, Cardone A, D'Ambrosio R, Paternoster B (2017) Numerical solution of time fractional diffusion systems. Appl Numer Math 116:82–94. https://doi.org/10.1016/j.apnum.2017.02.004
6. Califano G, Conte D (2018) Optimal Schwarz waveform relaxation for fractional diffusion-wave equations. Appl Numer Math 127:125–141. https://doi.org/10.1016/j.apnum.2018.01.002
7. Capobianco G, Conte D, Paternoster B (2017) Construction and implementation of two-step continuous methods for Volterra integral equations. Appl Numer Math 119:239–247. https://doi.org/10.1016/j.apnum.2017.02.008
8. Caputo M, Fabrizio M (2015) Damage and fatigue described by a fractional derivative model. J Comput Phys 293:400–408. https://doi.org/10.1016/j.jcp.2014.11.012
9. Cardone A, Conte D (2020) Stability analysis of spline collocation methods for fractional differential equations. Math Comput Simul (to appear). https://doi.org/10.1016/j.matcom.2020.07.004
10. Cardone A, Conte D, D'Ambrosio R, Paternoster B (2018) Collocation methods for Volterra integral and integro-differential equations: a review. Axioms 7(3). DOI https://doi.org/10.3390/axioms7030045
11. Cardone A, Conte D, Paternoster B (2009) A family of multistep collocation methods for Volterra integro-differential equations. AIP Conf Proc 1168(1):358–361. https://doi.org/10.1063/1.3241469
12. Cardone A, Conte D, Paternoster B (2018) Two-step collocation methods for fractional differential equations. Discrete Contin Dyn Syst Ser B 23(7):2709–2725. https://doi.org/10.3934/dcdsb.2018088
13. Cardone A, D'Ambrosio R, Paternoster B (2017) Exponentially fitted IMEX methods for advection-diffusion problems. J Comput Appl Math 316:100–108. https://doi.org/10.1016/j.cam.2016.08.025
14. Cardone A, D'Ambrosio R, Paternoster B (2017) High order exponentially fitted methods for Volterra integral equations with periodic solution. Appl Numer Math 114:18–29. https://doi.org/10.1016/j.apnum.2016.05.003
15. Cardone A, D'Ambrosio R, Paternoster B (2019) A spectral method for stochastic fractional differential equations. Appl Numer Math 139:115–119. https://doi.org/10.1016/j.apnum.2019.01.009
16. Cardone A, Ixaru LG, Paternoster B, Santomauro G (2015) Ef-Gaussian direct quadrature methods for Volterra integral equations with periodic solution. Math Comput Simul 110:125–143. https://doi.org/10.1016/j.matcom.2013.10.005

17. Cardone A, Jackiewicz Z, Mittelmann H (2012) Optimization-based search for Nordsieck methods of high order with quadratic stability polynomials. Math Model Anal 17(3):293–308. https://doi.org/10.3846/13926292.2012.685497
18. Cardone A, Jackiewicz Z, Verner JH, Welfert B (2015) Order conditions for general linear methods. J Comput Appl Math 290:44–64. https://doi.org/10.1016/j.cam.2015.04.042
19. Conte D, D'Ambrosio R, Jackiewicz Z, Paternoster B (2013) Numerical search for algebraically stable two-step almost collocation methods. J Comput Appl Math 239:304–321. https://doi.org/10.1016/j.cam.2012.08.012
20. Conte D, D'Ambrosio R, Moccaldi M, Paternoster B (2019) Adapted explicit two-step peer methods. J Numer Math 27(2):69–83. https://doi.org/10.1515/jnma-2017-0102
21. Conte D, Ixaru LG, Paternoster B, Santomauro G (2014) Exponentially-fitted Gauss-Laguerre quadrature rule for integrals over an unbounded interval. J Comput Appl Math 255:725–736. https://doi.org/10.1016/j.cam.2013.06.040
22. Conte D, Mohammadi F, Moradi L, Paternoster B (2020) Exponentially fitted two-step peer methods for oscillatory problems. Comput Appl Math 39(3). https://doi.org/10.1007/s40314-020-01202-x
23. Conte D, Paternoster B (2009) Multistep collocation methods for Volterra integral equations. Appl Numer Math 59(8):1721–1736. https://doi.org/10.1016/j.apnum.2009.01.001
24. Conte D, Paternoster B (2016) Modified Gauss-Laguerre exponential fitting based formulae. J Sci Comput 69(1):227–243. https://doi.org/10.1007/s10915-016-0190-0
25. Conte D, Paternoster B, Moradi L, Mohammadi F (2019) Construction of exponentially fitted explicit peer methods. Int J Circ Syst Signal Process 13:501–506
26. Cottone G, Di Paola M, Zingales M (2007) Dynamics of non-local systems handled by fractional calculus. In: Kartalopoulos S, Demiralp M, Mastorakis N, Soni R, Nassar H (eds) Selected topics on circuits, systems, electronics, control & signal processing, electrical and computer engineering series, pp 81–89. WSEAS (2007). Proceedings of the 6th WSEAS international conference on circuits, systems, electronics, control & signal processing (CSECS 07), Cairo, Egypt, Dec 29–31, 2007
27. Cuevas C, Sepúlveda A, Soto H (2011) Almost periodic and pseudo-almost periodic solutions to fractional differential and integro-differential equations. Appl Math Comput 218(5):1735–1745. https://doi.org/10.1016/j.amc.2011.06.054
28. Daftardar-Gejji V, Jafari H (2005) Adomian decomposition: a tool for solving a system of fractional differential equations. J Math Anal Appl 301(2):508–518. https://doi.org/10.1016/j.jmaa.2004.07.039
29. D'Ambrosio R, Paternoster B (2014) Exponentially fitted singly diagonally implicit Runge-Kutta methods. J Comput Appl Math 263:277–287. https://doi.org/10.1016/j.cam.2013.12.022
30. D'Ambrosio R, Paternoster B (2019) Multivalue collocation methods free from order reduction. J Comput Appl Math. https://doi.org/10.1016/j.cam.2019.112515
31. De Luca P, Galletti A, Ghehsareh H, Marcellino L, Raei M (2020) A GPU-CUDA framework for solving a two-dimensional inverse anomalous diffusion problem. Adv Parallel Comput 36:311–320. https://doi.org/10.3233/APC200056
32. Diethelm K (2007) Smoothness properties of solutions of Caputo-type fractional differential equations. Fract Calc Appl Anal 10(2):151–160
33. Diethelm K (2010) The analysis of fractional differential equations, vol 2004. Lecture Notes in Mathematics. Springer, Berlin
34. Diethelm K, Ford NJ, Freed AD (2002) A predictor-corrector approach for the numerical solution of fractional differential equations. Nonlinear Dyn 29(1–4):3–22. https://doi.org/10.1023/A:1016592219341
35. Diethelm K, Ford NJ, Freed AD (2004) Detailed error analysis for a fractional Adams method. Numer Algorithms 36(1):31–52. https://doi.org/10.1023/B:NUMA.0000027736.85078.be
36. Duan JS (2013) The periodic solution of fractional oscillation equation with periodic input. Adv Math Phys pp Art. ID 869,484, 6. https://doi.org/10.1155/2013/869484

37. Duan JS, Chaolu T, Rach R, Lu L (2013) The Adomian decomposition method with convergence acceleration techniques for nonlinear fractional differential equations. Comput Math Appl 66(5):728–736. https://doi.org/10.1016/j.camwa.2013.01.019
38. Garrappa R (2010) On linear stability of predictor-corrector algorithms for fractional differential equations. Int J Comput Math 87(10):2281–2290. https://doi.org/10.1080/00207160802624331
39. Garrappa R, Popolizio M (2011) On accurate product integration rules for linear fractional differential equations. J Comput Appl Math 235(5):1085–1097. https://doi.org/10.1016/j.cam.2010.07.008
40. Ixaru LG, Vanden Berghe G (2004) Exponential fitting, mathematics and its applications, vol 568. Kluwer Academic Publishers, Dordrecht. https://doi.org/10.1007/978-1-4020-2100-8. With 1 CD-ROM (Windows, Macintosh and UNIX)
41. Jackiewicz Z (2009) General linear methods for ordinary differential equations. Wiley, Hoboken, NJ. https://doi.org/10.1002/9780470522165
42. Khader MM (2011) On the numerical solutions for the fractional diffusion equation. Commun Nonlinear Sci Numer Simul 16(6):2535–2542. https://doi.org/10.1016/j.cnsns.2010.09.007
43. Khodabakhshi N, Vaezpour SM, Baleanu D (2014) Numerical solutions of the initial value problem for fractional differential equations by modification of the Adomian decomposition method. Fract Calc Appl Anal 17(2):382–400. https://doi.org/10.2478/s13540-014-0176-2
44. Kilbas AA, Srivastava HM, Trujillo JJ (2006) Theory and applications of fractional differential equations, North-Holland Mathematics Studies, vol 204. Elsevier Science B.V, Amsterdam
45. Kobelev V (2017) A proposal for unification of fatigue crack growth law. J Phys Conf Ser 843(1):012,022. https://doi.org/10.1088/1742-6596/843/1/012022
46. Li X (2012) Numerical solution of fractional differential equations using cubic B-spline wavelet collocation method. Commun Nonlinear Sci Numer Simul 17(10):3934–3946. https://doi.org/10.1016/j.cnsns.2012.02.009
47. Lubich C (1985) Fractional linear multistep methods for Abel-Volterra integral equations of the second kind. Math Comp 45(172):463–469. https://doi.org/10.2307/2008136
48. Mainardi F (2010) Fractional calculus and waves in linear viscoelasticity. Imperial College Press, London. https://doi.org/10.1142/9781848163300
49. Paternoster B (2000) A phase-fitted collocation-based Runge-Kutta-Nyström method. Appl Numer Math 35(4):339–355. https://doi.org/10.1016/S0168-9274(99)00143-9
50. Pedas A, Tamme E (2011) On the convergence of spline collocation methods for solving fractional differential equations. J Comput Appl Math 235(12):3502–3514. https://doi.org/10.1016/j.cam.2010.10.054
51. Pedas A, Tamme E (2014) Numerical solution of nonlinear fractional differential equations by spline collocation methods. J Comput Appl Math 255:216–230. https://doi.org/10.1016/j.cam.2013.04.049
52. Peng Y, Zhao J, Sepehrnoori K, Li Z (2020) Fractional model for simulating the viscoelastic behavior of artificial fracture in shale gas. Eng Fract Mech 228(106):892. https://doi.org/10.1016/j.engfracmech.2020.106892
53. Podlubny I (1999) Fractional differential equations, Mathematics in science and engineering, vol 198. Academic Press, Inc., San Diego, CA. An introduction to fractional derivatives, fractional differential equations, to methods of their solution and some of their applications
54. Rawashdeh EA (2006) Numerical solution of fractional integro-differential equations by collocation method. Appl Math Comput 176(1):1–6. https://doi.org/10.1016/j.amc.2005.09.059
55. Yang W, Chen Z (2019) Fractional single-phase lag heat conduction and transient thermal fracture in cracked viscoelastic materials. Acta Mech 230(10):3723–3740. https://doi.org/10.1007/s00707-019-02474-z
56. Zayernouri M, Karniadakis GE (2014) Exponentially accurate spectral and spectral element methods for fractional ODEs. J Comput Phys 257(part A):460–480. https://doi.org/10.1016/j.jcp.2013.09.039
57. Zayernouri M, Karniadakis GE (2014) Fractional spectral collocation method. SIAM J Sci Comput 36(1):A40–A62. https://doi.org/10.1137/130933216

Mode Shape Based Approach to Identify the Location of Crack and the Influence of Crack on Critical Speed of Gas Turbine Disc

Ranjan Kumar, Vinayak Ranjan, Saikat Chaterjee, and Sanjoy K. Ghoshal

Abstract In an aeroengine, the rotating gas turbine disc is susceptible to cracks in its service tenure due to its complex loadings and operating conditions. The appearance of crack in disc alters the stiffness near crack location which further affects vibration characteristics of the rotating disc and it is a serious reliability concern for aeroengine applications. Critical speed is one of the important design parameters which is also affected by the presence of crack on the disc. The early detection of crack is a challenging task and the variation in the vibrational behaviour under the influence of crack must be known. In the present work an attempt has been made to introduce an efficient procedure to locate the crack position on the disc and also to compare the vibrational characteristics of healthy and cracked disc under the factual operating condition specific to aeroengine. Critical speed of cracked discs is compared with that of the healthy disc to identify the influence of crack on the safe zone of operation within the permissible speed limit. In the present findings, it is appealing to note that the vibrational mode shape can be employed as an effective tool in detection of the location of the crack. The present work is expected to serve as guidelines during the design phase of gas turbine discs of aeroengine applications.

Keywords Crack · Rotating disc · Vibration · Critical speed

R. Kumar · S. K. Ghoshal
Indian Institute of Technology (Indian School of Mines), Dhanbad, India
e-mail: ranjansinha.k@gmail.com

V. Ranjan (✉)
Bennett University, Greater Noida, India
e-mail: Vinayak.ranjan@bennett.edu.in

R. Kumar · S. Chaterjee
Adamas University, Kolkata, India

M. Abdel Wahab (ed.), *Proceedings of the 8th International Conference on Fracture, Fatigue and Wear*, Lecture Notes in Mechanical Engineering,
https://doi.org/10.1007/978-981-15-9893-7_22

1 Introduction

Gas turbine disc is one of the crucial components in aero gas turbine engine due to its sophisticated working environment. Initially, individual discs on each stage are generally assumed to be free of any crack and defects however, it is not possible to ensure the same for discs life span. Crack may also arise during manufacturing, handling, mounting or maintenance operations of the discs. Crack in the disc causes dynamic instability and further affects the structural integrity of the disc which further causes in-service failure of the disc. No doubt, airworthiness authorities are very serious and the set norms is mandatory to be followed by the engine manufacturers to deal with cracking event. In general, alarming sensors are being installed in the engine to detect any malfunction including such cracks. However, it is irrefutable that all the cracking event that has occurred in past in connection to aeroengine turbine disc had been handled safely. Referring to the very recent accidental report [1], it is clearly stated that stage 2 turbine disc has failed because of low-cycle fatigue cracks and was the main reason of accident. The conventional methodology life to first crack (LTFC) approach which decides the service life of a disc based upon the engineering crack of approximately 0.75 mm length at its surface and recommends to retire the disc from service without being inspected and further reuse of disc is not advocated [2]. In the literature, various cracking issue in the rotating disc has been investigated and analyzed to detect the crack and to understand the vibrational characteristics. In pioneering works as available in open literature, it is reported that under the influence of crack, the structural member experiences a local flexibility that affects its vibration response [3, 4]. A novel study is presented by Sinha and Elbhbah [5] who conducted a composite spectrum and bispectrum analysis to identify the fault location in rotating disc. Shanyavskiy [6] discussed fatigue cracking of turbine discs based on crack growth analysis. Emphasizing on the severity of crack it is recommended, in-service inspection discs with non-destructive tests at certain flight hours needs to be followed to avoid early failure of disc due to crack. Aziz and Woike [7] presented a rotor disc health assessment study in their work. Spin test experiment is conducted on the healthy disc without any crack and the disc with artificially induced notched crack and rotated at a rotational speed of up to 10,000 rpm. Microwave and capacitive sensors, is used to identify flaws in the disc. Ebrahimi et al. [8] proposed a continuous model to detect flexural vibration of cracked rotors. Moreover, the effect of external excitation of gravity and unbalance forces on the rotor behavior has also been studied. Wang et al. [9] considered the transverse crack in a rotor and 3D finite element solution is presented to investigate the mode shape variation under influence of crack. Giannella et al. [10] modeled fatigue crack-growth simulation in a compressor disc of a gas turbine engine under the application of operating loads. Leao et al. [11] discussed vibration based structural health monitoring techniques. In particular, the study is focused for the detection of crack in the rotating discs. Boscato et al. [12] presented a review study on structural health monitoring through vibration-based approaches. It is discussed that vibration-based condition monitoring systems permits observation of the global response for a structure and it is a promising method for damage detection in turbomachinery's natural frequency of different modes of vibration and modal behaviour is important to analyze in order to serve the purpose. The review indicates that the literature is lacking in the vibrational analysis for rotating gas turbine disc specific to aero-engine application under plausible operating conditions. Further, there is lack of literature emphasisizing on the influence of crack on critical speed corresponding to adverse mode of vibration. In the present work the prestress natural frequency analysis of the gas turbine discs under combined rotational and thermal loadings has been carried out to present a comparative analysis of the critical speed for healthy and cracked discs. Attention has been given to the influence of cracks of various geometry and known location. The obtained mode shapes are used to corroborate the location of the crack, thus, establishing mode shape analysis as a technique to successfully predict the location of real-life crack on gas turbine discs.

2 Mathematical Formulation

Figure 1 illustrates an annular disc rotating in the XY plane. The disc rotates with a constant angular velocity ω with respect to the inertial coordinate system (X, Y, Z) fixed to the centre of the disc. The absolute velocity of a particle point A is given by

$$V = \dot{u} + \Omega(r + u) \tag{1}$$

where, $r = \{x_A \ y_A \ 0\}^T$ is the position vector of A.

The displacement vector $u = \{u_x \ u_y \ u_z\}^T$; the over dot denoting differential with respect to time; the superscript T denotes the transpose operator; the angular velocity of the disc Ω in the matrix form is given as

Fig. 1 Annular rotating disc

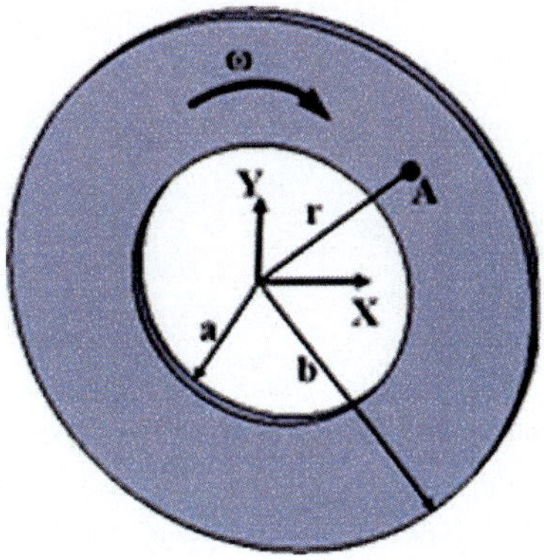

$$\Omega = \begin{bmatrix} 0 & -\omega & 0 \\ \omega & 0 & 0 \\ 0 & 0 & 0 \end{bmatrix} \tag{2}$$

The total kinetic energy of the disc is written as

$$T = \frac{1}{2}\int_{v} \rho V^T V dv \tag{3}$$

where, ρ is the density of the material; v is the volume of the disc.

Substituting Eq. (1) in Eq. (3), total kinetic energy is obtained as

$$T = \frac{1}{2}\int_{v} \rho(\dot{u}^T\dot{u} + 2u^T\Omega^T\dot{u} + 2\dot{u}^T\Omega r + 2u^T\Omega^T\Omega r + r^T\Omega^T\Omega r + u^T\Omega^T\Omega u)dv \tag{4}$$

The elastic strain energy can be written as

$$U_s = \frac{1}{2}\int_{v} \varepsilon_e^T \sigma dv \tag{5}$$

where, ε_e represents the elastic strain, and the corresponding stress σ is given by

$$\sigma = Y\varepsilon_e \tag{6}$$

where, Y is the material stiffness matrix. The total strain ε, including thermal strain ε_t, can be written as

$$\varepsilon = \varepsilon_e + \varepsilon_t \tag{7}$$

where, the thermal strain ε_t caused by a temperature change $\Delta T = T_r - T_o$ is expressed as

$$\varepsilon_t = \alpha(\Delta T) \tag{8}$$

where, α stands for the linear coefficient of thermal expansion vector, T_o is the reference temperature and T_r is the steady-state radial temperature distribution. The total strain can further be related to the displacement using the linear strain–displacement relationship

$$\varepsilon = Du \tag{9}$$

where, D is the strain–displacement matrix. Using Eqs. (6)–(9), Eq. (5) can be written as

$$U_s = \frac{1}{2}\int_v (u^T D^T Y D u - 2u^T D^T Y \alpha \Delta T + \Delta T \alpha^T Y \alpha \Delta T) dv \tag{10}$$

It may be noted that in the absence of any conservative external forces, the total potential energy corresponds to the total strain energy U_s. Moreover, the effect of the centrifugal loading due to the rotation of the disc has been taken into account by the fifth and sixth term of the Eq. (4). In order to articulate the dynamics of the rotating disc in the framework of finite element method (FEM), kinematic variables at the element level can be written as

$$r = Sp, \quad u = Sq, \quad \dot{u} = S\dot{q} \tag{11}$$

where, S is the shape function matrix of order $3 \times 3N_{nodes}$; p is the nodal coordinates vector of order $3N_{nodes}$; q and $\dot{q}$ are the nodal displacement and velocity vectors of order $3N_{nodes}$, respectively. N_{nodes} represent the number of element nodes.

Using Eq. (11), Eqs. (4) and (10) can be rewritten as,

$$T = \frac{1}{2}\int_v \rho(\dot{q}^T S^T S\dot{q} + 2q^T S^T \Omega^T S\dot{q} + 2\dot{q}^T S^T \Omega S p + 2q^T S^T \Omega^T \Omega S p + \\ p^T S^T \Omega^T \Omega S p + q^T S^T \Omega^T \Omega S q) dv \tag{12}$$

$$U_s = \frac{1}{2}\int_v (q^T S^T D^T Y D S q - 2q^T S^T D^T Y \alpha \Delta T + \Delta T \alpha^T Y \alpha \Delta T) dv \tag{13}$$

In the absence of any external load on the disc, the Lagrange's equation is expressed as

$$\frac{d}{dt}\left(\frac{\partial L}{\partial \dot{q}}\right) - \frac{\partial L}{\partial q} = 0 \tag{14}$$

where, the Lagrange's function

$$L = T - U_s \tag{15}$$

Substituting Eqs. (12) and (13) into Eq. (15), the equation of motion at the elemental level for rotating disc under the effect of thermal load can be obtained from Eq. (14) as

$$m_0\ddot{q} + c\dot{q} + (k_e + k_c)q + f_{cl} + f_T = 0 \tag{16}$$

where, m_0 is the elemental mass matrix; $\ddot{q}$ is the nodal acceleration vector; c is the Coriolis damping matrix; k_e is the element's structural stiffness matrix; k_c is the centrifugal stiffness matrix; f_{cl} is the element's centrifugal load vector; f_T is the element's thermal load vector. The matrices of Eq. (16) are of the order $3N_{nodes} \times 3N_{nodes}$, and vectors are of the order $3N_{nodes}$. The matrices and the vectors are obtained as

$$m_0 = \int\limits_v (\rho S^T S) dv \tag{17a}$$

$$c = \int\limits_v (2\rho S^T \Omega^T S) dv \tag{17b}$$

$$k_e = \int\limits_v (S^T D^T Y D S) dv \tag{17c}$$

$$k_c = -\int\limits_v (\rho S^T \Omega^T \Omega S) dv \tag{17d}$$

$$f_{cl} = -\int\limits_v (\rho S^T \Omega^T \Omega S p) dv \tag{17e}$$

$$f_T = -\int\limits_v (S^T D^T Y \alpha \Delta T) dv \tag{17f}$$

The equation of motion for the entire disc can be obtained by expressing Eq. (16) in its global form as

$$M_0 \ddot{Q} + C\dot{Q} + (K_e + K_c)Q + F_{cl} + F_T = 0 \tag{18}$$

where, M_0,C, K_e and K_c are the corresponding global matrices of the order.

$N_{dof} \times N_{dof}$, and the global vectors $\ddot{Q}$,$\dot{Q}$, Q, F_{cl} and F_T are the corresponding global vectors of the dimension N_{dof}. Here, N_{dof} represents the finite element mesh degrees of freedom. It is noted that F_{cl} and F introduces prestresses induced by rotational load and thermal load, respectively. Further, the effect of centrifugal loading incorporated by the centrifugal stiffness K_c apparently increases the overall stiffness of the disc. The eigenvalue problem with prestress defined by Eq. (18) is simulated in Ansys to obtain the natural frequencies f_{mn} and the corresponding mode shapes ϕ_{mn} where 'm' represents nodal circle and 'n' represents nodal diameter. The appearance of crack in disc is expected to affect the stiffness at the crack location which will be compensated by stiffness matrix terms of Eqs. (16) and (18). The following section highlights the considerations of crack affected stiffness through an elaborate modelling of the cracked disc.

3 Finite Element Simulation and Analysis

The Finite element analysis (FEA) is carried out in ANSYS as a coupled field analysis involving structural and thermal domains. The coupled field analysis considers the combined effect of thermal gradient and centrifugal load acting on the disc. The typical parametric geometric model [13] of the disc: the inner radius $a = 105$ mm, outer radius $b = 200$ mm while thickness h varying as *3.2*, *16* and *32* mm are considered to study the effect of disc thickness on the natural frequency of out-of-plane vibrational modes of interest. Inconel is the material for the disc and the relevant material properties are E = 209GPa, $\mu = 0.3$, $\rho = 8220$ kg/m^3 and $\alpha = 12.2$ e^{-6} (°C)$^{-1}$. The 3D solid model of the healthy disc is shown in Fig. 2a and cracked model of disc is shown in Fig. 2b. Considerable efforts have been inculcated in modelling the crack in the disc. From the reported literature it is realized that in case of rotating disc, radial cracks are more adverse and therefore only radially oriented crack is considered in this analysis. Two representative radial cracks of dimensionless length *(l/b = 0.1 and l/b = 0.25)* and depth throughout the thickness has been modeled. The V-notch type crack is considered whose angle is between the walls of crack and the crack opens up towards the rim portion of the disc. The parametric studies for both healthy and cracked disc are conducted to investigate the vibrational behaviour under the same rotational and thermal loadings. Finer mesh elements have been considered near crack locations and appropriate attention has been given to avoid singularity issue at crack tip.

The coupled field SOLID186 element is used for meshing the geometric model of the disc. The element is defined by 20-nodes having three degrees of freedom at each node: translations in the nodal x, y and z directions. The element exhibits quadratic displacement behavior, and it supports plasticity, stress stiffening, large deflection,

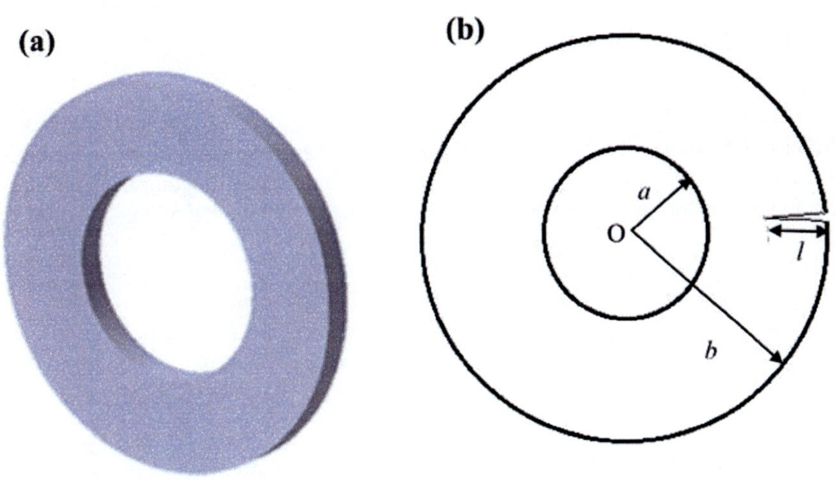

Fig. 2 Model of the disc generated in Ansys. **a** Healthy disc, **b** cracked disc

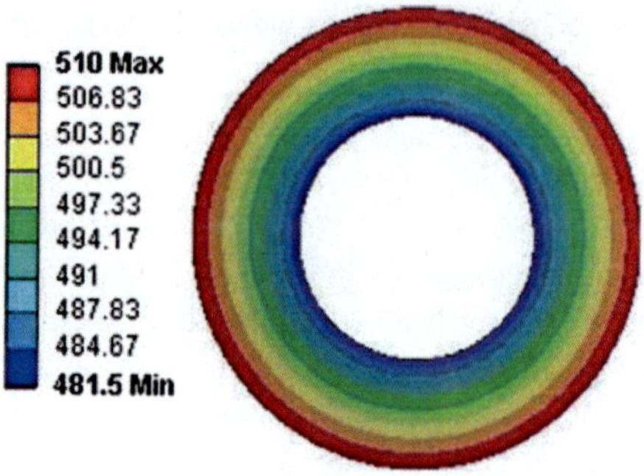

Fig. 3 Temperature distribution T_r (in °C) along the radius of the disc

large strain capabilities, and prestress effects due to thermal and centrifugal loads. The operating speed has been selected based upon the permissible speed identified in the author's earlier work [13, 14]. The boundary conditions correspond to fixed condition at the bore of the disc and free at the rim. A typical temperature of 481.5 °C is applied at the bore and a temperature of 510 °C is applied at the disc rim to get the temperature distribution along the disc radius, as shown in Fig. 3.

3.1 Validation of the Numerical Study

The solution technique using Ansys adopted in the present work is validated (Table 1) with the results reported in the open literature [2]. Firstly, a convergence validation for the numerical analysis is conducted by increasing the number of elements.

4 Results and Discussion

The main focus is to carry out critical speed analysis of rotating gas turbine discs. Initially, the healthy disc has been analyzed without any cracks and further the discs are analyzed in presence of cracks of known geometry and location. The results obtained from the analysis of healthy discs are compared with that of cracked discs in

Table 1 Natural frequency (Hz) for different vibrational modes for rotating disc

Bouboulas et al. [15]: disc radius ratio (a/b) = 0.1; $h = 2.413$ mm; b = 188.468 mm (E = 62.5GPa, $\rho = 2777$ kg/m^3, $\mu = 0.3$, rotational velocity = 5000 rpm)			
(m, n)	Bouboulas et al. [2]	Present study (FEM)	Error (%)
(0, 0)	95	94.384	0.648
(0, 1)	103.4	102.9	0.483
(0, 2)	154.5	154.54	0.349

terms of natural frequency, mode shapes and critical speed. Both healthy and cracked discs are analyzed under combined rotational and thermal loading for different modes of vibration of the discs, and the adverse modes of vibration are predicted. Further, a comparison of critical speeds has been presented.

4.1 Rotating Disc Subjected to Rotational and Thermal Loading

A comparison of natural frequency results are tabulated in Table 2a–c for healthy and cracked discs of the three representative thicknesses while considering two different dimensionless crack lengths ($l/b = 0.1$ and $l/b = 0.25$). f_h, f_C, f_C^* denotes the common twin mode frequency of healthy disc, frequency of cracked disc for symmetric mode, and frequency of cracked disc for antisymmetric mode, respectively. The natural frequencies increase with the speed of rotation as influenced by the combined rotational and thermal loading. For healthy as well as cracked discs of the considered thicknesses, it is noted that (0, 0) mode exhibits single mode of vibration, whereas, (0, 1), (0, 2) and (0, 3) modes exhibit twin modes of vibration. The remarkable point to note is that in case of healthy disc, the twin mode frequency values are of no striking difference and the variation in the corresponding mode shapes is of no practical significance. On the contrary, in case of cracked disc, the twin mode is characterized by significantly different symmetric and antisymmetric modes of vibration. The symmetric mode corresponds to in-phase vibration about the line of crack while the antisymmetric mode is out of phase. For the thin ($h = 3.2$ mm) disc, natural frequency (f_C) is minimum for $l/b = 0.25$, at each speed of rotation, for the (0, 1) mode of vibration. However, for the (0, 2) mode of vibration, (f_h) is observed to be minimum for the entire range of operating speed. The (0, 3) mode does not exist up to 12,000 rpm for both healthy and cracked discs, whereas, (0, 3) mode of vibration appears while increasing the thickness of the discs. For the discs with thickness $h = 16$ mm *and* 32 mm, it is observed that for (0, 0) mode of vibration of the disc, at each speed of rotation, natural frequency (f_C) of cracked disc exists only for $l/b = 0.1$. For the case with $l/b = 0.25$ (Table 2b, c), the single mode corresponding to order (0, 0) manifests to a symmetric mode of order (0, 0.5). The mixed type mode of order (0, 0.5) reveals a transition of the lower order mode to the present symmetric mode. The phenomenon of mixed type mode is not observed for the healthy discs. For the rest of the modes of vibration, as observed for the rotating discs without thermal loading, consideration of thermal load similarly affects the natural frequency (f_C^*) of the antisymmetric mode of vibration to be the lowest with significant decrease with increase in crack length. The overall differential effects observed in healthy and cracked discs may be attributed to the resultant variations in stiffness caused by variations in disc thickness, and, the interplay of stresses affected by the rotational speed, thermal loading effects, presence of nodal diameter and crack.

Table 2 **a** Natural frequency of rotating disc with h = 3.2 mm. **b** Natural frequency of rotating disc with h = 16 mm. **c** Natural frequency of rotating disc with h = 32 mm

Mode	Speed (rpm) 57257817017 Freq. (Hz)	−11997861584000	3000	6000	9000	12,000	15,000	18,000
(0, 0)	f_h	478.2	484.3	502.3	530.8	568.2	612.6	662.6
l/b = 0.1	f_C	473.5	480.1	498.51	527.43	565.22	610.14	660.49
l/b = s0.25	f_C	461.1	467.6	486.76	517.03	553.98	601.8	653.82
(0, 1)	f_h	425.7	433	454.1	487.1	529.7	579.7	635.2
l/b = 0.1	f_C^*	425.5	432.8	453.89	486.93	529.58	579.62	635.14
	f_C	419.4	427.0	448.67	482.36	525.69	576.54	632.41
l/b = 0.25	f_C^*	420.2	427.7	449.21	482.87	526.02	576.71	632.89
	f_C	388.8	397.7	423.45	461.0	496.26	556.65	619.94
(0, 2)	f_h	212.6	228.9	272.1	331.8	400.5	474.3	551.1
l/b = 0.1	f_C^*	222.5	238.2	279.84	337.95	405.44	478.11	554.09
	f_C	215.4	231.5	274.36	333.62	402.05	475.42	552.03
l/b = 0.25	f_C^*	267.4	281.6	321.40	360.26	425.69	495.93	561.35
	f_C	251.3	266.4	309.53	328.1	412.99	488.7	548.45
(0, 3)	f_h	–	–	–	–	–	250.5	400.6
l/b = 0.1	f_C^*	–	–	–	–	–	270.73	413.84
	f_C	–	–	–	–	–	266.29	410.17
l/b = 0.25	f_C^*	–	–	–	–	252.7	403.67	462.64
	f_C	–	–	–	–	229.33	380.57	459.45
(b)								
(0, 0)	f_h	1424.9	1426.8	1432.6	1442.1	1455.3	1472.1	1492.4
l/b = 0.1	f_C	1422.7	1424.6	1430.5	1440.2	1453.7	1470.7	1491.2
l/b = 0.25	f_C	–	–	–	–	–	–	–
(0, 0.5)	f_h	–	–	–	–	–	–	–
l/b = 0.1	f_C	–	–	–	–	–	–	–
l/b = 0.25	f_C	1402.9	1405.0	1411.4	1421.8	1436.3	1454.7	1476.8
(0, 1)	f_h	1427.0	1429.0	1435.1	1445.1	1459.0	1476.8	1498.1

(continued)

Table 2 (continued)

Mode	Speed (rpm) 57257817017 Freq. (Hz)	−11997861584000	3000	6000	9000	12,000	15,000	18,000
l/b = *0.1*	f_C^*	1426.7	1428.7	1434.7	1444.7	1458.6	1476.2	1497.6
	f_C	1427.1	1429.1	1435.2	1445.3	1459.2	1476.9	1498.3
l/b = *0.25*	f_C^*	1420.5	1422.6	1428.7	1438.9	1453.	1470.9	1492.5
	f_C	1425.6	1427.5	1433.5	1443.3	1457.	1474.5	1495.5
(0, 2)	f_h	1490.7	1493.0	1499.7	1510.9	1526.3	1546.0	1569.7
l/b = *0.1*	f_C^*	1490.5	1492.8	1499.5	1510.6	1526.	1545.6	1569.3
	f_C	1491.8	1494.1	1500.8	1511.9	1527.4	1547.	1570.7
l/b = *0.25*	f_C^*	1476.0	1478.2	1484.9	1495.9	1511.2	1530.6	1554.1
	f_C	1494.3	1496.6	1503.2	1514.3	1529.6	1549.1	1572.6
(0, 3)	f_h	1756.3	1758.7	1765.7	1777.3	1793.5	1814.1	1839.0
l/b = *0.1*	f_C^*	1754.5	1756.9	1763.8	1775.4	1791.5	1811.9	1836.6
	f_C	1756.2	1758.5	1765.5	1777.1	1793.2	1813.6	1838.3
l/b = *0.25*	f_C^*	1715.3	1717.6	1724.4	1735.7	1751.4	1771.3	1795.4
	f_C	1760.	1762.2	1768.9	1780.	1795.4	1815.	1838.6
(c)								
(0, 0)	f_h	2601.6	2602.5	2605.	2609.2	2615.1	2622.6	2631.8
l/b = *0.1*	f_C	2599.4	2600.3	2602.8	2607.1	2613.1	2620.7	2630.0
l/b = *0.25*	f_C	–	–	–	–	–	–	–
(0,0.5)	f_h	–	–	–	–	–	–	–
l/b = *0.1*	f_C	–	–	–	–	–	–	–
l/b = *0.25*	f_C	2580.0	2580.9	2583.7	2588.3	2594.7	2602.9	2612.9
(0, 1)	f_h	2603.9	2604.8	2607.5	2612.0	2618.3	2626.3	2636.1
l/b = *0.1*	f_C^*	2603.3	2604.1	2606.8	2611.3	2617.5	2625.5	2635.3
	f_C	2603.4	2604.3	2607.0	2611.5	2617.8	2625.8	2635.3
l/b = *0.25*	f_C^*	2596.0	2596.9	2599.6	2604.1	2610.4	2618.4	2628.3
	f_C	2602.6	2603.5	2606.1	2610.5	2616.6	2624.5	2634.0
(0, 2)	f_h	2737.6	2738.6	2741.6	2746.7	2753.7	2762.7	2773.6
l/b = *0.1*	f_C^*	2735.1	2736.1	2739.1	2744.1	2751.1	2760.0	2770.9
	f_C	2738.0	2739.0	2742.0	2747.0	2754.0	2762.9	2773.8

(continued)

Table 2 (continued)

Mode	Speed (rpm) 57257817017 Freq. (Hz)	−11997861584000	3000	6000	9000	12,000	15,000	18,000
l/b = *0.25*	f_C^*	2706.3	2707.3	2710.3	2715.2	2722.1	2730.9	2741.7
	f_C	2740.3	2741.3	2744.3	2749.2	2756.1	2765.	2775.8
(0, 3)	f_h	3278.3	3279.3	3282.4	3287.6	3294.8	3304.	3315.2
l/b = *0.1*	f_C^*	3270.	3271.	3274.1	3279.2	3286.4	3295.5	3306.7
	f_C	3272.7	3273.7	3276.8	3281.9	3289.	3298.2	3309.3
l/b = *0.25*	f_C^*	3179.5	3180.5	3183.5	3188.5	3195.4	3204.3	3215.2
	f_C	3257.3	3258.3	3261.3	3266.1	3273.	3281.7	3292.4

A major outcome of the present analysis is the determination of the location of the crack on the disc. It is observed that, except for the mixed type mode, the nodal diameter aligns with the crack at all rotational speed for the antisymmetric mode of vibration. It is about this nodal diameter, the disc segments separated by the crack vibrates in out of phase fashion. Thus, determination of the orientation of the nodal diameter and the points on the disc about this nodal diameter undergoing out of phase vibration will identify the orientation as well as the exact location of the crack. Whereas, in mixed type modes, the nodal diameter aligns with the crack for the symmetric mode of vibration. The proposed method of determination of crack location is further illustrated through the mode shapes of the disc ($h = 32$ mm) in Figs. 4, 5, 6 and 7. The present finding can serve as a guideline during experimental detection of crack location. Moreover, it is worth mentioning that this adverse out of phase motion may cause further propagation and crack growth leading to complete failure of the disc.

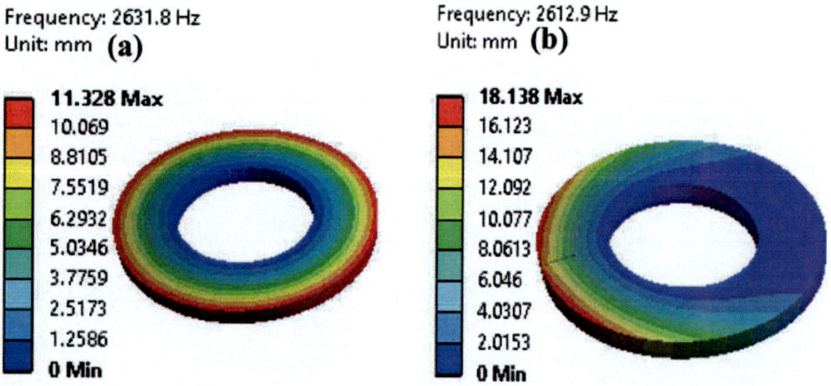

Fig. 4 Mode shape for out-of-plane vibrational modes of disc ($h = 32$ mm) rotating at 18,000 rpm. **a** Healthy disc (0, 0) umbrella mode. **b** Cracked disc, *l/b* = *0.25* (0, 0.5) mode

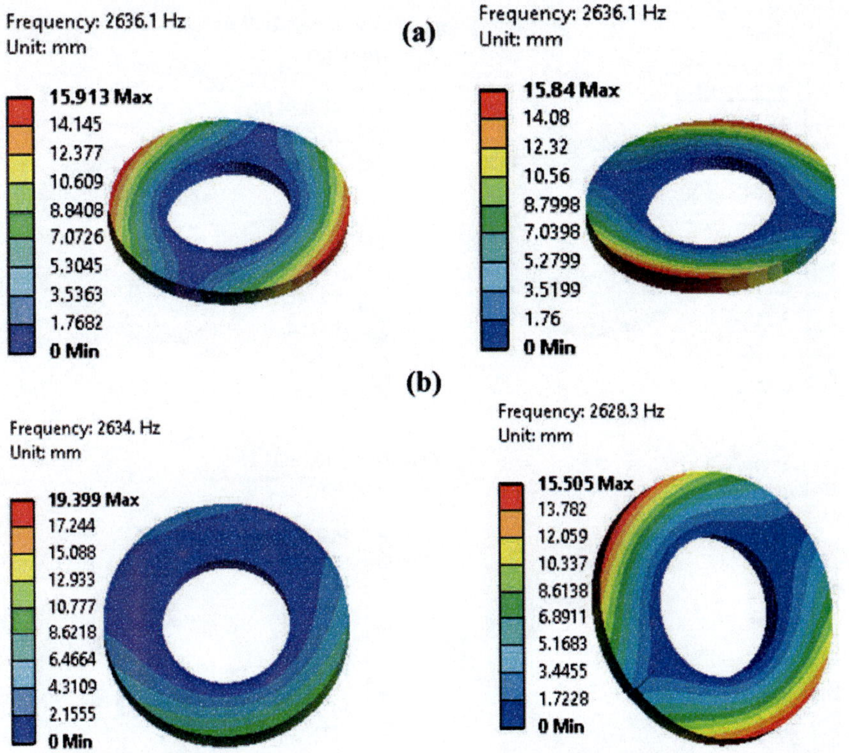

Fig. 5 Mode shape for out-of-plane vibrational modes (0, 1) of disc ($h = 32$ mm) rotating at 18,000 rpm. **a** Healthy disc twin modes. **b** Cracked disc ($l/b = 0.25$) symmetric and antisymmetric mode

4.2 *Critical Speed*

The identification of critical speed in connection to rotating machinery is always a design need in order to have safety against resonance situation. The consequences of resonance are well known amounting to loss of complete rotating machinery. The rotational speed corresponding to which resonance occurs is known as critical speed. In gas turbines, numerous sources of vibration excitation are present which may result in periodic excitations with excitation frequency components corresponding to the running speed, its harmonics and sub-harmonics. For a rotating disc, critical speed for a particular mode of vibration must be determined by the points of intersection of excitation order fan lines and the rotational speed dependent natural frequency for the considered mode of vibration. The critical speeds are determined as the speed N (rpm) corresponding to the point of intersection of the natural frequency line with the engine-order lines ($1 \times rpm$, $2 \times rpm$, $3 \times rpm$, and so on). In the present study,

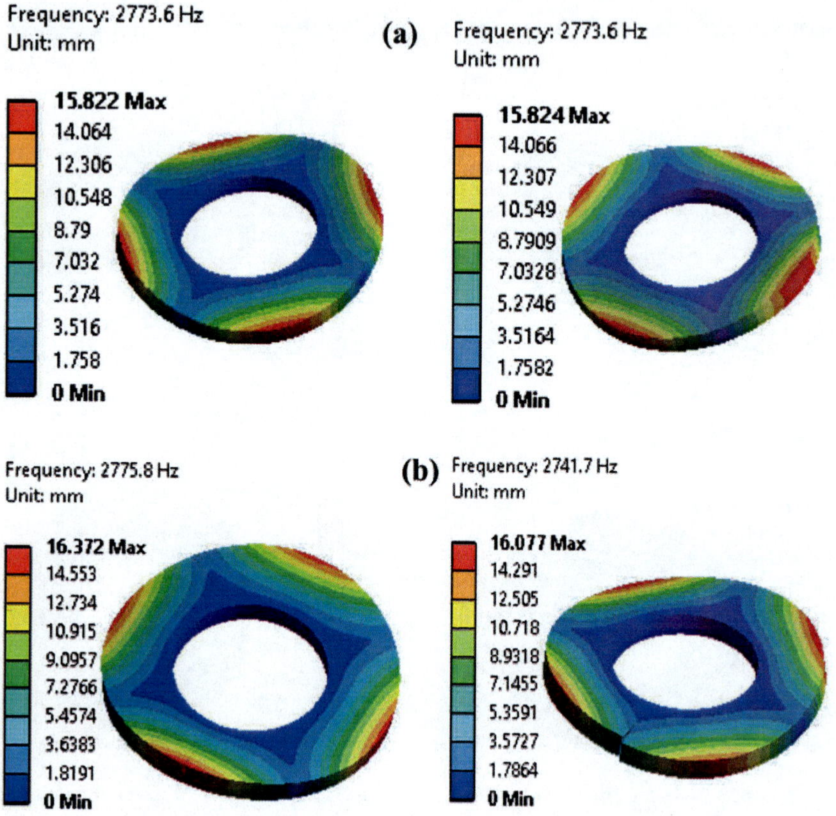

Fig. 6 Mode shape for out-of-plane vibrational modes (0, 2) of disc (h = *32* mm) rotating at 18,000 rpm. **a** Healthy disc twin modes. **b** Cracked disc ($l/b = 0.25$) symmetric and antisymmetric mode

engine-order lines up to the seventh (7 × rpm) order have been plotted (Figs. 8, 9, 10, 11, 12, 13 and 14).

For both healthy and cracked disc, three models ($a/b = 0.525$, and $h = 3.2$ mm, *16* mm and *32* mm) are analyzed and critical speed results are presented in Table 3a, b. $(N_{cr})_h$, $(N_{cr})_C$, $(N_{cr})_{C^*}$ denotes the critical speeds corresponding to healthy disc, symmetric mode of cracked disc and antisymmetric mode of cracked disc, respectively. For the disc with $h = 3.2$ mm, critical speed corresponding to $1 \times rpm$ line is obtained for the cracked disc only for the (0, 3) mode while the healthy disc depicted no critical speed for $1 \times rpm$ line. Critical speed decreased with the increase in the excitation order both for the cracked and the healthy discs. As shown in Table 3a, for the entire range of excitation order and mode order considered, critical speed follows a decreasing trend with increase in the number of nodal diameters. Moreover, at a given excitation order, critical speed decreased with the increase in crack length with the symmetric mode of vibration recording the lowest value. For the disc with h

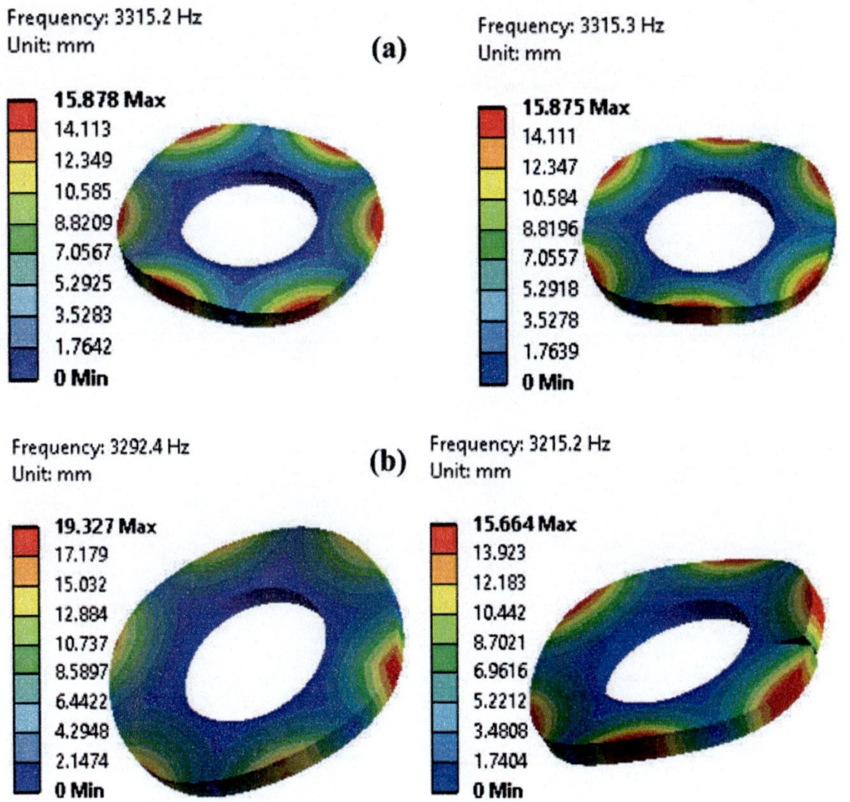

Fig. 7 Mode shape for out-of-plane vibrational modes (0, 3) of disc (h = *32* mm) rotating at 18,000 rpm. **a** Healthy disc twin modes. **b** Cracked disc ($l/b = 0.25$) symmetric and antisymmetric mode

$= 16$ mm, critical speed is obtained (Table 3b) only for the excitation orders $5 \times rpm$ and higher, whereas no critical speed is obtained for disc thickness $h = 32$ mm in the entire range of operating speed for both healthy and the cracked disc. This indicates that gas turbine discs of higher thickness have less chances of getting excited by the low order source excitations.

5 Conclusions

Under the rotating and thermal operating conditions, a comparative analysis of the out of plane vibrations and critical speed has been carried out for the healthy and cracked discs. Discs with radial crack of two representative non-dimensionless crack length ($l/b = 0.1$ *and* $l/b = 0.25$) have been considered in the present study. Prestress modal

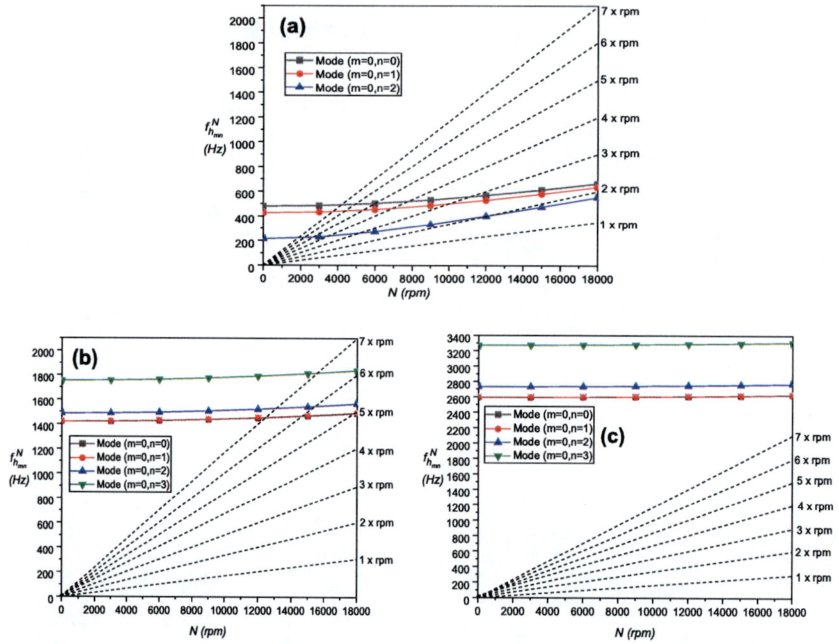

Fig. 8 Critical speeds for healthy discs **a** $h = 3.2$ mm, **b** $h = 16$ mm, **c** $h = 32$ mm

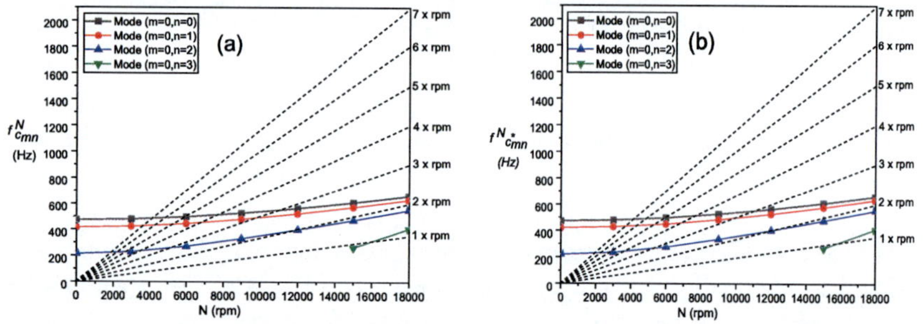

Fig. 9 Critical speeds for cracked disc ($h = 3.2$ mm, $l/b = 0.1$) (a) Symmetric (b) Antisymmetric

analysis has been performed with due consideration of the thermal and rotational prestresses. The variations in the mode shapes in the presence of crack elucidates a successful technique of the determination of the location of the crack on the gas turbine disc. At least one nodal diameter aligns with the crack for both general and mixed type modes. While this alignment is obtained for the antisymmetric mode of vibration in the general case, mixed type mode results in the alignment for the symmetric mode of vibration. Thus, the determined mode shape of the disc with emphasis on the node points can be used to infer the exact location of the crack.

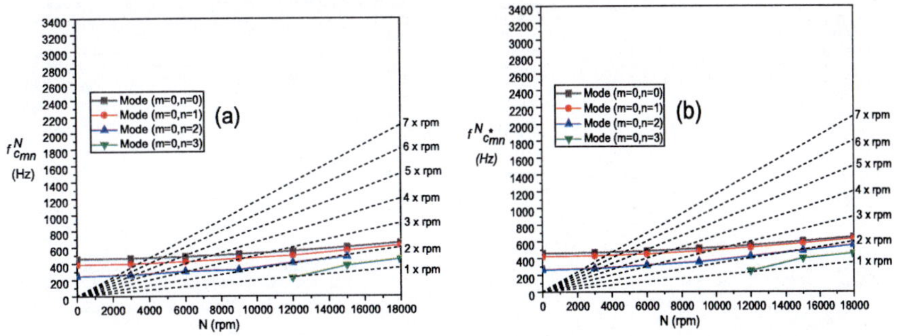

Fig. 10 Critical speeds for cracked disc ($h = 3.2$ mm, $l/b = 0.25$) (a) Symmetric (b) Antisymmetric

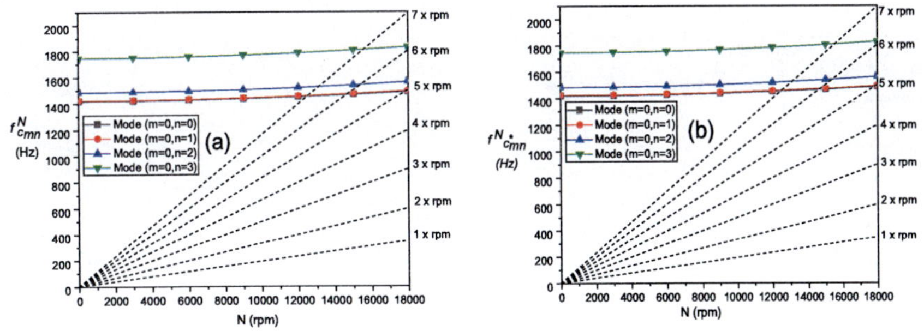

Fig. 11 Critical speeds for cracked disc ($h = 16$ mm, $l/b = 0.1$) (a) Symmetric (b) Antisymmetric

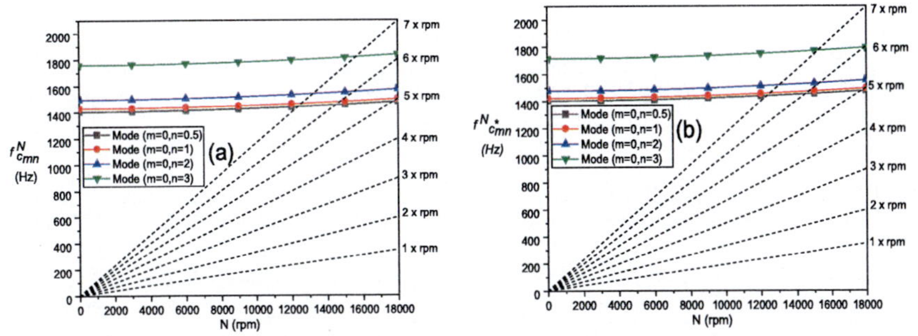

Fig. 12 Critical speeds for cracked disc ($h = 16$ mm, $l/b = 0.25$) (a) Symmetric (b) Antisymmetric

The present finding can serve as a guideline during experimental detection of crack location. It is concluded that in addition to rotational speed and temperature gradient, disc profile and thickness have a major influence on the critical speed. The present findings are expected to serve as guidelines during the design phase of gas turbine discs of aeroengine applications.

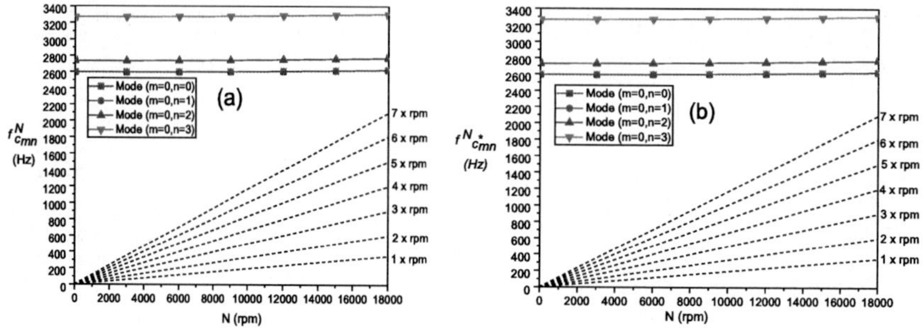

Fig. 13 Critical speeds for cracked disc ($h = 32$ mm, $l/b = 0.1$) (a) Symmetric (b) Antisymmetric

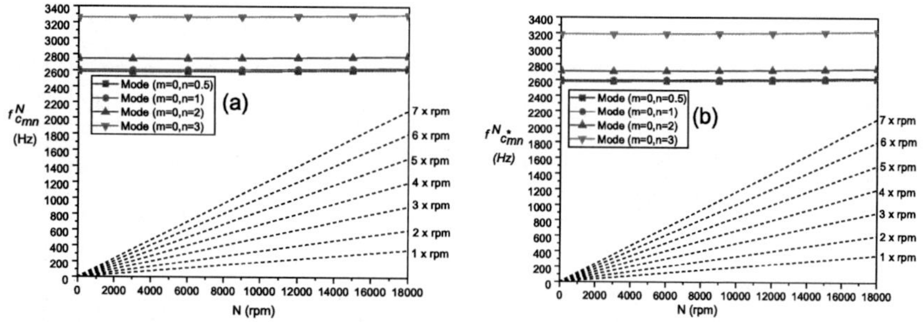

Fig. 14 Critical speeds for cracked disc ($h = 32$ mm, $l/b = 0.25$) (a) Symmetric (b) Antisymmetric

Table 3 **a** Critical speeds for healthy and cracked discs ($h = 3.2$ mm). **b** Critical speeds for healthy and cracked discs ($h = 16$ mm)

Mode	Excitation order / Critical speed		1 × rpm	2 × rpm	3 × rpm	4 × rpm	5 × rpm	6 × rpm	7 × rpm
(a)									
(0,0)		$(N_{cr})_h$	–	–	11,142	7801	6000	4871	4154
$l/b = 0.1$		$(N_{cr})_C$	–	–	11,000	7666	6000	4880	4160
$l/b = 0.25$		$(N_{cr})_{C^*}$	–	–	10,744	7522	5770	4822	4075
(0,1)		$(N_{cr})_h$	–	–	10,000	6956	5392	4350	3738
$l/b = 0.1$		$(N_{cr})_C$	–	–	9850	6800	5306	4358	3697
		$(N_{cr})_{C^*}$	–	–	10,080	7000	5392	4387	3725
$l/b = 0.25$		$(N_{cr})_C$	–	–	9000	6352	5000	4071	3467
		$(N_{cr})_{C^*}$	–	–	10,000	7000	5331	4317	3652
(0, 2)		$(N_{cr})_h$	–	12,000	5051	3504	2696	2212	1871
$l/b = 0.1$		$(N_{cr})_C$	–	12,000	5187	3550	2745	2257	2000
		$(N_{cr})_{C^*}$	–	12,150	5388	3664	2831	2343	2000
$l/b = 0.25$		$(N_{cr})_C$	–	13,440	6000	4329	3176	2659	2228
		$(N_{cr})_{C^*}$	–	15,000	6344	4490	3447	2782	2363
(0, 3)		$(N_{cr})_h$	–	–	–	–	–	–	–
$l/b = 0.1$		$(N_{cr})_C$	15,767	–	–	–	–	–	–
		$(N_{cr})_{C^*}$	15,570	–	–	–	–	–	–
$l/b = 0.25$		$(N_{cr})_C$	–	–	–	–	–	–	–
		$(N_{cr})_{C^*}$	12,000	–	–	–	–	–	–
(b)									

(continued)

Table 3 (continued)

Mode	Excitation order / Critical speed	1 × rpm	2 × rpm	3 × rpm	4 × rpm	5 × rpm	6 × rpm	7 × rpm
(0, 0)	$(N_{cr})_h$	–	–	–	–	17,900	14,725	12,521
l/b = 0.1	$(N_{cr})_C$	–	–	–	–	17,880	14,737	12,537
l/b = 0.25	$(N_{cr})_{C^*}$	–	–	–	–	–	–	–
(0, 0.5)	$(N_{cr})_h$	–	–	–	–	–	–	–
l/b = 0.1	$(N_{cr})_C$	–	–	–	–	–	–	–
l/b = 0.25	$(N_{cr})_{C^*}$	–	–	–	–	17,659	14,515	12,385
(0, 1)	$(N_{cr})_h$	–	–	–	–	18,000	14,725	12,521
l/b = 0.1	$(N_{cr})_C$	–	–	–	–	17,880	14,737	12,537
	$(N_{cr})_{C^*}$	–	–	–	–	17,880	14,737	12,537
l/b = 0.25	$(N_{cr})_C$	–	–	–	–	17,934	14,757	12,595
	$(N_{cr})_{C^*}$	–	–	–	–	17,901	14,157	12,521
(0, 2)	$(N_{cr})_h$	–	–	–	–	–	15,389	13,153
l/b = 0.1	$(N_{cr})_C$	–	–	–	–	–	15,435	13,153
	$(N_{cr})_{C^*}$	–	–	–	–	–	15,455	13,186
l/b = 0.25	$(N_{cr})_C$	–	–	–	–	–	15,492	13,186
	$(N_{cr})_{C^*}$	–	–	–	–	–	15,246	13,000
(0, 3)	$(N_{cr})_h$	–	–	–	–	–	–	15,632
l/b = 0.1	$(N_{cr})_C$	–	–	–	–	–	–	15,632
	$(N_{cr})_{C^*}$	–	–	–	–	–	–	15,529
l/b = 0.25	$(N_{cr})_C$	–	–	–	–	–	–	15,668

(continued)

Table 3 (continued)

Mode	Excitation order / Critical speed		1 × rpm	2 × rpm	3 × rpm	4 × rpm	5 × rpm	6 × rpm	7 × rpm
		$(N_{cr})_{C^*}$	–	–	–	–	–	–	15,176

References

1. Uncontained engine failure and subsequent fire (2018) US National transportation safety board, Report no. NTSB/AAR18-01
2. Claudio RA, Branco CM, Gomes EC, Byrne J (2002) Life prediction of a gas turbine disc using the finite element method. 8AS Jornadas de Fractura 8:31–144
3. Dimarogonas AD (1996) Vibration of cracked structures: a state of the art review. Eng Fract Mech 55:831–857
4. Huang CH, Ma CC (2000) Vibration of cracked circular plates at resonance frequencies. J Sound Vib 236:637–656
5. Sinha JK, Elbhbah K (2013) A future possibility of vibration-based condition monitoring of rotating machines. Mech Syst Signal Process 34:231–240
6. Shanyavskiy AA (2013) Fatigue crack propagation in turbine disks of EI698 superalloy. Frat Ed Integrita Strutt 24:13–25
7. Aziz AA, Woike M (2013) Turbine rotor disk health monitoring assessment based on sensor technology and spin tests data. Sci World J 41387:1–8
8. Ebrahimi A, Heydari M, Behzad M (2017) Forced vibration analysis of rotors with an open edge crack based on a continuous vibration theory. Arch Appl Mech
9. Wang S, Zi Y, Qian S, Zi B, Bi C (2018) Effects of unbalance on the nonlinear dynamics of rotors with transverse cracks. Nonlinear Dyn 91(4):2755–2772
10. Giannella V, Citarella R, Perrella M, Shlyannikov V (2019) Surface crack modelling in an engine compressor disc. Theor Appl Fract Mech 103:1–11
11. Leao LS, Cavalini AA, Morais TS, Melo GP, Steffen V (2019) Fault detection in rotating machinery by using the modal state observer approach. J Sound Vib 458:123–142
12. Boscato G, Fragonara LZ, Cecchi A, Reccia E, Baraldi D (2019) Structural health monitoring through vibration-based approaches. Shock Vib 2380616 1–5
13. Kumar R, Ranjan V, Kumar B, Ghoshal SK (2018) Finite element modelling and analysis of the burst margin of a gas turbine disc using an area weighted mean hoop stress method. Eng Fail Anal 90:425–433
14. Kumar R, Ghoshal SK, Ranjan V, Kumar B (2018) Burst strength analysis of gas turbine disc based on deformation characteristics. AIP Conf Proc 020016:1–7
15. Bouboulas AS, Anifantis NK (2011) Vibration analysis of a rotating disk with a crack. ISRN Mech Eng 1–13

Time-Delay Fractional Optimal Control Problems: A Survey Based on Methodology

Dajana Conte, Eslam Farsimadan, Leila Moradi, Francesco Palmieri, and Beatrice Paternoster

Abstract We survey some representative results on time-delay fractional differential optimal control problems. In this paper we provide a review of the techniques, developed in the last decade, for the numerical solution of time-delay fractional optimal control problems. In particular, Chebyshev and Chelyshkov wavelet methods, continuous and discrete Chebyshev polynomials methods are focused on this study.

Keywords Chelyshkov wavelets · Chebyshev wavelet · Discrete and continuous chebyshev polynomials · Fractional calculus · Time-delay fractional optimal control problems

1 Introduction

Many mechanical devices play an important role in our daily life. How to "optimally" design them while satisfying the required system and design constraints is an important engineering concern. The optimality of the design can be measured by using an objective function that evaluates for example construction cost or some other performance measure. If a particular design makes the objective function as high or low as possible, depending the objective, it is called an optimal design.

Mechanical systems can be modeled with mathematical equations such as ordinary differential equations. These systems change according to time or any other

D. Conte · L. Moradi (✉) · B. Paternoster
Department of Mathematics, University of Salerno, Fisciano, Italy
e-mail: lmoradi@unisa.it

D. Conte
e-mail: dajconte@unisa.it

B. Paternoster
e-mail: beapat@unisa.it

E. Farsimadan · F. Palmieri
Department of Computer Science, University of Salerno, Fisciano, Italy
e-mail: efarsimadan@unisa.it

F. Palmieri
e-mail: fpalmieri@unisa.it

M. Abdel Wahab (ed.), *Proceedings of the 8th International Conference on Fracture, Fatigue and Wear*, Lecture Notes in Mechanical Engineering,
https://doi.org/10.1007/978-981-15-9893-7_23

independent variable in the relationship. Often, by applying some kind of external force or control, these systems can be moved from one state to another predefined state. It may also be possible to do the same thing in different ways. If there is more than one way to do the task, choosing the best one may be the most desirable. The best way can be determined by measuring the performance of the system. This measure is usually called the cost function, target performance, or performance indicator. The external force applied to the system according to the best performance is called optimal control. The aim of optimal control is to maximize or minimize a performance indicator by identifying control and state trajectories. The important examples of controlled mechanical systems are electromechanical and mechanical plants such as transport systems, diverse mechanisms, robots and so on [53].

In factory applications such as distillation units, heat transfer, steel manufacturing, mining processes, often time delays are involved. Also in other contexts, control techniques have been developed taking into account delay features, for example, average approximation method [5], linear-quadratic [6], iterative dynamic programming [27], dynamic programming [31], Bezier curve method [34], recursive shooting method [39], FD method [40], Legendre multiwavelets [44], Walsh series [57] and line-up competition algorithm [63]. Time-delay fractional optimal control problems (TDFOCPs) are characterized by time-delay equations in the dynamical system and are frequently used in transportation systems, robotic, mechanical, biological and chemical systems [4, 8, 27, 31, 33, 34, 40–42, 47].

For some optimal mechanical control problems, analytic solutions can be obtained from the necessary and sufficient optimally conditions. As in most cases TDFOCPs are strongly nonlinear, and do not have analytical solutions, it is generally necessary to use numerical methods [27, 36, 38, 45, 51, 52].

Several kinds of methods have been developed and implemented in the last years for TDFOCPs: shifted Legendre polynomials [8], linear programming [41], Chebyshev wavelet functions [48], Continuos and Discrete Chebyshev polynomials [49], Chelyshkov wavelet functions [50], Boubaker functions [60], Bernoulli wavelet [61] and Bernstein polynomials [62]. This paper reviews in particular the proposed methods in [48–50].

2 Optimal Control Problem with Time-Delay Fractional Dynamical System

We at first recall the definitions of fractional derivative and integral [3, 55, 58, 64], which will be used in the following.

Definition 1 Let $\mathfrak{f} \in C[0, \infty)$. Then $\mathfrak{f} \in C_\mu[0, \infty)$, $\mu \in \mathbb{R}$ if it can be written as $\mathfrak{f}(t) = t^p \mathfrak{f}_1(t)$, $t \in [0, \infty)$, with $p \in \mathbb{R}$, $p > \mu$ and $\mathfrak{f}_1 \in C[0, \infty)$ [46, 59]. Moreover $\mathfrak{f} \in C_\mu^n[0, \infty)$, $n \in \mathbb{N}$ if its n-th derivative $\mathfrak{f}^{(n)} \in C_\mu[0, \infty)$.

Definition 2 Let $\mathfrak{f} \in C_1^n[0, \infty)$. The Caputo fractional derivative of the function $\mathfrak{f}$, of order $\alpha > 0$, is

$$\mathfrak{D}^{\alpha}\mathfrak{f}(t) = \begin{cases} \dfrac{1}{\Gamma(n-\alpha)}\displaystyle\int_0^t \frac{\mathfrak{f}^{(n)}(\tau)}{(t-\tau)^{\alpha-n+1}}d\tau, \; t>0, & 0 \leq n-1 < \alpha < n, \\ \dfrac{d^n\mathfrak{f}(t)}{dt^n}, & \alpha = n \in \mathbb{N}. \end{cases}$$

Definition 3 Let $\mathfrak{f} \in C_{\mu}, \mu \geq -1$. The Riemann-Liouville fractional integration of $\mathfrak{f}$, of order $\alpha \geq 0$, is

$$\mathcal{I}^{\alpha}\mathfrak{f}(t) = \begin{cases} \dfrac{1}{\Gamma(\alpha)}\displaystyle\int_0^t (t-\tau)^{\alpha-1}\mathfrak{f}(\tau)d\tau, & \alpha > 0, \\ \mathfrak{f}(t), & \alpha = 0. \end{cases}$$

The TDFOCP consistis in minimizing the functional

$$\mathcal{J} = \frac{1}{2}\int_0^2 g\,(t, x(t), u(t))dt, \tag{1}$$

where g is a positive function and, the state function $x(t)$ and the control function $u(t)$, satisfy the dynamical system

$$\begin{aligned} \mathfrak{D}^{\alpha}x\,(t) = h(t) &+ \sum_{i=1}^{m}\left[p_i(t)x(t) + q_i(t)x\,(t-\eta)\right] \\ &+ \sum_{j=1}^{n}\left[r_j(t)u(t) + s_j(t)u\,(t-\tau)\right], \\ & \qquad t \in [t_0, t_1], \;\; \alpha \in (0, 1], \end{aligned} \tag{2}$$

$$\begin{aligned} x(t) &= f_1(t), \quad -\eta \leqslant t \leqslant 0, \\ u(t) &= f_2(t), \quad -\tau \leqslant t \leqslant 0, \end{aligned} \tag{3}$$

where $\tau, \eta > 0$ are known constant delays, $p_i(t)$, $q_i(t)$, $r_j(t)$, $s_j(t)$ and $h(t)$ are continuous functions, $f_1(t)$ and $f_2(t)$ are functions on $[-\eta, 0]$ and $[-\tau, 0]$, respectively.

In general, two main different approaches have been proposed for the numerical solution of TDFOCPs: indirect approaches [2, 37] which are based on the construction of a Hamiltonian system, and direct approaches [8, 41, 48–50, 60–62], which will be further analyzed in this paper. Since the results presented in [8, 60–62] are not consistent with the exact solution reported in [5] (when $\alpha = 1$), we will not go into details about them.

In [41], one of the proposed algorithms attempts to propose an efficient linear programming (LP) formula for TDFOCPs. A multivariable case study is presented in [48]. The author introduced the numerical method by using Chebyshev wavelets functions.

Afterwards, very recently, in [49], authors analyzed classical Chebyshev polynomials (CSCPs), proposed new discrete shifted Chebyshev polynomials (DSCPs) for solving TDFOCPs and shown the superiority of DSCPs over CSCPs. With the use of DSCPs, a good accuracy can be obtained with a small number of shifted Chebyshev polynomials. Therefore the computational cost of DSCPs (and the corresponding CPU time) is much less than CSCPs, as we will show in the following section.

In a very recent time, in [50] a new numerical method based on the Chelyshkov wavelet functions (CWFs) has been introduced. This method reduces a TDFOCPs having a quadratic performance indicator, to a system of algebraic linear equations. The performed numerical simulations and the analysis of outstanding results show significant features of the new method compared to previous findings.

As TDFOCPs have only recently been analyzed in the literature, the paper considers publications after 2018. We use the following notation for the methods of papers [48–50]:

- CCWFs =Chebyshev wavelet functions [48];
- DSCPs=Discrete shifted chebyshev polynomials [1, 35, 49];
- CSCPs=continuous shifted chebyshev polinomials [49];
- CWFs=Chelyshkov wavelet functions [28, 50].

Method	Explicit formula	Conditions
CCWs	$\psi_{nm}(t) = \begin{cases} \sqrt{2\xi^{k-1}} c_m T\left(2^k t - n\right) & \frac{n}{\xi^k} \le t < \frac{n+1}{\xi^k} \\ 0 & otherwise, \end{cases}$	$m = 0, 1, ..., M$ $n = 0, 1, ..., \xi^{k-1}$
CWFs	$\rho_{nm}(t) = \begin{cases} \sqrt{2m+1} 2^{\frac{k}{2}} \rho_m C\left(2^k t - n\right) & \frac{n}{2^k} \le t < \frac{n+1}{2^k} \\ 0 & otherwise, \end{cases}$	$m = 0, 1, ..., M$ $n = 0, 1, ..., 2^{k-1}$
CSCPs	$T^*_{n+1}(t) = (4t-2)\, T^*_n(t) - T^*_{n-1}(t)$	$n = 1, 2, ...N$
DSCPs	$T_{n,N}(t) = \sum_{k=0}^{n} (-1)^k \binom{n+k}{n} \binom{N-k}{n-k} \binom{t}{k}$	$n = 0, 1, ..., N$

where in Table 1:

Table 1 Approximation of f (t), for $t \in [0, 1]$

Method	Function approximation	Conditions
CCWs	$f_{M,k}(t) = \sum_{i=1}^{\hat{s}} c_i \theta_i(t) = C^T \Theta(t)$	$C = [c_i]_{i=1}^{\hat{s}}$ $\Theta(t) = [\theta_i(t)]_{i=1}^{\hat{s}}$ $\hat{s} = \xi^k (M+1)$
CWFs	$f_{M,k}(t) = \sum_{i=1}^{\hat{m}} c_i \phi_i(t) = C^T \Phi(t)$	$C = [c_i]_{i=1}^{\hat{m}}$ $\Phi(t) = [\phi_i(t)]_{i=1}^{\hat{m}}$ $\hat{m} = 2^k (M+1)$
CSCPs	$f(t) \simeq \sum_{i=0}^{N} c_i T_i^*(t) = C^T \Upsilon(t)$	$C = [c_i]_{i=0}^{N}$ $\Upsilon(t) = \left[T_i^*(t)\right]_{i=0}^{N}$
DSCPs	$f(t) \simeq \sum_{i=0}^{N} c_i T_{i,N}(t) = C^T \Psi(t)$	$C = [c_i]_{i=0}^{N}$ $\Psi(t) = \left[T_{i,N}(t)\right]_{i=0}^{N}$

- $\theta_i(t) = \psi_{nm}(x), \quad c_i = c_{nm}, \quad i = n(M+1) + m + 1, \quad \hat{s} = \xi^k(M+1),$
- $\phi_i(t) = \rho_{nm}(x), \quad c_i = c_{nm}, \quad i = n(M+1) + m + 1, \quad \hat{m} = 2^k(M+1).$

In the following subsections, we briefly recall the concepts of operational matrices [32, 56] of the CCWs method [48], DSCPs method [49], CWFs method [50] and CSCPs method [7, 65]. For summarize, consider the polynomial vector $\Omega(t)$ (which is chosen among $\Theta(t)$, $\Phi(t)$, $\Upsilon(t)$ or $\Psi(t)$ from Table 1, depending on the method).

2.1 Operational Matrices of CWFs (or CCWs)

Theorem 1 *The fractional integral of order α of the CWFs (or CCWs) vector $\Omega(t)$ can be obtained as*

$$I^\alpha \Omega(t) \simeq P^{(\alpha)} \Omega(t), \tag{4}$$

where $P^{(\alpha)} \in \mathbb{R}^{\hat{m} \times M+1}$ (or $\mathbb{R}^{\hat{s} \times M+1}$) is the fractional integration operational matrix of CWFs (or CCWs).

Theorem 2 *Suppose $\Omega(t)$ be the CWFs (or CCWs) vector and $W \in \mathbb{R}^{\hat{m}}$ (or $\mathbb{R}^{\hat{s}}$) be an arbitrary vector. Then*

$$\Omega(t)\Omega^T(t)W = \tilde{W}\Omega(t), \tag{5}$$

where $\tilde{W}$ is the product operational matrix of the CWFs (or CCWs) vector $\Omega(t)$.

2.2 Operational Matrices of DSCPs (or CSCPs)

Theorem 3 *The fractional integral of order α of the DSCPs (or CSCPs) vector $\Omega(t)$ can be obtained as*

$$I^\alpha \Omega(t) = Q^{(\alpha)} \Omega(t) \tag{6}$$

where $Q^{(\alpha)}$ is the fractional integration operational matrix of DSCPs (or CSCPs).

Theorem 4 *Let $\Omega(t)$ be the DSCPs (or CSCPs) vector and $Z \in \mathbb{R}^{N+1}$ be an arbitrary vector. Then*

$$\Omega(t)\Omega(t)^T(t)Z = \tilde{Z}\Omega(t), \tag{7}$$

where $\tilde{Z}$ is the product operational matrix of DSCPs (or CSCPs) vector.

3 Description of the Numerical Methods

Now, we briefly review the algorithms of the SCWs, CWFs, CSCPs and DSCPs techniques.

Algorithm

Inputs: L (equal to $\hat{s}$, $\hat{m}$ or $N+1$, depending on the method), α, m, n, τ, η, the functions g, h, f_1, f_2, $p_i, q_i, i=1,...,m$ and $r_j, s_j, j=1,...,n$.
Step 1: Consider the polynomial vector $\Omega(t)$ (which is chosen among $\Theta(t)$, $\Phi(t)$, $\Upsilon(t)$ or $\Psi(t)$ from Table 1, depending on the method).
Step 2: Define a vector $X \in \mathbb{R}^L$ which is unknown and approximate

$$\mathfrak{D}^\alpha x(t) \simeq X^T \Omega(t)$$

and

$$u(t) \simeq U^T \Omega(t), \tag{7}$$

by using Table 1 (Function approximation).
Step 3: Determine the coefficient vectors $P_i, R_i, i=1,...,m, Q_j, S_j, j=1,...,n, F_1, F_2$ and H and approximate all known functions in dynamical system (2) and (3).

$$p_i(t) \simeq P_i^T \Omega(t), \; q_i(t) \simeq Q_i^T \Omega(t)$$

$$r_j(t) \simeq R_j^T \Omega(t), \; s_j(t) \simeq S_j^T \Omega(t)$$

$$h(t) \simeq H^T \Omega(t), \; f_1(t) \simeq F_1^T \Omega(t), \; f_2(t) \simeq F_2^T \Omega(t).$$

Step 4: Determine the matrix $\mathcal{M}^{(\alpha)}$ (which is chosen among $P^{(\alpha)}$ or $\mathcal{Q}^{(\alpha)}$ depending on the method), from Theorems 1, 3.
Step 5: Approximate $x(t)$, $x(t-\eta)$ and $u(t-\tau)$ as follows:

$$x(t) \simeq I^\alpha X^T \Omega(t) + h_1(0) = X^T \mathcal{M}^{(\alpha)} \Omega(t) + H_1^T \Omega(0),$$

$$x(t-\eta) = X^T \mathcal{M}^{(\alpha)} \Omega(t-\eta) + H_1^T \Omega(0), u(t-\tau) = U^T \Omega(t-\tau).$$

Step 6: Compute the vector U by substituting the above formulas in the system (2), derive unknown vector U as a function of X.
Step 7: Use the obtained vector U from Step 6 and approximate control function $u(t)$ in (7).
Step 8: Substitute the $x(t)$ and $u(t)$ in (1).
Step 9: Approximate the optimal performance indicator with application of the Gauss-Legendre quadrature formula

$$\mathcal{J}[x_i] \simeq \sum_{k=1}^{r} w_k \Lambda(t_k, X),$$

where $i=1,...,L$ (if method is CCWs or CWFs), $i=0,...L-1$ (if method is DSCPs or CSCPs)

and

$$\Lambda(t, X) = g\left(t, X^T \mathcal{M}^{(\alpha)} \Omega(t) + H_1^T \Omega(0), U^T \Omega(t)\right).$$

Step 10: Solve a system of algebraic equations to obtain the unknowm vector X

$$\frac{\partial \mathcal{J}}{\partial x_i} = 0, \qquad i=0,...L-1 \; (or \; i=1,...,L).$$

Outputs: The optimal performance indicator $\mathcal{J}$, state and control functions $x(t)$, $u(t)$.

3.1 Selected Numerical Example

Consider the following system [8, 41, 48–50, 60–62]: minimize the performance indicator

$$\mathcal{J} = \frac{1}{2}\int_0^2 \left(u^2(t) + x^2(t)\right)dt,$$

subjected to

$$\begin{aligned} &\mathfrak{D}^\alpha x(t) = q(t)x(t-1) + u(t), \quad t \in [0, 2], \quad \alpha \in (0, 1], \\ &x(t) = 1, \quad t \in [-1, 0]. \end{aligned}$$

See Figs. 1, 2, 3, 4 and Tables 2.

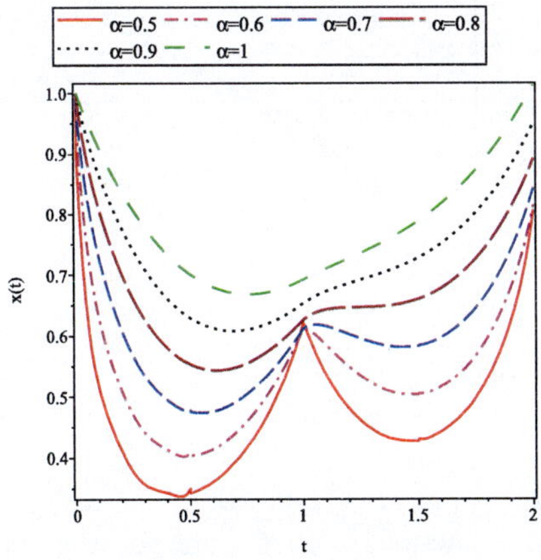

Fig. 1 Numerical solutions for $q(t) = 1, M = 6$ and $k = 2$, [50]

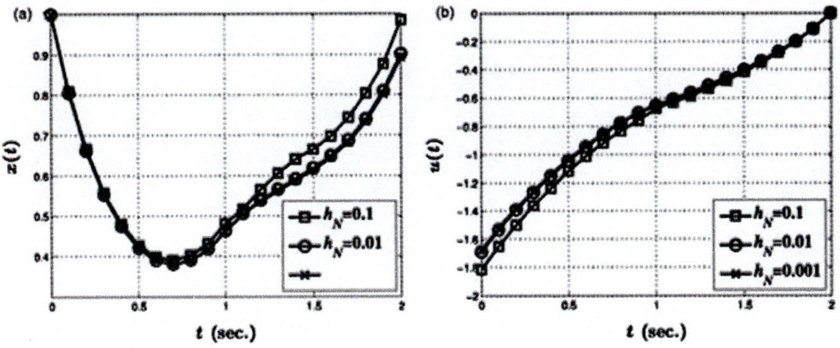

Fig. 2 Numerical solutions for $\alpha = 0.9$ with $q(t) = t$, [41]

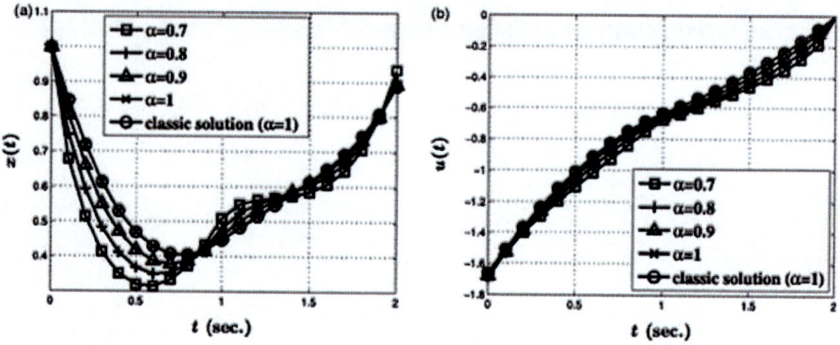

Fig. 3 Numerical solutions for $q(t) = t$ and different values of α, [41]

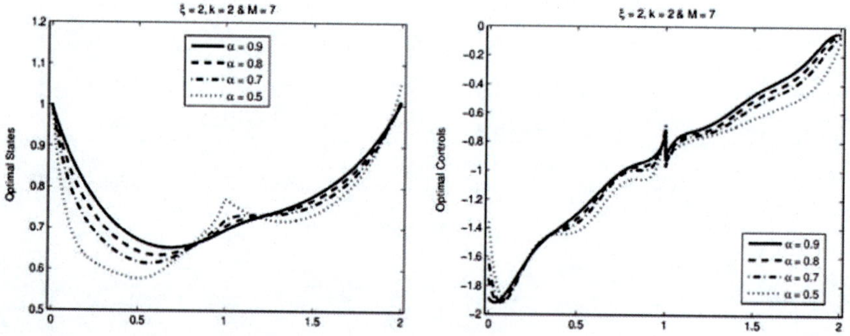

Fig. 4 Numerical results for $q(t) = 1$ and different values of α, [48]

From Table 3 we observe that numerical findings of [41, 48–50, 57] agree with the exact solution reported in [5] while the results of [8, 60–62] are not in good agreement with the exact solution (when $\alpha = 1$, $\mathcal{J} = 1.6419$) (Table 4).

Table 2 Optimal performance indicator $\mathcal{J}$ and CPU time for $q(t) = 1$ and various α, N [49]

	$\mathcal{J}$		CPU time	
	DSCPs	CSCPs	DSCPs	CSCPs
$N = 8$				
$\alpha = 0.8$	1.59078206	1.59307087	1.906	11.829
$\alpha = 0.9$	1.62390237	1.62504538	1.297	11.828
$\alpha = 0.95$	1.63681662	1.63763788	1.250	13.782
$\alpha = 1.0$	1.64731019	1.64793101	0.610	7.859
$N = 12$				
$\alpha = 0.8$	1.58623310	1.59279517	2.828	73.860
$\alpha = 0.9$	1.62194647	1.62493518	3.469	66.157
$\alpha = 0.95$	1.63574710	1.63757248	3.109	72.016
$\alpha = 1.0$	1.64685043	1.64789439	4.250	52.094
$N = 16$				
$\alpha = 0.8$	1.57924487	1.59272160	13.094	194.703
$\alpha = 0.9$	1.61735795	1.62490580	13.172	169.640
$\alpha = 0.95$	1.63254517	1.63755459	12.485	173.860
$\alpha = 1.0$	1.64487878	1.64788380	5.188	116.109

Table 3 Optimal performance indicator $\mathcal{J}$ for various M, k and α and $q(t) = 1$ [50]

	$\alpha = 0.7$	$\alpha = 0.8$	$\alpha = 0.9$	$\alpha = 0.95$	$\alpha = 0.99$
$(M = 3, k = 1)$	1.5520216	1.5926624	1.6248816	1.6375392	1.6459914
$(M = 6, k = 1)$	1.5519837	1.5926482	1.6248785	1.6375382	1.6459912
$(M = 6, k = 2)$	1.5519859	1.5926486	1.6248785	1.6375357	1.6459912

Table 4 Optimal performance indicator $\mathcal{J}$ by several methods for $N = 16$, $q(t) = 1$ and $\alpha = 1$ [49]

Method	Optimal performance indicator value $\mathcal{J}$
Average approximation method [5]	1.6419
Shifted Legendre polynomials [8]	0.4727464
Linear programming [41]	1.64886527
CCWFs method [48]	1.647874
DSCPs method [49]	1.64487878
CSCPs method [49]	1.64788380
CWFs method [50]	1.64787419
Walsh series [57]	1.6497
Boubaker functions [60]	0.00002674
Bernoulli wavelet [61]	0.3048
Bernstein polynomials [62]	0.6381

4 Conclusion

The review paper introduces the reader to recent advances with time delay fractional system on optimal control problems. We end this review by observing that TDFOCPs are spreading more and more in the applications in technology and science. This has led to many new advances within the broad field of fractional dynamic systems that have attracted the attention of scientists. Despite the large number of researches and results, there is still plenty of things to be done. In the area of numerical solution for FDEs, further advances can be carried out. One way of exploring other numerical methods for ODEs is to investigate acceptable extensions, for example peer methods and general linear methods [17, 18, 20, 22, 25]. As future works, exponential fitting techniques may be applied for FDEs with oscillatory behavior which are appropriate adapted numerical methods [13, 14, 16, 21, 24, 29]. Another field about numerical methods for FDEs, may include multi step collocation methods [9, 12, 14, 18, 21, 26, 30].

References

1. Asli BHS, Flusser J (2017) New discrete orthogonal moments for signal analysis. Sig Process 141:57–73
2. Bryson AE Jr, Ho YC (1975) Applied Optimal Control: Optimization, Estimation, and Control. Halsted Press book, Hemisphere Publishing Company
3. Baleanu D, Diethelm K, Scalas E, Trujillo JJ (2016) Fractional calculus: models and numerical methods. World Scientific, Singapore
4. Baleanu B, Maaraba T, Jarad F (2008) Fractional variational principles with delay. J Phys A Math Theor 41(31):315403
5. Banks HT, Burns JA (1978) Hereditary control problems: numerical methods based on averaging approximations. SIAM J Control Optim 16(2):169–208
6. Basin M, Rodriguez-Gonzalez J (2006) Optimal control for linear systems with multiple time delays in control input. IEEE Trans Autom Control 51(1):91–97
7. Bhrawy AH, Alofi AS (2013) The operational matrix of fractional integration for shifted Chebyshev polynomials. Appl Math Lett 26(1):25–31
8. Bhrawy AH, Ezz-Eldien SS (2016) A new Legendre operational technique for delay fractional optimal control problems. Calcolo 53(4):521–543
9. Cardone A, Conte D (2020) Stability analysis of spline collocation methods for fractional differential equations. Math Comput Simulation. (To appear) https://doi.org/10.1016/j.matcom.2020.07.004
10. Cardone A, Conte D, D'Ambrosio R, Paternoster B (2018) Collocation methods for volterra integral and integro-differential equations: a review. Axioms 7(3). DOI https://doi.org/10.3390/axioms7030045
11. Cardone A, Conte D, Paternoster B (2009) A family of multistep collocation methods for Volterra integrodifferential equations. AIP Conf Proc 1168(1):358–361. https://doi.org/10.1063/1.3241469
12. Cardone A, Conte D, Paternoster B (2018) Two-step collocation methods for fractional differential equations. Discrete Contin Dyn Syst Ser B 23(7):2709–2725. https://doi.org/10.3934/dcdsb.2018088

13. Cardone A, D'Ambrosio R, Paternoster B (2017) Exponentially fitted IMEX methods for advection-diffusion problems. J Comput Appl Math 316:100–108. https://doi.org/10.1016/j.cam.2016.08.025
14. Cardone A, D'Ambrosio R, Paternoster B (2017) High order exponentially fitted methods for Volterra integral equations with periodic solution. Appl Numer Math 114:18–29. https://doi.org/10.1016/j.apnum.2016.05.003
15. Cardone A, D'Ambrosio R, Paternoster B (2019) A spectral method for stochastic fractional differential equations. Appl Numer Math 139:115–119. https://doi.org/10.1016/j.apnum.2019.01.009
16. Cardone A, Ixaru LG, Paternoster B, Santomauro G (2015) Ef-Gaussian direct quadrature methods for Volterra integral equations with periodic solution. Math Comput Simul 110:125–143. https://doi.org/10.1016/j.matcom.2013.10.005
17. Cardone A, Jackiewicz Z, Mittelmann H (2012) Optimization-based search for Nordsieck methods of high order with quadratic stability polynomials. Math Model Anal 17(3):293–308. https://doi.org/10.3846/13926292.2012.685497
18. Cardone A, Jackiewicz Z, Verner JH, Welfert B (2015) Order conditions for general linear methods. J Comput Appl Math 290:44–64. https://doi.org/10.1016/j.cam.2015.04.042
19. Conte D, D'Ambrosio R, Jackiewicz Z, Paternoster B (2013) Numerical search for algebraically stable two-step almost collocation methods. J Comput Appl Math 239:304–321. https://doi.org/10.1016/j.cam.2012.08.012
20. Conte D, D'Ambrosio R, Moccaldi M, Paternoster B (2019) Adapted explicit two-step peer methods. J Numer Math 27(2):69–83. https://doi.org/10.1515/jnma-2017-0102
21. Conte D, Ixaru LG, Paternoster B, Santomauro G (2014) Exponentially-fitted Gauss-Laguerre quadrature rule for integrals over an unbounded interval. J Comput Appl Math 255:725–736. https://doi.org/10.1016/j.cam.2013.06.040
22. Conte D, Mohammadi F, Moradi L, Paternoster B (2020) Exponentially fitted twostep peer methods for oscillatory problems. Comput Appl Math 39(3). https://doi.org/10.1007/s40314-020-01202-x
23. Conte D, Paternoster B (2009) Multistep collocation methods for Volterra integral equations. Appl Numer Math 59(8):1721–1736. https://doi.org/10.1016/j.apnum.2009.01.001
24. Conte D, Paternoster B (2016) Modified Gauss-Laguerre exponential fitting based formulae. J Sci Comput 69(1):227–243. https://doi.org/10.1007/s10915-016-0190-0
25. Conte D, Paternoster B, Moradi L, Mohammadi F (2019) Construction of exponentially fitted explicit peer methods. Int J Circ Syst Signal Process 13:501–506
26. Canuto C, Hussaini M, Quarteroni A, Zang T (1988) Spectral Methods in fluid dynamics. Springer, Berlin
27. Chen CL, Sun DY, Chang CY (2000) Numerical solution of time-delayed optimal control problems by iterative dynamic programming. Opt Control ApplMethods 21(3):91–105
28. Chelyshkov VS (2006) Alternative orthogonal polynomials and quadratures. Electron Trans Numerical Anal 25(7):17–26
29. D'Ambrosio R, Paternoster B (2014) Exponentially fitted singly diagonally implicit Runge-Kutta methods. J Comput Appl Math 263:277–287. https://doi.org/10.1016/j.cam.2013.12.022
30. D'Ambrosio R, Paternoster B (2019) Multivalue collocation methods free from order reduction. J Comput Appl Math. https://doi.org/10.1016/j.cam.2019.112515
31. Dadebo S, Luus R (1992) Optimal control of time-delay systems by dynamic programming. Opt Control Appl Methods 13(1):29–41
32. Doha EH, Bhrawy AH, Ezz-Eldien SS (2011) A Chebyshev spectral method based on operational matrix for initial and boundary value problems of fractional order. Comput Math Appl 62(5):2364–2373
33. Eller D, Aggarwal J, Banks H (1969) Optimal control of linear time-delay systems. IEEE Trans Autom Control 14(6):678–687
34. Ghomanjani F, Farahi MH, Gachpazan M (2014) Optimal control of time-varying linear delay systems based on the Bezier curves. Comput Appl Math 33(3):687–715

35. Gogin N, Hirvensalo M (2017) On the generating function of Discrete Chebyshev polynomials. J Math Sci 224(2)
36. Haddadi N, Ordokhani Y, Razzaghi M (2012) Optimal control of delay systems by using a hybrid functions approximation. J Optim Theory Appl 153(2):338–356
37. Hestenes MR (1980) Calculus of variations and optimal control theory. Applied Mathematics series. R. E. Krieger Pub. Co.
38. Hwang C, Chen MY (1986) Analysis of time-delay systems using the Galerkin method. Int J Control 44(3):847–866
39. Jajarmi A, Hajipour M (2016) An efficient recursive shooting method for the optimal control of time-varying systems with state time-delay. Appl Math Model 40(4):2756–2769
40. Jajarmi A, Hajipour M (2017) An efficient finite difference method for the time-delay optimal control problems with time-varying delay. Asian J Control 19(2):1–10
41. Jajarmi A, Baleanu D (2017) Suboptimal control of fractional-order dynamic systems with delay argument. J Vib Control. https://doi.org/10.1177/1077546316687936
42. Jamshidi M, Wang CM (1984) A computational algorithm for large-scale nonlinear timedelay systems. IEEE Trans Syst Man Cybern 14:2–9
43. Kwakernaak H, Sivan R (1972) Linear optimal control systems. Wiley, Hoboken
44. Khellat F (2009) Optimal control of linear time-delayed systems by linear Legendre multiwavelets. J Optim Theory Appl 143(1):107–121
45. Khader MM, Hendy AS (2012) The approximate and exact solutions of the fractional-order delay differential equations using Legendre seudospectral Method. Int J Pure Appl Math 74(3):287–297
46. Kreyszig E (1989) Introductory functional analysis with applications. Wiley, New York
47. Malek-Zavarei M, Jamshidi M (1987) Time-delay systems: analysis, optimization and applications. Elsevier Science Ltd, New York
48. Malmir I (2019) A new fractional integration operational matrix of Chebyshev wavelets in fractional delay systems. Fractal Fract 3:46
49. Moradi L, Mohammadi F (2019) A comparative approach for time-delay fractional optimal control problems: discrete versus continuous Chebyshev polynomials. Asian J Control 21(6):1–13
50. Moradi L, Mohammadi F, Baleanu D (2018) A direct numerical solution of time-delay fractional optimal control problems by using Chelyshkov wavelets. J Vib Control, 1–15
51. Marzban HR, Razzaghi M (2004) Optimal control of linear delay systems via hybrid of block-pulse and Legendre polynomials. J Franklin Inst 341:279–293
52. Marzban HR (2017) Parameter identification of linear multi-delay systems via a hybrid of block-pulse functions and Taylor's polynomials. Int J Control 90(3):504–518
53. Nijmeijer H, van der Schaft A (1990) Nonlinear dynamical control systems. Springer, New York, NY
54. Nikiforov AF, Suslov SK, Uvarov VB (1991) Classical orthogonal polynomials of a discrete variable. Springer, Berlin, Heidelberg
55. Oldham KB, Spanier J (1974) The fractional calculus. Academic Press, New York
56. Patel VK, Singh S, Singh VK (2017) Two-dimensional shifted Legendre polynomial collocation method for electromagnetic waves in dielectric media via almost operational matrices. Math Methods Appl Sci 40(10):3698–3717
57. Palanisamy KR, Rao GP (1983) Optimal control of linear systems with delays in state and control via Wash functions. IEE Proc 130(6):300–312
58. Podlubny I (1999) Fractional differential equations: an introduction to fractional derivatives, fractional differential equations, to methods of their solution and some of their applications. Academic Press, New York
59. Riordan J (1980) An introduction to combinatorial analysis. Wiley, New York
60. Rabiei K, Ordokhani Y, Babolian E (2017) Fractional-order Boubaker functions and their applications in solving delay fractional optimal control problems. J Vib Control 1077546317705041
61. Rahimkhani P, Ordokhani Y, Babolian E (2016) An efficient approximate method for solving delay fractional optimal control problems. Nonlinear Dyn 86(3):1649–1661

62. Safaie E, Farahi MH, Farmani-Ardehaie M (2015) An approximate method for numerically solving multi-dimensional delay fractional optimal control problems by Bernstein polynomials. Comput Appl Math 34(3):831846
63. Sun DY, Huang TC (2010) The solutions of time-delayed optimal control problems by the use of modified line-up competition algorithm. J Taiwan Inst Chem Eng 41(1):54–64
64. Samko SG, Kilbas AA, Marichev OI (1993) Fractional integrals and derivatives: theory and applications. Gordon and Breach, Langhorne
65. Sedaghat S, Ordokhani Y, Dehghan M (2012) Numerical solution of the delay differential equations of pantograph type via Chebyshev polynomials. Commun Nonl Sci Numer Simul 17:4815–4830
66. Wang XT (2007) Numerical solutions of optimal control for time delay systems by hybrid of block-pulse functions and Legendre polynomials. Appl Math Comput 184(2):849–856

Failure Analysis of Strip Foundation on Layered Soil Under Static Loading

Masyitah Md Nujid, Fatimah Abdul Rahman, Ng Kok Shien, Noorsuhada Md Nor, Juraidah Ahmad, Nor Faizah Bawadi, Ali Akhbar Firoozi, and Eka Kusmawati Suparmanto

Abstract A soil underneath foundation structure must satisfy design criteria of bearing capacity analysis which is safety against failure and tolerable settlement. Failure analysis of bearing capacity shows differently for a strip foundation on a single layer of soil or layers of soil in a deposit. For a design criterion, the strip foundation on layer deposit can be analyzed easily by using analytical and experimental approaches. Besides, the strip foundation failure behaviour on single soil is also predictable. However, foundation failure on layered soil is difficult to analyze and predict for the complex behaviour of soil underneath the foundation by analytical or experimental method. Thus, a three-dimensional finite element of strip foundation failure is analyzed. The three-dimensional geometry model of layered soils incorporated in the geotechnical finite element of PLAXIS (V8) program with the Mohr-Coulomb soil failure criterion model were chosen in the analysis. The strip foundation bearing capacity on dense sand overlying soft clay with ratio foundation width (B) to the top layer thickness (H) subjected to vertical loading was evaluated. The shear failure mechanism of strip foundation on layered soils was investigated. The numerical analysis results were verified with theoretical analytical formula from published equations. It shows that the ultimate capacity of strip footing bearing on layered soils increase in width of footing (B). The general shear failure mechanism

M. Md Nujid (✉) · F. Abdul Rahman · N. K. Shien · N. Md Nor
Faculty of Civil Engineering, Universiti Teknologi MARA, Cawangan Pulau Pinang Permatang Pauh Campus, Permatang Pauh, Malaysia
e-mail: masyitahmn@uitm.edu.my

J. Ahmad
Faculty of Civil Engineering, Universiti Teknologi MARA, Cawangan Selangor 40450, Malaysia

N. F. Bawadi
School of Environmental Engineering, Universiti Malaysia Perlis, Kompleks Pusat Pengajian Jejawi 3, Arau 02600, Perlis, Malaysia

A. Akhbar Firoozi
Department of Civil Engineering, Faculty of Engineering & Technology, University of Botswana, Gaborone, Botswana, Iran

E. Kusmawati Suparmanto
Jabatan Kerja Raya (Cawangan Kejuruteraan Geoteknik), Kuala Lumpur 50400, Malaysia

M. Abdel Wahab (ed.), *Proceedings of the 8th International Conference on Fracture, Fatigue and Wear*, Lecture Notes in Mechanical Engineering,
https://doi.org/10.1007/978-981-15-9893-7_24

is observed for dense overlying soft clay in all cases except for B/H = 0.5 (B = 2 m) shows local shear failure mechanism.

Keywords Shear failure · Strip foundation · Static loading · Three-dimensional · Finite element method

1 Introduction

Strip foundation is a shallow foundation type used to provide continuous or linear superstructure loading such as wall which transfer the loading to the soil beneath the superstructure. The founding depth of the strip foundation is no greater than 3 m from the soil surface. The soil underneath footing structure must satisfy design criteria of bearing capacity analysis which is safety against failure, adequate depth and tolerable settlement [1].

Ultimate bearing capacity determination is important when designing a building. It is defined as the strength of bearing pressure at which the ground is estimated to fail in shear which failure will happen. Three modes of shear failure experienced by the foundation soil which is depending on the hardness of base and depth soil layers. There is general shear failure, local shear failure and punching shear failure. A general bearing capacity collapse of foundation occurs when footing loading triggers huge progress of the soil on a shear failure surface which expands away from the footing and up to the ground surface. The capacity of the general bearing of the foundation depends on the footing size and properties of layers soil.

The surface failure of ultimate load in the soil changes base on the soil layer's shear strength characteristics such as upper layer thickness, footing shape, size and embedment, a ratio of the upper layer thickness to a width of footing. Failure analysis of bearing capacity shows differently for a strip foundation on a single layer of soil or layers of soil in the deposit. The bearing stratum either softer or stiffer for layers soil than underlying stratum which varying strength properties. It is found increasing in ultimate bearing capacity with an increase in depth top layer thickness [2]. Also the thickness of the mechanism of collapse differs on the strength of the layered soil [3]. Therefore, it is important to determine and calculate the bearing capacity for soil profile accordingly.

In early years, the collapse shear failures of a strip over a two layers foundation clay were analyzed analytically from experimental owed to [4]. The bearing capacity of shallow footings on layers soil has been studied extensively. Study approaches are covered analytically [5], semi-empirical [6], empirical to finite element methods. Most recent research conducted on the ultimate bearing capacity of the layered soil chose finite element analysis for handle complex layered pattern and complex behaviour of soil for instances [7] studied plane-strain footings on sand overlaying clay soils and [8] evaluated the undrained soil on two-layered clays. Meanwhile [3] studied a square footing rested on the ground surface of two-layered clay soils. Most analysis assumptions were made on the general shear failure mechanism happen with small penetration before reaching the ultimate load. Unlikely to happen where a hard layer overlays a soft layer, a punching shear failure with large penetration occurs.

In the current study, the three-dimensional (3D) strip foundation model on dense sand overlying soft clay failure analysis was employed to investigate the failure of the foundation under static vertical loading of 500 kN/m^2 imposed. The bearing capacity analysis by the ratio of footing width to the top layer thickness (B/H) and size of footing width effects of layered soil by finite element method (FEM) was evaluated and verified by the semi-empirical method. The failure mechanism of strip foundation in two-layered soil also was investigated.

2 Failure Analysis of Layered Soil

The ultimate bearing capacity aims to determine the load which can be carried by the carried by the soil underneath the foundation before failing in shear. A re-evaluation of the existing solutions to predict failure analysis of shallow foundation on two layers of soil have been published in a few years back. For punching shear failure, semiempirical approaches have also been suggested based on experimental studies by [5, 6, 12] which also called punching shear models, since the sand layer is in a state of passive failure along vertical planes below the edges of the footing. The punching shear coefficient, K_s depends on the angle of internal friction of the sand layer [9] mobilized angle of friction (δ), the undrained shear strength of the clay (c_u), the angle of friction of the sand (ϕ), and the bearing capacity ratio (q_1 /q_1) based on Eq. 3 proposed by [6] for estimating a strip foundation's ultimate bearing capacity in the event of a punching shear failure [7]. Table 1 shows a formula to evaluate bearing capacity for a strip foundation resting on a dense sand that overlies a soft clay while Fig. 1 shows the punching shear coefficient, K_s of 36° friction angle with $q_2/q_1 = 0$.

Table 1 Formula for bearing capacity analysis of layered soil

q_u: ultimate bearing capacity	Term	Equation
$cNc+ \gamma H^2 \left(1+\left(\frac{2D}{H}\right) K_s \tan \frac{\varnothing}{B} + \gamma D\right)$	c: undrained cohesion of clay N_c: bearing capacity factor (5.14) $N\gamma$: unit weight factor D: depth of embedment H: thickness of top layer B: footing width γ: unit weight of top layer K_s: punching shear coefficient	(1) [5]
$cNc + \frac{(2C_a H_1)}{B} + \gamma D_f$	D_f: depth of foundation H_1: thickness of top layer C_a: soil adhesion, a function of c/c_1	(2) [12]
$\frac{q_1}{q_2} = \frac{cN_c}{0.5\gamma B\ N_\gamma}$		(3) [6]

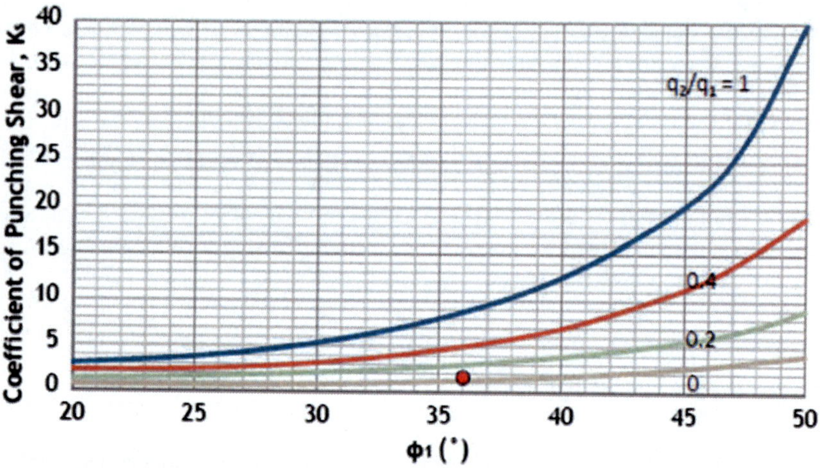

Fig. 1 Meyerhof and Hanna punching shear coefficient K_s Hanna and Meyerhof [6]

Burd and Frydmen [7] conducted a rigid two-dimensional plane-strain shallow foundation study which imposed on the ground surface consisting of a uniform layer of sand overlying dense clay uniform beds. The thickness of the sand layer is comparable to the footing width and the ground surface and the interface between the two soil layers are horizontal. It is believed that the layer of clay was undrained and that the layer of sand was drained. Yu et al. [10] simulated three-dimensional punching shear failure or the penetration responses of square footings in strong over soft clays. Increasing factors in soil layers strength ratio, relative top soil thickness and weight, as the peak foundation and penetration depth increase can reduce potential punching shear failure. The failure mechanism of foundation on dense sand overlying soft clay shows the potential for "punch-through" failure exists [10, 11] due to a sudden decrease in bearing resistance when the footing punches a block of sand into the underlying soft clay in an uncontrolled manner.

3 Methodology

3.1 Numerical Modelling

In the 3D incremental stress-strain analysis, a plane stain strip footing was considered as rigid and the footing was constructed on the embedded at various depth from the ground surface in the top strong soil layer. Vertical loading of 500 kN/m^2 was imposed until failure achieved. The vertical boundary was set at 4B from the ground surface. It was assumed that the stress-induced terminated at this depth. As the size footing,

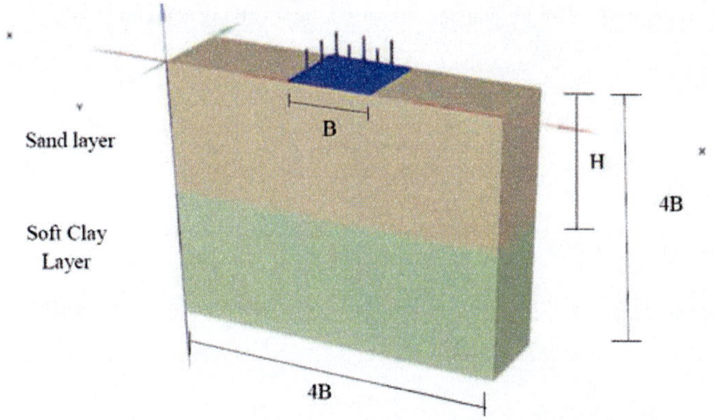

Fig. 2 Geometry of the plane strain strip foundation model on layered soil

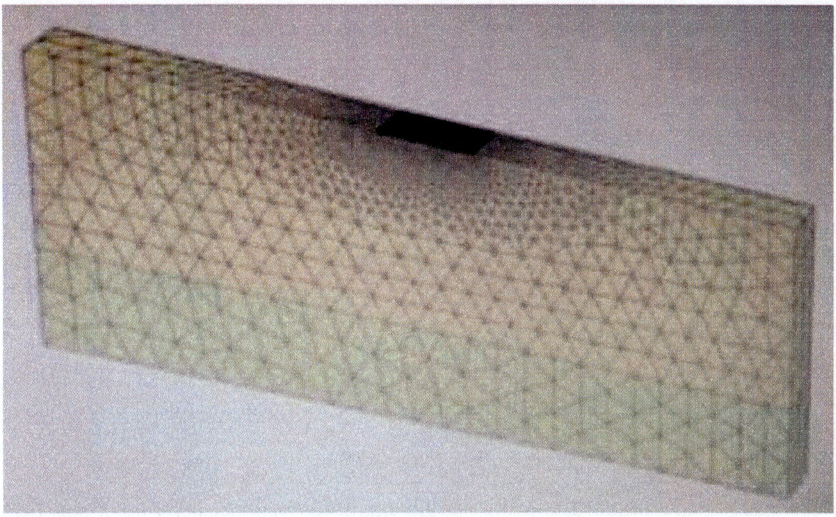

Fig. 3 Mesh assigning in the analysis

B increased, the model size increase, therefore the number of mesh will vary. Figure 2 shows the model geometry of the problem.

Figure 3 shows mesh configuration with fine mesh was assigned near the area below of strip foundation. Meanwhile, the medium mesh was used for the rest of layered soil due to avoid time-consuming while analyzing.

A PLAXIS 3D finite element program was used and Mohr-Coulomb soil failure criterion was chosen to evaluate failure analysis of the foundation dense overlying soft clay which foundation fail in a linear elastic. The effects of width footing to the top layer thickness ratio (B/H_1) and width of footing were studied with a unit weight of sand and clay were 20 kN/m^3 and 18 kN/m^3 respectively. The soil layers Poisson's

Table 2 Input parameters for foundation on dense sand overlying soft clay

Parameter	Sand	Clay	Footing
Material model	Mohr-Coulomb	Mohr-Coulomb	Linear elastic
Type of material behaviour	Drained	Undrained (B)	Non-porous
Soil unit weight, γ_{unsat}, kN/m^3	18	17	–
Soil unit weight, γ_{sat}, kN/m^3	20	18	–
Young's modulus, E, kN/m^2	2×10^4	3×10^4	1×10^{12}
Poisson's ratio, ν	0.3	0.3	0.15
Cohesion, c, kN/m^2	0.2	20	–
Friction angle, ϕ	35	–	–

ratio was set to v = 0.30 while the ratio of the elastic modulus was taken as E = 20 MPa. The value of internal friction angle for dense sand and undrained cohesion of soft clay was 35° and 20 kN/m^2 which were obtained experimentally from shear box test and unconsolidated undrained respectively. Table 2 shows a set of input parameters assigned in the finite element program. In the PLAXIS 3D program, the undrained B method was used as drainage type for the clay. This type of drainage is chosen for undrained behaviour or short-term material behaviour in which stiffness is defined in terms of effective properties whereas strength are defined as undrained shear strength.

3.2 *Numerical Modelling Scheme*

Table 3 shows the ratio of the width of foundation to the upper layer thickness B/H that was used in this study which are 0.5, 1.0, 1.5, and 2.0. The total thickness of layered soil was set as 4B as shown in Fig. 2. The value of the top layer thickness depends on the ratio of footing width to the top layer thickness, B/H which the upper sand layer thickness decreases as the ratio of B/H increase.

Table 3 Numerical modelling scheme

B/H	Top layer thickness, H(m)		
	$B = 1$	$B = 2$	$B = 3$
0.5	2.0	4.0	6.0
1.0	1.0	2.0	3.0
1.5	0.7	1.3	2.0
2.0	0.5	1.0	1.5

4 Results and Discussion

The finite element program adopted the technique which produced the curve of ΣM_{stage} through the calculating phase to determine the ultimate bearing capacity of the layered soil from numerical analyses.

The curve of ΣM_{stage} was used to determine the value of q_u where the ultimate load takes a position in the state of failure, it is acceptable to take into account the rate of yielding in the determination of the ultimate bearing capacity. The yielding rate of soil elements was believed in combination with other methods. An indicator of the ultimate bearing capacity, q_u is the point at which the curve experiences a distinctive change in the typical slope, showing accelerated plastic yielding. The value from ΣM_{stage} curve will multiply with the surface loads to assess the ultimate bearing capacity. In summary, there is condition to consider the failure of the footing as below:

1. If the M_{stage} curve exhibits distinct yield point, then that point is taken to indicate the q_u as shown in Fig. 4a.
2. If the curve shows in Fig. 4b with no distinct yield point.

Another technique proposed in the second method was easier compared to the first method. It is possible to assume that the final point will reduce by 20%. However, each method is used this study to check accuracy of ultimate bearing capacity, q_u.

Figure 5 shows the relationship between q_u and B to observe the effect of upper sand layer thickness through the ratio of B/H value. The ratio of B/H = 0.5 had the highest value of bearing capacity compared with another ratio of B/H. It can be seen the upper sand layer thickness decrease as the ratio of B/H increase. It has been analyzed that the two-layered soil bearing capacity decrease by incrementing the ratio of B/H. This is because of the ultimate bearing capacity, q_u decrease as the H (thickness of the upper dense sand layer) decrease from B/H = 0.5 to B/H = 2. The result shows the H increases, the q_u also increases. The bearing strength decreases from the peak initial value to the least for a base on a thick sand deposit.

Figure 6 shows the relationship between q_u and B/H to analyse the effect of B (footing size) on the qu ultimate bearing capacity through four difference values of B/H. It also shows that B = 3 had the highest value of q_u. It seems that the larger the value of B, the larger value of q_u. Thus, it can be summarized by increasing the footing width, B the q_u ultimate bearing capacity also increases.

The numerical results were then verified from the semi-empirical approach from [5, 6]. Table 4 shows differences in the ultimate bearing capacity results from both methods. For B = 2 m show smallest percentage in difference for B/H ratio except B/H = 2.0. Meanwhile, B = 1 m had the highest percentage in the difference between numerical and semi-empirical approaches compared to B = 2 and 3 m. The B = 2 and 3 m of strip foundation width can cater to the loading of 500 kPa. Most B sizes show percentage differences for both methods of analysis were below 20%.

The failure mechanism of strip foundation in two layers of soil using shear strain contours in term of shadings and contour line was investigated. Bearing capacity

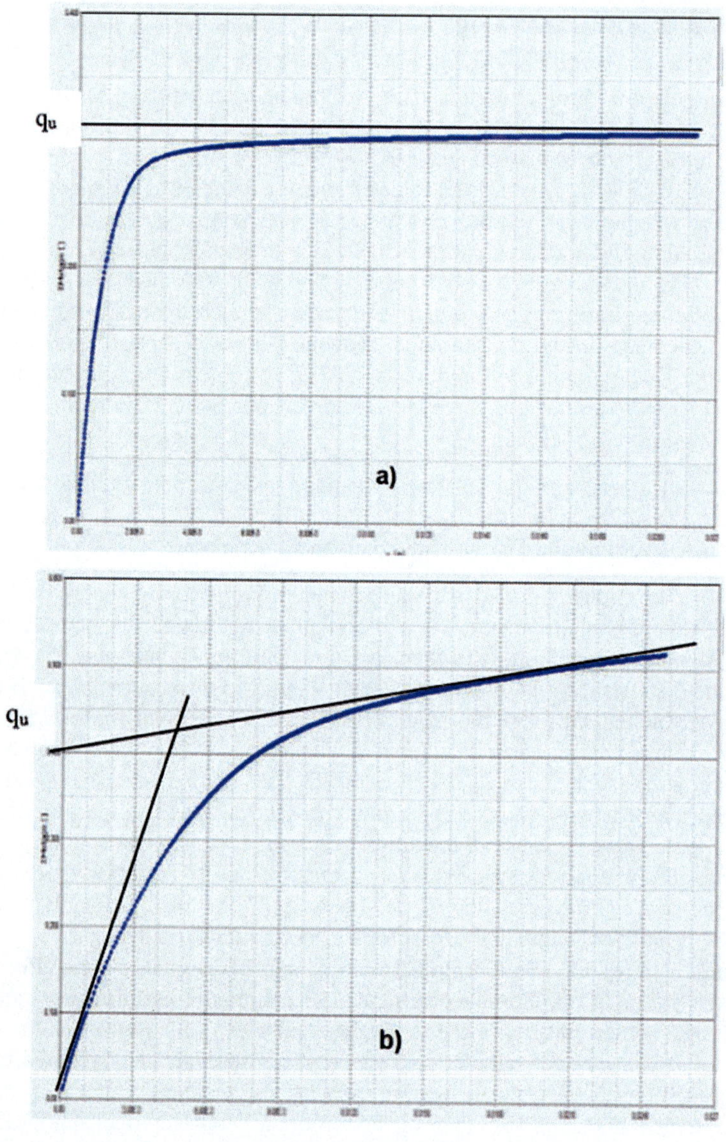

Fig. 4 **a** Distinct failure mode. **b** No distinct failure mode

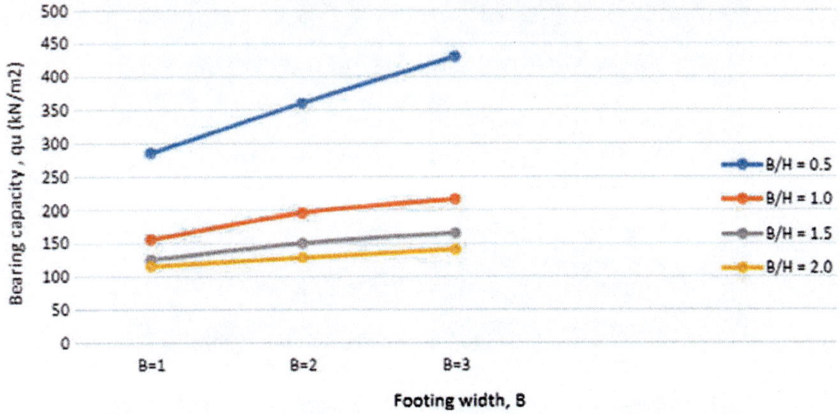

Fig. 5 The relationship between q_u and B

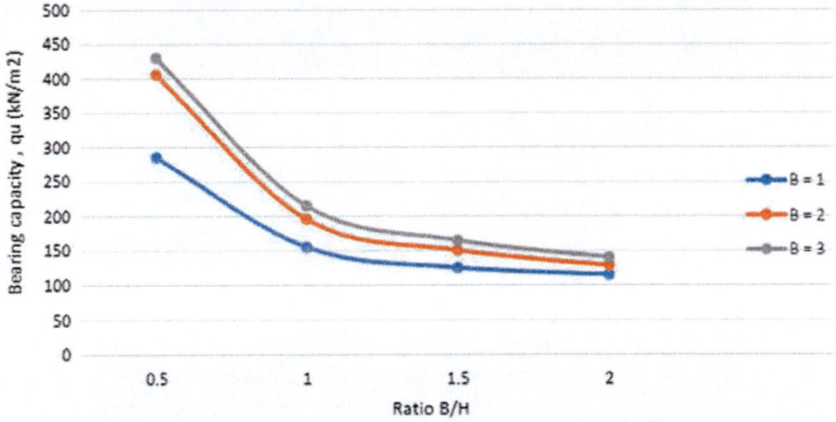

Fig. 6 The relationship between q_u and B/H

Table 4 Ultimate bearing capacity results from analytical and numerical methods

RatioB/H	B = 1 (m)			B = 2 (m)			B = 3 (m)		
	Semi empirical (kPa)	Numerical method (kPa)	Difference %	Semi empirical (kPa)	Numerical method (kPa)	Difference %	Semi empirical (kPa)	Numerical method (kPa)	Difference %
0.5	327.32	285	12.9	381.52	360	5.6	457.65	430	6.0
1.0	183.64	155	15.6	192.72	195	1.2	210.61	215	2.1
1.5	151.99	125	17.8	153.59	150	2.3	161.21	165	2.4
2.0	138.74	115	17.5	138.77	128	7.8	142.68	140	1.9

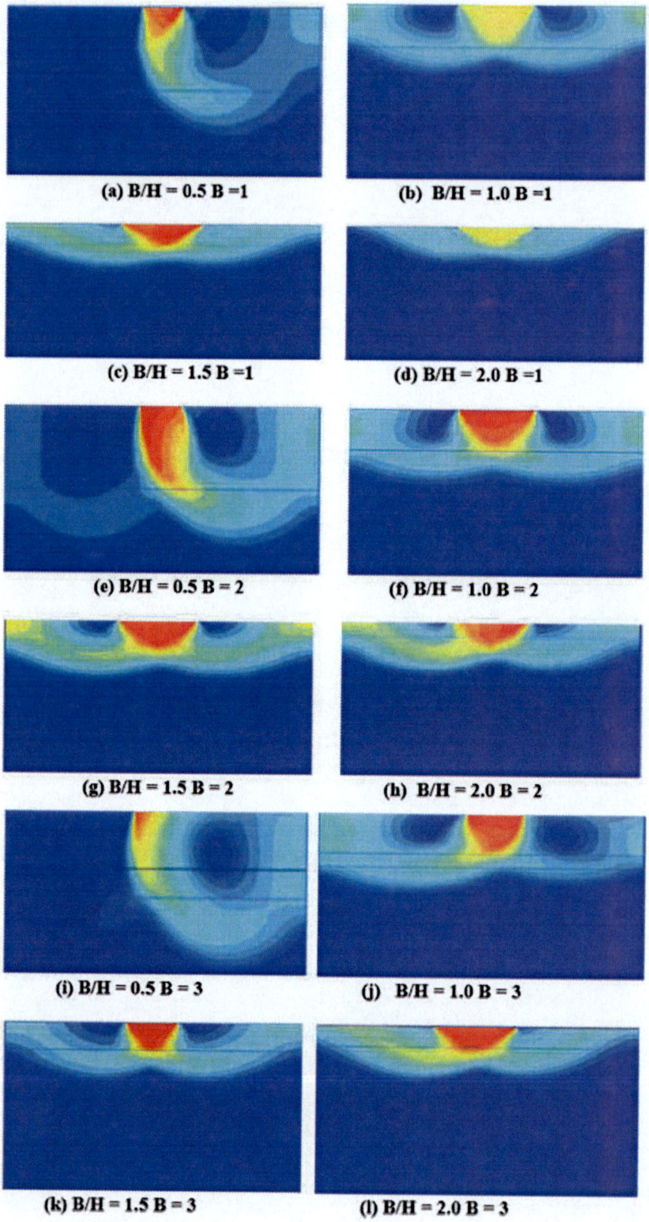

Fig. 7 Deformed shear contour

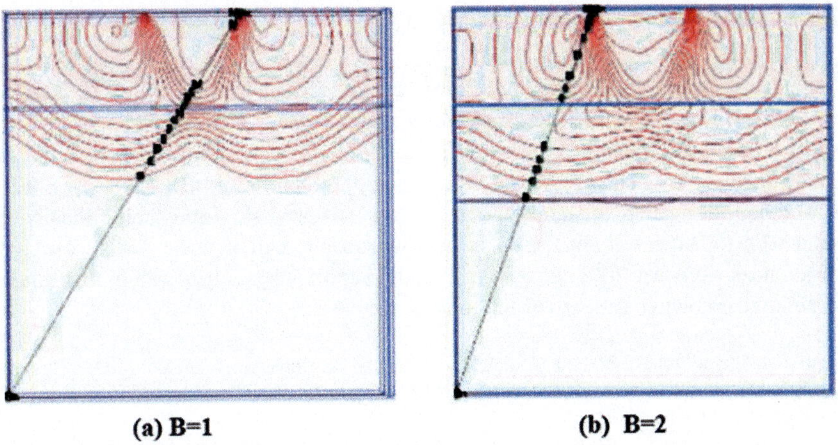

Fig. 8 Contour line for ratio B/H = 2.0

of foundation will change with various factors such as foundation width and thickness of top layer soil. These factors affect the failure mechanism of the foundation. Besides, there is three types of failure mechanism, which are the general shear failure, punching failure and local shear failure. Deformed shear contour is observed in total incremental displacements obtained from FEM given the failure mode. The failure mechanism of three different size of footing as shown in Fig. 7 for B/H = 0.5, 1.0, 1.5 and 2.0. The red line marked the boundary between the upper and bottom soil layer. The squeeze breakdown mode of dense sand overlying clay soils was observed. It was found that the depth of the failure mechanism decreases as the ratio of B/H increase. Figure 6 shows the type of failure for each case that was obtained from finite element analysis. Based on the results, general shear failure was observed for dense sand overlying soft clay in most of the case. General shear failure occurs as the layer of soft clay acts as a fairly rigid foundation. However, local shear failure became obvious in the case of B/H = 0.5 and B = 2 m. The soft layer of clay is relatively weaker than the thick layer of sand. Therefore, it can be concluded that the type of failure mechanism of layered soil depends on the thickness and width of footing.

Figure 8 shows the shear strain contours at failure in term of contour line for B/H = 2.0. From this figure, it is observed that the depth of the contour line of B = 2 was deeper compared to B = 1. As the footing size increase, the contour line produced will increase too. It can be concluded that the depth of the failure zone increases when the size of the footing increase. Hence, footing width is one of the contributing factors that affects the failure mode.

5 Conclusion

In the failure analysis, the ultimate bearing capacity of strip footing subjected to static vertical loading was investigated and verified with the semi-empirical approach. The shear failure mechanism was also observed. The ultimate bearing capacity of dense sand overlying soft clay decreases as the sand thickness ratio, B/H increases. The larger the footing size, the larger the bulb stress (stress zone). Therefore, if the dense sand in the top layer is thicker, then the bearing capacity will increase. The percentage differences were less 20% between two method approaches. The depth of the failure zone increases when the size of the footing increase.

Acknowledgements The authors would like to thanks to the Universiti Teknologi MARA Cawangan Pulau Pinang and Universiti Teknologi MARA Selangor for providing financial support for the publication of this manuscript.

References

1. Padmini D (2018) Review of methods to predict the bearing capacity of shallow foundations. Int Res J Eng Technol 5(11):217–226
2. Gupta A, Dutta RK, Shrivastava R, Khatri VN (2017) Ultimate bearing capacity of square/rectangular footing on layered soil. Indian Geotech J 47(3):303–313
3. Zhu M, Michalowski RL (2006) Bearing capacity of rectangular footings on two-layer clay Capacité porteuse des semelles de fondationrectangulaires sur deux couches argileuses, pp 997–1000
4. Button SJ (1953) The bearing capacity of footings on two-layer cohesive subsoil. In: Proceedings on 3rd International Conference Soil Mechanisms Foundation Engineering, vol 1. Zurich, pp 332–335
5. Meyehof G (1974) Ultimate bearing capacity of footings on sand layer overlying cay. Canadian Geotech J 11(2):223–229
6. Hanna AM, Meyerhof GG (1980) Design charts for ultimate bearing capacity of foundations on sand overlying soft clay. Canadian Geotech J 17(2):300–303
7. Burd HJ, Frydman S (1997) Bearing capacity of plane-strain footings on layered soils. Canadian Geotech J 34(2):241–253
8. Merifield R, Nguyen V (2006) Two- and three-dimensional bearing-capacity solutions for footings on two-layered clays. Geomech Geoeng Int J 1(2):151–162
9. Ramadan MI, Hussien MH (2015) Bearing capacity of sand overlying clay. Int J Sci Res (IJSR) 4(11):1852–1859
10. Yu L, Liu J, Kong X, Hu Y (2011) Three-dimensional large deformation FE analysis of square footings in two-layered clays. J Geotech Geoenviron Eng 137(1):52–58
11. Randolph MF, Lee KK, Cassidy MJ (2013) Bearing capacity on sand overlying clay soils: experimental and finite-element investigation of potential punch-through failure. Géotechnique 63(15):1271–1284
12. Meyerhof GG, Hanna AM (1978) Ultimate bearing capacity of foundations on layered soils under inclined load. Canadian Geotech J 15:565–572

Evaluation of the Residual Stress Behavior of 316L Stainless Steel in Electric Generator Blades Subjected to Different Hours of Service

A. Morales-Ortiz, J. C. Arango, and C. C. Palacio

Abstract This work studies the stability of residual stresses on cooling blades of electricity generators made of 316L stainless steel subjected to different hours of service. New blades made by stamping were analysed and residual stresses were measured in high-stress areas by finding magnitudes of up to 400 MPa. A finite element simulation was performed to determine the magnitude of the applied stress and measurements were made at the same points on blades of 13.000, 35.000 and 60.000 h of service. It is demonstrated how the location and direction of applied load generates that residual stress decrease while the number of cycles increase, but others, coincident with the points of high applied load the residual stresses increase over time.

Keywords Residual stress · Fatigue · X-ray diffraction

1 Introduction

A blade is a structural element that seeks to displace an specific amount of air when it moves, its geometry is far from simple and its manufacturing process is by stamping and die-cutting. Like any structural element, its mission beyond being aerodynamically effective is to have enough resistance to withstand the operating cycle to the rotor rotation conditions of a hydraulic generation turbo-group.

The behavior of the blades of electric generators is an issue that for many years has been studied in order to solve the problems that they exhibit. Several studies conducted by large companies and universities at the international level have revealed different phenomena associated with the problems of breaking these elements (fatigue, resonance to the natural frequency, corrosion by the ozone generation, etc.) and have made significant advances that have avoided new accidents in the stators of the power plants.

A. Morales-Ortiz (✉) · J. C. Arango · C. C. Palacio
EAFIT University, Medellín, Colombia
e-mail: amoraleso1@eafit.edu.co

M. Abdel Wahab (ed.), *Proceedings of the 8th International Conference on Fracture, Fatigue and Wear*, Lecture Notes in Mechanical Engineering,
https://doi.org/10.1007/978-981-15-9893-7_25

One of the main factors influencing reduction of service life of mechanical parts and engineering components, such as a blade, is the residual stresses that are present in its structure even when external forces or loads are absent [8, 10, 18]. Sometimes the presence of these stresses may be intentional to improve some of the characteristics of the material (as is the case of shot-peening on the turbine surface to improve its fatigue behavior and induce compression stresses in the material), but when residual stresses are not desired and are present, they can be harmful and significantly affect phenomena such as fracture, dimensional stability, residual life, crack propagation and critical load [9].

In practice, any manufactured part is unlikely to be completely free of residual stress [17], either because they are generated during processing or because they are induced during their work cycle, so when designing, manipulating and operating a component there is an implicit need to determine the presence of residual stresses as well as the applied load to which it will be used.

The stresses present in a structure can be of various types: traction, compression, shear, bending, buckling, torsion and fatigue. When a material is under external forces as bending, shearing or torsion it induce compressive or tensile stresses [13], which are considered as positive stress when a body is under the action of forces acting in the opposite direction tending to stretch it and are considered as negative stress when the body is under the action of forces that act in the same direction and tend to decrease its dimensions [5]. Cyclic loads and residual stresses affects the components service life promoting the occurrence of fatigue failures [16]. When the material has no residual stress, the range of stress represented by $\Delta\sigma = \sigma_{max} - \sigma_{min}$, is not affected (see Fig. 1) [7].

The presence of residual stresses in the material affects the magnitude of maximum and minimum stress, which makes the material more susceptible to failures (see Fig. 2) [7].

Currently there are many techniques for the quantification and location of residual stress in order to know the actual condition of the parts in service. However, only those categorized as non-destructive have real applicability in the industry, as they allow the evaluation of the component without damage it like the X-ray diffraction (XRD) [10]. XRD determine the elastic deformations that occur in a crystalline material, such as metals or ceramics, for correlate them with the residual stresses in its structure. When a deformation occurs the interplanar space of the crystalline

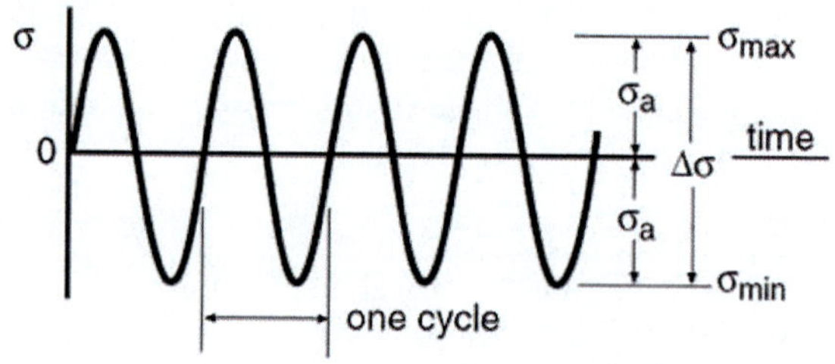

Fig. 1 Stress behavior in fully reversible cyclic loads on a material without the presence of residual stresses [7]

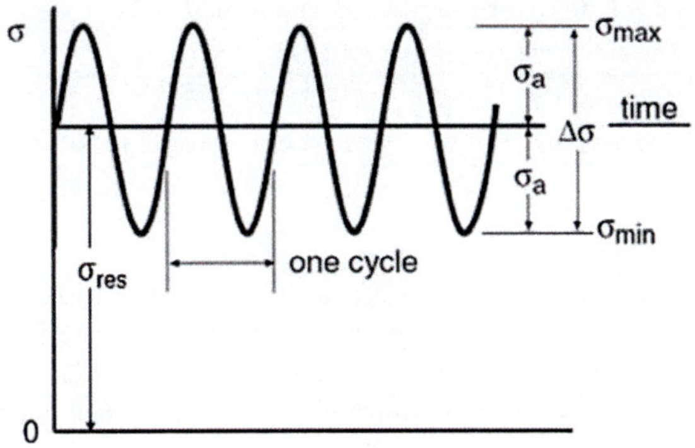

Fig. 2 Stress behavior in cyclic loads on a material with the presence of residual stresses

network changes from the normal stress-free state in a proportion that corresponds to the magnitude of the applied stress, as seen in the Fig. 3 [11, 12].

In this technique the sample is irradiated with X-rays and its physical behavior is described by Bragg's Law [6].

$$2dsen\theta = n\lambda \quad (1)$$

When the material has residual stresses, the diffraction peaks generated by the diffraction pattern can be altered from the normal position, as can be seen in the Fig. 4.

In this way, measurements can be made on parts, such as the blades of the generators, and thus determining the magnitude and direction of the residual stress.

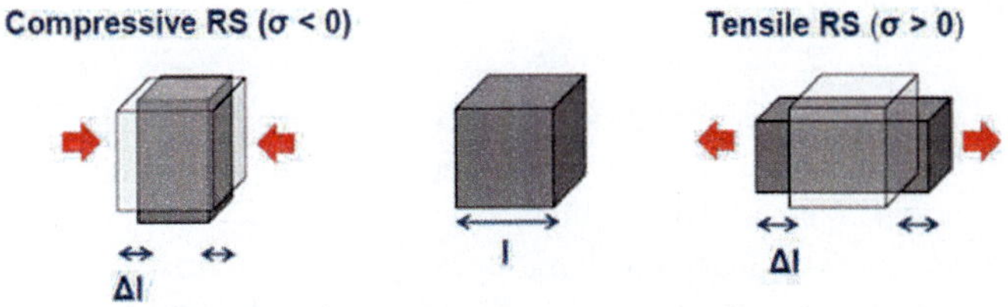

Fig. 3 Deformations in the interplanar space of a crystal

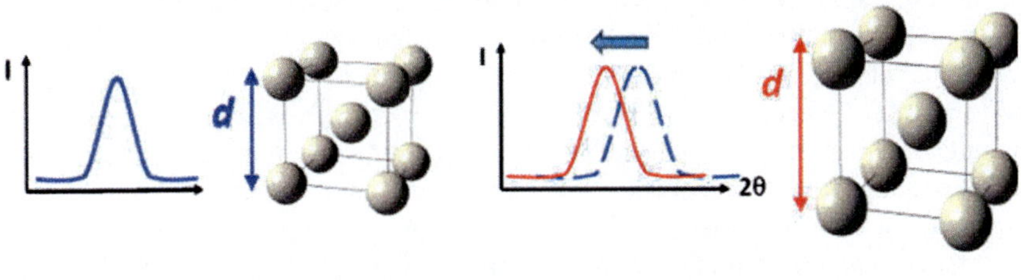

Fig. 4 Diffraction peak movement due to residual stresses present in the structure [1]

Table 1 Chemical composition of the AISI 316L steel

Chemical composition steel AISI 316L									
Alloy elements	%C	%Cr	%Ni	%Si	%Mn	%P	%S	%Mo	%Fe
Composition	0.02	17.1	10.1	0.48	1.76	0.031	0.001	2.06	68.4

On the other hand, finite element simulation, based on the finite element method (FEM), allows to generate an approximation of solutions to problems that can be very complex [4]. This technique allowed to simulate the behavior of the blades when they are put into service on the generators, in order to know the magnitude and location of the stress to which they are subjected during the operation and evaluate their behavior when residual stresses are present, letting evaluate their condition after different hours of service [2].

2 Materials and Methods

2.1 Material

The manufacturing material of the studied blades corresponds to a 316L stainless steel according to the AISI classification. Below the chemical composition is presented for the material of the blades (Table 1).

2.2 Metallography

A sample of the material was obtained by waterjet cut HBI of reference WAC 1313 at a speed of 0.15 m/min. The sample was subsequently prepared by immersion in resin and metallographic polishing in accordance with ASTM E407-07 [2]. The prepared samples were 3 and were taken one longitudinally, one in a transverse direction and another superficial, as it presented in (Fig. 5), in order to validate a possible influence of the sample orientation of properties on the material associated with the orientation of the grains in the microstructure.

Metallographic revealing was made on the surface of the 3 samples using a solution of 30 ml hydrochloric acid, 15 ml of nitric acid mixed with and 45 ml of glycerol, for further observation in optical microscope (Olympus MDS 5000d) using of 100, 500 and 5000X magnifications. The observed microstructure is presented in Fig. 6. A granular microstructure is identified, and it is according to the chemical composition of the material which promotes the formation of austenite grains, with the presence of intergranular phases that are typical of the processes of precipitate formation at the grain limits of stainless steels such as 316L [15]

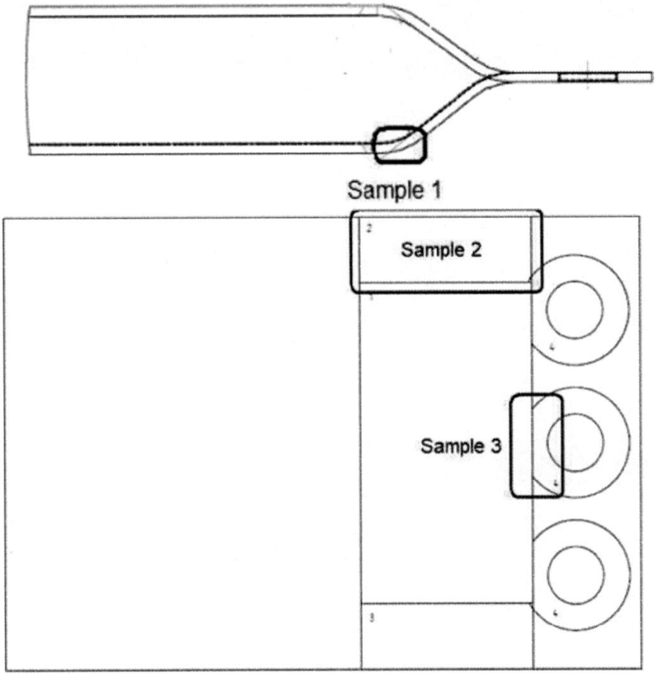

Fig. 5 Sampling points

Fig. 6 Microstructure of a blade after 30.000 hours of service

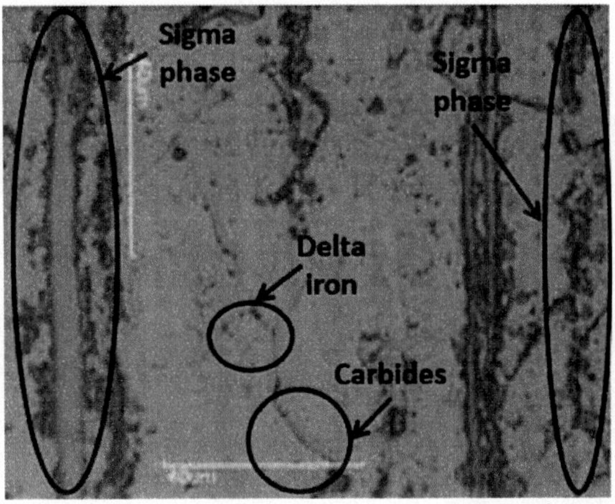

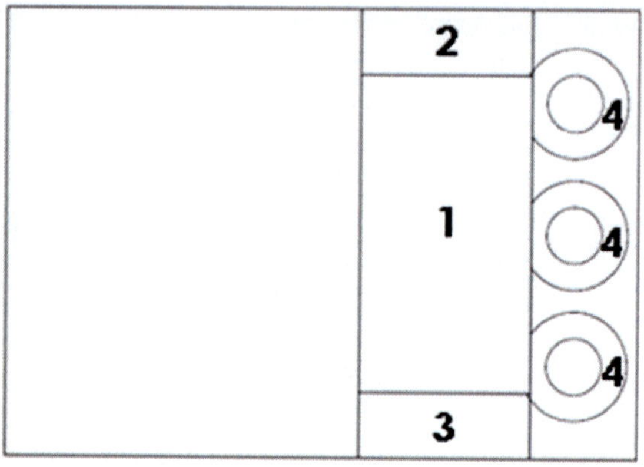

Fig. 7 Hardness measurement zones

Table 2 Brinell hardness results

Zone	Average Brinell hardness HBN
1	180
2	182
3	178
4	175

2.3 *Hardness*

Brinell hardness measurements were performed on different areas of the blades identified as it is illustrated in Fig. 7. A DUD-500 durometer was used, with a 2.5 mm diameter steel ball indenter and applying a load of 100 kg in accordance with ASTM E10-18 [3]. The results obtained are presented in Table 2.

2.4 *Residual Stress Determination by XRD*

Measurement of residual stresses was performed on the surface of areas identified as 2, 3 and 4 on the blades as shown in the Fig. 8a and on the points identified in Fig. 8b.

These measurements were performed with a portable X-ray diffractometer (Spider X 2018 model) with residual stress measurement module for austenitic stainless steels.

The condition of blades analysed were grouped as follows:

- 5 new upper blades.
- 5 new lower blades.
- 5 upper used blades with 35.000 h.
- 5 lower used blades with 35.000 h.

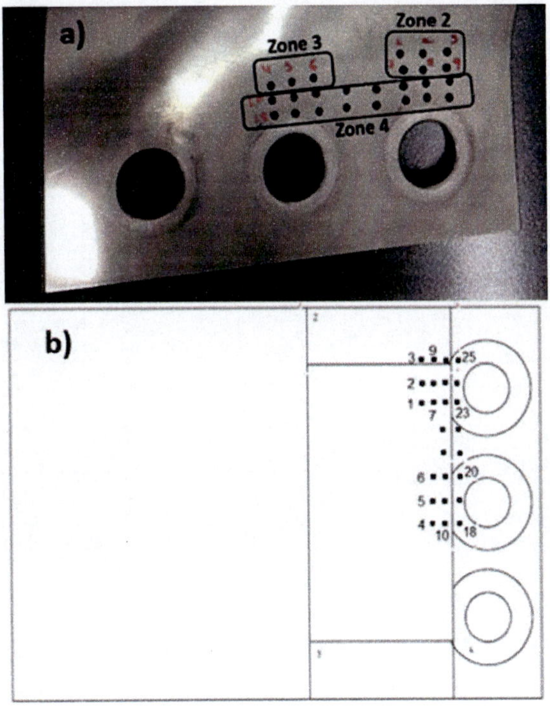

Fig. 8 Zones **a** and points **b** of measurement of residual stress

- 5 upper used blades with 63.000 h.
- 5 lower used blades with 63.000 h.

The location of these blades within the rotor can be seen in Fig. 9, where it is possible to identify the upper blades and the lower ones.

Figure 10 shows the blades used as samples of this study.

2.5 Finite Elements Simulation for the Stress Determination in the Blades

The determination of stresses acting on the blades of the study was carried out through simulations by FEM taking into account the loads and variables that act on the blade during the operation as it is presented below:

- Number and location of blades: 24 lower and 24 uppers
- Temperature: 35–40 °C.
- Turning speed: 300 rpm.
- Airflow through the air cooler: 14 m^3/s.
- Tangential blade speed: 80.8 m/s.

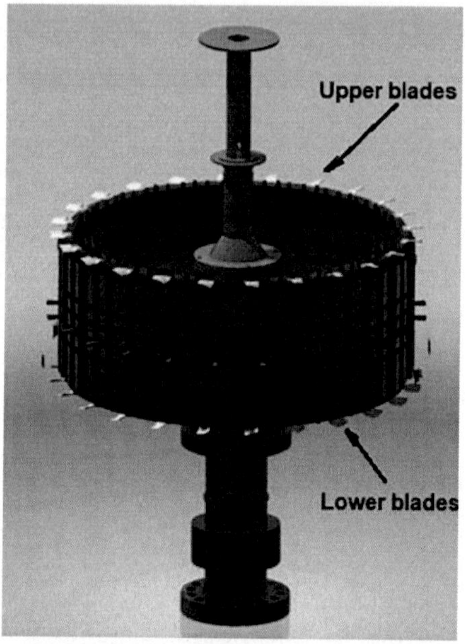

Fig. 9 Location of the blades within the rotor

Fig. 10 Blade under study

- Blade turning radius: 2.57 m.
- Air density: 1 kg/ m^3.
- Centrifugal force: 4824 N.
- Horizontal air force: 29.5 N.
- Vertical air force: 422.2 N.

The simulation process was carried out in two stages:

1. Simulation of the flow conditions that occur within the generator during its operation, considering the air speed and the rotation direction of the generator, the volume of air between the rotor and the stator, the volume of air in the radiators, the air pressure acting on the blades;
2. Simulation of mechanical response of the blades, before the action of the rotation and airflow conditions associated with the operation, to identify the magnitudes of Von-Mises stress, unitary and total strain, and shear stresses.

3 Discussion

Microstructure analysis show the presence of austenite grains, which corresponds to a typical microstructure of a 316L steel. The presence of segregations in grain boundaries was observed which is corresponding to this material and the operating conditions such as temperature and accumulated hours; these segregations correspond to chromium carbides and sigma phases which are characteristic phases of fatigued austenitic stainless steels [15].

Brinell hardness values shows an aging hardening derived from the phases and carbides present in the fatigued alloy. Furthermore, simulations showed that the blades exhibit stress concentration in the vicinity of the holes obtaining up to 61 MPa, as presented in the Fig. 11.

The maximum stress obtained from the simulation reach a value of 61 MPa, which shows that induced loads due to service do not represent a mechanical risk to the structure when the material strength (Su = 596 MPa) is taken in to account. However, the presence of precipitates in the grain limits of the microstructure reduces the capacity of the material to support load and can induce the nucleation and crack propagation [14].

XRD residual stress results show the presence of tensile stresses (identified with positive sign) and its magnitude depends of the zone of the blade measured and the service time of the blades. There is a clear inverse relationship between the magnitude of the stress and the number of hours of service, as can be seen in the Fig. 12. Contrary, in the neighborhood of the holes, residual stress tends to increase its magnitude directly with the number of cycles.

Figure 12 shows the residual stresses results in function of number of hours of service.

The upper blades residual stresses exhibit relaxation phenomenaes regarding to accumulated cycles except on the zones around the holes on which it was identified a

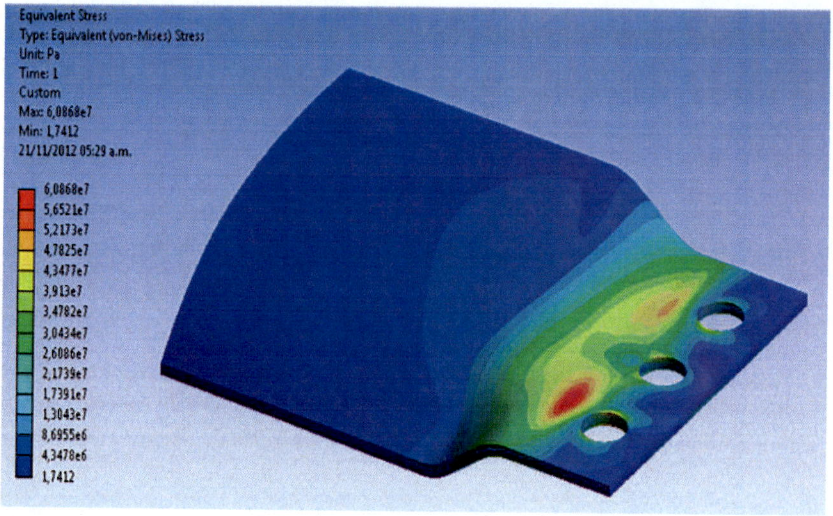

Fig. 11 Concentration of stress on a blade in service

competition between stress intensity factor associated to the presence of the holes and the stresses relaxation behavior; it was evidenced in the residual stress measurements for the points 16, 17, 22 & 25 located around the holes. It was noticed that for the blades zones under high loads the residual stresses tend to exhibit a weak stress relaxation phenomenae, as it can be seen in the Fig. 13.

Tensile residual stresses (positive magnitude in residual stresses) reduce the strength of the material as they are already pre-stressed being the most relevant factor both for the appearing and location of cracks in the presence of applied loads.

Residual stresses are not constant over the time, they have a dynamic behavior with two alternatives either their relax them to become neutral or they increase as the applied stress continue rising.

4 Conclusions

Residual stresses on components that have been manufactured through severe plastics deformation have high residual stress values, which limit the stress that can be applied as total stress that easily contributes to reach the limit values such as the yield strength.

Obtained results show that the residual stresses are not stable due to inducing alternating stress, on the contrary they have the ability to increase their magnitude if the direction of the applied stress match with the residual stress. On the other hand, it is possible to reduce the magnitude of residual stresses by applying alternating stress with opposite directions that allow residual stresses to reduce their magnitude.

Fig. 12 Residual stresses results in function of number of hours of service

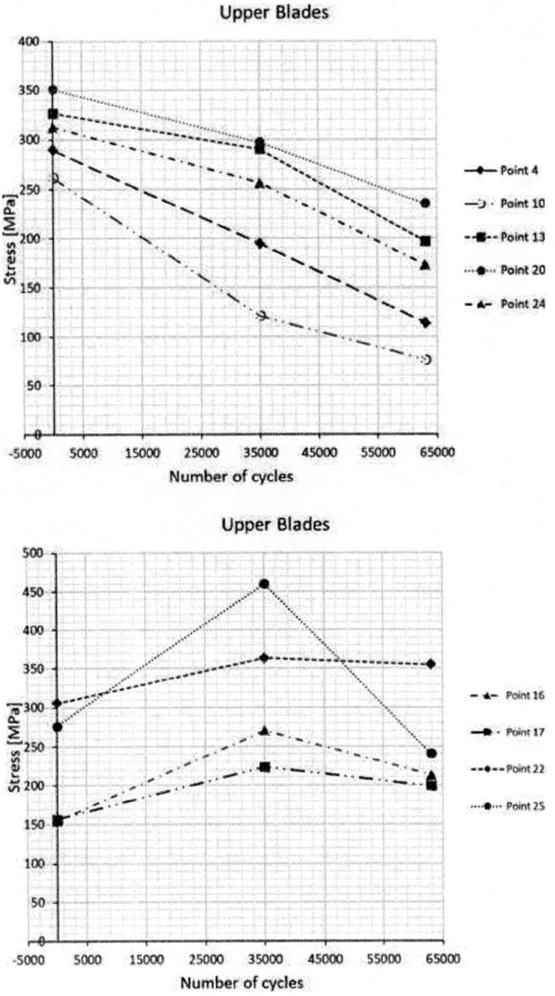

Fig. 13 Residual stress behavior on lower blades

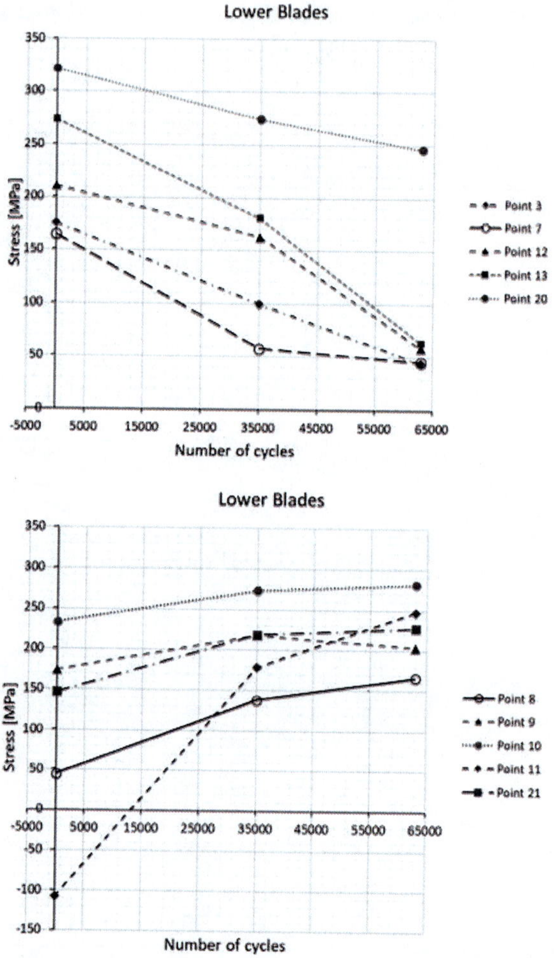

Acknowledgements To EAFIT University and the GEMA Research group for economical and technical support with internal projects that allow the realization of this research.

References

1. Analytical Instruments Group G SpiderX
2. ASTM E407 (2015) E407-07 Standard practice for microetching metals and alloys. ASTM Int. https://doi.org/10.1520/E0407-07R15E01.2
3. ASTM Standards (2018) Standard test method for brinell hardness of metallic materials E10−18. Annu B. ASTM Stand

4. Bathe K-J (2006) Finite element procedures: Klaus-Jurgen Bathe. Englewood Cliffs, NJ Klaus-Jürgen Bathe, cop
5. Capote Luna V (2011) Tipos de esfuerzos físicos
6. Cullity B, Stock S (2014) Elements of X-ray diffraction, 3rd edn. Pearson Education Limited, Edinburgh
7. Dowling NE, Prasad K, Narayanasamy R (2013) Mechanical behavior of materials engineering methods for deformation, fracture, and fatigue, 4th edn. Pearson Education Limited, Harlow
8. Epp J (2016) X-ray diffraction (XRD) techniques for materials characterization. Elsevier Ltd
9. García Jacobino JL, Álvarez Luna Macyuri (2006) Tensiones Residuales en la Soldadura. Métodos para su Atenuación-Monografias.com
10. Guinebretiere R (2007) X-ray diffraction by polycrystalline materials. ISTE
11. Kandil FA, Lord J, Fry A, Grant PV (2002) A review of residual stress measurement methods. Measurement of residual stress in components. NPL Report, UK
12. Kandil FA, Lord JD, Fry AT, Grant PV (2001) Review of residual stress measurement methods—a guide to technique selection. Crown 1:45
13. Nasir NSM, Razab MKAA, Mamat S, Ahmad MI (2016) Review on welding residual stress. ARPN J Eng Appl Sci 11:6166–6175
14. Obrtlík K, Polák J, Hájek M, Vašek A (1997) Short fatigue crack behaviour in 316L stainless steel. Int J Fatigue. https://doi.org/10.1016/S0142-1123(97)00005-4
15. Renk O, Hohenwarter A, Pippan R (2012) Cyclic deformation behavior of a 316L austenitic stainless steel processed by high pressure torsion. Adv Eng Mater. https://doi.org/10.1002/adem.201200015
16. Rossini NS, Dassisti M, Benyounis KY, Olabi AG (2012) Methods of measuring residual stresses in components. Mater Des 35:572–588. https://doi.org/10.1016/j.matdes.2011.08.022
17. Withers PJ, Bhadeshia HKDH (2001) Residual stress part 2—nature and origins
18. Yavuz Peter McIntyre M, Karaman Dennis O IL, Anderoglu O (2004) residual stress measurement using X-ray diffraction. Texas A&M Univ, pp 1–64

Influence of Microcracks on Strength of Diamond Wire Sawn Silicon Substrates

Florian Wallburg, Kevin Meyer, Michael Budnitzki, Meinhard Kuna, Felix Kaule, and Stephan Schoenfelder

Abstract So-called microcracks are crucial for material removal during diamond wire sawing of silicon. But they also reduce the strength of silicon substrates after sawing due to the remaining subsurface damage. In order to investigate the influence of microcracks on strength of diamond wire sawn silicon substrates, more than 180 specimens of {100}-silicon were prepared using a reproducible scratching procedure with a VICKERS indenter. Subsequently, 4-point bending tests were conducted parallel and perpendicular to the scratch grooves. The results indicate a clear anisotropy in the mechanical strength, which is mainly caused by median and also affected by radial cracking. The microstructural analysis of specimens with a single scratch or multiple scratches show results as a continuous as well as straight median crack propagation at a nearly constant damage depth over the complete length of the scratch groove. Additionally, in the case of multiple scratching, periodic crack kinking towards pre-damaged zones can be observed and, thus, further emphasises the importance of considering pre-existing damage in terms of material removal mechanisms and strength analyses.

Keywords Diamond wire sawing · Scratching · Silicon · Strength · Fracture

1 Introduction

Multi-wire sawing of high-purity silicon (Si) bricks is the dominant technology for the production of thin Si substrates, called wafers, for photovoltaic application [1]. Sawing with fixed abrasive grains bonded to the wire (diamond wire sawing, DWS)

F. Wallburg (✉) · F. Kaule · S. Schoenfelder
Faculty of Engineering, Leipzig University of Applied Sciences, 04277 Leipzig, Germany
e-mail: florian.wallburg@htwk-leipzig.de

F. Wallburg · K. Meyer · F. Kaule · S. Schoenfelder
Fraunhofer Center for Silicon Photovoltaics CSP, 06120 Halle, Saale, Germany

F. Wallburg · M. Budnitzki · M. Kuna
TU Bergakademie Freiberg, Institute of Mechanics and Fluid Dynamics, 09599 Freiberg, Germany

M. Abdel Wahab (ed.), *Proceedings of the 8th International Conference on Fracture, Fatigue and Wear*, Lecture Notes in Mechanical Engineering,
https://doi.org/10.1007/978-981-15-9893-7_26

shows considerable advantages over sawing with loose abrasive grains in a carrier fluid around the wire (loose abrasive sawing, LAS). Considering higher sawing velocities, a reduced kerf loss and the usage of environmentally friendly water-based coolants during DWS [2], it is not surprising that a rapid technology change from LAS to DWS has taken place under the enormous economic cost pressure in the photovoltaic industry [3, 4]. However, due to a change in the microscopic abrasive mechanism from indentation (in LAS) to scratching (in DWS), research results concerning subsurface damage (SSD) in the LAS process cannot be directly transferred to DWS as discussed in [5].

During LAS, material removal is achieved by a large number of single indentations [6]. Since loose abrasive particles required for the removal processes can move freely in the carrier fluid (three-body-abrasion) as depicted in Fig. 1a, the induced SSD is, from a macroscropic point of view, randomly distributed in its orientation. In terms of the substrate strength, this mechanism prevents the occurrence of anisotropic effects as reported in studies based on 4-point bending tests of over 2,200 specimens from 53 different batches by Kaule et al. [7]: Neither the type of material (monocrystalline or multicrystalline) nor the orientation of loading rollers in relation to the sawing direction (parallel $\parallel$ or perpendicular $\perp$) significantly influence the strength of Si substrates which is typically around 150 MPa as characteristic fracture stress value for LAS (156 mm × 156 mm) wafers in 4-point bending tests [7]. This isotropic value is in good agreement with the results of several studies carried out by other authors, see [8–12].

However, in contrast to substrates sawn by LAS, DWS substrates behave differently. No significant difference can be found between LAS and DWS substrates regarding testing the strength perpendicular to the saw marks for both monocrystalline and multicrystalline substrates. This is different for testing parallel to the saw marks, i.e. to the sawing direction. For monocrystalline and multicrystalline material, a decrease in DWS substrate strength of up to 50% and 70%, respectively, is observed by Kaule et al. [7]. These results are in excellent agreement with recent experiments carried out by Carton et al. [13]. It is conceivable that grain structure have an influence on strength of multicrystalline substrates since the cutting characteristics changes near grain boundaries [14, 15].

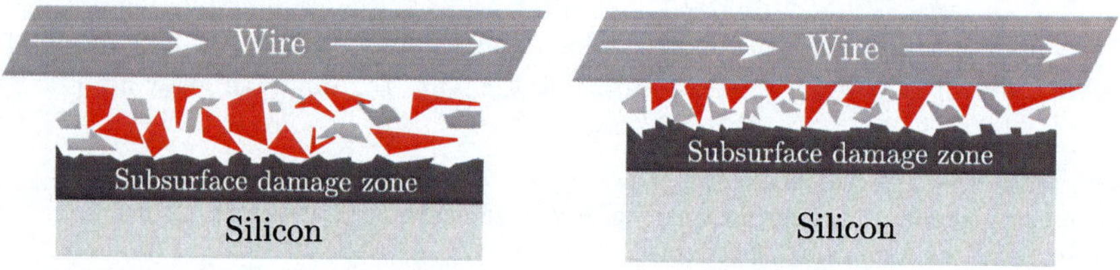

(a) Loose abrasive sawing (LAS). (b) Diamond wire sawing (DWS).

Fig. 1 Illustration of material removal in LAS (three-body abrasion) and DWS (two-body abrasion) with remaining subsurface damage zone (SSD) in the material after the sawing process

(a) Substrate surface after diamond wire sawing with common topology showing multiple scratch grooves and chipped material fragments.

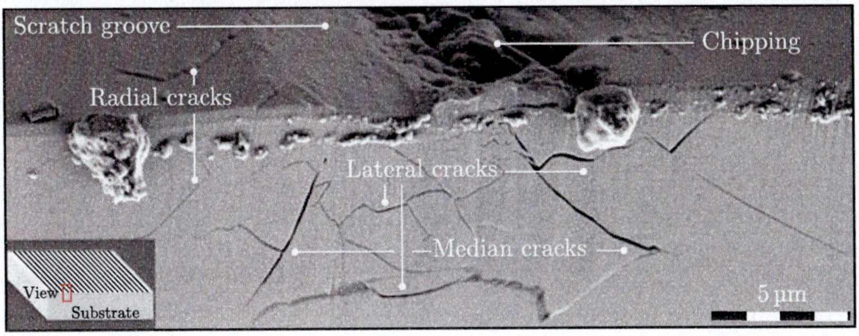

(b) Subsurface damage induced by single scribing in the ⟨100⟩ direction on a {100} Si surface showing median, radial and lateral cracks whereas the last two mentioned types of cracks are responsible for the characteristic abrasion mechanism, the so-called chipping.

Fig. 2 Cross-sections of silicon substrates processed by scratching with sharp abrasive particles

In general, the directional difference in strength can be assumed to be related to the different removal process. Since abrasive particles are bonded to the wire, SSD mechanisms are based on a so-called two-body-abrasion as depicted in Fig. 1b. As a consequence, abrasion processes are characterised by continuous scratch grooves with partially repeated SSD patterns consisting of median, radial and lateral cracks [16, 17]. A typical substrate surface with associated underlying SSD generated by DWS is shown in Fig. 2.

The investigations presented here are focused specifically on the micromechanical mechanisms for scratching-induced SSD in silicon. In this respect, the effects of different crack types on the substrate strength are studied and detailed microstructural analysis of SSD are carried out to provide a comprehensive understanding of the anisotropic strength of DWS substrates or wafers.

2 Methodology

In order to investigate the effect of different types of cracks on the substrate strength, rectangular monocrystalline {100}-Si microelectronic chips ($30 \times 7 \times 0.525\,mm^3$, with ⟨110⟩ direction of edges) were scratched at the chip center parallel and perpendicular with a diamond VICKERS indenter over a length of $s = 4.5$mm (cf. Fig. 3). The influence of the sawing force is represented by different applied normal scratching forces of $F_n = 40$ mN and $F_n = 150$ mN inducing different SSD. Scribing on these monocrystalline Si chips was performed in the ⟨110⟩ direction which forces the growth of only one single median crack directly located below the indenter. The ⟨110⟩ direction was chosen as the corresponding {110} plane is one of the two possible principal cleavage planes of the Si crystal [18]. Cracks in this {100}⟨110⟩ configuration propagate in depth approximately perpendicular to the surface as shown in [19]. Thus, it cause maximum damage depth in the material. After processing, all specimens were analysed by light microscopy to check if the scratch grooves had been successfully introduced, to verify that the scribes had the intended length as well as to study the corresponding substrate damage.

In total, 230 specimens were tested by 4-point bending: 100 thereof (50 ∥, 50 ⊥) were previously scratched using a normal force of $F_n = 40$ mN, another 80 specimens (40 ∥ and 40 ⊥) were processed using $F_n = 150$ mN and 50 unscratched specimens served as reference. Figure 3 shows the experimental setup of 4-point bending with the respective scratch groove orientation in relation to the loading rollers (∥ or ⊥). It should be noted here, that the parallel arrangement of loading rollers and saw marks corresponds to a fracture mode I loading of the scribes, tensile stress direction is perpendicular to the scribes. A perpendicular orientation of loading rollers and scribes to each other is equal to a parallel tensile load regarding the scribes. All tests were conducted on a universal testing machine ZWICK 005 using a load cell of 500 N. The deflection of the specimens was measured based on the machine position. The loading speed was set to 0.01 mm/s. During the tests, force and deflection were recorded for each specimen starting after an initial force of 1 N was reached and ending at fracture. Fracture stresses σ_i of the specimens during 4-point bending were calculated using a Finite–Element (FE) model taking into account non-linearities such as large deflections as well as contacts between the loading rollers and the Si specimen.

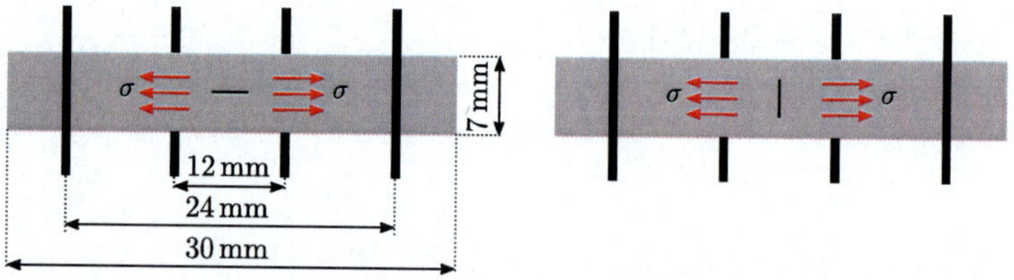

Fig. 3 Schematic representation of the 4-point bending setup (bottom view) for testing the influence of scratch grooves, oriented perpendicular (left) and parallel (right) to the loading rollers, on the substrate strength with indication of the resulting tensile stresses

The specimen (with linear-elastic material propertiesYOUNGS's modulus $E = 168.9$ GPa, POISSON's ratio $\nu = 0.278$) was modelled with SHELL93 elements in the FE-code ANSYS. Experimentally and numerically obtained force-displacements curves were matched by varying the coefficient of friction μ. As a result, a coefficient of $\mu = 0.15$ was determined.

The calculated fracture stresses σ_i were then used for statistical evaluation purposes by means of a two-parameter WEIBULL distribution [20, 21]:

$$P(\sigma_i) = 1 - \exp\left[-\left(\frac{\sigma_i}{\sigma_\theta}\right)^m\right], \tag{1}$$

where σ_θ denotes the characteristic fracture stress at which 63.2% of all specimens fail. The WEIBULL modulus m represents the scattering of σ_i whereas a low value is associated with a large scattering of fracture stresses and vice versa. The maximum likelihood estimation (MLE) was used for calculating the WEIBULL parameters following [22].

In order to investigate the scratching-induced subsurface damage (SSD) in more detail, two further $\{100\}\langle 110\rangle$-Si chips were prepared using a normal force of $F_\mathrm{n} = 150$ mN. This force value was chosen since a better defect characteristic and recognisability can be expected. In contrast to the specimens for 4-point bending analysis, these two specimens were prepared with several scratch grooves. One chip was only processed with single scribes while the other chip contains pairs of grooves at a distance of 10 or 5 µm to each other. In order to ensure that single scribes/pairs do not interact with each other, a distance of $\xi = 100$ µm was set between each scribe setup. A bevel polishing of the damaged specimens under an angle of $\alpha = 1°$ was applied for a more precise evaluation of the SSD depth profile and analysis of crack continuity over the complete length of the scratch groove (see Fig. 4). Furthermore, the specimens were slightly rotated around the surface normal axis (z-axis, $\langle 100\rangle$) so that the respective surface was not polished exactly parallel to the scratch

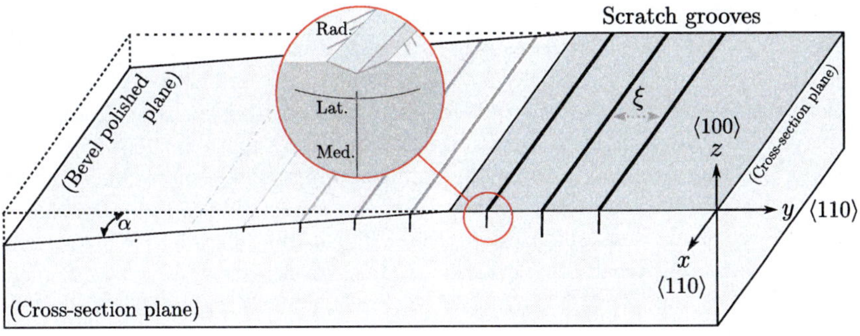

Fig. 4 Illustration of a specimen preparation for microstructural analysis of subsurface damage induced by scratching

grooves. This methodology allows a much better depth-sensitive evaluation along the scratching direction in addition to the analysis across the grooves.

Afterwards, the bevel polished specimens were etched isotropically with a SECCO etchant (HF, $K_2Cr_2O_7$, DI-water) for 3 s in order to achieve a better defect decoration. For precise information about the surface topology generated by means of bevel polishing, the topology of the prepared specimens were measured using an optical interferometer (SENTRONICS SEMDEX 301). Together with detailed information about the polishing depth provided by the topology measurements, SSD was evaluated using a scanning electron microscope (HITACHI SU- 70).

3 Results

The following chapter initially deals with strength analyses of parallel and perpendicular testing of the scribes. Subsequently, the results of the microstructural analysis are presented, which are intended to provide a better understanding of the influence of different crack types on strength.

3.1 Fracture Strength

In Table 1 the characteristic fracture stress σ_θ and the WEIBULL modulus m for each series are given. Furthermore, in Fig. 5 the probability plots for all tested specimens are shown.

It can be seen that the reference series has the highest strength (509.0 MPa). The value of σ_θ concerning the 40mN-$\perp$ series is only slightly lower than for the reference case (500.9 MPa), with a significantly smaller WEIBULL modulus m (9.2 < 17.2). It is overlapping with the reference series with competing defect distributions of the scribe and the edge defects of the chip which has to be kept in mind when comparing with other series.

As expected, strength decreases with increasing scribing forces: Considering the tested configurations in this article ($F_n = 40$ mN and $F_n = 150$ mN), it can be found that strength decreases by about 34% for the parallel testing scenario. In the case of perpendicular testing, strength is strongly reduced by approximately 65%. Furthermore, testing direction is important, parallel configurations show much smaller fracture stresses. The characteristic fracture stresses are 2.5-fold ($F_n = 150$ mN) and 5-fold ($F_n = 40$ mN) smaller than for perpendicular testing configuration of the same scribes. Considering the absolute values of σ_θ, the data also show that even a damage induced with $F_n = 150$ mN and subjected to perpendicular testing (150 mN-$\perp$) does not reduce strength as much as a damage induced with $F_n = 40$ mN and subjected

Table 1 Weibull parameters with 90% confidence bounds

Sample	Orientation	Characteristic fracture stress σ_θ (MPa)		Weibull modulus m (−)	
Reference	–	509.0	(500.8 ...517.1)	17.2	(13.9 ...20.9)
40 mN	⊥ perpendicular	500.9	(485.8 ...515.6)	9.2	(7.3 ...11.4)
	‖ parallel	107.8	(106.1 ...109.4)	27.7	(20.3 ...36.1)
150 mN	⊥ perpendicular	177.8	(171.3 ...184.4)	7.6	(6.1 ...9.2)
	‖ parallel	71.4	(70.5 ...72.2)	23.2	(19.1 ...27.5)

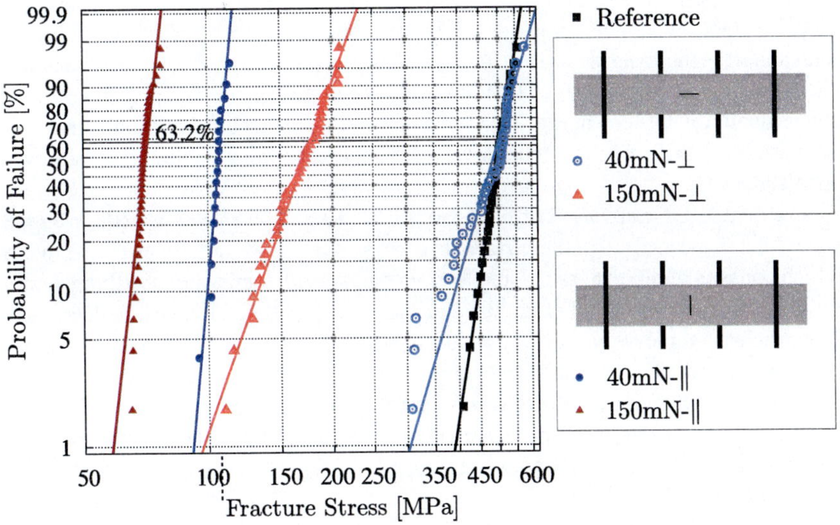

Fig. 5 Probability plots for specimens with scratch grooves aligned parallel and perpendicular to the loading rollers which have been damaged by two different scratching forces each ($F_n = 40$ mN, $F_n = 150$ mN) in contrast to the unscratched reference specimen

to parallel testing (40 mN-‖)[1]. It can also be seen that the WEIBULL modulus m within both perpendicular experiments is lower than in the parallel testing scenario indicating a higher scattering of σ_i.

3.2 *Microstructural Analysis of Single Scribes*

On the basis of the substrate damage that forms during scribing, different crack types can be observed which are responsible for different strength behaviour depending on the loading direction. Figure 6a shows a scratch groove after scribing with $F_n = 150$ mN. It can be seen that many radial cracks are located at the edges

[1]It should be noted that 21 specimens of the 40 mN-‖ configuration could not be evaluated for technical reasons. Nevertheless, the statistical results are very clear.

of the groove (often at an acute angle). In combination with lateral cracks this leads to chipping which is of interest for material removal. In contrast, no radial cracks can be detected on the surface when scribing with $F_n = 40$ mN. Also no chipping at the edges of the grooves takes place as shown in Fig. 6b. However, on the basis of surface analyses no predictions can be made regarding median cracks (i.e. depth, existence) and radial cracks, if existing, concerning their SSD depth. For this reason, the results from bevel polished single scribe experiments are evaluated. Figure 7a shows an image taken by light microscopy after bevel polishing. For total of 15 induced single scribes (R01–R15), only the first five (R01–R05) can still be detected on the surface since their scratch grooves were only partially polished. Particularly, with regard to the first two scribes R01 and R02, the induced substrate damage is still clearly visible at the beginning. Following R01 or R02 along the scribing direction, it can be qualitatively seen in Fig. 7a that material removal through bevel polishing was not parallel to the scribes which allows SSD analyses across and along the grooves as intended.

The qualitative evaluation of the scribes R01, R05, R06 as well as R08 in Fig. 8 shows that in the vicinity of the surface (Fig. 8a) severe substrate damage can be found which consists of median, radial and lateral cracks. With increasing polishing depth, the SSD in terms of lateral (Fig. 8b) and radial (Fig. 8d) crack quantity decreases. Nevertheless, some very well pronounced radial cracks can be detected at larger polishing depths. For example, considering R08 (Fig. 8d) a large radial crack can be identified at a depth of 9.8 μm while the SSD under scratch groove R06 (Fig. 8c) at 6.9 μm consists only of a median crack. Since the median crack type appears to be particularly deep and continuous in general, the following evaluation focuses on median cracks.

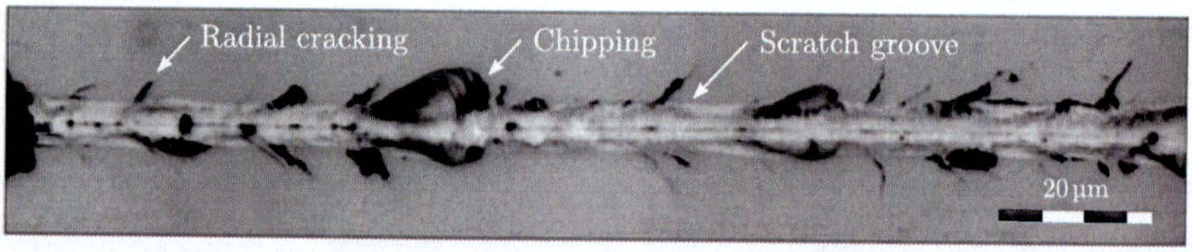

(a) Single scribe induced with $F_n = 150$ mN.

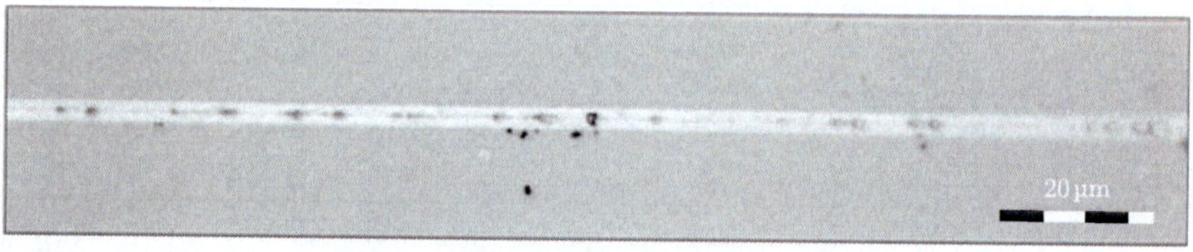

(b) Single scribe induced with $F_n = 40$ mN.

Fig. 6 Surface view taken by light microscopy of two specimens prepared with single scribes (which were afterwards used for fracture stress analyses) illustrating the increase in substrate damage at higher scribing forces

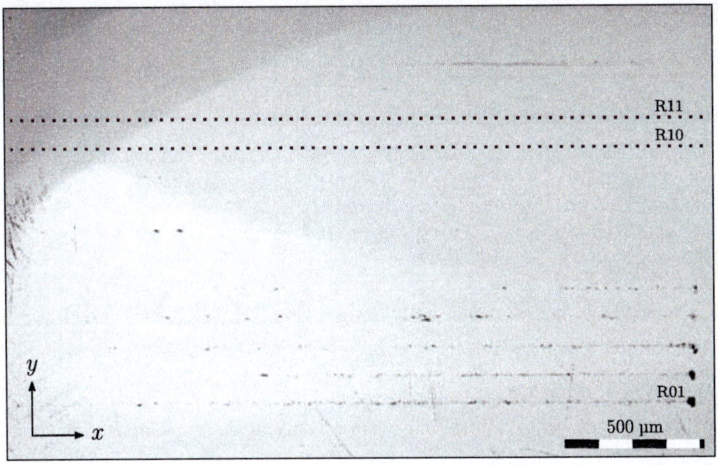

(a)

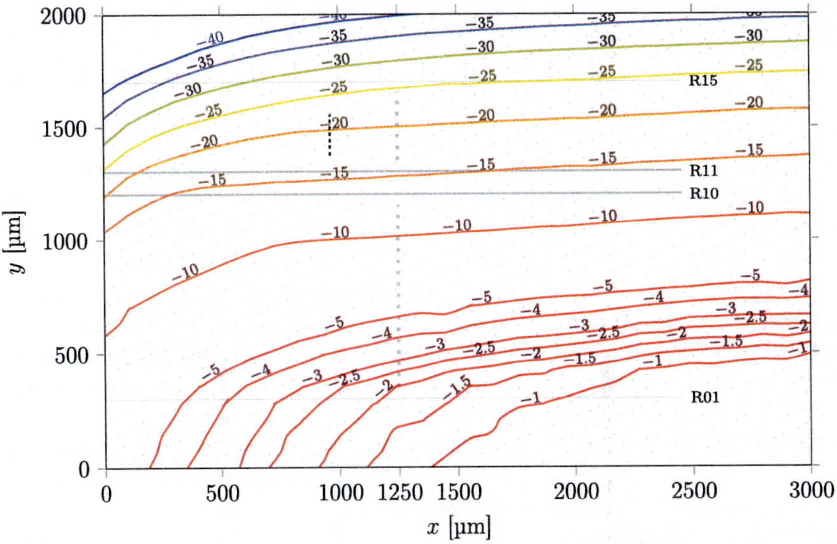

(b)

Fig. 7 Surface view of the bevel polished specimen processed by single scribing: **(a)** image of the surface taken by a light microscope, **(b)** surface topology represented by isolines

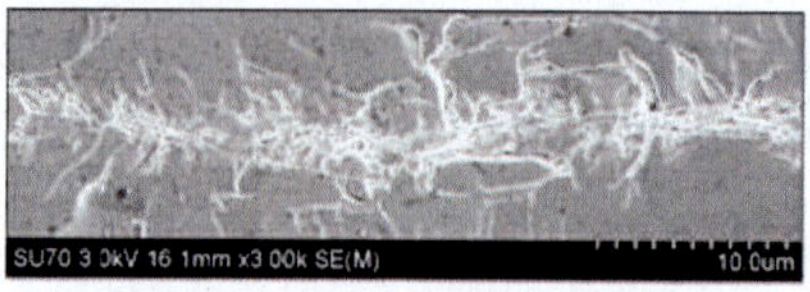

(a) R01 at 1.5 µm polishing depth.

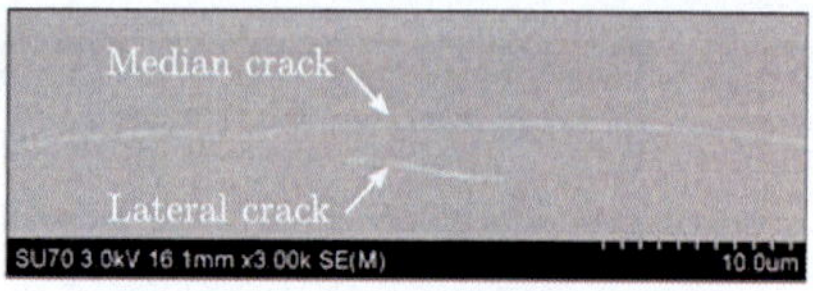

(b) R05 at 5.5 µm polishing depth.

(c) R06 at 6.9 µm polishing depth.

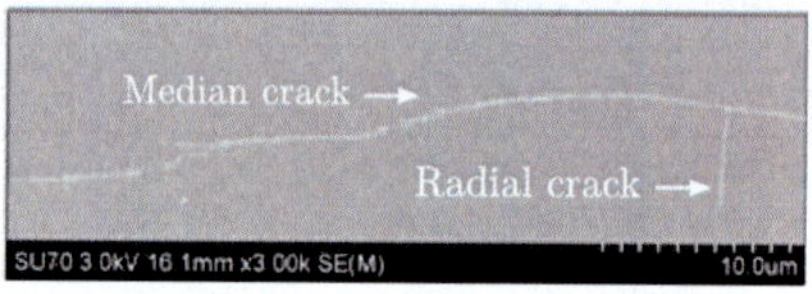

(d) R08 at 9.8 µm polishing depth.

Fig. 8 SEM images of the single scribed bevel polished specimens showing the decrease in SSD from median, radial and lateral cracking to almost only median cracking at higher polishing depths

In this respect, the SSD profile of the surface perpendicular to the scratch grooves is initially studied. The evaluation takes place approximately in the middle of the groove at $x = 1250\,\mu m$ (cf. Fig. 7b, dotted line). It is found that below each scratch groove there is one median crack as presumed for the crystal structure of the specimen. Each median crack shows a nearly continuous, uninterrupted crack path if it is detectable. Median cracking can be found below R01–R10. In the case of the scratching experiments R11–R15, however, no median cracks can be found. Between the last detectable (below R10) and the first undetectable crack (below R11) about 1.75 µm of material was removed by bevel polishing. The detailed removal profile is shown in Fig. 9. Thus, no strongly varying median crack profile can be observed within an uncertainty of 1.75 µm in the transition area between damaged and undamaged material. After the cross-sectional analysis of the scratch grooves, the single scratch R11 is analysed along its groove (x-direction in Fig. 7a) in order to determine the position from which the crack below R11 becomes fully visible again and, in this respect, within which limits a crack tip fluctuation can be observed. The results are shown in Fig. 10. In total 1.1 µm of the material is removed over a scratch length of 750 µm whereby the median crack tip is initially not detectable (Fig. 10a) but quickly begins to fluctuate within this interval (Fig. 10b, c) before the crack path can be regarded as continuous (Fig. 10d).

From these results not only a continuous path of the median crack type can finally be derived, it can also be shown that the crack tip fluctuates within a range (so-called transition zone) of only 1.1 µm. This value is more precise regarding the initial estimation based on two adjacent scribes (1.75 µm). Furthermore, the cracks themselves show only slight directional deviations below the grooves. Thus, the SSD characteristic in terms of median cracking is very reproducible.

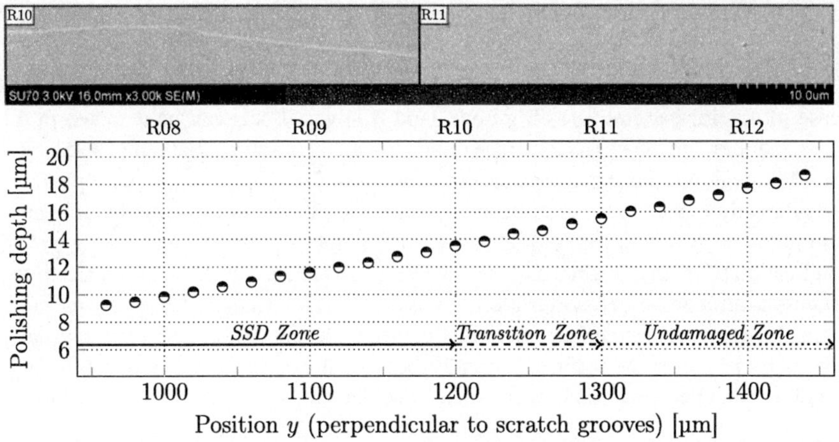

Fig. 9 Profile of the bevel polishing depth at position $x \approx 1250\,\mu m$ (see Fig. 7b) together with a SEM image of the last detectable SSD caused by median cracking under scribe R10 as well as an SEM image of the first undamaged material zone under scribe R11

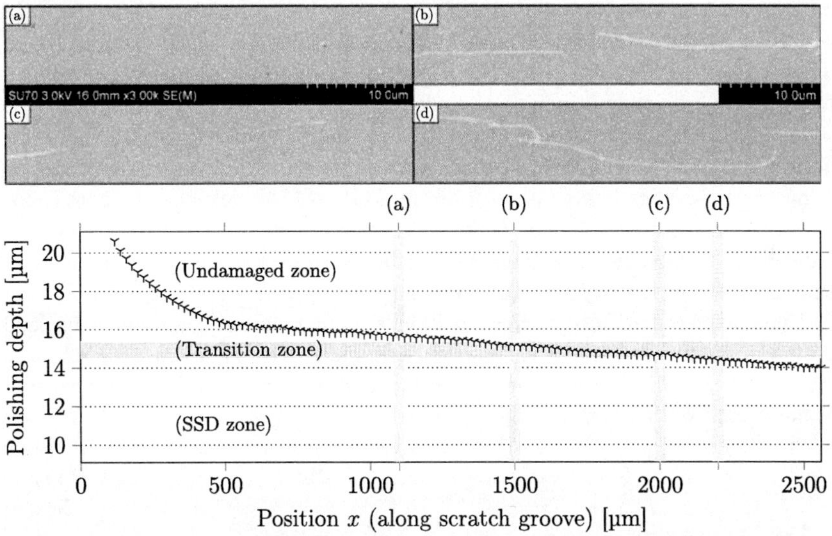

Fig. 10 Investigation of crack continuity along the bevel polished scribe R11 between damaged and undamaged substrate zone with SEM images of characteristic areas; the vertical markings correspond in their width to the respective SEM image section

3.3 Microstructural Analysis of Multiple Scribes

This analysis is intended to show which substrate damage can be expected in the case of several parallel scratch grooves and if novel effects can be detected in the crack pattern. On the basis of surface images it can be observed that radial and lateral cracks merge into large chipped areas as shown in Fig. 11. The material removal is significantly more pronounced than for a single scribe under the same loading condition ($F_n = 150$ mN, Fig. 6a). However, investigations concerning kinking and/or unification of median cracks and, as a result, possible derivations with regard to the strength-influencing behaviour are not possible from surface analyses for two reasons: Firstly, because median cracks are formed in deeper material layers and usually do not propagate to the surface. Secondly, because the damage caused by chipping is so severe that no deeper crack structure can be detected under the respective chipped segment.

As in the case of the investigation of single scribes, the analysis of SSD induced by multiple scribing is carried out using a bevel polished specimen. The results indicate a tendency for kinking of median cracks in the presence of pre-damaged areas. It can be observed that cleavage plane changing cracks occur at regular intervals, whereby a minimum distance of 5 µm can be determined. Merging of different crack planes (see Fig. 12a or b, respectively) as well as kinking cracks without crack tip contact to the other crack plane are possible (cf. Fig. 12c). The macroscopic angular relationship of the cracks to each other is in most cases defined by an average angle of 45°. But from a crystallographic point of view, there is no specific fracture plane at this angle in the lattice. This means that even if the crack macroscopically follows a certain direction, it must microscopically propagate in several {110}/{111} cleavage planes to follow the stress driven direction. The kinking exclusively takes place in the direction of the pre-damaged material. The already existing damage has a crack stopping effect since no crack grows beyond it (cf. Fig. 12b).

Underneath the scratch grooves, however, no remarkable fluctuations of the depth profile are detectable in the bevel polished specimens. Also in this case it can be seen that the crack path of first scribes is constant and only the direction of propagation of the second scribes can vary, cf. Fig. 12.

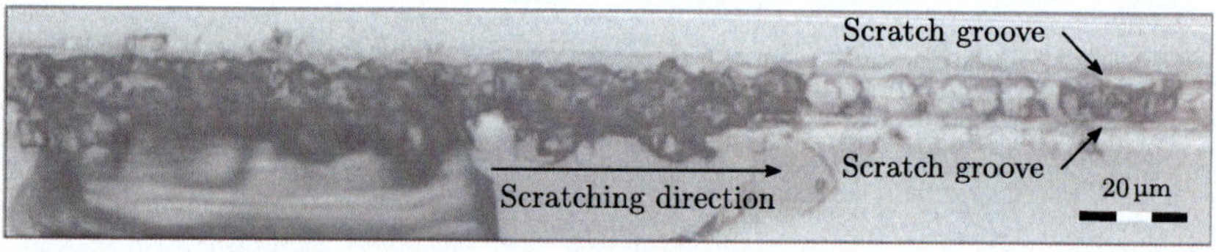

Fig. 11 Surface view of two parallel scratch grooves before bevel polishing (also etched) induced by a normal scribing force of $F_n = 150$ mN at a distance of 10 µm showing severe substrate damage

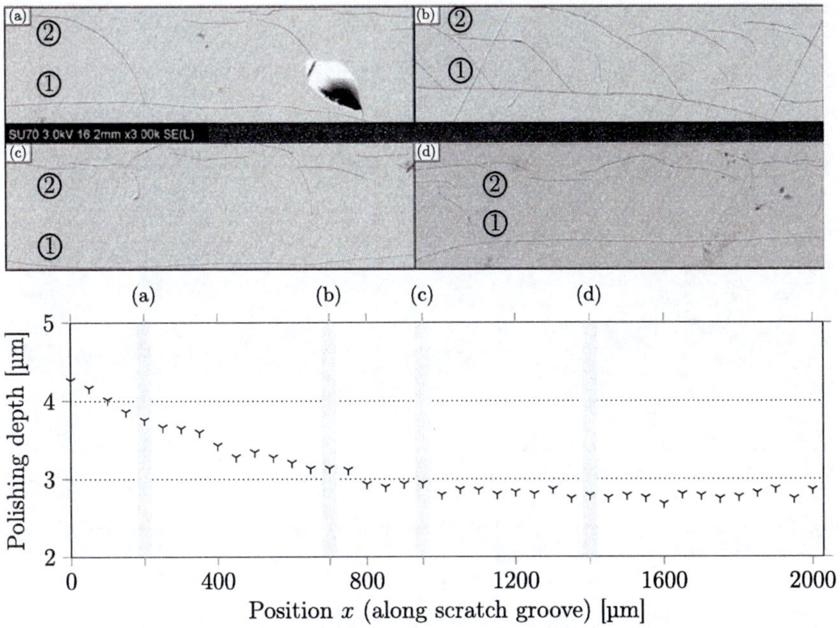

Fig. 12 Bevel polished scribes in a distance of $\delta = 10\,\mu m$ to each other revealing occasional crack kinking of the second (2) to the first (1) median crack, sometimes with coalescence of crack planes and crack stopping at the first induced crack plane, sometimes without coalescence of crack planes

4 Discussion

4.1 *Influence of Different Crack Types on Strength*

Based on the conducted microstructural investigations it can be seen that radial cracking occurs at the edges of the scratch grooves (see, e.g., Fig. 6a or Fig. 8a) while median cracking appears directly below the grooves (cf. Fig. 8b and Fig. 9 (R10)). Lateral cracks propagate nearly parallel to the surface. This means that median cracks are opened in fracture mode I during parallel testing of the scribes (or saw marks, cf. Fig. 13). Similarly, radial cracks are relevant for fracture during perpendicular testing (see Fig. 14). Lateral cracks are not relevant for strength, due to their orientation almost parallel to the surface. They are not or only slightly loaded. It can be assumed that they do not contribute to a possible reduction in strength.

In combination with the obtained fracture stresses, it can be therefore concluded that median cracks are mainly responsible for the failure of the 40 mN-|| and 150 mN-|| specimens. Likewise, the failure of the 150 mN-⊥ specimens can be attributed to radial cracking. In contrast to this experimental study with a scribing force of $F_n = 150$ mN, no radial cracks can be detected on the surface of the

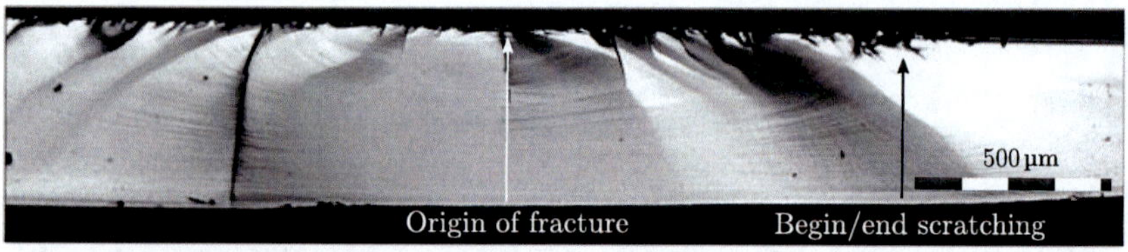

(a) Side view of a specimen with its origin of fracture.

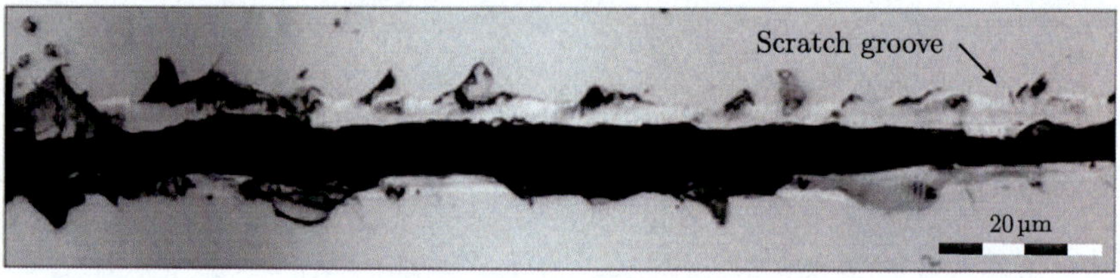

(b) Surface view of another specimen showing the separated scratch groove.

Fig. 13 Failure of a specimen ($F_n = 150$ mN) caused by parallel testing due to median crack growth directly under the scratch groove

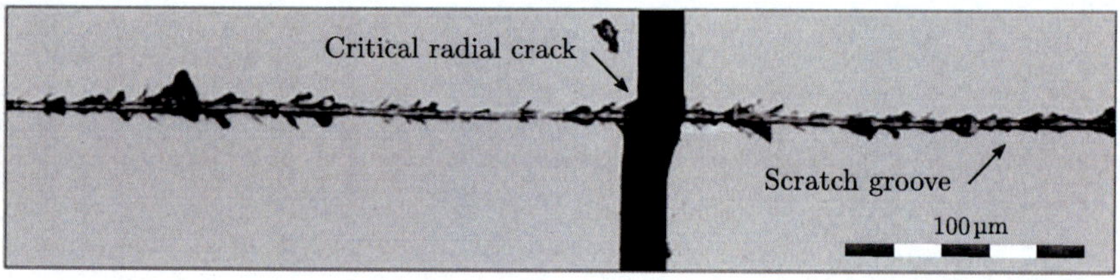

Fig. 14 Failure of a specimen ($F_n = 150$ mN) caused by perpendicular testing due to radial crack growth at the edges of the scratch groove

40 mN-⊥ specimens. However, the WEIBULL moduli m of the perpendicularly tested specimens ($m = 9.2$ for $F_n = 40$ mN and $m = 7.6$ for $F_n = 150$ mN) and specimens with a parallel scratch groove orientation ($m = 27.7$ for $F_n = 40$ mN and $m = 23.2$ for $F_n = 150$ mN) are very similar (cf. Table 1). Therefore, a partly formation of radial cracks can be assumed for the 40 mN-⊥ series as well.

Interestingly, a more detailed examination of the 40mN-⊥ series shows a bimodal behaviour, see Fig. 15. While some specimens fail due to radial cracks ($m = 4.8$, 40 mN-⊥-type-1), other specimens ($m = 25.4$, 40 mN-⊥-type-2) behave very similar to the reference series whose strength is determined by the sawn chip edges. In order to be able to identify radial cracking for this scribing force configuration, it might be necessary to perform further microstructural analyses of specimens processed with this force configuration including etching and following analysis of the corresponding surface with a SEM. Though, strength analysis indicates the existence of radial cracks being optical invisible at the surface.

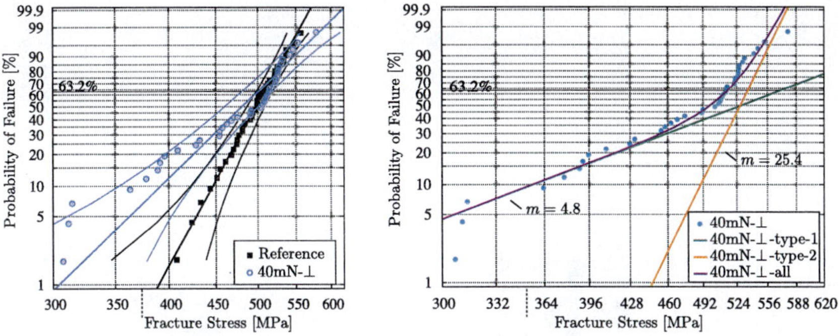

(a) Probability plot of the reference and 40mN-⊥ series with 90% confidence intervals

(b) Probability plot of the 40mN-⊥ series showing a bimodal behaviour of the WEIBULL parameter m

Fig. 15 Probability plots indicating different reasons for failure: while for some specimens the sawn chip edges are responsible for failure as is the case with the reference series (see confidence interval overlap in the upper range of (**a**) and the corresponding WEIBULL modulus of 40 mN-⊥-type-2 in (**b**)) other specimens fail due to radial cracking (see lower range in (**a**) and low WEIBULL modulus of 40 mN-⊥-type-1 in (**b**))

Generally exists a relationship between the size distribution function of defects and the WEIBULL modulus m [23]. If a PARETO distribution $f(a)$ of defect size a with characteristic parameter β is assumed, then the WEIBULL modulus reads as

$$f(a) = c\,a^{-\beta} \quad \rightarrow \quad m = 2(\beta - 1). \tag{2}$$

This means, the sharper the defect distribution function, the higher is the modulus m, the smaller the scattering of fracture stresses. Thus, failure due to median cracking is very well reproducible which can be seen by the large WEIBULL modulus (little scattering of the measured data and narrow defect size distribution) for parallel orientation of loading rollers and scratch grooves. In the case of radial cracking many cracks occur along the scratch groove which have a wider distribution of size resulting in a larger scattering (meaning lower WEIBULL modulus) for perpendicular orientation.

In general, it can be seen that the strength-reducing influence of radial cracks is smaller for both scribing force configurations than that of median cracks since the SSD depth based on radial cracks (compare, e.g., Fig. 8a with Fig. 8b) is considerably smaller compared to median cracks. For example, a specimen scratched with a scribing force of $F_n = 150$ mN which is subsequently tested perpendicular to the induced scratch groove (loading of radial cracks) achieve a, from a statistical point of view, 2.5 times higher strength in contrast to parallel testing based on the calculations of the characteristic WEIBULL stresses (cf. Fig. 5 and Table 1).

In comparison to the results for DWS Si wafers provided by Kaule et al. [7] it can be noticed that the strength for perpendicular testing (185 MPa) is twice as

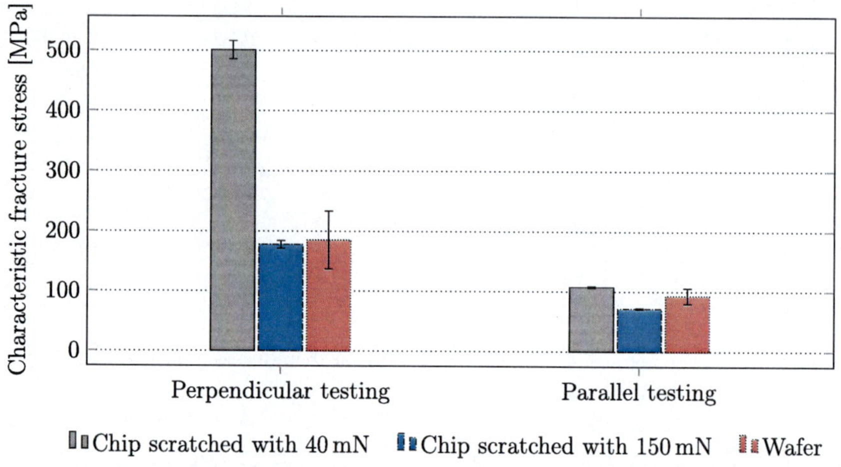

Fig. 16 Comparison of the influence of perpendicular and parallel testing on strength of different substrates (single scribed chips prepared with a normal scribing force of $F_n = 40$ mN as well as $F_n = 150$ mN in relation to as-cut wafers sawn with diamond wires, cf. Kaule et al. [7]); the error bars indicate the 90% confidence bounds

much higher compared to parallel testing (93 MPa, see Fig. 16). It can further be assumed that median and radial cracks induced with a normal scribing force of around $F_n = 150$ mN are responsible for failure of full wafers. Thus, the low strength of wafers in parallel orientation is caused by continuous median cracks on the wafer surface. The higher strength of perpendicular orientation is dominated by radial cracks. Experimental studies carried out by Moeller et al. [24] suggest an average sawing force of about 100 mN on wafer level, which fits fairly well to this work here. In the case of multiple scratching like in the sawing process, a possible influence of kinked median cracks on wafer strength appears to be low since they are not ideally oriented in loading direction and usually significantly smaller than single median cracks [19]. Based on the results of the microstructural analysis and strength evaluations, the median crack has the greatest influence on strength. It will be further discussed in the following.

4.2 *Aspects Concerning the Median Crack Continuity*

It can be seen from the microstructural investigations, that for single scribing median cracks are characterised by a continuous as well as straight crack propagation and an almost constant depth profile. Further, only slight directional deviations below the indenter can be determined. These findings differ from the analysis of Wuerzner et al. who found that vertical (=median) cracks induced by irregularly shaped abrasive particles show in a cross-section analysis no continuous behaviour for the same

{100}⟨110⟩-scribing combination. However, a continuity in the SSD depth has also been determined in the investigations presented here. On the other hand, the applied cross-section analysis in [16] was conducted parallel to the scratch groove (x-z-cross-section plane, cf. Fig. 4). Since the median crack path is characterised by slight fluctuations around its imaginary centre line, a cross-section through this plane always leads to a separation of the median crack structure. A further problem could be caused by the difficult differentiation to radial cracks in this type of analysis, which is why the bevel polishing technique is preferable to cross-section analysis.

4.3 Comparison of Median Crack Depths Obtained from Bevel Polished and Cross-Sectional Processed Specimens

Bevel polishing is very time-consuming insofar as a certain range of the polishing depth is targeted like in this work (over 20 h polishing time). From the bevel polished specimen as presented in Fig. 10, a median crack depth between 14.4 and 15.5 µm could be detected. Due to the continuous damage pattern, a cross section perpendicular to the scratch groove (y-z-cross-section plane, cf. Fig. 4) is representative for the median crack depth of the entire scribe. The range obtained by the bevel polished specimen is therefore in good agreement with already published studies [19], in which a median crack depth between 12.5 µm and 14.5 µm (95% confidence interval, $n = 120$ specimens) was obtained using cross-section analysis of scratch grooves with the same normal scribing force as well as the same indenter ($F_n = 150$ mN, VICKERS indenter).

5 Summary and Conclusions

The presented investigations have shown that median cracks due to scratching have a very large influence on wafer strength. Since median cracks directly occur below the process-related scratch grooves, this means that testing wafers parallel to the scribes (mode I loading of scribe/saw marks) is much more critical than perpendicular testing. In the case of a perpendicular orientation of loading rollers and scratch grooves to each other, failure is caused by radial cracks whose SSD depth is smaller in comparison to that of median cracks. Differences in strength of up to factor 5 have been found. Further microstructural investigations have shown that median cracks continuously propagate at an almost constant depth. This is valid for single as well as for multiple scribing. Concerning multiple scribing, it could be observed that median cracks tend to kink if other median cracks are in the vicinity. This can be seen as particularly interesting with regard to the material removal, whereas the impact on strength is considered to be minor.

The results of the microstructural investigations as well as strength analysis suggest that surfaces of DWS sawn wafers are currently processed with particle scribing forces of up to 150 mN. A reduction of this force could significantly increase the substrate strength, especially if the saw marks are under a pure mode I loading which means a parallel alignment of saw marks and loading rollers in terms of the conducted tests of this work. In any case, wafers should always be handled in such a way that loading is parallel to the saw marks and prevent loading of median cracks. Moreover, if it is possible to avoid large scribing forces on the wafer surfaces during sawing ($F_n < 40$ mN), high strength values can be achieved since radial crack growth is increasingly inhibited. One conceivable approach for reducing the sawing forces on the sides of the sawing channel could be a specific particle design on the wire so that the sawing channel boundaries, i.e. the future wafer surfaces, are only slightly processed. This should be subject of future research.

Acknowledgements This work has been financially supported by the European Union as part of the ESF-Program (Contract No. 100284335) and the German Federal Ministry of Education and Research (BMBF) within the research project ENOWA II (Contract No. 0325805I), which are gratefully acknowledged. The authors thank Carola Klute from Fraunhofer-Center for Silicon Photovoltaics CSP in Halle (Saale) for etching the samples and creating the scanning electron images.

References

1. Moeller HJ, Funke C, Rinio M, Scholz S (2005) Multicrystalline silicon for solar cells. Thin Solid Films 487(1–2):179–187
2. Watanabe N, Kondo Y, Ide D, Matsuki T, Takato H, Sakata I (2010) Characterization of polycrystalline silicon wafers for solar cells sliced with novel fixed-abrasive wire. Progr Photovolt Res Appl 18:485–490
3. ITRPV (2017) International technology roadmap for photovoltaics eighth edition: 2016 results
4. ITRPV (2019) International technology roadmap for photovoltaics tenth edition: 2018 results
5. Wu H (2016) Wire sawing technology: a state-of-the-art review. Precis Eng 43:1–9
6. Moeller HJ (2004) Basic mechanisms and models of multi-wire sawing. Adv Eng Mater 6(7):501–513
7. Kaule F, Oswald M, Koepge R, Klute C, Schoenfelder S (2015) Comprehensive analysis of strength and reliability of silicon wafers and solar cells regarding their manufacturing processes. Photovolt Int 29 Edition:22–29
8. Wu H, Melkote SN, Danyluk S (2012) Mechanical strength of silicon wafers cut by loose abrasive slurry and fixed abrasive diamond wire sawing. Adv Eng Mater 14(5):342–348
9. Yang C, Wu H, Melkote SN, Danyluk S (2013) Comparative analysis of fracture strength of slurry and diamond wire sawn multicrystalline silicon solar wafers. Adv Eng Mater 15(5):358–365
10. Bidiville A, Heiber J, Wasmer K, Habegger S, Assi F (2010) Diamond wire wafering: wafer morphology in comparison to slurry sawn wafers. In: 25th European photovoltaic solar energy conference and exhibition/5th world conference on photovoltaic energy conversion
11. Buchwald R, Froehlich K, Wuerzner S, Lehmann T, Sunder K, Moeller HJ (2013) Analysis of the sub-surface damage of mc- and Cz-Si wafers sawn with diamond-plated wire. Energy Proc 38:901–909
12. Funke C, Wolf S, Stoyan D (2009) Modeling the tensile strength and crack length of wire-sawn silicon wafers. J Solar Energy Eng 131(1):011012

13. Carton L, Riva R, Nelias D, Fourmeau M, Coustier F, Chabli A (2019) Comparative analysis of mechanical strength of diamond-sawn silicon wafers depending on saw mark orientation, crystalline nature and thickness. Solar Energy Mater Solar Cells 201:110068
14. Kumar A, Melkote SN (2018) Diamond wire sawing of solar silicon wafers: a sustainable manufacturing alternative to loose abrasive slurry sawing. Proc Manuf 21:549–566
15. Wu H, Melkote SN (2013) Effect of crystal defects on mechanical properties relevant to cutting of multicrystalline solar silicon. Mater Sci Semicond Process 16(6):1416–1421
16. Wuerzner S, Buchwald R, Moeller HJ (2015) Surface damage and mechanical strength of silicon wafers. Phys Status Solidi (C) Current Topics Solid State Phys 12(8):1119–1122
17. Swain MV (1979) Microfracture about scratches in brittle solids. Proc R Soc A Math Phys Eng Sci 366(1727):575–597
18. Pérez R, Gumbsch P (2000) Directional anisotropy in the cleavage fracture of silicon. Phys Rev Lett 84(23):5347–5350
19. Wallburg F, Kuna M, Budnitzki M, Schoenfelder S (2020) Experimental and numerical analysis of scratching induced damage during diamond wire sawing of silicon. Wear 454–455:203328
20. Weibull W (1939) A statistical theory of the strength of materials. Stockholm: Generalstabens litografiska anstalts förlag 15:1–45
21. Weibull W (1951) A statistical distribution function of wide applicability. J Appl Mech 18:293–297
22. DIN EN 843. DIN EN 843-5:2007-03, Hochleistungskeramik-Mechanische Eigenschaften monolithischer Keramik bei Raumtemperatur-Teil 5: Statistische Auswertung; Deutsche Fassung EN_843–845:2006. NORM, Beuth Verlag GmbH
23. Munz D, Fett T (2001) Ceramics: mechanical properties, failure behaviour, materials selection. Springer, Berlin, corr. 2. print edition
24. Moeller HJ, Buchwald R, Winstanley S (2012) Basic mechanism of diamond wire sawing. In: 6th international workshop on crystalline silicon for solar cells, Aix-les-bains, France

Residual Stress Behavior on Welding Joints in Different Steels Using X-Ray Diffraction

A. Morales-Ortiz, J. C. Arango, A. S. Marulanda, and C. C. Palacio

Abstract Residual stresses are one of the main conditioning factors of the mechanical behavior of components subjected to repair by welding. The magnitudes, directions, location and homogeneity of the residual stresses change with the chemical composition of the base material, the welding process, among other variables. Although it is inevitable to induce residual stresses during the welding process, it is possible to determinate magnitudes and directions using characterization techniques such as X-ray diffraction. In this work, the residual stress distribution measurement is carried out using a portable X-ray diffractometer in the welds of two types of steel (304 stainless steel and A36 carbon steel). On 304 stainless steel, the presence of residual compressive stresses on the base material and the thermally affected zone (HAZ) and residual stresses on tension on the weld seam were identified. For its part, for A36 carbon steel, a distribution of residual stresses in tension and random compression was evidenced in all areas of the sample due to the welding process and material formation. It is concluded that the heterogeneity in the type of residual stresses distributed over the material favors the weakening of the propagation of possible cracks that appear in the areas with residual compressive stresses and the strong influence of the chemical composition of the filler material of welding in the distribution and type of residual stresses that are induced in as a result of the joining process.

Keywords Residual stress · X-ray diffraction · Welding

1 Introduction

Residual stresses are present in the crystalline structure of the material even when external forces or loads are not being applied [3, 4, 24]. A manufactured part is unlikely to be free of residual stress [22], either because they are generated during processing, because they are induced during their service cycle or because of repair

A. Morales-Ortiz (✉) · J. C. Arango · A. S. Marulanda · C. C. Palacio
EAFIT University, Medellín, Colombia
e-mail: amoraleso1@eafit.edu.co

M. Abdel Wahab (ed.), *Proceedings of the 8th International Conference on Fracture, Fatigue and Wear*, Lecture Notes in Mechanical Engineering,
https://doi.org/10.1007/978-981-15-9893-7_27

processes such as welding [20]. Currently there are several techniques for determination of residual stresses, such as the hole drilling, the curvature method and some methods based on diffraction (neutron or X-ray) in order to know the residual stress magnitude of components. However, some of these techniques are destructive (hole drilling and curvature) or difficult to access (synchrotron, neutron diffraction), which means that X-ray diffraction is one of the most widely used techniques, due to recent developments in portable field measurement equipment [9, 10, 17, 23, 24]. This technique is based on the measurement of the strains that occur in the structure of a crystalline material due to the existing proportional relationship between strains and the magnitude of the applied load, making it possible of being measured and characterized [7, 8].

In welding processes, the residual stress are the result of the solidification process, heat transfer and the thermal expansion that occur when the material expands and contracts due to changes in temperature [2]. These stresses are unavoidable, but can be determined using techniques such as X-ray diffraction. [5, 6, 12–14, 16, 21, 25]. Several studies have reported the behavior of residual stresses evaluated by XRD in materials with welding joints, conducting to the identification of the distribution of compressive stresses on the areas in the weld toes, as well as the presence of tensile stresses on the weld [16]; the identification of tensile stresses on the weld followed by lower magnitudes of tensile residual stresses over the near area and in contrast to the presence of compressive residual stresses on the base metal of stainless steel plates with approximately the same amount of austenite and ferrite phases [13]; the compressive stresses around the last weld layer as well as tensile stresses near the low temperature zone in stainless steel samples with 13% Cr–4% [21]; the behavior of residual stresses compared to relaxation processes applied to pipes fabricated of S355J2H + N ferritic-pearlitic structural steel and welded circumferentially, which showed that before performing the relief, the samples had tensile stresses in the weld and their proximities, but after the heat treatment process they became compressive or even tensile ones but with lower magnitues than before; the effect of the presence of austenite in the microstructure on the distribution of residual stresses on pieces joined by welding, as well as its origin was evaluated experimentally and by numerical simulations [6].

Establishing theoretical distributions of residual stresses on welds is possible from experimental studies of their structure, however, these distributions depend on multiple factors specific to the process that not in all circumstances can be considered and make more complex the understanding of its origin and behavior on welded metal structures. Due to this, this work evaluates the distribution of the residual stresses generated by welding processes on two types of steel (stainless steel 304 and carbon steel A36) and which were determined using the X-ray diffraction technique that allows the characterization of components quickly, non-destructively, non-invasively and without modifying the existing residual stress field on the surface during measurements, in order to identify the effect of the type of material on the distribution and kind of residual stresses.

Table 1 Chemical composition of 304 stainless steel and A36 carbon steel

Chemical composition									
Base metal	Cr %	Ni %	S %	P %	Si %	Mn %	C %	N %	Fe %
ASTM 304	17.5–19.5	8–10.5	0–0.03	0–0.05	0–1.00	0–2	0–0.07	0–0.11	Balance
ASTM A36	0.20	0.00	0.05	0.04	0.28	1.03	0.25–0.29	0.00	Balance

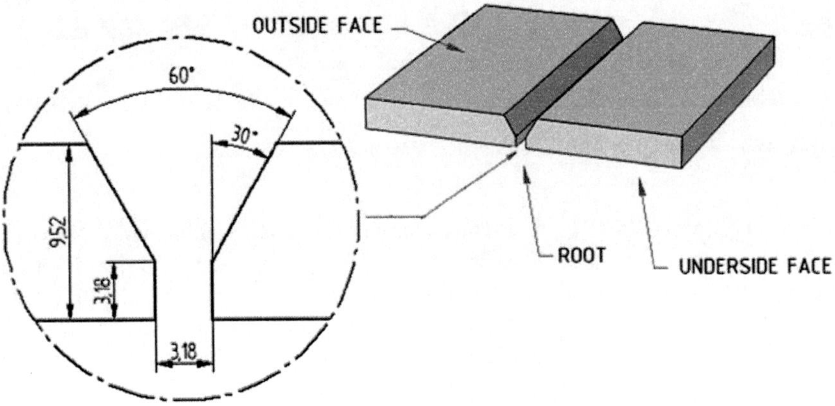

Fig. 1 Welding joint geometrical parameters

2 Materials and Methods

2.1 Materials

For this study, samples made of two different types of steel were prepared in order to compare the stress distribution due to the welding processes in each one. The steels used for the studied samples were: 304 stainless steel and A36 carbon steel, both identified using the ASTM classification. The chemical composition of these steels is shown in Table 1.

The plates of each material were prepared by machining with bevel cooling to 60 ° as shown in Fig. 1.

2.2 Welding Process

Two methods were used for the welding process: TIG (Tungsten Inert Gas) and SMAW (TIG for the 304 stainless steel and SMAW for A36 steel), both performed

Fig. 2 Welded samples: **a** ASTM 304 stainless steel. **b** ASTM A36 carbon steel

on a Miller Maxstar 150 STL unit. The plates were joined by welding with 3.175 mm OD 308L electrodes for the ASTM 304 stainless steel sample and 7018 electrodes for the ASTM A36 steel sample. A total of 5 weld passes were applied in the following sequence: the first pass-string bead at the root to initiate the joint, the following 3 weave bead passes to full-fill the filet and the last one at the root from the underside face to generate full penetration and structural function.

The ASTM 304 stainless steel sample was welded using the TIG process with a current of 105 A and a voltage of 220 V and the ASTM A36 steel sample was welded by the SMAW process with a current of 90 A and a voltage of 220 V. Figure 2 shows the two manufactured samples.

2.3 *Residual Stress Measurements*

Residual stress measurements were carried out with the portable X-ray diffractometer Spider X (model 2018) operating at a voltage of 30 kV, current of 84 μA and scanning at diffraction angles between 115 ° and 150 ° for the ASTM 304 stainless steel sample; and at a voltage of 30 kV, current of 84 μA and scanning at diffraction angles between 140 ° and 150 ° for the ASTM A36 steel sample.

The measurement process is executed according to the following stages: (1) identification of measurement points (Fig. 3), (2) cleaning the area for measurements with a soft cloth to remove impurities, dust and grease. The measured areas correspond to the weld bead (filet), the area close to the weld and weld toes (heat affected zone or HAZ) and the base material. For each piece the same areas and the same number of points were taken. The stresses were measured over the samples going from the base metal, passing on the first HAZ, the weld bead, the second HAZ and finally to finish on the opposite side of the base metal. Residual stresses magnitudes at each

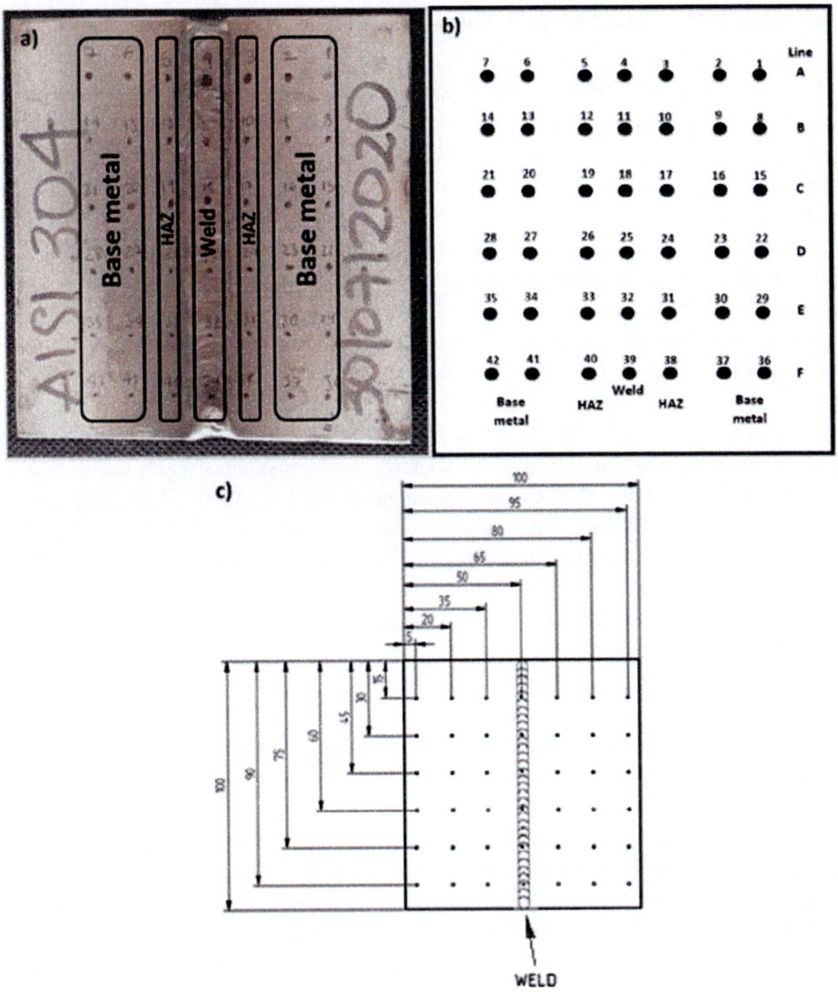

Fig. 3 Identification of the measurement points of the residual stresses on the samples: **a** measurement zones, **b** measurement points, **c** points location (mm)

point were measured through the scan of 9 incidence angles by the goniometer of the diffractometer.

3 Discussion

The results and analysis of the measurement of residual stresses by XRD for each sample are presented below.

3.1 Results of ASTM 304 Stainless Steel Sample

The results obtained after the residual stresses measured on the ASTM 304 stainless steel sample are shown in Table 2. It is identified the presence of compressive stresses (named by its negative sign) on both the base metal and the HAZ noting that the magnitudes are greater on the HAZ than on the base metal. Otherwise on the weld bead the residual stresses are in tension. This behavior in the residual stresses on welded joints and their adjacent surfaces has been reported by Nasir et al. [15] Rossini et al. [16] for samples made of ASTM 304 stainless steel and using TIG welding processes.

Figure 4 shows the behavior of the residual stresses on the ASTM 304 stainless steel sample. Each line in the graph presents the same behavior with a peak of residual stress for the points located on the weld. All the points corresponding to the weld have residual stresses in tension while all the points on both the HAZ and the base metal are in compression. It is also remarkable that the residual stresses for the points on the HAZ are greater than the ones located on the base metal. This behavior of high values in the residual compressive stresses measured on the HAZ is an indicator of the high load that the filler material confers on the weld bead and the restrictions to the expansion that the HAZ undergoes by the base metal, which makes the HAZ is constricted at the end of the process. For its part, the weld bead that is formed by the contribution of metal in the molten state, during the cooling and solidification process, experiences internal contraction loads that translate into a tensile residual stress on the surface.

The appearance of cracks in this steel can be favored by the high magnitudes of the residual stresses on the weld, which are found in tension throughout its length and in magnitudes that are close to the tensile strength for this type of steels (Su = 510 MPa) [11]. This fact provides risk of fracture for the welded joint, for which the content of delta ferrite or FN (ferrite number) can be modulated based on the composition of the filler material electrode; in the case of the stainless steel sample (ASTM 304) welded with 308L electrode the FN is between 3 and 7%, which is adequate given the recommended maximum limit of 8% [18, 19].These results indicate the sensitivity to cracking of 308L steel welds due to its chemical composition and is supported

Table 2 Residual stress results of ASTM 304 stainless steel sample

Residual stress [MPa]							
Line	Base metal		HAZ	Weld	HAZ	Base metal	
A	−286	−359	−711	523	−674	−265	−266
B	−266	−290	−763	502	−485	−194	−375
C	−339	−155	−485	469	−491	−174	−365
D	−314	−325	−426	615	−281	−430	−170
E	−303	−341	−285	540	−361	−234	−348
F	−302	−398	−363	456	−558	−368	−298

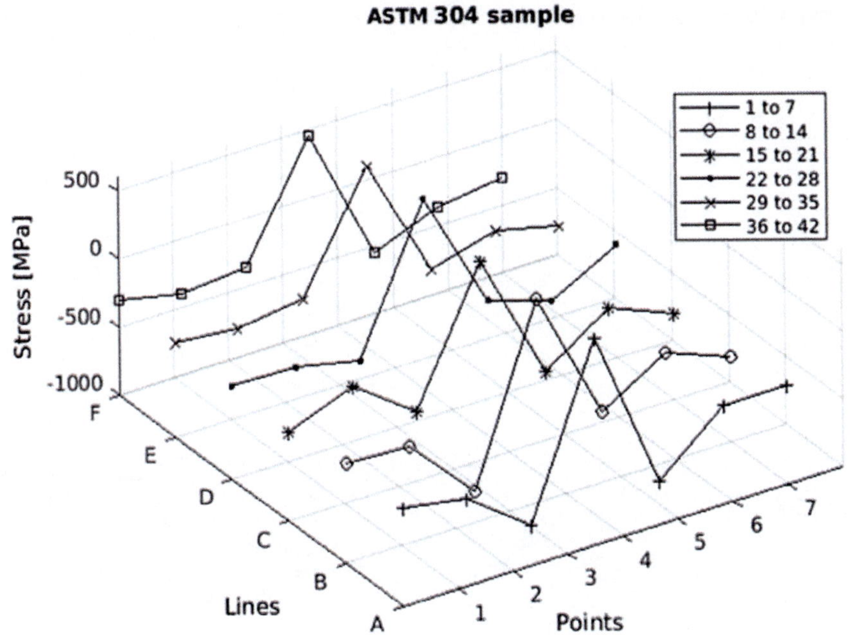

Fig. 4 Results of residual stresses of ASTM 304 stainless steel

by the magnitudes and type of residual stresses identified for this sample as it was reported in the Table 2 and Fig. 4.

3.2 *Results of ASTM A36 Carbon Steel*

This sample presents a heterogeneous distribution about the type of residual stresses remarking that there is not preference in compressive and tensile residual stresses appearance regarding to the zones on the sample as it is shown in Table 3. It is possible to notice the presence of residual stresses in compression and in tension both on the base metal and on the weld. Regarding the HAZ zone, compressive residual stresses are identified in most of the points analyzed, however, it is noticed the presence of some points with low tensile residual stresses surrounded by compressive residual stresses. For the residual stresses on the weld, both tensile and compressive stresses are also obtained, noting that the residual stresses in tension have lower magnitudes than the tensile strength for this type of steels (Su = 485 MPa) [1, 11].

Figure 5 evidences the residual stresses behavior of the ASTM A36 sample at each of the measured points. In the results of Table 3 and Fig. 5 the global stress range (difference between the maximum value in tension and the maximum in compression)

Table 3 Residual stress results of ASTM A36 sample

Stress [MPa]							
Line	Base metal		HAZ	Weld	HAZ	Base metal	
A	−38	−215	−312	−122	−277	−218	−83
B	−32	−178	−137	49	−122	−91	−47
C	−112	80	−151	144	−17	57	−182
D	−57	123	−89	−45	−83	−53	−150
E	137	78	77	−173	27	64	97
F	−140	−244	−277	37	−119	−239	−193

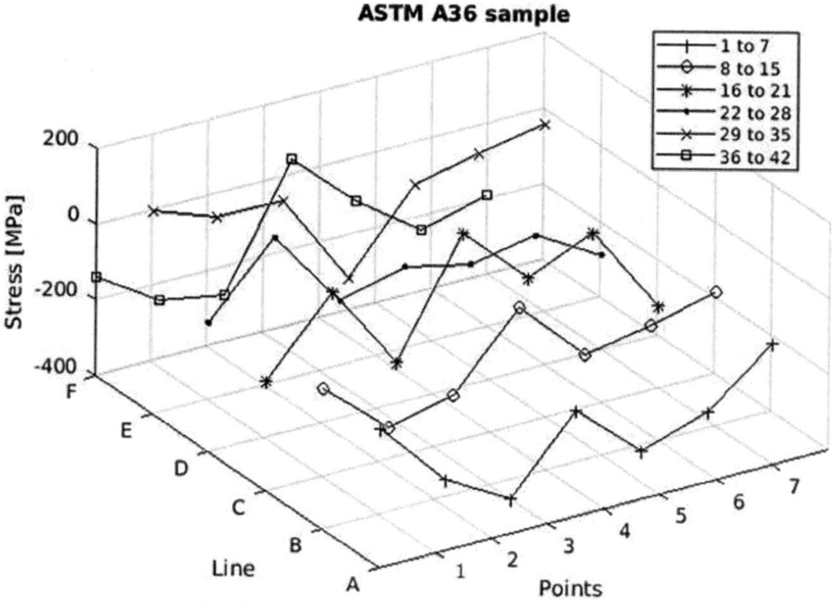

Fig. 5 Residual stress results of ASTM A36

is less than 500 MPa for the sample of A36, which represents the half of the global residual stress range observed in the 304 stainless steel sample which is 1000 MPa.

Despite the heterogeneity in the A36 residual stresses distribution, the welding process in this material reveals the following advantages: (i) the maximum tensile residual stress identified in the A36 sample is 4 times below than the one obtained for the 304 sample (144 MPa versus 615 MPa), this fact makes the last material more sensitive to fracture as consequence of the welding process since the residual stress approaches to the ultimate tensile strength (Su) of the material; (ii) the tensile/compression stress delta of A36 sample is less than 500 MPa while in 304 it is 1 GPa, being the ratio 1:2 of A36:304 which evidences the heterogeneity in residual

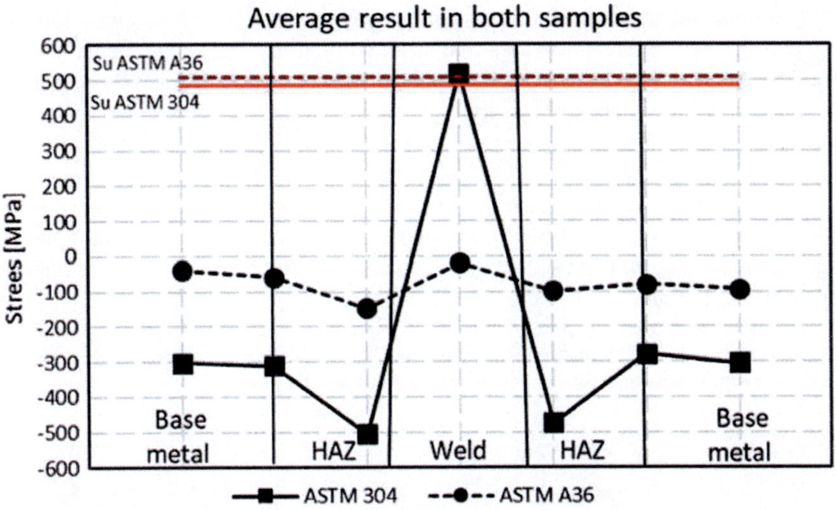

Fig. 6 Average result in both samples for ASTM 304 and ASTM A34 steels

stresses is in fact more marked in the 304 sample than in the A36 sample; (iii) in the A36 steel sample, the 42 points on which the residual stresses were measured, 12 are in tension and the rest in compression and it is identified that the data that are in compression are randomly immersed and surrounded by points with magnitudes of compressive residual stresses, which configures colonies of tensile zones surrounded by a field of compressive residual stresses, a fact that favors the weakening for flaws propagation if they appear within the areas in tension.

Figure 6 shows a comparison between the average results in both samples and the magnitudes of tensile strength for ASTM 304 and ASTM A36 steels (Su = 485 MPa and Su = 510 MPa, respectively).

4 Conclusion

When ASTM 304 stainless steel plates are joined by welding processes like the TIG technique, a distribution of residual stresses is obtained according to the theoretical distributions: residual stresses in tension on the weld bead and residual stresses in compression on the HAZ and the base metal. For its part, the A36 carbon steel sample joined by SMAW electrode welding process shows that the stress distribution is not apparently homogeneous because a random allocation of points in tension surrounded by points in compression which configures colonies of compressive residual stresses confining small fields of tensile residual stresses. This situation can be favorable to control the propagation of cracks (mechanism of arresting cracks in welded joints and consequently improving the fracture toughness in welds).

The presence of residual stresses that represent a risk for the welded joint is represented by the tensile stresses that are identified in higher magnitudes on the weld bead of the 304 stainless steel sample than in the A36. This shows the influence of the chemical composition on the appearance of residual tensile stress on the weld. The high Cr and Ni contents of 304 stainless steel that are responsible for the formation of delta ferrite dendrites can have an effect on this behavior.

The measurement of residual stresses by XRD in welds makes it possible to diagnose and project the risks of the appearance of cracks both in the weld and in the base metal, as well as to identify the parameters that may influence the quality of the welded joints.

Acknowledgements To the EAFIT University and the GEMA group for the support for the realization of internal projects that allow the realization of this research.

References

1. ASTM a 36/a 36 M-05 (2005) Standard Specification for Carbon Structural Steel. Standards. https://doi.org/10.1520/A0036
2. DeWald A (2016) Residual Stress in Welding—Hill Engineering
3. Epp J (2016) X-Ray Diffraction (XRD) Techniques for materials characterization. Elsevier Ltd
4. Guinebretiere R (2007) X-ray diffraction by polycrystalline materials. ISTE
5. Hempel N, Bunn JR, Nitschke-Pagel T, Payzant EA, Dilger K (2017) Study on the residual stress relaxation in girth-welded steel pipes under bending load using diffraction methods. Mater Sci Eng A. https://doi.org/10.1016/j.msea.2017.02.005
6. Hensel J, Nitschke-Pagel T, Dilger K (2014) On the effects of austenite phase transformation on welding residual stresses in non-load carrying longitudinal welds. Weld World. https://doi.org/10.1007/s40194-014-0190-3
7. Kandil FA, Lord J, Fry A, Grant PV (2002) A review of residual stress measurement methods. Measurement of Residual Stress in Components, NPL Report, UK
8. Kandil FA, Lord JD, Fry AT, Grant P V (2001) Review of residual stress measurement methods -a guide to technique selection. Crown 1:45
9. Lu J, (U.S.) S for EM (1996) Handbook of measurement of residual stresses. Fairmont Press , Distributed by Prentice Hall PTR, Lilburn, GA, Upper Saddle River, NJ
10. Martins JA, Cardoso LP, Fraymann JA, Button ST (2006) Analyses of residual stresses on stamped valves by X-ray diffraction and finite elements method. J Mater Process Technol. https://doi.org/10.1016/j.jmatprotec.2006.03.072
11. Milad M, Zreiba N, Elhalouani F, Baradai C (2008) The effect of cold work on structure and properties of AISI 304 stainless steel. J Mater Process Technol. https://doi.org/10.1016/j.jmatprotec.2007.09.080
12. Monin VI, Gurova T, Castello X, Estefen SF (2009) Analysis of residual stress state in welded steel plates by X-ray diffraction method. Rev Adv Mater Sci 20:172–175
13. Monin VI, Lopes RT, Turibus SN, Payão Filho JC, de Assis JT (2014) X-Ray diffraction technique applied to study of residual stresses after welding of duplex stainless steel plates. Mater Res 17:64–69. https://doi.org/10.1590/S1516-14392014005000047
14. Monine VI, Da Cruz Payão Filho J, Gonzaga RS, Passos EKD, De Assis JT (2018) X-Ray diffraction technique for residual stress measurement in NiCrMo alloy weld metal. Adv Mater Sci Eng. https://doi.org/10.1155/2018/8986423

15. Nasir NSM, Razab MKAA, Mamat S, Ahmad MI (2016) Review on welding residual stress. ARPN J Eng Appl Sci 11:6166–6175
16. Rossini NS, Dassisti M, Benyounis KY, Olabi AG (2012) Methods of measuring residual stresses in components. Mater Des 35:572–588. https://doi.org/10.1016/j.matdes.2011.08.022
17. Ruud CO (2000) Residual stress measurements. Mech Test Eval 8:0
18. Saluja R, Moeed K (2015) Formation quantification and significance of delta ferrite for 300 series stainless steel weldments. Int J Eng Technol Manag Appl Sci
19. Saluja R, Moeed KM (2016) Assessment of delta ferrite for SA 240 type 304L austenitic weld metal using different filler materials. 7:493–498
20. Smith WF (2010) Fundamentos de la ciencia e ingeniería de materiales, 4th ed. Ciudad de México
21. Thibault D, Bocher P, Thomas M (2009) Residual stress and microstructure in welds of 13%Cr-4%Ni martensitic stainless steel. J Mater Process Technol 209:2195–2202. https://doi.org/10.1016/j.jmatprotec.2008.05.005
22. Withers PJ, Bhadeshia HKDH (2001) Residual stress Part 2—Nature and origins
23. Yadroitsev I, Yadroitsava I (2015) Evaluation of residual stress in stainless steel 316L and Ti6Al4V samples produced by selective laser melting. Virtual Phys Prototyp 10:67–76. https://doi.org/10.1080/17452759.2015.1026045
24. Yavuz Peter McIntyre M, Karaman Dennis O IL, Anderoglu O (2004) Residual stress measurement using X-Ray diffraction. Texas A&M University, pp 1–64
25. Zhang Y, Liu J, Zhao E, Xu X, Cheng H (2009) Elimination and evaluation of welding residual stress in sterilizers melded by combined process. Hanjie Xuebao/Transactions China Weld Inst

Effect of Specimen Size on Localization using Digital Image Correlation

Rupesh K. Verma, Giang D. Nguyen, Ha H. Bui, and Murat Karakus

Abstract Failure of structures made of brittle materials like rock and concrete exhibits deterministic size effect on post-localization mechanical behavior. The application of advanced image-based instrumentation, along with conventional testing can provide full-field strain and its evolution that can be useful for the analysis of localized behavior of brittle materials. In this view, this study collaborates the *Digital Image Correlation technique* (DIC) with lateral strain-controlled uniaxial compression tests on Hawkesbury sandstone for insights into size effect on localized behavior. Full-field strain obtained from DIC is synchronized with macroscopic responses to investigate the effect of specimen size on strain localization mechanism. Results from this study demonstrate that the increase in specimen size intensifies the local scale deformation, accelerates the localization process, and increases both strain gradients across the shear band and shear band thickness. The shear band evolution analysis in this study also provides rich quantitative data useful for the development of constitutive models to capture size effect on the behavior of rocks.

Keywords Size effect · Softening · Localization · Shear band

1 Introduction

Strain localization is usually associated with softening behavior and loss of load-carrying capacity of materials. It has been well known that deterministic size effect is a consequence of softening and localized failure, leading to the dependence of mechanical behavior on specimen size [1, 2]. Similar effects have also been observed in strain burst experiments, where the specimen's height/width (H/W) ratio affects the burst characteristics of rock [3–5]. Nevertheless, lack of details on the evolution of

R. K. Verma · G. D. Nguyen (✉) · M. Karakus
School of Civil Environmental and Mining Engineering, University of Adelaide, Adelaide, Australia
e-mail: g.nguyen@adelaide.edu.au; giang.nguyen@trinity.oxon.org

H. H. Bui
Department of Civil Engineering, Monash University, Clayton, Australia

M. Abdel Wahab (ed.), *Proceedings of the 8th International Conference on Fracture, Fatigue and Wear*, Lecture Notes in Mechanical Engineering,
https://doi.org/10.1007/978-981-15-9893-7_28

Table 1 Sample specification and key material properties

Sr. No.	Diameter (mm)	Aspect ratio (H/D)	Peak load (kN)	Elastic modulus (GPa)	UCS (MPa)	Fracture Angle
1	42	2.5	48.4	8.06	34.9	67.9^0
2	42	3.5	46.8	8.07	33.8	61.4^0

localized failure and its correlation with macro responses of the specimens prevents obtaining insights into the effects of specimen size on localized failure and producing quantitative results for advanced constitutive models [6].

In this direction, *Digital Image Correlation* (DIC) can offer benefits on top of the traditional measurements of macro responses. It is not only cost-effective and faster compared to X-ray tomography but also provides non-contact full-field kinematics of planar or non-planar surfaces [7]. It enables us to obtain much more data than possible via conventional strain gauges. As a result, it has been widely used to explore the behavior of brittle geomaterials under different loading conditions. For instance, Tran et al. [8] utilized the DIC technique to investigate the mechanism of shear band formation in rock-like material formed by bonded rigid particles. Verma et al. [9] utilized the three-dimensional DIC technique to analyze the mechanism of localization and shear band formation in the Hawkesbury sandstone rock under uniaxial compression loading. Xing et al. [10] presented a review of the fundamentals and current state of the art of high-speed imaging applications in geomechanics. Mazela et al. [11] extended DIC applications to investigate the mechanical properties of tablets. Similarly, DIC has been used to investigate the mechanical behavior of a wide range of materials, including rock, concrete, magnets, graphene, and ceramics [12–19]. Despite the above, limited attempts were made to provide in depth analysis of rich data obtained from the DIC applications and utilize it further for the development of constitutive models, particularly size effect on the initiation and development of shear bands.

This study presents an innovative approach to analyzing the effect of specimen size on the localization mechanism and shear band formation in rocks using a three-dimensional DIC technique. For this purpose, lateral strain-controlled uniaxial compression tests were conducted on cylindrical specimens of Hawkesbury sandstone with different aspect ratios (height/diameter) of 2.5 and 3.5. The indirect control using lateral strain allows us to not only acquire the complete load-displacement response that exhibit snap-back but also enable the efficient application DIC in analyzing local-scale behavior. In this view, Sect. 2 briefs the experimental setup followed by analysis of the effect of specimen size on macro response in Sect. 3. Section 4 analyzes the DIC data to evaluate the effect of specimen size on the localization mechanism and formation of shear band.

2 Experimental Setup

This study presents the results from the lateral strain-controlled uniaxial compression experiments on Hawkesbury sandstone. Cylindrical samples of 42 mm diameter with two different aspect ratios (i.e., height/diameter) of 2.5 and 3.5 were prepared for these experiments. Five samples were tested for each aspect ratio. These experiments were performed on the universal compression loading machine—INSTRON. The specimens were compressed via moving top platen while keeping the bottom platen fixed. The lateral deformation rate (i.e., obtained from the chain extensometer) of the specimen was maintained constant and used as feedback to the loading frame. The axial deformation was recorded continuously via two Linear Variable Differential Transformers (LVDTs). The above setup allows capturing snapback (or class II) responses in brittle materials.

In order to obtain local scale deformation, this study utilizes the three-dimensional DIC technique. For this purpose, each cylindrical specimen was exposed to black speckles on white paint. These randomly distributed speckles provided a unique pattern essential for better correlation. Additionally, two CCD (charged coupled device) cameras were used to capture the specimen images continuously throughout the test with a pre-defined rate. These cameras were placed under sufficient lighting conditions in front of the specimen with an inclination of 40^0 approximately. For post-processing, this study utilizes the Correlation Solutions—VIC 3D, i.e., commercial software for image processing [20]. It enabled us to acquire a good correlation between the pixel intensity array on the concerned image with the reference image of the undeformed specimen.

3 Specimen Size Effect on Sample Scale Response

The sample scale load-displacement responses enabled us to acquire the essential properties, including elastic modulus (E), uniaxial compressive strength (UCS) along with the fracture characteristics (i.e., shear band formation) and angle. Table 1 presents a summary of the specimen's geometric specification along with key macroscopic properties.

The above results shows a slight variation of material strength (less than 4%) and E (less than 0.2%) due to the inhomogeneity of the material.

Figure 1 presents the macroscopic stress-strain response from the rock specimens with the above two aspect ratios. It shows that, for the same rock type, an increase in aspect ratio from 2.5 to 3.5 induces more severe snapback in the post-peak regime, demonstrating the non-uniqueness of stress-strain behavior. The failure of both specimens with aspect ratios (AR) of 2.5 and 3.5 show clear and distinct shear bands of 68^0 and 61^0 inclinations, respectively. These macroscopic observations indicate that the sample scale strength reduction, in principle, is the consequence of localized material damage resulting in the formation of shear band in both specimens. Apart

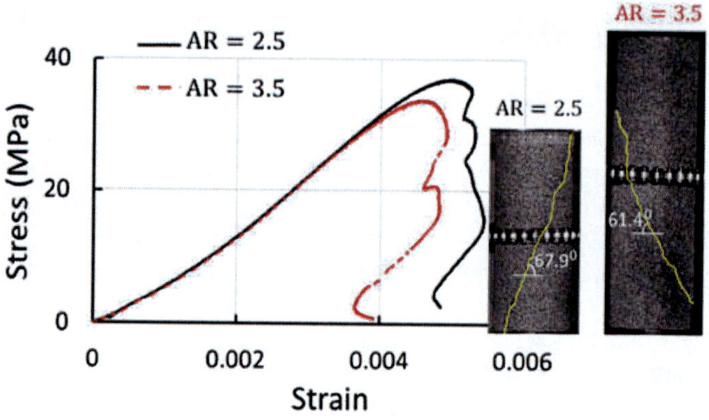

Fig. 1 Stress-strain response and effect of aspect ratio (AR)

from this, without details on the evolution of shear band that DIC can provide, it is not possible to acquire much insight into the localization mechanism and its variations with specimen size.

4 Specimen Size Effect on Localization

This Section collaborates the effect of local-scale responses obtained from DIC data with sample scale load variation with time. The overall results are analyzed separately in pre- and post-peak regimes.

4.1 Pre-Peak Regime

In the pre-peak regime, three points on load versus time plot (i.e., 'a' at ~20%, 'b' at ~65% of peak and 'c' at peak load) are considered to synchronize this sample scale response with the local scale behavior. Figure 2 presents the contours of full-field Von-Mises strain evolution over the area of interest (AOI) in both specimens. Inspection lines (i.e., section 'x–x') are considered to extract strain evolution data on the surface of the specimens. The position of these inspection lines is adjusted to minimize the interference from the chain extensometer and acquire good strain evolution data across the shear band.

At initial stages 'a' and 'b' in Fig. 2, the above two specimens undergo elastic deformation with relative homogeneity in full-field strain contours. The overall strain and strain increments (i.e., from stage 'a' to 'b') across the vertical sections in both specimens are also comparable at both samples (i.e., Fig. 1) and local (i.e., Fig. 2) scales. The minor difference one can observe here is that the local scale responses

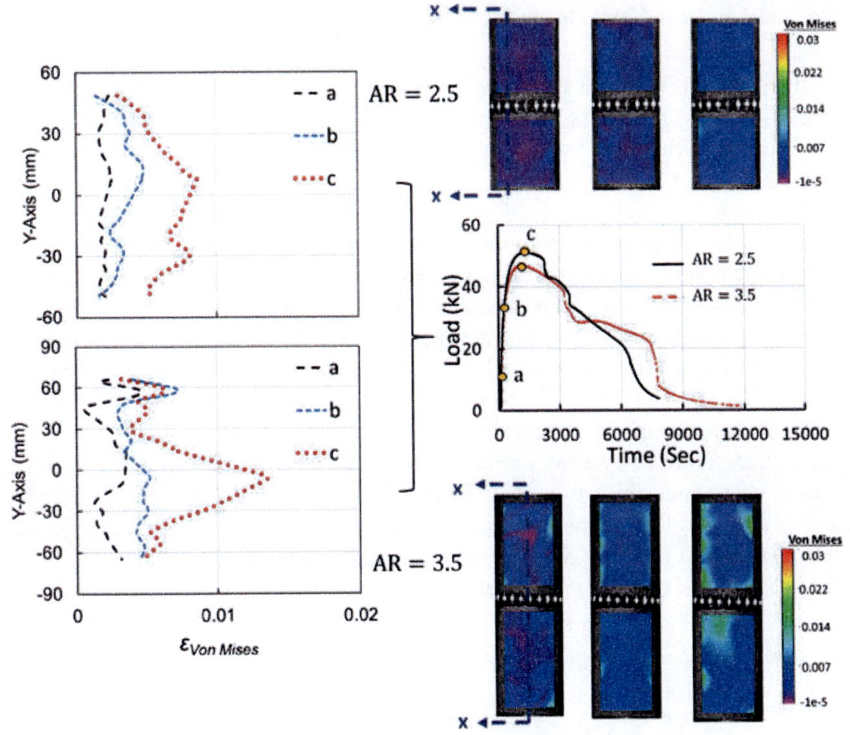

Fig. 2 Pre-peak regime: strain evolution with load–reduction

are fluctuating around an averaged one, reflecting the inhomogeneity of the material below the macro continuum scale. Overall, these results show that the pre-peak elastic regime are under relatively uniform compression, and the influence of specimen size on the local scale responses is insignificant.

As the load approaches the peak (i.e., point ‘c’), the effect of specimen size on the local scale responses becomes evident. For the specimen with aspect ratio 3.5, one can observe the beginning of strain localization with maximum strain increment of ~80% on the ‘x–x’ section from stage ‘b’ to ‘c.’ The full-field strain contours at stage ‘c’ also indicate the strain intensification in localized regions, with the start of a localization band. On the other hand, specimen with aspect ratio of 2.5 demonstrates relatively uniform distribution with maximum local strain increment observed to be ~60% smaller than the more slender specimen (aspect ratio of 3.5) at this stage. Overall, despite having a lower peak load, specimen with a higher aspect ratio undergoes quicker intensification of localization with a higher magnitude of strain in the localized region.

4.2 Post-Peak Regime

Figure 3 presents the local strain evolution data in two specimens at different stress states in the post-peak regime, i.e., stage 'd,' 'e,' and 'f.' Localized strain along the 'x–x' section in the more slender specimen intensifies with a maximum increase of ~55% at stage 'd' and ~140% at stage 'e' as compared to the peak loading stage. During this localization process, the remaining portions of the specimen outside this localized zone experience negligible strain increment. At stage 'f,' the maximum strain on the 'x–x' section increases by ~300% (i.e., compared to the peak loading stage), followed by strain relaxation in the regions outside the localization zone. It indicates strong localization prior to cracking (i.e., visible as inclined shear band crack), causing the release of stored strain energy and triggering strain relaxation in the zone outside the localization band.

For the specimen with aspect ratio of 2.5, stage 'd' is also marked with strain localization followed by strain jumps in the localized region at stages 'e' and 'f', causing shear band prior to complete failure. Despite having a higher peak load, the magnitude of localized strain on the 'x–x' section in all post-peak stages of this specimen (aspect ratio of 2.5) is significantly smaller than what observed in the more slender specimen of the same diameter (aspect ratio of 3.5).

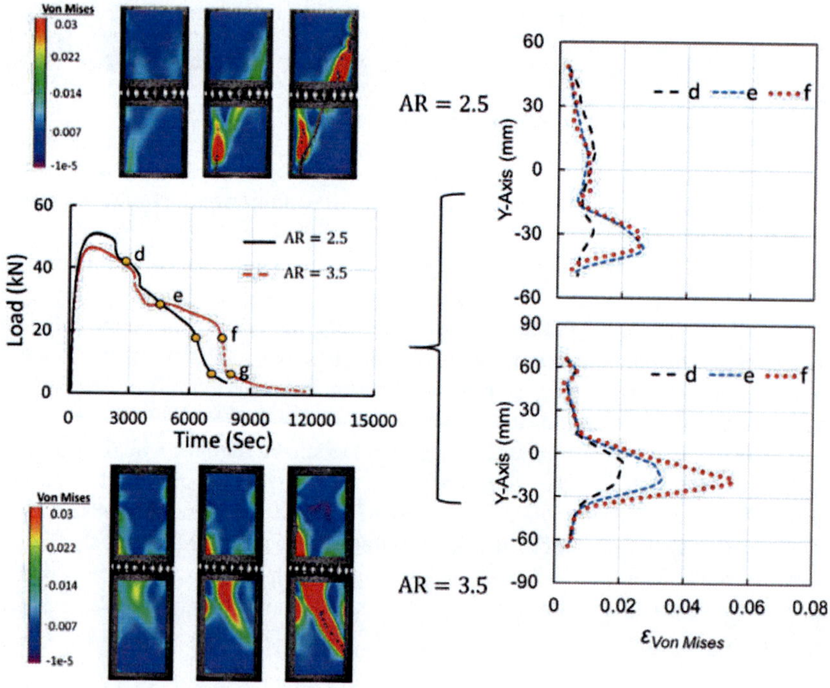

Fig. 3 Post-peak regime: strain evolution with load-reduction

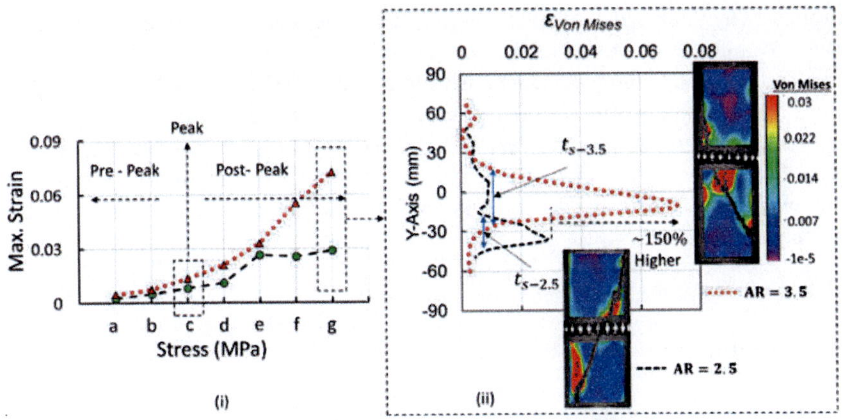

Fig. 4 (i) Variation of maximum strain along 'x–x' section with loading stages (i.e., stress states); (ii) strain variation and shear band estimations

Figure 4a demonstrates the explicit comparison of these magnitudes of maximum strains at different stress states (i.e., from stage 'a' to 'g'). It shows that the local scale responses of both specimens are similar up to stage 'b.'. Beyond this point 'b', the maximum strain starts bifurcating and differs significantly after stage 'e' in the post-peak regime. The strain distribution on the 'x–x' section at the failure of the more slender specimen is ~150% higher than what observed in the other one. Overall, the speed of localization and the magnitude of strain jump in the localized region is significantly higher in specimen with higher aspect ratio of 3.5. Figure 4b presents an estimation of shear band thickness, i.e., $t_{s-3.5}$ and $t_{s-2.5}$ in specimens with 3.5 and 2.5 aspect ratios. It shows that the thickness of shear bands depend on the slenderness of specimen (i.e., $t_{s-3.5}$ ~ two times of $t_{s-2.5}$). However further studies on different types of brittle materials are needed to reinforce this observation.

This effect of specimen size on the local scale responses can be explained via strain energy storing and dissipation mechanism. As the cylindrical specimen deforms, it stores elastic strain energy inside the specimen. For the same stress level, magnitude of elastic strain energy stored is governed by the specimen volume. Thus, the increase in specimen aspect ratio while keeping the diameter constant results in the increased specimen volume (i.e., ~1.4 times in the present case). On the other hand, the energy dissipation depends upon the crack surface, i.e., energy dissipated to create a new surface. Unlike the volume, increase in crack surface area with specimen size is comparatively insignificant (theoretically, it remains the same, given the same stress condition and same diameter). It shows that for higher aspect ratio, the increase in stored elastic strain energy at peak load is significant, while the dissipation energy capacity theoretically remains the same. This leads to much higher localized strain along the 'x–x' section of the more slender specimen after stage 'b' (start of localization) and at subsequent stages in the post-peak regime, as shown in Figs. 3 and 4. This increase in excess strain energy at the peak load with specimen size further explains why thicker shear band was observed in the specimen with a higher aspect ratio. Overall, the above results and analysis show the significant effects of specimen size on the localization mechanism and the evolution of shear bands in rocks. The shear band thickness prior to fracture has been shown to be significant compared to the specimen size, despite the very brittle response of the specimen (snapback, or class II behavior). This is the basis for theoretical development taking into account the evolution of shear band thickness [6] towards better constitutive models for geomaterials.

5 Conclusions

This study synchronizes the local scale strain evolution data obtained from the image processing technique with the macroscopic response obtained from lateral strain-controlled UCS tests on Hawkesbury sandstone. This allows insights into the effect of specimen size on the localization mechanism and evolution of the shear band. The presented results and corresponding analysis demonstrate that, even in the controlled cracking environment, an increase in specimen size accelerates the localization process, increases the strain gradients across the shear bands, and also shear band thickness. These effects specimen dimension is attributed to the excess strain energy, which increases with specimen size and influences the localization characteristics. The shear band evolution analysis in this study also provides more quantitative data useful for the development of constitutive models intended to capture size effect on the behavior of rocks [6, 21–24].

Acknowledgment Support from the Australian Research Council to Giang D. Nguyen, Ha H. Bui and Murat Karakus (projects FT140100408, LP150100539, DP170103793, FT200100884) is acknowledged. The authors also thank honours project students P. Dawidoswki, C. Dullona, Z. Naimi and M. Zamparelli under Giang D. Nguyen's supervision for their support in carrying out the experiments.

References

1. Bažant ZP (2000) Size Effect. Int J Solids Struct 37:69–80
2. Bažant ZP (1999) Size effect on structural strength: a review. Arch Appl Mech 69:703–725. https://doi.org/10.1007/s004190050252
3. Tuncay E, Hasancebi N (2009) The effect of length to diameter ratio of test specimens on the uniaxial compressive strength of rock. Bull Eng Geol Environ 68:491–497. https://doi.org/10.1007/s10064-009-0227-9
4. Li D, Li CC, Li X (2011) Influence of sample height-to-width ratios on failure mode for rectangular prism samples of hard rock loaded in uniaxial compression. Rock Mech Rock Eng 44:253–267. https://doi.org/10.1007/s00603-010-0127-0

5. Zhao XG, Cai M (2015) Influence of specimen height-to-width ratio on the strainburst characteristics of Tianhu granite under true-triaxial unloading conditions. Can Geotech J 52:890–902
6. Nguyen GD, Bui HH (2020) A thermodynamics-and mechanism-based framework for constitutive models with evolving thickness of localisation band. Int J Solids Struct 187:100–120. https://doi.org/10.1016/j.ijsolstr.2019.05.022
7. Nguyen TB, Mcnamara S, Crassous J, Amon A (2017) Experimental study of shear bands formation in a granular material. 10001:1–5
8. Tran TPH, Bouissou S, Chemenda A et al (2019) Initiation and evolution of a network of deformation bands in a rock analogue material at Brittle-Ductile transition. Rock Mech Rock Eng 52:737–752. https://doi.org/10.1007/s00603-018-1641-8
9. Verma RK, Pour AF, Dawidowski P et al (2019) Analysing localisation behaviour of rocks using digital image correlation technique. In: 13th International Conference on Mechanical Behaviour of Materials 11–14
10. Xing HZ, Zhang QB, Braithwaite CH et al (2017) High-speed photography and digital optical measurement techniques for geomaterials: fundamentals and applications. Springer Vienna
11. Mazel V, Guerard S, Croquelois B et al (2016) Reevaluation of the diametral compression test for tablets using the flattened disc geometry 513:669–677
12. Jonsén P, Häggblad H, Sommer K (2007) Tensile strength and fracture energy of pressed metal powder by diametral compression test. Powder Technol 176:148–155. https://doi.org/10.1016/j.powtec.2007.02.030
13. Zhang X, Yi Y, Zhu H et al (2018) Measurement of tensile strength of nuclear graphite based on ring compression test. J Nucl Mater 511:134–140
14. Nath F, Mokhtari M (2018) Optical visualization of strain development and fracture propagation in laminated rocks. J Pet Sci Eng 167:354–365
15. Cao R, Lin H, Cao P (2018) Strength and failure characteristics of brittle jointed rock-like specimens under uniaxial compression : digital speckle technology and a particle mechanics approach. Int J Min Sci Technol 28:669–677
16. Xing HZ, Zhang QB, Ruan D et al (2018) Full-field measurement and fracture characterisations of rocks under dynamic loads using high-speed three-dimensional digital image correlation. Int J Impact Eng 113:61–72
17. He W, Hayatdavoudi A (2018) A comprehensive analysis of fracture initiation and propagation in sandstones based on micro-level observation and digital imaging correlation. J Pet Sci Eng 164:75–86. https://doi.org/10.1016/j.petrol.2018.01.041
18. Candamano S, Sgambitterra E, Lamuta C et al (2019) Graphene nanoplatelets in geopolymeric systems : A new dimension of nanocomposites. 1–10
19. Fallah Pour A, Nguyen GD, Vincent T, Ozbakkaloglu T (2020) Investigation of the compressive behavior and failure modes of unconfined and FRP-confined concrete using digital image correlation. Compos Struct 252:112642. https://doi.org/10.1016/j.compstruct.2020.112642
20. Correlation Solutions (2018) VIC-3D Reference Manual–Correlation Solutions
21. Le LA, Nguyen GD, Bui HH et al (2018) Localised failure mechanism as the basis for constitutive modelling of geomaterials. Int J Eng Sci 133:284–310. https://doi.org/10.1016/j.ijengsci.2018.09.004
22. Nguyen GD, Nguyen CT, Nguyen VP et al (2016) A size-dependent constitutive modelling framework for localised failure analysis. Comput Mech 58:257–280. https://doi.org/10.1007/s00466-016-1293-z
23. Nguyen GD, Korsunksky AM, Einav I (2014) A constitutive modelling framework featuring two scales of behaviour: fundamentals and applications to quasi-brittle failure. Eng Fract Mech 115:221–240. https://doi.org/10.1037/0033-2909.I26.1.78
24. Nguyen GD, Einav I, Korsunsky AM (2012) How to connect two scales of behaviour in constitutive modelling of geomaterials. Géotechnique Lett 2:129–134. https://doi.org/10.1680/geolett.12.00030

Fatigue

Prognosis of Damage Intensity on Reinforced Concrete Beam Under Cyclic Loading

Noorsuhada Md Nor, Soffian Noor Mat Saliah, Shahrum Abdullah, Norrul Azmi Yahya, and Masyitah Md Nujid

Abstract Active damage such as crack initiation owing to cyclic loading in reinforced concrete beams is a pivotal global phenomenon. This damage is because the crack initiation is slowly generated in the structure and cannot be observed visually. As the cyclic load progresses, the cracks need to be classified. This paper presents the prognosis of damage intensity on reinforced concrete beams under cyclic loading. Twelve singly reinforced concrete beams were prepared for the purpose of testing. The beams were subjected to three point loading with six phases of maximum cyclic loading. The acoustic emission technique was utilised for monitoring the crack progression. For cyclic loading, the 1 Hz frequency and sinusoidal wave mode were utilised. For each phase of loading, 5000 cycles were applied. A zone intensity chart was used for classifying the cracks that occurred in the beams. The classification was constructed based on the acoustic emission signals collected from the located events and on a channel basis. The located events were concentrated on the critical sections of SEC A and SEC B. Meanwhile, for the channel basis, the closest sensor to critical sections was used from CH3. It was found that the prognosis of the active crack intensity in the beam can be identified, which closely corresponds to the load increment. This study is useful for understanding crack behaviour in a beam subjected to cyclic loading.

Keywords Acoustic emission · Damage intensity · Cyclic loading · Reinforced concrete

N. Md Nor (✉) · S. N. Mat Saliah · M. M. Nujid
Faculty of Civil Engineering, Universiti Teknologi MARA, Cawangan Pulau Pinang, Pulau Pinang, Malaysia
e-mail: ida_nsn@uitm.edu.my

S. Abdullah
Department of Mechanical and Manufacturing Engineering, Faculty of Engineering and Built Environment, Universiti Kebangsaan Malaysia, Bangi, Malaysia

N. A. Yahya
Faculty of Civil Engineering, Universiti Teknologi MARA, Shah Alam, Malaysia

M. Abdel Wahab (ed.), *Proceedings of the 8th International Conference on Fracture, Fatigue and Wear*, Lecture Notes in Mechanical Engineering,
https://doi.org/10.1007/978-981-15-9893-7_29

1 Introduction

Acoustic emission is a technique for detecting and monitoring the rapid release of energy from localised sources within a material deforming under stress [1, 2]. It is a non-destructive technique to assess the condition of a reinforced concrete structure and is a useful technique for structural health monitoring. It is normally applied to monitor the condition of bridges, dams, and buildings. The technique relies on acoustic emission sources such as crack growth or any deformation that happens in the material or structure, which needs to be assessed. The assessment is significantly vital for detecting crack progression in the structure before it becomes worse. The cracks have to be categorised so that the prognosis of the crack behaviour can be classified into a suitable zone intensity.

Recently, a review of crack inspection on beams of actual size using the acoustic emission technique has been extensively reported by Noorsuhada [3]. The crack mode that correlated to the acoustic emission parameter was investigated by Md Nor et al. [4]. The average frequency—rise angle (RA) value was analysed to determine the two types of cracks that occurred in the beam i.e. tensile and shear cracks. Acoustic emis-sion has been used to monitor the fracture of high-strength steel subjected to constant loading and tensile by Toubal et al. [5]. Based on scanning electron microscopy images, they found a good correlation between the fracture surfaces and recorded acoustic emission signatures.

Acoustic emission has been utilised to determine the bond strength of a corroded large-scale reinforced concrete beam using three sensors attached at specific distances on the beam [6]. They found that the acoustic emission parameters examined through the cumulative signal strength, hits, historical index, and severity have a reasonable correlation with the corrosion propagation levels in the beam. An in situ assessment of a reinforced concrete slab was carried out by Di Benedetti and Nanni [7] using the acoustic emission technique. A load test was performed on the slab to estimate the acoustic emission parameters related to the damage. They affirmed that acoustic emission provides useful information for real-time monitoring of in situ reinforced concrete slabs. Muhammad Sauki et al. [8] investigated the acoustic emission activity in RC slabs at different thicknesses. They found that the acoustic emission activities for a thicker slab are higher than for a thinner slab. A study carried out by Larosche et al. [9] on an actual sized pile connection to bent cap subjected to cyclic displacement found that the acoustic emission can assess the integrity of the specimen. At the same time, they inferred that the acoustic emission intensity analysis is apt for structural damage classification, especially damage that occurs between the sensors. The review indicates that an intensity analysis of the acoustic emission signal is significantly useful in classifying the damage zone. It provides a measurable assessment of the damage zones based on the historical index and severity [10–13]. The historical index is a measurement related to the alteration of the acoustic emission signal strength as the load phase is applied.

Meanwhile, the severity is determined as the average signal strength among the most significant numerical values of the signal [13]. In the present study, a reinforced

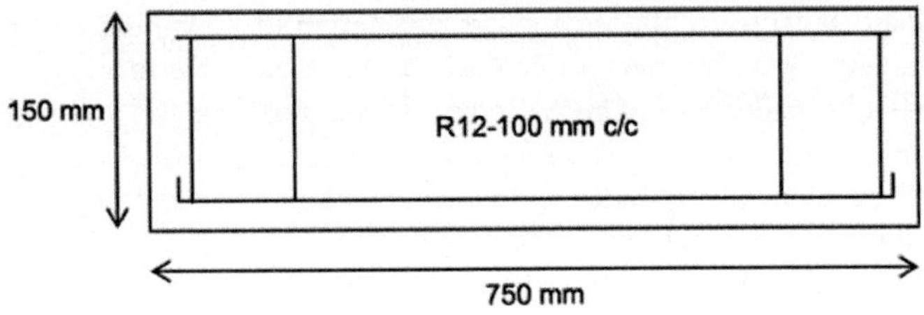

Fig. 1 Detail of the reinforcement (side view)

Fig. 2 Cross-sectional details of the reinforced concrete beam

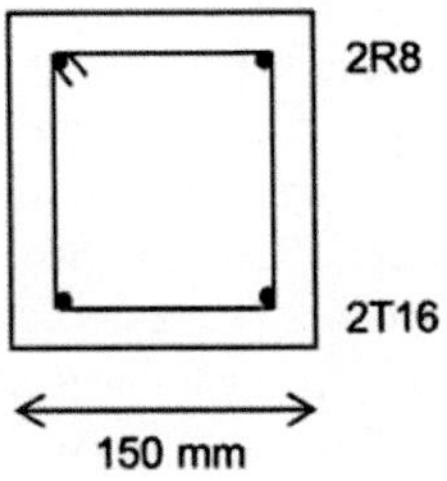

concrete beam subjected to cyclic loading concurrent with acoustic emission monitoring for the prognosis of active crack zones was investigated. The intensity analysis of the acoustic emission signal was used to categorise the cracks into zones.

2 Methodology

2.1 Specimen Preparation

A total of twelve singly reinforced concrete beams were constructed to evaluate the crack propagation when the beam is subjected to increasing fatigue loading. Each beam was 150 mm wide × 150 mm high × 750 mm long. Two high yield 16 mm steel bars were used as tensile reinforcement while two mild yield 8 mm steel bars were used on the compression part as hanger bars. For the links, mild yield steel bars of 12 mm diameter with 100 mm spacing as shown in Fig. 1 was used. Figure 2 shows the detail of the reinforcements used in this study.

2.2 Cyclic Loading and Acoustic Emission Monitoring

Six phases of maximum cyclic loading concurrent with the acoustic emission monitoring were adapted for the present experimental work. The first load crack (Pcr) and ultimate load (Pult) were utilised for the application of maximum cyclic loading, which was taken from the monotonic test. For the monotonic loading, five beams were constantly loaded to failure to determine the Pult and the average Pult was found to be 158.85 kN. The Pcr was identified from the monotonic test on the same sized beams, which corresponded to the first crack that showed on the surface of the beam during testing. The Pcr was found to be 24 kN. The six load phases are identified in Table 1. Figure 3 shows the specimen set up for the acoustic emission monitoring. For the cyclic test, constant parameters such as type of load cycle and frequency were used. The load cycle type of sine wave and 1 Hz frequency were applied throughout the cyclic test. For each load phase, 5000 cycles were applied to the beam.

For the acoustic emission monitoring, the VS75-V acoustic emission sensor was used. A total of four acoustic emission sensors were fixed on the beam at specific locations. Two sensors were fixed at 200 mm from the centre of the beam on both sides. The other two sensors were coupled at the centre of the beam on its cross-section. The connectivity of all the sensors was then verified using pencil lead break (PLB) to gain a good sensitivity between the beam and sensor surfaces. Prior to performing the acoustic emission monitoring, some parameters were applied in the acoustic emission hardware as presented in Table 2. In this study, AMSYS-6 software was used. A 45-dB constant threshold level was used throughout the test, and wave

Table 1 The six phases of maximum cyclic loading

Maximum cyclic load based on P_{cr}			Maximum cyclic load based on P_{ult}		
Load phases	Load ratio	P_{max} (kN)	Load phases	Load ratio	P_{max} (kN)
Phase 1 (P1)	0.5	12	Phase 4 (P4)	0.2	32
Phase 2 (P2)	0.8	19	Phase 5 (P5)	0.5	63
Phase 3 (P3)	1.0	24	Phase 6 (P6)	0.6	79

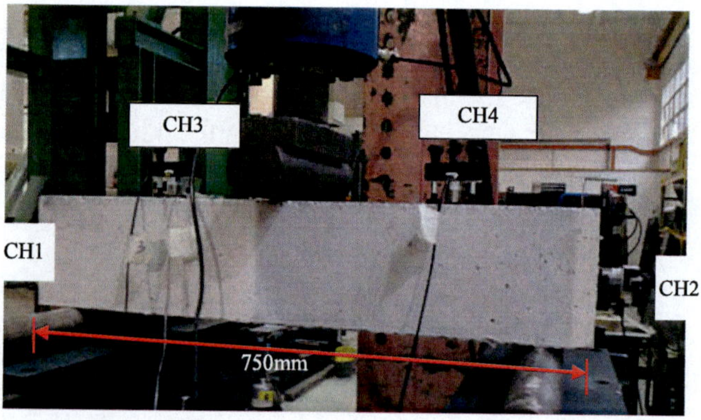

Fig. 3 Setup of the three-point bending test and the acoustic emission monitoring applied to the beam

Table 2 Acoustic emission parameters set on the hardware

Acoustic emission parameters	Value
Sampling rate	10 MHz
Rearm time	1.62 ms
Duration discrimination time	400 μs
Pre-trigger samples	200
Digital setting	25 to 850 kHz

velocity of 4000 m/s was applied. The results from the acoustic emission monitoring during cyclic loading for each phase were stored and visualised on an acoustic emission visual. The cracks that occurred in each load phase were marked on the beam's surface during the loading pause. The cracks were marked as 1 to 6 to indicate that they occurred during load Phase 1–Phase 6.

2.3 Intensity Analysis

Intensity analysis is a quantitative measurement for the deterioration classification of a structure, which corresponds to the time changes during testing. In this analysis, two indices were used namely severity, *Sr* and historical index, *HI*. The indices were extracted from the acoustic emission signal strength during testing. Historical index and severity can be calculated using the following Eqs. 1 and 2, respectively [11]:

$$HI = \frac{N}{N-K}\left(\frac{\sum_{i=K+1}^{N} S_{oi}}{\sum_{i=1}^{N} S_{oi}}\right) \tag{1}$$

$$S_r = \frac{1}{J}\left(\sum_{m=1}^{J} S_{om}\right) \tag{2}$$

where; N is defined as the amount of acoustic emission hits up to time t; S_{oi} is acoustic emission signal strength of the ith event; K and J are factual extracted constants based on material properties; and S_{om} is acoustic emission signal strength of the mth hit, where the order of m is based on signal strength amplitude. The K constant generally used for concrete is based on the N value. If N is less than or equal to 50, K becomes 0. If N is between 51 and 200, K becomes N – 30. If N is between 201 and 500, K becomes 0.85 N. If N is greater than or equal to 501, K becomes N–75. Meanwhile, for the J constant, if N is less than 50, J is 0. If N is greater than or equal to 50, J is 50 [10].

From the plots of historical index and severity in the intensity chart, as shown in Fig. 4, Gostautas et al. [11] recommended five intensity zones. Zone A is no significant emission, which is no follow up required. Zone B is minor damage, which is noted for reference in future tests. Zone B damage is normally due to

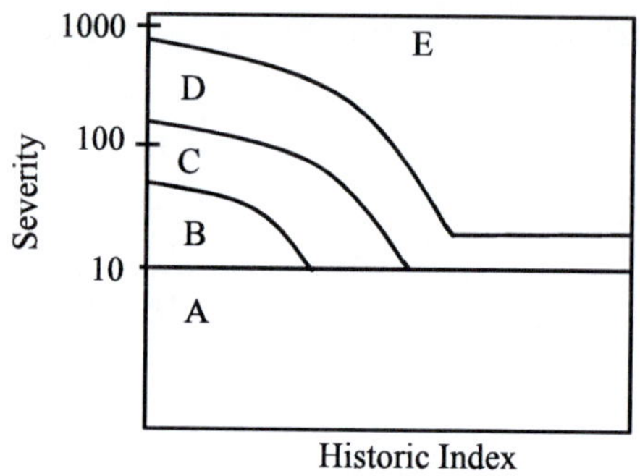

Fig. 4 Intensity chart for classification of crack zones [11]

minor surface damage such as corrosion and cracked attachment welds. Zone C is intermediate damage where a follow-up evaluation is required. Zone D is follow-up, which is significant damage that requires a further follow-up inspection. Finally, Zone E is major damage, which requires an immediate shutdown and frequent follow-up inspections of the structure.

In this study, the crack intensity identification of the beam was based on the acoustic emission signals attained in two sections and one channel. Two sections known as SEC A and SEC B were selected with distances of 100–300 mm and 310–440 mm from the beam edge, respectively. The selection of these sections was based on the critical part of the beam, which received the highest stress concentration when subjected to monotonic loading. Since Channel 3 (CH 3) was the closest sensor to these selected sections which was exposed to crack formation and received more acoustic emission activities compared to the other sensors, the acoustic emission signal strength collected from CH 3 was utilised to evaluate the crack.

3 Results and Discussion

The acoustic emission monitoring was carried out for six phases of different ranges of maximum cyclic loading. Figure 5 shows the crack formation that appeared on one part of the beam's surface. Figures 6, 7 and 8 show the summaries of intensity plots which were attained at SEC A, SEC B, and CH 3 in the beam subjected to cyclic loading in all phases. Intensity analysis provides a quantitative assessment of the crack zones using plots of the historical index, HI, and severity index, Sr [11]. The intensity chart is utilised for classification of crack levels in the beam specimens when subjected to six phases of maximum cyclic loading. Five intensity zones i.e. Zone A, Zone B, Zone C, Zone D, and Zone E as stated in the preceding section were adopted to classify the damage.

When the maximum cyclic load of 0.5 P_{cr} (12 kN) was applied to the beam, a vertical crack appeared from the bottom part to beyond the neutral axis of the beam, as shown in Fig. 5a. This crack was identified as a flexural crack. This first

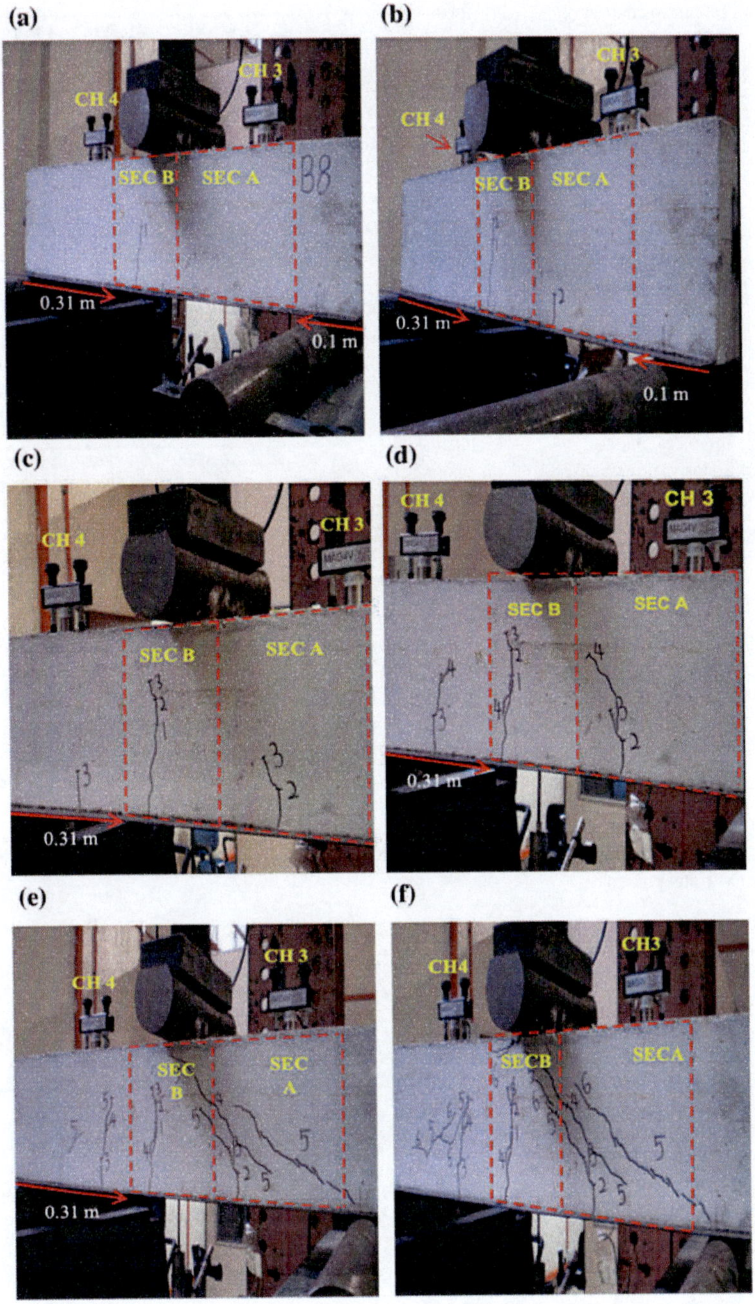

Fig. 5 Crack modes for beam subjected to **a** 0.5 P_{cr} **b** 0.8 P_{cr} **c** 1.0 P_{cr} **d** 0.2 P_{ult} **e** 0.5 P_{ult} and **f** 0.6 P_{ult}

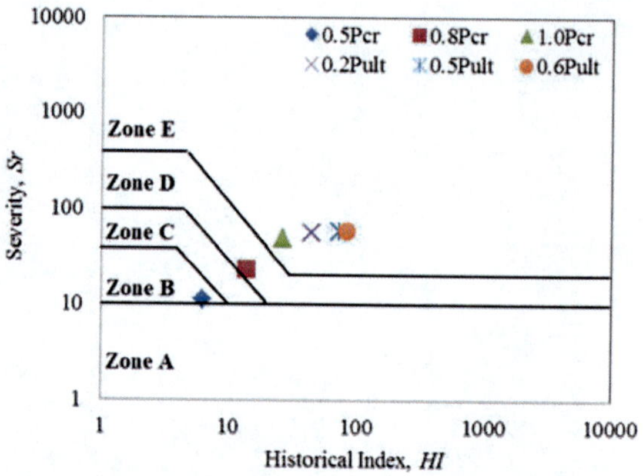

Fig. 6 Intensity chart at SEC A under various phases of cyclic load

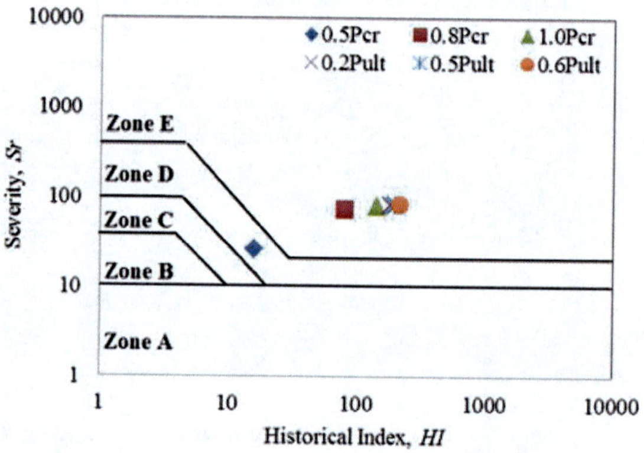

Fig. 7 Intensity chart at SEC B under various phases of cyclic load

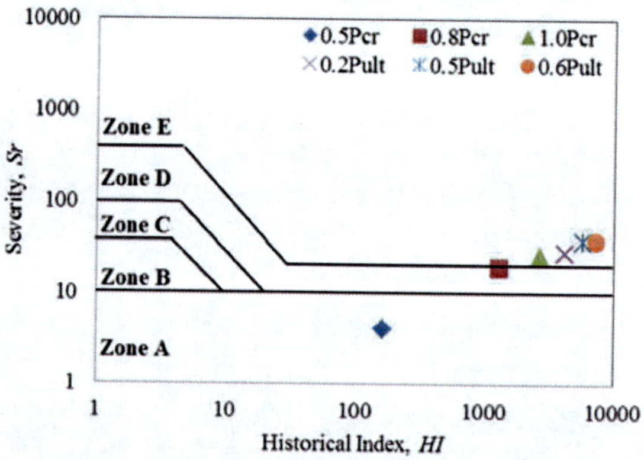

Fig. 8 Intensity chart at CH 3 under various phases of cyclic load

crack formation was marked as 1 to denote the cyclic load applied under Phase 1 (0.5 P_{cr}). In the 0.5 P_{cr} load phase, it could be clearly seen that no crack formed at SEC A, while this first crack occurred in the SEC B area. From the intensity chart as presented in Figs. 6 and 7, the plots indicate that SEC A and SEC B fall under Zones B and D, respectively. It indicates that as there was no crack at SEC A, the beam was experiencing a minor crack. From this plot, the crack development in the beam can be detected earlier prior to being visually observed. However, the crack that appeared on the beam's surface at SEC B and the Zone D intensity plot were indications that the beam was experiencing significant damage. This means that the beam required further follow-up inspections periodically before the crack became worse. Hence, the plot on SEC B gave useful information on the condition of the beam. It contradicts with the plot extracted from CH 3, where the beam falls under Zone A, as shown in Fig. 8. This plot shows that no crack occurred in the beam, and the actual condition of the beam was not well mapped. Consequently, it is noted that the acoustic emission intensity plot at CH 3 underestimated the actual condition of the beam, which showed the appearance of a flexural crack at the centre of the beam.

As the maximum cyclic load increased to 0.8 Pcr (19 kN), the first crack at SEC A was detected, as shown in Fig. 5b. Due to the crack, the acoustic emission intensity plot indicates that SEC A fell in the middle of intensity Zones C and D as shown in Fig. 6. Zone C shows that an intermediate crack has occurred in the beam and requires a follow-up evaluation. It signifies that the prognosis of the intensity zone extracted from SEC A is able to predict the crack occurrence as the maximum cyclic loading of Phase 1 is imposed on the beam. Thus, the plot in the intensity chart when the beam is subjected to the maximum cyclic load of Phase 2 can be utilised to diagnose the future beam condition as the Phase 3 load is imposed on the beam. In order to obtain a better prediction, Zone D is preferred to show the condition of the beam so that earlier precautions can be taken.

As the crack continuously propagated from the tip of the previous crack on both sides of SEC B when Phase 1 loading was applied to the beam, the plot in the intensity chart indicates that the beam falls in Zone E (Fig. 7). This implies that severe damage has occurred in the beam as it underwent a tensile crack [14]. If the acoustic emission intensity collected from SEC B is utilised for a prognosis of the beam condition, it closely reflects that the beam requires a rapid shut down and a generic inspection must be performed.

However, the intensity plot extracted from CH 3 as shown in Fig. 8 demonstrates that the crack intensity falls between Zones D and E. If Zone E is selected for predicting the beam condition, an early precaution can be taken. Hence, as the maximum cyclic load of 0.8 P_{cr} was applied to the beam, CH 3 represented a more acceptable plot for predicting the true beam condition. It is more reasonable since the crack at the mid-span of the beam propagated beyond the neutral axis. At the same time, a new crack at SEC A was detected.

As the maximum cyclic loading increased to 1.0 Pcr to 0.6 Pult, more cracks were generated in the areas of SEC A and SEC B as shown in Fig. 5c and d. From the illustration in the intensity chart, SEC A and SEC B both fell under intensity Zone E. This was caused by the acoustic emission activity such as signal strength that increased as the load increased, which was reflected in the determination of the plot in the intensity chart. At the same time, it resulted from the progression of the crack in the beam. Hence, the plot in the intensity chart progressed to the upper right area [11].

As the maximum cyclic load increased to 1.0 Pcr (24 kN), more crack formations were induced in the beam which boosted the high acoustic emission activities. The high acoustic emission activities are related to the presence of a high stress concentration due to the increasing load in the beam. This condition is reflected in the plots of intensity zones for SEC A, SEC B, and CH 3, where the plots are in intensity Zone E. The intensity zones presented for SEC A, SEC B, and CH 3 are well matched with the crack pattern that appeared on the beam surface as shown in Fig. 5c. Hence, it is reasonable to state that the beam was under severe damage.

Similar intensity zones were found at SEC A, SEC B, and CH 3 when the beam was subjected to maximum cyclic loading of 0.2–0.6 Pult i.e. they fell under Zone E. This matched well with the crack pattern as depicted in Fig. 5d–f, where excessive cracks were visually observed on the beam surface through the presented photograph. Hence, it is acceptable to state that the intensity zones presented from data derivation of located events and channel basis are able to predict the condition of the beam. At the same time, the intensity zones at SEC A and SEC B can indicate more precisely the condition of the beam at earlier stages. Moreover, the acoustic emission intensity plots in the charts of previous phases are well matched with the occurrence of cracks in the latter cyclic load phases.

It can be surmised that as the maximum cyclic load applied to the beam increased to 0.2 Pult (32 kN), excessive cracks were formed. Both tensile and shear cracks were noticed on the beam surface. If these cracks occurred, the beam durability would re-duce, affecting the integrity and condition of the beam, and would reflect major dam-age. Intensity Zone E was recorded based on the plots at SEC A, SEC B, and CH 3. The intensity zones are well-matched with the crack pattern as presented in Fig. 5d.

More shear cracks were noticed as the maximum cyclic load increased to 0.5 P_{ult} and 0.6 P_{ult} as presented in Fig. 5e and f, respectively. The shear cracks were inclined at about 45° from the plane indicating that the beam was continuously experiencing severe damage. The intensity plots at SEC A, SEC B, and CH 3 for both load phases were well matched with the crack pattern on the beam with the intensity falling under Zone E. It indicates that the beam has experienced major damage. Hence, the beam structure is no longer acceptable for use, and an immediate shut down should be taken into consideration. A strict follow-up inspection of the beam condition is recommended.

Hence, it can be inferred that the plots of intensity zones derived from SEC A and SEC B, which is based on located events in the initial phase can be utilised as a useful tool for predicting the beam condition in the future. The crack pattern for each load phase was well matched with the intensity plots. However, at the earlier load phases, the intensity plots on the CH 3 were not well-matched with the crack pattern that was visually observed on the beam surface. As the maximum cyclic load increased, the intensity plots at CH 3 were well matched with the crack patterns

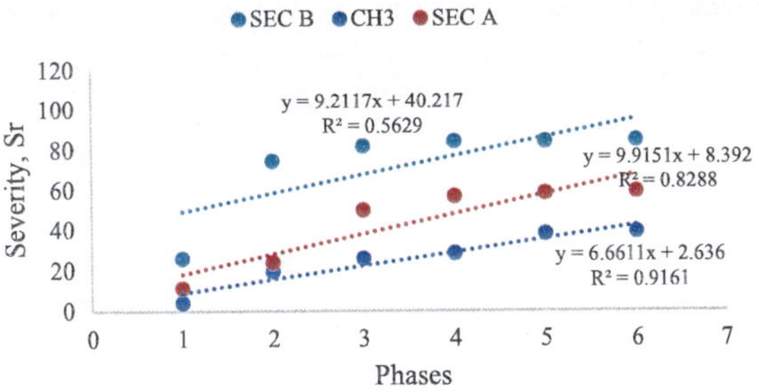

Fig. 9 Relationship between severity and load phases

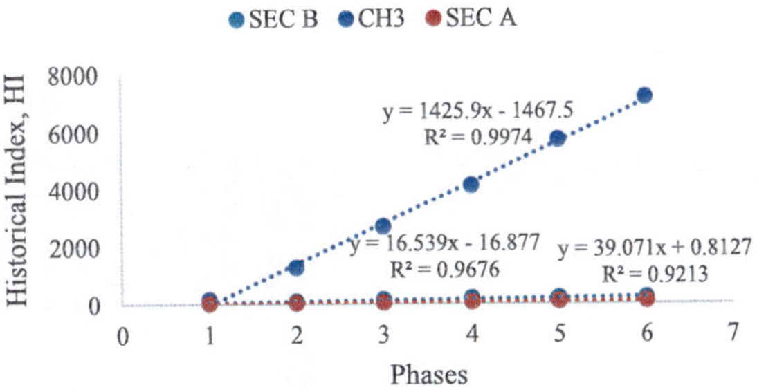

Fig. 10 Relationship between the historical index and load phases

and gave valuable information in the prediction of the beam condition. Although the plot in the intensity zone for CH 3 is less effective for predicting future cracks in the beam, it presents a good relationship between severity and historical index and load phases, as shown in Figs. 9 and 10, respectively. The squared R for CH3 represents the highest relationship with the value of 0.9161 and 0.9974 for severity and historical index compared to SEC A and SEC B.

However, the relationship between severity and load phases for SEC B, where the crack progressively occurred is low with the squared R of 0.5629. It indicates that the location which was exposed to excessive stress concentration as the load was applied caused an unpredictable crack severity.

Table 3 shows the summary of intensity zones corresponding to acoustic emission signals obtained from SEC A, SEC B, and CH 3. Since SEC B was the critical location for crack concentration, it produced an early warning for follow-up inspections even when a minimum load in Phase 1 (0.5 P_{cr}) was applied. This was followed by a major defect when the load increased in Phase 2 (0.8 P_{cr}). In SEC A, in the first phase of loading, a minor defect was detected although the crack was invisible. Then, an intermediate-follow-up defect was detected when the first crack appeared on the beam surface. It was then followed by a major defect when the load range increased. The intensity zone derived from CH 3 shows the beam was in intact condition as the Phase 1 load was applied. It seems that the acoustic emission signal produced by the crack in SEC B was delayed in its arrival at the sensor at CH 3 since the closest sensor was at CH 4. Hence, at this phase, a lower acoustic emission activity was produced, leading to insignificant emission. It is deduced that the intensity plots based on

Table 3 Summary of intensity zones in SEC A, SEC B and CH 3

Phase	Load (kN)	Intensity zone		
		Located event		Channel basis
		SEC A	SEC B	CH 3
Phase 1	12	Minor	Follow-up	Insignificant
Phase 2	19	Follow-up	Major	Major
Phase 3	24	Major	Major	Major
Phase 4	32	Major	Major	Major
Phase 5	63	Major	Major	Major
Phase 6	79	Major	Major	Major

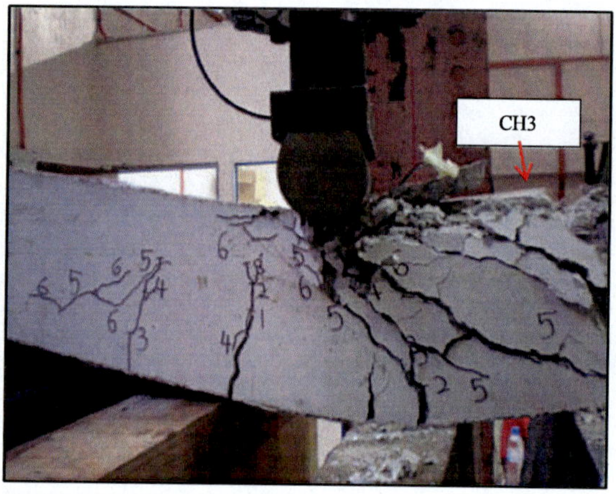

Fig. 11 Failure condition of the beam after application of different phases of loading

located events are more suitable to present the beam condition and correspond better to visual crack patterns than those captured on a channel basis. Figure 11 shows the beam was monotonically loaded to failure and the crushing of the beam is close to the sensor at CH 3.

4 Conclusion

In a nutshell, the classification of active cracks in a reinforced concrete beam subjected to various cyclic loadings can be performed using the intensity analysis. It is found that the plots on SEC A and SEC B are well-matched with the crack occurrence in the beam when subjected to various cyclic loads ranging from 0.5 P_{cr} up to 0.6 P_{ult}. The intensity plots at CH 3 do not reflect the occurrence of cracks in the beam in earlier phases. However, as the cyclic load increases, the intensity plots on CH 3 also represents an informative classification for prognosticating the condition of the reinforced concrete beam. It is inferred that the prognosis of intensity zones based on the data obtained from located events are more presentable and better correspond to the crack modes appearing on the beam surface than those obtained at CH 3.

Acknowledgements The authors would like to express their thanks to Universiti Teknologi MARA, Cawangan Pulau Pinang and Universiti Teknologi MARA, Malaysia for providing financial support towards the publication of this manuscript.

References

1. ASTM (2005) Standard Guide for Determining the reproducibility of acoustic emission sensor response. Non Destructive Testing Standard
2. Kaphle M (2011) Review: acoustic emission technique—opportunities, challenges and current work at QUT. In: eddBE2011 Proceedings, pp 312–317
3. Md Nor N (2016) An overview on fatigue damage assessment of reinforced concrete structures with the aid of acoustic emission technique. Construct Build Mater
4. Md Nor N, Ibrahim A, Muhamad Bunnori N, Mohd Saman H (2013) Acoustic emission signal for fatigue crack classification on reinforced concrete beam. Construct Build Mater 49:583–590
5. Toubal L, Chaabouni H, Bocher P, Jianqiang C (2020) Monitoring fracture of high-strength steel under tensile andconstant loading using acoustic emission analysis. Eng Failure Anal 108
6. Abouhussien AA, Hassan AAA (2018) Acoustic emission monitoring of corrosion damage propagation in large-scale reinforced concrete beams. J Perform Constructed Facil 32(2)
7. Di Benedetti M, Nanni A (2014) Acoustic emission intensity analysis for in situ evaluation of reinforced concrete slabs. J Mater Civil Eng 26(1):6–13
8. Muhammad Sauki MIH, Md Nor N, Juhari N, Endur MZ (2019) Acoustic emission characteristic for different thicknesses of RC slabs. J Phys Conf Ser 1349:1–8
9. Larosche A, Ziehl P, Mangual J, ElBatanouny MK (2015) Damage evaluation of prestressed piles to cast in place bent cap connections with acoustic emission. Eng Struct 84:184–194
10. Noorsuhada MN (2018) Structural health monitoring through acoustic emission, in Eco-Efficient repair and rehabilitation of concrete infrastructures. In: Pacheco-Torgal F, Melchers RE, Shi X, De Belie N, Tittelboom KV, Sáez A (eds) Woodhead Publishing Series in Civil and Structural Engineering, pp 123–146
11. Gostautas RS, Ramirez G, Peterman RJ, Meggers D (2005) Acoustic emission monitoring and analysis of glass fiber-reinforced composites bridge decks. J Bridge Eng 10:713–721
12. Mat Saliah SN, Md Nor N, Kamaruzzaman KN, Mohd Mesron MA, Abd Rahman N (2019) Intensity crack zones of reinforced concrete beam under monotonic loading. J Phys Conf Ser 1349
13. Nair A, Cai CS (2010) Acoustic emission monitoring of bridges: review and case studies. Eng Struct 32:1704–1714
14. Mc Cormac JC, Nelson JK (2005) Design of reinforced concrete ACI 318-05 Code Edition. Wiley and Sons, Inc

An Experimental Investigation into the Fatigue Behavior of Spot Welded Tensile Shear (TS) Specimens

Ahmet H. Ertas and Mustafa Akbulut

Abstract The resistance spot welds (RSWs) are extensively used in many industries especially automotive industry as a joining process of sheet metals due to its flexibility and adaptability to automation. In addition to easy automation capability, there are many more advantages that this joining process offers in the attachments of thin sheet body components, such as effectiveness, low cost and reliability. Despite all the favorable properties of this manufacturing process, the spot welded sheet metals are prone to mechanical fatigue failure especially under cyclic sinusoidal loadings. Therefore, understanding and elucidating the fatigue phenomenon of the RSWs are crucial during the design phase, in terms of estimating and preventing undesired failure conditions. However, in the design procedure, there exist considerable amount of challenges to be overcome to analysis and accordingly predict the fatigue failure in the sheet metals connected by spot welds, among them it can be counted electrode wear-down, non-uniform sheet metal forms, defected or incomplete weld nuggets. Many of these faults cause non-uniform stress distributions inside the sheet metal and make it difficult to estimate fatigue failure. Thereof experimental work, in most cases, shines as a viable option in the fatigue analysis. Within the scope of this work, an attempt is made to examine the fatigue phenomenon of the spot welded sheet metals experimentally. For this purpose, tensile shear (TS) type spot welded test specimens were utilized. A series of fatigue life tests were conducted on the previously prepared samples and the deformed test specimens were examined. The experimental results were compared with those of similar research studies available in the literature. It was observed that the outcomes of the present work are reasonably consistent with the literature results.

Keywords Fatigue · Crack Initiation · Crack Propagation · Fatigue Fracture · Tensile Shear (TS) Specimen · Spot Weld

---

A. H. Ertas (✉)
Department of Mechanical Engineering, Faculty of Engineering and Natural Sciences, Bursa Technical University, Bursa 16330, Turkey
e-mail: ahmeth.ertas@btu.edu.tr

M. Akbulut
RUTE, TUBITAK, Kocaeli 41470, Turkey

M. Abdel Wahab (ed.), *Proceedings of the 8th International Conference on Fracture, Fatigue and Wear*, Lecture Notes in Mechanical Engineering,
https://doi.org/10.1007/978-981-15-9893-7_30

1 Introduction

Resistance Spot Weld (RSW) has been in use to assemble sheet metals [1–4]. There are sometimes thousands of RSWs in some structures which make the manufacturing spot welded structures important [5–20]. Among the parameters influencing the safety, fatigue is considered to be one of the most widespread failure mechanism for the spot welded joints undergoing cyclic loadings. In this respect, understanding fatigue response of RSWs are important.

2 Materials and Methods

It is well known that the failure mechanism of RSWs and accordingly spot welded structures, as being in other structures, is mostly related to the stress values. In addition, spot weld nugget diameter, metal sheet thicknesses and metal type are also crucial and play important roles in deformation of spot welds under fatigue loading [5, 11, 12, 21].

In this study, since it is the most commonly used type in industries, e.g. automobile, aerospace, ship etc., and also its critical loading mode is fatigue; the two spot-welded TS type specimens shown in Fig. 1 consisting of sheet metals connected by spot welds subjected to fatigue loading have been investigated experimentally. Accordingly, deformation mechanisms for TS type spot-welded structures have been examined closely. Investigating deformation mechanisms in this type of specimens makes it possible for engineers and scientists to take necessary steps for the prevention of catastrophic damages. Additionally, in order to make reliable predictions on structures numerically, it is necessary to make experimental studies and investigate deformation mechanisms.

The mechanical properties of DIN 1623 steel [5] used are listed in Table 1.

A typical TS type spot welded specimen examined is shown in Figure 1. The thickness of the metal sheets and spot weld diameters were chosen as 1 mm and 4 mm, respectively obeying the equation shown (Eq. 1) to conform the standards of AWS and SAE. In Eq. 1 d and t represent diameter of the spot weld and thickness of

Fig. 1 A typical TS type spot welded specimen

Table 1 Mechanical properties of DIN 1623 steel [5]

Average tensile strength (MPa)	319.64
Average yield strength (MPa)	217.41
Average reduction of area, %	35.59
Average Brinell hardness, kgf/mm^2	105.5

Table 2 Welding parameters [5]

Electrode force (kp)	150
Welding current (Amps)	10
Weld time (cycles @ 60–80 Hz)	8.5
Electrode diameter (mm)	4.45

the metal sheet, respectively. On the other hand, the welding parameters used when preparing spot welded Tensile Shear (TS) specimens are given in Table 2 [5].

$$d \approx 4\sqrt{t} \quad (1)$$

For the fatigue testing, MTS (Material Testing System) having 100 kN capacity was used. MTS testing machine has closed loop servo-hydraulic testing system, and also contains a digital controller. Experiments were done in air at the room temperature. The operating frequency was chosen between 1 and 5 Hz.

Some of the tested TS specimens were examined under optical microscope. Figure 2 shows both optical micrographs of the cross section of a partially failed

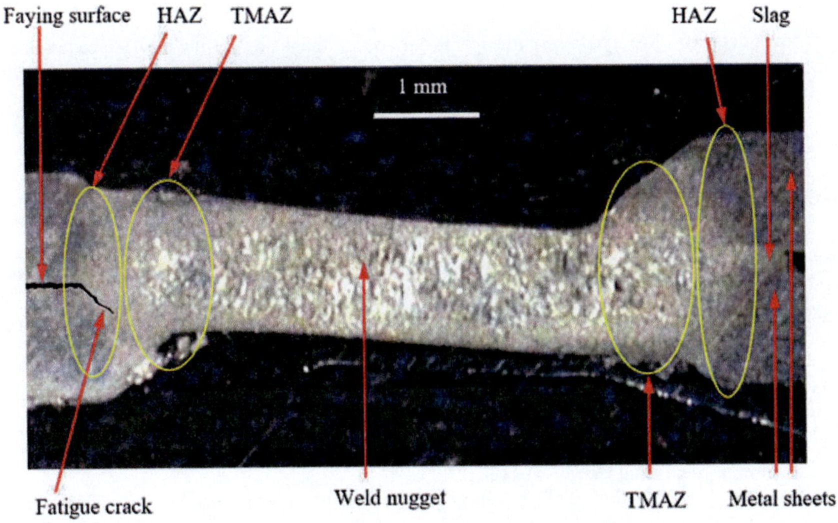

Fig. 2 A partially failed spot welded Tensile Shear (TS) specimen

spot welded TS specimen and also different regions on the investigated cross section which are Thermal-Mechanical Affected Zones (TMAZs) and Heat Affected Zones (HAZs).

3 Results

In this study many untested and tested specimens have been investigated. Comparing the results with those of similar research studies available in the literature [5, 11, 12] it was observed that the outcomes of the present work are reasonably consistent with the literature results in terms of crack patterns of TS specimens under fatigue loading. The cracks which propagate through the sheet thickness and lead to fracture were identified in Figs. 2 [5, 11, 12, 22–26]. This identification is known as typical failure mode for TS specimens.

4 Conclusions

In this work, the fatigue failure of TS samples was investigated experimentally. Crack development and accordingly deformation pattern for TS type spot welded specimens were investigated. Examination of deformed specimens showed that cracks leading to fatigue fracture occurred most generally on the peripheries of the nuggets on the inner surfaces of the plates, which conforms to the fatigue crack initiation sites observed in the experiments conducted previously in references [5, 11, 12, 23–28].

Acknowledgement We would like to express our special thanks of gratitude to Mercedes-Benz Turk A.S. for their help and support during preparation and testing of the specimens.

References

1. Ertas AH, Sonmez FO (2008) A parametric study on fatigue strength of spot-weld joints. Fatigue Fract Eng Mater Struct 31(9):766–776
2. Ertas AH, Yilmazi Y, Baykara C (2008) An investigation of the effect of the gap values between the overlap portions of the spot-welded pieces on fatigue life. Proc Inst Mech Eng Part C-J Mech Eng Sci 222(6):881–890
3. Ertas AH (2015) Design optimization of spot welded structures to attain maximum strength. Steel Compos Struct 19(4):995–1009
4. Ertas AH, Sonmez FO (2011) Design optimization of spot-welded plates for maximum fatigue life. Finite Elem Anal Des 47(4):413–423
5. Ertas AH, Vardar O, Sonmez FO, Solim Z (2009) Measurement and assessment of fatigue life of spot-weld joints. J Eng Mater Technol Trans ASME 131(1), Article Number:011011
6. Ertas AH, Sonmez FO (2009) Optimization of spot-weld joints. Proc Inst Mech Eng Part C-J Mech Eng Sci 223(3):545–555

7. Ertas AH, Sonmez FO (2007) A parametric study on fatigue life behavior of spot welded joints. In: 8th International fracture conference, Istanbul-Turkey, pp 498-508
8. Ertas AH, Vardar O (2005) Fatigue behavior of spot welds in modified tensile shear specimens using experimental data. In: 9th International research/expert conference "trends in the development of machinery and associated technology" TMT 2005, Antalya/Turkey, pp 243-246
9. Ertas AH (2006) Fatigue behavior of spot welds in modified tensile shear specimens using numerical data. In: 10th International research/expert conference "trends in the development of machinery and associated technology" TMT 2006, Barcelona-Lloret de Mar/Spain, pp 777–778
10. Ertas AH, Sonmez FO (2006) Optimization of spot weld joints. In: The 12th international conference on machine design and production, Kusadasi/Turkey, pp 883-893
11. Pan N (2000) Fatigue life study of spot welds. PhD Thesis, Stanford University
12. Ertas AH (2004) Fatigue behavior of spot welds. MSc Thesis, Bogazici University
13. Barkey ME, Kang H, Lee YL (2001) Failure modes of single resistance spot welded joints subjected to combined fatigue loading. Int J Mater Product Technol 16(6/7):510–527
14. Rathburn RW, Matlock DK, Speer JG (2003) Fatigue behavior of spot-welded high-strength sheet steels. Weld J 82(8):207–218
15. Chao YJ (2003) Failure mode of spot welds: interfacial versus pullout. Sci Technol Weld Join 8(2):133–137
16. Deng X, Chen W, Shi G (2000) Three-dimensional finite element analysis of the mechanical behavior of spot welds. Finite Elem Anal Des 35(1):17–39
17. Hou Z, Wang Y, Li C, Chen C (2006) An analysis of resistance spot welding. Weld J 85(3):36–40
18. Barkey ME, Kang H, Lee YL (2000) Evaluation of multiaxial spot weld fatigue parameters for proportional loading. Int J Fatigue 22(8):691–702
19. Di Fant-Jaeckels H, Galtier A (2000) Fatigue lifetime prediction model for spot welded structures. Revue de Metallurgie-Cahiers Inf Tech 97(1):83–95
20. Salvini P, Scardecchia E, Demofonti G (1997) A procedure for fatigue life prediction of spot welded joints. Fatigue Fract Eng Mater Struct 20(8):1117–1128
21. Davidson JA, Imhof EJ (1983) A fracture mechanics and system-stiffness approach to fatigue performance of spot-welded sheet steels. In: Proceedings of the SAE international congress and exposition, USA, pp 19–38
22. Lin PC, Su ZM, He RY, Lin ZL (2012) Failure modes and fatigue life estimations of spot friction welds in cross-section specimens of aluminum 6061-T6 sheets. Int J Fatigue 38:25–35
23. Fujii T, Tohgo K, Suzuki Y, Yamamoto T, Shimamura Y (2016) Fatigue strength and fatigue fracture mechanism of three-sheet spot weld-bonded joints under tensile-shear loading. Int J Fatigue 87:424–434
24. Xiao L, Liu L, Chen DL, Esmaeili S, Zhou Y (2011) Resistance spot weld fatigue behavior and dislocation substructures in two different heats of AZ31 magnesium alloy. Mater Sci Eng A-Struct Mater Prop Microstruct Process 529:81–87
25. Radakovic DJ, Tumuluru M (2008) Predicting resistance spot weld failure modes in shear tension tests of advanced high-strength automotive steels. Weld J 87(4):96S–105S
26. Nakayama E, Fukumoto M, Miyahara M, Okamura K, Fujimoto H, Fukui K, Kitamura T (2010) Evaluation of local fatigue strength in spot weld by small specimen. Fatigue Fract Eng Mater Struct 33(5):267–275
27. Ong JH (2003) An improved technique for the prediction of axial fatigue life from tensile data. Int J Fatigue 15(3):213–219
28. Lee YL, Pan J, Hathaway RB, Barkey ME (2005) Fatigue Test Anal, 1st edn. Elsevier, USA

A Fatigue–Reliability Analysis of Spot Welded Modified Tensile Shear (MTS) Specimens

Ahmet H. Ertas and Mustafa Akbulut

Abstract In sheet metal structures, generally the resistance spot welds (RSWs) are used as a joining method especially for its adaptability to automation, effectiveness, robustness and low cost. On the other hand, the spot-welded sheet metals operate, most of the time, under cyclic loading conditions, which lead to fatigue fracture in the structure. In order to predict and prevent fatigue damage in the spot-welded sheets properly, many factors which affect the quality of welds in the design phase are to be considered such as arbitrary shapes of the sheet metal, missing or incomplete spot welds, and abrasion of the welding electrodes etc. All these disadvantages cause deviations in the stress analysis and make it very hard to obtain uniform stress distributions. Hence deterministic computational approaches, even pure experimental investigations of spot welded structures, do not always result in reliable outputs. In that case, probabilistic approach appears to be a feasible alternative in the fatigue assessment of RSWs. In this study, a probabilistic fatigue analysis is carried out on spot welded modified tensile-shear (MTS) test specimens. In the probabilistic analysis a g-function was introduced for each case study. The results obtained through aforementioned method were shown to be in good agreement with the related studies in literature.

Keywords Fatigue · Reliability analysis · Modified tensile shear (MTS) specimen · Spot weld

1 Introduction

Reliability and risk data analysis are especially important when the risks are high. Accordingly, modern approach aims to give protection when the risks are felt

A. H. Ertas (✉)
Department of Mechanical Engineering, Faculty of Engineering and Natural Sciences, Bursa Technical University, Bursa 16330, Turkey
e-mail: ahmeth.ertas@btu.edu.tr

M. Akbulut
RUTE, TUBITAK, Gebze 41470, Kocaeli, Turkey

M. Abdel Wahab (ed.), *Proceedings of the 8th International Conference on Fracture, Fatigue and Wear*, Lecture Notes in Mechanical Engineering,
https://doi.org/10.1007/978-981-15-9893-7_31

to be high via probability based analysis [1, 2]. Probabilistic design techniques are being used in a wide spectrum of application areas in industry including aerospace, automotive, mechanical, civil, chemical, electrical, and manufacturing industries.[1–3].

In general, the reliability of a product is directly related to many factors like the type of manufacturing process, environment, external load, cost etc. which must be considered in design [4, 5]. Today during especially vehicle development process, engineering based effort is spent [6, 7]. Probabilistic design, on the other hand, provides an alternative approach to these studies.

Naturally there are many uncertainties which have to be considered in design. Hence, it is necessary to use safety factors in a deterministic design. However, in probabilistic design, statistical methods are used [7–9].

Being a design parameter, the fatigue phenomenon is a slow process, so it introduces numerous uncertainties [10–37]. Therefore, as it is explained, in addition to the numerical and experimental analysis, a probabilistic analysis is also necessary to treat the problem completely.

For spot welded structures fatigue analysis is done using different fatigue life prediction models—the nominal stress, the notch stress, the notch strain, the fracture mechanics, and volumetric fatigue approaches etc. Since most of the existing numerical analyses are still limited [38], reliability analysis is also necessary for reliable failure predictions. Hence in this study, in addition to experimental study, a reliability analysis was also done.

2 Materials and Methods

2.1 *Comparison of Probabilistic and Deterministic Design Methodologies*

Deterministic approaches to the engineering problems may result in inconsistent products especially in the future because of the limitations in design, service, material, manufacturing etc. [38, 39]. On the other hand, there is no single method available for solving all engineering problems efficiently. As a result, many probabilistic design methods have been created for solving engineering problems [3, 39, 40].

In deterministic design it is necessary to use Factor of Safety (FS) to consider afore mentioned uncertainties. Selection of factor of safety generally depends upon the engineer's knowledge or experience [3, 39, 40].

In the probabilistic design, on the other hand, the performance function which originated from the limit state theory is expressed mathematically as [39]

$$g(x) = R - P \tag{1}$$

where R and P represent resistance and load, respectively; x is the vector of basic random variables, $g(x) < 0$ and $g(x) > 0$ are the failure region and the safe region, respectively. The Probability of Failure (POF) can be evaluated via Eq. (2) [3]

$$\mathrm{POF} = \int_{g(x)\langle 0} \ldots \int f_x(x)dx \tag{2}$$

If the limit state function combines more than two statistically independent variables, x_i the variance of response $g(x_i)$ can be approximated by Taylor expansion with eliminating higher order terms. In this case, the variance for a function, σ_g, can be expressed analytically via Eq. (3), where σ and μ are the variable standard deviation and its mean, respectively [39]

$$\sigma_g^2 \approx \sum \left(\frac{\delta g}{\delta x_i}\right)^2 \sigma_{x_i}^2 \tag{3}$$

The standard normal variant—which is known as First Order Reliability Method (FORM)—and then The Probability of Failure (POF) can be calculated by Eqs. (4), and (5), respectively [39]

$$z = -\frac{\mu_R - \mu_P}{\sigma_g} \tag{4}$$

$$\mathrm{POF} = \phi(z) \tag{5}$$

In the standard normal variant method, the limit state is linearized around the “design point”. Accordingly, The POF is given by Eq. (6) [39]

$$\mathrm{POF} = \varphi(-\beta) \tag{6}$$

2.2 A Probabilistic Study on Fatigue Strength of Spot Weld Joints

Reliability of a machine component is very important. Because depending on the reliability, a component will be unsafe under certain circumstances. Hence reliability linked to probability can be defined as the probability that failure will not occur [7–9].

The probabilistic fatigue analysis used here is based on the work of Pan [4] and Ertas [3]. In this type of analysis, the variability of the material yield strengths, which are DQSK and HSLA steels, is taken into account via using nearly normal distribution. In many cases, because the loading exhibits sinusoidal behavior it is necessary to use plus or minus tolerance, L, and the standard deviation, σ, shown in

Table 1 Fundamental formulas

Function	Mean (μ_Z)	Standard Deviation (σ_Z)
$Z = a$	a	0
$Z = Xa$	$a\,\mu_X$	$a\,\sigma_X$
$Z = X + a$	$a + \mu_X$	σ_X
$Z = X \pm Y$	$\mu_X + \mu_Y$	$\sqrt{\sigma_X^2 + \sigma_Y^2}$
$Z = X$	$\mu_X^2 + \sigma_X^2$	$\sqrt{4\mu_X^2\sigma_X^2 + 2\sigma_Y^4}$
$Z = 1/X$	$1/\mu_X$, nearly	σ_X/μ_X^2, nearly
$Z = X/Y$	μ_X/μ_Y	$\sqrt{4\mu_X^2\sigma_Y^2 + \mu_Y^2\sigma_X^2/\mu_Y^2}$, nearly

Eqs. (7) and (8), respectively [10, 41] which are valid for all normal distribution:

$$L = 3\sigma \tag{7}$$

$$\sigma = L/3 \tag{8}$$

There are many mathematics of random variables for different mathematical expressions which can be found in text books [10, 41]. Table 1, for instance, shows the fundamental formulas. In this table, "*X*" and "*Y*" represent independent random variables, on the other hand, "Z" and "*a*" represent the result and any constant, respectively.

Case 1: Fatigue Life Analysis: In this case, simply an attempt is made to solve the problem using conventional solution technique. For a typical spot welded Modified Tensile Shear (MTS) specimen shown in Fig. 1, the calculation was done.

Conventionally fatigue life of a spot welded plate can be calculated using the following formula which is called as "Morrow's mean stress equation".

$$\frac{\Delta\varepsilon}{2} = \varepsilon_a = \frac{\Delta\varepsilon_e}{2} + \frac{\Delta\varepsilon_p}{2} = \frac{\sigma'_f - \sigma_m}{E}\left(2N_f\right)^b + \varepsilon'_f\left(2N_f\right)^c \tag{9}$$

where $\Delta\varepsilon$, σ_m, N_f, c, E, σ'_f, ε'_f, b, $\Delta\varepsilon_e$, $\Delta\varepsilon_p$ represent local strain at the notch, mean stress, cyclic life of the specimen, material constant determined experimentally, modulus of elasticity, the fatigue strength coefficient, the fatigue ductility coefficient, material constant determined experimentally, alternating true elastic strain and alternating true plastic strain, respectively.

In this case shown in Fig. 1 let's assume that the spot welds have to carry 9700 lb. In addition, the spot weld diameters, actually the heat affected zone (HAZ) diameters, change between 0.926–0.928 in. for the outer diameter and 0.850–0.852 in. for the inner diameter. Here it is known that spot welds consist of two different parts which are spot weld oneself (the center of spot weld) and heat affected zone (HAZ). Because the deformation occurs inside of the HAZ, it is necessary to examine HAZ instead

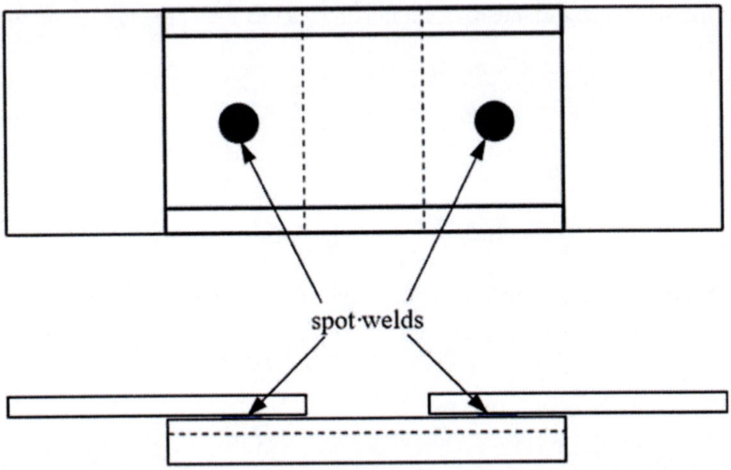

Fig. 1 A typical geometry of the two spot-welded modified tensile shear (MTS) type specimen (top and side views)

of spot weld itself. So we choose the diameters according to HAZ and because of welding mechanism it was chosen the diameters with tolerance values.

Under this loading case, a tensile stress occurs and this stress value can be calculated using Eq. (10):

$$S = \frac{P}{(\pi/4)(D^2 - d^2)} \tag{10}$$

where D, d, S, P represent mean outside diameter (inches, random variable), mean inside diameter (inches, random variable), unit stress (psi), and applied force (lb).

Using aforementioned data, unit stress is calculated and found as 91,396 psi. Assuming tensile yield strength as 130,000 psi and ultimate tensile strength as 135,000 psi, the safety factors can be calculated by using Eq. (11):

$$S = \frac{9700}{(\pi/4)(Z)} = \frac{12350{,}4}{Z} \tag{11}$$

where $Z = (D^2 - d^2)$ is a random variable. In this application, the outer diameter and inner diameters are 0.927 ± 0.001 inches and 0.851 ± 0.001 inches, respectively. Standard deviation is taken as tolerance/3. Random variables of outer and inner diameters are:

$$D = (\mu_D, \sigma_D) = (0.927, 0.00033)\text{ in.}$$

$$D = (\mu_d, \sigma_d) = (0.851, 0.00033)\text{ in.}$$

The mean value and standard deviation of Z can be found easily and summarized as the following,

$$Z = (\mu_Z, \sigma_Z) = (0.13513, 0.000831)\ \text{in.}^2$$

Hence, stress value became;

$$S = (12350.4/Z) = (12350.4/(0.13513, 0.000831)) \quad (12)$$

Likewise, the mean value and standard deviation can be found as follows;

$$\mu_S = (12350.4/\mu_Z) = 91396\ \text{psi}$$

$$\sigma_S = (12350.4)\frac{\sigma_Z}{\mu_Z^2} = 562\ \text{psi}$$

$$S = (\mu_S, \sigma_S) = (91396, 562)\ \text{psi}$$

If difference distribution, $W = T - S$, is taken into account, the following values are obtained. Here:

$$T = \text{ tensile yield strength } = (\mu_T, \sigma_T) = (130000, 7500)\ \text{psi},$$

$$S = \text{ calculated stress } = (\mu_S, \sigma_S)$$

$$\mu_W = 130000 - 91396 = 38604\ \text{psi and}$$

$$\sigma_W = \sqrt{7500^2 + 562^2} = 7521\ \text{psi}$$

Hence difference distribution become;

$$W = (\mu_W, \sigma_W) = (38604, 7521)$$

In order to apply W to standard distribution, $N = (0, 1)$, the following formula can be used:

$$X = \frac{0 - \mu_W}{\sigma_W} = ((0 - 38{,}604) - 7521) = -5.133$$

where 0 and X represent the horizontal coordinate in the difference distribution and corresponding coordinate in the standard distribution table, respectively.

Using a suitably expanded table of standard normal distribution, it can be found that the area to the right of this X coordinate is equal to 0.9999 or probability of failure equal to 0.0001. The corresponding reliability of 0.9999 can be designated by the shorthand expression as:

$$R = 0.9999$$

Here using the given function (tensile stress formula, that is Eq. (10)) and NESSUS the probability of failure and corresponding reliability are seen the same as found by hand calculation.

Secondly let's look at the effect of variable load, which results in fatigue failure in components. Hence, the loading, that is the mean value, and the corresponding standard deviation will be as follows:

$$P = (9700, 9700 * 10\% of 9700) = (9700, 970)\ \text{lbs},$$

Hence tensile stress equation (our "G" function) became;

$$S = \frac{(\frac{4}{\pi})P}{Z} \tag{13}$$

where

$$Z = D^2 - d^2 = (0.13513, 0.000831)\ \text{in}^2$$

Using Table 1,

$$\left(\frac{4}{\pi}\right)P = (12350.4, 1235)\ \text{lb and}$$

$$S = ((12350.4, 1235)/(0.13513, 0.000831)) = (91396, 9159)\ \text{psi}$$

For this new situation:

$$T = (130000, 7500),\ \text{given}$$

$$\mu_W = 130000 - 91396 = 38604\ \text{psi},\ \text{unchanged}$$

$$\sigma_W = \sqrt{7500^2 + 9159^2} = 11838\ \text{psi},\ \text{changed: so}$$

$$W = (\mu_W, \sigma_W) = (38, 604, 11, 838)\ \text{psi},$$

Finally to apply W to standard normal distribution, $N = (0, 1)$, it can be used the following formula,

$$X = \frac{0 - \mu_W}{\sigma_W} = ((0 - 38,604)/11,838) = -3.26$$

Using a suitably expanded table of standard normal distribution, it can be found that the area to the right of this X coordinate to be equal to 0.99944 or probability of failure equal to 0.00056. The corresponding reliability of 0.99944 can be designated by the shorthand expression as:

$$R = 0.99944$$

Here using the given function (tensile stress formula, that is Eq. (10)) and NESSUS the probability of failure and corresponding reliability are seen the same as found by hand calculation.

This steps can be expanded and used for calculation of the probability of failure for fatigue life, that is for the load applications of N_f, using Eq. (9). For this case the reliability function is defined as [42]:

$$L_N(n) = P(N\rangle n) \tag{14}$$

and the probability of failure in the interval (from first to the nth application of load) $[1, n]$ is [42]:

$$F_N(n) = 1 - L_N(n) = P(N \leq n) \tag{15}$$

Finally let's look at the fatigue behavior of spot welded TS specimen overall considering Eq. (9). There are some constants and material properties. The so called values for a material (DQSK or DIN 1623 steel material) and loading case (cycling between 3700 and 200 N) using previous studies, [20, 21], can be summarized as the following:

$$\sigma_m = \left(\mu_{\sigma_m}, \sigma_{\sigma_m}\right) = (159.7, 22.933)\ \text{MPa}$$

N_f = for life it was wanted 1,000,000 (infinite life) with an anticipated normal distribution over the fatigue life of the spot welded TS specimen such that its standard deviation is 10% of its mean value, then the mean value and standard deviation, that is the case, for N_f will be:

$$N_f = \left(\mu_{N_f}, \sigma_{N_f}\right) = (1000000,\ 100000),$$

$$c = -0.4,$$

$$E = 207000\ \text{MPa},$$

$$\sigma_f^{'} = 499 \text{ MPa},$$

$$\varepsilon_f^{'} = 0.104,$$

$$b = -0.06,$$

$$\varepsilon_a = \left(\mu_{\varepsilon_a}, \sigma_{\varepsilon_a}\right) = (0.000684, 0.00075),$$

Using Eq. (9) with the aforementioned values, the value of the function becomes;

$$G = \frac{\sigma_f^{\prime} - \sigma_m}{E}\left(2N_f\right)^b + \varepsilon_f^{\prime}\left(2N_f\right)^c - \varepsilon_a$$

$$G = \frac{499 - \sigma_m}{207,000}\left(2N_f\right)^{-0.06} + 0.104\left(2N_f\right)^{-0.4} - \varepsilon_a \quad (16)$$

where

$$N_f = \left(\mu_{N_f}, \sigma_{N_f}\right) = (1,500,000, 150,000),$$

$$\sigma_m = \left(\mu_{\sigma_m}, \sigma_{\sigma_m}\right) = (159.7, 22.933), \text{ and}$$

$$\varepsilon_a = \left(\mu_{\varepsilon_a}, \sigma_{\varepsilon_a}\right) = (0.000684, 0.00075).$$

It is clear that the probability of failure or the corresponding reliability of the spot welded MTS specimen obtained by using the last formula (Eq. (16)) will be much lower than the computed values, since the resulting mean value and standard deviation is taken into account along with each component with their mean values and standard deviations. Here it was considered only one component of the loading and in this single load two cases, which are the tensile load itself and diameter of the HAZ, was examined as an example. In short, using the "G" function, that is Eq. (16) with NESSUS, it is found that the probability of failure is 0.337, which is an expected value.

Case 2: Correlations among applied load and experimental fatigue life data: In fatigue analysis, the uncertainty is originated from two main subjects. The first one is physical properties based uncertainties and the second one is modeling uncertainty which is related to analysis assumptions for models used. In order to figure out whether the stress versus fatigue life (S–N) curve of fatigue tests have a considerable scatter, the Coefficient of Variation (COV) of fatigue life (N) can be checked. Typically fatigue life data used to define fatigue strength of welded joint is characterized by a COV equals to 0.50 [43]. In this study, the experimental fatigue life data for every loading case have lower COV values as seen in Table 2 [14] which shows the

Table 2 The reliability data for MTS type spot welded specimens [14]

Life range	Mean value	Standard deviation	COV
13,280–13,450	13,370	80.932	0.006
58,660–62,970	61,190	1851.4	0.030
90,750–95,000	93,480	1875.26	0.020
159,250–172,300	167,100	6916.83	0.041
265,700–300,600	283,150	24,678.03	0.087
633,250	633,250	-	-

Table 3 Regression analysis (scatter diagram) results for MTS type spot welded specimens

Life range	R^2
13,280–13,450	0.8926
58,660–62,970	0.9341
90,750–95,000	0.8109
159,250–172,300	0.8899
265,700–300,600	1.0000
633,250	–

reliability of the data, in other words the uncertainty of the experimental fatigue life data have a considerable scatter.

In addition to checking COV value, it can also be looked at the regression coefficients values via the regression analysis, or scatter diagram, of the experimental data. Table 3 shows the coefficient of determination (R^2) values of the regression analysis or the scatter diagrams of MTS type spot weld data.

The result of the regression analysis (scatter diagram), which is the estimated coefficient of determination, given in Table 3 indicate that the predictability of the regression is quite high and the general applicability of the experimental fatigue life data are good.

Secondly, to find a distribution from observed (experimental) data, as an example for selected loading case (3700–200 N), and then whether the assumed distribution is acceptable for a significance level (α), i.e. for $\alpha = 1\%$, Kolmogorov–Smirnov (K-S) statistical test was used. Before doing K-S test, the plots, experimental fatigue life data versus Cumulative Distribution Function (CDF) and Probability Density Function (PDF) values were obtained using a commercial software, NESSUS (Southwest Research Institute) [44], and shown in Fig. 2.

The maximum difference between CDFs of the ordered data can be estimated using Eq. (17) [45]:

$$D_n = \max |F_X(x_i) - S_n(x_i)| \tag{17}$$

$F_X(x_i)$ and $S_n(x_i)$ can be estimated via Eq. (18) [45]:

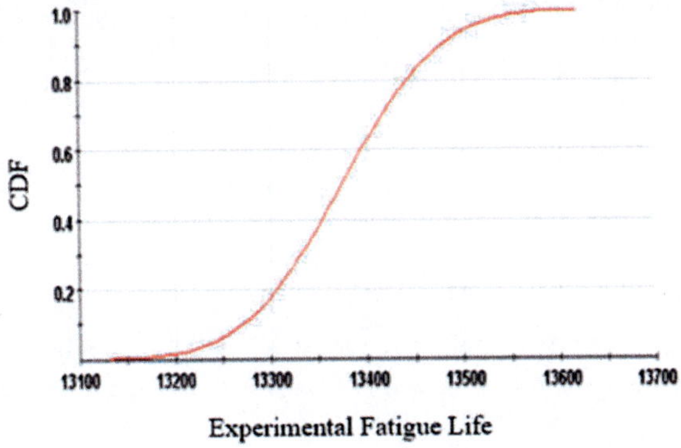

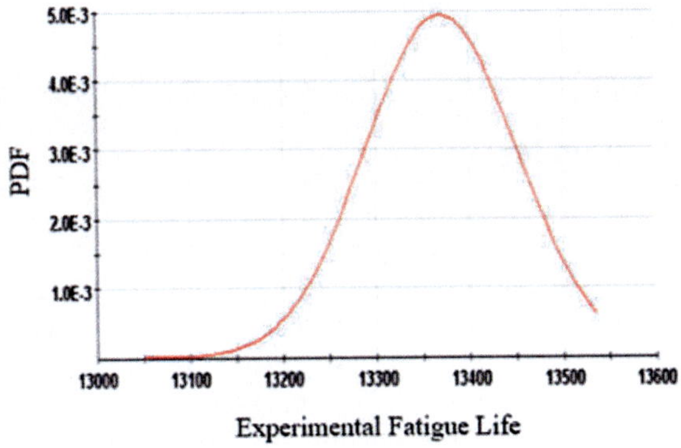

Fig. 2 CDF and PDF values for experimental fatigue life data of MTS type spot welded specimen under 3700–200 N loading case

$$F_X(x_i) = \Phi\left[\frac{\ln x_i - \lambda}{\xi}\right] \tag{18}$$

where λ and ξ are the two parameters of the lognormal distribution.

$$S_n(x_i) = \begin{Bmatrix} 0, & x \langle x_1 \\ \frac{m}{n}, & x_m \le x \le x_{m+1} \\ 1, & x \ge x_n \end{Bmatrix} \tag{19}$$

The CDF of D_n^α can be related to the significance level α as [45];

$$P\left(D_n \le D_n^\alpha\right) = 1 - \alpha \tag{20}$$

and D_n^α values at various significance levels α can be obtained from a standard mathematical table as shown in reference [45]. Table 4 shows the K-S test results of the experimental fatigue life data for MTS type spot welded specimen taking into account the loading case of 3700–200 N.

Table 4 K-S test results for experimental results of MTS type spot welded specimens under the loading of 3700–200 N

(x_i)	$F_X(x_i)$	$S_n(x_i)$	D_n
13,280	0.4364	0.20	0.2364
13,290	0.4841	0.40	0.0841
13,390	0.5832	0.60	0.0169
13,440	0.7968	0.80	0.0032
13,450	0.8291	1.00	0.1710

Using 1% significance level $D_n^\alpha = D_5^{0.01}$ can be obtained from a standard mathematical table [45], which is K-S test, as 0.669. Because all the D_n values founded are less than from $D_n^\alpha = D_5^{0.01} = 0.669$, the assumed lognormal distribution is acceptable for 1% significance level and K-S test and this shows the reliability of the data.

3 Conclusions

In this work, the fatigue failure behavior of MTS specimens was investigated probabilistically and the following points were determined:

The anticipated failure probability, for the function of Eq. (11), of "one in ten million" indicates a very safe design; this is the case in spite of the comparatively low safety factor of 1.42, based on the mean values of yield strength and stress.

The stress-strength reliability value of 0.9994 (for the function of Eq. (13)) corresponds to 6 failures per 10,000. This is a relatively lower reliability than that obtained in first case (from Eq. (11)) for the constant load condition, and would be considered unacceptable in most applications.

For the final case (for the function of Eq. (16)), the reliability of the spot welded MTS specimen under fluctuating loading condition, that is in terms of fatigue loading, is low because the system was considered overall and it is clear that failure will occur. This value also shows why fatigue loading is one of the most important deformation mechanisms.

Acknowledgements We would like to express our special thanks of gratitude to Mercedes-Benz Turk A.S. for their help and support during preparation and testing of the specimens. We also acknowledge great respect for the assistance with the application of NESSUS package program, which we have received from Mechanical Engineering Computer Laboratory of Bogazici University.

References

1. Fragola JR (1996) Reliability and risk analysis data base development: a historical perspective. Reliab Eng Syst Saf 51:125–136

2. Vrijling JK, Hengel WV, Houben RJ (1998) Acceptable risk as a basis for design. Reliab Eng Syst Saf 59:141–150
3. Goh YM, Booker J, McMahon C (2004) A comparison of methods in probabilistic design based on computational and modeling issues. In: 5th international conference on integrated design and manufacturing in mechanical engineering, University of Bath England
4. Mahadevan S, Ni K (2004) Fatigue crack growth analysis of spot-welded joints. Fatigue Fract Eng Mater Struct 27:473–480
5. Murty ASR, Naikan VNA (1997) Machinery selection - process capability and product reliability dependence. Int J Qual Reliab Manag 14(4):381–390
6. Guede Z, Sudret B (2005) Probabilistic assessment of thermal fatigue in nuclear components. Nucl Eng Des 235:1819–1835
7. Kloess A, Mourelatos ZP, Meernik PR (2004) Probabilistic analysis of an automotive body-door system. Int J Veh Des 34(2):101–125
8. Penny RK, Weber MA (1991) Probabilistic stress analysis methods. Int J Reliab Stress Anal Fail Prev 30:21–27
9. Swift KG, Raines M, Booker JD (2000) Case studies in probabilistic design. J Eng Des 11(4):299–316
10. Ertas AH, Sonmez FO (2008) A parametric study on fatigue strength of spot-weld joints. Fatigue Fract Eng Mater Struct 31(9):766–776
11. Ertas AH, Yilmazi Y, Baykara C (2008) An investigation of the effect of the gap values between the overlap portions of the spot-welded pieces on fatigue life. Proc Inst Mech Eng Part C-J Mech Eng Sci 222(6):881–890
12. Ertas AH (2015) Design optimization of spot welded structures to attain maximum strength. Steel Compos Struct 19(4):995–1009
13. Ertas AH, Sonmez FO (2011) Design optimization of spot-welded plates for maximum fatigue life. Finite Elem Anal Des 47(4):413–423
14. Ertas AH, Vardar O, Sonmez FO, Solim Z (2009) Measurement and assessment of fatigue life of spot-weld joints. J Eng Mater Technol-Trans ASME 131(1), Article Number: 011011
15. Ertas AH, Sonmez FO (2009) Optimization of spot-weld joints. Proc Inst Mech Eng C-J Mech Eng Sci 223(3):545–555
16. Ertas AH, Sonmez FO (2007) A parametric study on fatigue life behavior of spot welded joints. In: 8th international fracture conference, Istanbul-Turkey, pp. 498–508.
17. Ertas AH, Vardar O (2005) Fatigue behavior of spot welds in modified tensile shear specimens using experimental data. In: 9th international research/expert conference "Trends in the development of machinery and associated technology" TMT 2005, Antalya/Turkey, pp. 243–246
18. Ertas AH (2006) Fatigue behavior of spot welds in modified tensile shear specimens using numerical data. In: 10th international research/expert conference "Trends in the development of machinery and associated technology" TMT 2006, Barcelona-Lloret de Mar/Spain, pp. 777–778
19. Ertas AH, Sonmez FO (2006) Optimization of spot weld joints. In: The 12th international conference on machine design and production, Kusadasi/Turkey, pp. 883–893
20. Pan N (2000) Fatigue life study of spot welds. PhD Thesis, Stanford University
21. Ertas AH (2004) Fatigue behavior of spot welds. MSc Thesis, Bogazici University
22. Barkey ME, Kang H (2001) Lee YL (2001) Failure modes of single resistance spot welded joints subjected to combined fatigue loading. Int J Mater Prod Technol 16(6/7):510–527
23. Rathburn RW, Matlock DK, Speer JG (2003) Fatigue behavior of spot-welded high-strength sheet steels. Weld J 82(8):207–218
24. Chao YJ (2003) Failure mode of spot welds: interfacial versus pullout. Sci Technol Weld Joining 8(2):133–137
25. Deng X, Chen W, Shi G (2000) Three-dimensional finite element analysis of the mechanical behavior of spot welds. Finite Elem Anal Des 35(1):17–39
26. Hou Z, Wang Y, Li C, Chen C (2006) An analysis of resistance spot welding. Weld J 85(3):36–40

27. Barkey ME, Kang H, Lee YL (2000) Evaluation of multiaxial spot weld fatigue parameters for proportional loading. Int J Fatigue 22(8):691–702
28. Di Fant-Jaeckels H, Galtier A (2000) Fatigue lifetime prediction model for spot welded structures. Revue Metall-Cahiers Inf Tech 97(1):83–95
29. Salvini P, Scardecchia E, Demofonti G (1997) A procedure for fatigue life prediction of spot welded joints. Fatigue Fract Eng Mater Struct 20(8):1117–1128
30. Davidson JA, Imhof EJ (1983) A fracture mechanics and system-stiffness approach to fatigue performance of spot-welded sheet steels. In: Proceedings of the SAE international congress and exposition, USA, pp. 19–38
31. Ong JH (2003) An improved technique for the prediction of axial fatigue life from tensile data. Int J Fatigue 15(3):213–219
32. Lee YL, Pan J, Hathaway RB, Barkey ME (2005) Fatigue testing and analysis, 1st edn. Elsevier, USA
33. Lin PC, Su ZM, He RY, Lin ZL (2012) Failure modes and fatigue life estimations of spot friction welds in cross-section specimens of aluminum 6061–T6 sheets. Int J Fatigue 38:25–35
34. Fujii T, Tohgo K, Suzuki Y, Yamamoto T, Shimamura Y (2016) Fatigue strength and fatigue fracture mechanism of three-sheet spot weld-bonded joints under tensile-shear loading. Int J Fatigue 87:424–434
35. Xiao L, Liu L, Chen DL, Esmaeili S, Zhou Y (2011) Resistance spot weld fatigue behavior and dislocation substructures in two different heats of AZ31 magnesium alloy. Mater Sci Eng A-Struct Mater Prop Microstruct Process 529:81–87
36. Radakovic DJ, Tumuluru M (2008) Predicting resistance spot weld failure modes in shear tension tests of advanced high-strength automotive steels. Weld J 87(4):96S-105S
37. Nakayama E, Fukumoto M, Miyahara M, Okamura K, Fujimoto H, Fukui K, Kitamura T (2010) Evaluation of local fatigue strength in spot weld by small specimen. Fatigue Fract Eng Mater Struct 33(5):267–275
38. Lee YL, Lu MW (1994) Fatigue-reliability analysis of resistance spot-welds. In: Proceedings of annual reliability and maintainability symposium, Anaheim-California USA, pp. 178–184
39. Bramley A, Brissaud D, Coutellier D, McMahon C (2005) Advances in integrated design and manufacturing in mechanical engineering. Springer, Netherlands
40. Mahadevan S, Haldar A (2000) Probability, reliability and statistical methods in engineering design. Wiley, USA
41. Welling M, Lynch J (1985) Probabilistic design techniques applied to mechanical elements. Naval Eng J 97(4):116–123
42. Freudenthal AM, Garrelts JM, Shinozuka M (1966) The analysis of structural safety. J Struct Div ASCE 92:267–325
43. Stepsen RI, Fatemi A, Stepsen RR, Fuchs HO (2000) Metal fatigue in engineering, 2nd edn. Wiley, USA
44. NESSUS (2010) NESSUS user's manual, version 9.3, San Antonia, Texas (USA).
45. Sankaran M, Achintya H (2000) Probability and statistical methods in engineering design. Wiley, USA

A Comparative Study on Fatigue Life Prediction of Spot Welded Coach Peel (CP) and Modified Coach Peel (MCP) Type Test Specimens

Mustafa Akbulut and Ahmet H. Ertas

Abstract Besides friction, ultrasonic and laser welding, resistance spot welding is extensively employed for joining sheet metal panels in automotive, railroad and aerospace industry as a reliable, cost-effective, and flexible manufacturing process. Structural panels attached through the spot welding usually undergo cyclic loadings in service, which cause fatigue fracture damage. In order to predict and prevent this type of failure, an in-depth assessment of the fatigue phenomenon is inevitable. In this study, a strain based model known, as Coffin-Manson is adapted to estimate the S–N diagram of various materials, for the first time, on Coach Peel (CP) and Modified Coach Peel (MCP) Type Test Specimens. The strain values used in the model are computed through the FEM analysis. The results obtained in the present work are compared with those of alternative models and experimental data available in literature. Consistencies and discrepancies among the results are presented and interpreted.

Keywords Fatigue · Spot weld · FEM · Coach peel (CP) and modified coach peel (MCP) · Coffin-Manson

1 Introduction

In automotive, railroad, aerospace and many other engineering applications, sheet metals are joined together by means of various welding techniques like ultrasonic, laser and resistance spot welding. Due to its flexibility and adaptability to automation as well as cost effectiveness, spot welding is preferred much more frequently. Most of the time, the sheet metals attached together through this method undergo harsh mechanical conditions such as cyclic mechanical stresses. The alternating mechanical

M. Akbulut (✉)
MAM, TUBITAK, Gebze 41470, Kocaeli, Turkey
e-mail: mustafa.akbulut@tubitak.gov.tr

A. H. Ertas
Department of Mechanical Engineering, Faculty of Engineering and Natural Sciences, Bursa Technical University, Bursa 16330, Turkey

M. Abdel Wahab (ed.), *Proceedings of the 8th International Conference on Fracture, Fatigue and Wear*, Lecture Notes in Mechanical Engineering,
https://doi.org/10.1007/978-981-15-9893-7_32

stresses cause fatigue fracture inside the sheet metals after a period. In order to avoid undesired mechanical failure because of fatigue, the conditions leading to that phenomenon are to be understood very well and necessary precautions must be taken to prevent it. In this respect, fatigue life prediction plays an important role in the design of sheet metals joined by resistance spot welding.

There have been many studies [1–27] dealing with the fatigue behavior of resistance spot welding. Some of them [1, 2] are parametric studies, some others [3, 4] are optimization studies to increase the life span of the spot welded sheet metals. In the fatigue analysis of spot weld, various approaches have been employed such as structural stress [5], nominal stress [6], notch stress approaches [6–9], notch strain approach [5, 9–12], volumetric fatigue approach [5, 13] and crack propagation approach [5, 14–18].

In the fatigue life assessment of the resistance spot welds, certain type of standard test specimens are employed such as Tensile Shear [1, 2, 5] (TS), Modifies Tensile Shear [1, 2, 5] (MTS), Coach Peel [19] (CP) and Modified Coach Peel [5] (MCP). Since it is most commonly encountered in real applications and more conservative in industrial applications, TS and MTS [16, 20, 22–24] type specimens have been studied much more frequently compared to CP and MCP type specimens. As a result, there is a lack of information in literature on experimental and numerical behavior of these pieces. In this study, a comparative approach was adapted to investigate the fatigue life performance of CP and MCP test samples both through experimental results and numerical computations.

2 Models for Fatigue Life Estimation

As discussed in literature section of the present work, there are numerous approaches for fatigue life prediction of spot welded joints. Here, within the scope of this study, notch strain approaches are taken into consideration and a utile one known as Coffin-Manson is expressed.

2.1 *Strain-Based Approaches*

Coffin-Manson is a strain based approach that can yield remarkable agreement with experimental results and predict the fatigue life of spot welds considerably. This model [9, 11, 16, 21–27] offers an expression, which relates alternating true elastic and plastic strains, $\Delta\varepsilon_e$ and $\Delta\varepsilon_p$, to fatigue life, that is number of cycles to failure ($2N_f$), represented by:

$$\frac{\Delta\varepsilon}{2} = \varepsilon_a = \varepsilon_e + \varepsilon_p = \frac{\Delta\varepsilon_e}{2} + \frac{\Delta\varepsilon_p}{2} = \frac{\sigma_f'}{E}\left(2N_f\right)^b + \varepsilon_f'\left(2N_f\right)^c \quad (1)$$

where, ε_e is the elastic component of the cyclic strain amplitude, ε_p the plastic component of the cyclic strain amplitude, ε'_f the regression intercept called 'the fatigue ductility coefficient, c the regression slope called 'the fatigue ductility exponent', σ'_f the regression intercept called 'the fatigue strength coefficient', N_f the number of cycles to failure, and b the regression slope called 'the fatigue strength exponent'.

In addition to Coffin-Manson, there are two more universal relations which are used in fatigue life calculations proposed in the literature based on experimental data [11, 21]. These relations are "Muralidharan & Manson" and "Brinell hardness" equations which are expressed in Eq. (2) and Eq. (3) respectively. In this study, Brinell hardness equation was also employed in the fatigue life predictions in order to make a comparison between experimental and Coffin-Manson results.

$$\frac{\Delta\varepsilon}{2} = \varepsilon_a = 0.623\left(\frac{S_{ut}}{E}\right)^{0.823}\left(2N_f\right)^{-0.09} + 0.0196\left(\varepsilon_f\right)^{0.155}\left(\frac{S_{ut}}{E}\right)^{-0.53}\left(2N_f\right)^{-0.56} \tag{2}$$

$$\frac{\Delta\varepsilon}{2} = \frac{4.25(HB) + 225}{E}\left(2N_f\right)^{-0.09} + \frac{0.32(HB)^2 - 487(HB) + 191000}{E}\left(2N_f\right)^{-0.56} \tag{3}$$

3 Numerical Analysis

In the fatigue-life estimation model defined above, plastic and elastic strains need to be computed. In order to calculate these values pertaining to the CP and MCP test specimens, a finite element model (FEM) is built and numerical solutions are found out.

3.1 Finite Element Model Description

In the FEM analysis, a structural module of ANSYS 17.2 workbench was employed. A mapped mesh strategy using solid elements was applied in such a way that stress concentration locations were meshed substantially fine in order to obtain sensitive results. Contact properties between the sheet metals and the nugget-sheet metal interface were created automatically in the workbench. The necessary coefficients ensuring the reliable analysis outputs were determined through parametric studies. Finite Element Mesh configurations pertaining to CP and MCP specimens are illustrated in Figs. 1 and 2 as follows.

In spot welding process, heat input through the electrodes does not affect the mechanical properties of the materials. Therefore, the material properties pertaining

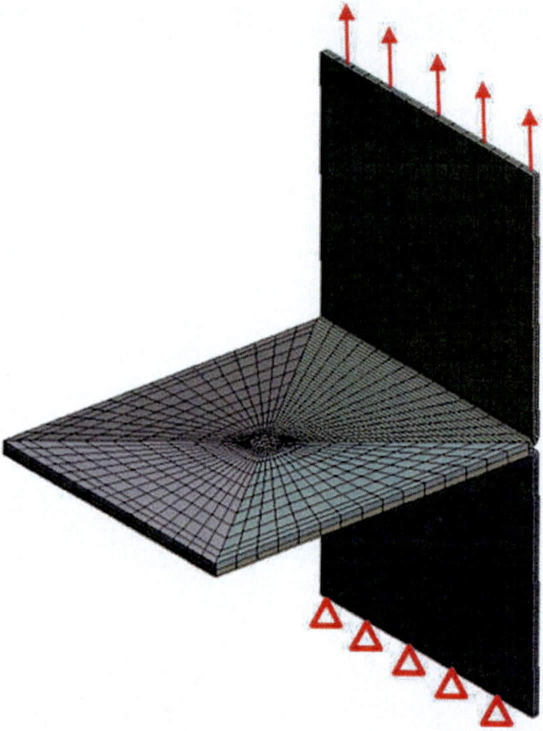

Fig. 1 Finite element mesh configuration of coach peel (CP) specimen

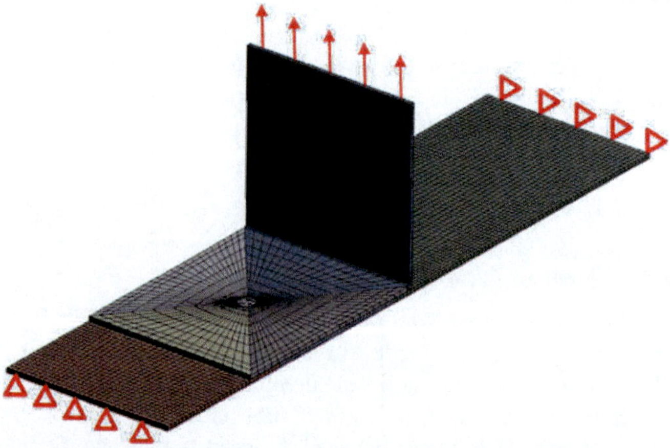

Fig. 2 Finite element mesh configuration of modified coach peel (MCP) specimen

Table 1 Mechanical properties of the materials to be used

	Yield strength (MPa)	Ultimate tensile strength (MPa)	Elasticity (MPa)	Average elongation (%)
HSLA [5]	340	400	2.06×10^5	33
DQSK [5]	174	250	2.06×10^5	23
DP780GI [19]	450	850	1.90×10^5	19.5

Table 2 Dimensions of CP and MCP specimens

$L_1 = 25$ mm	$H_1 = 80$ mm	$W_1 = 38$ mm	$D_1 = 7$ mm	$t_1 = 1.61$ mm
$L_2 = 177.8$ mm	$H_2 = 31.75$ mm	$W_2 = 34.93$ mm	$D_2 = 6.35$ mm	$t_2 = 0.9$–1.5 mm

to the nugget and base metal sheets were used without being modified. In Table 1 the mechanical properties of the materials to be used in the analysis were presented.

In the analysis, two main specimens CP and MCP were investigated and the dimensions pertaining to both parts are listed in Table 2.

Figures 3 and 4 show the geometry and dimensions of both specimens.

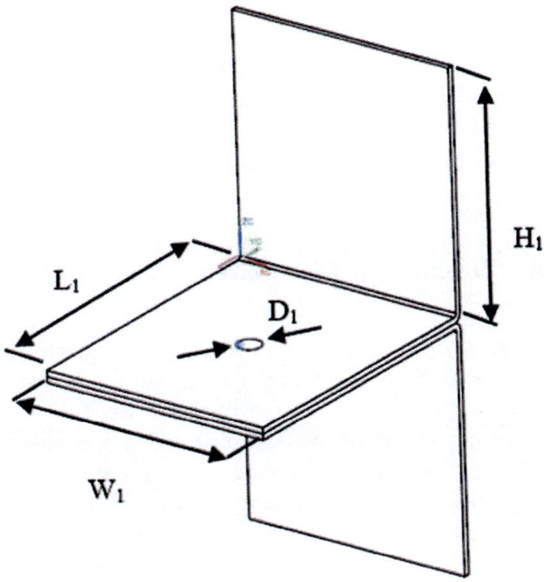

Fig. 3 Coach peel (CP) specimen

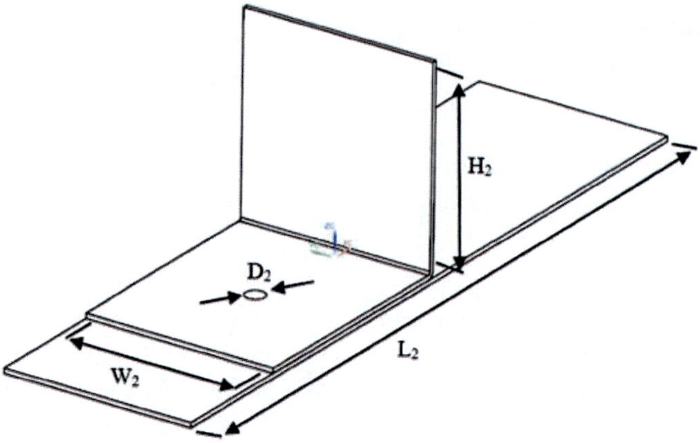

Fig. 4 Modified coach peel (MCP) specimen

3.2 *Boundary Conditions*

In the finite element models of CM and MCP specimens, the force application points are illustrated by red arrows (Figs. 1 and 2). All displacement and rotation degrees of freedom of the nodes on the surface of the bottom part of CP and left and right hand side of the MCP were fixed. Out of plane displacements on the surface on which the force is applied is also restricted. A tensile load is applied in both parts. Since loading ratios® are chosen as "0", maximum value Fmax is applied to read the principal elastic and plastic strains.

4 Results and Discussions

In this study, it is aimed to investigate the fatigue life prediction of MCP and CP type specimens. Since many of the researches on the spot welding are related to Tensile Shear (TS) and Modified Tensile Shear (MTS), the number of experimental studies regarding the CP and MCP are quiet scarce compared to the former. Within the scope of the present work, the experimental outputs of Pan [5] for MCP and Guanghan et al. [19] for CP were utilized to make comparisons between the numerical studies and real engineering data. In the work of Guanghan et al. [19] S–N life curve pertaining to DP780GI material for only one type of geometry with a single thickness value is presented. In Fig. 5, the elastic strain distribution of this geometry is illustrated. Contrary to TS and MTS specimens, there are two critical regions, one is the nugget surrounding and the other is the corner of L-shaped extension. In both regions, elastic and plastic strains gain the highest values.

As to the study of Pan [5], two different materials HSLA and DQSK were used for different thickness values of MCP base and L-shape piece. The examples used

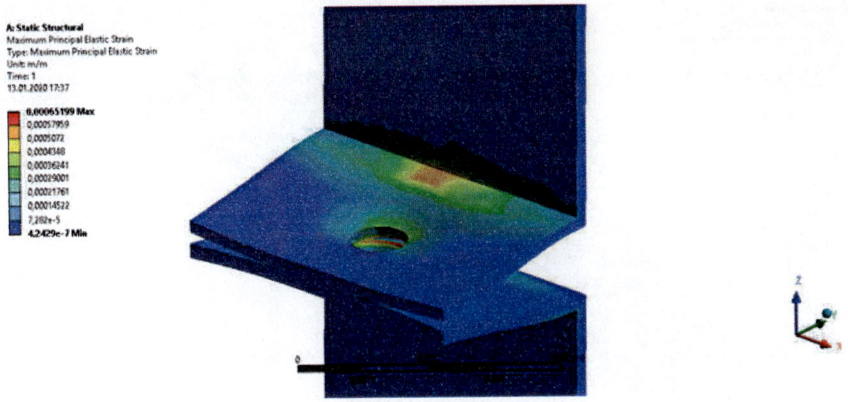

Fig. 5 Finite element analysis of coach peel (CP) specimen

in the present work are analyzed and the elastic strain distributions were presented in Figs. 6, 7 and 8.

In Fig. 6, both pieces are 1.04 mm, in Fig. 7 both pieces 1.5 mm. In the last analysis, the thicknesses of L-shape and base plate are different, in Fig. 8, L-shape has 0.9 mm, while base plate has 1.5 mm.

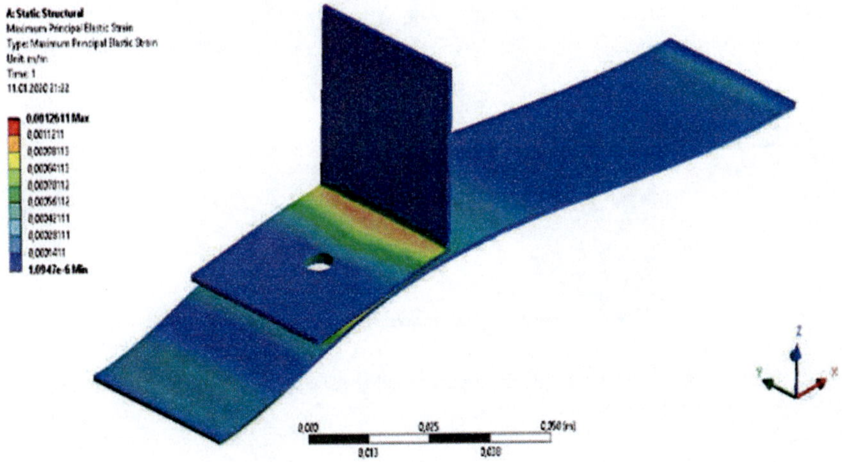

Fig. 6 Finite element analysis of modified coach peel (MCP) specimen (L-shape = 1.4 mm, base plate = 1.4 mm)

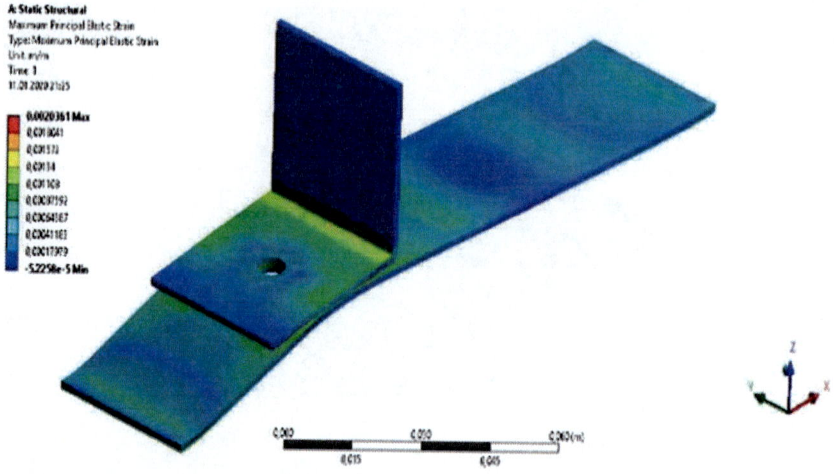

Fig. 7 Finite element analysis of modified coach peel (MCP) specimen (L-shape = 1.5 mm, base plate = 1.5 mm)

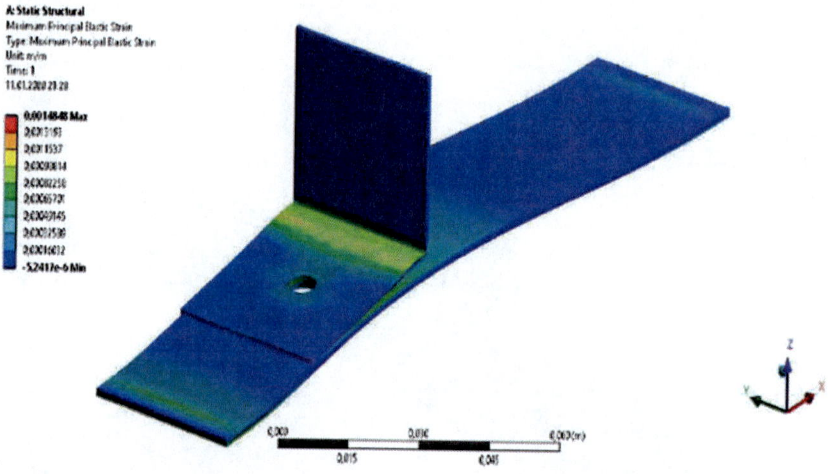

Fig. 8 Finite element analysis of modified coach peel (MCP) specimen (L-shape = 0.9 mm and base plate = 1.5 mm)

4.1 Comparisons of Experimental and Numerical Data

As mentioned earlier, for CP type specimen, the experimental data of a previous study [19] was utilized. S–N curve for DP780GI were plotted on the basis of numerical data calculated from Coffin-Manson and Brinell-Hardness expressions discussed in

Sect. 2. Both Coffin-Manson and Brinell-Hardness empiric relations yielded predictions smaller than the experimental data as illustrated in Fig. 9. The deviation of Brinell-Hardness results from the experimental data is much larger than that of Coffin-Manson.

In Figs. 10 and 11 the experimental and computational predictions of MCP specimens of different thicknesses are illustrated. For $t = 1.5$ mm sample, real cycle to

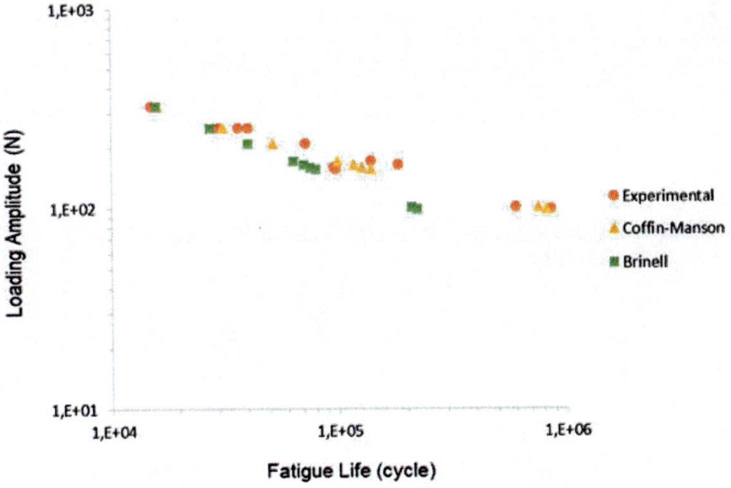

Fig. 9 Experimental data and numerical predictions for coach peel (CP) specimen

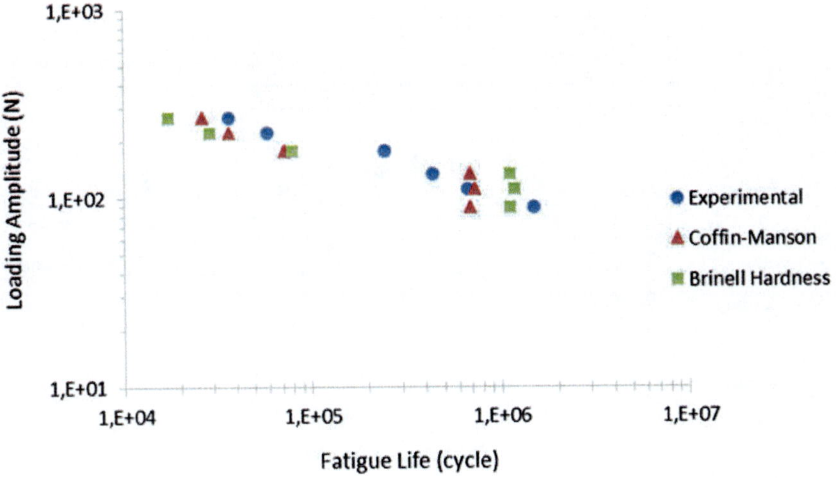

Fig. 10 Experimental data and numerical predictions for modified coach peel (MCP) specimen ($R = 0$, $t = 1.04$, HSLA steel)

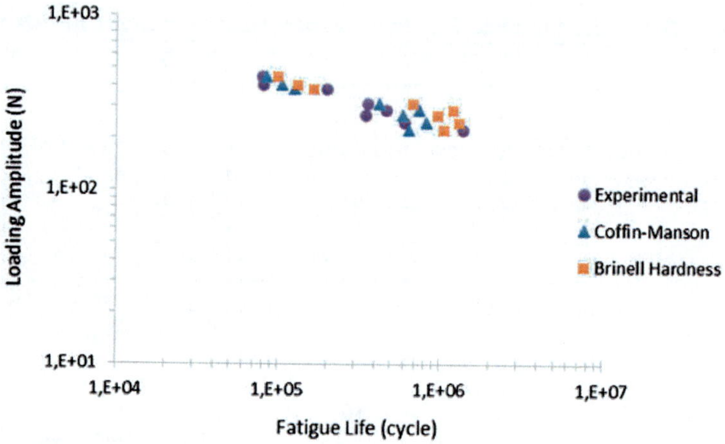

Fig. 11 Experimental data and numerical predictions for modified coach peel (MCP) specimen ($R = 0$, $t = 1.5$, HSLA steel)

failure and empiric estimations are relatively higher than $t = 1.04$, as the strain values are quite smaller. In both cases, for higher cycles, predictions of Coffin-Manson and Brinell-Hardness deviate from the experimental data substantially.

In Fig. 12, MCP consists of two different thicknesses, L-shape is 1.5 mm and base plate is 0.9 and both empiric expressions resulted in conservative predictions, even though Coffin-Manson is much safer for higher number of cycles.

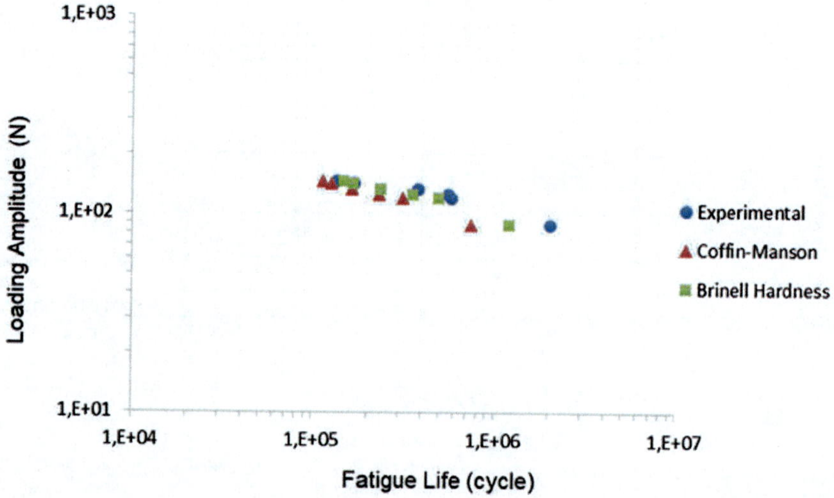

Fig. 12 Experimental data and numerical predictions for modified coach peel (MCP) specimen ($R = 0$, L ($t = 1.5$ mm) 0.9 mm, DQSK steel)

5 Conclusions

In this study, fatigue life values of CP and MCP type spot weld test specimens were computed through Coffin-Manson and Brinell-Hardness empiric relations. Experimental data of previously conducted studies were utilized in the comparisons. Both empiric expressions gave conservative results for CP specimen, which is made of a material with quite high ultimate strength. Both experimental and computational values for MCP specimen with higher thickness resulted in augmented fatigue life as expected and they are in good agreement with experimental data especially for low cycle values. As the fatigue cycle gets increased, both numeric approaches deviate from the measured values considerably.

References

1. Ertas AH, Sonmez FO (2008) A parametric study on fatigue strength of spot-weld joints. Fatigue Fract Eng Mater Struct 31(9):766–776
2. Ertas AH, Sonmez FO (2007) A parametric study on fatigue life behavior of spot welded joints. In: 8th international fracture conference, Istanbul-Turkey, pp. 498–508
3. Ertas AH (2015) Design optimization of spot welded structures to attain maximum strength. Steel Compos Struct 19(4):995–1009
4. Ertas AH, Sonmez FO (2013) Design optimization of spot-welded plates for maximum fatigue life. Finite Elem Anal Des 47(4):413–423
5. Pan N (2000) Fatigue life study of spot welds. PhD Thesis, Stanford University
6. Ertas AH (2004) Fatigue behavior of spot welds. MSc Thesis, Bogazici University
7. Radaj D (1996) Theory of forces and stresses in spot welded overlap joints. Arch Appl Mech 67:22–34
8. Zhang S (1997) Stress intensities at spot welds. Int J Fract 88:167–185
9. Socie DF (1977) Fatigue-life prediction using local stress strain concepts. Exp Mech 2:50–56
10. Peeker E, Niemi E (1999) Fatigue crack propogation model based on a local strain approach. J Constr Steel Res 49:139–155
11. Roessle ML, Fatemi A (2000) Strain-controlled fatigue properties of steels and some simple aproximations. Int J Fatigue 22:495–511
12. Roessle ML, Fatemi A, Khosrovaneh AK (1999) Variation in cyclic deformation and strain-controlled fatigue properties using different curve fitting and measurement techniques. In: Proceedings of the SAE international congress and exposition, Detroit, Michigan, USA
13. Adib H, Gilgert J, Pluvinage G (2004) Fatigue life duration prediction for welded spots by volumetric method. Int J Fatigue 26:81–94
14. Henrysson HF (2002) Variable amplitude fatigue of spot welds. Fatigue Fract Eng Mater Struct 25:1187–1197
15. Zhang Y, Taylor D (2007) Optimization of spot welded structures. Finite Elem Anal Des 37:1013–1022
16. Newman JA, Dowling NE (1998) A crack growth approach to life prediction of spot-welded lap joints. Fatigue Fract Eng Mater Struct 21:1123–1132
17. Davidson JA, Imhof EJ (1983) A fracture mechanics and system-stiffness approach to fatigue performance of spot-welded sheet steels. In: Proceedings of the SAE international congress and exposition, Detroit, Michigan, USA
18. Cooper JF, Smith RA (1985) The measurement of fatigue cracks at spot welds. Int J Fatigue 7:137–140

19. Guanghan W, Dayong L, Xuming S, Yinghong P, Yandong S, Li H, Shiyao H, Weiqin T (2017) Experiment and modeling on fatigue of the DP780GI spot welded joint. Int J Fatigue 103:73–85
20. Chao YJ (2003) Failure mode of spot welds: interfacial versus pullout. Sci Technol Weld Joint 8(2):133–137
21. Stephens RI, Fatemi A, Stephens RR, Fuchs HO (2001) Metal fatigue in engineering. Wiley Inter science Publication, New York, USA
22. Deng X, Chen W, Shi G (2000) Three-dimensional finite element analysis of the mechanical behavior of spot welds. Finite Elem Anal Des 35:17–39
23. Hou Z, Wang Y, Li C, Chen C (2006) An analysis of resistance spot welding. Weld J 85:36–40
24. Wondseok J, Bae D, Sohn I (2004) Fatigue design of various type spot welded lap joints using the maximum stress. KSME Int J 18(1):106–113
25. Suresh S (2004) Fatigue of Materials. Cambridge University Press, Cambridge, UK
26. Cui W (2002) A state of the art review on fatigue life prediction methods for metal structures. J Mar Sci Technol 7:43–56
27. Ong JH (1993) An evaluation of existing methods for the prediction of axial fatigue life from tensile data. Int J Fatigue 15(1):13–19

Numerical Methodology to Predict Subsurface Crack Initiation from Non-metallic Inclusions Due to Rolling Contact Fatigue

Gopalakrishnan Ravi, Wim De Waele, and Stijn Hertelé

Abstract Bearings used in the intermediate and high speed stages of offshore wind turbine gearboxes run in harsh working conditions and may fail prematurely due to rolling contact fatigue (RCF). The microstructural alterations linked with such premature bearing failures are often described as white etching crack failures (WEC). Understanding these white etching bearing failures requires the study of RCF phenomenon at multiple scales ranging from macroscopic to microscopic and at different stages such as crack initiation and propagation. This paper presents a 2D numerical framework to evaluate subsurface RCF crack initiation originating from non-metallic inclusions in bearings. The global finite element (FE) model simulates parts of the contact bodies such as roller elements and raceway to represent the contact zone. A submodel containing a non-metallic inclusion is derived from the global FE model of a rolling contact. The inclusion is elliptical in shape and its position, dimensions and orientation can be changed. A moving Hertzian pressure distribution is modeled to simulate the rolling pass and the stress history around the inclusion/matrix interface. A multi-axial critical plane approach is adopted to calculate fatigue initiation damage. After explaining the numerical framework, this paper highlights the functionalities based on a number of case studies, focusing on effects of normal load and surface traction between the roller and inner raceway, and on inclusion characteristics. The results give us an insight into the underlying physics behind the mechanism of subsurface initiated RCF. Investigation of the link between RCF and white etching cracking is ongoing.

Keywords Rolling contact fatigue · Crack initiation · Fatigue damage modelling · Non-metallic inclusion · Critical plane approach

G. Ravi (✉) · W. De Waele · S. Hertelé
Department EMSME, Soete Laboratory, Ghent University, Ghent, Belgium
e-mail: gopalakrishnan.ravi@ugent.be

G. Ravi
SIM vzw, Technologiepark, 48, 9052 Zwijnaarde, Ghent, Belgium

M. Abdel Wahab (ed.), *Proceedings of the 8th International Conference on Fracture, Fatigue and Wear*, Lecture Notes in Mechanical Engineering,
https://doi.org/10.1007/978-981-15-9893-7_33

1 Introduction

Roller bearings are amongst the most commonly used engineering components in a variety of machinery such as turbines, automotive engines, transmissions, drills etc. They permit rotary motion and are designed to sustain significantly heavy loads. Figure 1 represents a common rolling element bearing (REB) and its corresponding components. It consists of an inner and an outer ring, a number of rolling elements, a cage to isolate the rolling elements and optional seals to avoid lubrication seepages or external contamination. The ever increasing demand for higher power density from modern machinery during the past 50 years puts a higher mandate on load carrying capacity and reliability of bearings [1]. Harsh bearing working environments can be damaging for the material. These damages can be affected by contact pressure, rotational speed, sliding, and environmental factors. The components, although manufactured from high strength steels, are liable to rolling contact fatigue (RCF) failure mode, resulting in material flaking from the surface of bearings or raceways [2]. In the current study, we focus our attention towards computational modelling of RCF crack initiation triggered by the presence of non-metallic inclusions, as the crack initiation tends to occupy 80–90% of the total RCF lifetime [3]. Therefore the RCF crack initiation methodology is profoundly important and thus the goal of this work is to design a simple and robust framework to predict RCF crack initiation based on continuum mechanics, cyclic contact loading and a multi-axial critical plane fatigue damage approach. The modeling approach helps to understand the stress–strain response around non-metallic inclusions present in the bearings under the effect of normal load and surface traction between the roller and raceway. Simulations carried out in this study aim to provide an in-depth understanding towards the physical mechanism of subsurface initiated damage from inclusions due to RCF.

1.1 *Premature Bearing Failures: Inclusions as Crack Initiators*

In the past decade, nearly 27,000 wind turbines, varying from 500 to 5 MW, for both onshore and offshore applications required frequent gearbox repair, replacement or maintenance well before their service life [4]. Premature failures often happen in

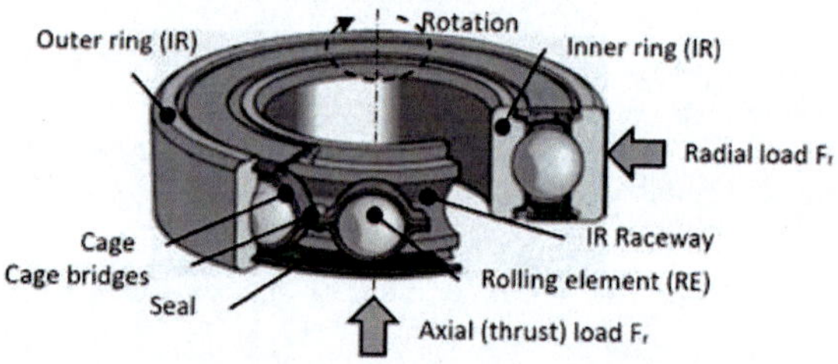

Fig. 1 Typical rolling element bearing (REB) structure (courtesy of SKF)

inner raceways of the high-speed parallel-stage rolling element bearings, leading to high deadtime in wind turbines and implying significant maintenance costs for wind plant operators and original equipment manufacturers [4] (see Fig. 2). The bulk of all gearbox failures initiates at bearings, and the principal failure mode in these bearings is associated with local microstructural alterations. The failure of these bearings is understood to be often due to the presence of inhomogeneities and non-metallic inclusions [5], which act as crack initiation sites under RCF conditions. From previous studies, it has been observed that material defects such as manganese sulphide (MnS) and Alumina (Al_2O_3) inclusions are found to be the primary white etching crack (WEC) damage initiators [6, 7]. The cracks usually initiate around these inclusions, where the material exhibits high stress concentrations and typical butterfly formations are observed, refer Fig. 3 (left) [7]. Therefrom, WECs may originate and propagate as seen in Fig. 3 (right) [8]. In order to simulate this phenomenon, modeling

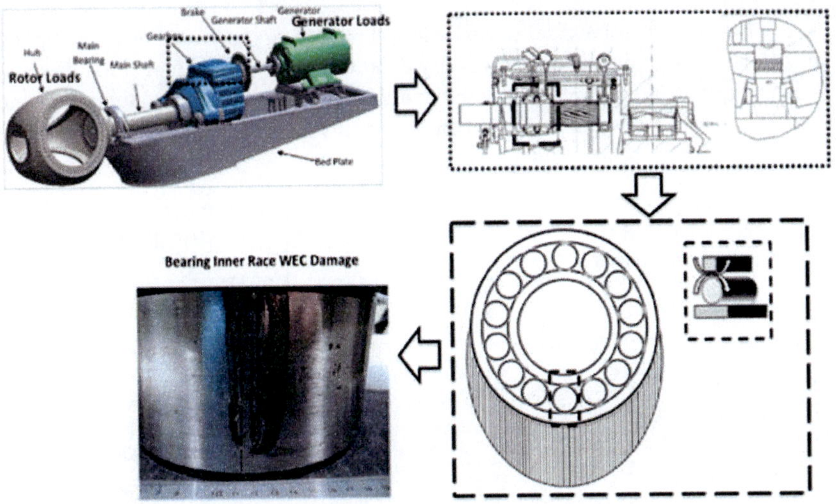

Fig. 2 Illustration of WEC failure in wind turbine gearbox bearings [9]

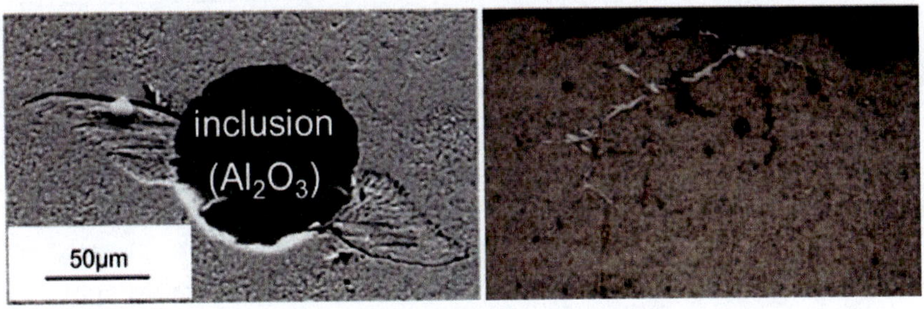

Fig. 3 Butterfly formation around inclusions (Left) [7], White etching cracks (Right) [8]

rolling contact fatigue seems to be the common starting point. Using this RCF model as a foundation, different aspects such as inclusion modeling, butterfly simulation, phase transformation, hydrogen pileup, crack initiation, propagation, branching and spalling can be included and extended for WEC analysis. As part of such wider research, this paper investigates the crack initiation hotspots from bearing inclusions due to RCF.

2 A Strategy to Predict RCF Crack Initiation from Inclusions

Figure 4 shows the strategy developed for analysis of RCF crack initiation from a subsurface inclusions. The approach is divided into two parts: pre-processing and post-processing.

Pre-processing involves identifying relevant parameters such as bearing roller and raceway geometry, dimensions, boundary conditions, meshing, contact loading and modeling this in a finite element (FE) analysis environment. The fatigue damage analysis is part of post-processing, where the results from the FE analysis are used to

Fig. 4 An approach to predict RCF crack initiation from non-metallic inclusions

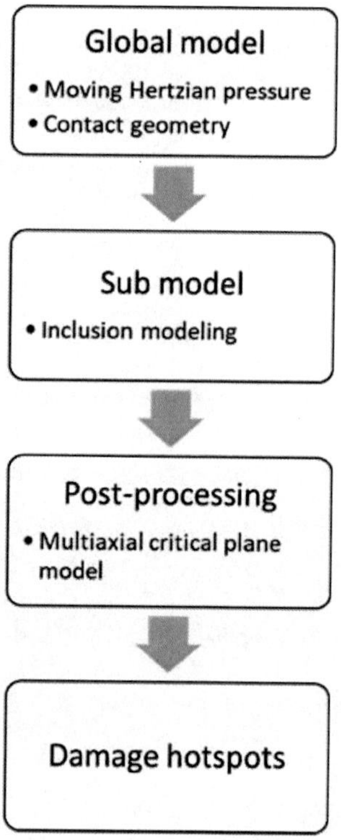

identify the most critical point at the inclusion-matrix interface and the corresponding critical plane for crack initiation. The critical plane is determined using a search algorithm that accounts for multi-axial, non-proportional loading together with a damage parameter for fatigue crack initiation around the inclusions. This gives us the most vulnerable regions (hotspots) with respect to fatigue damage initiation.

3 Numerical RCF Crack Initiation Modeling Framework

The authors propose a 2D plane strain model that is implemented in finite element software ABAQUS/standard version 2019, to evaluate the RCF crack initiation at non-metallic inclusions within bearing steels. This model mimics a portion of an inner raceway in the circumferential direction. The analysis methodology is subdivided as described below: FE based global model and sub model, and post-processing using a multi-axial critical plane approach which calculates a fatigue initiation indicator.

3.1 Global Model

The global model consists of a homogenous bulk material representing bearing steel (Fig. 5). This is done to realistically estimate the linear-elastic stress loading cycle for each observed material point at the surface or subsurface. The mechanical properties for the bearing steel are taken as: Poisson ratio $\nu = 0.3$, Young's modulus $E = 200$ GPa.

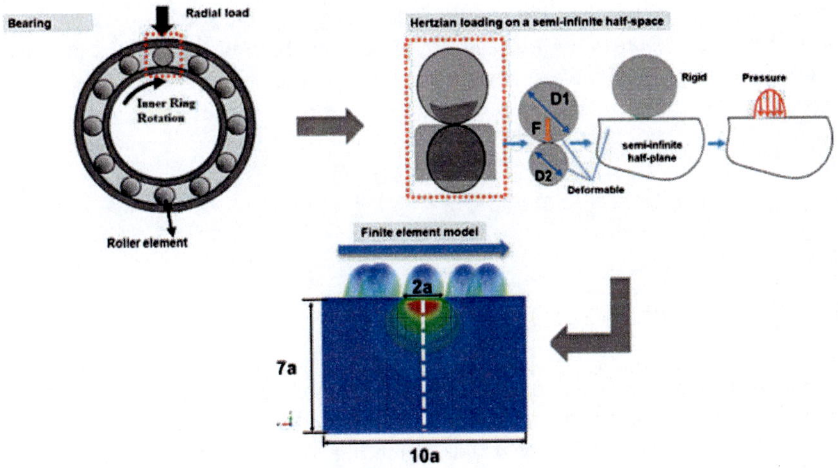

Fig. 5 Modeling of Hertzian contact pressure

3.1.1 Contact Load Modeling

As per contact mechanics theory, under the assumption of plane strain conditions, the contact of a roller bearing on a raceway at any location can be implied as two spherical bodies with different radius (R_1, R_2) contacting each other, as shown in Fig. 5. This can be further simplified to a circular body with equivalent radius R (see Eq. (1)) on a semi-infinite half plane [9].

$$\frac{1}{R} = \frac{1}{R_1} + \frac{1}{R_2} \tag{1}$$

This contact loading can be further denoted as Hertzian pressure loading on a semi-infinite half-space (Fig. 5). These pressure distributions were applied as non-uniform, general surface traction vectors in ABAQUS/Standard using FORTRAN user subroutines. Normal pressure load was modeled through a DLOAD subroutine and the surface tractions were applied using UTRACLOAD subroutine. Equations (2) and (3) represent the component of the Hertzian pressure profile applied through DLOAD and UTRACLOAD subroutines for point and line contact loading, respectively [10].

$$p(x, y) = P_{max}\sqrt{1 - \frac{(x - x_c)^2}{a^2} - \frac{(y - y_c)^2}{b^2}} \tag{2}$$

$$p(x, y) = P_{max}\sqrt{1 - \frac{(x - x_c)^2}{a^2}} \tag{3}$$

where (x_c, y_c) is the center of the Hertzian pressure profile, (x, y) represents the position of the pressure distribution coordinates and $p_{\max}$ refers to the peak pressure obtained at the center (x_c, y_c). a and b represent the half contact widths in x and y directions respectively. Shear surface traction $t_x(x)$ along the rolling direction is proportional to the normal pressure according to the friction coefficient μ_s (Eq. (4)).

$$t_x(x) = \mu_s |p(x, y)| \tag{4}$$

3.2 *Sub Model*

An inclusion was introduced in a sub model, where the macroscopic stresses calculated from the global model were applied as a boundary condition to transfer the contact stresses towards the inclusions, refer Fig. 6. In this work, two types of inclusion (Al_2O_3 and MnS) were studied. The inclusion was positioned where the

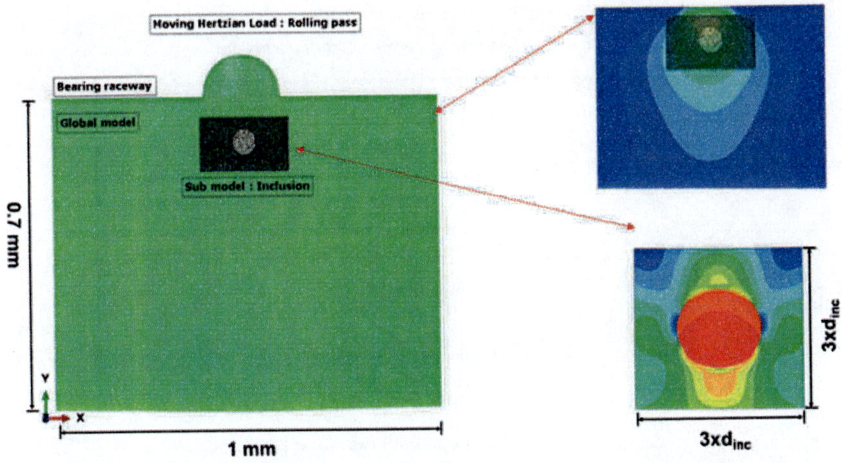

Fig. 6 Finite element model of rolling pass: global model and sub model with inclusion

maximum subsurface shear stress occurs (τ_{max}) according to Eq. (5). Here, z represents the depth below the raceway and a refers to the half-contact width. The sub model size was chosen as a square with a length of 3 times the diameter of the inclusions (d_{inc}) [8].

$$\tau_{max} = 0.30p_o, z = 0.78a \tag{5}$$

The resulting stress distributions under the surface were validated using the analytical solutions for both line and circular contact (see Fig. 7).

3.3 Multi-axial Critical Plane Approach

To take non-proportionality and the time dependent multiaxial stress state into account during the rolling contact process, different multiaxial fatigue damage criteria have been reported in literature among which the critical plane based fatigue prediction approach is widely accepted for RCF initiated bearing failures [11, 12]. Fatemi and Socie (F–S) developed a shear dominated multi-axial fatigue model for both high cycle fatigue (HCF) and low cycle fatigue (LCF) failure types [13]. As RCF failures are predominantly shear based [12], this F–S approach has been used in this study to investigate the fatigue crack initiation hotspots around the inclusions at the continuum scale.

The Fatemi–Socie multiaxial fatigue model can be described as per below Eq. (6)

$$FS(\theta) = \left(1 + k.\frac{\sigma_n(\theta)}{\sigma_{ys}}\right).\frac{\Delta\gamma(\theta)}{2} \tag{6}$$

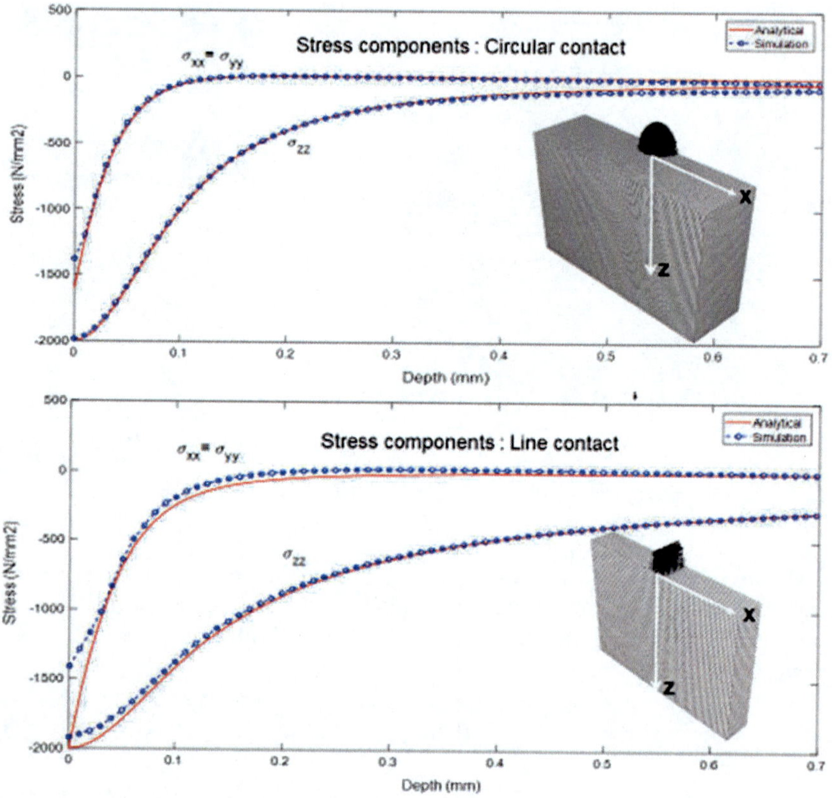

Fig. 7 Validation of FE global model for circular and line contact (without traction)

where k is a material parameter, σ_{ys} represents the yield strength, $\sigma_n(\theta)$ represents the maximum normal stress perpendicular to the material plane and $\Delta\gamma(\theta)$ is the shear strain range on the material plane. The plane orientation is represented by its angle θ with respect to the x-axis. The critical plane corresponds with a maximum value for $\Delta\gamma(\theta)$.

From the Fatemi–Socie model, we understand that the damage tends to initiate on the plane with maximum shear strain range and the corresponding normal stress acting on the maximum shear strain plane accelerates the damage process by facilitating the crack opening. The maximum normal stress is normalized by the yield strength as seen in the equation. The material constant k is further arbitrarily taken as unity [12].

For this study, a script was developed to calculate the fatigue hotspots around the inclusion-matrix interface using the F-S model described above. The script uses the stress results from the FE model to calculate shear strain and normal stress evolution along all possible planes using 2-D transformation laws (Mohr's circle). Then, the script uses a search algorithm to identify the plane with highest value of shear strain

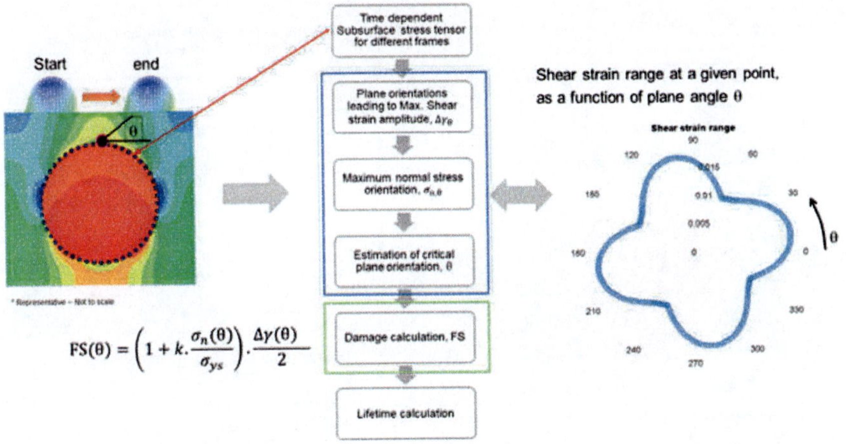

Fig. 8 Fatemi–Socie critical plane algorithm for subsurface inclusions

range and its corresponding normal stress value. This process is repeated around the inclusion-matrix interface to find the vulnerable regions for crack initiation. A flow chart explaining this approach is shown in Fig. 8.

An other method which has been reported in literature to predict crack initiation from non-metallic inclusions for rolling contact in bearings is the stress-based Dang Van (DV) multiaxial failure criterion. This method incorporates the time dependent maximum shear stress amplitude ($\tau_{max}(\theta, t)$) which is also a function of plane angle, and instantaneous hydrostatic stress tensor ($\sigma_H(t)$). This method relates the change in stress state at a point to a damage parameter which is generally a function of fatigue limits in pure torsion and pure bending. The underlying idea behind this DV approach is that a fatigue crack initiates when a critical value of the Dang Van parameter is exceeded. More details about this method can be found in [14, 15]. In this study, the results obtained from the F–S model are compared with the DV criteria as benchmark.

The shear stress amplitude (τ_a) is expressed as,

$$\tau_a(\theta, t) = |\tau(\theta, t) - \tau_m| \tag{7}$$

where τ represents shear stress on any plane (stress transformation laws) and τ_m is defined as half the sum of minimum and maximum shear stress on a plane.

The Dang Van criterion used in this study can be expressed as

$$\mathrm{DV}(\theta_c, t) = \tau_{max}(\theta_c, t) + \rho.\sigma_H(t) \tag{8}$$

where θ_c is the critical plane angle according to where τ_a reaches its maximum. ρ is a material parameter referred from [15].

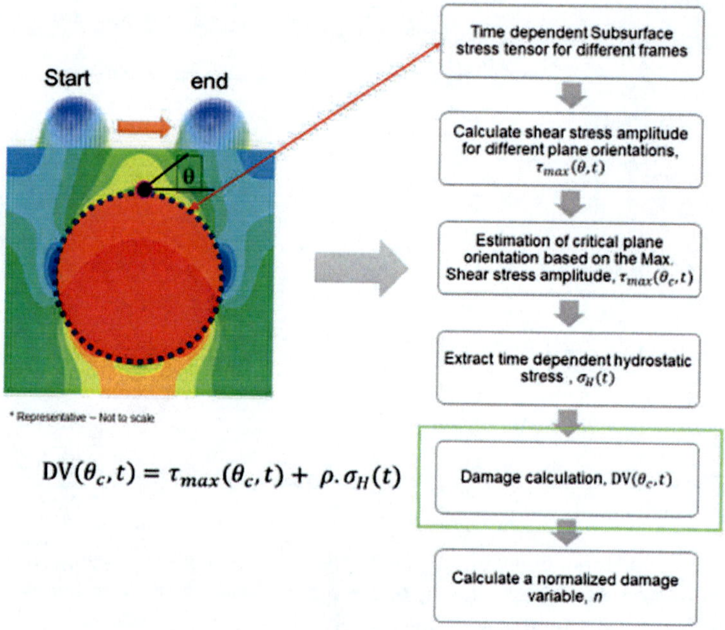

Fig. 9 Dang Van critical plane algorithm for subsurface inclusions

For this study, a script to calculate a normalized damage variable n [15] around the inclusion-matrix interface was developed to identify the fatigue hotspots around the inclusions. The flowchart explaining the implementation process is shown in Fig. 9.

4 Case Studies

As discussed higher, RCF crack nucleation is assumed to start at the subsurface inclusions due to the cyclic contact load acting between the roller and raceway. In this case study two different types of harmful inclusions (Al_2O_3 and MnS) have been studied (refer Fig. 10). The first type is a circular alumina inclusion with 0.01 mm radius and $\nu = 0.3$, $E = 380$ GPa. The second type focuses on elliptical MnS inclusions inclined 45° to the rolling direction; its dimensions are 0.0096 mm (semi-minor length) and 0.028 mm (semi-major length) [5]. The mechanical properties for the MnS are taken as $\nu = 0.25$, $E = 108$ GPa. A 3 μm size mesh and CPE4R element type is used for the presented case. The matrix-inclusion interfaces were assumed to be perfectly bonded.

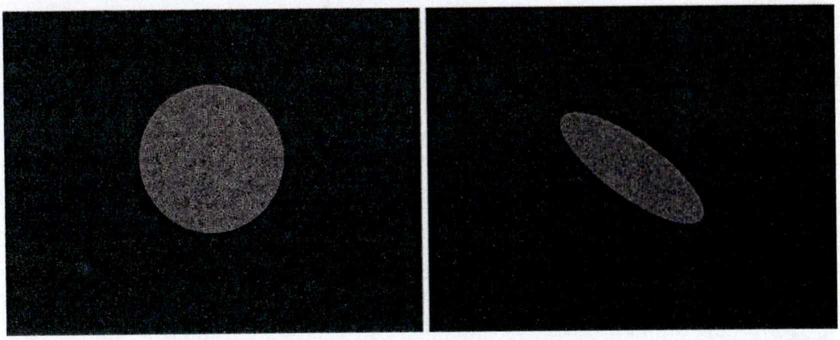

Fig. 10 Alumina inclusion (Left), Manganese sulfide inclusion (Right)

4.1 *Effect of Inclusion Shape on Damage Profile*

4.1.1 Damage Initiation Around Circular Inclusion

Figure 11 shows the F–S damage contour around the circular alumina inclusion. As the failure is expected to take place in the steel matrix rather than in the inclusion, these damage plots are taken from the steel matrix close to the matrix-inclusion interface. The highest damage for this intact perfectly bonded inclusion is reached at the four poles (top, bottom, left and right). These results indicate that crack initiation is most likely to happen at these hot spots. As the Fatemi–Socie model uses the shear strain range as primary factor and the maximum normal stress as secondary, these values are extracted around the inclusion and plotted in Fig. 11.

The F–S damage regions predicted from this study are found to be in agreement with studies from literature [15]. As benchmark, stress based multiaxial Dang Van

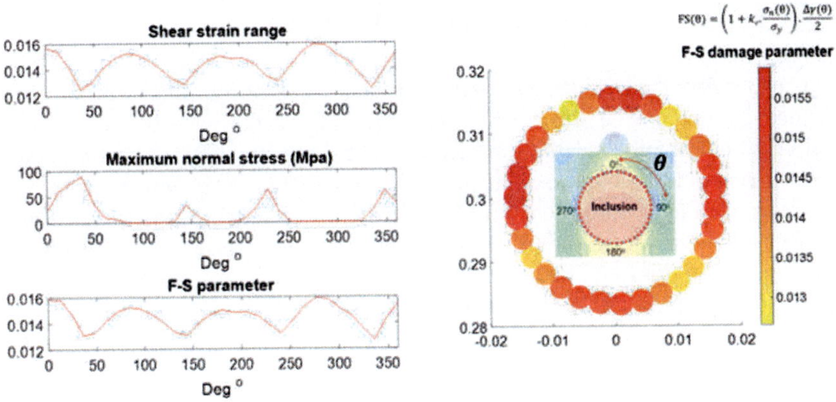

Fig. 11 F–S damage profile around the circular inclusion

damage criterion is implemented. Figure 12b shows the normalized multi-axial Dang Van (DV) damage factor (*n*) for material points around the matrix-inclusion interface for the case of a circular alumina inclusion under Hertzian loading with a maximum contact pressure of 1 GPa. All material properties and fatigue limits used in this DV damage calculation were referred from work of Cerullo [15]. The DV damage factor *n* at a material point around the interface is calculated based on the time dependent shear stress amplitude and hydrostatic stresses extracted from the submodel. From the results (Fig. 12a and b) we observe that the highest damage factor in both cases are predicted to occur at the four poles of the inclusion. The results predicted from this study are in qualitative agreement with the previous work from Cerullo (see Fig. 12c). This supports the observation that a fatigue crack is likely to initiate from one of these poles for the case of a circular alumina inclusion. Despite being different multiaxial approaches, the similarity between the results of the F–S and DV analyses can be motivated due to the predominant influence of shear stress/strain amplitude. As can be seen in Fig. 11, the F–S damage factor is dominated by shear strain range as the normal stress reaches close to zero at these regions, which makes the damage factor independent of normal stress and purely shear dominated. Similar patterns can be observed from Fig. 12b, where the DV damage factor *n* is shear stress dominated with less dependency on hydrostatic stress. Note that, for the presented case study, F–S and DV predict very similar damage hotspots, and that this is logical because both criteria are dominated by shear stress/strain range.

Figure 12d is cited from the work of Grabulov [7], illustrating the butterfly crack initiation formed at an artificially introduced alumina inclusion with a diameter of 0.045 mm and located at a depth of 0.040 mm from the raceway. From the figure, we see that the initial crack nucleates from one of the damage spots (right pole in this case) predicted by the numerical model before the damage shifts 45° towards the raceway. The crack initiates at the matrix-inclusion interface which is in line with our previous assumption that failure occurs in the matrix rather than in the inclusion.

4.1.2 Damage Initiation Around Elliptical Inclusion

Figure 13 shows the F–S damage contour around the elliptical MnS inclusion. The results show a significant increase in stress concentration at the inclusion tips. Accordingly, the highest fatigue damage is predicted at the tips (top left and bottom right), which indicate the possible crack initiation sites. Figure 14a and b shows the evidence of subsurface cracks initiated from MnS inclusions when the inclusion is viewed axially and circumferentially [6]. This damage prediction correlates well with the experimental findings where cracks initiating from MnS inclusion tips were identified to be one of the potential WEC damage initiators.

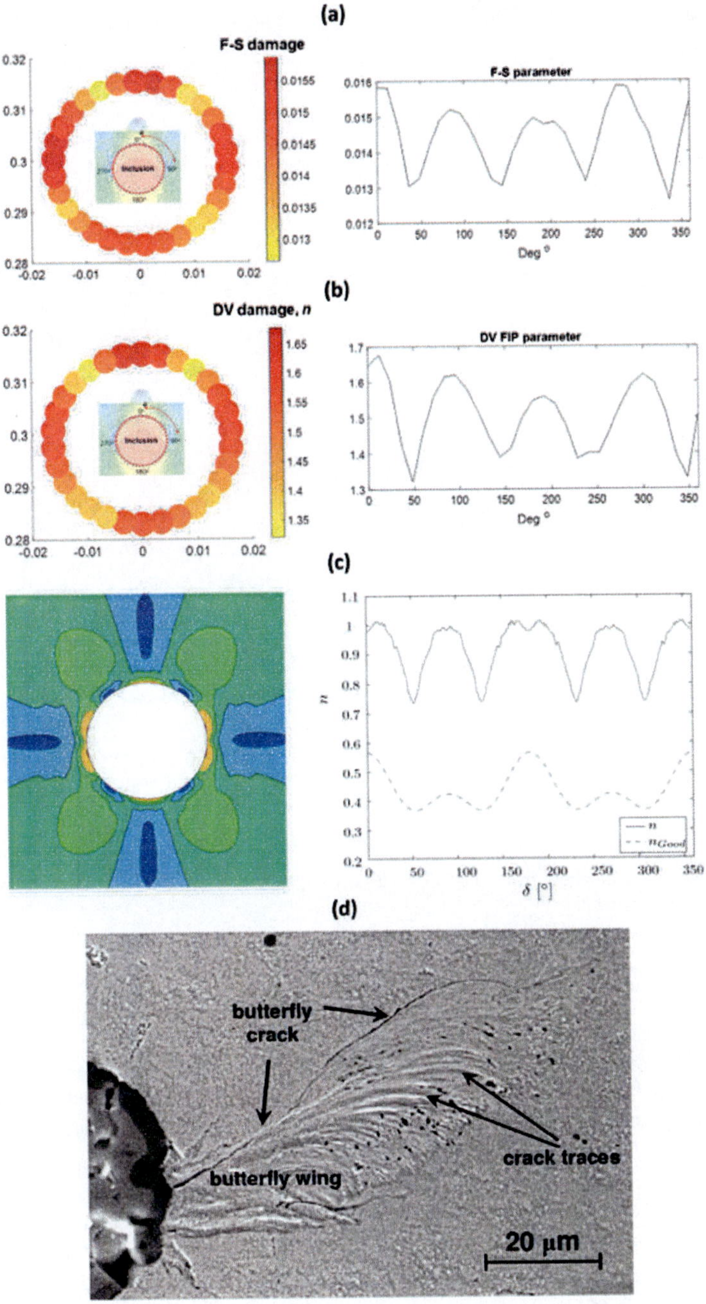

Fig. 12 **a** F–S damage distribution around the alumina inclusion, **b** DV damage distribution around the alumina inclusion, **c** Dang Van and Goodman damage distribution around the alumina inclusion [15], **d** butterfly crack initiating at a spherical alumina inclusion from a damage hotspot [7]

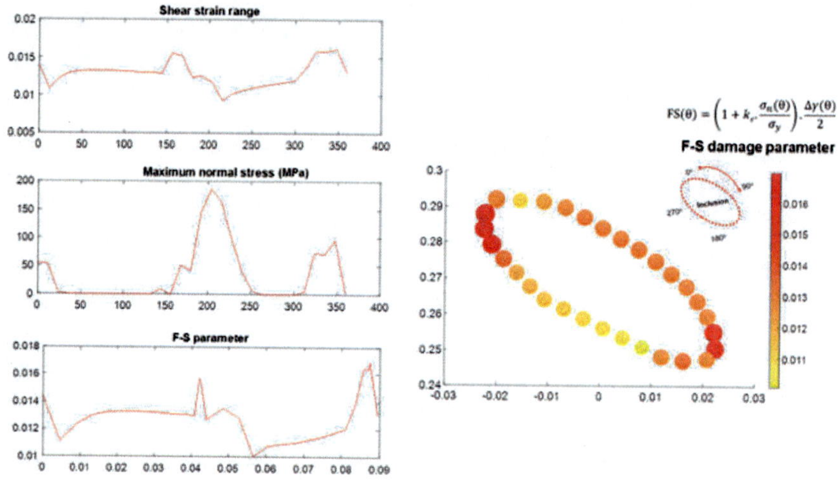

Fig. 13 F–S damage profile around the elliptical inclusion

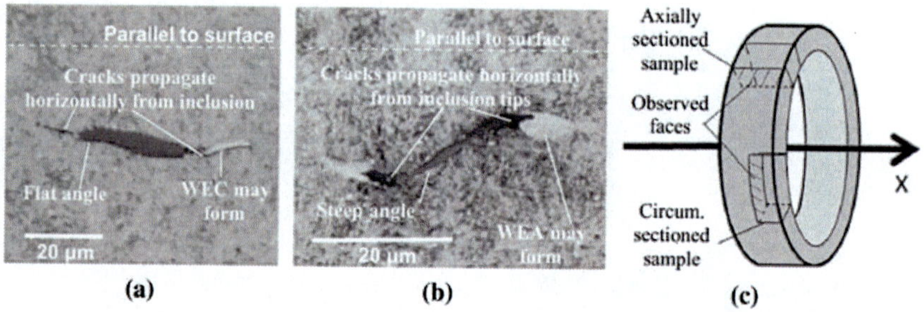

Fig. 14 Crack nucleation from a MnS inclusion [6]. **a** Axial view. **b** Circumferential view. **c** Specimen representation

4.2 Effect of Surface Traction

It has been observed that surface traction plays a significant role in premature bearing failures [16]. As reported from previous studies, significant sliding happens during transient operating conditions and torque reversals take place during dynamic events such as emergency braking, startup and shutdown [17]. This introduces unfavorable loading conditions over the entire wind turbine drive train. It has been reported that, this surface friction has a major influence on micro-cracks initiation at the inclusion interface leading to surface spalling [18]. In order to know the impact of this phenomenon, the surface traction was varied by changing the friction coefficient between 0 and 0.3 [5] and its influence on the damage initiation was studied.

From the results, it is seen that increasing friction between roller and raceway slightly increases fatigue damage and shifts the location of the hot spots. This means

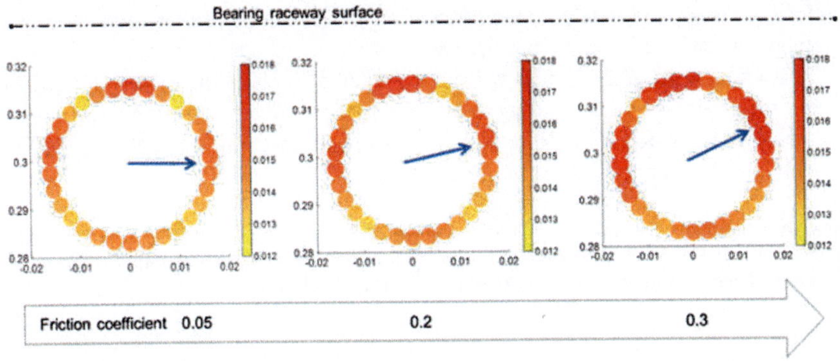

Fig. 15 Impact of surface traction on F–S damage profile around alumina inclusion (all at same depth below the surface)

that the cracks that initiated at shallow depths have a higher chance of leading to surface damage. Also, introduction of surface traction tilts the maximum damage value towards the raceway as shown by the arrow, see Fig. 15 similar result was observed by Bruce for a void under the effect of surface traction [16].

5 Conclusion

This work presented a simplified 2D numerical framework to predict RCF crack initiation from inhomogeneities present in wind turbine bearings. First, a global FE model was developed that uses a moving contact pressure to represent the rolling pass and to realistically estimate the subsurface stresses under elastic plane strain conditions. These stress- strain responses are fed as displacement boundary condition to a submodel containing an inclusion. The stress results of that submodel are used in conjunction with continuum based damage to predict the fatigue hotspots around the inclusions. The damage models used in this case are the Fatemi–Socie and Dang Van multi-axial, non-proportional critical plane models. The results of both models have also been compared with different numerical method and experimental results from the literature. From this study, it can be concluded that the developed method is capable of predicting fatigue crack initiation on a continuum basis using the bulk material properties.

Numerical case studies were carried out to predict the location of crack initiation around both circular and elliptical different inclusion shapes (Al_2O_3 and MnS). The effect of surface traction on the damage initiation was also studied, which shows that the damage value increased and tilted towards the raceway signaling premature surface damage. A further advantage of this framework is that in addition to information on damage initiation, it also provides information on critical regions around the inclusion, and on the critical plane orientation which can be further used for crack propagation and total lifetime predictions.

For future studies, the model will be extended to include the effect of residual stress, matrix-inclusion debonding and probabilistic analysis of the effects of inclusion size and position. This RCF framework will be coupled to a statistical WEC modeling framework as part of future scope.

Acknowledgements The authors gratefully acknowledge the financial support via the MaDurOS program from VLAIO (Flemish Agency for Innovation and Entrepreneurship) and SIM (Strategic Initiative Materials) through project SBO MaSiWEC (HBC.2017.0606).

References

1. Lai J, Stadler K (2016) Investigation on the mechanisms of white etching crack (WEC) formation in rolling contact fatigue and identification of a root cause for bearing premature failure. Wear 364–365:244–256. https://doi.org/10.1016/j.wear.2016.08.001
2. Slack TS, Raje N, Arakere NK, Sadeghi F (2009) A review of rolling contact fatigue. J Tribol 131:041403. https://doi.org/10.1115/1.3209132
3. Özden UA, Giang NA, Chen G, Bezold A, Broeckmann C (2014) A FEM model for lifetime prediction of Ancorsteel1000C in rolling contact fatigue. PAMM 14:167–168. https://doi.org/10.1002/pamm.201410071
4. Sheng S, Link H, LaCava W, van Dam J, McNiff B, Veers P, Keller J, Butterfield S, Oyague F (2011) Wind turbine drivetrain condition monitoring during grc phase 1 and phase 2 testing. https://doi.org/10.2172/1027157
5. Al-Tameemi HA, Long H (2019) Finite element simulation of subsurface initiated damage from non-metallic inclusions in wind turbine gearbox bearings. Int J Fatigue 105347. https://doi.org/10.1016/j.ijfatigue.2019.105347
6. Bruce T, Rounding E, Long H, Dwyer-Joyce RS (2015) Characterisation of white etching crack damage in wind turbine gearbox bearings. Wear 338–339:164–177. https://doi.org/10.1016/j.wear.2015.06.008
7. Grabulov A (2010) Fundamentals of rolling contact fatigue. Technische Universiteit Delft
8. Moghaddam SM (2016) Analytical and experimental investigation of microstructural alterations in bearing steel in rolling contact fatigue. Purdue University
9. Gazizulin D, Klein R, Bortman J (2017) Towards efficient spall generation simulation in rolling element bearing. Fatigue Fract Eng Mater Struct 40:1389–1405. https://doi.org/10.1111/ffe.12580
10. Bomidi JAR, Weinzapfel N, Sadeghi F, Liebel A, Weber J (2013) An improved approach for 3d rolling contact fatigue simulations with microstructure topology. Tribol Trans 56:385–399. https://doi.org/10.1080/10402004.2012.754072
11. Basan R, Marohnić T (2018) Multiaxial fatigue life calculation model for engineering components in rolling-sliding contact. MATEC Web Conf 188:02016. https://doi.org/10.1051/matecconf/201818802016
12. Sauvage P (2018) On an extension of the fatemi and socie equation for rolling contact in rolling bearings. In: FFW: Proceedings of the 7th international conference on fracture fatigue and wear, pp. 438–457
13. Fatemi A, Socie DF (1988) A critical plane approach to multiaxial fatigue damage including out-of-phase loading. Fatigue Fract Eng Mater Struct 11:149–165. https://doi.org/10.1111/j.1460-2695.1988.tb01169.x

14. Liu H, Liu H, Zhu C, He H, Wei P (2018) Evaluation of contact fatigue life of a wind turbine gear pair considering residual stress. J Tribol 140:041102. https://doi.org/10.1115/1.4039164
15. Cerullo M, Tvergaard V (2015) Micromechanical study of the effect of inclusions on fatigue failure in a roller bearing. Int J Struct Integrity 6:124–141. https://doi.org/10.1108/IJSI-04-2014-0020
16. Bruce T (2016) Analysis of the premature failure of wind turbine gearbox bearings. The University of Sheffield
17. Verstraeten T, Nowe A, Keller J, Guo Y, Sheng S, Helsen J (2019) Fleetwide data-enabled reliability improvement of wind turbines. Renew Sustain Energy Rev 109:428–437
18. Bruce T, Long H, Dwyer-Joyce RS (2018) Threshold maps for inclusion-initiated micro-cracks and white etching areas in bearing steel: the role of impact loading and surface sliding. Tribology Lett 66

Fatigue of Multiaxially Loaded Shaft-Hub Connection Under Different Load Parameters

Lukáš Suchý and Alexander Hasse

Abstract In addition to free-surfaced components, which have predictable fatigue life, the forces in drive train systems inevitably also pass through contacting surfaces. Failures, that occur at shaft-hub connections, are caused by high stress concentration and tribological loading. In the present paper, a test rig developed in-house for combined rotating bending load and alternating torsion is introduced. Experimental results obtained from the test rig on high-cycle fatigue (10^7 cycles) of press-fit connections are presented. To eliminate the tribological effects, a notched specimen from the same material is examined with corresponding parameter. The phase shift as well as amplitude ratio were investigated on normalised steel 1045. Compared to the calculation standards, which are frequently used in the industry, lower fatigue limits are obtained from experimental results. For a better consideration of the multiaxial fatigue, different estimation calculations are performed. Using numerical investigations, the calculation approaches based on integral methods are applied in combination with a critical distance approach. The smallest calculation error is reached at the critical distance of 0.096 mm.

Keywords Press fit · Shaft-hub connection · High-cycle fatigue · Contact · Experimental · Multiaxial fatigue · Rotating bending · Torsion · Integral fatigue approach

1 Introduction

Along the force flow in drivetrains, primary and secondary loads lead to complex stress states in joined components. In the case of these rotating components, the combined loading consists of the alternating torque and the rotating bending moment. Additionally, friction type connections exhibit a normal pressure and contact shear

L. Suchý (✉) · A. Hasse
Technische Universität Chemnitz, 09107 Chemnitz, Germany
e-mail: lukas.suchy@mb.tu-chemnitz.de

Institute of Design Engineering and Drive Technology, Chemnitz, Germany

M. Abdel Wahab (ed.), *Proceedings of the 8th International Conference on Fracture, Fatigue and Wear*, Lecture Notes in Mechanical Engineering,
https://doi.org/10.1007/978-981-15-9893-7_34

stresses. These secondary parameters have a constant initial value but vary with introduced loads. As a result, the multiaxial stress state lead to failure of shaft-hub connections.

To improve the fatigue life, of shaft-hub-connections, local time-dependent stress components at failure points are input for the high-cycle-fatigue strength prediction. Several approaches are used for the multiaxial fatigue of specimens on free surfaces. Depending on the loading situation, stress-based calculations using the critical plane approach (CPA) or integral approach (IA) provide acceptable results for the fatigue limit of plain specimen. An extensive overview is given in a survey by Papuga [1]. 17 different methods compared with a series of 407 experiments. In this study, Papuga introduced also his own criteria, integral-plane (PI) and critical-plane (PCR) types [2]. In separated study [3], in which only integral approaches are compared, a method from LIU and Zenner [4] and the extended PI criterion (PIN) [3] results in smallest deviations. Further investigated integral methods in [3] were developed by Papadopoulos [14], and Fogue/Kenmeugne [15].

In contrast to the fatigue of free surfaces, estimation of fretting fatigue for contact components is more challenging owing to additional parameters. Some authors [5–7] propose critical-plane methods for fatigue limit calculations of basic uniaxial tensile testing with fretting pads. The calibration occurs on a tensile specimen with additional tribological load. In the experimental process, fretting pads provide a slip movement under defined normal forces during the tensile testing. Several pad contours and contact types are used to calibrate such models. The position of the evaluation point depends on theory of critical distance (TCD) [8]. By using an elastic stress field in vicinity of the notch, TCD considers the gradient effect. The fracture mechanical property of threshold value, point, line, or area method can be used, to improve the calculation accuracy. According to Araújo [5, 9], half of the maximum non-propagating crack length (MNC) is used for the calculation and leads to better results than the hot-spot approach. For assessing this material specific length, El Haddad [10] proposed the relation.

$$a_0 = 1/\pi \cdot (\Delta K_{th}/\Delta\sigma_0)^2. \tag{1}$$

ΔK_{th} corresponds to the range of the threshold value of the stress intensity factor. $\Delta\sigma_0$ represents the range of uniaxial plain fatigue in reverse mode.

Atzori provides in [11] an overview of necessary material fatigue properties for different steel and an improvement for assessing the MNC for steels with.

$$a_{0,Atzori} = \left(\frac{36200}{\sigma_{UTS} \cdot \Delta\sigma_0}\right)^{1.17}, \tag{2}$$

where ultimate tensile stress σ_{UTS} is also taken into account. The corresponding position for evaluating the stress state is at

$$L = a_0/2 \tag{3}$$

Using his experiments, Nesladek [6] introduced an optical slip measurement system and compared in [12] experimental fretting fatigue with critical plane based criteria from [1]. His results show that McDiarmid critical-plane method was the most accurate for the investigated contact cases. Authors in [13] obtain the same result. Integral approaches were not considered in these investigations.

During the design phase, the fatigue limit of a shrink fit is typically calculated in a nominal approach using experimentally estimated notch factors. Calculation standards and guidelines [14, 15] show material-dependent notch factors for limited geometry definition for fully reversing torsional and bending load. Additionally, the calculation of the notch factor for shrink fit as a function of diameter and interference is introduced in [15]. For the strength assessment of combined loads, the calculation is based on Gough's ellipse formulation [16]

$$\frac{\sigma_a^2}{f_{-1}^2} + \frac{\tau_a^2}{t_{-1}^2} = 1. \tag{4}$$

Here σ_a and τ_a represents the stress amplitudes, and f_{-1} and t_{-1} the plain-reversed-fatigue limit for axial load and bending.

Within this framework further effects on fatigue are considered in a nominal approach in mentioned standards, including effects of size, support, or stress gradient, in the numerator term instead of plain fatigue values in Eq. (4). It is remarked that the validity of the calculation according to the standards is limited to proportional stress states. With these complex stress states in component scenarios, especially in shaft-hub connections, merging the sole load notch factors can lead to abbreviations in fatigue limit calculations.

In general, by combining circumferential bending and alternating torsion several parameters can be considered (Table 1). The ratio of the load frequencies and amplitudes of each load is described by

$$V_f = f_{\sigma_b}/f_\tau \tag{5}$$

and

$$V_A = \sigma_{ba}/\tau_{ta}, \tag{6}$$

Table 1 Multi-channel load parameters

Frequency ratio	Amplitude ratio	Torsion stress ratio	Phase shift
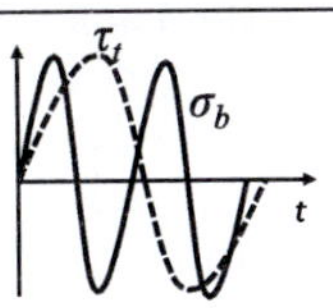	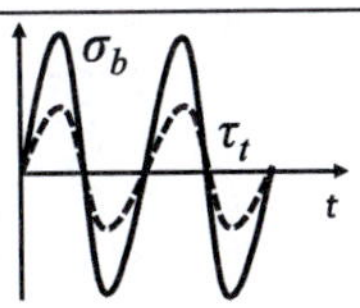	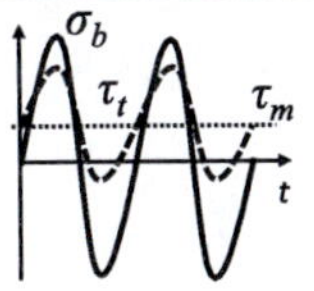	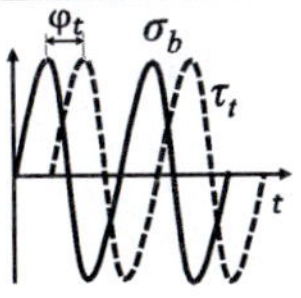

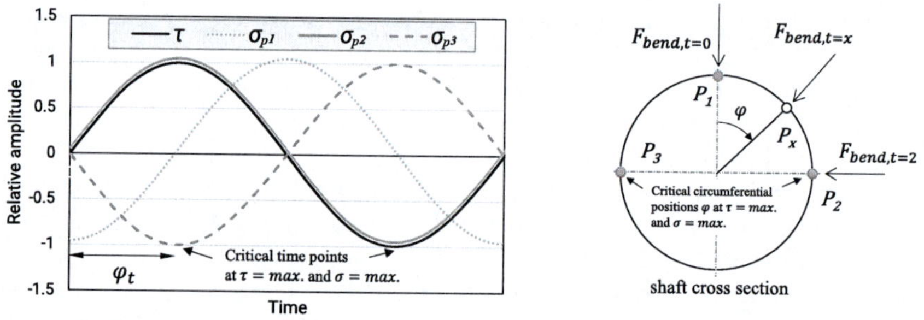

Fig. 1 Evolution of circumference position coordinate from time dependent course of bending and torque load at equal frequency and amplitude

σ_{ba} and τ_{ta} correspond to the nominal bending and torsion amplitudes, while f represents their frequencies.

The torsional stress ratio is calculated using

$$R_{\tau} = \tau_{\max}/\tau_{\min}, \tag{7}$$

where the maximum $\tau_{\max}$ and minimum $\tau_{\min}$ values represent the range of stress in the time domain. Using described loads, the phase shift parameter is by nature connected to the spatial φ coordinate on the circumference of shaft specimen. Figure 1 depicts the relation of the load phase shift in the time domain and the circumference position in the distance/path domain.

In the case of equal frequencies, two zones of maximal equivalent stress (von Mises) are achieved at the phase shifts of $\varphi_t = 90°$ and 270°, where both load channels reach their maximum values. As the bending load is omnipresent and only changes its circumferential position, the points of highest stress (P_2, P_3) correspond to the position of the bending force when torque load peaks at ¼ and ¾ of the time (Fig. 1). Considering this relationship, every phase shift value is present on circumference of the shaft during the test.

In this paper, the integral approaches are investigated within the scope of shaft-hub connection. In a comparing study [3], criteria from the shear stress intensity hypothesis (SIH) by Liu/Zenner [13] and the PIN method by Papuga [2, 3], resulted in the smallest error indices among the methods with integral approach owing to the consideration of shear mean stress for each evaluated stress plane.

Using these criteria, experimental results are input to the finite element analysis (FEA) of the stress state. Regarding to the presented literature, the evaluation positions vary in the corresponding range. Subsequently, the stress state should be evaluated over the shaft circumference.

2 Experimental Setup

2.1 Specimen Characterization

The material properties in Table 2 were achieved in tensile testing in terms of [17]. Table 3 shows the geometry of the shrink-fit connection. The shaft and hub were produced from the same normalised material 1045.

Relative interference ξw is the quotient of diametric interference I and the shaft diameter D_F according to DIN 7190 [18]. It results in a nominal contact pressure of 122.4 MPa. The exact interference was measured for each connection (see Table 4). The hub edge is sharp cornered. The joining process was performed in a dry and cleaned state at room temperature of 20 °C by shaft temperature of −30 °C and hub temperature of 250 °C. Rest period was at least 24 h at room temperature and normal air moisture of 52% ± 10%.

Additionally, a fillet shaft produced from same material was investigated. The corresponding roughness was $Rz = 2.5$ μm.

Table 2 Material properties of 1045 + N

	Young's modulus	Yield stress	Tensile strength	Elongation
	E (GPa)	$R_{p0.2}$ (MPa)	σ_{UTS} (MPa)	A (%)
Shaft	190	313	622	25.5
Hub	200	352	645	21.2

Table 3 Specimen geometries

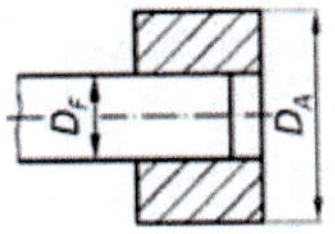

Interference fit connection

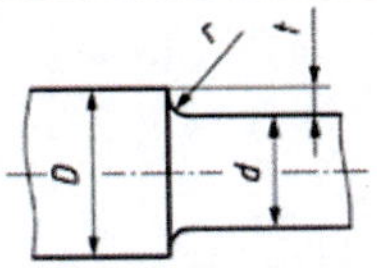

Filleted shaft

Shaft diameter	D_F	40 mm	*Shaft diameter*	d	40 mm
Hub thickness Q_A	D_F/D_A	0.5	*Fillet height*	t	4 mm
Relative interference ξ_w	I/D_F	1.7 ‰	*Radius*	r	0.4 mm
Relative joining length l_{rel}	l_F/D_F	0.9	*Fillet diameter*	D	48 mm
Material	1045+N (hub + shaft)		Material		1045+N

2.2 Test Performance

For the purpose of superimposing the presented loads, a test bench for combined loads was developed (Fig. 2).The rotating bending load was applied in four-point bending mode by magnetic forces. A hydraulic drive provided the alternating torque.

The maximum number of load cycles is defined at 10^7 load cycles for a shrink-fit connection and $5 \cdot 10^6$ cycles for a notched shaft. Due to security issues, the switch-off criterion of each test was defined by loss of stiffness due to crack-related reduction of the specimen cross section, measured by the radial displacement.

The investigated parameter was the amplitude ratio $V_A = 1$ and 0.5 for each specimen type as well as the phase shift along the specimen circumference.

3 Experimental Results

The fatigue limit σ_{ba} is calculated according to the staircase method of Hück [19], which corresponds to 50% reliability and is calculated by.

$$\sigma_{ba} = \sigma_{a0} \cdot d^{A/F}, \tag{8}$$

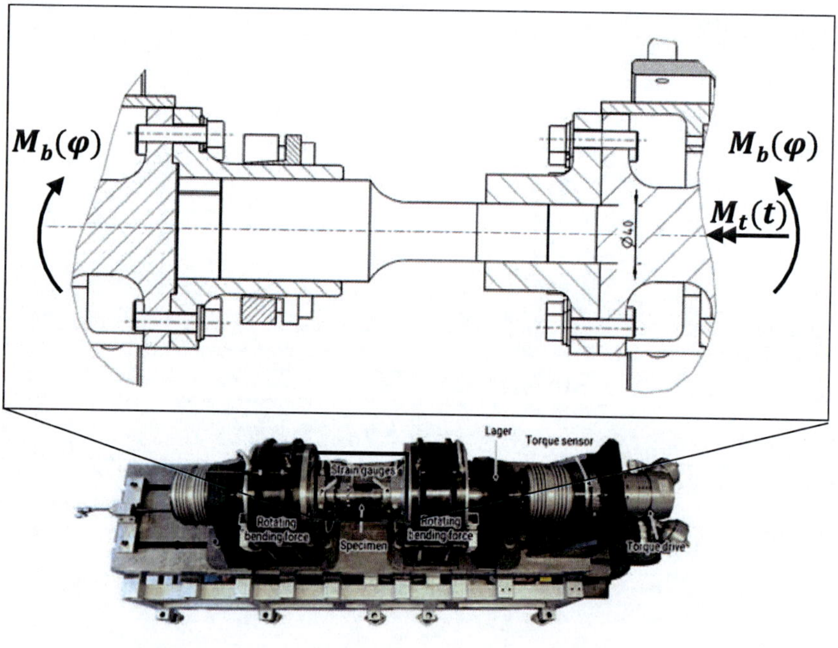

Fig. 2 Test bench for combined rotating bending and alternating torsion

σ_{a0} corresponds to the lowest level of the load staircase. d is the step interval. Values F and A summarise all events at evaluated staircase levels according to [19] and [20].

For this purpose, each event on the repeated stress level is considered independent of the result (failure or run-out). Therefore, the consideration of one fictive result on level of potential last step can be taken into account. Considering Eq. (6) the fatigue limit τ_{ta} can be derived from σ_{ba}.Tables 4 and 5 shows the experimental results for each specimen at each step.

A constant step size was chosen for the test following [20]. In addition, the measured interference between the shaft and hub for each connection was outlined. The bending and torsional loads are denoted as nominal. The corresponding standard deviation of real load resulted in 2.14% for bending and 0.46% for torsion.

Table 4 Results of fatigue experiments for shrink fit

Test	Nominal bending stress	Nominal torsion stress	Stress ratio	Radial interference	Number of cycles
	σ_n (MPa)	τ_n (MPa)	V_A	U (mm)	N (cycles)
1	100	100	1.0	6.5E−02	1,462,461
2	80	80	1.0	6.8E−02	7,601,235
3	75	75	1.0	6.8E−02	8,218,955
4	70	70	1.0	6.45E−02	10,000,000
5	75	75	1.0	7.0E−02	7,234,998
6	70	70	1.0	6.75E−02	10,000,000
7	75	75	1.0	6.85E−02	6,033,139
8	70	70	1.0	6.95E−02	5,012,603
9	65	65	1.0	6.65E−02	10,000,000
10	100	100	0.5	6.8E−02	3,450,123
11	95	95	0.5	6.85E−02	7,028,817
12	90	90	0.5	6.7E−02	10,000,000
13	95	95	0.5	7.05E−02	9,121,256
14	90	90	0.5	6.9E−02	6,273,407
15	85	85	0.5	6.95E−02	10,000,000
16	90	90	0.5	6.6E−02	8,372,210
17	85	85	0.5	7.0E−02	5,230,150
18	80	80	0.5	6.8E−02	10,00,0000

4 Fatigue Calculation

4.1 Finite Element Analysis

FEA of the specimen geometry and load case was performed to achieve stress tensor as input for further calculations using the software version 14.4 of. The specimen geometries were modelled in full scale by simplificating the geometry in undesired areas (e.g. bolt holes). In contrast to this, Fig. 3 shows the evaluation areas, which are final meshed finely according to investigation from [21]. An minimal element edge of length of 0.008 mm was used in the area of interest. C3D8I (8-node linear brick) elements were used in the analysis.

For the contact definition, a penalty friction formulation with 1.25% elastic slip definition and a friction coefficient of 0.8 were chosen, which corresponds to the realistic value of developed friction due to the fretting mechanism according to [22]. Normal contact behaviour is set to "hard" contact. An average interference value of 0.0675 mm from Table 4 was used in contact between shaft and hub.

The linear-elastic analysis was performed in three steps. First, frictionless interference was applied, the friction is activated in next step, and, finally, loads are applied on the shaft. The load values were derived by Eq. (7) and are listed in Table 6. The load definition of rotating bending momentum as well as the alternating torque was defined by harmonic functions. Corresponding to the four-point-bending test setup (Fig. 3), the distribution of the bending momentum is constant in the axial direction.

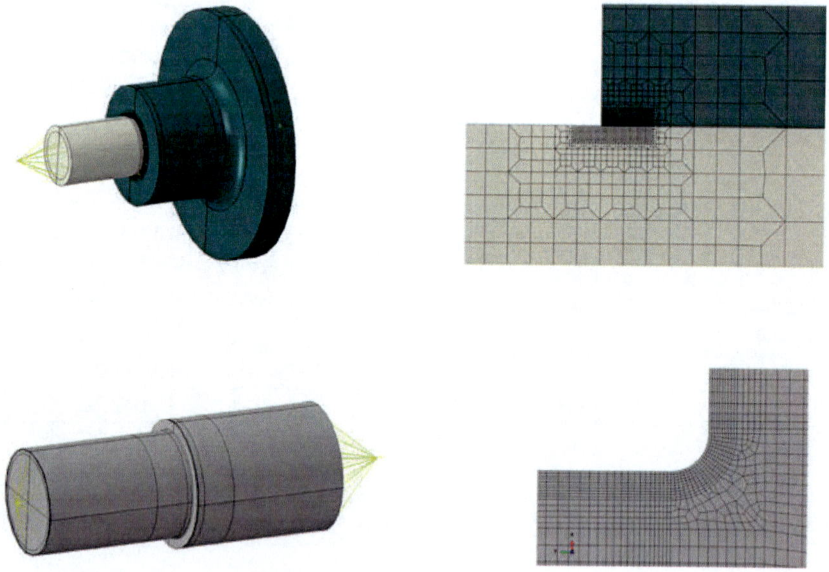

Fig. 3 Model of shrink-fit assembly and fillet shaft

Table 5 Results of fatigue experiments for filled shaft

Test	Nominal bending stress	Nominal torsion stress	Stress ratio	Radial interference	Number of cycles
	σ_n (MPa)	τ_n (MPa	V_A	U (mm)	N
17	80	80	1.0	(free surf.)	1,300,053
18	75	75	1.0	(free surf.)	1,951,201
19	70	70	1.0	(free surf.)	2,327,938
20	65	65	1.0	(free surf.)	4,050,232
21	60	60	1.0	(free surf.)	5,000,000
22	65	65	1.0	(free surf.)	5,000,000
23	70	70	1.0	(free surf.)	4,525,162
24	65	65	1.0	(free surf.)	3,632,624
25	60	60	1.0	(free surf.)	5,000,000
26	70	70	1.0	(free surf.)	2,360,095
26	75	75	0.5	(free surf.)	5,000,000
27	80	80	0.5	(free surf.)	3,486,601
28	75	75	0.5	(free surf.)	5,000,000
29	80	80	0.5	(free surf.)	3,531,400
30	75	75	0.5	(free surf.)	5,000,000
31	80	80	0.5	(free surf.)	5,000,000
32	85	85	0.5	(free surf.)	2,414,506
33	80	80	0.5	(free surf.)	5,000,000

Table 6 Input of failure criteria from experiment and literature

Type	Amplitude ratio	Specimen fatigue limit Ø40 mm			
	V_A	σ_{ba}	τ_{ta}	M_b	M_t
	(–)	(MPa)	(MPa)	(Nm)	(Nm)
Shrink fit	1	71.3	71.3	447.991	895.982
Shrink fit	0.5	88.3	44.15	554.805	1 109.610
Fillet	1	65.0	65.00	408.407	816.814
Fillet	0.5	78.5	39.25	493.230	986.460

The fixing point at the hub and loading point at the shaft were provided by a kinematic coupling constraint. All degrees of freedom were fixed in the hub coupling point during the last step.

4.2 Integral Approach at Critical Distance

Both investigated fatigue criteria, SIH in Eq. (9) and PIN in Eq. (10) are of similar design but of different arrangement and consider all four stress components acting on the cutting plane: shear stress amplitude C_a, mean stress C_m, normal stress amplitude N_a, and mean stress N_m. Each stress component is weighted by material parameters $a, b, c, d,$, which are derived from reversed and repeated torsion and axial fatigue limits and were taken from [4] and [3].

$$\sqrt{\frac{15}{8\pi}\int_{\varphi=0}^{2\pi}\int_{\psi=0}^{\pi}\left[a_{SIH}C_a^2\left(1+c_{SIH}C_m^2\right)+b_{SIH}N_a^2(1+d_{SIH}N_m))\right]\sin\psi d\psi d\varphi} \leq f_{-1} \tag{9}$$

$$\sqrt{\frac{1}{4\pi}\int_{\varphi=0}^{2\pi}\int_{\psi=0}^{\pi}[a_{PIN}C_a(C_a+c_{PIN}C_m)+b_{PIN}(N_a+d_{PIN}N_m)]\sin\psi d\psi d\varphi} \leq f_{-1} \tag{10}$$

Using the formula from Atzori et al. [11] for the non-propagating short crack length a_0, the evaluation position is defined at half distance $L = a_0/2 = 0.031$ mm according to Araújo [9]. Comparing to the calculation of El Haddad [10], the threshold crack value ΔK_{th} range is reviewed in the literature. For presently examined material, values between 8.07 MPa$\sqrt{m}$ and 13.6 MPa$\sqrt{m}$ at $R = -1$ are found in [11] and [23] and result according to Eq. (1) in $L = 0.0268$–0.0761 mm. Therefore, the presented criteria are performed in this range.

The integrals in Eq. (9) and (10) are solved numerically by sum of part solutions of each cutting plane. The estimation of the shear components is realised by the smallest circumscribed circle of the shear path (see also [24]).

4.3 Calculation Results

For each FEA, the equivalent stress is calculated by presented integral approaches using fatigue limits in Table 7. The repeated plain fatigue limit ($R = 0$) is obtained by using the formula from FKM-Guideline [15]

$$f_0 = 2 \cdot f_{-1}/\left(1 + 0.35 \times 10^{-3} \cdot \sigma_{UTS}/\text{MPa} - 0.1\right) \tag{11}$$

Table 7 Fatigue values 1045 + N

	f_{-1} (MPa)	f_0 (MPa)	t_{-1} (MPa)	t_0 (MPa)
1045 + N	292.34	480.15	168.78	304.43

For the repeated torsional fatigue limit, the approach

$$t_0 = 4 \cdot \frac{t_{-1}}{2 \cdot \frac{f_{-1}}{f_0 + 1}} \tag{12}$$

proposed by Liu and Zenner [4] is used.

For each calculation point, the accuracy is estimated using the formula

$$S = f_{-1}/\sigma_{criterion} \tag{13}$$

according to the current criterion. A value of $S = 1$ represents an exact prediction of the fatigue limit. A value less than $S = 1$ represents an optimistic prediction.

First, SIH and PIN are plotted for the fillet shaft surface over the circumference angle φ in Fig. 4. The evaluation was performed at critical distances $L = 0 - 0.09$ mm at radius position of highest maximal principal stress. Surprisingly, both criteria show very similar results for equivalent stress. With increasing evaluation distance L, the safety factor drifted between $V_A = 0.5$ and $V_A = 1$.

Evaluation of stresses in shaft of the shrink fit connection is performed at critical distances $L = 0$–0.096 mm. The evaluation position ranges over the crack threshold stress intensity ΔK_{th} values listed in the literature. Furthermore, the nearest mesh point to the evaluation according to Eq. (2) by Atzori [11] at $L = 0.024$ mm is

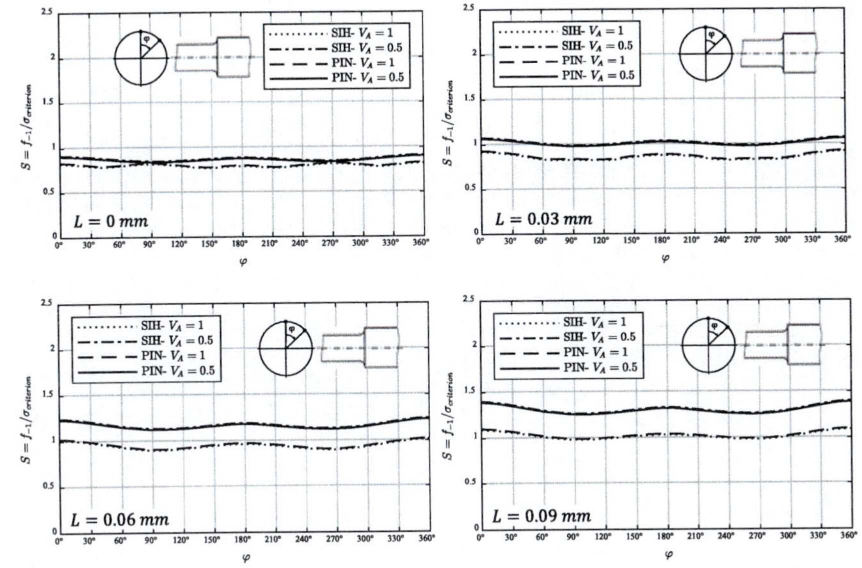

Fig. 4 Safety factors of SIH and PIN at combined bending and torque of fillet shaft in angle domain φ for $V_A = 1, 0.5$ at critical distances $L = 0 - 0.09$ mm

considered. The coordinate L runs radially inward from the shaft surface at the axial position of the hub corner.

For the smallest evaluation distance of $L = 0.024$ mm in Fig. 4, PIN and SIH predict failure for all circumference points similar to the evaluation on the shaft surface at $L = 0$ mm. The equivalent stresses of both criteria are too high. Increasing the critical distance, values of S rise. The only case reaching values of $S = 1$ was for $V_A = 0.5$ at $L = 0.040$ mm (Fig. 5). The overall results show optimistic estimations of press-fit fatigue limits. The stress state at further critical distances of up to $L = 0.096$ mm were also evaluated (Fig. 6). At these points the lowest average

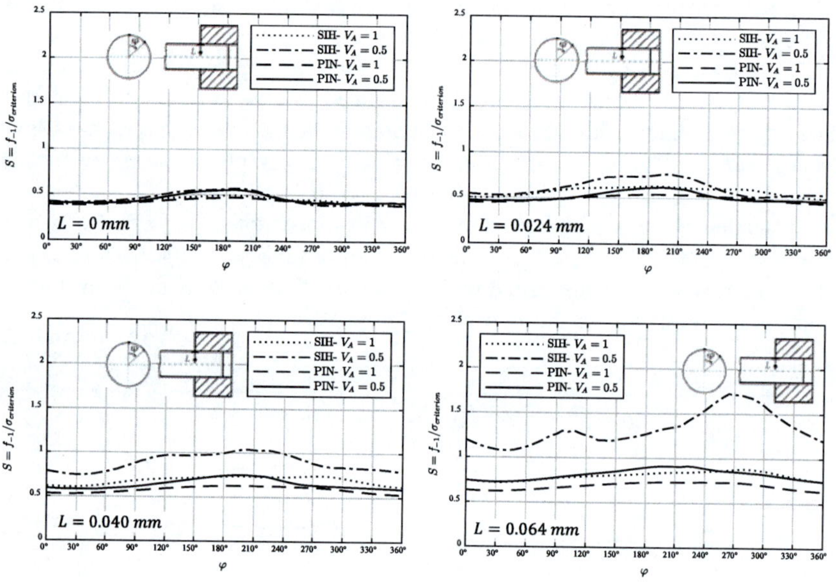

Fig. 5 IH and PIN at combined bending and torque of shrink fit with plain shaft in angle domain φ for $V_A = 1, 0.5$; $D_F = 40$ mm; $\xi_w = 1.7‰$; $L = 0$–0.064 mm

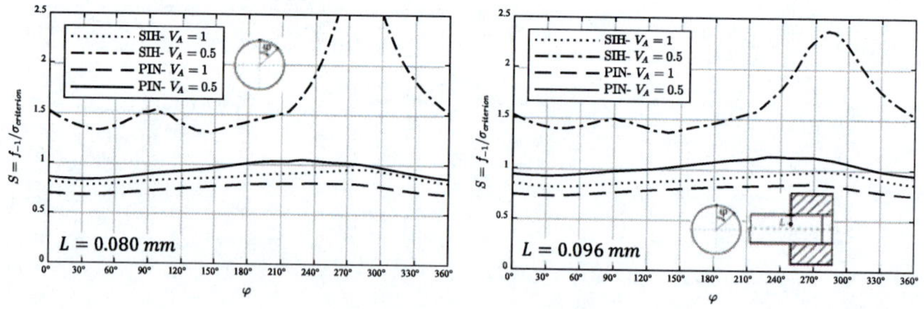

Fig. 6 IH and PIN at combined bending and torque of shrink fit with plain shaft in angle domain φ for $V_A = 1, 0.5$; $D_F = 40$ mm; $\xi_w = 1.7‰$; $L = 0.080$ mm and $L = 0.096$ mm

error of S was obtained. With the exception of S_{SIH} at $V_A = 0.5$, both criteria seem to commute around $S = 1$. The error range of PIN is from -27% to 13%. The error range of SIH is between -18% and 138%.

At critical distances starting at $L = 0.024$ mm, the SIH curves for $V_A = 0.5$ and $V_A = 1$ drift further apart.

5 Summary and Conclusion

In the present paper, the multi-channel high cycle fatigue of a press-fitted specimen and a fillet shaft from normalised 1045 was investigated. By varying of the load ratio of rotating bending and alternating torsion the fatigue limit was reached for each specimen type. This load combination results in different phase shift conditions along the specimen circumference.

Using the estimated results, an FEA was performed to obtain local stresses. According to the literature preview, the two best methods of integral type SIH and PIN were used to calculate equivalent stress over the circumference of the specimen. In the case of a shrink fit, the evaluation distance underneath the shaft surface was varied. Following the approach of El Haddad and Atzori, the fatigue evaluation points were estimated from material properties.

The results of both strength criteria show a sensitivity to phase shift. The best predictions for the fatigue limit are obtained out of the range of plausible distance according to the literature. The corresponding crack growth threshold value ΔK_{th} at this point is high in comparison to reviewed literature. While both criteria show similarly accurate for a notched specimen, PIN exhibits a smaller calculation error for a shrink-fit connection than SIH at a critical distance of $L = 0.096$ mm.

In conclusion, the available integral methods are calibrated by plain fatigue experiments when only considering reversed biaxial loading and stress ratio $R = 0$. In contrast, the contact components show a multiaxial stress state at failure position. Furthermore, the shrinking of the hub on the shaft leads to compressive mean stresses with stress ratio $R > 1$. The mean stress sensitivity in this range is not considered in any investigated criteria. Nevertheless, a notable accuracy was reached at verified evaluation points.

In following steps, investigations on different frequency ratios as well as load conditions are of interest. Also, a comparison of evaluation methods with critical plane methods is recommended.

Acknowledgements The research project "Shaft-hub connections under combined loads", on which the results are based, was supported by the "Forschungsvereinigung Antriebstechnik e.V." research association in Germany.

References

1. Papuga J (2011) A survey on evaluating the fatigue limit under multiaxial loading. Int J Fatigue 33:153–165. https://doi.org/10.1016/j.ijfatigue.2010.08.001
2. Papuga J, Růžička M (2008) Two new multiaxial criteria for high cycle fatigue computation. Int J Fatigue 30:58–66. https://doi.org/10.1016/j.ijfatigue.2007.02.015
3. Papuga J, Suchý L, Růžička M (2019) Mean shear stress effect built into the multiaxial fatigue strength estimation method of an integral type. MATEC Web Conf 300:16010. https://doi.org/10.1051/matecconf/201930016010
4. Liu J, Zenner H (2003) Fatigue limit of ductile metals under multiaxial loading. In: Carpinteri A, de Freitas M, Spagnoli A (eds) European structural integrity society. Elsevier, pp 147–164
5. Araújo JA, Susmel L, Taylor D et al (2008) On the prediction of high-cycle fretting fatigue strength: Theory of critical distances vs. hot-spot approach. Eng Fract Mech 75:1763–1778. https://doi.org/10.1016/j.engfracmech.2007.03.026
6. Nesládek M, Španiel M, Jurenka J et al (2012) Fretting fatigue – Experimental and numerical approaches. Int J Fatigue 44:61–73. https://doi.org/10.1016/j.ijfatigue.2012.05.015
7. Bhatti NA, Abdel Wahab M (2017) A numerical investigation on critical plane orientation and initiation lifetimes in fretting fatigue under out of phase loading conditions. Tribol Int 115:307–318. https://doi.org/10.1016/j.triboint.2017.05.036
8. Taylor D (2008) The theory of critical distances. Eng Fract Mech 75:1696–1705. https://doi.org/10.1016/j.engfracmech.2007.04.007
9. Araújo JA, Susmel L, Taylor D et al (2007) On the use of the theory of critical distances and the modified Wöhler curve method to estimate fretting fatigue strength of cylindrical contacts. Int J Fatigue 1:95–107. https://doi.org/10.1016/j.ijfatigue.2006.02.041
10. El Haddad MH, Smith KN, Topper TH (1979) Fatigue Crack Propagation of Short Cracks. J Eng Mater Technol 101:42–46. https://doi.org/10.1115/1.3443647
11. Atzori B, Meneghetti G, Susmel L (2005) Material fatigue properties for assessing mechanical components weakened by notches and defects. Fatigue Fract Eng Mater Struct 28:83–97. https://doi.org/10.1111/j.1460-2695.2004.00862.x
12. Nesládek M, Španiel M, Kuželka J et al (2017) A fretting damage correction factor applicable to the McDiarmid criterion of plain high-cycle fatigue. Fatigue Fract Eng Mater Struct 40:27–44. https://doi.org/10.1111/ffe.12470
13. Gandiolle C, Garcin S, Fouvry S (2017) A non-collinear fretting-fatigue experiment to compare multiaxial fatigue criteria: critical shear plane strategy is better than invariant formulations. Tribology International C:57–68. https://doi.org/10.1016/j.triboint.2016.09.011
14. DIN 743-1:2012-12, Calculation of load capacity of shafts and axles. Beuth Verlag GmbH
15. Rennert R, Kullig E, Vormwald M et al (2012) Analytical strength assessment of components in mechanical engineering: FKM-guideline, 6, rev. VDMA Verl, Frankfurt am Main
16. Gough HJ, Pollard HV, Clenshaw WJ (1952) Some experiments on the resistance of metals to fatigue under combined stresses part I and part II. Reports and Memoranda No. 2522. Aeronautical Research Council (Great Britain)
17. DIN EN ISO 6892-1:2017-02, Metallic materials - Tensile testing - Part 1: Method of test at room temperature (ISO_6892-1:2016). Beuth Verlag GmbH
18. DIN 7190-1:2017-02, Interference fits - Part 1: Calculation and design rules for cylindrical self-locking pressfits. Beuth Verlag GmbH
19. Hück M (1983) Ein verbessertes Verfahren für die Auswertung von Treppenstufenversuchen. Materialwiss Werkstofftech 14:406–417. https://doi.org/10.1002/mawe.19830141207
20. DIN 50100:2016-12, Load controlled fatigue testing - Execution and evaluation of cyclic tests at constant load amplitudes on metallic specimens and components. Beuth Verlag GmbH
21. Hofmann S (2016) Untersuchungen zur Ermüdungsfestigkeit von Pressverbindungen. Universitätsverlag Chemnitz
22. Hauschild S (2020) Reibdauerbeanspruchte Stahl-Kontakte - Auslegung und Bewertung mittels systemspezifischer Reibkorrosionsfaktoren. Universitätsverlag Chemnitz

23. Susmel L (2004) A unifying approach to estimate the high-cycle fatigue strength of notched components subjected to both uniaxial and multiaxial cyclic loadings. Fatigue Fract Eng Mater Struct 27:391–411. https://doi.org/10.1111/j.1460-2695.2004.00759.x
24. Sunde SL, Berto F, Haugen B (2020) Efficient implementation of critical plane for 3D stress histories using triangular elements. Int J Fatigue 134:105448. https://doi.org/10.1016/j.ijfatigue.2019.105448

Evaluation of Fatigue Strength Characteristics of Al–Mg Oxide Dispersed Aluminum Composite

Yuki Ueno, Koichiro Nambu, and Masahiro Okumiya

Abstract Aluminum is a lightweight metal that is soft and highly malleable; thus, it offers good workability. In addition, it also exhibits corrosion resistance due to the oxide film formed on its surface. Therefore it can be adopted for various applications such as engine cylinders and, the frames and rims in a bicycle. Typically, aluminum alloys are widely used. Recently, metal matrix composites, which are composites of a base metal (such as aluminum alloys) and a reinforcing material (such as ceramics), have been employed as they are cheap and feature excellent properties that cannot be obtained via other materials. In this regard, this study aims to develop a novel composite material wherein Al–Mg oxide is dispersed in Al alloy. Fatigue tests were employed to evaluate the fatigue strength of this novel composite, and the results were compared with those of the Al/SiC composite. It was found that the Al/SiC composite material exhibited higher fatigue strength for a lower number of cycles, whereas the proposed material achieved greater fatigue strength for a higher number of cycles. Furthermore, scanning electron microscopy images of the fracture surfaces of these materials indicated that Al/SiC primarily failed due to an intergranular fracture, whereas the proposed composite failed due to a transgranular fracture. It was also observed that the fractures originated at segregated Si crystals near the surface.

Keywords Aluminum · Al–Mg oxide · Composite · Fatigue · Fracture surface

1 Introduction

In recent years, from an environmental perspective, it has become essential to reduce the increasing emission of CO_2 from automobiles. As a solution to this problem, it is necessary to improve the fuel efficiency of automobiles. This can be achieved by reducing the weight of the vehicle by employing lightweight materials with improved

Y. Ueno (✉) · K. Nambu · M. Okumiya
Materials Processing Laboratory, Toyota Technological Institute, 2-12-1 Hisakata, Tempaku-ku, Nagoya, Japan
e-mail: sd16016@toyota-ti.ac.jp

M. Abdel Wahab (ed.), *Proceedings of the 8th International Conference on Fracture, Fatigue and Wear*, Lecture Notes in Mechanical Engineering,
https://doi.org/10.1007/978-981-15-9893-7_35

rigidity for maintaining structural integrity. Composite materials, wherein an inorganic material, such as ceramics, is dispersed in an aluminum alloy (hereinafter referred to as aluminum-based composite materials), can be used to meet these requirements. Aluminum-based composite materials offer improved tensile strength, heat resistance, and wear resistance as well as superior rigidity, as compared to conventional aluminum alloys [1].

Thus far, applications of aluminum-based composite materials include pistons, for which aluminum is pressure-infiltrated into alumina fibers or SiC whiskers, and disk rotors, for which SiC particles are dispersed in aluminum [2]. However, the application of this material to automobile components is associated with severe issues in terms of productivity and cost; thus, aluminum-based composites have not been employed in automobiles since 2000. Currently, SiC particle reinforced aluminum composites (Al/SiC) are primarily used for industrial machinery.

In this study, we investigate the feasibility of using oxide particles as an alternative to the conventionally used SiC particles. Thus, Al–Mg oxide was selected and dispersed in molten aluminum to obtain the Al–Mg oxide-dispersed aluminum composite material (Al/Al–Mg oxide). Furthermore, the structure, hardness, tensile strength, and fatigue strength of this material were evaluated, in comparison with the corresponding values of Al/SiC reported in literature [3]. Finally, the specimens were subjected to fatigue failure, and its fracture surface was analyzed.

2 Specimen and Methods

Specimens. Al–Mg oxide with an average particle size of 10–50 μm was dispersed in molten aluminum. Dimensions of the Al/Al–Mg oxide specimens prepared for the fatigue test are shown in Fig. 1.

Fatigue Tests. Fatigue tests were performed at room temperature using a quadruple rotary bending fatigue testing machine (R-ratio of -1 and rotational speed of 3000 rpm). These tests were terminated when the sample remained intact even

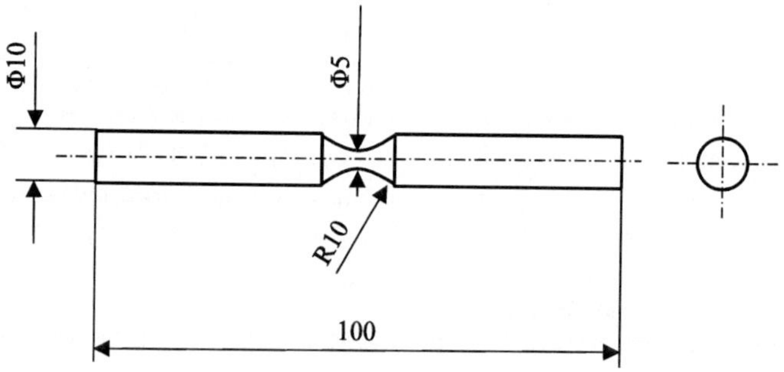

Fig. 1 Fatigue test piece

when the number of cycles exceeded 10^7 times. In other words, the stress that the samples could withstand 10^7 cycles without breaking was defined as the fatigue limit.

Microstructure, Hardness, and Tensile Strength Measurement. Microstructure observation and hardness measurement were cut from the fractured specimen and polished. Thereafter, the microstructures were observed using an optical microscope (OM), and a micro Vickers hardness test machine was employed to determine the hardness (HV) of the samples. As the tensile strength of this material has been evaluated in previous studies, these tests were not performed. The results obtained via the characterizations were compared with the values reported in literature.

Fractography. The fracture surface, obtained through the fatigue test, was observed using scanning electron microscopy (SEM) equipped with energy dispersive X-ray spectroscopy (EDS).

3 Results and Discussion

Fatigue Tests. Figure 2 shows the S–N curves of Al/Al–Mg oxide composite and Al/SiC composite. The arrow in the figure indicates that the specimens did not break even if the number of cycles exceeded 10^7 cycles for the load. From this figure, Al/SiC showed higher fatigue strength on the low cycle side, but higher value on Al/Al–Mg oxide on the high cycle side. The fatigue limit of Al/Al–Mg oxide and Al/SiC were 151 MPa and 140 MPa, respectively.

Microstructure, Hardness, and Tensile Strength. Figure 3 presents an OM image of Al–Mg oxide. In this figure, the spherical black portion is the Al–Mg oxide, the white portion is the primary α-Al phase, and the gray portion is the eutectic Si [4]. For a composite material produced via casting, Al–Mg oxide is dispersed in the

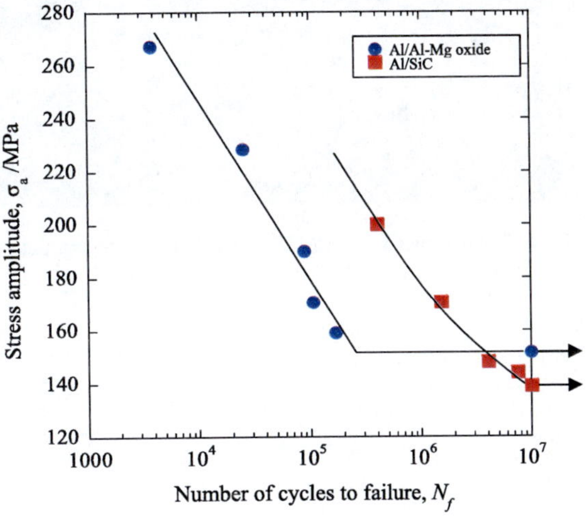

Fig. 2 Stress amplitude with respect to the number of cycles for the fatigue failure of Al/Al–Mg oxide and Al/SiC

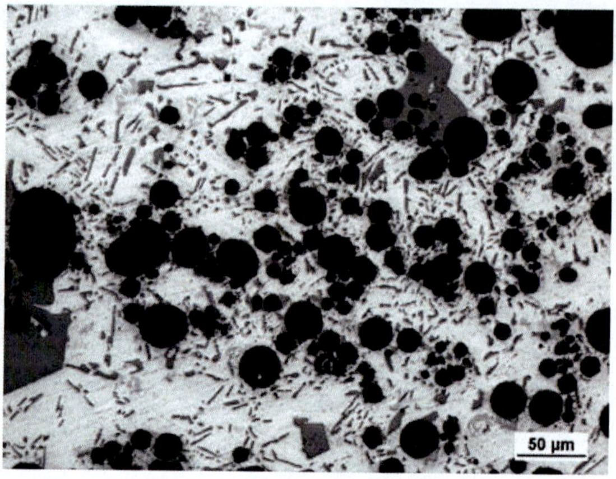

Fig. 3 OM images of Al/Al–Mg oxide composite

primary α-Al phase boundaries, because the primary α-Al phase grows in a dendritic state during solidification [5]. This figure indicated that a majority of the Al–Mg oxides are completely dispersed.

On the contrary, as shown in Fig. 4, segregated portions were also confirmed. Thus, it was concluded that future efforts are required to prevent this segregation.

Table 1 lists the hardness (HV) and tensile strength values of Al/Al–Mg oxide and Al/SiC. The hardness of these materials was almost identical; however, the tensile strength of Al/Al–Mg oxide was 119 MPa lower than that of Al/SiC.

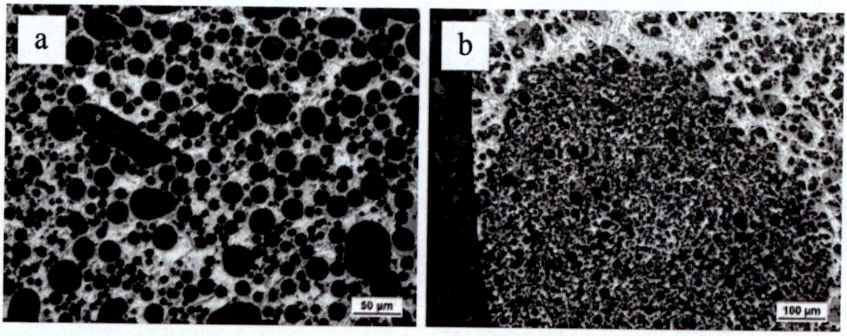

Fig. 4 Optical images of Al/Al–Mg oxide indicating particle segregation: **a** high magnification; **b** low magnification

Table 1 Micro Vickers hardness and tensile strength of Al/Al–Mg oxide and Al/SiC

Materials	HV	Tensile strength (MPa)
Al/Al–Mg oxide	174	272
Al/SiC	182	391

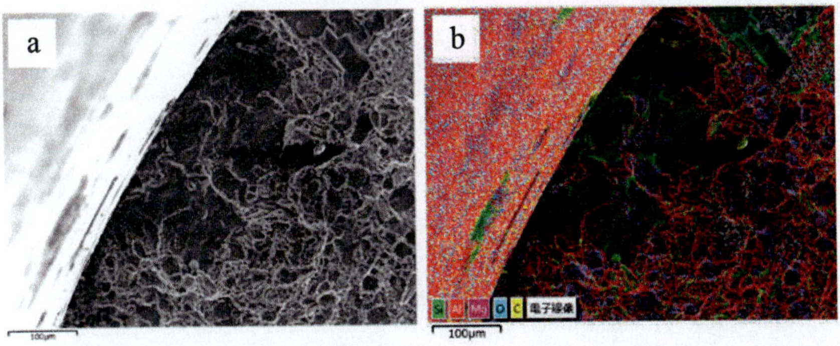

Fig. 5 Fracture surfaces of Al/Al–Mg oxide (which failed at a stress of 267 MPa after 2780 cycles): **a** SEM image and **b** EDS image

Fractography. Figure 5 depicts the SEM and EDS images of the origin of the fracture in Al/Al–Mg oxide, which underwent failure at a stress of 267 MPa after 2780 cycles. The images indicate that the fracture origin lies near the segregated Si crystals. An analysis of the fracture surface via SEM revealed that Al/SiC failed due to an intergranular fracture, whereas Al/Al–Mg oxide failed due to a transgranular fracture. This difference in the fracture morphologies of the two materials is attributed to the high fatigue limit of the proposed Al/Al–Mg oxide.

4 Conclusions

- Al/SiC exhibited higher fatigue strength for a lower number of cycles, whereas Al/Al–Mg oxide exhibited greater fatigue strength for a higher number of cycles. The fatigue limits of Al/Al–Mg oxide and Al/SiC were 151 MPa and 140 MPa, respectively.
- Based on the OM images, a majority of the particles were completely dispersed. However, a few segregated portions were also observed.
- The hardness of the two composites was almost identical; however, the tensile strength of Al/Al–Mg oxide was 119 MPa lower than that of Al/SiC.
- The fractures in the materials were found to originate near the segregated Si crystals. In addition, Al/SiC underwent failure due to an intergranular fracture, whereas Al/Al–Mg oxide failed due to a transgranular fracture.

References

1. Yusaku M, Naoya M, Hirofumi M, Keisaku O (2006) Influence of notch size and the solidification microstructure on the rotating bending fatigue characteristics of SiC particle-reinforced AC4B alloy composites. J Japan Inst Metals 70(7):554–561
2. Toshio Y (2004) Recent manufacturing technology of metal matrix composite for automobiles. J JFS 76(12):1014–1021
3. Tateoki I, Qiubao O (2013) Strength and fatigue properties of SiC particle-reinforced AC4C based aluminum alloy composites fabricated by melt stirring-gravity casting. Japan Inst Light Metals 63(11):400–405
4. Sanji K, Chozo F, Akihiko K (1988) Aluminum-silicon alloys. Japan Inst Light Metals 38(7):427
5. Toshiro K (2001) Aluminum alloy strength. Uchida Rokakuho Publishing Co. Ltd, Japan

Assessment of a Thermal Fatigue Test Conducted Under Cyclic Non-proportional Loading Using Open Source CAE and Finite Element Analysis Methods

Shosuke Miyahira and Terutaka Fujioka

Abstract Thermal fatigue assessment is often required for high temperature components such as those in thermal power and chemical plants. The procedures used to evaluate fatigue life consumption of a component subjected to cyclic non-steady thermal loading are rather complicated and, in many cases, difficult to carry out. In addition, experimental validation of a thermal fatigue assessment by using FEM is rarely carried out due to the high cost and length of time required to perform tests that subjected a specimen to cyclic thermal loading. The authors herein propose simplified fatigue assessment methods and undertake their experimental validation by comparing the FEA assessment with the result of a previously reported thermal fatigue test performed by JAEA (Japan Atomic Energy Agency) on a tapered cylindrical specimen. In the test considered, the loading was almost proportional, and thus the validity of these methods for general non-proportional loading remained to be solved. This paper describes the fatigue assessment of components subjected to cyclic thermal loading, in which simulation was undertaken of a previously conducted thermal fatigue test that used an axisymmetric specimen with three different types of structural discontinuities. The test was also performed by JAEA in their sodium flow test facility. The specimen was subjected to cyclic thermal loading by alternating liquid metal flows injection at different temperatures. In this study finite element analyses, non-steady-state thermal transient and stress analyses were performed using commercial FEA software (MSC.Marc) and open source FEA software (Salome-Meca, Code-Aster) to evaluate change in thermal stress, and the two FEAs were compared. The FEA results indicate non-proportional stress histories. The simplified assessment methods previously proposed for determining the pair of time steps that define the maximum stress range were applied and subsequently compared. The simplified assessments made using the methods in this study are consistent with those made based on ASME Section III Division 5.

S. Miyahira (✉) · T. Fujioka
Department of Mechanical Engineering, Faculty of Science & Engineering, Toyo University, Saitama, Japan
e-mail: s36a02000154@toyo.jp

M. Abdel Wahab (ed.), *Proceedings of the 8th International Conference on Fracture, Fatigue and Wear*, Lecture Notes in Mechanical Engineering,
https://doi.org/10.1007/978-981-15-9893-7_36

Keywords Thermal fatigue · Verification and validation · Finite element analysis · Open source CAE · Stress range · Non-proportional loading

1 Introduction

1.1 Objective

An assessment of thermal fatigue is often required in the design and/or maintenance of high temperature components, such as thermal power plant components. Low-cycle fatigue failure of these components may occur when the components are subjected to cyclic thermal stress caused by the temperature changes of frequent start-up and stop operation. Such stress changes can be estimated by performing heat transfer and stress analyses with FEA software. The fatigue life consumption evaluation procedures are given in ASME Section III Division 5 [1], which describes inelastic FEA-route assessment procedures for determining the maximum equivalent strain range. Japanese fast breeder reactor design code [2] describes an elastic FEA-route method that is equivalent to those in ASME code. These methods require the analyst to specify the pair of time steps in a given thermal cycle that maximizes the stress or strain range at a specific location. However, this task is troublesome and requires a large computer memory when surveying a broad section of the component surface.

The lack of experimental verification and validation data represents another difficulty in thermal fatigue assessment. Generating high stress changes requires a large specimen and therefore the test becomes costly and time-consuming. Some experimental studies have been carried out at the Japan Atomic Energy Agency (JAEA) using their sodium loop test facilities. The test data obtained there were costly but very valuable, and should be used as a reference for analysts to evaluate their skills and expertise. In a previous paper [3], one of the sodium loop tests was evaluated by various elastic and inelastic FEA-route fatigue assessment procedures, which ultimately produced equivalent results to those obtained by the ASME method. The previous study employed simplified methods, in which the determination procedures for the time step pair requires less memory and a shorter computation time. The specimen evaluated in [3] was simple in shape and the stress history was mostly proportional. However, validation of these methods under more complex, non-proportional loading remains to be solved.

The objectives of this paper are to validate the previously proposed simplified methods by simulating the thermal fatigue test of a specimen with structural discontinuities that is subjected to non-proportional loadings. The test considered was also performed by JAEA [4]. Unlike in the design codes [1, 2], this study employs the von Mises' equivalent stress and the separately studied stress redistribution locus (SRL) methods, as simplified elastic–plastic strain range estimates, without performing stress categorizations.

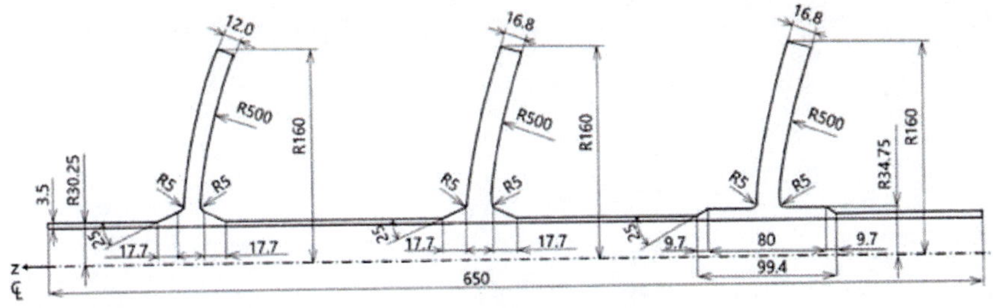

Fig. 1 Dimensions and configuration of the axisymmetric specimen [4]

2 Thermal Transient Test Methods

2.1 Previously Conducted Thermal Transient Test Overview

This study considers a previously performed thermal fatigue test that used an axisymmetric type 304stenless steel (304SS) specimen with structural discontinuities [4]. The test was performed by JAEA in their sodium flow test facilities. The specimen configuration and dimensions are shown in Fig. 1. The test specimen can be divided into three different umbrella-like curved disks connected by a cylinder, denoted Model 1, Model 2, and Model 3, respectively. Model 1 and Model 2 had the same cylinder connection section but differed in disk thickness, whereas the disks in Model 2 and Model 3 had the same thickness but differed in their cylinder connection component. The inner diameter of the cylinder was 53.5 mm and the wall thickness was 3.5 mm.

Thermal cycles were applied to the specimen by alternately injecting hot (550 °C) and cold (300 °C) sodium flows into the specimen. The hot sodium flow was maintained for one hour while the cold sodium flowed for five hours, with maximum flow rate of 200 L/min. This thermal transient was repeated for 1700 cycles. After completing the test, the specimen was cut, and the initiated cracks were observed.

The crack observations reported in [4] are shown in Fig. 2, and the measured crack sizes are summarized in Table 1. The images in Fig. 2 were digitally processed from the photographs in [4] (Fig. 3).

3 Assessment of Thermal Fatigue Using Simplified Methods

3.1 Finite Element Analysis

FEA analyses, consisting of thermal transient and elastic stress analyses, were performed by using commercial FEA software (MSC.Marc) [5] and open source FEA software (Salome-Meca, Code-Aster [6]) to evaluate the change in thermal

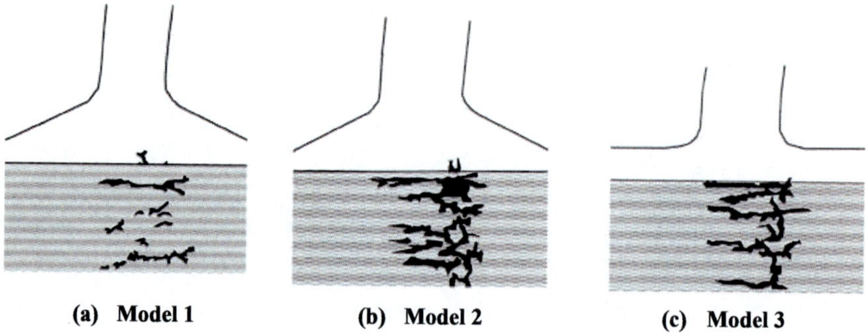

Fig. 2 Initiated thermal cracks on the inner surface of the specimen after the test [4]

Table 1 Observed crack size (dimensions in mm) [4]

Model type	Crack length		Crack depth
	Upstream	Downstream	Center
Model 1	8	11	4.07
Model 2	9	15	5.95
Model 3	8	15	5.62

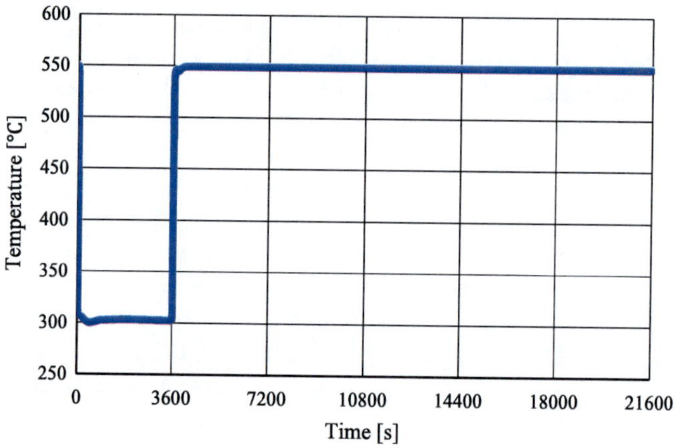

Fig. 3 Measured temperature history of sodium during one thermal cycle [4]

stress. The same finite element mesh models were used for both FEAs. First, thermal transient analysis was performed to examine the temperature distribution in the analysis models. Next, elastic stress analysis was conducted using the temperature distribution results. The material properties of 304SS and the heat transfer coefficient between sodium and the specimen employed in the analysis of [4] were used in the FEAs in this study. The analysis models used in the simulation and the mechanical boundary conditions are shown in Fig. 4. This simulation used the same analysis

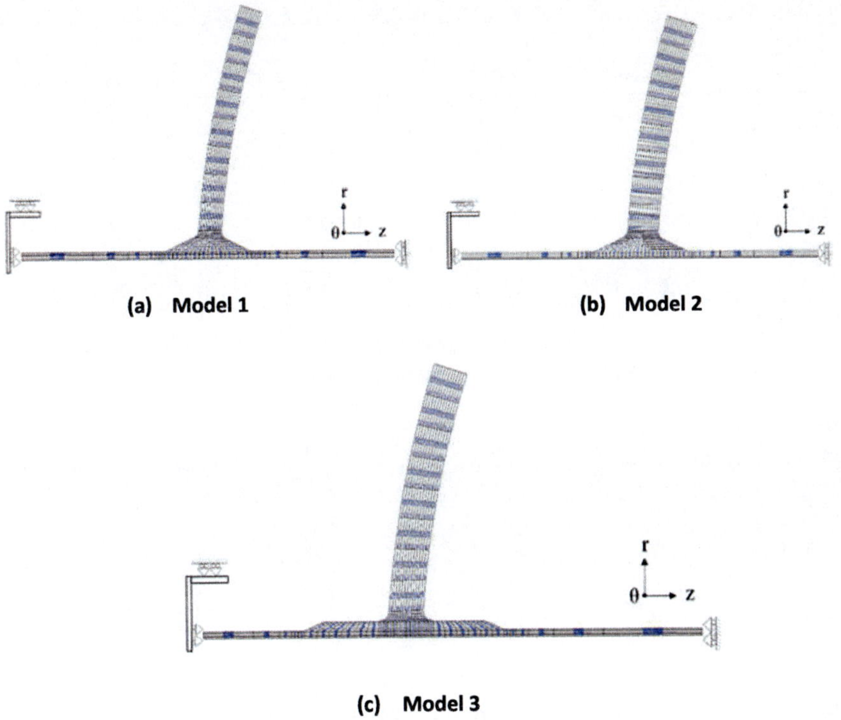

Fig. 4 Analysis models and mechanical boundary conditions

models to compare the simulation results of MSC.Marc and Salome-Meca Characteristics. Characteristics of the analysis models are shown in Table 2 along with the element type of each model. Mechanical and material properties are summarized in Tables 3 and 4.

Table 2 Mesh characteristics of the analysis

	Number of nodes	Number of elements	Element type
Model 1	1027	312	8-node axisymmetric quadratic element
Model 2	1177	360	8-node axisymmetric quadratic element
Model 3	1577	486	8-node axisymmetric quadratic element

Table 3 Material properties of 304SS and heat transfer coefficient of sodium used in heat transfer analysis

T (°C)	ρ (kg/mm^3)	λ (W/m K)	C_p (J/kg °C)	h (W/m °C)
300	7.90E–09	18.28	539.7	13.098
350	7.88E–09	19.08	548.1	12.709
400	7.86E–09	19.67	552.3	12.335
450	7.83E–09	20.33	556.5	11.976
500	7.81E–09	21.00	560.7	11.633
550	7.79E–09	21.67	564.8	11.303

Table 4 Material properties of 304SS used in elastic analysis

T (°C)	ν	α (1/°C)	E (MPa)	S_m (MPa)
300	0.287	1.88E–05	176,519.7	114.7
350	0.291	1.92E–05	172,597.7	110.7
400	0.295	1.96E–05	168,674.38	106.8
450	0.298	1.99E–05	163,771.06	101.9
500	0.302	2.03E–05	158,867.73	98.0
550	0.306	2.06E–05	153,964.41	95.1

3.2 *Simplified and ASME-Based Methods for Stress Range Estimation*

The previous study proposed three types of simplified methods for calculating stress range that do not consume a large memory and long calculation time. The calculation of the maximum stress range required to determine the pair of time steps can be estimated using the following von Mises equivalent stress:

$$S_p = \frac{1}{\sqrt{2}}\left\{(\Delta\sigma_z - \Delta\sigma_r)^2 + (\Delta\sigma_r - \Delta\sigma_\theta)^2 + (\Delta\sigma_\theta - \Delta\sigma_z)^2 + 6\Delta\tau_{zr}^2\right\}^{1/2} \quad (1)$$

$$\Delta\sigma_z = \sigma_{z\_t1} - \sigma_{z\_t2}$$

$$\Delta\sigma_r = \sigma_{r\_t1} - \sigma_{r\_t2}$$

$$\Delta\sigma_\theta = \sigma_{\theta\_t1} - \sigma_{\theta\_t2}$$

$$\Delta\tau_{zr} = \tau_{zr\_t1} - \tau_{zr\_t2}$$

The evaluation procedures in the ASME-defined FEA-route method are rather complicated and difficult to carry out because the pair of time steps, t_1 and t_2, must be chosen from all possible combinations of time step in a load cycle. In addition, the calculated stress components and temperatures must be stored while the FEA is being performed. However, in Method 1, time step t_1 is determined first, when the instantaneous equivalent stress reaches its largest in the 2nd cycle, and subsequently time step t_2 is determined at the point that produces the maximum equivalent stress range in the 3rd cycle. These procedures are undertaken in Method 1 for representative evaluation points.

In Method 2, a pair of time steps t_1 and t_2 is determined for each stress component, after which the identified pair for the component with the largest stress range is selected for estimating the equivalent stress range. In Method 3, the equivalent stress range is defined as one-half of the stress history path. Method 3 does utilize a pair of time steps, similarly to the previously proposed method for strain range estimation for a multi-axial fatigue test conducted under non-proportional loading [7]. Conceptual schematics of the three simplified methods and the ASME-based method are shown in Fig. 5. In Fig. 5, the stress paths and time step definitions are expressed in an ideally simplified principal stress space. In cases with uniform mechanical properties and

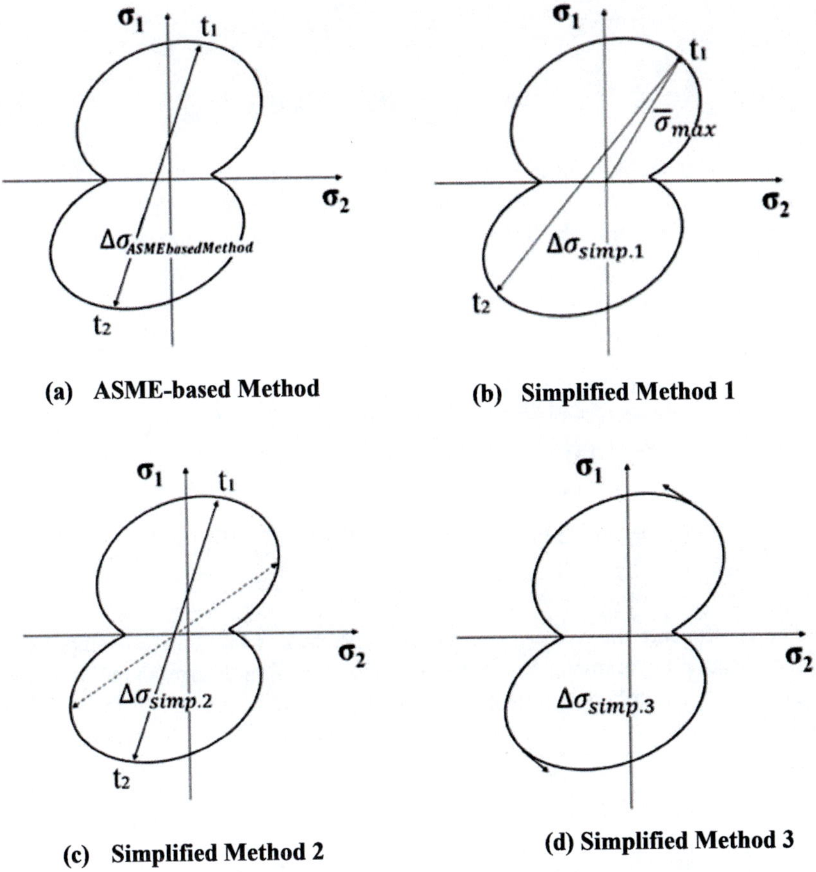

Fig. 5 Conceptual schematics of the ASME-based method and the three simplified methods

perfectly proportional bi-axial loading, the three methods lead to the same result of stress range.

3.3 Fatigue Life Consumption

The elastic–plastic strain range is required to calculate the fatigue life consumption N/N_f, where N is the number of thermal loading cycles and N_f is the corresponding fatigue life. In this study, stress categorization was not performed, and the simple criteria given in Eqs. 2 and 3 were employed for the elastic limit. In the following equations, $3S_m$ represents the elastic domain limit, S_p the maximum equivalent stress range at an evaluation point, and E the Young's modulus.

$$\Delta\varepsilon = \frac{S_P}{E}\left(S_p \leq 3S_m\right) \tag{2}$$

$$\Delta\varepsilon = \frac{S_P^2}{3S_m E}(S_p > 3S_m) \tag{3}$$

The best-fit fatigue failure curve of 304SS estimates the allowable number of cycles N_f, and thus calculates the fatigue life consumption N/N_f. The best-fit fatigue failure curve is expressed by the following equations, given in the fast reactor design code [2]:

$$\left(\log_{10} N_f\right)^{-\frac{1}{2}} = A_0 + A_1\Delta\log_{10}\Delta\varepsilon + A_2\Delta\left(\log_{10}\Delta\varepsilon\right)^2 + A_3\Delta\left(\log_{10}\Delta\varepsilon\right)^4 \tag{4}$$

$$A_0 = 1.621827 - 0.4567850 \times 10^7 \times T^2 \times R$$
$$A_1 = 1.131346 + 0.8665061 \times 10^8 \times T^2$$
$$A_2 = 0.3439663$$
$$A_3 = -0.1374387 \times 10^{-1} + 0.4910723 \times 10^{-4} \times R$$
$$R = \log_{10}\dot{\varepsilon} = -3$$

The coefficients A_0, A_1, A_2, and A_3 are expressed as functions of temperature and the strain rate parameter R. the metal temperature T is determined in Method 3 based on the highest temperature at the evaluation point during one cycle. The other methods, for conservativeness, use the higher of the two temperatures at time steps t_1 and t_2.

4 Assessment of Thermal Fatigue

4.1 Thermal Transient and Elastic Stress Analyses

A comparison was made of the thermal transient analysis results obtained with commercial and open source FEA software. The temperature distributions at 46,860 s, during the 3rd hot shock is show in Fig. 6 for each model. Figure 6 indicates that as the thickness of the analysis model increases, the temperature difference also becomes large.

The changes in stress components are shown in Figs. 7, 8, and 9. The results of the analyses suggest that dominant stress components are hoop and axial stresses.

The stress ranges calculated using the obtained stress components, are shown in Figs. 10, 11, and 12. The thicker models were subjected to higher stress ranges than their thinner counterparts. Method 1 resulted in almost the same assessment as that obtained by the ASME-based method. The results also confirm that some differences exist between the results given by the Salome-Meca and MSC.Marc software. Method 2 appears to lead to non-conservative assessment under non-proportional loading. On the other hand, Method 3 always leads to overestimate of stress range under the non-proportional loading.

The stress paths of the dominant stress components (hoop and axial) obtained using the FEA methods are shown in Figs. 13, 14, and 15, for Model 1, 2 and 3 respectively.

4.2 Calculation of Fatigue Life Consumption

The fatigue life consumption was calculated from the estimated strain range using an elastic FEA-route assessment. Crack initiation should be observed at the location where N/N_f (fatigue life consumption)$\geq$ 1. The results of fatigue life consumption obtained by each method and using each FEA software are shown in Figs. 16, 17, and 18. Method 1 and the ASME-based method resulted in almost the same consumption rate. Although some differences were revealed between the result given by Salome-Meca and MSC.Marc software, the crack initiation limit ($N/N_f = 1$) occurred at approximately the same z coordinate. As with the stress range evaluations, Method 2 lead to slightly non-conservative assessments and Method 3 to conservative ones. The underestimate given by Method 2 is attributable to the difference in the pair of time steps and reference temperatures. In addition to the obvious differences in stress ranges, the reference temperatures obtained by Method 3 were higher than those obtained in any other methods.

The ranges where the fatigue life-consumption rate exceeds one are shown in Figs. 19, 20, and 21, along with a comparison to the crack observations reported by JAEA. These figures confirm that the fatigue assessments made in this study agree well with the test observations.

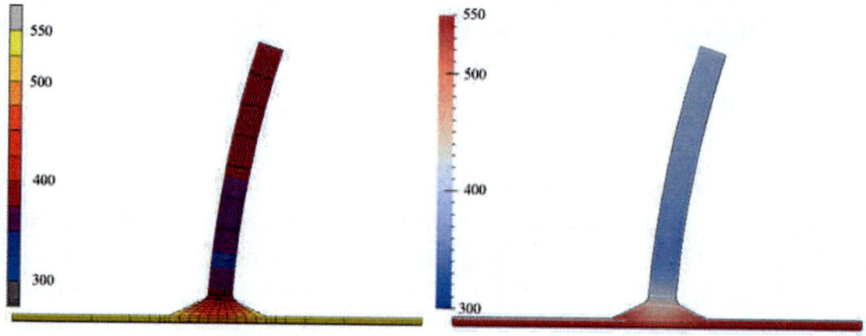

(a) Model 1 (Left: MSC.Marc, Right: Salome-Meca)

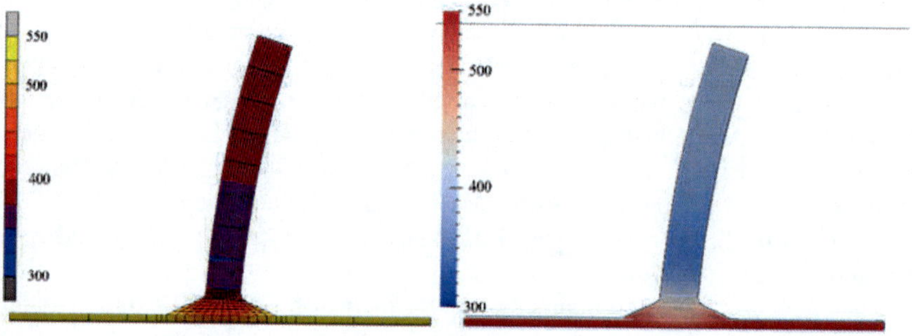

(b) Model 2 (Left: MSC.Marc, Right: Salome-Meca)

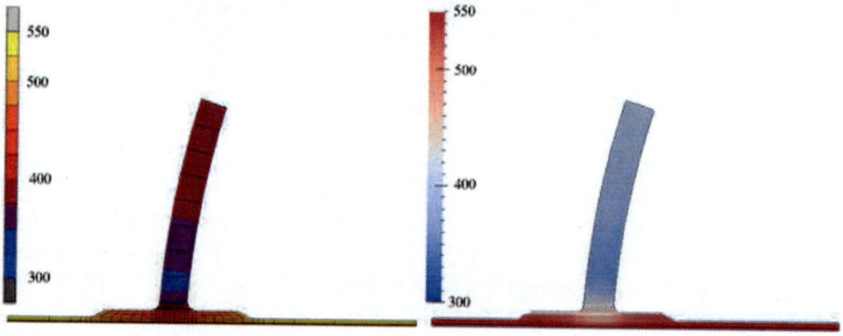

(c) Model 3 (Left: MSC.Marc, Right: Salome-Meca)

Fig. 6 Comparison of thermal transient analyses obtained at 46,860 s (during the 3rd hot shock) MSC.Marc and Salome-Meca

5 Conclusion

In this paper, assessments of thermal fatigue life consumption were made by referencing the thermal fatigue test performed by JAEA. FEAs were conducted using two different types of FEA software, MSC.Marc and Salome-Meca, and the results were compared. One of the previously proposed simplified methods, Method 1, effectively

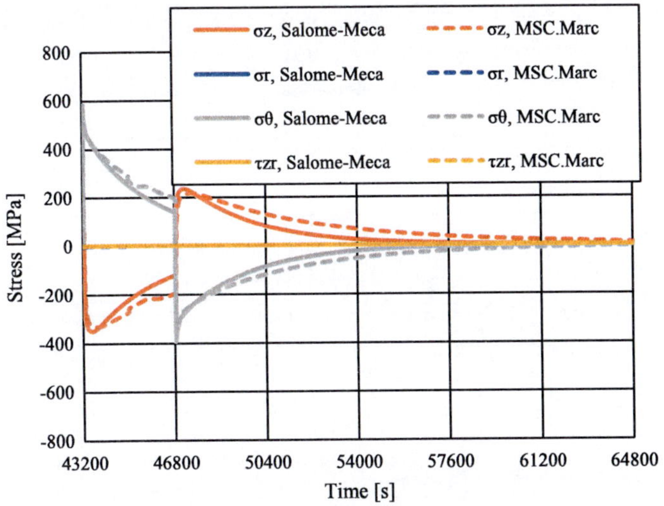

Fig. 7 Comparison of elastic stress analyses obtained using MSC.Marc and Salome-Meca for Model 1

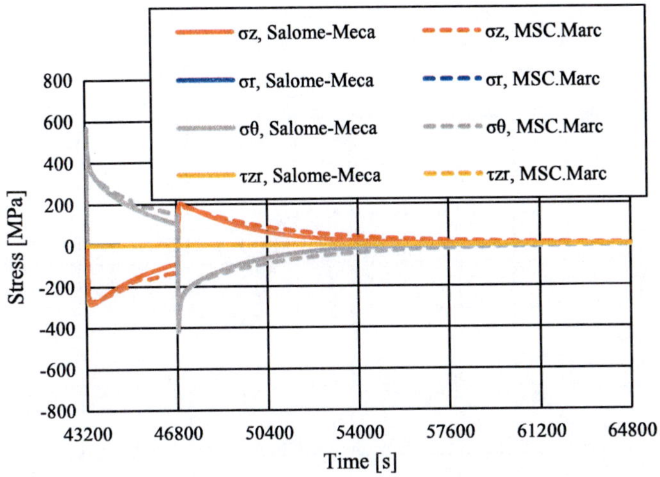

Fig. 8 Comparison of elastic stress analyses obtained using MSC.Marc and Salome-Meca for Model 2

achieved an assessment equivalent to that of the ASME-based method for defining the time step pair. Method 2 may result in a non-conservative assessment, while Method 3 can be employed for simple and conservative fatigue assessment results.

Slight differences were observed in the results obtained using the MSC.Marc and Salome-Meca (Code-Aster) software programs.

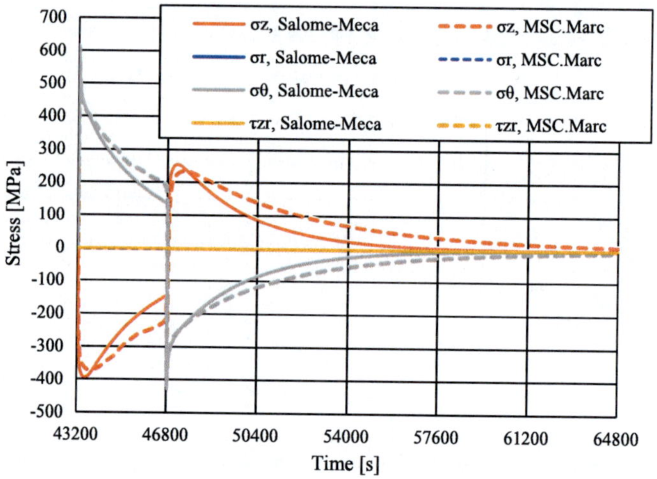

Fig. 9 Comparison of elastic stress analyses obtained using MSC.Marc and Salome-Meca for Model 3

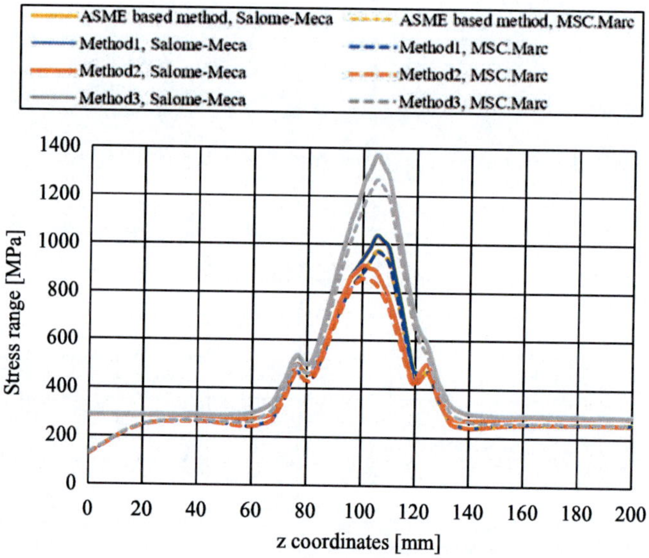

Fig. 10 Comparison of stress range obtained using MSC.Marc and Salome-Meca for Model 1

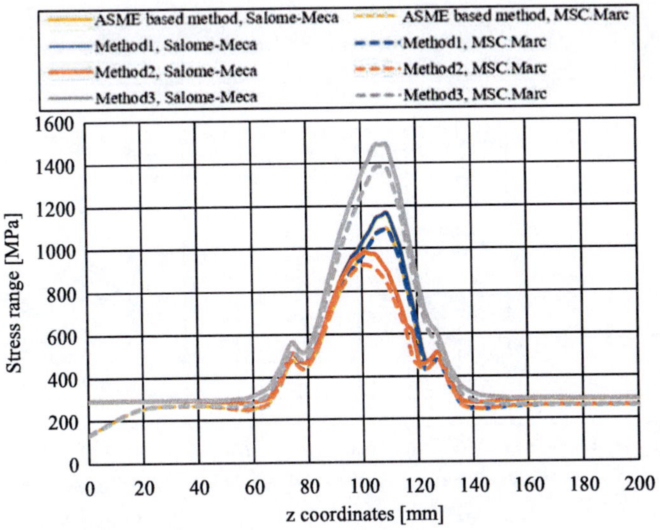

Fig. 11 Comparison of stress range obtained using MSC.Marc and Salome-Meca for Model 2

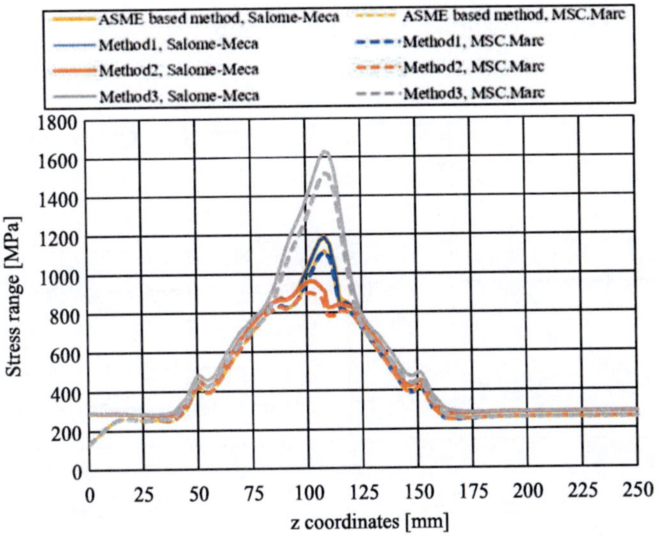

Fig. 12 Comparison of stress range obtained using MSC.Marc and Salome-Meca for Model 3

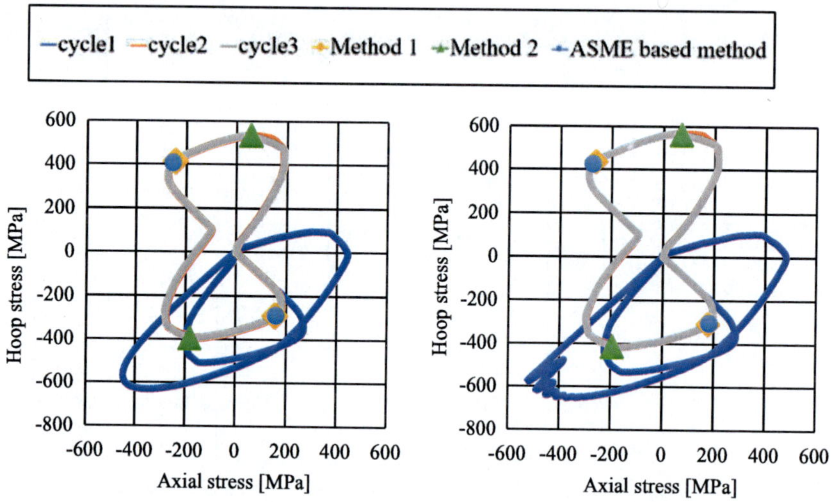

(a) MSC.Marc (b) Salome-Meca

Fig. 13 Stress paths comparison of MSC.Marc and Salome-Meca results for Model 1

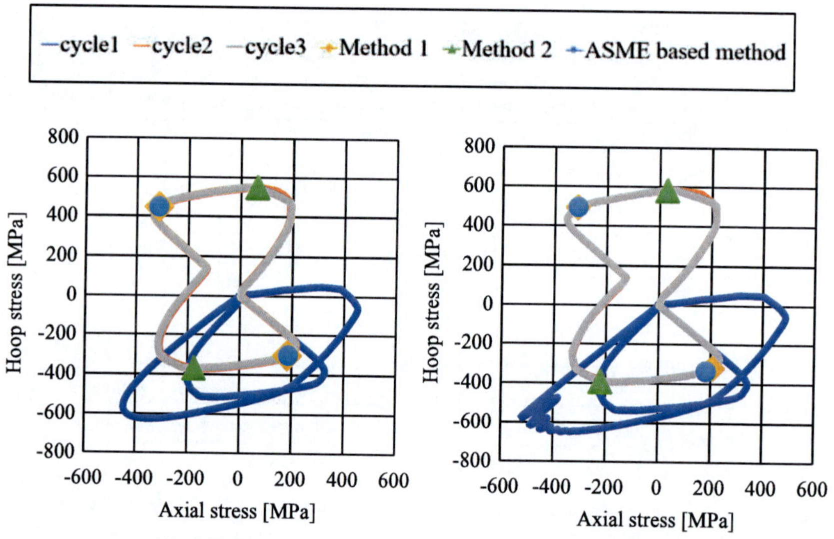

(a) MSC.Marc (b) Salome-Meca

Fig. 14 Stress paths comparison of MSC.Marc and Salome-Meca results for Model 2

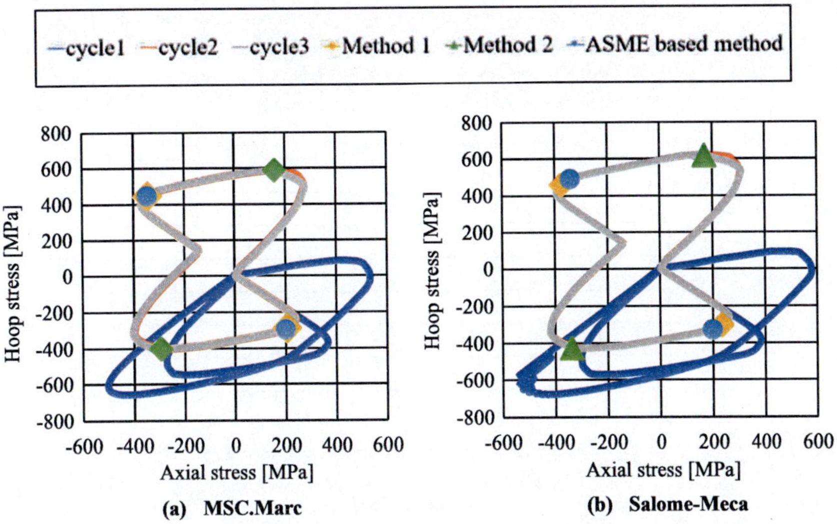

Fig. 15 Stress paths comparison of MSC.Marc and Salome-Meca results for Model 3

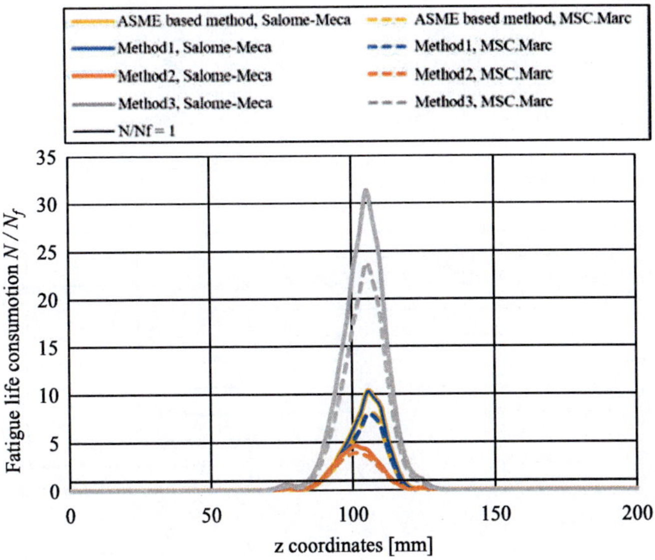

Fig. 16 Fatigue life consumption comparison of MSC.Marc and Salome-Meca results for Model 1

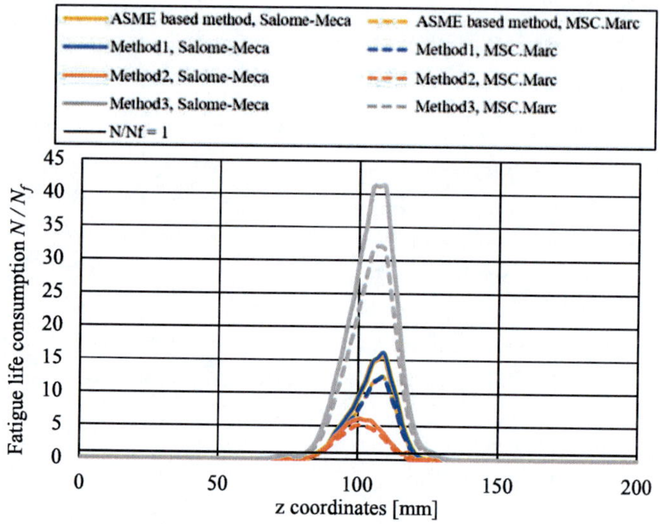

Fig. 17 Fatigue life consumption comparison of MSC.Marc and Salome-Meca results for Model 2

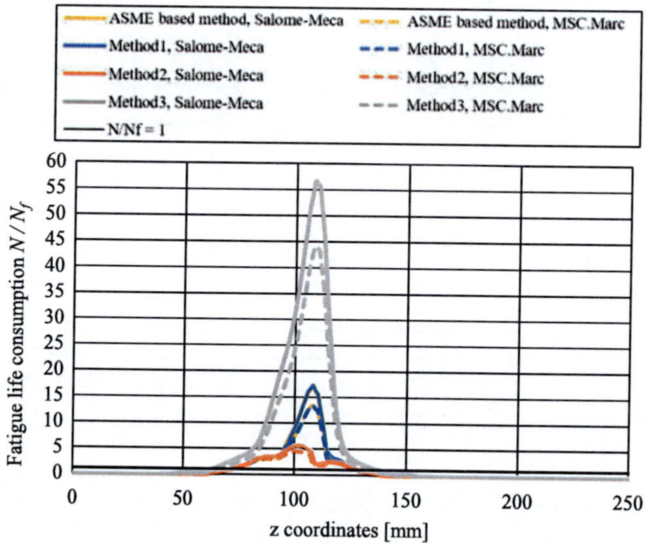

Fig. 18 Fatigue life consumption comparison of MSC.Marc and Salome-Meca results for Model 3

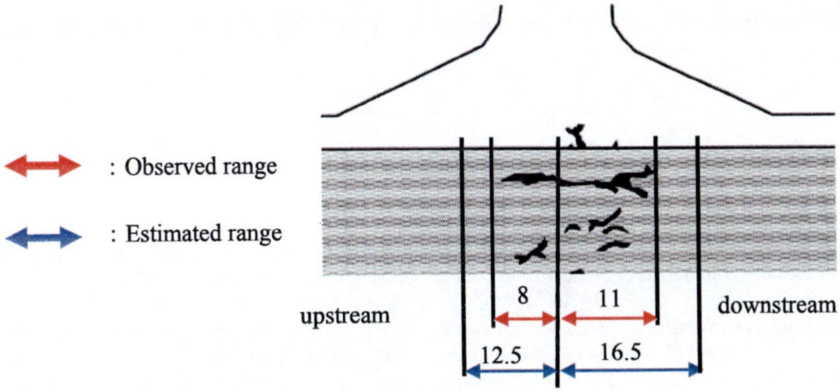

Fig. 19 Comparison of the observed and estimated crack ranges for Model 1

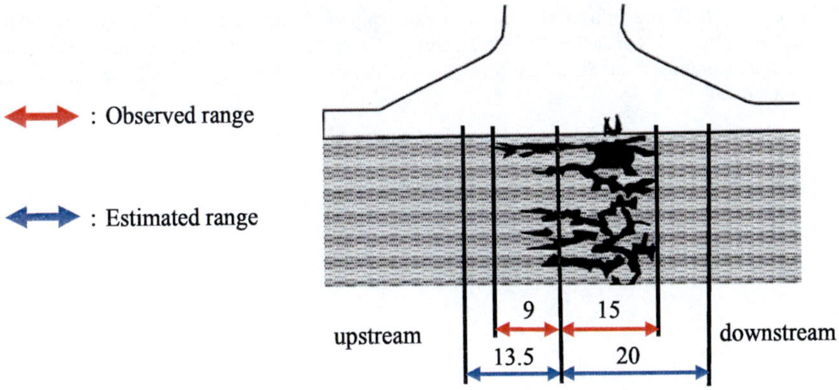

Fig. 20 Comparison of the observed and estimated crack ranges for Model 2

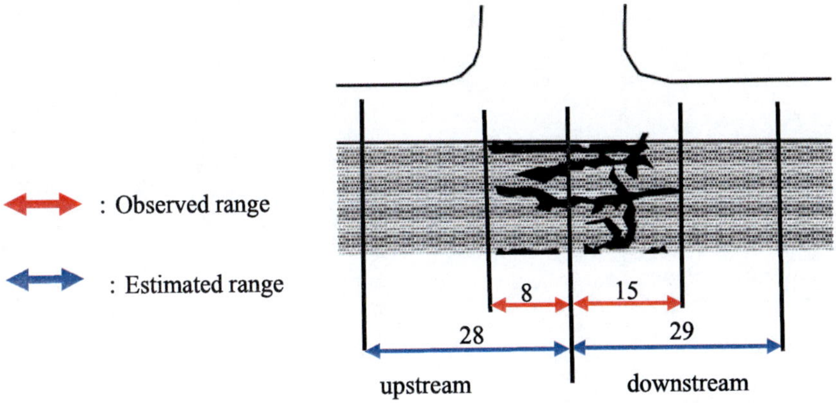

Fig. 21 Comparison of the observed and estimated crack ranges for Model 3

Acknowledgements This work was supported by JSPS KAKENHI Grant Number JP18K02963.

References

1. ASME, Boiler and Pressure Vessel Code, Section III, Division5, 2019
2. JAEA, Structural Design Guide for Class 1 Components of Prototype Fast Breeder Reactor for Elevated Temperature; Standard for Strength of Material, PNC TN9421 85–08, p.253. Japan Atomic Energy Agency, Japan, (1985), (in Japanese).
3. Ishizaki K, Watashi K, Takahashi N, Iwata K (1989) Thermal Fatigue of SUS304, PNC TN 9410 84 101, pp.1–112. Japan Atomic Energy Agency (1989) (in Japanese)
4. Tanaka N, Ishizaki K, Kikuchi M, Watashi K, Iwata K (1992) Creep-Fatigue Test of Nozzle Models under Thermal Loadings, PNC TN9410 92–284, pp 1–53. Japan Atomic Energy Agency, Japan, (in Japanese)
5. MSC Software. https://www.mscsoftware.com/. Last accessed 7 Jan 2020
6. Electricity de France. https://www.code-aster.org/spip.php?rubrique2. Last accessed 7 Jan 2020
7. Sakon T, Wada H, Asada Y (1987) Procedure of Creep-Fatigue Life Consumption Applied to Inelastic Conference on Structural Mechanics in Reactor Technology, SMiRT-9, pp 267–272, Switzerland

Depreciation Accounting in Longevity Evaluation of Complicated Systems

B. Avotyn', A. Smirnov, and B. Belobragin

Abstract Consideration is given to the systems, which consist of nonrepairable expendable items and the restorable items operating in a cyclic mode. Durability of the latter, along with the no-failure operation, maintenance ability and storage qualities is characterized by longevity, i.e. the object property to ensure safe operation until offset of the ultimate limit state providing usage of the established maintenance and repair system. Service durability also directly depends on: human errors resulting in the undeliberate result; dependent failures caused by the system latent fault, etc. The most useful technique is the operation life setting, i.e. the number of operating cycles, which the object should perform over the period of its service life. At that, the comprehensive description is given of the operation cycle, environment conditions, and qualifications of personnel. In making of the theoretical assessments to describe the materials fatigue resistance and depletion, the Weibull-Gnedenko distribution is used. In carrying out of tests, in case of revealing of the component part excessive wear (due to the safety criteria) to be checked is the deterioration rate of the devices and units mostly subjected to wear, along with the maintenance works in compliance with the article maintenance documentation. Based on all the information and analysis, the composition of the SPTA sets shall be determined proceeding from the specified values of the system reliability value. Matters related to the depreciation accounting in the complex systems longevity evaluation have been addressed. Flagship approaches to solving of the emerging problems have been outlined.

Keywords Durability · Wearability · Wear out

1 Introduction

This paper deals with the problematic points of the systems component parts wear out model, the study of which allows ensuring the sample service durability parameters. The system durability is prescribed to be a requirement to the number of the working

B. Avotyn' (✉) · A. Smirnov · B. Belobragin
JSC SPLAV SPA named after A. Ganichev, RF 33, Shcheglovskaya zaseka, Tula 300004, Russia
e-mail: bavotyn@mail.ru

M. Abdel Wahab (ed.), *Proceedings of the 8th International Conference on Fracture, Fatigue and Wear*, Lecture Notes in Mechanical Engineering,
https://doi.org/10.1007/978-981-15-9893-7_37

cycles, which the system shall perform without interfering of the repair services. Durability depends on the equipment reliability and wear out of the instruments work surfaces. The system durability can be increased by means of: the system elements redundancy and improvement of the quality of manufacture; providing the system with the necessary quantity of the spare parts, tools and accessories (SPTA sets [1]). The limitation of the first approach is the item cost. The limitation of the second approach is the design value of the availability function taking into account the periods of routine maintenance. Above all others, the designer using the first approach, tries to maximize the durability. Having used up all his possibilities, he uses the second approach.

2 Methodology

Obvious problem of the first approach is establishing the connection between the value of durability, influencing parameters, wearability parameters of the particular system constituent with its specific dimensions, characteristics and limitations. The bases for establishing connection are the results of the academic research on wear out of specific materials under the specified conditions. Let us consider the typical academic methodology of the given type [2] in which there are described the test bench of centrifugal type and test methodology for erosive wear of materials in model medium, which is an analog to formation fluid of oil well and containing water, abrasive particles, corrosive elements, is described. The test results for eight types of powder materials on the basis of iron and copper, alloyed by Ni, Cr, Mo and having ferrite-pearlite, pearlite or ferrite structure, in water with abrasive are presented. The parameters of the experiment are: fluid jet velocity 11–16 m/s; attack angle 45°, 90°; quartz and corundum particles of various dispersion F100, F40, F24 used as abrasive. The linear dependence of wear intensity on test time at various attack angles is revealed. Wear out of powder materials in water with corundum particles of F100 size is approximately five times higher, than with use of quartz particles of the same size. The material No 1 (*designation in Russian according to GOST* (*State Standard*) 28378-89 "*Iron-based powder structural materials. Types*" see [2]—*author's note*) has maximum wear resistance at action of small particles of F100 size. Comprehensive description is given of the samples and test conditions (sorry about this long citation but this work can serve as an example of a thorough presentation of the research results): "…Proof samples were manufactured by "Keramet" company in the form of square blocks with the size of 25×15×3 mm made of four types of iron-based powder materials. The blanks were obtained by powder metallurgical technique: pressing of mixture of powders and dissociated ammonia-sintering with the simultaneous copper infiltration. The samples roughness (R_a)—0.14 µm…"

For optimization of the methodology, the wear out kinetics was investigated, based on which the time of the experiment was chosen. The tests of the sample made of material No 2 (*designation in Russian according to GOST* (*State Standard*) 28378-89 "*Iron-based powder structural materials. Types*" *see* [2]—*author's note*) were carried out in the water with quartz at the jet velocity of 16 m/s and the samples angle m/s and the samples angle of setting 45° and 90°. After each experiment with duration of 160 min, the sample was weighed, the profilographs of the wear out area was recorded, the wear out intensity was determined, and the sample was set again in a like manner. The test results have revealed that the wear out kinetics has a linear character, i.e. with the increase of the experiment duration, the wear out intensity **I** practically does not change. At this experiment duration, significant values of wear out can be obtained both for the least wearproof and for the more wearproof materials, with the acceptable repeatability errors. Under given conditions, the wear out intensity average value comes to $2.11\ -10^{-6}$ g/g, scatter of readings, within limits of 7%. Worn-out surface profilometry under given conditions has revealed that the wear out pattern depends on the angle of setting of sample …

Based on the results obtained, the following main conclusions can be made. The major parameter influencing the powder materials erosion resistance at the specified jet velocity is the type of the abrasive material. Wear out of the powder materials in the water with corundum F100 is about 5 times more than in the water with quartz of the similar fraction. The abrasive material size also influences the wear out intensity of the powder materials. With the increase of the corundum size from F100 to F24, the wear out intensity of material No 2 (*designation in Russian according to GOST* (*State Standard*) 28378-89 "*Iron-based powder structural materials. Types*" *see* [2]—*author's note*) has increased as a factor of 1.8 higher. Under given conditions, hardness of powder materials has a little influence on their erosion resistance. Material No 1 (*designation in Russian according to GOST* (*State Standard*) 28378-89 «*Iron-based powder structural materials. Types*» *see* [2]—*author's note*) has the maximum wear out resistance when exposed to the abrasive particles with the size of F100. In carrying out of the erosive wear out tests, it is necessary to take into account the nonuniformity of wear out of the sample surface.

In the design of a specific gas separator screw under the specified requirements of the technical design specification, it is necessary to answer the questions, as follows:

1. In which way does the life ratio of the sample subject to confirmation to conformance to the requirements of the technical design specification is relevant to the times given in this article. If it exceeds the ones cited in this article, it should be established if the type of the wear out function observed within the experiment described in this publication was preserved within the restricted time.
2. Whether the gas separator operating conditions defined in the technical design specification confirm to those referred to in this work.
3. What is the allowable wear out of the unit component parts.
4. What are the criteria of malfunction, their connection with the wear out, etc.

Usually, the academic works on the wear out are only the source for the test methodology of the real samples, the results of which are analyzed by a specially appointed committee. At that, the major problem here is the choice of the reliability index inspection plan subject to confirmation by tests. A distinction is made between:

- check tests on determination of the system sample reliability index (to be defined during development);
- evaluation tests of the system sample production model (within the frames of assessment of the quality of manufacture).

A distinction is made between the quality control based on quantitative criterion, quality criterion or alternative criterion. Under the control based on the quantitative criterion, defined shall be the numerical values of one or several indices, which are to be compared with the standard indicators. Under the control based on the quality criterion, each tested unit of product is referred to a definite group. A particular case of the control based on the quality criterion is the control based on the alternative criterion, when the production is referred to one of two groups: fit/unfit. The problem is confirmation of the quality control requirements under the conditions of the restricted number of the operating cycles at the development stage. Some samples are accepted for usage together with the reliability enhancement schedule and achievement of the planned in the technical design specification reliability factor within a certain period of the trial operation. Example of the standard work schedule see [3]. The reliability growth-planning curve is important for development of the general strategy of carrying out of tests and assessment, defining of the necessary resource support. For the feature-based inspection of characteristics of the systems and their component parts during serial production based on the criterion of quality of manufacture (taking into account the reliability attained during the sample development) stringent branch standards are in force based on GOST (State Standard) P 50779.52-95 [4]. The number of the articles in the test samples is defined in the Technical Specifications and pushes up the cost of the delivered lot from 2…3% or as a factor of two higher. GOST [5] if compared to GOST P 50779.52-95, sets less overloaded 8-level inspection plans for a disciplined supplier. Amounts of sampling and the acceptance numbers specified by this standard ensure high probability of acceptance of batches of good products but have, as a consequence, high risk for the consumer. It is worth to note the timeliness of development of the new methodology for checking of the production component reliability index, especially for the component parts of the system with high characteristics of reliability and quality, because of necessity of high amount of sampling to confirm the specifications stated. Of all the plans with the total amount of control **n**, the minimum probability (the most favorable for the consumer risk) of the batch acceptance with the number of off-spec products equal to the rejectable quality level of defects per hundred is achieved with the single sampling plan (**n**, 0). In accordance with this plan, the production is accepted when and only when from **n** of the checked units of output all are considered fit. Operational characteristic for the plan (**n**, 0) is a curve circumscriptive from below all possible operational characteristics of the inspection plans. Let's take a usual presupposition that the consumer's risk shall not exceed 0.2. As is known, rejectable quality level of defects p_d for the plan (**n**, 0) and binomial distribution of the defective items is calculated from the condition:

$$(1 - p_d)^n = 0.2 \tag{1}$$

Based on which, the necessary amount of sampling shall be:

$$\mathbf{n} = 1.6/p_d. \tag{2}$$

For example, let's take $p_d = 0.01$; i.e. let's demand that not less than 80% of batches in which the number of defective items is in excess of 1 from 100, were rejected as defective. Then, volume of inspection must come to not less than **n** = 160. Further on, if a sufficiently high defect level has been attained, such as the customer should receive not more than 1 defective item out of 10,000, i.e. $p = 0.0001$, the volume of inspection must come to not less than **n** = 160,000 (the example is taken from [6]). Solution of this problem is found in implementation of the task-orientated methodologies of rapid testing and statistical quality control techniques.

3 Results

Stated above problems for the system unrestorable component parts will be found in the works [7, 8, 10]. For example, in [7] it is shown that in estimating of the expendable unrestorable items reliability factor:

1. Usage of biased or unbiased estimate of parameters dispersion does not influence the result of comparison of small samples in the range covered.
2. Method of comparison of abnormal samples by t test works well.
3. In comparing of the nondefective samples along with the sampling mean, it is necessary to carry out comparative test of the sampling variances.
4. Existing normative and methodological procedures of finding of the confidence span lower limit for integrated reliability indicators in the field of the minute volumes of full-scale testing can bring about the logically incorrect results.
5. It is being suggested a simple way of tentative estimate of the confidence span lower limit for the reliability indices of the expendable unrestorable items in the field of the minute volumes of full-scale testing.

In [8] it is shown that:

1. The narrowest confidence interval without accounting for the prior information is provided by the equation
$$\widehat{R} = 1 - \frac{0,5}{n+2}. \tag{3}$$
2. Estimates of the reliability factor taking prior information into account allow specifying a predicted value significantly (towards increasing) for a limited number of tests, when substantiated assumptions about the quantitative level of the reliability of an object are formulated.
3. Simultaneous consideration of the methods of accounting for prior information in the case of highly reliable devices allows for establishing the priority of selection of a method based on preliminary discussions (*a* and *b*): (*a*) first,

when the results of earlier tests with similar devices by the binomial scheme in identical conditions are known. The number of tests is summed up and used in equations like $\underline{P} = 0.2^{\frac{1}{n}}$. (*b*) Another method with theoretically experimentally substantiated assumptions about the quantitative level of reliability of an object is formulated based on prior information. As a result, the estimate interval is truncated ($[R_{lower}; 1]$ instead of [0, 1]).

In this way it can be concluded that unification of the results of a limited sample with prior (also insufficient) information by truncation of the interval of finding a reliability estimate may be inefficient. This conclusion is in agreement with [9], p. 393, where it is noted "...the method of fiducially probabilities becomes more effective with an increasing number of failures...," i.e., not in the case considered. In [10] it is shown that with the availability of the sufficient amount of priory information (for example, established batch production), possibility emerges to estimate efficiency of the earlier set inspection plans, starting out from the achieved testing parameter input level. With the achieved high quality level of batch production, the initially set inspection plan can be in a well-argued manner decreased because the expenditures it involves, practically, do not change the defects per hundred. In the example given, the two-fold reduction in the amount of sampling **n** (from 8 to 4) will reduce the average outgoing quality level by 0.5% (from 0.999750 to 0.999701). The standard method of the operational capability limits defining is the test measurement of the system-monitored parameters after each real working cycle in the course of tests. Owing to the limitations on the number of the cycles performed by the system in the course of its development and optimization, in most cases, no reaching of limit state in the course of the tests takes place. The results thus obtained, serve as a basis for the projection curve plotting reflecting the component part under investigation degradation till the limit tolerance state. The RF reference documentation on assurance of reliability of the samples included into the Uniform Codifier of Issue Item [11],[1] prescribe taking a decision on the number of the required SPTA sets on the lower bound of confidence interval of the testing parameter. Thus, the number of the SPTA sets can initially be over and above what is needed, however, based on the system practical application results, this value is corrected decreasingly. Based on the results of such truncated tests, the relationships for the lower bound of confidence interval of the normally distributed quantity under study possess a classic view:

$$\underline{M}_{gen} = M_{sample} - t - m_M, \tag{4}$$

where $\underline{M}_{gen}$ is the lower interval limit containing the true value of the quantity under study (for example, wear parameter of the contact blocks after three working cycles); M_{sample} is the mean value of the testing parameter after a series of measurements (for example, average value of the wear parameter of contact blocks of the sample five identical component parts having operated during three working cycles [2]; m_M is

[1]The reference is given to the document of general applicability.

[2]In practical work, the parameter possesses a more complicated form taking into account the changes in the wear process in dependence to the external influencing factors (ambient temperature, dust

Table 1 Dependence of the confidence criterion **t** on the probability of the error-free estimate **P** (at **n** > 30)

Degree of possibility of the error-free estimate (**P**%)	Confidence criterion **t**
95	1.96
99	2.6
99.9	3.3

Table 2 **t** criterion value for three *P* probabilities and the number of measurements **n** < 30

P	The degree of possibility of the error-free calculation (**P**, %)		
n−1	0.95	0.99	0.999
1	12.7	63.7	37.0
2	4.3	9.9	31.6
3	3.2	5.8	12.9
4	2.8	4.6	8.6
5	2.6	4.0	6.9
6	2.4	3.7	6.0
7	2.4	3.5	5.3
8	2.3	3.4	5.0
9	2.3	3.3	4.8
10	2.2	3.2	4.6
11	2.2	3.1	4.4
12	2.2	3.1	4.3
13	2.3	3.0	4.1
14–15	2.1	3.0	4.1
16–17	2.1	2.9	4.0
18–20	2.1	2.9	3.9
21–24	2.1	2.8	3.8
25–29	2.0	2.8	3.7

non-sampling error (measurement error); **t** is the confidence criterion. The value of **t** criterion is connected to error-free estimate (**P**) (Table 1 at the number of observations well in excess of 30).

To find the **t** criterion at the number of observations **n** < less 30, use can be made of a special table (sometimes called N. A. Plokhinski Table[3] in the domestic publications) (Table 2).

content, spread of the weight-and-dimensional characteristics of the interacting component parts, configuration of the subsystems).

[3]Based on the Student's **t** test boundary value tables.

Complexity in estimation of the necessary quantity of the SPTA sets is connected with the selection of its sufficiency criterion, based on which the SPTA sets will be composed on the basis of the technical design specification requirements to the system reliability factor. Extremely labour intensive is the SPTA sets optimization based on criterion of the minimum aggregate expenditures involved in meeting the technical design specification requirements to the reliability factor (primal problem of optimization) or optimization based on criterion of the maximal reliability factor with restrictions to the aggregate expenditures (inverse problem). On the one side, the SPTA set is completed to ensure reliability, keeping operational (failure-free performance, operational readiness), meeting the reliability requirements. On the other side, it is necessary for maintainability assurance, performance of maintenance in the interests of keeping operational. There is another criterial function in defining the SPTA set: average delay in fulfillment of the request. Comparing repair and replacement of the failed module for a spare part from the SPTA set, it is validly to ask a question on the possibilities and equivalence of these methods for restoring to the working condition. It is clear that the possibility of repair does not increase probability of no-failure operation, but makes it possible to increase the article availability factor. If time for replacement of the failed module for the available spare part is comparable to the time of restoration by repair, then, both methods are equivalent. If in compliance with the article intended use and the properties of the processes of functioning, in which the article takes part, some inaction period necessary for restoration by repair or replacement is allowed, then it can be said about availability of the time reserve. If there is no time reserve, then the SPTA set does not improve the reliability factors. In such case, for the corresponding component parts of the article structural redundancy is necessary, which ensures usage of the spare parts to improve the reliability factor. The question of importance of the time reserve is not just a question of measurement of time. This can also be the design problem [12]. To determine value t_{extra}, it is necessary to simulate the behavior of the device where the failed component is found under the conditions of the temporary disability. In some cases, new design solutions can become necessary, for example, some temporal change of functioning condition necessary to slow the unfavorable processes and to increase t_{extra} to the acceptable value. In the different purpose systems special conditions can be created, which will result in availability of the time reserve used to replace the failed module for a spare part. It means that to improve the reliability factor, the SPTA sets can satisfactorily be used even in a nonredundant system. The complexity and high cost of the presented approach, which initially leads to obtaining quite a rough information, intensifies the analytical evaluation of the wear parameters and development of methods for evaluation of the degradation of characteristics by simulating of wear based on mathematical models of the interaction of the studied component parts. For example, in the case under consideration, you can use a typical scientific work [13] (see Fig 1).

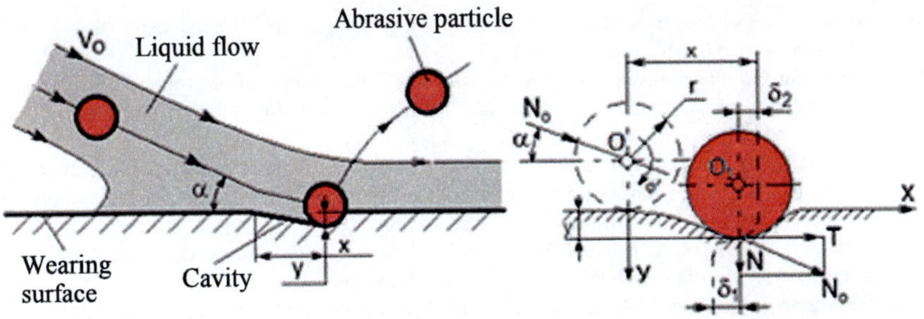

Fig. 1 Structural model for the interaction of an abrasive particle with the surface of the blade during erosion wear-out

In most cases, a simulated evaluation is characterized by high labour intensity and low degree of applicability in engineering calculations, since it is based on a number of assumptions concerning dominant phenomena and ignorance of the others that are considered to be less important. It is especially difficult to build an adequate mathematical model under the conditions of creating of the new samples and absence of the prior information.

4 Conclusions

Thus, the problematic points of determining the required amount of SPTA sets to ensure a given system reliability based on the wear model construction which allows to set the limits of the components functional ability and the periods of routine maintenance, are the high cost of works (both experimental and theoretical) and large confidence interval of the statistics obtained.

References

1. Interstate standard GOST 2.601-2013 (Unified system for design documentation. Exploitative documents) M.: Standartinform (2014)
2. Smirnov NI, Yagovkina A, Prozhega M, Smirnov NN (2017) Development of methods for materials erosion tests. M J Eng Industr Eng Educ 2:60–67
3. Growth Reliability (2014) Enhancing defense system reliability. National Academies Press, Washington
4. GOST P 50779.52-95 Statistical methods (1996) Acceptance sampling by attributes
5. GOST P ISO 2859-1-2007 Statistical methods (2007) Sampling procedures for inspection by attributes. Part 1: sampling schemes indexed by acceptance quality limit for lot-by-lot inspection ISO 2859-1:1999 sampling procedures for inspection by attributes—part 1: sampling schemes indexed by acceptance quality limit (AQL) for lot-by-lot inspection (IDT)

6. Orlov AI (2004) Mathematical basis of sampling quality control (Decision-making theory). Publishing house MART, Learning aid.-M
7. Denezhkin GA, Belobragin BA, Avotyn' BA (2017) Estimators of the reliability index of a nonrepairable one-shot bibitem by small statistical samples. J Mach Manuf Reliab 46(2):154–160
8. Belobragin BA, Avotyn' BA, Ustinkin AI (2019) Comparative efficiency of the equations of conditional distributions for estimation of the reliability factor of highly reliable unmaintainable single-use devices. J Mach Manuf Reliab 46(2):136–140
9. Ushakov IA (ed) (1985) EdNadezhnost' tekhnicheskikh ystem. Spravochnik (Reliability of technical systems, the handbook). Radio Svyaz', Moscow
10. Belobragin BA, Avotyn' A, Ustinki AI (2020) Accounting of the priory information in a batch production of the unmaintainable single-use disposable devices for optimization of the inspection plan parameters. J Eng Autom Probl 1:22–27
11. P 50.5.002-2001 Recommendation on cataloguing. Products cataloguing for Federal Government needs. Uniform codifier of issue article and the order of development and maintaining of the directory partitions of the Federal product catalogue for Federal Government needs. M.: State Committee for the Russian Federation for Standardization and Metrology (2007)
12. Tcherkesov UN (2010) Concerning the reliability calculation of the repairable systems upon availability of the spare parts. Part 1. M.: Reliability 3:29–9
13. Litvinenko K (2016) Predictive modeling of the technical state of ECPU (electric centrifugal pump units—autom.) under conditions of intensive removal of solids. Abstract of dissertation for the degree of Candidate of Technical Sciences. Federal State Budgetary Educational Institution of Higher Education, Ufa State Petroleum Technological University, Ufa city

Fatigue Assessment of Aged Steel Specimens Under Uniaxial Cyclic Loading

A. Yosri, A. Zayed, S. Saad-Eldeen, and H. Leheta

Abstract Corrosion degrades mechanical properties and fatigue life of structures considerably; hence reliable numerical fatigue analysis is essential for assessing the remaining life of aged steel structures. In this paper, the influence of corrosion topology and the minimum thickness location on the stress concentration over corroded specimens and, in consequence, their fatigue life was comprehensively studied. The corroded surfaces were modelled using random spatial distributions approach. The results of the numerical simulation have been compared with published fatigue test results. The comparison showed good agreement between the experimental results and the adopted approaches for fatigue life assessment.

Keywords Stress concentration · Corrosion modelling · Fatigue life · Corrosion fatigue · S–N curve

1 Introduction

Ships and offshore structures operate in an aggressive and complex environment with high humidity and corrosive atmosphere. Those structures are subjected to highly fluctuated and complex loading conditions. The combination of tensile stresses and the corrosive environment culminates with environmental cracking. Static tensile stresses provoke stress corrosion cracking and hydrogen embrittlement or hydrogen-induced cracking, while the cyclic tensile stresses spur corrosion fatigue cracking [1]. Those cracks may finally lead to leakage and unstable failures. Therefore, corrosion can be considered to be the most aggressive age-related structural degradation factor.

A. Yosri (✉) · A. Zayed · H. Leheta
Faculty of Engineering, Department of Naval Architecture & Marine Engineering, Alexandria University, Alexandria, Egypt
e-mail: ahmed.yosri@alexu.edu.eg

S. Saad-Eldeen
Faculty of Engineering, Department of Naval Architecture & Marine Engineering, Port Said University, Port Said, Egypt

M. Abdel Wahab (ed.), *Proceedings of the 8th International Conference on Fracture, Fatigue and Wear*, Lecture Notes in Mechanical Engineering,
https://doi.org/10.1007/978-981-15-9893-7_38

Garbatov et al. [2] performed fatigue test on 11 corroded specimens which were cut from a corroded box girder. The experiments showed that there is an excessive degradation of the corroded specimens fatigue life. The negligible levels of residual stresses in the test specimens during corrosion reduced the possibility of environmental cracking interference with the results of the fatigue tests. The corroded surface topology was captured by 3D scanner and the principal characteristics of the corroded specimens were reported [2].

Moreover, Shan-hua et al. [3] conducted fatigue test on 21 corroded specimens of different corrosion ratios and compared them with the fatigue life of 2 intact specimens. It was found that the fatigue life of the corroded specimens was about 3% of the intact ones.

The corroded specimens were subjected to two forms of corrosion. The dominant one is general (uniform) corrosion and the second form is localized pitting corrosion. The localized corrosion is more dangerous than uniform corrosion because it is more difficult to detect, predict and design against [1]. It causes stress concentration at pits tip where fatigue cracks always initiate. Therefore, recently, considerable attention has been paid to the effect of pitting corrosion on the fatigue life of steel structures.

Ho Ahn et al. [4] studied the correlation between pit depth and the fatigue life of single-pitted specimens. It was demonstrated that the fatigue life decreases apparently with increasing pit depth. DuQuesnay et al. [5] investigated experimentally the influence of corrosion damage on the fatigue life of corroded aluminium coupons. It was stated that the depth of the corrosion pit is a suitable parameter for charachterizing the corrosion damage and for predicting fatigue life of the corroded coupons. It has to be stated that excessive reduction of fatigue life due to corrosion has been reported [6, 7], however, there is no straightforward equation that can predict the fatigue life of corroded structures precisely.

In this paper, the major factors that affect the stress concentration and in consequence the fatigue life of naturally corroded specimens is comprehensively studied using finite element method (FEM) and fatigue analysis codes. The corrosion topology is modelled using random spatial distribution of the corrosion wastage. The estimated fatigue lives of the corroded FE models are compared with the published fatigue test results.

2 S–N Approach

The fatigue strength of structures is highly influenced by the material type and the ultimate tensile stress. For example, the fatigue life of ship structures made of steel is 2.5 times higher than those made of Aluminum alloy [8]. In addition, for a group of similar alloys, the fatigue strength depends on the ultimate tensile stress. For instance, the fatigue strength of an un-notched specimen can be enhanced by increasing its ultimate strength by heat treatment or changing the alloy composition [9].

The fatigue life can be assessed by S–N approach or the fracture mechanics approach. However, S–N approach is widely used for fatigue design purposes and also it is often the most suitable [10]. In this paper, S–N approach has been adopted to investigate the fatigue life of corroded specimens.

The fatigue properties of materials can be described using S–N curve, also called Wöhler curve, and fatigue diagram. Those curves can be evaluated experimentally by constant amplitude fatigue tests on un-notched specimens or can be estimated using various approaches. For high cycle fatigue, the S–N curve can be described by 2 points. The first point corresponding to 1000 cycle fatigue strength (S_{1000}) which is approximately equal to 90% of ultimate strength S_U for bending test and 75% of S_U for tensile test. The second point corresponding to the endurance limit at 10^6 cycle S_f, for steel materials S_f is around half of the ultimate strength [9, 11].

Przemyslaw et al. [12] investigated the errors of the different S–N curve estimation approaches. Based on qualitative and quantitative verification, they found that the errors of those estimated S–N curves may be up to three times less or more than the experimental fatigue life. They proposed a hybrid method combines experimental and analytical fatigue data to reduce the errors of the analytical S–N curves.

In this paper, the S–N curve of Carbon Steel SAE1015 [13] has been adopted to study numerically the fatigue life of the corroded specimens tested by Garbatov et al. [2]. The material has the same mechanical properties ($S_U = 400$ MPa, $S_{yield} = 235$) and approximately the same chemical composition as the corroded specimens. The ratios $S_{1000}/S_U = 84\%$, S_f (at 10^6 cycle)/$S_U = 43.8\%$ and the stress ratio $= -1$.

3 Corrosion Finite Element Modelling

The aim of the finite element modelling of the corroded specimens is to predict the stress concentration over the corroded surface which is essential for fatigue life estimation. Stress concentration is usually defined by Stress Concentration Factor, SCF (K_t) which is defined as the ratio between the peak stress and the nominal stress.

$$K_t = \frac{\sigma_{peak}}{\sigma_{nominal}} \tag{1}$$

Three methods can be used to determine the SCF. The most reliable method is experimental measurement, using strain gauges; parametric equations, as Peterson [14] and Roark's [15] formulas; and finite element analysis, FEA, which is adopted in this study.

Various approaches in the literature have been adopted to represent corrosion wastage and pits in finite element analysis. The uniform reduction of thickness which is adopted by the classification society guidelines [16] is useful in global FEA, while it is imprecise for local FEA. In order to model the localized corrosion, different corrosion pit shapes such as cylindrical, conical, hemispherical, square and rectangular were adopted in literature [17–21]. Teixeira and Guedes Soares [22] modelled the corrosion wastage using random spatial distribution approach to study the collapse strength of corroded plates. Converting the thickness measurements by

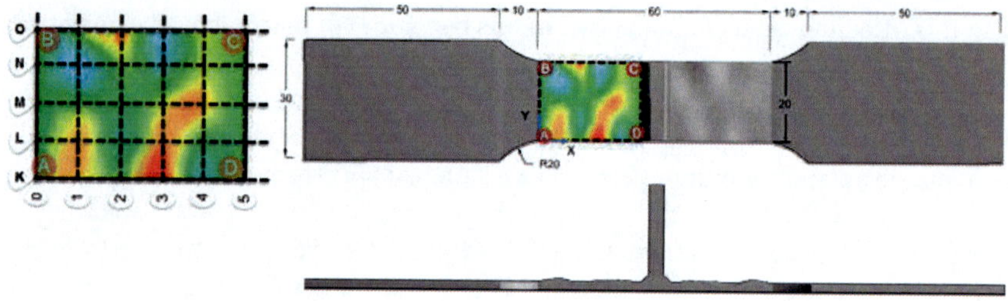

Fig. 1 Finite element solid model of corroded specimen

Table 1 Principal characteristics of corroded specimens [2]

Specimen	A6	A7	A8	A10	B9	B14	B17	C4	C11	C12	C13
t_{max} “mm”	3.08	3.07	3.23	3.07	3.22	3.23	3.14	3.28	3.29	3.09	2.98
t_{min} “mm”	1.5	1.52	1.62	1.48	2.13	2.1	2.09	2.04	1.83	1.96	2.02
t_{mean} “mm”	2.44	2.39	2.53	2.4	2.72	2.69	2.6	2.53	2.57	2.61	2.52
Standard deviation “mm”	0.39	0.28	0.32	0.19	0.36	0.11	0.17	0.19	0.23	0.22	0.16

laser or photogrammetric techniques into 3D finite element model was also adopted in [23].

For the current study, the 11 corroded specimens tested by Garbatov et al. [2] were modelled using random spatial distribution approach. The corroded surface was generated using a grid of 30 points, 5 rows (K:O) × 6 columns (0:5), as indicated in Fig. 1, where red contours represent max thicknesses and blue contours represent minimum thicknesses in mm. The distribution of random thickness and its values was estimated rationally to meet the principal characteristics, minimum thickness (t_{min}); maximum thickness (t_{max}); mean thickness (t_{mean}) and volume loss, of the corresponding real corroded specimen, as given in Table 1.

The accuracy of the calculated stress concentration factor using FEA usually depends on the element size [24]. In the subsequent analysis, the same finite element size and type are adopted in the evaluation of stress concentration factors arisen by corrosion; whereas the elements size may cause slight fluctuation of K_t that can be ignored. Figure 2 demonstrates a typical FE mesh employed in the present study.

4 Stress Concentration Factor

In order to investigate the major factors that affect stress concentration over the corroded surfaces of the tested specimens by Garbatov et al. [2], A series of FE models; 125 models, in total, were modelled. The categorization of the models are as

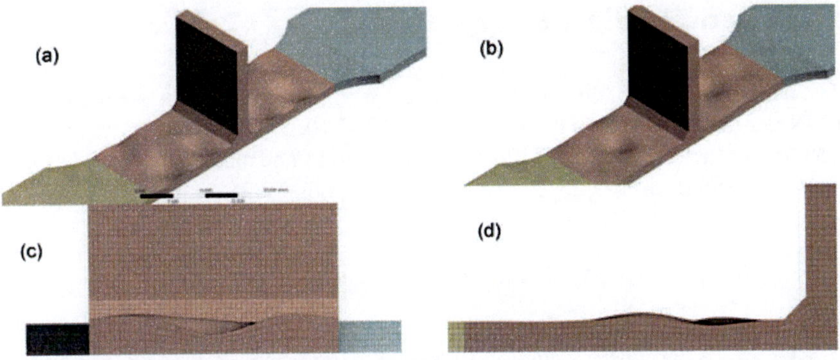

Fig. 2 Typical FE mesh employed in the present study; **a, b** FE models of the corroded specimens; **c** cross-section; **d** longitudinal section

follow, 10 FE models for each real corroded specimen and 15 models for investigating the influence of the location of minimum thickness on the fatigue life.

Figure 3 (left) illustrates the correlation between stress concentration factor K_t and $(t_{mean} - t_{min})$ and Fig. 3 (right) for cross-sectional area at minimum thickness A_{tm}. The stress concentration factor fluctuates considerably for small thicknesses, while it becomes steadier for thick specimens. Moreover, SCF decreases with increasing cross-sectional area of the corroded specimens.

Based on the linear regression analysis, the stress concentration factor may be calculated using Eq. 2. This equation can be used only for thickness measurements close to the reported above and over an area of 500 mm^2 (20 mm × 25 mm).

$$K_t = 1 + 1.0783(t_{mean} - t_{min}) \tag{2}$$

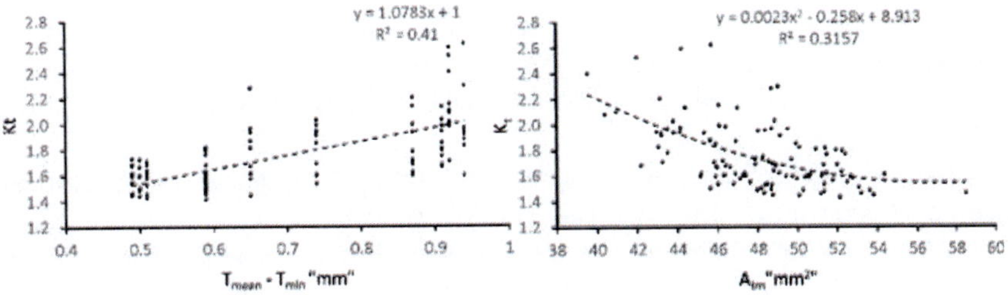

Fig. 3 Correlation between stress concentration factor K_t and $(t_{mean} - t_{min})$ (left) and cross-sectional area at minimum thickness A_{tm} 110 model (right)

5 Numerical Fatigue Life Assessment

The numerical fatigue life calculation has been conducted using the simple method (S–N approach) [9, 11]. The S–N curve of SAE1015 [13] has been modified to account for the corrosion and different loading setup. The modification due to mean stress correction was calculated using Gerber Parabola:

$$S_N = \left(1 - \left(\frac{S_m}{S_u}\right)^2\right) \times (S_N)_{S_m=0} \tag{3}$$

where S_m is the experiment mean stress and S_N is the fatigue strength for a certain fatigue life N.

The micro-roughness is assumed to be 250 μm and its effect on low cycle fatigue is ignored [9]. To account for the deterioration of the material mechanical properties and the high scatter of the stress concentration factor, a modification factor $C_R = 0.8$ is introduced which refers to a reliability level of 0.99 [11]. The estimated SCF in the previous section will have a considerable effect on high cycle fatigue, while in low cycle fatigue it will have a small effect [9]. Therefore, in low cycle fatigue, its effect was reduced by a suggested factor $\varnothing = 0.8$, whereas $\varnothing \bullet K_t > 1$. Equations 4 and 5 combine the contribution of all these parameters which define the modified S–N curve.

$$S_{afk} = \left(1 - \left(\frac{S_{ma}}{S_U}\right)^2\right) \frac{S_{f(S_m=0)}}{K_t} C_R \gamma \tag{4}$$

$$S_{a1000k} = \left(1 - \left(\frac{S_{ma}}{S_U}\right)^2\right) \frac{S_{1000(S_m=0)}}{\varnothing K_t} C_R \tag{5}$$

where $\gamma = 0.8$ is the surface roughness reduction factor corresponding to 250 μm micro-roughness [9], S_{afk} is fatigue strength of the corroded specimens at 10^6 cycle, S_{a1000k} is fatigue strength of the corroded specimens at 10^3 cycle and $S_{1000(Sm=0)}$ and $S_{f(Sm=0)}$ are the fatigue strengths of the material at 10^3 and 10^6 cycle and mean stress = 0.

Group A was tested under stress range $S_r = 120$ MPa ($S_{max} = 126$ MPa), the stress range of Group B is $S_r = 155$ MPa ($S_{max} = 163$ MPa) and Group C $S_r = 200$ MPa ($S_{max} = 210$ MPa). Figure 4 illustrates the correlation between the numerical fatigue life of the corroded specimens, experimental fatigue test results of Garbatov et al. [2] and different fatigue categories [16, 25]. It can be observed that the fatigue category of the welded joint is degraded from fatigue class 80 to fatigue class 40 comparing with the numerical results and fatigue class 50 taking into account only the fatigue experiment. It was observed experimentally and numerically that all failures occurred away from the welded web. The locations of the numerical failures were investigated using ncode (commercial fatigue design code). Figure 5 shows the fatigue life contours

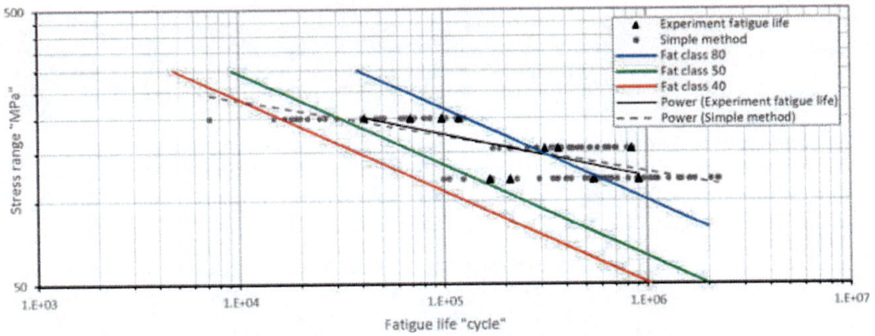

Fig. 4 Comparison between fatigue test results by Garbatov et al. [2], different fatigue categories and simple method

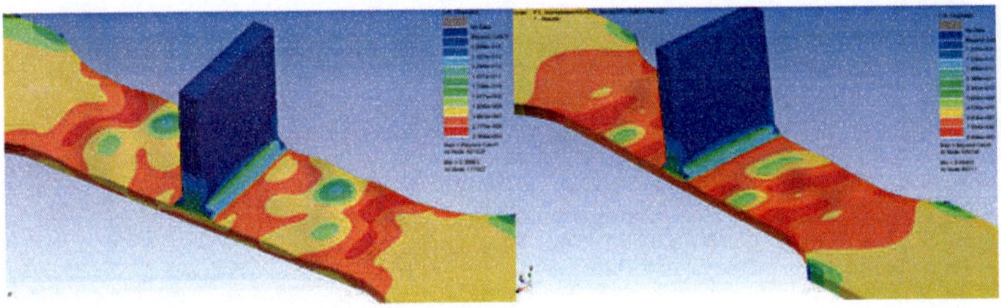

Fig. 5 Fatigue life contours show the expected location of fatigue failure (ncode)

of the expected locations of fatigue failure (ncode). It is found that the numerical failures occurred at or near the minimum thickness location [16].

5.1 *Influence of Minimum Thickness Location on Fatigue Life*

In order to explicate the high fluctuation of fatigue life of the corroded specimens, 15 FE models were created for A7 corroded specimen, Garbatov et al. [2]. All the 15 models have the same t_{min}, t_{mean} and volume loss of the real corroded specimen; while the minimum thickness location has been altered between rows and columns indicated in Fig. 6. The fatigue life was estimated using ncode. Figure 7 illustrates the correlation between minimum thickness location and the corresponding fatigue life. It is noticeable that the minimum fatigue life occurs when the minimum thickness location is at row M and column 1, near the geometrical notch, which is due to the combined effect of corrosion damage and geometrical notch. The fatigue life increases with shifting the minimum thickness towards the welded web and reduces at column 4 before increasing again when it is located near the welded web. The

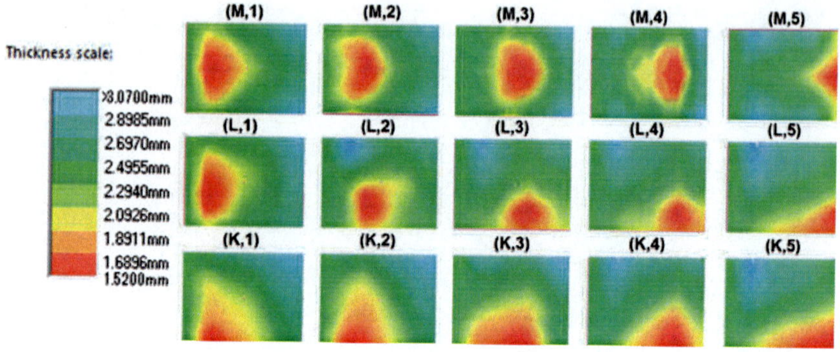

Fig. 6 Different minimum thickness locations of the corroded specimen A7 (15 models)

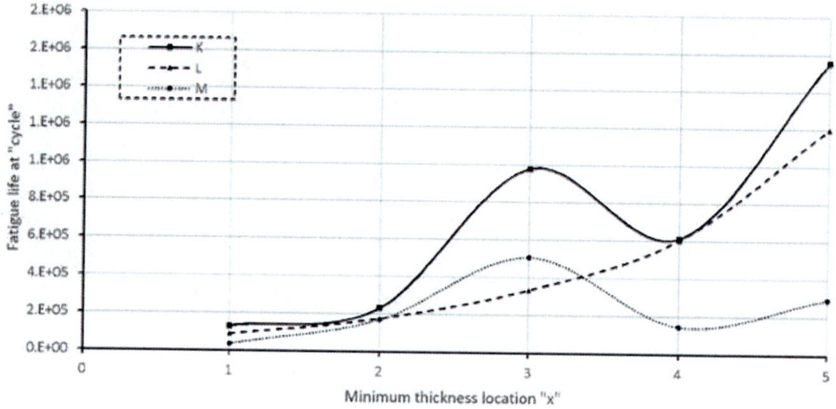

Fig. 7 Minimum thickness location and fatigue life of specimen A7 at stress range = 120 MPa

increment of fatigue life at column 4 may be due to the thickness reduction at the weld toe which may act as burr grinding (technique used to reduce stress concentration at weld toes) that improves fatigue life [16].

It has to be stressed that, stress concentration factors within thin regions of corroded plates/panels have to be taken into account; otherwise, fatigue cracks may initiate prematurely as described earlier. These cracks may propagate to thicker regions or to the adjacent welded joints causing serious damage to the structure.

The present study shows a good agreement between the numerical results and the published experimental fatigue test results of the corroded specimens. It has to be stated that further numerical analysis and fatigue tests are still essential to study the behavior of thick corroded plates ($t_{mean} > 4$ mm) under various mean stresses.

6 Conclusion

The present study dealt with different approaches that have been adopted to estimate the fatigue life of aged ship structural. The main causes of deterioration and highly scattered fatigue life have been studied. The results of the numerical simulation have been compared with fatigue test results.

The random spatial distribution approach, which was adopted to represent the corrosion wastage, gives a reliable assessment of stress concentration over the surface of corroded structures. Therefore, it can be adopted in numerical investigations of SCF for corroded ship structural details.

A regression equation was proposed to estimate the maximum stress concentration factor over a small patch of a corroded plate.

The effect of location of the deepest corrosion pit or minimum thickness on fatigue life is investigated. It was found that the location of minimum corroded thickness relative to the neck, edges and the welded web has a significant effect on the fatigue life.

The outcomes of this paper can be useful to estimate the S–N curve of corroded structures numerically, which is essential for variable amplitude loading fatigue analysis.

References

1. Roberge PR (2000) Handbook of corrosion engineering. McGraw-Hill, McGraw-Hill Education, New York
2. Garbatov Y, Guedes Soares C, Parunov JJ (2014) Fatigue strength experiments of corroded small scale steel specimens. Int J Fatigue 59:137–144
3. Shan-hua X, Qiu BJMS, E. A, Experimental study on fatigue behavior of corroded steel 584:163–169
4. Seung-Ho Ahn S, Lawrence Jr F, Metzger M (1992) Corrosion fatigue of an HSLA steel. Fatigue Fract Eng Mater Struct 15(7):625–642
5. DuQuesnay DL, Underhill PR, Britt HJ (2003) Fatigue crack growth from corrosion damage in 7075-T6511 aluminium alloy under aircraft loading. Int J Fatigue 25(5):371–377
6. Medved J, Breton M, Irving PE (2004) Corrosion pit size distributions and fatigue lives—a study of the EIFS technique for fatigue design in the presence of corrosion 26(1):71–80
7. Pidaparti RM, Patel RRJML (2008) Correlation between corrosion pits and stresses in Al alloys 62(30):4497–4499
8. Fricke W (2017), Fatigue and fracture of ship structures. Encyclopedia of Maritime and Offshore Engineering pp 1–12
9. Schijve J (2009) Fatigue of structures and materials, 2nd ed. Springer, Berlin
10. Bai Y, Jin W-L (2015) Marine structural design. Butterworth-Heinemann
11. Lee Y, Pan J, Hathaway R, Barkey M (2005) Fatigue testing and analysis, 1st ed. Butterworth-Heinemann
12. Strzelecki P, Sempruch J (2016) Experimental verification of analytical models of the SN curve within limited fatigue life 54(1):63–73
13. Boardman BE (1982) Crack initiation fatigue—data, analysis, trends and estimation. In: Fatigue conference, United States
14. Pilkey WD (1997) Peterson's stress concentration factors, 2nd ed. Wiley

15. Young WC, Budynas RG (2002) Roark's formulas for stress and strain. McGraw-Hill
16. DNV.GL, Fatigue assessment of ship structures. In: Guideline C (ed) Rules for classification of ships. Det Norske Veritas
17. Ahmmad MM, Sumi Y (2010) Strength and deformability of corroded steel plates under quasi-static tensile load. J Mar Sci Technol 15:1–15
18. Huang Y, Zhang Y, Liu G, Zhang Q (2010) Ultimate strength assessment of hull structural plate with pitting corrosion damnification under biaxial compression. Ocean Eng 37:1503–1512
19. Rahbar-Ranji A (2012) Ultimate strength of corroded steel plates with irregular surfaces under in-plane compression. Ocean Eng 54:261–269
20. Shi X, Zhang J, Guedes Soares C (2018) Numerical assessment of experiments on the ultimate strength of stiffened panels with pitting corrosion under compression. Thin-Walled Struct 133:52–70
21. Zhang Y, Huang Y, Meng F (2017) Ultimate strength of hull structural stiffened plate with pitting corrosion damage under unaxial compression. Marine Struct 56:117–136
22. Teixeira AP, Guedes Soares C (2008) Ultimate strength of plates with random fields of corrosion. Struct Infrastruct Eng 4(5):363–370
23. Gathimba N, Kitane Y (2018) Static ductility evaluation of corroded steel plates considering surface roughness characteristics. The 2018 Structures Congress (Structures18), Songdo Convensia, Incheon, Korea
24. Méndez G, Cuamatzi-Meléndez R, Hernández AA, Capula-Colindres SI, Angeles-Herrera D, Velázquez JC, Vazquez-Hernández O (2017) Correlation of stress concentration factors for T-welded connections – finite element simulations and fatigue behavior. Soldagem Inspeção 22(2):194–206
25. Hobbacher A (2015) Recommendations for fatigue design of welded joints and components. Springer, Berlin

Estimation of Steam Turbine Shafts Fatigue Damage Caused by Torsional Vibrations

O. Chernousenko, V. Peshko, B. Marisyuk, and A. Bovsunovsky

Abstract In Ukraine, 74 power units with a capacity of 150–800 MW of thermal power plants out of 102 are on the verge of exceeding the park resource (220 thousand hours). Of these, 17 power units are close to the end the park resource, and 11 units have reached their estimated resource (100 thousand hours). Statistics of power plants accidents in different countries indicates that most of them are caused by the long term accumulation of fatigue damage. Among the main causes of fatigue damage are the pulsating of high-temperature steam flows in the turbine cylinders, transverse and torsional vibrations of shafts during long-term operation etc. This causes the development of computational and experimental methods for the determination of fatigue damage of turbine units' shafts under dynamic loads. The evaluation of fatigue damage of steam turbine K-200-130 shaft is based on the 3D finite element model. The calculations take into account experimentally determined fatigue properties of rotor steel. The fatigue damage as a result of torsional vibrations of turbine shaft caused by the abnormal operation of turbine generator is calculated. Zones of stress concentration in the rotating elements of the steam turbine type K-200-130 are established for various operating modes. The procedure to predict the effect of fatigue damage on the estimated resource of turbine unit is proposed.

Keywords Turbine shaft · 3D model · Finite element method · Fatigue damage · Torsional vibrations

O. Chernousenko · V. Peshko (✉) · B. Marisyuk
Department of Cogeneration Installations of Thermal and Nuclear Power Plants, National Technical University of Ukraine "Igor Sikorsky Kyiv Polytechnic Institute", 37 Prosp. Peremohy, Kyiv 03056, Ukraine
e-mail: vapeshko@gmail.com

A. Bovsunovsky
Department of Mechanical Engineering, Standardization and Certification of Equipment, National University of Food Technologies, 68 Volodymyrska Str., Kyiv, Ukraine

M. Abdel Wahab (ed.), *Proceedings of the 8th International Conference on Fracture, Fatigue and Wear*, Lecture Notes in Mechanical Engineering,
https://doi.org/10.1007/978-981-15-9893-7_39

1 Introduction

In the world energy sector, cases of brittle destruction of turbine units due to the fatigue damage are rare. However, they lead to catastrophic consequences with significant material losses. An example of such destruction is the accident at the Galatin power station (USA), when the rotor of the medium pressure cylinders of the turbine unit with a capacity of 225 MW was broken into 30 parts [1]. A similar accident occurred in the United States in Pittsburgh [2]. The reason of these accidents was long-term accumulation of fatigue damage. For a similar reason, the turbine shaft of the Kashira thermal station with a capacity of 300 MW fell to pieces in 2004 [3].

The fatigue damage of steam turbine rotating elements during operation is dependent on the dynamic regimes of loading, which can cause transverse and torsional vibrations of the shaft. A lot of investigations were devoted to transverse vibrations of turbine shafts. At the same time, torsional vibrations are insufficiently studied. In general case, the reasons of turbine shafts torsional vibrations can be the reactive torque that occurs due to a short-circuit on generator, short circuit in power lines, connection of generator to power network with insufficient (rough) synchronization, touching of rotating parts to the turbine shell, all possible electromagnetic effects on generator [4–6]. As a result, the magnitude of the reactive torque can be by several times bigger than the magnitude of the torque at stationary regime of turbine operation.

The turbine generator short-circuit is very rare event, but under certain conditions it can cause the catastrophic failure of turbine shaft [5, 6]. At the same time, the long-term accumulation of fatigue damage in turbine shaft is not possible because of this reason. However, the fracture analysis of fatigue cracks in the emergency destroyed shafts [1–3] revealed that cracks growth lasted several years.

The accumulation of fatigue damage in turbine rotating elements can be attributed to the frequent start-ups, which are usual for thermal power stations. The estimated number of such launches during the operation time of turbine units is nearly 2000. At each start-up at least one networking of turbine generator happens. The short-term reactive electromagnetic shaft's torque arises by reason of the difference of the phase shift angle between the vector of turbine generator electromotive force and the vector of power network voltage [7]. Thus, torsional vibrations of turbine shafts occur due to this reactive torque. In some cases, the intensity of torsional vibrations is high enough to be the reason of fatigue damage [8]. The number of cycles of vibration with maximal stress amplitude exceeding the fatigue limit of rotor steel at each start-up of turbine can be in the range from tens to thousands. As a consequence, during the life-time of turbine operation fatigue damage gradually accumulates.

This work is the continuation of the research [7]. Unlike the research [7], the level of damping of shafts torsional vibrations were taken from the experimental investigation [9], which makes prediction of turbine shaft fatigue damage more realistic.

Given the above, it is evident that the study of influence of turbine shafts torsional vibrations, caused by the electromagnetic and mechanical processes in generator, on the stress–strain state (SSS) and fatigue damage of shafts is important. The object under investigation was steam turbine K-200-130.

2 The Computational Model of Turbine Rotors in Three-Dimensional Setting

Steam turbine of K-200–130 type consists of three rotors (high pressure, middle pressure and low pressure, respectively, that is HPR, MPR and LPR) and generator (TG). The three-dimensional model of the shaft line was created using the graphic editor SolidWorks. The model consisted of 177,538 finite elements, which are interconnected by 303,260 nodes (Fig. 1).

The geometric model of the shaft is developed in accordance with the drawings of the manufacturer and takes into account the main structural rotating elements of the turbine. At the same time, the geometric design of the blades and bandage connections was replaced by toroidal discs to reduce the time of numerical calculation. The sizes of these disks are specially selected in such way that their mass is equal to the mass of all blades and its mountings of this stage. Also, the outer diameter of the torus corresponds to the similar diameter of the blades. The studies conducted by the authors [10] showed that such a replacement is justified, since the error is about 5%, and the calculation term is reduced significantly.

The shaft is loaded by static torques T_{hp}, T_{mp}, T_{lp}, caused by the pressure of steam respectively on HPR, MPR and LPR (Fig. 1). Torsional vibrations are excited by the

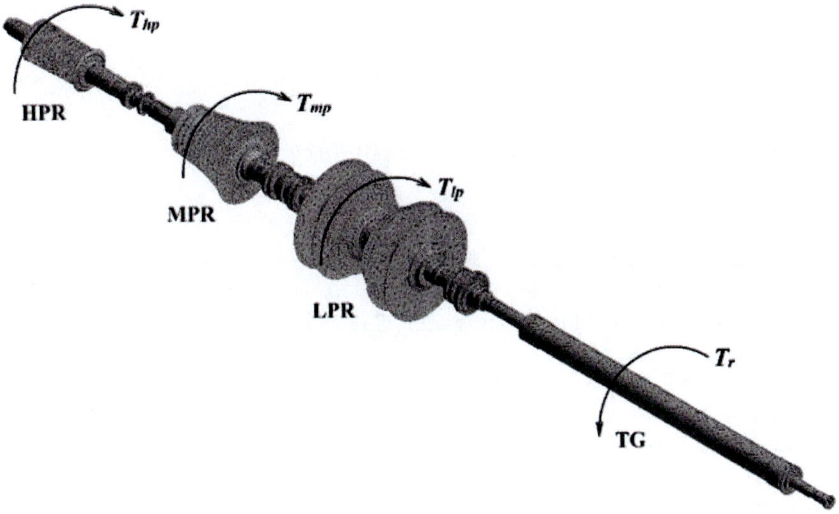

Fig. 1 3-D mesh of turbine K-200-130 shaft

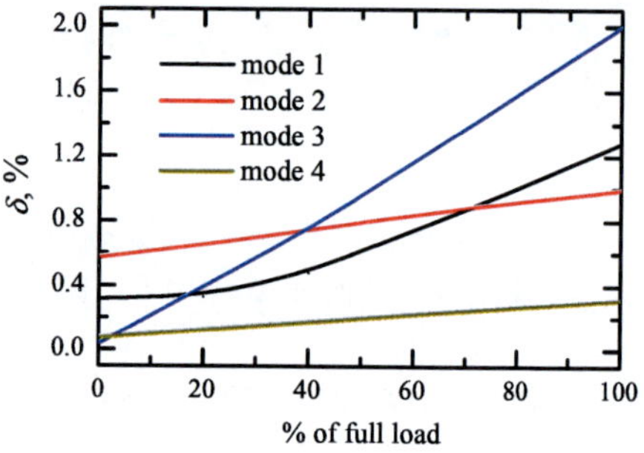

Fig. 2 Generator power dependence of damping characteristic of four turbine shaft torsional modes of vibration [9]

reactive torque (T_r), which can be of electromagnetic or mechanical nature. Total static torque is the sum of static torques acting on rotors.

$$T_{st} = T_{hp} + T_{mp} + T_{lp}. \tag{1}$$

The reactive torque was assumed to be Π-shaped function [6] with short duration, which can simulate unsuccessful connection of generator to power network or mechanical interaction of turbine structural elements with the turbine shell. Its amplitude was related to the static torque in the following way [6]

$$T_r/T_{st} = 3. \tag{2}$$

Damping characteristic of shaft was experimentally determined for different mode shapes (Fig. 2) [9]. In calculations damping in terms of logarithmic decrement there was accepted to be $\delta_1 = 1.28\%$; $\delta_2 = 1.0\%$; $\delta_3 = 2.0\%$; $\delta_4 = 0.32\%$, where subscript means number of mode shape.

3 Calculation of shaft's SSS at Stationary Mode of Operation

This calculation did not take into account the pulsations of the steam flow, voltage fluctuations on the generator and in the electric network, because the influence of these factors on the SSS of the shaft is negligibly small.

The values of torque on rotors were assumed to be $T_{hp} = 0.196$ MNm, $T_{mp} = 0.291$ MNm and $T_{lp} = 0.163$ MNm. The sum of these moments is the nominal static torque $T_{st} = 0.65$ MNm acting in the area of coupling between the shaft of the steam turbine and the generator.

Figure 3 demonstrates the equivalent stresses by Mises in the stationary mode of operation of steam turbine that produces 200 MW of electric power. As can be seen, the highest stresses are observed between the low pressure rotor and turbine generator. They are in the range of 150–170 MPa. The stresses that occur in the

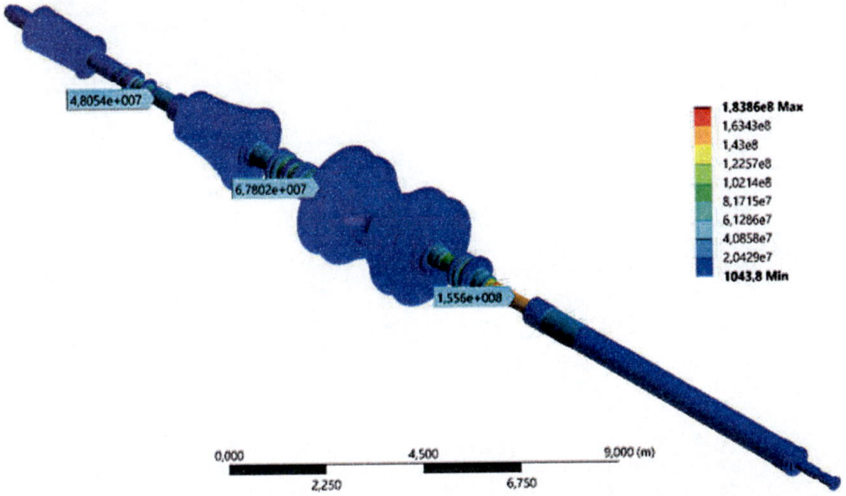

Fig. 3 Equivalent stresses of the shaft line of the K-200-130 turbine unit according to Mises at stationary mode of operation

area of the semi-rigid coupling between the MPR and LPR are in the range 1.4–72 MPa, and in the area between the HPR and MPR are in the range 3.4–53 MPa. The obtained stresses do not exceed the strength limit for 25Cr1Mo1V steel, so the stationary operation does not lead to the significant accumulation of damage in the turbine shaft.

4 Calculation of shaft's SSS Under Reactive Torque Loading

The shaft's SSS was calculated for the forced vibrations caused by the short-term reactive torque and then—for the free vibrations, the reactive torque disappears. As it was mentioned above, the reactive torque was accepted to be $T_r = 3T_{st}$. The duration of reactive torque was chosen so that the intensity of torsional vibrations was the highest, namely 0.02 s.

The most dangerous areas of the turbine shaft are the coupling between HPR and MPR, as well as the sections between MPR and LPR and between LPR and TG. Figure 4 demonstrates the time dependence of equivalent stresses amplitudes for these areas as a result of action of the reactive torque.

The maximal amplitude of equivalent stresses between HPR and MPR is observed after 0.071 s from the beginning of the reactive torque and reaches 515 MPa. The maximal stress amplitude between MPR and LPR is observed on the fourth cycle of vibration at about the same time and reaches 495 MPa. Equivalent stress amplitude

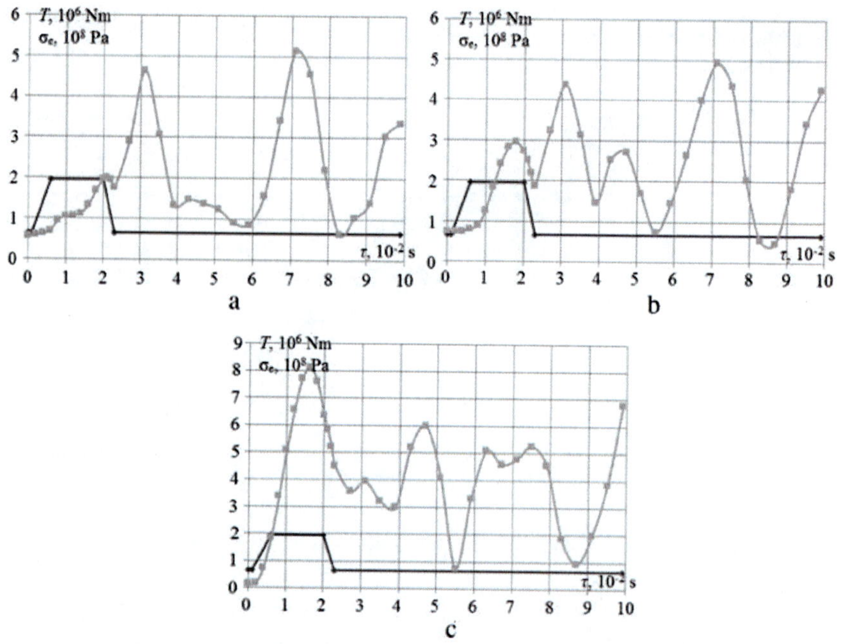

Fig. 4 Variation dynamics of the reactive moment (black line) and equivalent stresses (gray line) between HPR and MPR (**a**), MPR and LPR (**b**), LPR and TG (**c**) for the turbine K-200-130

between LPR and TG reach their maximal value after 0.016 s after the arising of torque and reaches 811 MPa. This value somewhat exceeds the ultimate strength of rotor steel 25Cr1Mo1V. Approximately in 4 s after cessation of the reactive torque, the vibrations completely damp.

5 Effect of Reactive Torque on Fatigue Damage of Shaft

The nature of reactive torque is very different. One of them, connected with mechanical problems, is not frequent event and hardly can be the reason of fatigue damage. Electromagnetic processes on generator happens every time during the networking. Taking into account that every networking of generator can cause intensive torsional vibrations of the steam turbine rotors, the amplitude of which exceeds the fatigue limit of rotor steel, several thousand of networking during lifetime of turbine unit causes long-term accumulation of fatigue damage.

The fatigue damage of the shaft in its most stressed areas (see Sect. 4) was estimated with usage of the Palmgren–Miner hypothesis [11, 12]

$$D = \sum_{i=1}^{n} \frac{1}{N_{id}}, \tag{3}$$

where n is the number of cycles with stress amplitude which exceeds the fatigue limit of rotor steel; N_{id} is the ultimate number of cycles till destruction.

Calculations of fatigue damage of the shaft was calculated with usage of the fatigue curve for the rotor steel 25Cr1Mo1V [7]. The experiment was carried out at symmetric loading cycle and temperature 20 °C. The effect of high temperature, scale factor and asymmetry of loading on the fatigue characteristics of rotor steel was taken into consideration by means of experimental coefficients. Woehler diagram in this case has the appearance.

$$\tau_{max} = K_s K_a K_t \left[\eta_0 + \eta_v (N)^c + \eta_u (N)^b\right], \tag{4}$$

where $\tau_{\max}$ is the amplitude of cyclic stress; $\eta_0 = 208{,}3$; $\eta_v = 3176{,}2$; $\eta_u = 1{,}86 \cdot 10^5$; $c = -0.3114$; $b = -0.8348$; N is the number of cycles; K_s is the scale coefficient; K_a is the asymmetry coefficient; K_t is the temperature coefficient.

Scale and temperature coefficients were taken from the experimental study [13], that is $K_s = 0.58$ and $K_t = 0.78$. The asymmetry coefficient was calculated with the equation [14]

$$K_a = \frac{\sqrt{\tau_{-1}^2 - \kappa^2 \psi_\tau^2 \tau_m^2}}{\tau_{-1}}, \tag{5}$$

where τ_{-1} is the fatigue limit (for the rotor steel 25Cr1Mo1V $\tau_{-1} = 230$ MPa); $\kappa = 0.922$; $\psi_\tau = 0.505$ [15]; τ_m is the mean stress.

The turbine shaft fatigue damage which is accumulated during operating time was calculated as a result of 2000 actions of the reactive torque (see Introduction), that is $D_{ac} = 2000D$. It is necessary to note, that failure takes place in case of $D_{ac} = 1$.

The estimation of fatigue damage was based on the supposition that the reactive torque t_{el} lasting is distributed uniformly in the range of time $\Delta t_{el} = t_{el,K} - t_{el,1}$ ($t_{el,1}$ and $t_{el,K}$ are the range limits). The averaged over this range fatigue damage was calculated according the following equation

$$D_{av} = \frac{\sum_{i=1}^{K} D_{ac,i}(t_{el,i})}{\Delta t_{el}}, \tag{6}$$

where $D_{ac,i}$ is the accumulated fatigue damage during the action of the reactive torque ($t_{el,i}$).

Averaged fatigue damage was calculated only for the areas of the turbine shaft between HPR and MPR, and between MPR and LPR because for the considered parameters of the reactive torque the area between LPR and TG most probably destructs immediately. As can be seen from Fig. 5, the exhaustion of the turbine

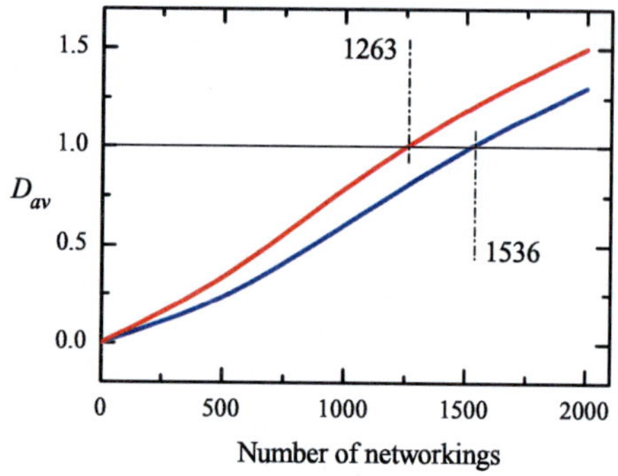

Fig. 5 Number of networkings dependences of averaged fatigue damage of turbine shaft in the area between HPR and MPR (red line) and between MPR and LPR (blue line)

resource due to fatigue damage to the shaft occurs in 1536 actions of the reactive torque in the area between HPR and MPR and in 1263 actions of the reactive torque in the area between MPR and LPR.

It should be noted, that the above estimation of fatigue damage is somewhat simplified since it suggests that all 2000 actions of the reactive torque have the same parameters. Of course, real scenarios of loading are much more complex. The developed procedure can be applied for the evaluation of turbine shaft fatigue damage if the parameters of the reactive torque will be measured in operation of turbine units.

6 Conclusions

Long-term exploitation of turbine equipment of thermal power stations which undergoing to frequent start-ups and shut-downs gradually accumulates fatigue damage in the most stressed areas of turbine shaft. The reason of fatigue damage is the intensive transverse or torsional vibrations of the turbine rotating elements of different nature (see Introduction).

The experimentally-analytical procedure to estimate the fatigue damage of turbine rotors because of torsional vibration was developed. To apply this procedure in practice it is needed to know the intensity of turbine shaft torsional vibrations in operation of turbine unit. Modern methods of vibration measurement made it possible to solve this problem [16].

For the reactive moment with the parameters, which were assigned as an example to show the idea of the procedure, the most stressed area of the turbine shaft is the area between LPR and TG. The least stressed area is the area between MPR and LPR. In any case under the action of the reactive moment with the chosen parameters the resource of turbine unit substantially drops.

In general case the intensity of turbine shaft torsional vibrations is dependent on the magnitude, duration and shape of the reactive moment time function.

References

1. Dorfman LS, Trubelja M (1999) Torsional monitoring of turbine-generators for incipient failure detection. In: Proceedings of the sixth EPRI steam turbine/generator workshop, pp 1–6. St. Louis, MO
2. DeForest DR, Grobel LP, Schabtach C, Seguin BR (1957) Investigation of the generator rotor burst at the Pittsburg station of the pacific gas and electric company. ASME Paper No. 57-PWR-12, 3-14
3. Zagretdinov ISh, Kostyuk AG, Trukhnii AD, Dolzhanskii PR (2004) Failure of the 300 MW turbine unit of the state district power station at Kashira: causes, consequences and conclusions. Teploenergetika 5:5–15
4. Walker DN, Placek RJ, Bowler CEJ, White JC, Edmonds JS (1984) Turbine generator shaft torsional fatigue and monitoring. In: CIGRE session, pp 11–17
5. Barella S, Bellogini M, Boniardi M, Cincera S (2011) Failure analysis of a steam turbine rotor. Eng Fail Anal 18(6):1511–1519
6. Bovsunovskii AP, Chernousenko OYu, Shtefan EV, Bashta DA (2010) Fatigue damage and failure of steam turbine rotors by torsional vibrations. Strength Mater 42(1):108–113
7. Bovsunovsky AP (2015) Fatigue damage of steam turbine shaft at asynchronous connections of turbine generator to electrical network. Phys: Conf Ser 628
8. Kima C-K, Jang G (2007) Effect of an excitation system on turbine-generator torsional stress in an HVDC power system. Electr Power Syst Res 77:926–935
9. Steam turbine-generator torsional vibration interaction with the electrical network. Final Report. Electric power research institute. Palo Alto, California (2005)
10. Chernousenko O, Peshko V, Marisyuk B (2019) Rationalization of the approaches of numerical simulation of the geometrically complex rotating equipment of electrical stations. Power Eng: Econ Tech Ecol 3:30–36
11. Palmgren A (1924) Life of ball bearings. Z Ver Dtsch Ing 68:339–341
12. Miner MA (1945) Cumulative damage in fatigue. J Appl Mech 12:159–164
13. Kudryavtsev IV, Naumchenkov NE (1978) Fatigue strength characteristics of steel 25KhNZMFA in relation to absolute dimensions and stress concentrations. Strength Mater 10(4):386–392
14. Troshchenko VT (1969) A “scale effect” in tests of long-term strength. Strength Mater 1(1):22–26
15. Troshchenko VT, Shestopal LR (1975) Criteria of torsional fatigue strength of metals. Strength Mater 7(2):127–131
16. Vaclavik J, Chvojan J (2017) Torsion vibrations monitoring of turbine shafts. In: Francesco I (ed) Procedia Structural Integrity 5. Elsevier, Amsterdam, pp 1349–1354

Crystal Orientation Effect on the Fretting Fatigue Behavior in Ni-Based Single Crystal Superalloys

Qinan Han, Jianwen Fang, and Haitao Cui

Abstract The crystal orientation effect on the fretting fatigue behavior in Ni-based single crystal superalloys (NBSX) was studied in this paper using crystal plasticity finite element (CPFE) simulation. The crystal plasticity constitutive model considering back stress was applied in the CPFE simulation, and four different crystal orientations were calculated to study the effect of crystal orientation. The CPFE simulation results showed significant differences between the four crystal orientations. The Mises stress and the cumulative shear strain under different crystal orientations was obtained. The change of stress and strain with time at different locations showed obvious difference under four crystal orientations. The simulation results showed obvious stress concentration near the contact edge, but the peak locations of cumulative shear strain would change under different crystal orientations.

Keywords Fretting fatigue · Superalloy · Crystal orientation · CPFE simulation

1 Introduction

Fretting fatigue refers to the phenomena when contact surfaces are subjected to alternating loads. Micron-scale relative slip occurs at the contact edge and crack initiates, leading to the material and structure failure. The dovetail contact between turbine disk and blade of aeroengines is a typical fretting fatigue condition. According to statistic data from the US Air Force, 1/6 of the fatigue related failures of the aeroengines are caused by fretting fatigue [1].

Q. Han · J. Fang · H. Cui (✉)
Aero-Engine Thermal Environment and Structure Key Laboratory of Ministry of Industry and Information Technology, College of Energy and Power Engineering, Nanjing University of Aeronautics and Astronautics, Nanjing 210016, China
e-mail: cuiht@nuaa.edu.cn

M. Abdel Wahab (ed.), *Proceedings of the 8th International Conference on Fracture, Fatigue and Wear*, Lecture Notes in Mechanical Engineering,
https://doi.org/10.1007/978-981-15-9893-7_40

Ni-based single crystal superalloys (NBSX) have excellent high-temperature strength, toughness and resistance to degradation in corrosive or oxidizing environments [2] and are widely used in the manufacturing of turbine blades. The face-centered cubic (FCC) crystallographic structure of NBSX causes its anisotropy, and different crystal orientations can show different mechanical properties.

Some previous research has shown that crystal orientation can significantly affect the fretting fatigue properties of materials. Matlik et al. conducted fretting fatigue experiments of NBSX, and showed that fretting contact stress led to the nucleation and propagation of cracks. It was observed that the crystallographic cracks initiated under fretting load and then propagated along the (111) octahedral plane [3]. Fleury et al. conducted high-temperature fretting fatigue tests of single crystal and polycrystalline alloy, and the influence of crystal orientation on the fretting fatigue life was studied. It's also shown that fretting stress was a function of crystal orientation [4]. Huang et al. carried out fretting experiments in different directions on the (100) crystal surface of NBSX, and found the slip trace, plastic and friction response was affected by the crystal orientation [5].

Many scholars have used finite element method to study the fretting fatigue behavior of materials. Zhang et al. established a finite element model of contact fretting fatigue between a rigid cylinder and an infinite half space, and the influence of microstructure on the fretting fatigue behavior of biphasic Ti–6Al–4V was studied [6]. Goh et al. conducted the two-dimensional crystal plasticity finite element (CPFE) simulation of cylinder-plane-contact fretting fatigue of Ti–Al–4V. The grain dispersion, crystal surface texture was considered, and the simulation results were consistent with the experiments [7]. Fleury et al. conducted finite element analysis to study the stress distribution near the contact edge under different crystal orientation [4]. However, the CPFE simulation considering back stress of the fretting fatigue behavior of NBSX under different crystal orientations is still lacked.

In this paper, the CPFE model of fretting fatigue of NBSX was developed. The crystal plasticity constitutive model considering back stress was applied in the finite element model. CPFE simulations under different crystal orientations were calculated to analyze the difference of stress and strain distribution, peak position.

2 Establishment of Finite Element Model

2.1 *Set-Up of Finite Element Model*

The single clamp model is established for simulation. The model includes a dog-bone shaped specimen and a pair of fretting pads. The shape and size of the specimen and fretting pad are shown in Figs. 1 and 2. The thickness of both the specimen and the fretting pad is 2 mm.

Figure 3 shows the schematic diagram of fretting model. The specimen is fixed at the left end, with a normal load applied at the top of the fretting pad. The tangential

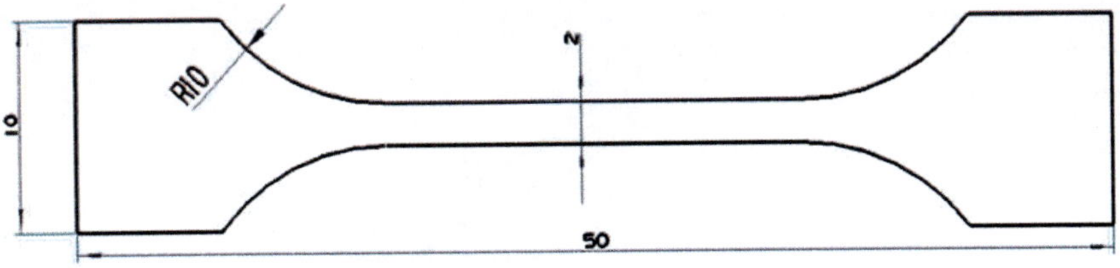

Fig. 1 Geometric dimensions of the specimen (unit: mm)

Fig. 2 Geometric dimensions of the fretting pad (unit: mm)

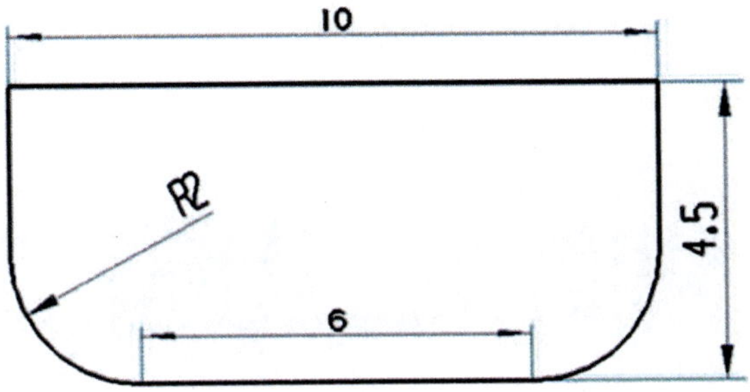

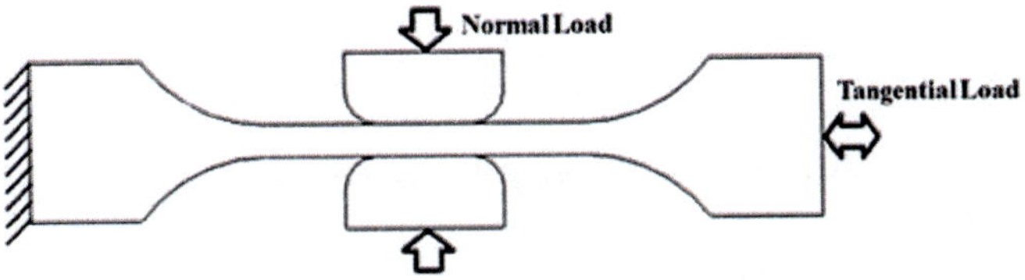

Fig. 3 Schematic diagram of fretting model

load is applied at the right end of the specimen. The finite element model is shown in Fig. 4. Considering the symmetry of the model, the semi-model is used for analysis to reduce the calculation cost.

The surface to surface contact is established between the specimen and the fretting pad, and friction coefficient of 0.3 [8] is used. Figure 5 shows the boundary conditions of finite element model. The boundary condition YSYMM ($UX = URY = URZ = 0$) is applied to the symmetry plane of the fretting model. The left end face of the specimen is set to be fixed, and the displacement constraint $UX = UZ = URX = URZ = 0$ is set on the upper surface of the fretting pad. The loading of the finite element model is divided into three steps, each of which takes 1 s. In step 1, the pre-applied force $P_0 = 100$ N is applied on the upper surface of the fretting pad to establish contact. Then in step 2 the normal load of $P = 1000$ N is applied, and the pre-applied force in step 1 is deactivated. In step 3, the normal load P is maintained and a tangential fatigue load $Q = 200 \sin 8\pi t$ N in the X direction is applied to

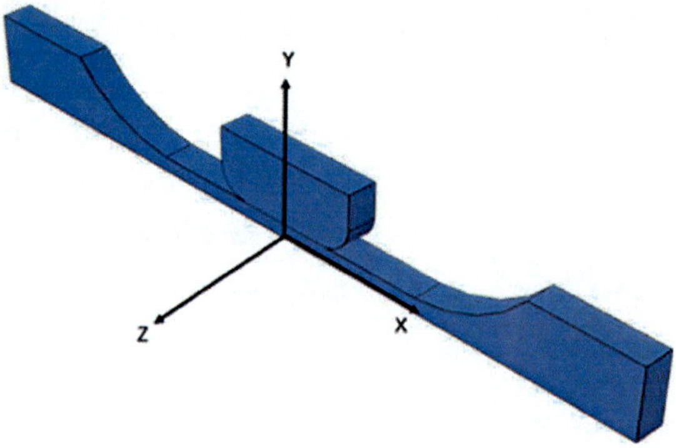

Fig. 4 The shape of the semi-model

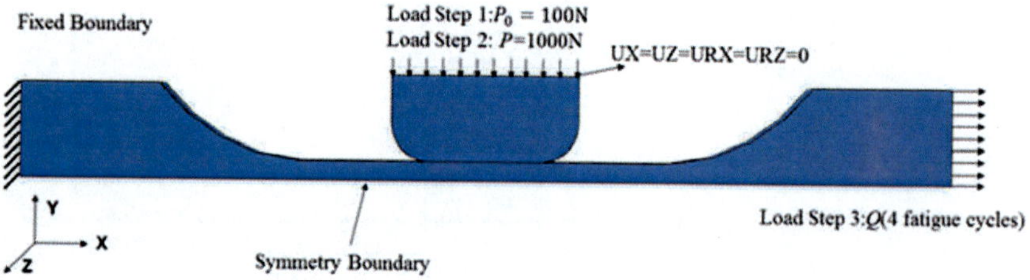

Fig. 5 The load and boundary conditions of the finite element model

the right end of the specimen. Considering the calculation cost, the fatigue load of 4 cycles is calculated in this study.

The mesh near the contact area is refined (see Fig. 6). The criterion for grid convergence is that the difference of peak stress is less than 10% after the grid

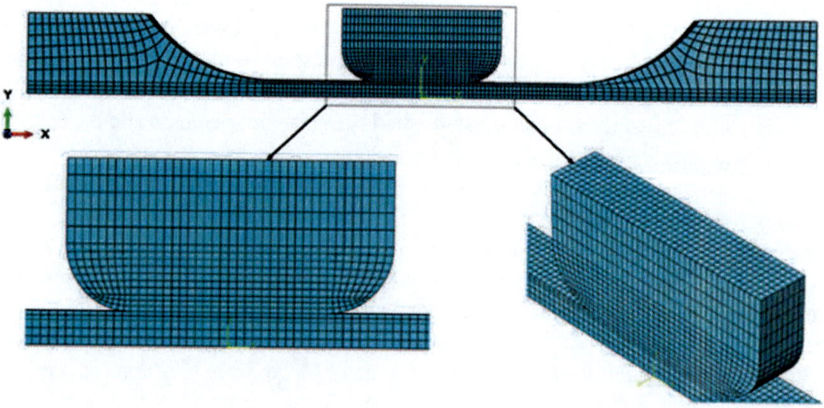

Fig. 6 Mesh of the finite element model

density is doubled. The element type is 8-node hexahedral (C3D8I). The model is segmented and structured meshing is adopted.

2.2 *Crystal Plasticity Constitutive Law*

The crystal plasticity constitutive theory summarized by Huang et al. [9] is briefly introduced here. The plastic deformation is assumed to be caused by crystallographic dislocation slip, and the resolved shear stress on the slip system is assumed to be the driving force of slip [9]. In this paper, 12 octahedral slip systems {111} <110> are considered.

The total deformation gradient can be decomposed into elastic part ($\mathbf{F}^{e}$) and plastic part ($\mathbf{F}^{p}$):

$$\mathbf{F} = \mathbf{F}^{e} \cdot \mathbf{F}^{p} \tag{1}$$

where the elastic part represents the lattice stretching and rotation, while the plastic part represents crystallographic slip.

$\mathbf{F}^{p}$ can be associated with the slip rate on the activated slip system:

$$\dot{\mathbf{F}}^{p} \cdot \mathbf{F}^{p\,-1} = \sum_{\alpha} \dot{\gamma}^{\alpha} \mathbf{s}^{\alpha} \mathbf{m}^{\alpha} \tag{2}$$

In the formula, $\dot{\gamma}^{(\alpha)}$ represents the slip rate of the α slip system, and unit vectors **s** and **m** represent the slip direction and normal direction of the slip plane. The slip rate $\dot{\gamma}^{(\alpha)}$ of the α slip system is determined by the resolved shear stress:

$$\dot{\gamma}^{(\alpha)} = \dot{\gamma}_0 \mathrm{sgn}(\tau^{(\alpha)} - X^{(\alpha)}) \left(\frac{\tau^{(\alpha)} - X^{(\alpha)}}{g^{(\alpha)}} \right)^{n} \tag{3}$$

where $\dot{\gamma}_0$ is relative strain rate, $g^{(\alpha)}$ represents the current strength of the α slip system, n is the exponent on the rate, $\tau^{(\alpha)}$ is the response to the resolved shear stress of the slip system:

$$\boldsymbol{\tau}^{\alpha} = \boldsymbol{\sigma}(\mathbf{s}^{\alpha} \otimes \mathbf{m}^{\alpha}) \tag{4}$$

The back stress is given by the Chaboche model, which can reflect the cyclic stress–strain response of metal materials:

$$\dot{X}^{(\alpha)} = \zeta^{(\alpha)} \left(r^{(\alpha)} \dot{\gamma}^{(\alpha)} - X^{(\alpha)} \left| \dot{\gamma}^{(\alpha)} \right| \right) \tag{5}$$

where $\zeta^{(\alpha)}$ and $r^{(\alpha)}$ are material dependent constants, $r^{(\alpha)}$ represents the relationship between back stress rate and slip rate $\dot{\gamma}^{(\alpha)}$, and $\zeta^{(\alpha)}$ represents the relationship between back stress rate and current back stress.

The evolution of the current strength is as follows:

$$\dot{g}^{(\alpha)}(\gamma) = \sum_{\beta}^{n} h_{\alpha\beta}(\gamma)\left|\dot{\gamma}^{(\beta)}\right| \tag{6}$$

$$h_{\alpha\beta}(\gamma) = h(\gamma)[q + (1 - q)\delta_{\alpha\beta}] \tag{7}$$

where $h_{\alpha\beta}$ is the potential hardening modulus, which represents the influence of slip rate on the current strength. q is the potential hardening parameter, and $h(\gamma)$ is the self-hardening modulus and is given by the following relation:

$$h(\gamma) = h_0 \mathrm{sech}^2\left(\frac{h_0\gamma}{\tau_s - \tau_0}\right) \tag{8}$$

where h_0 is the initial hardening modulus, τ_0 and τ_s represent the initial and saturation slip resistance, γ is the Taylor cumulative shear strain of all slip systems:

$$\gamma = \int \sum_{\beta=1}^{n} \left|d\gamma^{(\beta)}\right| \tag{9}$$

Material parameters that are applied in the CPFE simulation are shown in Table 1. The parameters describing the potential hardening law and the back stress hardening parameters are obtained from the work of Ma et al. [10]. The CPFE simulation is carried out by ABAQUS with a user material (UMAT) subroutine.

Table 1 Material parameters in CPFE simulation

Parameters	Value
C_{11}	187.85 GPa
C_{12}	99.64 GPa
C_{44}	129.49 GPa
n	3.60
h_0	548.00 MPa
τ_0	60.64 MPa
τ_s	100.64 MPa

3 Results and Discussions

3.1 CPFE Simulation Results

The crystal orientation is shown in Fig. 7, the X direction (length direction) of the global system is set as [001], and the Y direction (contact direction) of the global system is set as [110], and the crystal orientation denoted as crystal orientation A. To facilitate subsequent demonstration, individual locations on the model were named. The point with coordinates (−10, 1, 0) is denoted as point A, the point with coordinates (−3, 1, 0) is B, the point with coordinates (3, 1, 0) is C, and the point with coordinates (10, 1, 0) is D, and path 1 is defined as going from point A to point D.

Figure 8 shows the Mises stress map obtained after loading. The stress at the contact edge of the specimen and the fretting pad is large, and the maximum value of Mises stress is 141.7 MPa. Figure 9 shows the Mises stress curve along path 1, showing the peak of Mises stress is located at the contact edge. This indicates that stress concentration will occur at the contact edge in the fretting condition.

Figures 10 and 11 shows the change curve of the Mises stress of contact edge and the cumulative shear strain of all slip systems of point B and C over time, respectively. As shown in Fig. 10, the Mises stress of Point B first increases with time, then as the cyclic load is applied, enters a fluctuation state. The cumulative shear strain keeps increasing with time. When the stress goes down, the cumulative shear strain keeps increasing, but the increase amplitude becomes smaller, which makes the cumulative shear strain curve a step-rising shape. As shown in Fig. 11, Mises stress curve of point C is different from point B after t = 2 s, namely when applying tension at the right end of the specimen, the contact stress at point B is increased, while the contact stress at point C is reduced. Besides, the cumulative shear strain curve at point C has no obvious step-rising shape compared with point B.

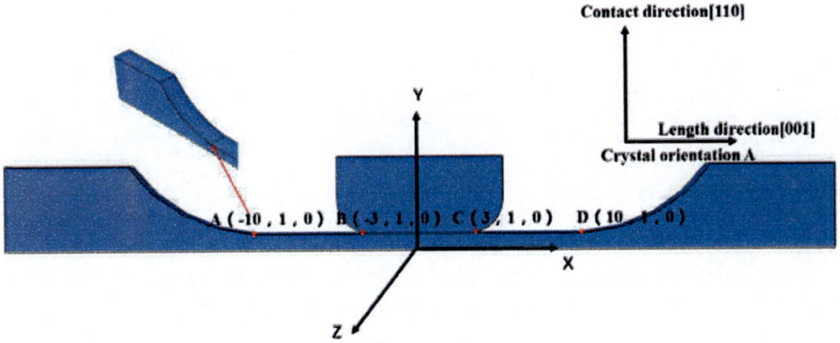

Fig. 7 The crystal orientation of the model

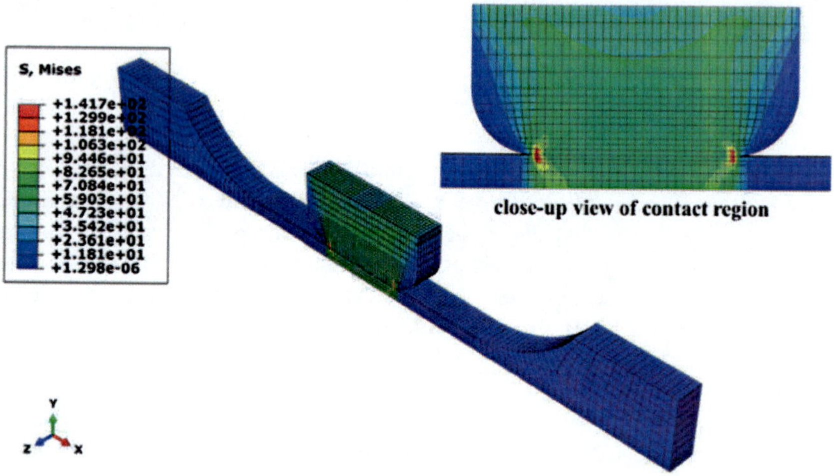

Fig. 8 Mises stress map

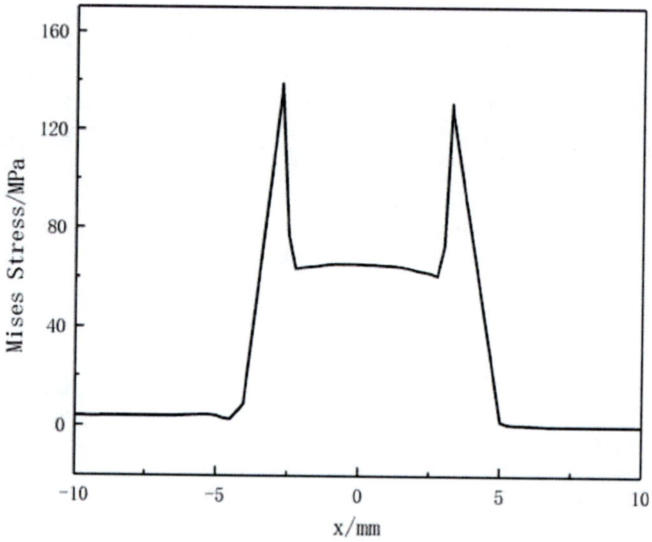

Fig. 9 The Mises stress curve along path 1

The simulation results also show the position of maximum cumulative shear strain does not coincide with the position of maximum Mises stress, Fig. 12 is the cumulative shear strain map at t = 3 s of the crystal orientation A, the maximum value occurs about 0.125 mm from the contact surface of the specimen, which locates at subsurface. This is different from the Mises stress map (Fig. 8) where the peak stress occurs at the contact edge and on the contact surface.

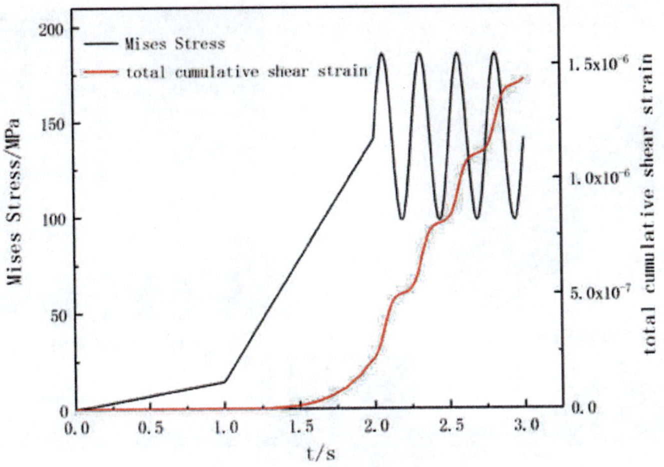

Fig. 10 The Mises stress curve and the cumulative shear strain curve at point B over time

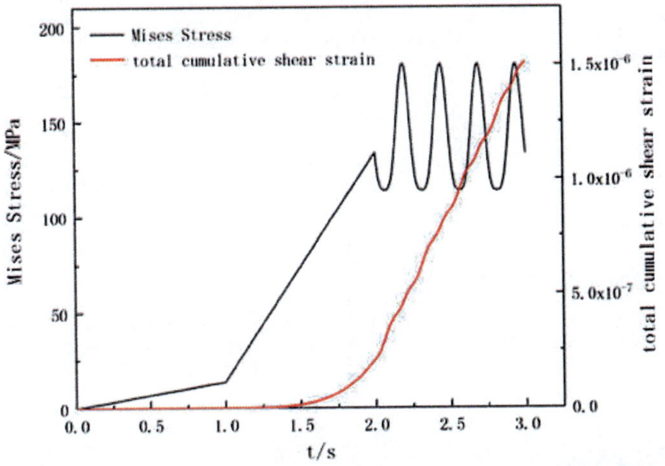

Fig. 11 The Mises stress curve and the cumulative shear strain curve at point C over time

3.2 Comparison of Different Crystal Orientations

To study the influence of crystal orientations on the fretting fatigue behavior, a total of 4 different crystal orientations are calculated including crystal orientation A ([001]/[110]) in previous section. The set-up of four crystal orientations are shown in Table 2 and denoted as crystal orientation B, C and D respectively. As shown in Fig. 13, the crystal orientations of these four groups along the X-axis are consistent and in the direction of [001].

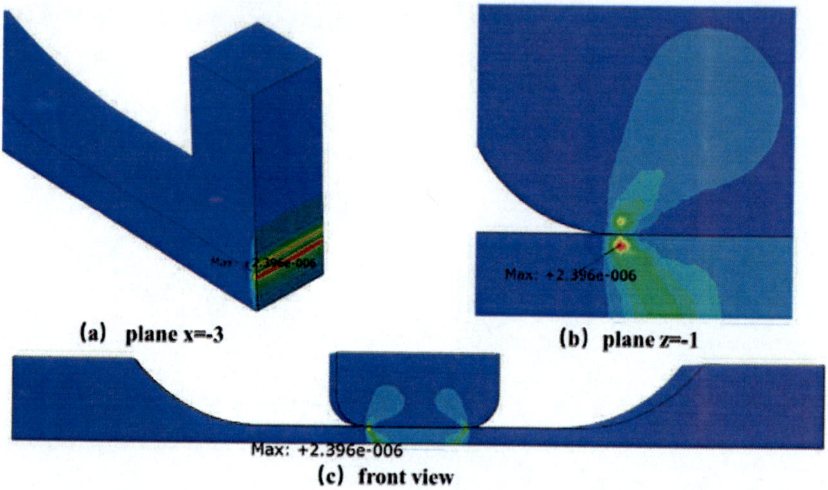

Fig. 12 Cumulative shear strain map (**a**) plane x = −3 (**b**) plane Z = −1 (**c**) front view

Table 2 Set-up of four crystal orientations

Crystal orientation	Direction of length	Direction of contact
A	[001]	[110]
B	[001]	[010]
C	[001]	[1$\bar{1}$0]
D	[001]	[100]

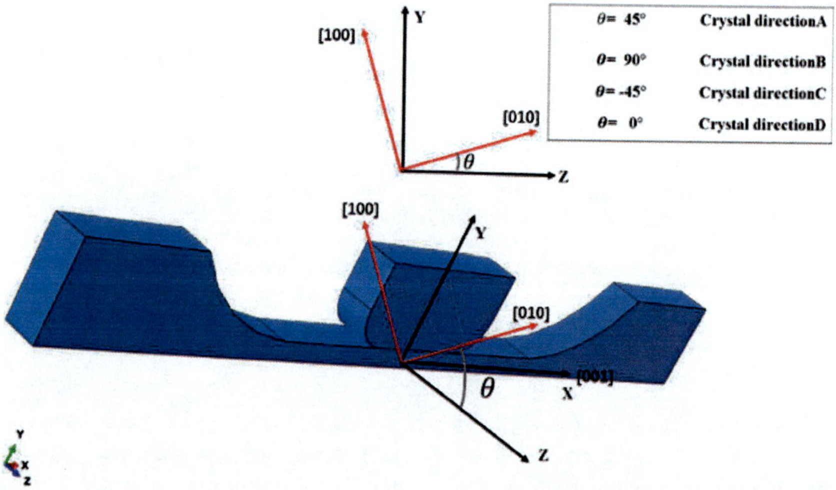

Fig. 13 Schematic diagram of four crystal orientations (θ is the angle between the crystal orientation [010] and the Z-axis)

The calculated results of four crystal orientations are summarized in Table 3. The maximum values of Mises stress and the maximum values of cumulative shear strain of all slip systems on different crystal orientations were compared. The table shows a modest difference in the Mises stress peaks, but a significant difference in cumulative shear strain, where the cumulative shear strain of crystal orientation D is about 8 times of crystal orientation A.

Figure 14 shows the change curve of the stress at point B of contact edge with the loading process of four crystal orientations. The Mises stresses at point B with different crystal orientations are not significantly different from each other during normal loading (before t = 2 s). However, under the action of sinusoidal fatigue load,

Table 3 Calculation results of four crystal orientations

Crystal orientation	Maximum of Mises stress (MPa)	Maximum cumulative shear strain	The position of the maximum cumulative shear strain
Crystal orientation A [001]/[110]	141.7	2.396e−6	The left side of the specimen
Crystal orientation B [001]/[010]	139.8	5.800e−6	The left side of the specimen
Crystal orientation C [001]/[1$\bar{1}$0]	146.6	9.396e−6	The right side of the fretting pad
Crystal orientation D [001]/[100]	146.2	1.809e−5	The right side of the fretting pad

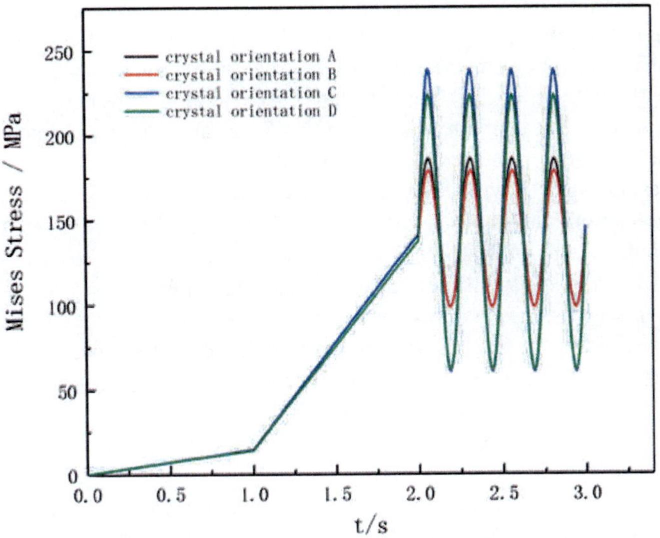

Fig. 14 The Mises stress change curve of Point B with different crystal orientations

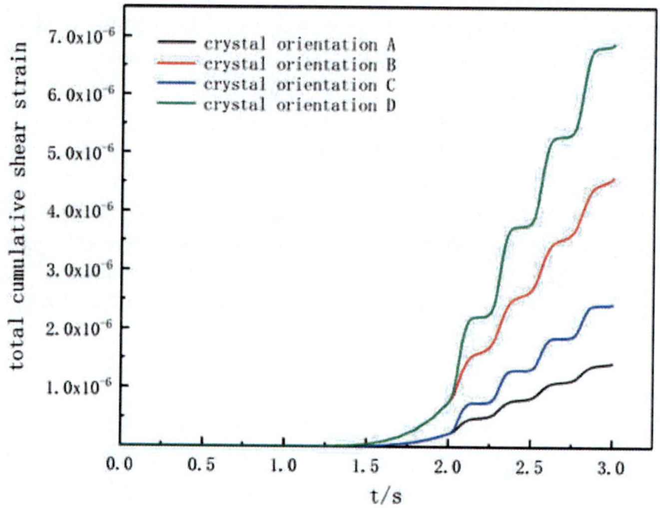

Fig. 15 The curve of cumulative shear strain change curve of Point B with different crystal orientations

the stress curve is obviously different. Specifically, the stress curve approximately maintains the sinusoidal shape and the stress mean is approximately equal, but the stress amplitude is greatly different under different crystal orientations. The stress amplitude of crystal orientation C and D is larger, while that of crystal orientation A and B is smaller.

Figure 15 shows the curve of the cumulative shear strain at point B with the loading process under different crystal orientations. The rate of change of cumulative shear strain under different crystal orientation with time is different. When the normal load increases gradually (t = 1.5 s), the cumulative shear strain begins to increase at point B. The growth rate of cumulative shear strain of crystal orientation D is the highest, while that of crystal orientation A is the lowest.

Comparing Figs. 14 with 15 it can be seen that under fretting fatigue load, the Mises stress amplitude is not strictly correlated with the cumulative shear strain. The stress amplitude at point B is the lowest, but the cumulative shear strain at point B is large, which is obviously different from isotropic materials.

For all four crystal orientations, the positions of maximum Mises stress are on the contact edge, however, the positions of maximum cumulative shear strain are at a certain distance from the contact surface. Take crystal orientation A as an example, as shown in Fig. 16, the maximum cumulative shear strain of crystal orientation A is located near the left contact edge of the specimen, the position coordinate is (−3, 0.875, −1) and it is denoted as point M.

In Table 3, for the crystal orientation A, the maximum cumulative shear strain of the entire model appears near the contact edge on the left of the specimen (as shown in point M in Fig. 16). However, as shown in Fig. 17, the position of maximum value

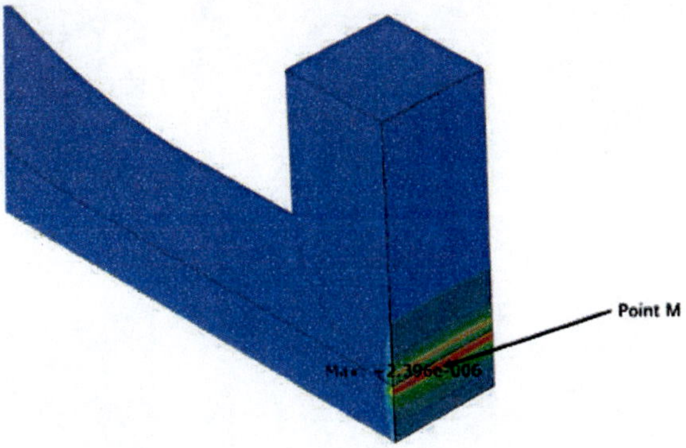

Fig. 16 The position of maximum cumulative shear strain of crystal orientation A

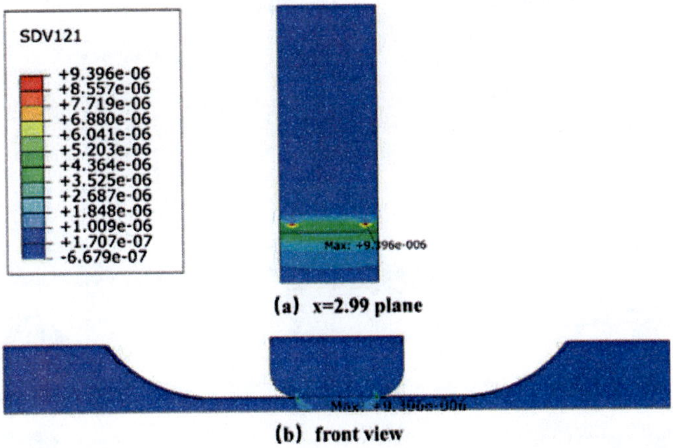

Fig. 17 The cumulative shear strain contour of crystal orientation C after loading **a** plane x = 2.99 **b** front view

of crystal orientation C is near the contact edge on the right side of the fretting pad, which indicates the potential failure location will change with the crystal orientation.

The cumulative shear strain maps of the specimen and the fretting pad with different crystal orientations are shown in Figs. 18 and 19, respectively. It can be seen that the cumulative shear strain of the fretting pad and the specimen of crystal orientation A is the smallest, while that of the fretting pad and the specimen of crystal orientation D is the largest and is 10 and 4 times larger than that of crystal orientation A, respectively.

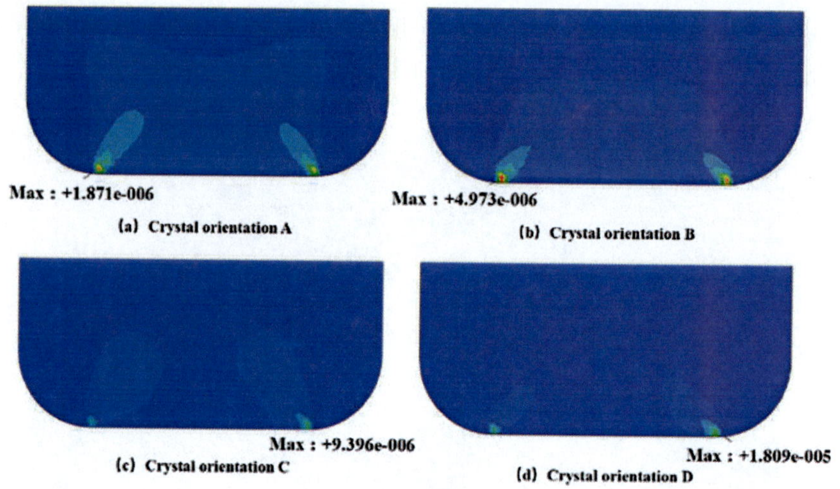

Fig. 18 Cumulative shear strain maps of fretting pads (**a**) crystal orientation A (**b**) crystal orientation B (**c**) crystal orientation C (**d**) crystal orientation D

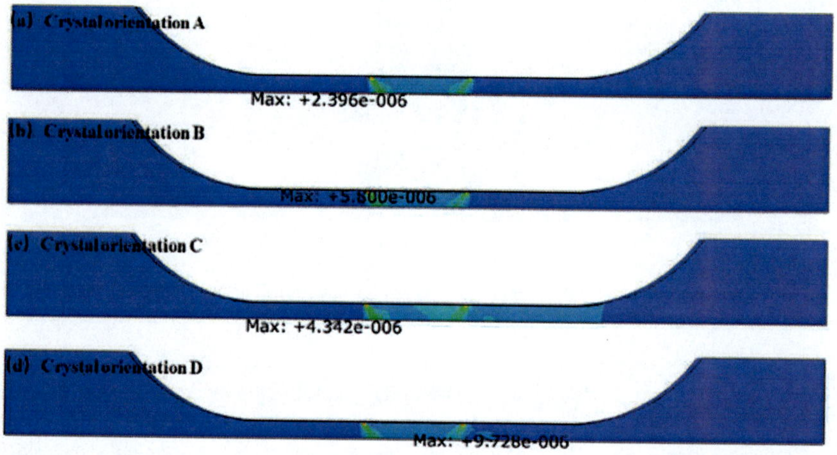

Fig. 19 Cumulative shear strain maps of specimens (**a**) crystal orientation A (**b**) crystal orientation B (**c**) crystal orientation C (**d**) crystal orientation D

4 Conclusions

In this paper, the influence of crystal orientation on the fretting fatigue behavior of NBSX is studied using CPFE simulation. Four crystal orientations are calculated and the Mises stress and cumulative shear strain are compared.

Under the fretting condition, the pad and the specimen have obvious stress concentration at the contact edge, and the plastic deformation is mainly concentrated on the near surface. The change of crystal orientation will affect the fretting fatigue behavior of NBSX. The cumulative shear strain of four crystal orientations can differ by about 8 times. CPFE simulation results also show difference of peak positions of cumulative shear strain under four crystal orientations, which indicates the potential failure location can change with the crystal orientation.

Acknowledgements This work is financially supported by the National Natural Science Foundation of China (Nos. 91860101 and 11632010).

References

1. Ciavarella M, Demelio G (2001) A review of analytical aspects of fretting fatigue, with extension to damage parameters, and application to dovetail joints. Int J Solids Struct 38(10):1791–1811
2. Pollock TM, Tin S (2006) Nickel-based superalloys for advanced turbine engines: chemistry, microstructure and properties. J Propul Power 22(2):361–374
3. Matlik J, Farris T, Haake F, Swanson G, Duke G (2006) High-frequency, high-temperature fretting-fatigue experiments. Wear 261(11–12):1367–1382
4. Fleury R, Paynter R, Nowell D (2014) The influence of contacting Ni-based single-crystal superalloys on fretting fatigue of Ni-based polycrystalline superalloys at high temperature. Tribol Int 76:63–72
5. Huang X, Gibson T, Zhang M, Neu R (2009) Fretting on the cubic face of a single-crystal Ni-base superalloy at room temperature. Tribol Int 42(6):875–885
6. Zhang M, McDowell D, Neu R (2009) Microstructure sensitivity of fretting fatigue based on computational crystal plasticity. Tribol Int 42(9):1286–1296
7. Goh C-H, Wallace J, Neu R, McDowell D (2001) Polycrystal plasticity simulations of fretting fatigue. Int J Fatigue 23:423–435
8. Matlik J, Farris T, Haynes J, Swanson G, Ham-Battista G (2009) Prediction of fretting crack location and orientation in a single crystal nickel alloy. Mech Mater 41(10):1133–1151
9. Huang Y (1991) A user-material subroutine incorporating single crystal plasticity in the Abaqus finite element program
10. Ma X, Shi H, Gu J, Wang Z, Harders H, Malow T (2008) Temperature effect on low-cycle fatigue behavior of nickel-based single crystalline superalloy. Acta Mech Solida Sin 21(4):289–297

Three-Dimensional Numerical Analysis of Shrink-Fitted Shafts Under Rotating Bending Fretting Fatigue Condition

Jingchen Wang and Yukui Gao

Abstract The cantilever rotating bending fretting fatigue (RBFF) damage of GH4169 was investigated in this work. With the help of ABAQUS and FRANC3D, the stress distribution and the contact status in the fretting zone were studied for the different fatigue loads. Furthermore, the critical plane shear stress amplitude model, Smith-Watson—Topper (SWT) model and Ruiz-Chen model were implemented and compared for fretting crack nucleation locations. The results showed that with the increase of the fatigue loads, the contact pressures decreased and the contact area reduced at the tension side. The stick zone, slip zone and open zone were formed due to fretting contact. At the compression side, the percent of slip zone area reached 4, 15 and 60% along the top surface axis of the shaft with the fatigue load of 500, 600 and 700 MPa. However, the slip zone area reduced with larger fatigue load and were almost the same for the three fatigue loads at the tension side. Moreover, the stress distribution results showed that the negative impact of fretting were weakened with the fatigue load increasing. Finally, the critical plane shear stress amplitude model and SWT model were determined to be more accurate in crack location prediction than the Ruiz-Chen model.

Keywords Fretting fatigue · Crack nucleation · Numerical simulation

J. Wang
School of Aerospace Engineering and Applied Mechanics, Tongji University, No. 100 Zhangwu Rood, Shanghai 200092, China

Y. Gao (✉)
School of Materials Science and Engineering, Tongji University, No. 4800 Caoan Road, Shanghai 201804, China
e-mail: ykgao12088@126.com

Shanghai Key Laboratory of R&D for Metallic Function Materials, No. 4800 Caoan Road 201804, Shanghai, China

M. Abdel Wahab (ed.), *Proceedings of the 8th International Conference on Fracture, Fatigue and Wear*, Lecture Notes in Mechanical Engineering,
https://doi.org/10.1007/978-981-15-9893-7_41

1 Introduction

Fretting damage usually occurs at the surfaces of two contacting mechanical components, which experience an oscillatory movement of small amplitude. Particularly, the fretting damage becomes more detrimental and increases the tensile and shear stresses when one or both contacting components undergo fluctuating loading. This process is termed as "fretting fatigue" [1]. Fretting fatigue damage is a common issue in aviation field, such as structural lap joints [2], dovetail blade/disk type attachments [3, 4], aircraft turbine shafts and so on. As an important transmission component in aero-engines, the turbine shaft bears the bending force that is generated by the turbine blades. With the impact of the high-speed rotation and airflow changes, the axis deflection may result in RBFF of the turbine shaft and the eventual failure of the aero-engine.

The RBFF of press-fitted assembly was investigated mostly on the railway materials by the combination of experimental and simulation methods. The influence factors were carried out and the damage mechanisms of the fretting fatigue were studied. Hoger et al. pointed out that clamping pressure, shape or geometry of assembly, residual stress were important factors by experiments of full-size press-fitted assemblies [5]. Nishioka et al. conducted a systematic research on the effect of fillet on the improvement of fretting fatigue limit [6]. In terms of fretting fatigue crack nucleation, Alfredsson reported the application of the Findley and Ruiz multiaxial criterions on the RBFF crack nucleation prediction [7]. Erena et al. investigated the influences of the geometry modification on the improvement of fretting fatigue resistance [8]. The SWT multiaxial fatigue parameter was successfully conducted to evaluate the extent of the improvement. Moreover, several scholars studied the fretting fatigue by the three-dimensional fracture mechanics-based software FRANC3D. Carter et al. introduced the approach for integrating existing fretting fatigue crack nucleation models and discrete crack growth simulation capabilities to compute total fatigue life by FRANC3D [9]. Experimental fretting data was comparable to the simulation results and an example for realistic Ti-6Al-4V turbine blade/disk predictions were carried out. Furthermore, the fretting fatigue of an aircraft engine fan dovetail attachment made of Ti-6Al-4V alloy was investigated by Mangardicha using FRANC3D [10]. The effect of coefficient of friction between contact surfaces were given and the classical plain fatigue SWT model was compared with the modified SWT model proposed by Hirsch [11].

However, the fretting fatigue was mostly studied on the Ti-6Al-4V materials. The details of the fretting fatigue properties of the superalloys have not been well considered, except for some Ni-based single crystal superalloys [12]. With the help of the ABAQUS and FRANC3D, the stress field of the fretting zone and the crack nucleation were investigated for the cantilever RBFF of GH4169 (modified Inconel 718 alloy), which was a common nickel-based superalloy in aeronautical applications. Different fretting fatigue nucleation criterions were implemented and the predicted nucleation locations were compared for different fatigue loads.

2 Numerical Method

The analysis was conducted using the general-purpose nonlinear code ABAQUS combined with the fracture analysis code FRANC3D. The FRANC3D is a software that allows to model cracks and compute the value of SIF relying on finite element (FE) codes (e.g. ANSYS, ABAQUS, NASTRAN) [13]. The results of stresses, displacements or temperature calculated by FE software are read by FRANC3D. The fretting fatigue crack nucleation analysis could be employed with the existing nucleation models.

2.1 ABAQUS Modeling

The FE assembly model for the cantilever RBFF comprises two parts: the shaft as the specimen and the hub as the fretting pad. The assembly model and the dimensions are illustrated in Fig. 1. The two parts are both considered as GH4169 and the elastic–plastic material with linear kinematic hardening is applied. Since the superalloy plays a vital role in the high temperature application, node temperature of 650 °C is employed to the model and the material expansion is taken into account. The material parameters at 650 °C are shown in Table1. Furthermore, the eight-node linear 3D element C3D8 is used and the mesh is refined in the fretting zone, which is also the contact edge of the shaft and the hub. The minimum size of the element is 10 μm along the z-axis.

The hub inner surface is considered as the master surface and the shaft surface is considered as the slave surface, using the master–slave contact approach. The interference fit of 10 μm is applied. The coefficient of friction between the contact surfaces is chosen as 0.4. The rotatory bending moment is applied by reference point forces of $F_x = F_0\sin(2\pi t)$ and $F_y = F_0\cos(2\pi t)$ at the free end of the shaft. The

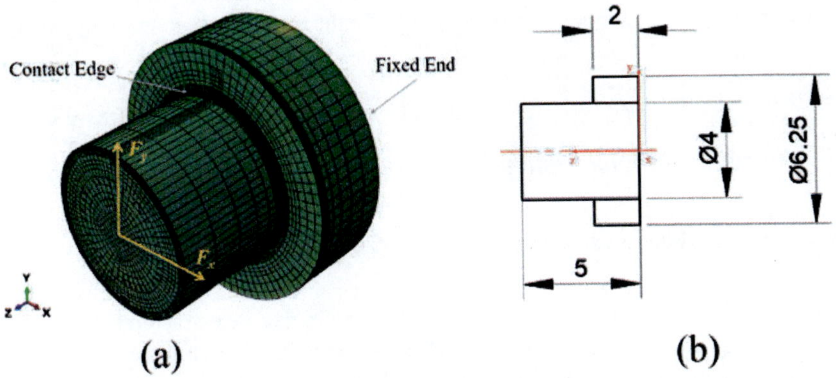

Fig. 1 **a** The shrink-fit assembly FE model. **b** The dimension of the shrink-fit (in mm)

Table 1 Material properties of GH4169 superalloy at temperature of 650 °C

Properties	Values
Density $\boldsymbol{\rho}$	8.24×10^3 kg/m^3
Elastic modulus $\boldsymbol{E}$	142 GPa
Poisson ratio $\boldsymbol{\nu}$	0.361
Yield stress $\boldsymbol{\sigma}_{0.2}$	1010 MPa
Hardening modulus $\boldsymbol{E}^T$	6 GPa
Thermal conductivity $\boldsymbol{\lambda}$	21.2 W/(m·°C)
Linear expansivity $\boldsymbol{\alpha}$	14.8×10^{-6} °C^{-1}

fatigue stress ratio $R = -1$ is implemented using this method. The calculation results of the applied forces are used that the final fatigue load of 500, 600 and 700 MPa are obtained at the location of z = 2 mm for the plain fatigue (where the contact edge is for fretting fatigue). Moreover, the displacements are avoided by a constraint for the z-displacements at z = 0 whereas the material expanding along the radial direction is allowed. Three cycles are considered.

2.2 *FRANC3D Modeling*

The result file of the FE model could be imported into FRANC3D. The main result for the fretting analysis is the constant normal load (P) provided by the interference fit of the hub and the cyclic fatigue load (S_{max} and S_{min}) of the shaft. The stress results at the moment when $F_y = F_0$ and $F_y = -F_0$ are chosen to be the maximum and minimum fatigue load.

Five fretting nucleation models are included in the software to predict the initial crack location. The five nucleation models are the equivalent stress [14], the critical plane shear stress amplitude [15], the critical plane SWT [16], the Ruiz-Chen model [17] and the modified SWT model [11]. The critical plane shear stress amplitude, SWT and Ruiz-Chen models are implemented for the fretting parameters comparison in this work.

3 Results and Discussions

3.1 *Stress Analysis in the Fretting Zone*

The contact stresses, bending stresses and slip amplitudes are extracted from the FE results along a z-axis path at the top surface of the shaft. The ultimate status when $F_y = F_0$ and $F_y = -F_0$ are considered. Therefore, the condition that S_{max} and S_{min} are achieved represent the compression and tension of the top of the shaft.

The contour plots of the Mises stress σ_{Mises} of the press-fitted assembly model with the fatigue loads of 500, 600 and 700 MPa are depicted in Fig. 2. With the increase of the fatigue load, it is easy to understand the stress concentration increasing at the contact edge, which is the most dangerous zone. The contact pressures $p(x)$ under the compression and tension conditions are illustrated in Fig. 3 The horizontal axis represents the contact position, which is the ratio of the coordinate of z to the radius r = 2 mm. Under the compression condition, $p(x)$ reaches highest level at almost the contact edge. However, $p(x)$ decreases near the contact edge at the tension side. The contact pressures become zero at the locations of z/r = 0.975, 0.956 and 0.923 with the fatigue loads of 500, 600 and 700 MPa, respectively. With the higher fatigue load, the higher bending stress leads to larger deformation and reduction of the contact area. Furthermore, the higher fatigue load results in lower contact pressures and then lower fretting damage, as introduced by Gutkin [18].

For the fretting contact status, the slip region is identified by the frictional shear traction $q(x)$ and the product of coefficient of friction μ and $p(x)$. The contact part is in the slip status when $q(x) = \mu p(x)$ and in the stick status when $q(x) < \mu p(x)$ [18].

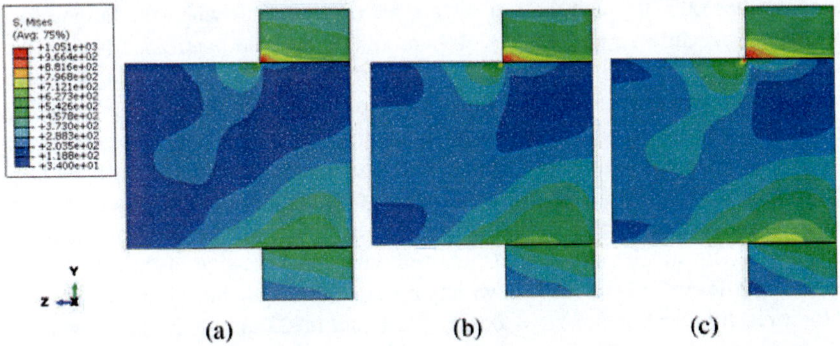

Fig. 2 σ_{Mises} contour plots with the fatigue loads of **a** 500 MPa, **b** 600 MPa and **c** 700 MPa

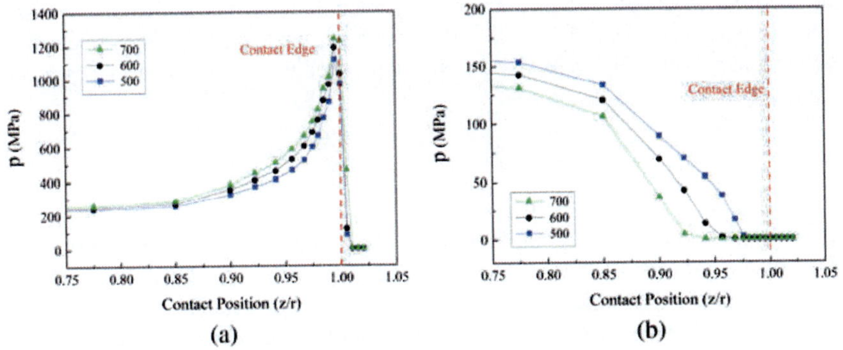

Fig. 3 $p(x)$ under **a** the compression and **b** the tension conditions

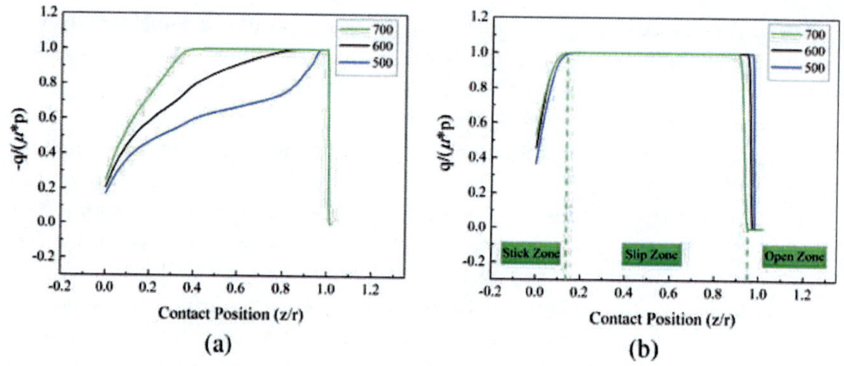

Fig. 4 Contact status under **a** the compression and **b** the tension conditions

In Fig. 4, the ratio $q(x)/\mu p(x)$ is plotted for the compression and tension conditions with different fatigue loads. The $-q(x)/\mu p(x)$ for the compression condition means the reverse movement of the shaft and hub. The percent of the reversed slip zone area reaches 60% through the top surface axis of the shaft when the fatigue load is 700 MPa, while 15 and 4% for 600 and 500 MPa. The open zone is achieved out of the contact edge for all fatigue loads. Although the values of the contact pressures are almost the same for the three loads under compression condition, the shear tractions are in quite different trend that result in diverse contact status, which may cause different wear scars and then stress redistribution in the experiment [19]. Nevertheless, the slip zone occupies the majority for all three loads at the tension side. The zones are regularly divided into stick zone, slip zone and open zone, as illustrated in Fig. 4b. The smaller slip zone is obtained with the increasing of the fatigue load for the cantilever RBFF, which shows an opposite trend with the compression status. Moreover, the locations of z/r = 0.975, 0.956 and 0.923 are the boundaries of the slip zones and the open zones, consistent with the distribution of $p(x)$.

The bending stress σ_{zz} is an important value as a key crack propagation driving force for the future mode I SIF calculation [13]. Figure 5a depicts the σ_{zz} values for different fatigue load cases under the tension condition. The σ_{zz} value increases with the increase of the fatigue loads. For each case, the maximum stress appears at the boundary of the slip zone and the open zone (see Fig. 4). For plain fatigue, the bending stress σ_{z0} is calculated as:

$$\sigma_{z0} = 32 \cdot F_0 \cdot (5 - z)/\left(\pi \cdot d^3\right) \quad (1)$$

where σ_{z0} is the bending stress, F_0 is the force at the free end of the specimen, d is the diameter of the specimen. The value of σ_{zz}/σ_{z0} represents the ratio of fretting fatigue stress to the plain fatigue bending stress at the same location of the specimen. Figure 5 plots the results of the σ_{zz}/σ_{z0} for the three fatigue loads. It could be seen that the lower fatigue loads achieve higher stress ratio near the contact edge. For the

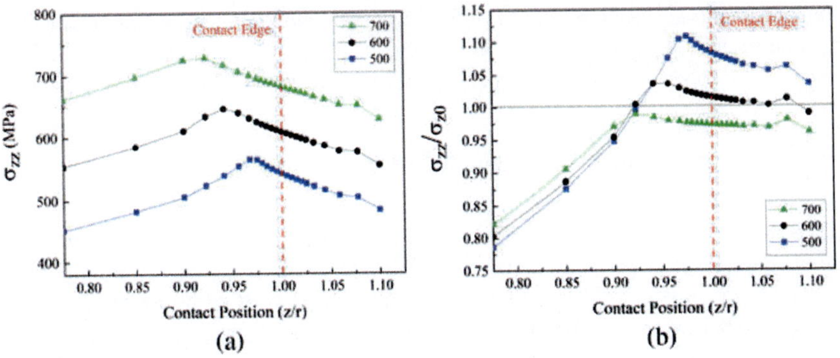

Fig. 5 σ_{zz} and σ_{zz}/σ_{z0} at the tension states

fatigue loads of 500 and 600 MPa, the contact pressures induced by fretting contact of the hub intensify the negative influence of fretting fatigue. However, the value of σ_{zz}/σ_{z0} is less than 1.00 for the fatigue load of 700 MPa, which means fretting fatigue weakens the crack propagation driving force of the original bending stress for the plain fatigue. Therefore, under the condition of the cantilever RBFF, the negative impact of fretting fatigue reduces with the increasing of the fatigue load, regardless of the effect of slip amplitudes and wear.

The relative slip amplitudes are another vital impact factor for the fretting fatigue nucleation evaluation. The results in Fig. 6 shows that the slip amplitudes increase with the increasing of the fatigue loads. Spink has stated that the increase in slip amplitude induces a reduction in fatigue life by an increase of the fretting damage [20]. From the discussions above, the increase of the fatigue load results in lower contact pressures and then slower crack nucleation. However, it causes the larger slip amplitudes leading to faster crack nucleation, either. Therefore, the fretting crack

Fig. 6 Relative slip amplitudes with the fatigue loads of 500, 600 and 700 MPa

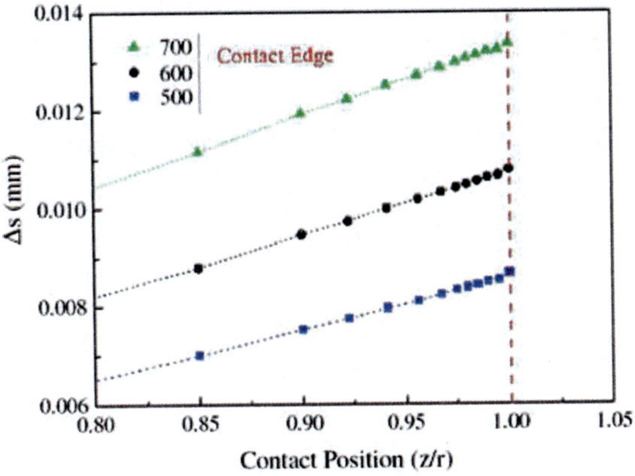

nucleation models are important for predicting the crack nucleation damage, which will be discussed in the next section.

3.2 Fretting Crack Nucleation Prediction

Three nucleation models are implemented in this work. Lykins has utilized the shear stress-based parameter to predict the fretting fatigue crack nucleation life, location and orientation [9, 15]. The maximum shear stress is verified to govern the crack nucleation behavior of fretting fatigue. Moreover, the significant effect of mean stress is taken into account by Walker [21]. The critical plane shear stress amplitude model is defined as followed:

$$\gamma_{crit} = \tau_{\max}/G(1 - R_\tau)^m \tag{2}$$

where $\tau_{\max}$ is the maximum shear stress on the critical plane, R_τ is the shear stress ratio i.e. $\tau_{\min}/\tau_{\max}$, G is the shear modulus and m is a material dependent fitting parameter.

SWT model is one of the most popular strain energy-based criterions that is used to determine the crack nucleation location and life [22]. Szolwinski and Farris made the modification and reported that the crack nucleation would occur on the plane where the product of the strain amplitude and maximum stress normal to the plane was the largest value [23]. The SWT model is defined by:

$$\Gamma_{crit} = \sigma_{\max}(\Delta\varepsilon/2) \tag{3}$$

where $\sigma_{\max}$ is the maximum normal stress on the critical plane and $\Delta\varepsilon$ represents the normal strain amplitude on the critical plane.

Ruiz et al. have studied the fretting fatigue of dovetail joints and proposed two parameters as a basis for the life assessment, of which the fretting parameter is known as:

$$F_2 = \sigma\tau\delta \tag{4}$$

where τ is the shear stress, δ is the relative slip and σ is the bulk normal stress. However, the effect of contact pressure on the fretting fatigue life is unable to be explained by the Ruiz-Chen parameter [22]. Lykins et al. have found that the F_2 parameter could be applied only when considering the effects of cyclic range or mean stress [16].

The FE results at the moment $F_y = F_0$ and $F_y = -F_0$ are imported into the FRANC3D and used for the $S_{\max}$ and $S_{\min}$ analysis. Therefore, the shaft along the positive y-direction is the compression side.

The contours of the three nucleation models are illustrated in Fig. 7. The parameters increase with the fatigue loads increasing, which represents the decrease of the nucleation life according to the nucleation life fitting models: γ_{crit}, Γ_{crit}, $F_2 = aN_i^b + cN_i^d$ [9]. In spite of the lack of the material fitting constants *a*, *b*, *c* and *d* for the GH4169 superalloys, the relationship between the nucleation life and nucleation parameters of Ti-6Al-4V could be referenced [16].

The definite locations of the crack nucleation are given by the nucleation parameter analysis. Although the critical plane shear stress amplitude model predicts that the crack nucleation locations are both at the contact edge (z = 2 mm) for fatigue loads of 600 and 700 MPa, they are at the coordinate of x = −0.467 mm and x = 0.157 mm, respectively. The SWT model predicts that crack nucleation locations are at the same position, which is the top center of the contact edge (x = 0 mm, z = 2 mm). The cracks predicted by the shear stress model and SWT model both locate at the pressure side. However, the Ruiz-Chen model gives the prediction on the tension side of the shaft, as shown in Fig. 7e, f. In addition, they are located at the inner of the contact surface with a distance of 0.45 mm to the contact edge.

As the RBRR tests of Ti-6Al-4V shown by Bertini et al. [24], the cracks are supposed to initiate at the contact edge. Since the model in this work is based on the RBFF, each point at the contact edge experiences the same load condition in the one cycle. The shear critical plane shear stress amplitude model and SWT model are both more accurate than the Ruiz-Chen model, which are all at the contact edge. Nevertheless, the validity of the two approaches needs to be further checked by experiments and nucleation life prediction studies.

4 Conclusions

In this work, the cantilever RBFF performances of GH4169 have been investigated through the simulation methods by ABAQUS and FRANC3D. The stress distribution and the fretting status in the fretting zone have been studied for the different fatigue loads. Furthermore, three fretting crack nucleation models have been implemented and compared, which are the critical plane shear stress amplitude, SWT and Ruiz-Chen models. The main conclusions are:

(1) The contact pressures become zero at the locations of z/r = 0.975, 0.956 and 0.923 with the fatigue loads of 500, 600 and 700 MPa, respectively. With the increase of the fatigue loads, the contact pressures decrease at the tension side.
(2) The stick zone, slip zone and open zone are formed due to fretting contact. With the fatigue load increasing, the area of slip zone at the compression side increases with different trends, while decreases at the tension side.
(3) Although the z-axial stress increases with the fatigue load increasing, the negative impact of fretting reduces.
(4) The locations of the crack nucleation predicted by the critical plane shear stress amplitude model and SWT model are more reasonable than the Ruiz-Chen

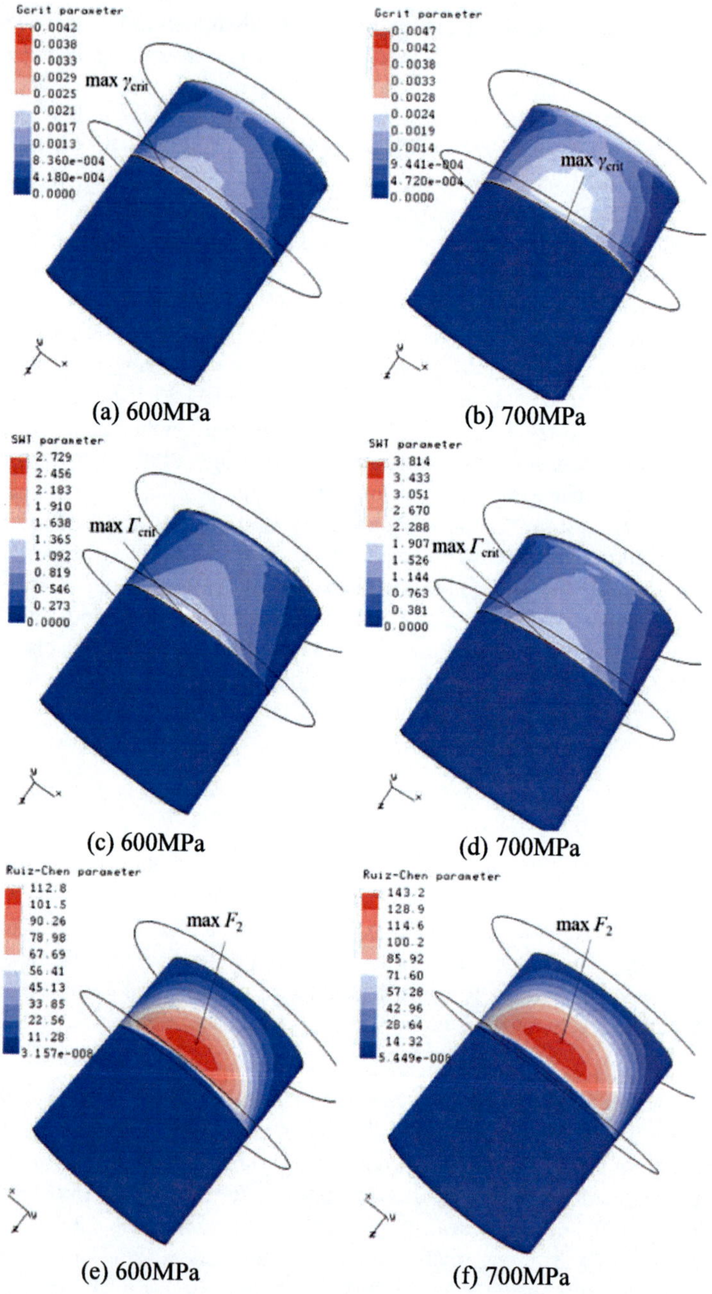

Fig. 7 Contours of parameter **a** and **b** γ_{crit}; **c** and **d** Γ_{crit}; **e** and **f** F_2

model, which are all at the contact edge. However, further experiments and nucleation life prediction studies are needed to evaluate the accuracy of the shear stress amplitude model and SWT model.

References

1. Waterhouse RB (1996) Mechanics of fretting fatigue. Tribol Int 29(2):175–176
2. Houghton D, Wavish PM, Williams EJ, Leen SB (2009) Multiaxial fretting fatigue testing and prediction for splined couplings. Int J Fatigue 31(11–12):1805–1815
3. Wei DS, Yuan SH, Wang YR (2012) Failure analysis of dovetail assemblies under fretting load. Eng Fail Anal 26:381–396
4. Golden PJ, Nicholas T (2005) The effect of angle on dovetail fretting experiments in Ti-6Al-4V. Fatigue Fract Eng Mater Struct 28(12):1169–1175
5. Oscar H (1999) Influence of fretting corrosion on the fatigue strength of fitted members. In: DeVilliers T (eds) Symposium on fretting corrosion. ASTM International, West Conshohocken
6. Nishioka K, Komatsu H, Morita Y (1971) Researches on increasing the fatigue strength of press-fit axle: 2nd report press-fitted shaft assembly two dimensional photoelasticity. Bulletin JSME 14:629–635
7. Alfredsson B (2009) Fretting fatigue of a shrink-fit pin subjected to rotating bending: experiments and simulations. Int J Fatigue 31(10):1559–1570
8. Erena D, Vázquez J, Navarro C, Domínguez J (2020) Numerical analysis of toroidal voids as stress relievers in shrink-fitted shafts. Tribol Int 143:105996
9. Carter BJ, Schenck EC, Wawrzynek PA, Ingraffea AR, Barlow KW (2012) Three-dimensional simulation of fretting crack nucleation and growth. Eng Fract Mech 96:447–460
10. Mangardich D, Abrari F, Fawaz Z (2019) A fracture mechanics based approach for the fretting fatigue of aircraft engine fan dovetail attachments. Int J Fatigue 129:105213
11. Hirsch M (2013) Temperature dependent fretting damage modeling of AISI 301 stainless steel. PhD thesis, Georgia Institute of Technology
12. Su Y, Han QN, Zhang CC, Shi HJ, Niu LS, Deng GJ, Rui SS (2019) Effects of secondary orientation and temperature on the fretting fatigue behaviors of Ni-based single crystal superalloys. Tribol Int 130:9–18
13. Wang J, Gao Y (2019) The stress intensity factor calculation for combined sliding wear and fatigue of GH4169 superalloy based on three-dimensional simulation. Wear 436–437:203012
14. Garcia DB, Grandt AF (2007) Application of a total life prediction model for fretting fatigue in Ti-6Al-4V. Int J Fatigue 29(7):1311–1318
15. Lykins CD, Mall S, Jain V (2001) A shear stress-based parameter for fretting fatigue crack initiation. Fatigue Fract Eng Mater Struct 24(7):461–473
16. Lykins CD, Mall S, Jain V (2000) An evaluation of parameters for predicting fretting fatigue crack initiation. Int J Fatigue 22(8):703–716
17. Ruiz C, Chen KC (1986) Life assessment of dovetail joints between blades and discs in aeroengines. I Mech E Conference Publications (Institution of Mechanical Engineers) 1:187–194
18. Gutkin R, Alfredsson B (2008) Growth of fretting fatigue cracks in a shrink-fitted joint subjected to rotating bending. Eng Fail Anal 15(5):582–596
19. Zhang YB, Lu LT, Zou L, Zeng DF, Zhang JW (2018) Finite element simulation of the influence of fretting wear on fretting crack initiation in press-fitted shaft under rotating bending. Wear 400–401:177–183
20. Spink GM (1990) Fretting fatigue of a 212%NiCrMoV low pressure turbine shaft steel—the effect of different contact pad materials and of variable slip amplitude. Wear 136(2):281–297

21. Walker EK (1970) The effect of stress ratio during crack propagation and fatigue for 2024-T3 and 7075-T6 aluminum. In: Rosenfeld T (eds) Effects of environment and complex load history on fatigue life. ASTM International, West Conshohocken
22. Bhatti NA, Abdel Wahab M (2018) Fretting fatigue crack nucleation: a review. Tribol Int 121:121–138
23. Szolwinski MP, Farris TN (1996) Mechanics of fretting fatigue crack formation. Wear 198:93–107
24. Bertini L, Santus C (2015) Fretting fatigue tests on shrink-fit specimens and investigations into the strength enhancement induced by deep rolling. Int J Fatigue 81:179–190

Wear

A Model for Prediction and Optimization of Flank Wear in End Milling of AISI 316 Stainless Steel

Peter Babatunde Odedeyi and Khaled Abou-El-Hossein

Abstract Controlling Tool wear is a significant issue related with material removal process in machining. This current study focuses on flank wear end milling of AISI 316 stainless steel using PVD a high-performance TiAlN multilayer coated carbide tool. Three factors (cutting speed, feed and depth of cut) and three-level factorial experiment designs with Box Behnken and statistical analysis of variance were performed in order to investigate the effect of the cutting parameters on the tool and workpiece in terms of flank wear. The results show that flank wear is statistically significant influenced by feed and cutting speed. The model for the experiment explains 75% of the variation in the tool wear and the predicted effect of speed on the tool wear. Feed rate has shown to have highest effect on the tool wear. Accordingly, ANOVA and Multiple Regression were used to develop mathematical model for response, alongside with diverse diagnostic tests were also performed to test the validity and efficacy of the proposed model.

Keywords Flank wear · Stainless steel · Surface roughness · ANOVA · Cutting tool

1 Introduction

Manufacturing process is often influenced by a lot of limitations such as tool wear, Tool life, surface roughness, surface location error, and machining stability [1]. As a result of abrasiveness due to physical contact between the work piece and machine tool, Tool wear has posed to be the most significant phenomenon that accounts for major limitation to machining processes. It has drawn a lot of attention among researchers, many of these researchers revealed that, when Tool wear is properly managed, it can prevent damages of machine tools and workpieces because manufacturing enterprises always demand for higher Quality and productivity [1]. Tool wear

P. B. Odedeyi · K. Abou-El-Hossein (✉)
Precision Engineering Laboratory, Mechatronics Department, School of Engineering, Nelson Mandela University, Port Elizabeth, South Africa
e-mail: Khaled.Abou-El-Hossein@mandela.ac.za

M. Abdel Wahab (ed.), *Proceedings of the 8th International Conference on Fracture, Fatigue and Wear*, Lecture Notes in Mechanical Engineering,
https://doi.org/10.1007/978-981-15-9893-7_42

usually occur in combination with the principal wear mode which depends on cutting conditions, tool insert geometry, work piece and tool material. When we combine work piece and cutting condition together, primarily for an end milling process the focus for this study, the cutting speed of milling operation and chip dimension-thickness may be exclusively important together with other wear factors. It is well known that cutting conditions are used for prediction of tool wear in machining [2]. According to recent literatures [2–6], standard static linear prediction models are often found used for tool wear modelling. The simplest and most commonly used is the linear multiple regression model (MRM) [7]. This prediction models have being used to optimized both micro machining and precision machining of high-quality parts.

2 Literatures

A close look at literatures revealed that researchers have used different techniques such as mathematical modeling and experimental modeling to predict Tool wear during manufacturing process [8]. Kaye et al. [9] work established a mathematical model using a response surface method to foresee flank wear by changing spindle speed. Tsao [10] adapted Grey-Taguchi technique to enhance the milling technique parameters of aluminium alloy having employed more than one overall performance attributes, they determined that Flank put on are reduced as excessive as 0.177–0.067 mm. Recently, Garhewal et al. [11] carried out review on tool wear varying operation parameters and concluded that in order to improve productivity in terms of surface integrity, flank wear and cutting region temperature, variations in tool geometry is one of the principal parameters to be considered. Ambilkar et al. [12] presented a review studies on performance of multilayer coated tool at a high speed in turning when machining AISI 4340 steel, they determined out that the impact of feed, velocity and depth of cut on cutting forces and surface finish has been analyzed however a whole lot still needs to be completed on Tool wear [13]. Dogra et al. [14] did research under dry cutting conditions on AISI H11 steel using exceptional coated carbide cutting tools together with TiN, TiCN, CBN and untreated carbide insert tools [15, 16]. These researchers pronounced that free surface roughness within the CBN tools was lower than in the other equipment all accounted for by Tool wear state.

They had been capable to discover that the wear rate of the carbide equipment as the machining progresses was near to the damage values achieved in the CBN tools. It can therefore be concluded that carbide tools had been akin to the CBN in terms of tool life at low reducing speeds and feed rates. At low temperature, the tool life of a prepared coated carbide inserts changed into greater from 16 - 23% in distinctive instances when as compared with the untreated covered carbide insert [11, 14, 17]. Furthermore, through literatures numerous researchers have also done research on mixed modeling techniques to determine progressive Tool wear in end milling of AISI steel. In Chakraborty et al. [18] studies, a wear progressive mix effect model was developed, under semi-dry cutting condition, the lowest flank wear was recorded at a cutting pace of 138 m/min with a feed rate value of 0.1 mm. at a higher rate of cutting speed of about 229 m/min, a lower tool life that comes with high wear progression was clearly recorded. Therefore, they concluded that reducing speed has the most considerable impact on flank wear progression whilst as compared with feed rate and depth of cut. Dos Santos et al. [19] Dos Santos et al. [18] in their research worked on how to determine the best combination of system parameters so one can yield quickest convergence coefficients via using extended Taylor Tool life equation. It was recorded in their research that the mean percent error in Tool life estimates of AISI 304 stainless became 46% compared to 10% AISI 1045 steel under the equal dry cutting condition.

3 Relationship Between Surface Roughness and Wear

Austenitic and duplex stainless steels are often difficult to cut while machining process consequently, this pose an irregular tool wear and a built-up edge (BUE) in machining operations [20]. In Wilkinson et al. study, [21] they argued that the key indicators of tool wear are the precision and surface shape of the machined profile of the machined surface [21]. Matsunmura et al., [22]] have recommended that when considering tool wear on and surface roughness at the same time, to decrease cost, machining operations are advanced with the aid of predicting flank wear the use of analytical technique on metal cut-ting principle and the usage of predicting surface roughness with the usage of a neural Network. With this approach, flank wear will consider the initial wear and surface roughness hence Tool wear may be modeled with the use of adaptive prediction [22]. Goller et al., [23] established that the surface roughness produced during turning operation, is affected by a number of issues and it is therefore a challenge to predict in reality [23]. They further stated that the progressive dynamism that was experienced on the tool peripheral surface can be largely attributed to the roughness of the machine surfaces which in turn is based on the wear resistance of the tool materials [24]. However, Komaraiah et al., found that the harder the tool material, the lower the surface roughness [24]. On the contrary, Wada et al., suggested that the finished surface roughness of the workpiece may be affected if the wear occurs on the cutting edge [25, 26]. But Yang et al., state that the change in visible surface roughness is essentially prompted through cutting flank wear [27].

According to Lee et al., emphasis was placed on the surface roughness and profile in high-speed end milling, they presented a method using simulation of machine surface where accelerated signal was used rather than cutting force [28]. They provided an argument that the vibration encountered by high speed of the spindle suddenly depreciates the geometric accuracy of the machined surface. In the end, they concluded that the generated surface data, surface roughness could be calculated, and profile be plotted [29]. Ehmann et al., in their study introduced a method to represent the surface generation process, they called it "surface-shaping" the method comprises of two parts, the first part modelled the machine tool kinematics and the second part modelled the cutting tool geometry. The latter descried the intersection of the tools face and the flank surface along with the respective angles [30].

4 Experimental Set Up

The research work was carried out on an industrial five axis Deckel Maho DMU 40 CNC machine. The DMU 40 CNC machine belongs to the set of innovative mono BLOCK series of CNC machining centers produced by Deckel Maho. This ultra-precision machine has an improved performance when it comes to dynamics, high precision, higher machined surface, high quality and minima space which is required for uprunning and operations [31]. The motor spindle speed of DMU 40 is up to 12,000 rpm [31, 32]. This machine is ideal for this test because of its extensive large range of expansion options, advanced CNC control and numerous software features capabilities.

There are such a lot of innovative capabilities of mono-block design such as higher dynamics, more precision, extra sturdiness, lower space requirement and high surface quality. For this machine to produce brilliant floor quality, irrespective of the application height, the desk may be organized in a transverse way simplest in the vertical directions when mixed with big linear manual methods which automatically ends in a better robustness. Some highlights of this machine are: optimal access and greater visibility into the work area, a 19" CNC screen, Machine foot print of 4.2 m^2, accelerate up to 0.8 g and a rapid travers of 30 m/min, scraper-belt chip conveyor, 250-l coolant or lubrication tank, geometric temperature compensation, fully enclosed safety cabin, energy supply integrated into the milling head and a trailing cable installation in the base, has a maximum load up to 1100 kg with a fixed table up to 800 g, with the NC rotary table, chain magazine with 60 tool pockets as an option, HSC motor spindle ranges from 12,000(optional), 24,000–42,000 rpm, for production package with 600-l coolant tank. These features make this machine suitable for this study.

5 Work Piece

It known that stainless steels have high resistance to corrosion. However, due to some motives which includes low warmness conductivity, high BUE tendency and high deformation hardening, its machinability becomes more tough than the alternative alloy steels [33, 34]. Among the stainless steel family, austenitic stainless steels are characterized by using an excessive work hardening rate, low thermal conductivity and resistance to corrosion [34, 35]. Recently, several efforts are been made to enhance the machinability of austenitic stainless steels [34, 36, 37].

The workpiece used in this experiment is stainless steel 316 grade. This form of stainless steel belongs to the family of austenitic stainless steel 300-series which

Table 1 General chemical properties of 316 stainless steel (%)

	C	Mn	P	S	Si	Cr	Ni	N	Mo
316	0.08	2	0.045	0.03	0.07	16–18	10–14	0.1	2–3

can be generally hard to machine. It is an austenitic chromium nickel chrome steel containing molybdenum [38]. The metal in Table 1 contains high percentage of Nickel than 304 stainless steel. The resultant composition of this material gives these steels much improved corrosion resistance in many harsh weather conditions. The addition of Molybdenum gives greater resistance to pitting and cavy's corrosion in Chloride-containing media, sea water and chemical environments along with sulfuric acid compounds, phosphoric and acetic acids. 316 stainless steel gives good strength and creep resistance. It also possesses first-rate mechanical and corrosion-resistant properties at sub-zero temperatures. Meanwhile, 316L is a low carbon amendment of 316.

Unlike other austenitic steels, when machining 316 stainless steels, it alloys organization machines with a stringy swarf, rough, rigidly supported tool with as heavy a cut as feasible which is used to prevent glazing. Typical programs of 316 SS consists of Jet engine parts, furnace parts, warmness exchanger, exhaust manifold photographical and pharmaceutical equipment, digester, valve and pump trim, chemical equipment, evaporators, pulp, paper and textile processing equipment, parts exposed to marine structure and tubing [39]. Due to its Physical, chemical—Table 1 and business characteristics, it draws attention for the workpiece choice in this experiment even as Fig. 1, represent the workpiece dimension and microstructure of 316 stainless steel.

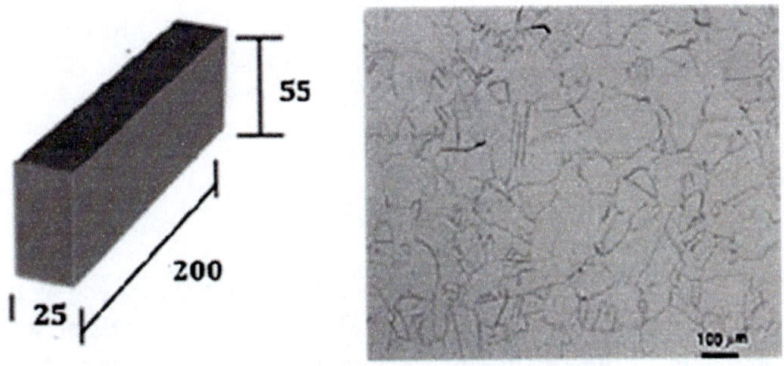

(a) (b)

Fig. 1 **a** The dimension of the work piece is (200 × 55 × 25) mm, **b** microstructure of 316 SS

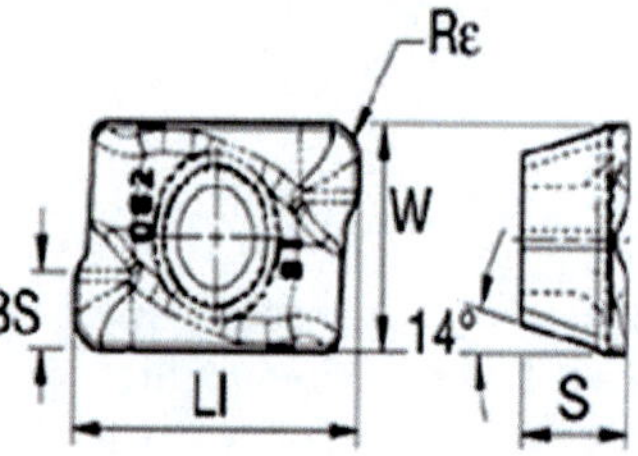

Fig. 2 Cutting tool geometry

Table 2 Inserts identification

Inserts cat No.	Insert shapes	Clearance angle	Size	Insert hand	Rake angle
ADPT150516ERHD	Parellelogram	7°	06	R	7°

6 Cutting Tool

End milling cutting experiments had been carried out on this study. The cutting tests have been performed on the usage of a 16 mm diameter Indexable end-mill with two reducing edges. Figure 2 shows a pattern of the tool from Kennametal with properties in Table 2. KC725M Grade Inserts is a multilayer compose of carbide grade with a TiAIN coatings. A high-performance TiAlN-PVD-coated carbide grade for milling steel [40], stainless metallic and ductile solid iron. KC725M Inserts can be recognized with their shapes as highlighted in Table 2. The purpose why this grade is an ideal desire for moist and dry machining in this experiment is due to the good thermal shock resistance of the substrate. KC725M is in most cases to be used in standard and heavy machining.

7 Analysis

According to ISO 1993, ISO 3685 [3], dictates that a tool wear such as flank wear has exceeded its useful life if the Vb is above 0.3 mm hence considered worn out. In this study, a total of 17 machining runs were carried out to evaluate and validate the modelling approach There are three machining parameters that have been considered, which are Speed (2900–5200) (rpm), feed rate (0.02–0.1)(mm/rev) and depth of cut (1–3)(mm) The output measured is Flank tool wear (mm) as shown in Table 3.

Table 3 Experimental design of the experiment and results (using Box-Behnken design)

Run	Speed (rpm)	Feed (mm/rev)	Feed rate (mm/min	DOC (mm)	TW (mm)
1	2900	0.02	58	2	0.43
2	5200	0.1	520	2	0.29
3	2900	0.1	290	2	0.18
4	4050	0.1	405	1	0.17
5	5200	0.02	104	2	0.29
6	5200	0.06	312	3	0.32
7	5200	0.06	312	1	0.2
8	2900	0.06	174	1	0.1
9	4050	0.02	81	1	0.33
10	4050	0.1	405	3	0.23
11	4050	0.06	243	2	0.3
12	2900	0.06	174	3	0.29
13	4050	0.06	243	2	0.26
14	4050	0.02	81	3	0.29
15	4050	0.06	243	2	0.36
16	4050	0.06	243	2	0.3
17	4050	0.06	243	2	0.28

8 ANOVA and Regression Analysis

8.1 Tool Wear Model

ANOVA [41–43] was employed to carry out the analysis using design expert software, the design of experiment of a 3 × 3 Box Behnken design with 3 center points will reflect 10 coefficients in a quadratic and 2FI model as follows:

$$\mathrm{TW} = \mathrm{A_0} + \mathrm{A_1D} + \mathrm{A_2F} + \mathrm{A_3S} + \mathrm{A_4D^2} + \mathrm{A_5F^2} + \mathrm{A_6S^2} + \mathrm{A_7\dot{D}F} + \mathrm{A_8\dot{D}S} + \mathrm{A_9\dot{F}S} \tag{1}$$

where:

$\mathrm{TW} = \mathrm{ToolWear}$
$\mathrm{A_0} = \mathrm{Intercept}$, $\mathrm{D} = \mathrm{Depthofcut}$,$\mathrm{F} = \mathrm{Feed}$, $\mathrm{S} = \mathrm{Speed}$,$\mathrm{F} = \mathrm{Feed}$
$\mathrm{A_1} - \mathrm{A_3} = \mathrm{Coeefficientoffirstordertermsinthemodel}$
$\mathrm{A_4} - \mathrm{A_6} = \mathrm{Coeefficientofsecondordertermsinthemodel}$
$\mathrm{A_7} - \mathrm{A_9} = \mathrm{Coeefficientofcrosstermsinthemodel}$ (Fig. 3).

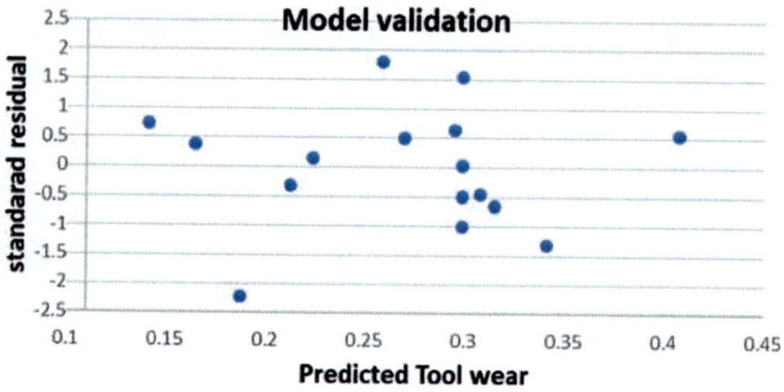

Fig. 3 Statistical model validation

8.2 *Predicted Model*

$$TW = 0.40 - 0.00007 * S - 0.70 * F + 0.27 * D + 0.0014 * S * F - 0.057 * D^2 \quad (2)$$

Figure 4 shows the comparison between the actual Flank wear data measured and the predicted flank wear (Table 4).

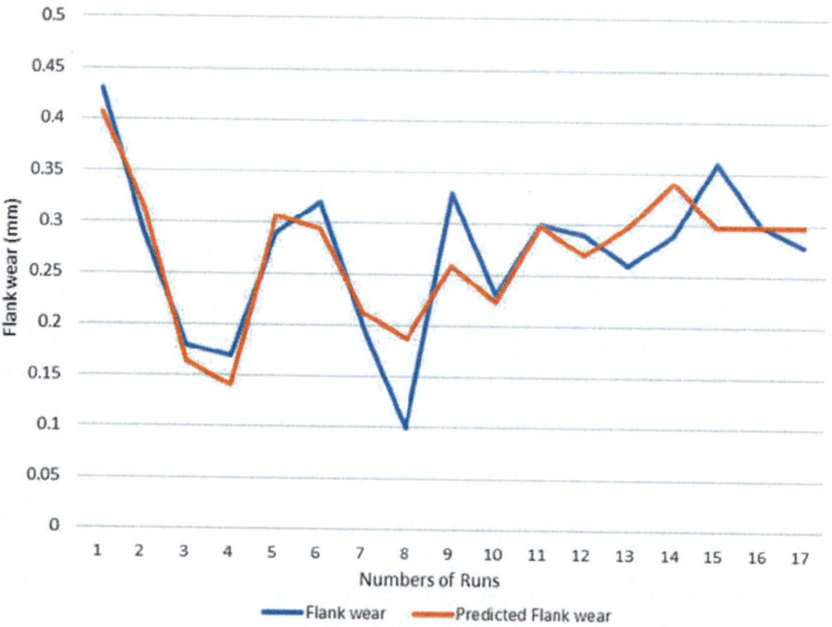

Fig. 4 Comparison between experimental data and predicted data

Table 4 Regression table

Summary output						
Regression statistics						
Multiple R	0.86325112					
R square	**7,452,025**					
Adjusted R Square	0.62938545					
Standard Error	0.04736357					
Observations	17					
ANOVA						
	df	*SS*	*MS*	*F*	*Significance F*	
Regression	5	0.07217067	0.01443	6.43431	0.0049631	
Residual	**11**	0.024676389	0.00224			
Total	16	0.096847059				
	Coefficients	*Standard Error*	*t Stat*	***P-value***	*Lower 95%*	*Upper 95%*
Intercept	0.36009964	0.166164953	2.16712	**0.05304**	−0.005627	0.72582623
Speed	−7.065E−05	3.41494E−05	−2.0689	**0.06289**	−0.0001458	4.5101E−06
Feed	−6.9714674	2.126639484	−3.2782	**0.00736**	−11.652169	−2.2907654
Dept	0.27180556	0.093568867	2.90487	**0.01432**	0.0658619	0.47774924
D^2	−0.0576389	0.02301456	−2.5045	**0.02927**	−0.1082936	−0.0069842
SF	0.0013587	0.000514821	2.63916	**0.02303**	0.0002256	0.00249181

In conclusion, the model explains 75% of the variation in the tool wear and the predicted impact of cutting speed on the tool wear. This may be more suitable by increasing Feed. However, in Table 5, Pred R-Squared of 0.3530 is not as close to the Adj R-Squared of 0.6294 as one might usually expect. This shows a huge block effect due to variation reduction and elimination of outliers. Adeq Precision measures the signal to noise ratio in which in step with literature, a ratio greater than the value of 4 is desirable. In this research work, a ratio of 9.467 shows an adequate signal therefore, this variation may be used to navigate the layout space.

9 Modelling and Optimization Plots

Figure 5 represents interaction plot which also can be called Perturbation plot, it allows to have a look at and compare the outcomes of all of the machining parameters (Feed, speed and depth of cut) at a point within the layout space. A tremendously flat line indicates insensitivity to change in that thing which we do not have in case

Table 5 Analysis of variance table

Source	Sum of Squares	df	Mean Square	F Value	p-value Prob > F
Model	0.072	5	0.014	6.43	0.0050 significant
A-speed	1.250E−003	1	1.250E−003	0.56	0.4710
B-feed	0.028	1	0.028	12.31	0.0049
C-depth	0.014	1	0.014	6.07	0.0315
AB	0.016	1	0.016	6.97	0.0230
C^2	0.014	1	0.014	6.27	0.0293
Residual	0.025	11	2.243E−003		
Lack of fit	0.019	7	2.725E−003	1.95	0.2711 not significant
Pure error	5.600E−003	4	1.400E−003		
Cor total	0.097	16			

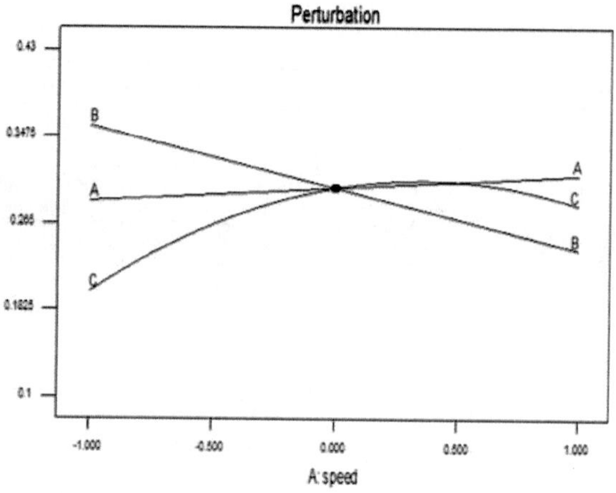

Fig. 5 Interaction plots

of this experiment. The perturbation plot might be used to locate those factors that maximum influence the Tool wear response. From Fig. 4, A = Feed, B = Speed and C = Depth of cut. Looking closely, it shows that the higher the speed the better or the slower the progressive tool wear, it appears that the increase in feed rate, the higher impact we have on the progressive tool wear. The depth of cut seems to have an interesting effect as it can be optimized. Initially, the larger the depth of cut during machining process, in as much as we desire to safe time, cost and maximize material removal volume the more damage we cause for the tool. But this condition can be salvage as the depth of cut keep increasing by lowering the speed and keep the feed a bit constant. Figures 6, 7 and 8 show contour plot, optimization plot and desirability plot respectively.

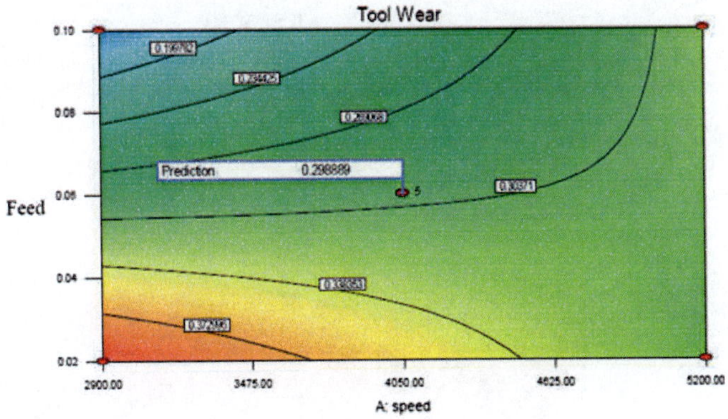

Fig. 6 3-D contour plot

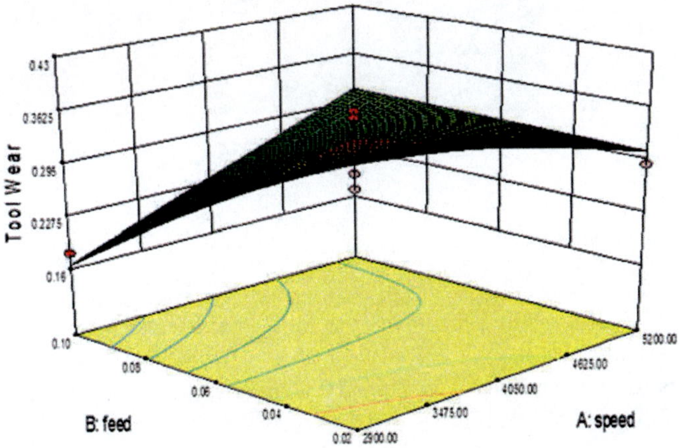

Fig. 7 3-D optimization plot

10 Conclusions

Present work emphasis on analysis of abrasive wear caused by flank wear in end milling of Austenitic AISI 316 grade stainless steel using coated carbide insert, each experiment was carried out with a new insert which was used until worn out i.e. minimum of flank wear 0.3 mm. The subsequent conclusions are drawn.

- The model for the experiment explains 75% of the variation in the tool wear and the predicted effect of speed on the tool wear. Where feed rate has shown the highest effect on the tool wear.
- For optimization, A minimum tool wear value of 0.174 mm was achieved through optimization at low values of feed, speed and depth of cut. Hence, better Tool

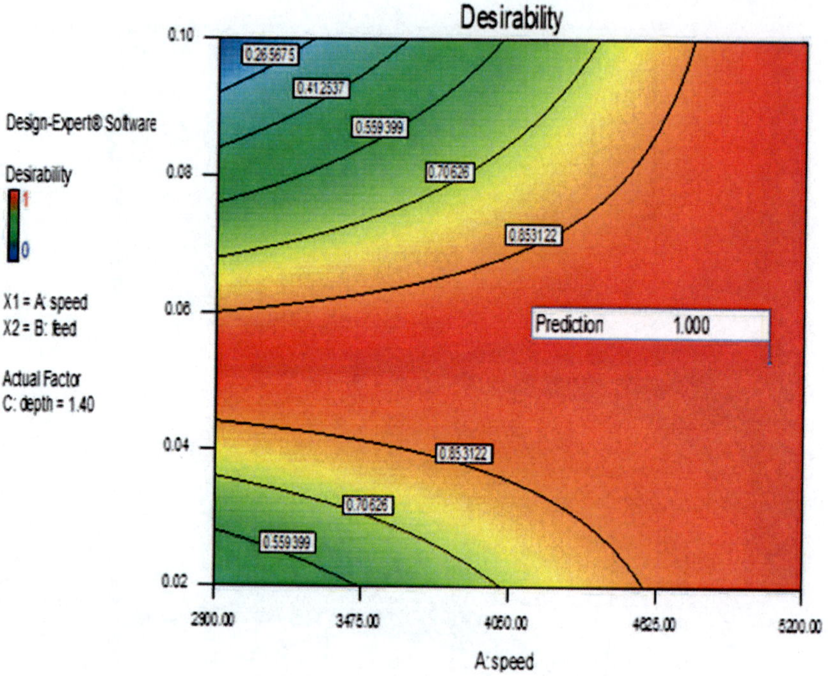

Fig. 8 Desirability plot

wear can be achieved by using lower machining parameters. When working with stainless steel 316, a maximum Tool wear value of 0.296 mm was achieved through optimization at low values of feed about 0.06 mm/rev, speed of 4050 mm/min and depth of cut about 2 mm.

Acknowledgements The authors would like to acknowledge and thank the National Research Foundation (NRF) of South Africa and Research Capacity Development (RCD) Department at Nelson Mandela University for their Financial supports RCD No: s215391292.

References

1. Drouillet C, Karandikar J, Nath C, Journeaux A-C, El Mansori M, Kurfess T (2016) Tool life predictions in milling using spindle power with the neural network technique. J Manuf Process 22:161–168
2. Zhou J-H, Pang CK, Zhong Z-W, Lewis FL (2011) Tool wear monitoring using acoustic emissions by dominant-feature identification. IEEE Trans Instrum Meas 60:547–559
3. Palanisamy P, Rajendran I, Shanmugasundaram S (2008) Prediction of tool wear using regression and ANN models in end-milling operation. Int J Adv Manuf Technol 37:29–41

4. Chen Z, Zhang X (2005) Monitoring of tool wear using feature vector selection and linear regression. In: International conference on natural computation, pp 1–6
5. Khan A, Maity K (2019) Statistical modelling and machinability assessment of commercially pure titanium (CP-Ti) grade II: an experimental investigation. Measurement
6. Kumar KA, Ratnam C, Rao KV, Murthy B (2019) Experimental studies of machining parameters on surface roughness, flank wear, cutting forces and work piece vibration in boring of AISI 4340 steels: modelling and optimization approach. SN Appl Sci 1:26
7. Pang CK, Zhou J-H, Zhong Z-W, Lewis FL (2010) Tool wear forecast using dominant feature identification of acoustic emissions. In: 2010 IEEE international conference on control applications, pp 1063–1068
8. Sivasakthivel P, Murugan VV, Sudhakaran R (2010) Prediction of tool wear from machining parameters by response surface methodology in end milling. Int J Eng Sci Technol 2:1780–1789
9. Kaye J, Yan D-H, Popplewell N, Balakrishnan S (1995) Predicting tool flank wear using spindle speed change. Int J Mach Tools Manuf 35:1309–1320
10. Tsao C (2009) Grey–Taguchi method to optimize the milling parameters of aluminum alloy. Int J Adv Manuf Technol 40:41–48
11. Garhewal G, Katiyar V (2018) Review on tool wear of varying turning operation parameters. J Sci Res Allied Sci 4:111–123
12. Ambilkar P, Kumari TP, Shrivas S (2017) Optimization of turning parameter for AISI 4140 by grey relation analysis. J Sci Res A Sci 3:141–147
13. Zareena A, Veldhuis S (2012) Tool wear mechanisms and tool life enhancement in ultra-precision machining of titanium. J Mater Process Technol 212:560–570
14. Dogra M, Sharma VS, Sachdeva A, Suri NM, Dureja JS (2011) Performance evaluation of CBN, coated carbide, cryogenically treated uncoated/coated carbide inserts in finish-turning of hardened steel. Int J Adv Manuf Technol 57:541–553
15. Ning L, Veldhuis S, Yamamoto K (2008) Investigation of wear behavior and chip formation for cutting tools with nano-multilayered TiAlCrN/NbN PVD coating. Int J Mach Tools Manuf 48:656–665
16. Vereschaka AA, Grigoriev SN, Sitnikov NN, Oganyan GV, Batako A (2017) Working efficiency of cutting tools with multilayer nano-structured Ti-TiCN-(Ti, Al) CN and Ti-TiCN-(Ti, Al, Cr) CN coatings: analysis of cutting properties, wear mechanism and diffusion processes. Surf Coat Technol 332:198–213
17. Science and Engineering Research. Available https://www.sersc.org/
18. Chakraborty P, Asfour S, Cho S, Onar A, Lynn M (2008) Modeling tool wear progression by using mixed effects modeling technique when end-milling AISI 4340 steel. J Mater Process Technol 205:190–202
19. Dos Santos A, Duarte M, Abrao A, Machado A (1999) An optimisation procedure to determine the coefficients of the extended Taylor's equation in machining. Int J Mach Tools Manuf 39:17–31
20. Krolczyk G, Nieslony P, Legutko S (2015) Determination of tool life and research wear during duplex stainless steel turning. Arch Civil Mech Eng 15:347–354
21. Wilkinson P, Reuben R, Jones J, Barton J, Hand D, Carolan T et al (1997) Surface finish parameters as diagnostics of tool wear in face milling. Wear 205:47–54
22. Matsumura T, Obikawa T, Shirakashi T, Usui E (1993) Autonomous turning operation planning with adaptive prediction of tool wear and surface roughness. J Manuf Syst 12:253–262
23. Goller J, Barrow G (1995) The prediction of surface finish in turning operations. In: Proceedings of the thirty-first international matador conference, pp 229–237
24. Komaraiah M, Reddy PN (1993) Relative performance of tool materials in ultrasonic machining. Wear 161:1–10
25. Wada R, Kodama H, Nakamura K, Mizutani Y, Shimura Y, Takenaka N (1980) Wear characteristics of single crystal diamond tool. CIRP Ann 29:47–52
26. Boud FM (1998) An investigation of the relationship between temperature, forces and tool wear in turning and drilling

27. Yang K, Jeang A (1994) Statistical surface roughness checking procedure based on a cutting tool wear model. J Manuf Syst 13:1–8
28. Lee KY, Kang MC, Jeong YH, Lee DW, Kim JS (2001) Simulation of surface roughness and profile in high-speed end milling. J Mater Process Technol 113:410–415
29. Benardos P, Vosniakos G-C (2003) Predicting surface roughness in machining: a review. Int J Mach Tools Manuf 43:833–844
30. Ehmann K, Hong M (1994) A generalized model of the surface generation process in metal cutting. CIRP Ann 43:483–486
31. Odedeyi PB, Abou-El-Hossein K, Liman MM, Jumare AI, Lukman AN (2017) Acoustic emission in stainless steel end milling with carbide tools. In: Key engineering materials, pp 206–210
32. Saimen www.siamintermach.com
33. Kopač J, Šali S (2001) Tool wear monitoring during the turning process. J Mater Process Technol 113:312–316
34. Xavior MA, Adithan M (2009) Determining the influence of cutting fluids on tool wear and surface roughness during turning of AISI 304 austenitic stainless steel. J Mater Process Technol 209:900–909
35. Grover MP, Groover MP (1996) Fundamentals of modern manufacturing, Wiley
36. O'Sullivan D, Cotterell M (2002) Machinability of austenitic stainless steel SS303. J Mater Process Technol 124:153–159
37. Fox-Rabinovich G, Kovalev A, Aguirre M, Beake B, Yamamoto K, Veldhuis S et al (2009) Design and performance of AlTiN and TiAlCrN PVD coatings for machining of hard to cut materials. Surf Coat Technol 204:489–496
38. Odedeyi P, Abou-El-Hossein K, Liman M (2017) An experimental study of flank wear in the end milling of AISI 316 stainless steel with coated carbide inserts. J Phys Conf Ser, 012058
39. SAB
40. Kulkarni AP, Sargade VG (2015) Characterization and performance of AlTiN, AlTiCrN, TiN/TiAlN PVD coated carbide tools while turning SS 304. Mater Manuf Process 30:748–755
41. Gopalsamy BM, Mondal B, Ghosh S (2009) Taguchi method and ANOVA: an approach for process parameters optimization of hard machining while machining hardened steel
42. Huo D, Lin C, Dalgarno K (2014) An experimental investigation on micro machining of fine-grained graphite. Int J Adv Manuf Technol 72:943–953
43. Periyanan P, Natarajan U, Yang S (2011) A study on the machining parameters optimization of micro-end milling process. Int J Eng Sci Technol 3:237–246

Study of Sliding Wear in Rail and Wheel Steels: Effect of Hardness Ratio and Normal Load in Pin on Disc Test

Thiago Gomes Viana, Gustavo Tressia, and Amilton Sinatora

Abstract Rails and wheels are one of the most expensive assets in the railways, and their life needs to be increased by existing methods such as using the optimum rail/wheel hardness ratio. This study aims to verify the synergistic effect of hardness ratio and normal load on the sliding wear resistance. For this, it was used as model system the test pin against disc. The pins were extracted from a class C steel cast wheel (AAR Standard) and the discs were extracted from premium and intermediate grade steel rails (AREMA Standard). Rails are pearlitic and the bainitic wheel is in its first life. Both are hypereutectoid steels with hardness between 321 and 392 HB. The results of the unlubricated sliding tests made possible to compare the sliding wear resistance as a function of the three normal loads (40, 80 and 120 N). The sliding wear behavior for some of the singly body wear rate cases here studied served to validate the field and laboratory tests presented in the literature. From a crossing between the results of the literature and the present study, it is recommended aiming reduction of the wear of rail and/or wheel in a unlubricated condition, to adopt the optimal hardness ratio between rail/wheel $R_{Optimal} \geq 1.0$, proposed in the field study (Steele and Reiff 1982), and/or $1.0 \leq R_{Optimal} \leq 1.1$ proposed by the laboratory test and mathematical model (Razhkovskiy et al. 2015).

Keywords Wheel-rail system · Sliding wear · Pin on disc test · Hardness ratio · Normal load

---

T. G. Viana (✉)
Department of Railway Track Engineering, VLI Logistics, São Paulo, Brazil
e-mail: thiago.viana@vli-logistica.com.br

T. G. Viana · G. Tressia · A. Sinatora
Department of Mechanical Engineering, University of São Paulo, São Paulo, Brazil

G. Tressia · A. Sinatora
Instituto Tecnológico Vale, Belém, Brazil

M. Abdel Wahab (ed.), *Proceedings of the 8th International Conference on Fracture, Fatigue and Wear*, Lecture Notes in Mechanical Engineering,
https://doi.org/10.1007/978-981-15-9893-7_43

1 Introduction

The criteria for dimensional, metallurgical and mechanical characteristics selection of rails and wheels in Railway companies are based on factors such as ton per axle of locomotives and wagons, track layout (inclination of ramps and radius of curves), trains traffic density and materials with high hardness.

Railway engineers work daily with the problem of high wear rate on rails and wheels. Efforts to solve this problem by increasing wheel and rail hardness have been made by researchers for many years. This decision has sparked controversy and discussion around the world, not being resolved and finalized [1].

Surface interactions between rail and wheel results in mass loss of the rail, with the latter being the critical component of the tribological system, representing one of the primary modes of degradation resulting in component replacement and is particularly dominant in sharp curves where obtained wheel-rail contact occurs in the gauge corner of railhead [2].

In the practice observed in daily inspection of rails and wheels, it is observed in Fig. 1 that the most accentuated type of wear, focus of this study, is in the gauge corner and lateral of the railhead and in the wheel flange. This is because the wheel flange is sliding in intense contact with the gauge corner on the high rail of curves.

According to the literature [3], reported studies of some field test which were used to propose wear prediction of the system depending on the wheel hardness to wheel hardness ratio (H_{Rail}/H_{Wheel}) [4]. Field results (track) of wear rate in rails and wheels at TTCI (Transportation Technology Center Incorporation—Pueblo, CO, USA), using different rail classes with hardness between 314 and 405 HB and different wheel classes with hardness between 249 and 317 HB, showed that the optimal hardness ratio between rail and wheel (H_{Rail}/H_{Wheel}) must be greater or equal than 1.0 [4]. These tests were performed on a closed-loop Railway called FAST (Facility for Accelerated Service Testing), where the train was traveling with a speed of 65 km/h,

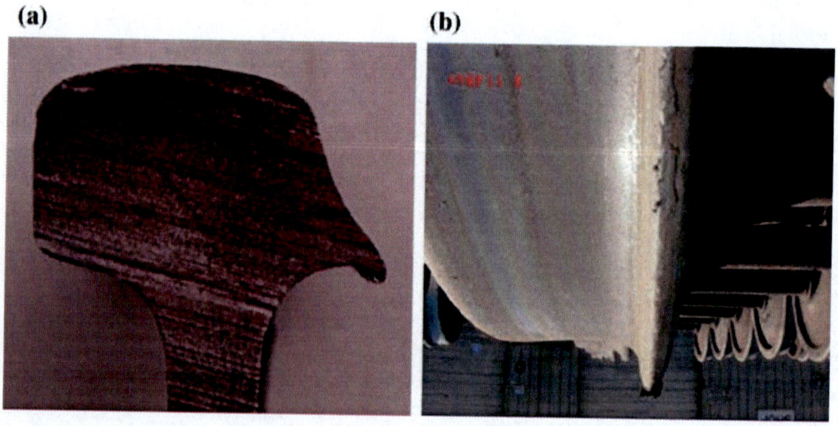

Fig. 1 Gauge corner and lateral wear of rail head (a) and wheel flange wear (b)

9,264 km/day (1 million gross tons/day), a curve radius of 350 m, unlubricated, inclination of ramps between 0.9 and 2% and train consisting of 4 locomotives plus 198 wagons.

These results (wear rate) were combined by wheel and rail wear rate data from field tests at TTCI in a manner proposed [5] and some conclusions can be doing about the wheel-rail system, such as when the hardness ratio is above 1.0 the wheel wear would stabilize and the rail wear would be reduced, as shown in Fig. 2.

In another study [1] laboratory experiments were conducted using the unlubricated vertical disc (wheel material C74GW-T-B) against horizontal bar (rail material M76TT) model (test rig), with 2% sliding (Fig. 3). The rail hardness samples bar was

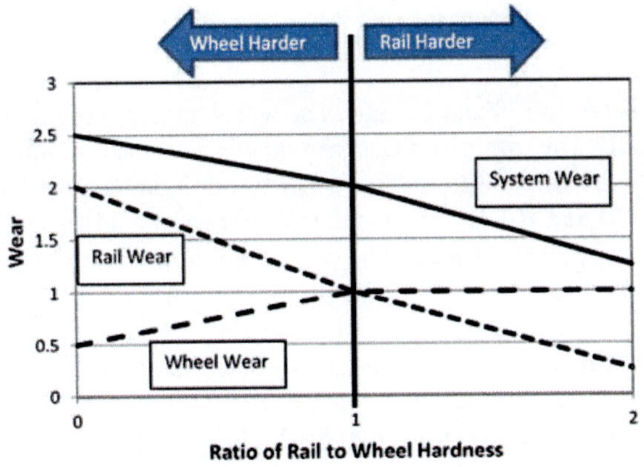

Fig. 2 Rail/wheel hardness optimal ratio [3, 4]

Fig. 3 Test rig for studying wear resistance of wheel and rail: **a** wheel model, **b** fork, **c** rail model, **d** plate, and **e** loading unit [1]

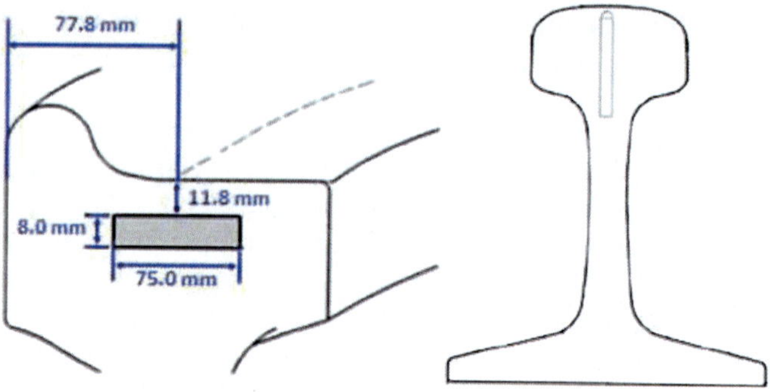

Fig. 4 Regions of the wheels and rail from where the discs and pins were removed (out of scale). Dimensions in mm [6]

between 321 and 401 HB. The wheel hardness samples disc was between 293 to 363 HB. The results of mass loss showed that the optimum hardness ratio (H_{Rail}/H_{Wheel}) was between 1.03 and 1.09, that is, the recommended values for the rails were between 363 and 401 HB and for the wheel between 361 and 388 HB.

In a recent study [6] pin on discs test were performed, where the pins were extracted from a 370 HB pearlitic rail while the disc were fabricated from a AAR class C wheel (Fig. 4). The unlubricated sliding test conditions include end pin rounded, a constant normal load and a sliding velocity range from 0.1 to 0.9 m/s. Pin wear rate and disc wear rate were higher (two to three times) in the condition with discs extracted from forged wheels (335 HB). This difference was justified by the authors due to the presence of degenerated pearlite on forged samples.

Friction management (rail lubrication) as an isolated solution provides a certain level of benefits for the Railway operator. However, in order to maximize the benefits of friction management, it is necessary to view this maintenance practice as one of the factors that contribute to keeping the wheel-rail system in good operating condition. Friction management interacts strongly with three other factors such as rail and wheel metallurgy, rail and wheel profile (grinding/milling) and track geometry as a function of actual train speed [7]. At VLI Logistics company, the Railway track engineering department entered a fifth factor about infrastructure (track and rolling stock), and Fig. 5 called the "magic pentagon" of the wheel-rail contact was created and, also, note that in this figure there is a specific factor that is associated to rail/wheel hardness ratio.

The objective of this work is to evaluate the influence of the hardness ratio of the rail/wheel steels and the normal load in sliding wear tests using the pin on disc configuration.

Fig. 5 "Magic pentagon" of wheel-rail interface. The red arrow highlights the hardness ratio parameter

2 Materials and Experimental Methods

2.1 Materials

Samples for the wear tests were extracted from two classes of rail steels (premium and intermediate) to fabricate the discs and from a class C Wheel to fabricate pins. The chemical compositions (mass percentage) are contained in Table 1 for the premium and intermediate class rail is that specified by the AREMA Standard [8] and the class C wheel is specified by the AAR Standard [9]. The hardness specified for these steels are shown in Table 2. These classes of steels have been heat treated by the manufacturers at their steel mill that performs normalization using compressed air or cool bath immersion directly on the rail head or on wheel thread/flange.

Table 1 Chemical composition (mass percentage) [8, 9]

Material/ (%)	C	Mn	P	S	Si
Premium and intermediate rail	0.74–0.86	0.75–1.25	0.020 max	0.020 max	0.10–0.60
Class C wheel	0.67–0.77	0.60–0.90	0.030 max	0.005–0.040	0.15–1.00

Table 2 Surface hardness [8, 9]

Material (steel class)	Hardness (HB)	
	Minimum	Maximum
Premium rail	370	–
Intermediate rail	321	–
Class C wheel	321	363

2.2 Experimental Methods

The microstructure of the samples was analyzed by scanning electron microscope (SEM). To prepare the sample a metallographic polishing was applied and a chemical etching using 3% Nital reagent.

Hardness tests on pin and discs were performed, being analyzed the two ends of the pin: the end that represents the wheel surface (that meets the disc) and the end that represents the wheel core. For discs hardness tests were performed on both the top and bottom.

The methodology for pin and disc extraction from wheels and/or rails, as well as the methodology for conducting the pin on disc tests are similar to those used in literature [6].

The sliding wear tests were conducted at the LFS (Surface Phenomena Laboratory)/USP Polytechnic School (University of São Paulo) using a pin on disc tribometer (Plint TE67). The test consists of a rotating horizontal disc and a fixed pin, contacting the same track and pressed with a constant normal load. The pin with 3 mm of diameter and the discs with 60 mm of diameter. To ensure homogeneity of parallelism and initial roughness for all tests, all pins (with flat tip) were subjected to the same ground process, passing #220, #400 and #600 sandpaper for approximately 30 s and the discs were rectified. The disc, with a constant rotation of 70 rpm, receives pin contact within a radius of 25 mm, resulting in a tangential velocity of 0.18 m/s. The unlubricated tests were performed for 60 min (Fig. 6).

Normal load varied between 40, 80 and 120 N. For each condition, at least three repetitions were performed. The friction force (tangential load) was monitored during the test using a load cell. The coefficient of friction was determined by the ratio between the friction force and the normal load. Mass loss was obtained by measuring the mass before and after tests and, after this, the wear rate was calculated.

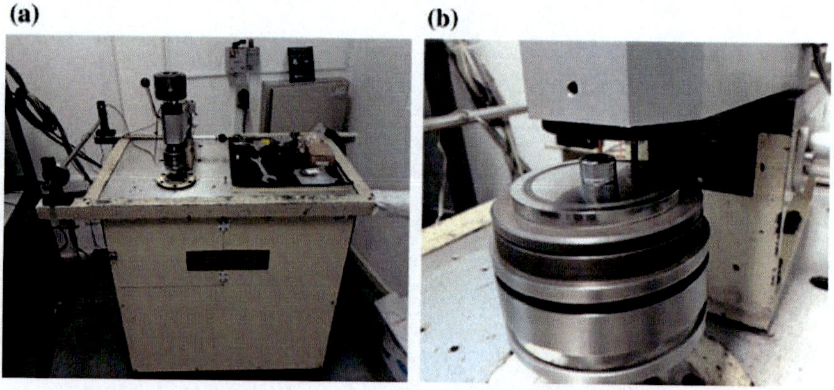

Fig. 6 **a** Tribometer pin on disc (Plint TE67) and **b** unlubricated sliding wear test

Table 3 Chemical composition: rails and wheel

Material (%)	C	Mn	P	S	Si
Premium rail	0.84	0.63	0.02	0.01	0.22
Intermediate rail	0.78	0.79	0.02	0.01	0.22
Class C cast wheel	0.72	0.82	0.02	0.01	0.74
Material (%)	Ni	Cr	Mo	V	Cu
Premium rail	0.01	0.20	0.00	0.00	0.01
Intermediate rail	0.01	0.09	0.00	0.00	0.00
Class C cast wheel	0.01	0.06	0.02	0.00	0.02
Material (%)	C (equivalent)				
Premium rail	1.03				
Intermediate rail	0.97				
Class C cast wheel	1.00				

3 Results and Discussion

3.1 *Chemical Composition*

The results of chemical composition of the samples are shown in Table 3, which shows that all three steels are hypereutectoid. Rails is in accordance with AREMA Standard [8] for chemical composition except that the Manganese content in the premium rail was softly below specified. The chemical composition of the wheel is in agree with AAR Standard [9].

3.2 *Microstructure*

Microstructure of premium and intermediate discs are fully pearlitic (Fig. 7). There is a higher refinement in the grain size and interlamellar spacing of the premium disc. This is due to the higher severity in the existing heat treatment for the premium rail head, where initially the railhead is heat up 700 °C and after is cooled rapidly to obtain a fine lamellar structure [10].

The pin showed heterogeneous microstructure as shown in Fig. 8, with the outer side of the pin (contact with the disc) being formed by bainite (Fig. 8b) and the inner side of the pin formed by pearlite (Fig. 8a). This microstructural difference is due to the variation of heat extraction along the wheel, being that, in regions where the cooling rate is high (external regions), the formation of bainitic structure occurs. The pin od disc tests were made in a bainitic pin's microstructure.

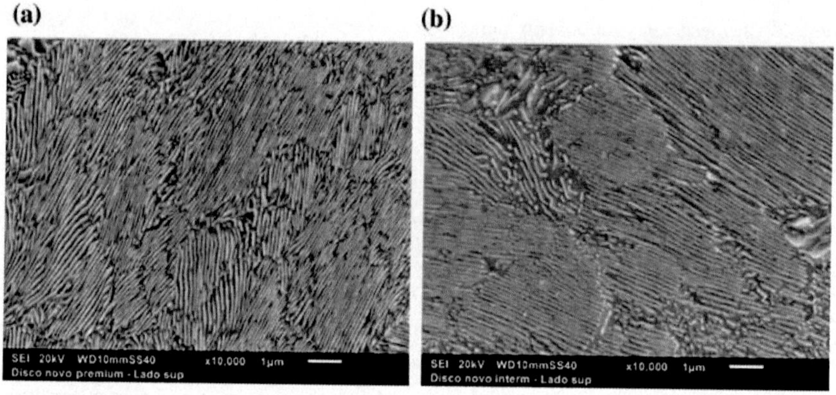

Fig. 7 Pearlite microstructure verified on **a** premium and **b** intermediate discs. Magnification: 10,000×

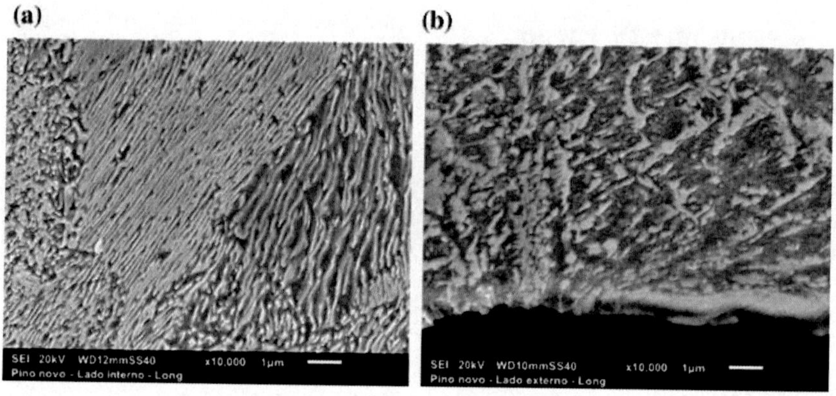

Fig. 8 Pearlitic microstructure verified on the **a** inner side of the pin and **b** bainitic microstructure checked on the outer side of the pin. Magnification: 10,000×

3.3 *Hardness*

Figure 9 illustrates the hardness values of discs and pin. The different disc materials showed significantly different hardness, indicating that the differences observed in the pearlitic refinement (heat treatment severity) evidencing the influence in the materials hardness.

It is noted that for sliding wear tests, premium rail hardness is 20% higher than that observed on class C cast wheel pin (R = 1.2). Intermediate rail disc hardness is 10% higher than that observed on class C cast wheel pin (R = 1.1). All results are conforming to hardness specification for rail and wheel materials.

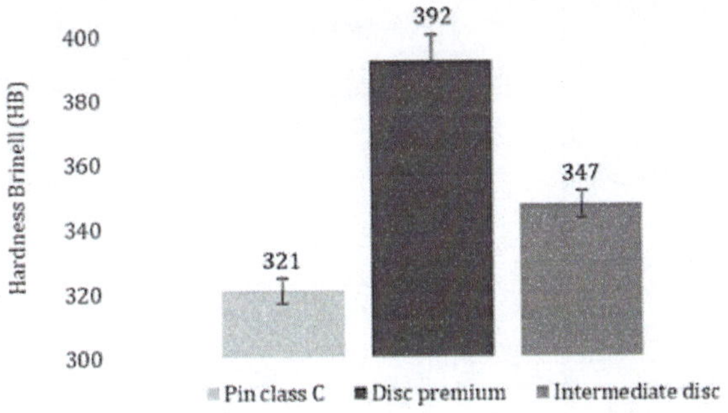

Fig. 9 Hardness Brinell values of pin and discs

3.4 *Wear Rate*

Unlubricated tests in premium disc (Fig. 10a) and intermediate disc tests (Fig. 10b) is showed in Fig. 8 being evidenced the behavior of wear rate in function of normal load.

In unlubricated tests, with the lowest normal load of 40 N, pins of wheel class C had similar wear rates when slid against discs of rails premium or rails intermediate. In the case of a disc intermediate, the wear rate of the pins was equal to the wear rate of the discs, and in the case of discs premium, the wear rate of the pins was greater than the wear rate of the discs. The results found for wear rate of premium disc steel versus class C pin are consistent with literature [4] for the normal load of 40 N. That is, for hardness ratio equal to 1.2 (premium rail disc versus class C wheel pin), wear rate of the wheel is greater than that of the rail. However, the results found for wear rate of intermediate disc steel versus class C pin are not in accordance with literature [4], as they show that for hardness ratio equal to 1.1, wear rate of both bodies are the same.

When the load increase from 40 to 80 N, the wear rate of the pins of wheel class C increased with the increase of the load for the tests against discs premium and intermediate. The wear rate of the pin was statistically equal than the wear rate of the disc premium and intermediate. The wear rate of the pin was similar when tested against the intermediate disc and against the premium disc. The wear rate of the intermediate disc was similar to the wear rate of the premium disc.

With the increase in the normal load from 80 to 120 N, the wear rate of the pins of class C wheels increased with the increase of the load for the tests against discs of the premium rail and intermediate rail. Pin wear rate was statistically equal to wear rate of the premium disc and greater than the intermediate disc. Pin wear rate was similar when tested on premium and intermediate discs. Wear rate of the intermediate disc was similar to that of the premium disc. Wear rate results for intermediate disc steel versus class C pin are consistent with literature [4] for normal load of 120 N. That is, for hardness ratio equal to 1.1 (intermediate rail disc versus class C wheel pin), wear rate of the wheel is greater than that of the rail. However, the results found for wear

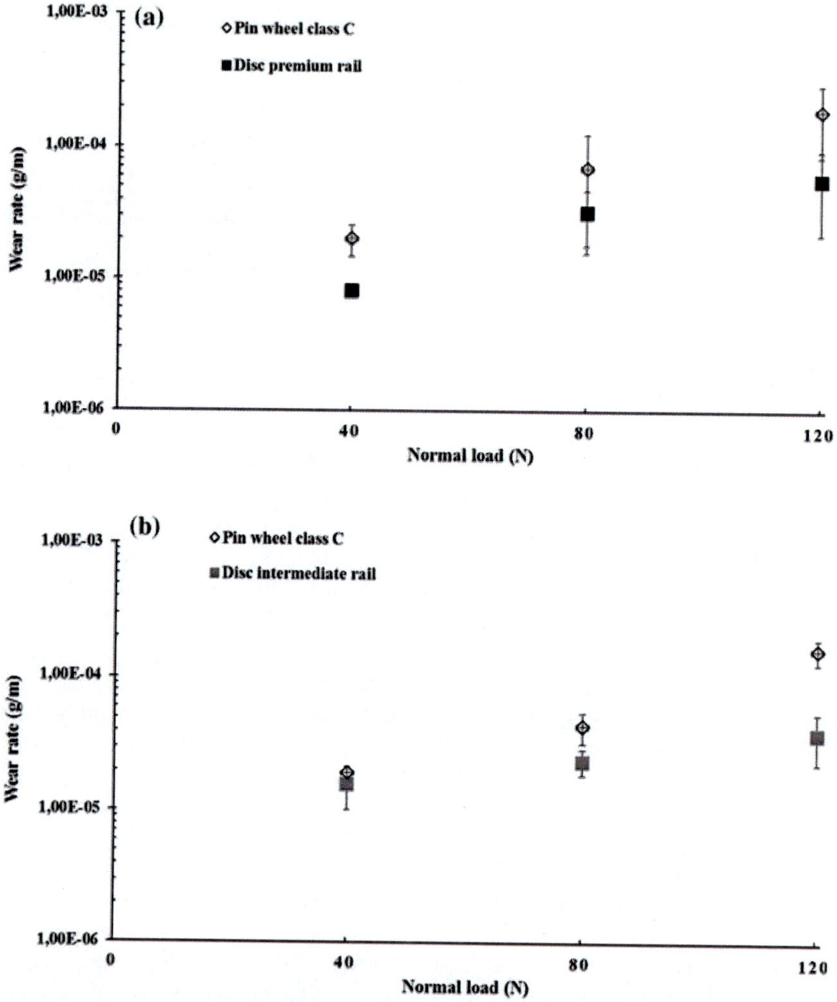

Fig. 10 Wear rate results as a function of the normal load applied to **a** the pin wheel class C above the disc premium rail in unlubricated condition and **b** the pin wheel class C above disc intermediate rail in unlubricated condition

rate of premium disc steel versus class C pin are not in accordance with literature [4], as they show that for hardness ratio equal to 1.2, wear of both bodies are the same.

The influence of normal load on wear rate for pin on disc sliding wear tests is a variable that complements the studies in Railway [11], and it is possible to correlate the normal load intensity with axle load of wagons or locomotives, type of layout of the track (curve radius).

The steel grades adopted in this laboratory study followed the peer recommendation that has the lower wear ratio for the wheel-rail system provided it is greater or equal than 1.0 [4] and/or between 1.0 and 1.1 [1] and/or greater than 1.0 [12, 13]. Another recommendation [1] followed in this study was to adopt, to reduce wear rate, approximately the same hardness ratio for steels, i.e., to move away from the ends of the lowest or highest hardness, such as how avoid using standard rail versus class C wheel or superpremium rail versus class D wheel.

3.5 Friction

The coefficient of friction (COF) as a function of the test time for each one normal load test conditions is in Fig. 11a, b.

The COF average results as a function of the normal load for all materials is presented in Fig. 11c.

It was possible to note two stages of COF along the test for all conditions. Initially, there is an increase in the COF as a function of the test time. In this period, the running-in, several interactions occur between the asperities of the surfaces, accompanied by deformation and wear of these asperities and work hardening of the surfaces [14–16]. The duration of this period depends of the test condition reaching, approximately, 1100 s. In the second period of the COF, steady regime, the COF was more stable, and

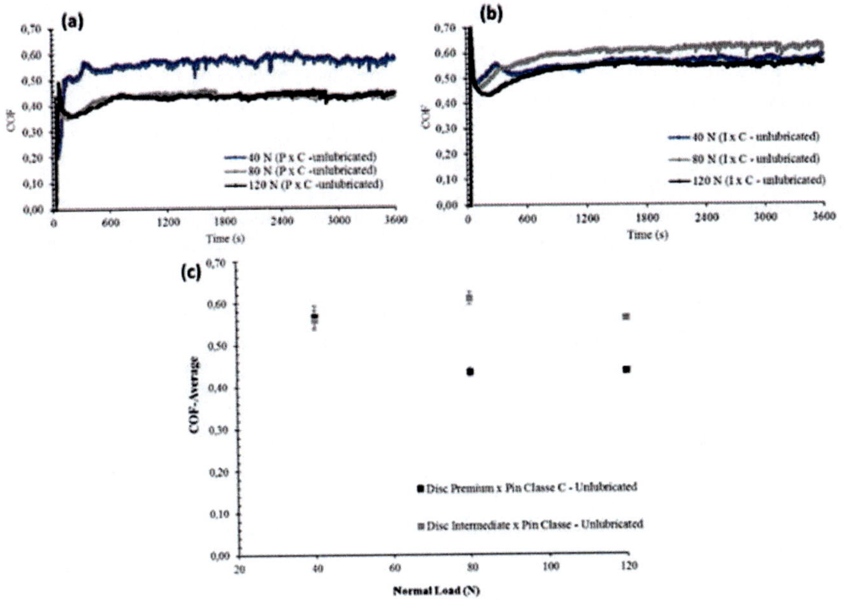

Fig. 11 Coefficient of friction as function of test time and normal load for: **a** disc premium rail unlubricated, **b** disc intermediate rail unlubricated and **c** the COF average on steady regime for all conditions

no significant variation occurred. At this stage, different behaviors were found for the tested conditions. Figure 10c shows the COF average for all conditions. This value was obtained as the average of the three tests of each condition, considering only the steady state of the test. For discs premium, the average value of COF decreased with the normal load, ranging from 0.57 to 0.43 when the normal load increased from 40 to 80 and/or 120 N. For disc intermediate, no significant variation on the COF average was noted, and the value remains about 0.58.

The reduction of the COF with an increase of the normal load for the condition with disc premium can be justified due to the increase in the temperature at the interface, thus increasing the reactivities of the surfaces and oxidation [17]. This, associated with greater hardness of the disc, providing greater support to the oxide layer, preventing the metal–metal contact, thus reducing the interaction between surfaces and consequently reducing the friction coefficient [16].

4 Conclusion

For the present study and the tribosystem here studied, body wear rate singly shows that in some cases the results are in accordance with literature and, however, in other cases are not in accordance.

For the cases that are in accordance with literature singly body wear rate show that:

- Under unlubricated test conditions with normal load of 40 N, the steel of premium rail discs wears less than the material of class C wheel pins (R = 1.2).
- For normal load of 120 N, the steel of intermediate rail discs wears less than the material of class C wheel pins (R = 1.1).

From this, as contributions to the railway aiming reduction of the wear of rail and/or wheel in a unlubricated condition, from a crossing between the results of the literature and the present study it is recommended to adopt the ratio between rail/wheel hardness $R_{Optimal} \geq 1.0$ proposed in the field study [4] and/or $1.0 \leq R_{Optimal} \leq 1.1$ proposed by the laboratory test and mathematical model [1].

5 Future Works

Aiming to study more aspects of sliding wear and its implications for the railroad, we have as future works:

- Total system wear rate (wheel-rail) in unlubricated condition;
- Sliding wear tests in lubricated condition;
- Sliding wear tests in unlubricated and with debris removal condition.

References

1. Razhkovskiy AA, Bunkova TG, Petrakova AG, Gateluk OV (2015) Optimization of hardness ratio in rail-wheel friction pair, Omsk State Railway University, Russia. ISSN 1068-3666. J Friction Wear 36(4):334–341. https://doi.org/10.3103/S1068366615040157
2. Zarembski AM (2005) The art and science of rail grinding, USA
3. Lewis R, Wang W, Burstow M, Lewis S (2016) Investigation of the influence of rail hardness on the wear of rail and wheel materials under dry conditions. In: Third international conference on railway technology: research development and maintenance, at paper 151, Volume: "proceedings of the third international conference on railway technology: research, development and maintenance", Civil-Comp Press, Stirlingshire, UK, Paper 151
4. Steele R, Reiff R (1982) Rail—it's behaviour and relationship to total system wear. In: Proceedings of the 2nd heavy haul conference, Colorado Springs
5. Rabinowicz E (1976) Friction and wear of materials, 1st edn. EUA
6. Prates L, Goldenstein H, Bozzi AC, Scandian C, Falqueto LE Study of sliding wear of the wheel flange—rail gauge corner contact conditions: comparative between cast and forged steel wheel materials. Wear. https://doi.org/10.1016/j.wear.2019.05.009
7. Roney M (2015) Introduction and discussion of guidelines. In: Guidelines to best practices for heavy haul railway operations. Management of the wheel and rail interface, International Heavy Haul Association (IHHA), pp 1.1–1.7
8. American Railway Engineering and Maintenance-of-Way Association (AREMA) (2013), Chap. 4 Rail
9. Association of American Railways (AAR) (2013), M-107/M-208 Wheels and axles
10. Heller W, Flugge J, Ratz G, Heller W, Flugge J, Ratz G (1989) Method for pearlitic rail steels, CA1256004 A
11. Zarembski AM, Palese JM, Martens JH (1991) The effect of improved rail manufacturing process on rail fatigue life. Am Railway Assoc Bulletin 733, 92
12. Rigney DA (1997) Comments on the sliding wear of metals. Tribol Int 30(5):361–367
13. Kato K, Hokkirigawa K (1985) Abrasive wear diagram. In: Proceedings 4th international congress on trihology (Eurotrib '851, Ecully, 1985). Elsevier, Amsterdan, paper 3
14. Blau PJ (1981) Mechanisms for transition friction and wear behavior of sliding metals. Wear, 72
15. Tressia G, Sinatora A, Goldenstein H, Masoumi M (2020) Improvement in the wear resistance of a hypereutectoid rail via heat treatment. Wear, 442–443
16. Viáfara C, Sinatora A (2009) Influence of hardness of the harder body on wear regime transition in a sliding pair of steels. Wear, 267
17. Kennedy FE (1984) Thermal and thermomechanical effects in dry sliding. Wear 100:453–4676

Friction Films and Their Influence on the Antifriction Properties of New High-Temperature Nickel Composites

Krzysztof Jamroziak and Tetiana Roik

Abstract The article studies the tribological properties formation of new composite antifriction materials based on nickel with solid lubricant CaF_2 for operation at temperatures 550–600 °C in air. It was shown the high antifriction properties are achieved due to the formation of homogeneous antiscoring friction films on contact surfaces under working conditions without liquid lubricant. It was analyzed the formed friction films are in the balance between the wear rate and the rate of new film's areas formation at temperatures 550–600 °C. This phenomenon is explained by the formation of the oxides certain amount in combination with pure chemical elements and CaF_2 solid lubricant. The contact surfaces are protected from abrasive phenomena and internal oxidation with such a chemical composition of films at 550–600 °C. Self-lubricating homogeneous antiscoring films provide the high tribological parameters. The research results allowed recommending the rational exploitation modes for new high-temperature nickel antifriction composite.

Keywords Composite antifriction material · Friction film · Tribological properties

1 Introduction

The durability and stable operation of the equipment depends on the resistance of parts to wear at contact interaction. Antifriction materials occupy one of the main places in such conditions. Among them, antifriction materials for severe operating conditions occupy a special place. These are materials based on nickel and its alloys for operation at high and elevated temperatures (500–600 °C) and loads

K. Jamroziak (✉)
Department of Mechanics, Materials and Biomedical Engineering, Wroclaw University of Science and Technology, Smoluchowskiego 25, 50370 Wroclaw, Poland
e-mail: krzysztof.jamroziak@pwr.edu.pl

T. Roik
National Technical University of Ukraine "Igor Sikorsky Kyiv Polytechnic Institute", 37 Peremogy ave., Kyiv 03057, Ukraine
e-mail: roik2011@gmail.com

M. Abdel Wahab (ed.), *Proceedings of the 8th International Conference on Fracture, Fatigue and Wear*, Lecture Notes in Mechanical Engineering,
https://doi.org/10.1007/978-981-15-9893-7_44

(5.0–8.0 MPa) in air [1–6]. The known cast antifriction materials based on nickel have high mechanical characteristics, high heat resistance, but don't satisfy operational requirements due to the high friction coefficient and wear rate at temperatures 500–600 °C in air [7–10].

Powder metallurgy technology is fundamental to solving this problem. Composites for working in hard friction conditions can combine alloying elements and various in nature additives that are not combined with use the traditional technology of manufacturing materials. However, powder technology is used insufficiently, especially at the creation of new antifriction materials. One of the main barriers is the limited information about the processes on the contact surfaces, especially under heavy friction conditions at high temperatures, loads and corrosive environments, as well as their effect on the wear resistance of materials. Therefore, the development of new antifriction composites is impossible without a comprehensive analysis of surface phenomena during high-temperature friction in air. The absence of such data limits the ability to reasonably choose the initial components, predict functional properties and establish rational modes of the composite's operation.

At high friction temperatures (above 500 °C), complex physical, thermal, chemical and deformation processes take place on the contact surfaces. Under these conditions, any liquid lubricant is inoperable and solid lubricant is added to the initial mixture. Surface processes lead to significant changes of part's functional properties [3, 7, 11–13]. It should be noted, the materials wear resistance is determined not only by their initial characteristics, but also by the properties of surface friction films that are formed on the contact surfaces in self-lubrication mode [11, 14, 15]. Therefore, information about the surface layers state is the biggest interest, especially in severe friction conditions.

As a rule, the known regularities of film formation processes in the self-lubrication mode are limited by measuring of composites' tribological properties. Such regularities are valid only in the framework of a specific experiment [11, 16]. It was established [11, 15, 17], the nature and properties of friction films, the so-called secondary structures, determine the operability of the antifriction composite and the friction unit as a whole.

However, until now, the behavior of friction films of new nickel antifriction composites with solid lubricant remains unexplored at different high temperatures in air. In addition, the behavior of chemical elements in the formed secondary structures is also not studied at high temperatures. Some nickel-based antifriction composites have been developed for high temperature operation (above 550 °C) [3, 8, 12, 18–20]. Chemically and thermally stable calcium fluoride (CaF_2) was added to the mixture as a solid lubricant. The effect of CaF_2 solid lubricant on the formation of antiscoring layers was studied in the publications [11–14, 19–23].

The chemical composition of friction films was studied for antifriction composites used for a work at high speeds or loads [11–14, 19–23]. It was shown the chemical elements from both contact materials present in such friction films, as well as oxygen as one of the main components of high-speed or high-loaded friction in air [11, 14, 19, 23].

However, there is still no information about the change in the chemical composition of friction films under various temperature operating conditions. This doesn't allow predicting the composite's frictional behavior and choosing rational temperature operating conditions, when the material demonstrates consistently high antifriction properties.

These arguments were the basis for studying the frictional behavior of new antifriction composite materials based on nickel during friction in the self-lubrication mode at temperatures 550–700 °C in air.

The obtained results will make it possible to select rationally the material's initial composition and its operating modes to ensure the high and stable functional properties at high temperatures.

The objective of this study is to research the friction films' chemical composition of the self-lubricating antifriction nickel composite material with CaF_2 solid lubricant and their effect on tribological properties at high temperatures.

2 Experimental Procedure

Antifriction films were studied using raster electron microscope; calcium fluoride in the matrix was identified using scanning electron microscopy (SEM), chemical elements distribution inside friction films were researched using Auger spectral analysis method on an Auger spectrometer with cylindrical mirror-type analyzer. Tribological tests were performed on a VMT-1 friction testing machine (temperature 550–650 °C, sliding speed V = 1.0 m/s and load P = 3.0–5.0 MPa), the counterface is made of high-speed heat-resistant steel R18 (DSTU, standard of Ukraine). This steel is equivalent to European standard steel EN X75WCrV18-4-1 or EN HSv18-0-1, and British standard steel BS 3355 or BS BT1. The counterface of such steel corresponded to the material of the real shafts in the high-temperature friction units of metallurgical equipment. The steel of counterface has the following chemical composition, wt.%: 0.73–0.83 carbon, 0.20–0.50 silicon, 0.20–0.50 manganese, 3.80–4.40 chromium, to 0.60 nickel, 17.00–18.50 tungsten, to 1.00 molybdenum, 1.008–1.40 vanadium, to 0.03 sulfur, to 0.03 phosphorus, and iron as the base.

The study was focused on the new antifriction composite materials based on nickel. Chemical composition of materials has been presented in Table 1 [24].

As it can be seen from Table 1, original mixture contains powder of calcium fluoride as a solid lubricant. This solid lubricant is thermal and chemically stable substance, and this fact guarantees the calcium fluoride presenting in non-changed state after sintering and further high temperature exploitation. It is known CaF_2 solid

Table 1 Chemical composition of the materials based on nickel, wt.%

Mo	W	Cr	CaF_2	Ni
12.0–15.0	12.0–15.0	14.0–18.0	6.0–12.0	Basis

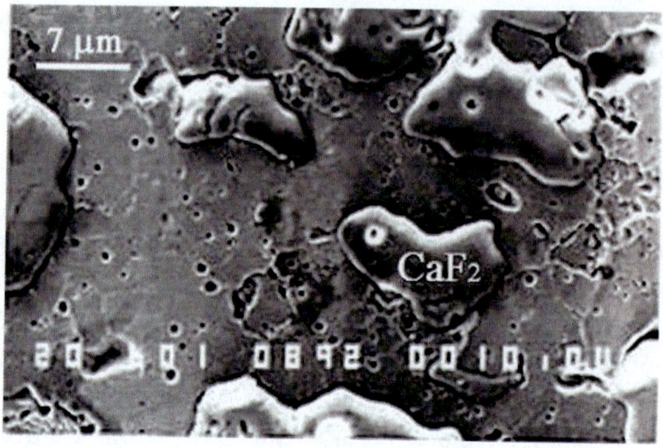

Fig. 1 SEM micrographs showing structure of material Ni + Mo + W + Cr + CaF_2 (compositional contrast mode)

lubricant is effective at elevated and high temperatures and retains its properties up to 1300 °C [7, 11, 13, 25, 26].

So we have researched antifriction composites—Ni + Mo + W + Cr + CaF_2.

The materials were obtained using powder metallurgy methods by mixing, pressing at a pressure 700–900 MPa and sintering at a temperature 1200 °C in hydrogen for 2 h. The porosity was 10–12% after sintering [25]. The material's structure is a metal matrix with CaF_2 solid lubricant inclusions (see Fig. 1).

The metal matrix is a γ solid solution of molybdenum, tungsten and chromium in nickel basis, as well as Ni_3Mo, Ni_3W, Ni_2Cr intermetallic strengthening phases detected by X-ray diffraction.

3 Results and Discussion

Antifriction properties of the examined materials have been presented in Table 2.

As can be seen from the Table 2, the studied composite demonstrates high antifriction properties in the temperature range 550–600 °C. This is ensured by the formation of smooth and homogeneous friction films. The friction surfaces have a smoothed microrelief, without areas of mechanical damage and fatigue failure up to 600 °C (see Fig. 2). To study this phenomenon, the Auger spectral analysis was performed both from the surface of the studied materials and from the counterface's surface after tribological tests at different temperatures. The Auger spectral analysis results are presented in Fig. 3 from dark and light areas of the composite and counterface's surfaces. The distribution of the main elements in friction films was evaluated at a depth to 100 nm. Auger electrons are excited by primary electrons with energy 5 keV. The sample surface was sprayed with argon ions (Ar^+) with energy Ei(Ar^+) = 600 eV and current density i = 5 $\mu A/cm^2$. This provided the friction films etching rate ~0.5 nm/min [12, 20].

Typical Auger electron spectra from surface selected areas of the composite (Ni–Mo–W–Cr) + 10% CaF_2 and counterface have been presented on Fig. 3.

Table 2 Antifriction properties of the materials at high temperatures

Chemical composition (wt.%)					Fraction coefficient (f) and wear rate (I), μm/km, at temperature (°C)						Maximum allowable T (°C)
Mo	Cr	W	CaF_2	Ni	550		600		650		
					f	I	f	I	f	I	
12.0	14.0	12.0	6.0	Rest	0.16	34	0.17	50	0.19	273	650
13.5	16.0	13.5	10.0	Rest	0.15	32	0.16	48	0.18	271	650
15.0	18.0	15.0	12.0	Rest	0.16	33	0.18	52	0.20	274	650

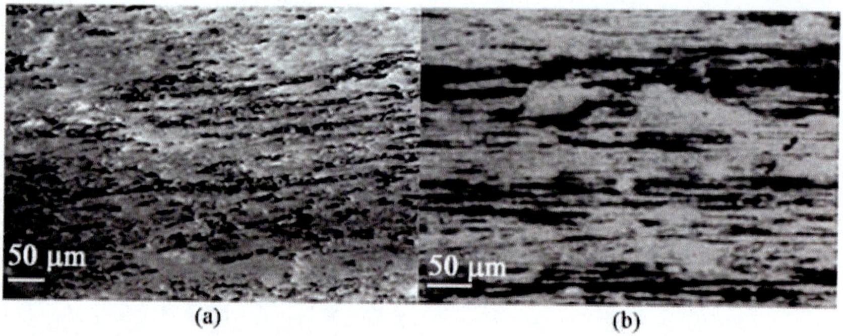

Fig. 2 Friction surfaces after testing at a temperature 600 °C: **a** composite; **b** high-speed steel counterface

It can be seen (see Fig. 3a, curves 2 and 3), nickel, molybdenum, tungsten, chromium, oxygen, sulfur, calcium and carbon are present in the Auger spectra of the composite's surface layer. The appearance of the carbon and sulfur Auger spectra is probably due to their presence in the material of the steel counterface. The concentration ratio of nickel (32.8 at.%), molybdenum and tungsten (4.9 at.% each) and chromium (6.1 at.%) indicates the formation of the alloyed nickel-based γ-solid solution. High intensity Cr, F, and Ca peaks were found in the Auger spectrum of the material's light areas (see Fig. 3a, curve 1). The increased intensity of chromium Auger peak is associated with its large number, both in the studied composite and counterface. This confirms the phenomenon of mass transfer under friction process. The results of the Auger analysis indicate the localization of fluoride particles in the light areas inside the friction film (see Fig. 3a, curve 1). The fluorine and calcium presence in the Auger spectra shows the direct participation of solid lubricant CaF_2 in the tribological process. Wear products (Fig. 3a, curves 4, and 5) contain chemical elements from both contact materials. Auger spectra of the counterface's surface from high-speed steel are shown in Fig. 3b. It was found (see Fig. 3b, curve 1) the light zone in Fig. 2 contains intense the Auger electrons peaks of Mo, Cr, Ca, O, S, Fe, and Ni. Dark areas (see Fig. 3b, curve 2) contain a carbon significant amount. Wear products (Fig. 3b, curve 3) contain a mixture of chemical elements C, O, W, S, Fe, and Ni from both materials of the contact pair. The friction surface's microrelief is characterized by a high dispersion degree of the solid lubricant CaF_2 particles (≤ 0.5 μm). This may indicate the formation of the friction film's fine structure and the maximum localization of surface deformation in film during friction process. An intense oxygen peak shows its significant amount in the friction film (see Fig. 3). This confirms the oxidative wear has occurred at high temperatures. During the high-temperature friction process, oxygen penetrates into the deformed volumes and increases their mobility that contributes to easy removal of wear products from the friction zone. Formed antifriction smooth and homogeneous friction films (see Fig. 2) not only provide high tribological properties, and

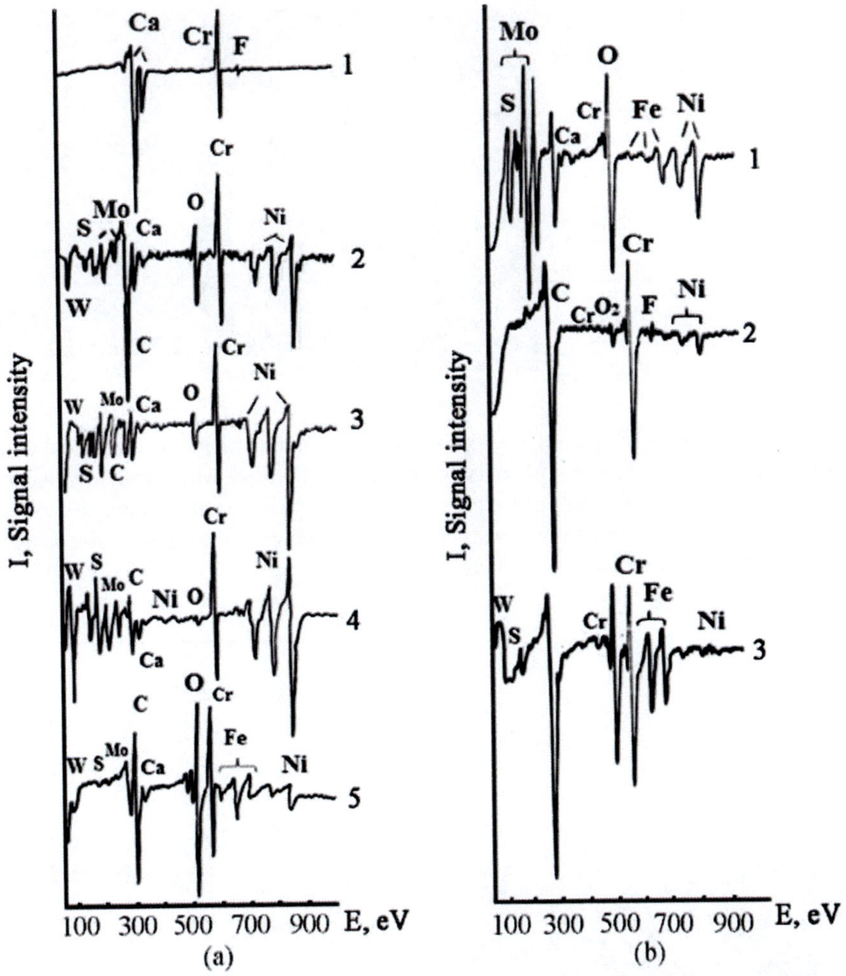

Fig. 3 Auger spectra of the surface layer after tribological tests at a temperature 550–600 °C: **a** composite (Ni–Mo–W–Cr) + 10% CaF_2 (1—light areas, 2 and 3—dark areas, 4 and 5—wear products); **b** high-speed steel counterface (1—light areas, 2—dark areas, 3—wear products)

also protect porous composite from internal oxidation at 550–600 °C. The fracture study of the composite has been shown in Fig. 4 after testing at 600 °C.

As can be seen (see Fig. 4), the fracture surface is a nickel γ-solid solution and particles of CaF_2 solid lubricant along the composite's grain boundaries without oxidized zones. X-ray diffraction analysis confirmed the presence of calcium fluoride at the grain boundaries in the initial state and didn't detect the formation of oxides. This confirms the absence of the composite's internal oxidation in the presence of a protective friction film on the surface. After tribological tests of the studied composite

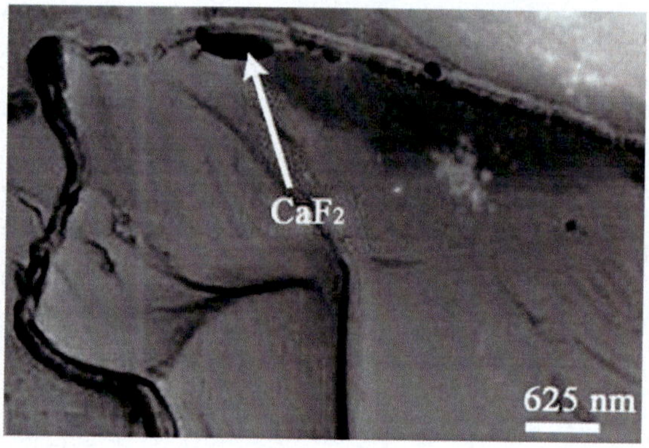

Fig. 4 SEM image of fracture surface for composite (Ni–Mo–W–Cr) + 10% CaF_2 after tribological test at 600 °C

at a temperature 650 °C, there is a significant increase in the friction coefficient and wear (see Table 2). The areas of coarse scratches and abrasive wear zones are visible on the composite's friction surface (see Fig. 5a). The friction film contains a lot of wear products with different dispersity. In this case, the wear of the composite occurs as a result of the individual surface areas peeling.

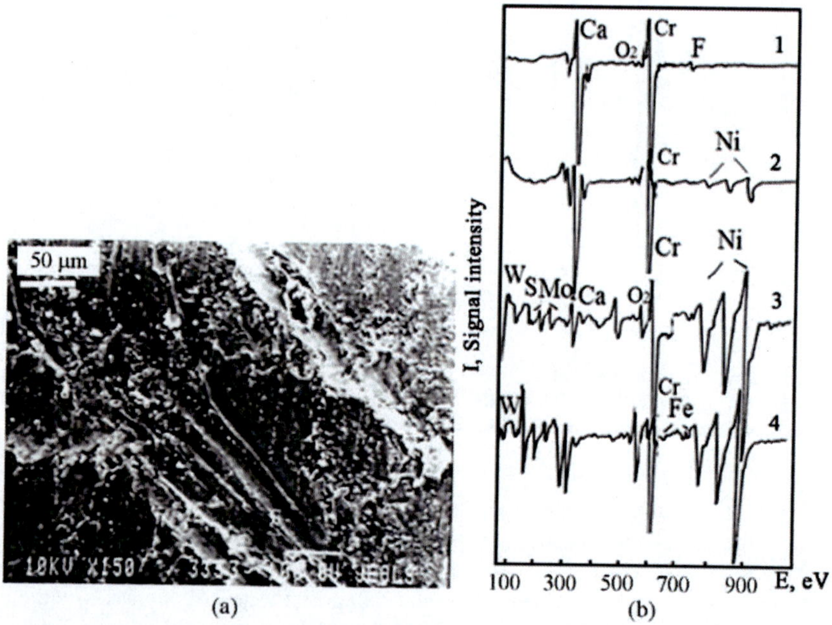

Fig. 5 Friction surface of composite (Ni–Mo–W–Cr) + 10% CaF_2 after tribological tests at 650 °C: **a** abrasive wear fragments; **b** Auger spectra (1—dark area, 2 and 3—light areas, 4—wear products)

A significant increase in the intensity of chromium Auger electrons takes place both in the light and dark regions of the friction track (see Fig. 5b). This is probably due not only to the high chromium concentration in the composite, but also to its high affinity to oxygen. The bifurcation and high intensity of Cr Auger electrons peaks confirms the formation of chromium oxides. In addition to chromium, an intense layer of Ca and F is also observed in the Auger spectra obtained from light regions (see Fig. 5b, curve 1). These elements probably screen all other chemical elements and don't allow the Auger electrons of other elements to be registered by the electron detector. Thus, it can be reasoned that the friction film's light areas are localization places of oxide and fluoride particles. After tribological tests at 650 °C, the composite's friction surface has a coarse heterogeneous structure (see Fig. 5a). The friction film is discrete with intermittent and chaotic localization of individual areas. The main components of the film are probably molybdenum, tungsten, chromium and iron high-oxide compounds. This indicates the oxidative nature of the composite's wear at such temperature. In this case, the main reason for increased wear is the high intensity of the oxides formation processes on the studied composite's surface at 650 °C in air. An imbalance occurs between the film's wear rate and the rate of its new areas formation, when such a film wears out faster than a new one is formed. As a result, the individual zones of the contact surfaces are not covered with a friction film and not protected, that leads to an increase in wear. The high oxide brittle layer cannot protect the contact surfaces from coarse failure appearance and abrasive wear. Analyzing the results of the friction films Auger spectral analysis at 550–600 and 650 °C, it is clear that the films contain chemical elements from both contact materials and oxygen of air in both considered cases. However, these elements amount is different in friction films depending on the temperature. This is indicated by Auger electrons peaks of different intensities.

The bifurcated high intensity peaks of W, Mo, Ni, Fe, and Cr indicate the formation of these elements oxides a large number at 650 °C. In such conditions, the friction films are unstable and discrete, unable to realize the self-lubrication mode of the composite. This is accompanied by the friction coefficient and wear rate significant increase up to 274 microns (see Table 2). However, this value is not critical under such difficult operating conditions. Therefore, we assume that the studied composite can work for a short time at this temperature. Thus, the studies have shown the cause of high antifriction properties for the new composites at 550–600 °C. In this temperature range, the formed friction films have a favorable ratio between the chemical elements of the contact pair and atmospheric oxygen. The balance takes place between friction film's wear rate and the rate of its new areas formation in such conditions. As a result, smooth and continuous friction films are formed on the contact surfaces without failure areas or abrasive damages. The tribological properties analysis made it possible to determine the rational operation modes of new composites: temperature 550–600 °C, loads 3.0–5.0 MPa, sliding speed 1.0 m/s, in air. Such materials can be recommended for a work in friction units of power equipment.

4 Conclusion

We have studied the antifriction properties of a new nickel-based composite for operation at high friction temperatures in self-lubrication mode. The study showed the Ni–Mo–W–Cr–CaF_2 composite has high antifriction properties at friction temperatures 550–600 °C in air. Such properties are achieved due to the formation of smooth and homogeneous friction films on the contact surfaces under operating conditions without liquid lubricant, when solid lubricant is added to the initial mixture.

It was shown, the formed antiscoring self-lubricating friction films are in balance between the wear rate and the rate of new film's areas formation at 550–600 °C. This phenomenon is explained by the formation of the oxides certain amount in combination with pure chemical elements and CaF_2 solid lubricant. The contact pair's surfaces are protected from abrasive phenomena and internal oxidation at such chemical composition of the film. This fact was confirmed by fractographic research.

When the temperature rises to 650 °C and above, the friction films lose their lubricating effect due to the oxidative processes high intensity. As a result, an imbalance is observed between the film's wear rate and the rate of its new areas formation, when such a film wears out faster than a new one is formed. This leads to the discreteness of the friction film and its inability to protect the contact surfaces from intense wear. High friction coefficient and wear rate are the result of this circumstance (see Table 2).

Further research will be devoted to the study of the phase composition and the phases quantitative ratio in the friction film. This will allow argumentatively predicting the functional properties of antifriction self-lubricating nickel-based composites by friction films quantitative analysis. Moreover, it becomes possible to control the antifriction properties by changing the choice of initial powders introduced into the charge for specific working high-temperature friction units.

References

1. Cui G, Liu Y, Gao G, Liu H, Kou Z (2020) Microstructure and high-temperature wear performance of FeCr matrix self-lubricating composites from room temperature to 800 °C. Materials 13(1):51. https://doi.org/10.3390/ma13010051
2. Kvapilova M, Dvorak J, Kral P, Hrbacek K, Sklenicka V (2019) Creep behaviour and life assessment of a cast nickel—base superalloy MAR—M247. High Temp Mater Process (London) 38:590–600
3. Jamroziak K, Roik T (2019) Structure and Properties of the new antifriction composite materials for high-temperature friction units. In: Abdel Wahab M (eds) Proceedings of the 7th international conference on fracture fatigue and wear, FFW-2018, LNME. Springer, Singapore, pp 628–637
4. Kumar A, Patnaik A, Bhat IK (2017) Investigation of nickel metal powder on tribological and mechanical properties of Al-7075 alloy composites for gear materials. Powder Metall 60:371–383

5. Zhen J, Li F, Zhu S, Ma J, Qiao Z, Liu W, Yang J (2014) Friction and wear behavior of nickel-alloy-based high temperature self-lubricating composites against Si_3N_4 and Inconel 718. Tribol Int 75:1–9
6. Kotkowiak M, Piasecki A, Kulka M (2019) The influence of solid lubricant on tribological properties of sintered Ni–20%CaF_2 composite material. Ceram Int 45:17103–17113
7. Jamroziak K, Roik T, Gavrish O, Vitsiuk I, Lesiuk G, Correia JAFO, De Jesus A (2018) Improved manufacturing performance of a new antifriction composite parts based on copper. Eng Fail Anal 91:225–233
8. Bin C, Tan Y, Tu Y, Wang X, Tan H (2011) Tribological properties of Ni-base alloy composite coating modified by both graphite and TiC particles. Trans Nonferrous Metals Soc China 21(11):2426–2432
9. Wang Z, Wang H, Liu G, Huang H, Zhang H, Shao Y, Hu B (2019) Effect of Ta on the microstructure of high performance Ni-based powder metallurgy superalloys. Sci China Technol Sci 62(11):1961–1967
10. Zielinska M, Zagula-Yavorska M, Poreba M, Sieniawski J (2010) Thermal properties of cast nickel based superalloys. Arch Mater Sci Eng 44(1):35–38
11. Roik TA, Shevchuk YF (2006) Development of predictable secondary structures in materials for high-temperature bearings. Powder Metall Met Ceram 45:531–539
12. Li JL, Xiong DX (2008) Tribological properties of nickel-based self-lubricating composite at elevated temperature and counterface material selection. Wear 265:533–539
13. Jamroziak K, Roik T (2019) New antifriction composite materials based on tool steel grinding waste. WIT Trans Eng Sci 124:151–159
14. Roik TA, Gavrish AP, Kirichok PA, Vitsyuk YY (2015) Effect of secondary structures on the functional properties of high-speed sintered bearings for printing machines. Powder Metal Ind Manager Econ 54:119–127
15. Cherepova T, Dmitrieva G, Tisov O, Dukhota O, Kindrachuk M (2019) Research on the properties of Co-TiC and Ni-TiC hip-sintered alloys. Acta Mechanica Et Automatica 13(1):57–67
16. Jozwiak P, Polkowski W, Bojar Z (2015) Applications of Ni_3Al based intermetallic alloys—current stage and potential perceptivities. Materials 8:2537–2568
17. Jianxin D, Tongkun C (2007) Self-lubricating mechanisms via the in situ formed tribofilm of sintered ceramics with CaF_2 additions when sliding against hardened steel. Int J Refract Metal Hard Mater 25(2):189–197
18. Dangsheng X (2001) Lubrication behavior of Ni–Cr-based alloys containing MoS_2 at high temperature. Wear 251(1–12):1074–1079
19. Zuomin L, Childs THC (2004) The study of wear characteristics of sintered high speed steels containing CaF_2, MnS and TiC additives at elevated temperature. Wear 257:435–440
20. Chen Z, Guo N, Ji L, Xu C (2019) Synthesis of CaF_2 nanoparticles coated by SiO_2 for improved Al_2O_3/TiC self-lubricating ceramic composites. Nanomaterials 9:1–11
21. Zhu S, Cheng J, Qiao Z, Yang J (2019) High temperature solid-lubricating materials: a review. Tribol Int 133:206–223
22. Pint BA, DiStefano JR, Wright IG (2006) Oxidation resistance: one barrier to moving beyond Ni-base superalloys. Mater Sci Eng, a 415:255–263
23. Yu J, Wang Y, Zhao X, Li Q, Qiao Q, Zhao J, Zhai S (2019) Wear resistance of Ni-based alloy coatings. Adv Mater Sci Eng ID 2548285. https://doi.org/10.1155/2019/2548285
24. Roik TA, Gavrish AP, Vitsyuk YY, Gavrish OA Powder wear-resistant material based on nickel. Patent UA104946 (Ukraine) 25.02.2009
25. Roik TA, Kholyavko VV, Vitsyuk YY, Melnyk OO (2009) Influence of mechanism of secondary-structures tribo-synthesis on properties of antifriction nickel-based composite materials. Metallophys Adv Technol 31(8):1111–1123
26. Kurzawa A, Roik T, Gavrysh O, Vitsiuk Iu, Bocian M, Pyka D, Zajac P, Jamroziak K (2020) Friction mechanism features of the nickel-based composite antifriction materials at high temperatures. Coatings 10(5):454. https://doi.org/10.3390/coatings10050454

Proposal for New Hardness Concept Using Herbert Hardness Tester

Masaaki Matsubara, Masayoshi Nakamura, and Ryosuke Suzuki

Abstract A Herbert hardness tester uses an inverted pendulum and performs measurements under loading conditions different from those used in conventional indentation and rebound hardness tests. Hardness is conventionally measured as a resistance to monotonic deformation. In contrast, a Herbert hardness tester oscillates on a test specimen, and provides information on resistance to cyclic deformation. The original Herbert hardness tester yielded three kinds of hardness data: (1) Scale hardness (indicated by the position of the bubble in a spirit level at the end of the first swing); (2) Time hardness (based on the pendulum swing time measured using a stopwatch); and (3) Flow hardness (defined as the scale hardness/time hardness ratio, and which is an indicator of work hardening). However, this test method did not become popular due to the time and effort involved compared with other techniques. Nevertheless, the present authors have continued to work towards making improvements in hardness measurement methods based on freely damped oscillations. These improvements included the use of rotary potentiometers, laser displacement meters, and image processing. And a new concept of "Damping hardness" has also been proposed. This paper covers the following topics: (1) A new hardness measure given by the natural angular frequency/damping factor; and (2) Development of a new Herbert hardness tester incorporating a wireless acceleration sensor.

Keywords Herbert hardness · Hardness tester · Hardness testing · Inverted pendulum · Freely damped oscillation

M. Matsubara (✉) · R. Suzuki
Graduate School of Science and Technology, Gunma University, Tenjincho 1-5-1, Kiryu, Gunma 376-8515, Japan
e-mail: m.matsubara@gunma-u.ac.jp

M. Nakamura
Forensic Science Laboratory, Metropolitan Police Department, Kasumigaseki 2-1-1, Chiyoda-ku, Tokyo 100-8929, Japan

M. Abdel Wahab (ed.), *Proceedings of the 8th International Conference on Fracture, Fatigue and Wear*, Lecture Notes in Mechanical Engineering,
https://doi.org/10.1007/978-981-15-9893-7_45

1 Introduction

Hardness measurement is typically performed by pressing an indenter into the test specimen. The indenter is usually harder than the material being tested. Hardness measurements are roughly divided into the following two types:

(1) Measurement where the size of the indentation determines the indentation hardness; and
(2) Measurement where the rebound height determines the rebound hardness.

Both types of hardness imply resistance to monotonic deformation. The loading type in pendulum hardness testing differs from other types of hardness tests, such as indentation hardness and rebound hardness. The tester swings like a pendulum.

Persoz and König hardness testing uses a suspended pendulum type tester to measure the hardness of coatings and films [1]. Such testing has been adopted by certain standards. In the Persoz and König test method, the hardness of any object is given by the number of oscillations made by the pendulum within a specified amplitude limit recorded by an electronic counter.

Herbert hardness testing uses an inverted pendulum type tester and performs measurements under loading conditions different from those used in conventional indentation and rebound hardness testing [2]. A Herbert hardness tester oscillates on a test specimen, and provides information on resistance to cyclic deformation. The original Herbert hardness tester provided three kinds of hardness data (see Fig. 1):

(1) Scale hardness (S_1) as indicated by the position of the bubble in a curved bubble tube at the end of the first swing. A curved tube with an air bubble and a scale graduated from 0 to 100 are mounted on top of the original Herbert hardness instrument. Scale hardness refers to the extent of tester tilting at the end of the first swing.
(2) Time hardness (T) based on the time (in seconds) needed to make ten single swings of the pendulum on any substance as measured using a stopwatch.
(3) Flow hardness (S_1/T) defined as the scale hardness/time hardness ratio, which is an indicator of resistance to flow.

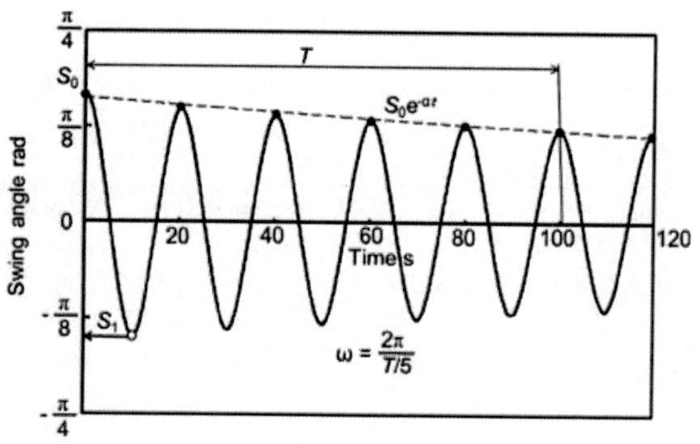

Fig. 1 Freely damped vibration waveform observed during Herbert hardness testing

Flow hardness could be used to predict the work-hardening properties of materials. However, this test method did not become popular due to the time and effort involved compared with other techniques. In the present study, the Herbert hardness tester modified by us freely oscillates with damping on a test specimen as shown in Fig. 1, although the original tester might not do so. The present authors have continued to work towards making improvements in hardness measurement methods based on freely damped oscillations. These improvements included the use of rotary potentiometers, laser displacement meters, and image processing. Such techniques allow us to observe freely damped vibrations of the tester on a test specimen.

Figure 1 shows an observed freely damped vibration waveform. The envelope line connecting successive amplitudes S(t) is given by

$$S(t) = S_0 e^{-\alpha t}, \tag{1}$$

where S_0 is the initial amplitude and α is a damping factor that indicates the amount of damping in the testing system. As α increases, the swinging motion diminishes more quickly. Thus, α is a newly defined hardness called the damping hardness, because it can be considered to represent hardness [3].

This chapter covers the following topics:

(1) A new hardness measure called the natural angular frequency/damping factor; and
(2) Development of a new Herbert hardness tester incorporating a wireless acceleration sensor.

2 Herbert Hardness Tester

The Herbert hardness tester is made of SUS304 austenitic steel and weighs 7.5 kg. The tester is provided at the center on the underside with an incomplete cylindrical shaped groove holding a cylindrical indenter made of carbide, 2 mm in diameter and 12 mm in length.

The cylindrical indenter contacting the test specimen is used for preventing three-dimensional pendulum movement, rather than having a ball- or cone-type form. The three weights are made of brass. Two side weights are operated by means of leveling screws to adjust the tester's center of gravity to coincide with the center of the cylindrical indenter. Thus, the weights can remain in neutral equilibrium. Vertically above the indenter is a screw-operated top weight. By moving the weight vertically, the center of gravity of the entire tester can be raised or lowered above the center of the cylindrical indenter. For hardness testing, the vertical position of the center of gravity is adjusted so that a single time period of device oscillation is 20 s on an artificial sapphire plate. Before conducting a test, the surfaces of the test specimen and the indenter are cleaned using acetone. The device is then tilted horizontally up to 30°. A solenoid switch maintains this initial tilt angle. Then the switch is released by

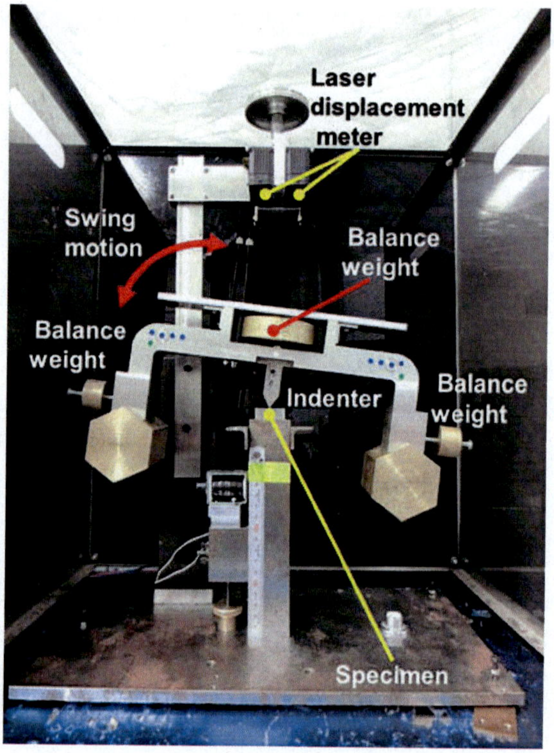

Fig. 2 Herbert hardness tester using laser displacement meters

retracting the plunger relative to the main body of the solenoid. The device then starts to freely undergo damped vibrations on the test specimen. Two laser displacement meters are set over the Herbert hardness tester to enable electrical measurement of the swing motion, as shown in Fig. 2.

Figure 3 shows an overview of the tester's measurement system. An electric power supply applies a voltage of 5 V to the laser displacement meters. The measurement range and resolution are 50 mm and 0.004 mm, respectively. The repeat accuracy is 0.004 mm. A personal computer (PC) stores the voltage change data from the laser displacement meter via a measurement unit. The computer indicates the swing angle as a change in voltage.

Figure 4 shows a schematic view of the swing angle measurement principle. The swing angle θ can be expressed as follows:

$$\theta = \tan^{-1}\frac{L_2 - L_1}{d}, \tag{2}$$

where d is the horizontal distance between the laser displacement meters, and L_1 and L_2 are the vertical distances between the tester and the laser displacement meters.

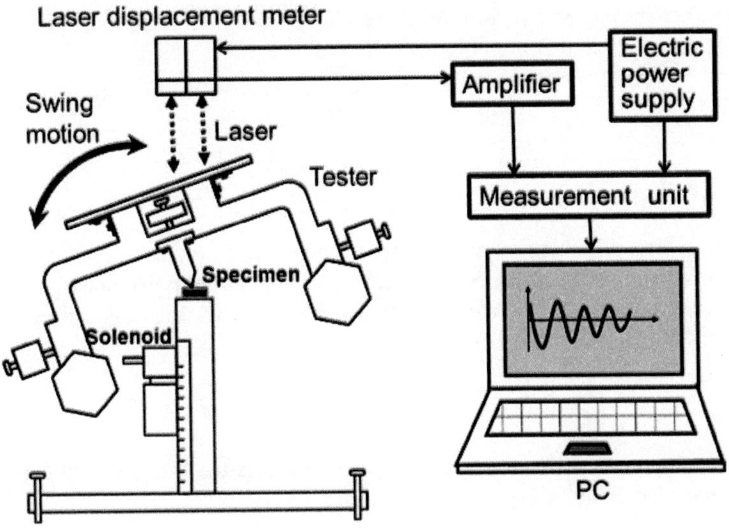

Fig. 3 Herbert hardness testing system

Fig. 4 Swing angle measurement principle

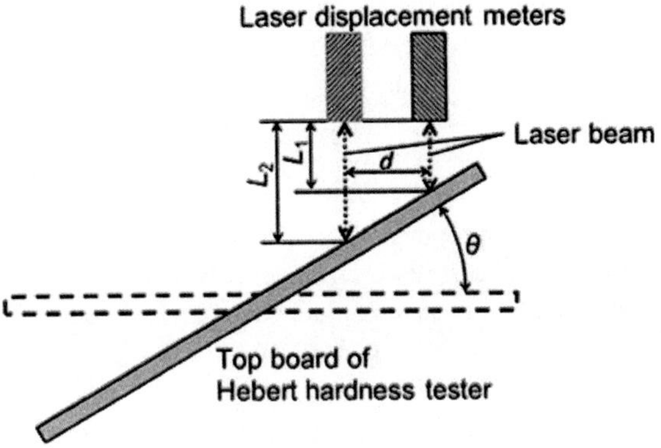

The angle θ can be precisely measured, and the measurement parameters d, L_1 and L_2 do not depend on the method of installation.

The test specimens used are standard test blocks for Brinell hardness (HBW) from 100 to 600 in steps of 50, made by Yamamoto Scientific Tool Laboratory Co. Ltd.

3 Experimental Results and Discussion

The past study on the original Herbert hardness testing clarified the following:

(1) Herbert proposed two formulae expressing the relationship between the time hardness (*T*) and Brinell hardness (*HBW*) [3].

$$HBW = 0.3T^2 \quad (3)$$

$$HBW = 10T(HBW > 333) \quad (4)$$

Therefore, *HBW* is proportional to the square of *T*.
For substances having a *HBW* greater than 333, *HBW* is 10 times *T*.

(2) Benedicks et al. proposed the following formula for the connection between *HBW* and *T* [4].

$$T = 1.7\sqrt{HBW} + 4.7 \times 10^{-5} HBW^2 \quad (5)$$

(3) The change from *T* to S_1 not only altered the relative hardness, but high speed steels also altered the sequence of hardness [2].

T values obtained from the original Herbert hardness test varied from about 10 to 70 with *HBW* ranging from about 50 to 700 [4].

Figure 5 shows the relationship between *T* and *HBW* obtained by the present Herbert hardness testing. The difference between the maximum and minimum of *T* is about 1.2 for almost the same *HBW* range as the results from the original Herbert hardness test. The following equation expresses the relationship.

$$T = 0.127\sqrt{HBW} + 98.0 \quad (6)$$

Here, *T* is proportional to the square root of *HBW*. The result is consistent with the results of past research as described in (2). Figure 6 plots *T* and S_1 in the order of harder *T*. S_1 is out of sequence for several items. These results are consistent with those of past research as described in (3).

Herbert defined the following [3]:

Fig. 5 Plot of time hardness versus Brinell hardness

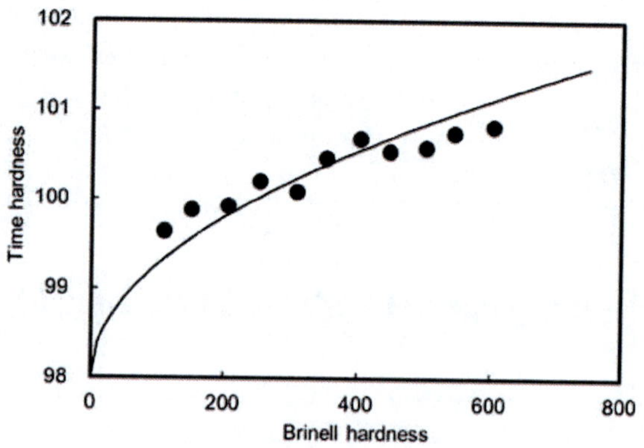

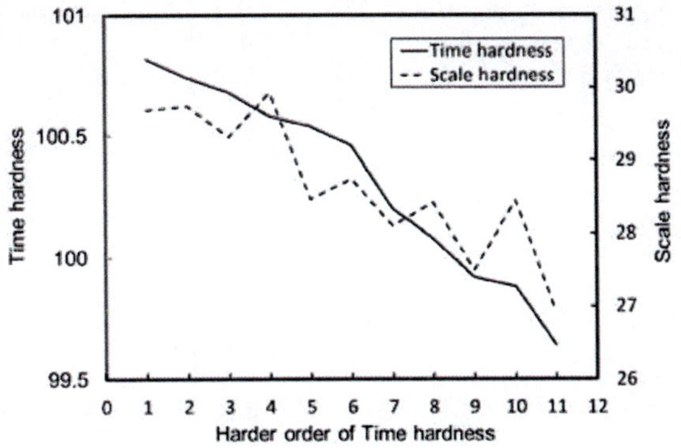

Fig. 6 Time hardness and scale hardness in the order of increasing time hardness

(1) S_1 means resistance to working, or work hardness.
(2) T means resistance to penetration, or indentation hardness.
(3) S_1/T means resistance to flow.

The three types of hardness above apparently do not express the swing motion of the Herbert tester on a test specimen, and their physical meaning is unknown. The present tester undergoes freely damped vibrations on a test specimen during hardness measurement as shown Fig. 1. The angle amplitude $S(t)$ of the vibrations decreases exponentially with time as expressed by Eq. (1). One of the authors defined α as the damping hardness because α is considered a physical quantity related to hardness [5]. α correlates with HBW as is evident from Fig. 7, and is considered to mean resistance to working, or work hardness. The relationship is a negative correlation, which means the harder the material, the smaller the value of α. A harder material would result in poorer accuracy for α as suggested in Fig. 7.

The angle $\theta(t)$ of free damping vibration is expressed by

$$\theta(t) = S_0 e^{-\alpha t} \cos\omega t \tag{7}$$

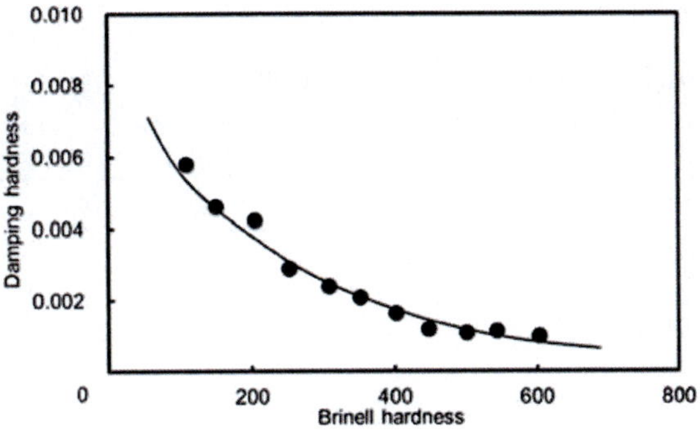

Fig. 7 Plot of damping hardness versus Brinell hardness

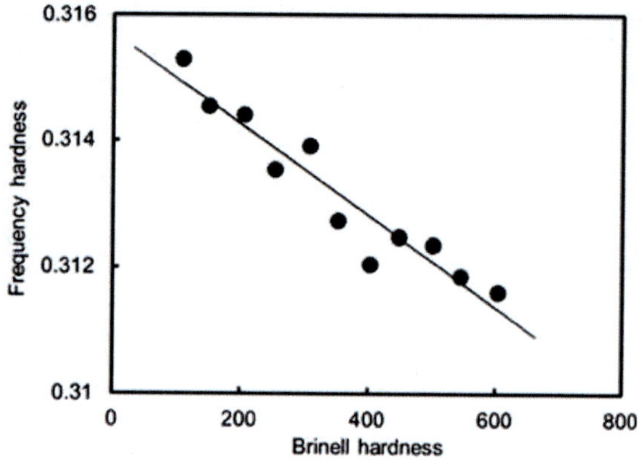

Fig. 8 Plot of frequency hardness versus Brinell hardness

where ω is the natural angular frequency. Not only α but also ω is related to hardness shown in Fig. 8. Although ω should be called the "Frequency hardness", it decreases with increasing *HBW* as in the case of α. However, the values of α and ω differ significantly from the typical *HBW* hardness values.

The ratio ω/α is considered a measure of hardness. Figure 9 shows the relationship between ω/α and HBW. ω/α increases with increasing HBW and has a stronger tie to HBW than ω. Because the numeric value of ω/α is also the about the same as the typical hardness value, ω/α is apparently a more appropriate hardness indicator than α or ω. The authors newly define ω/α as the "Matsubara hardness". It includes work hardness, indentation hardness, and resistance to flow.

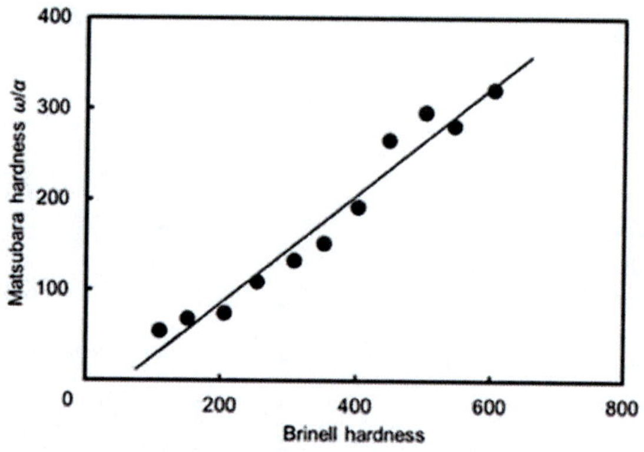

Fig. 9 A plot of Matsubara hardness versus Brinell hardness

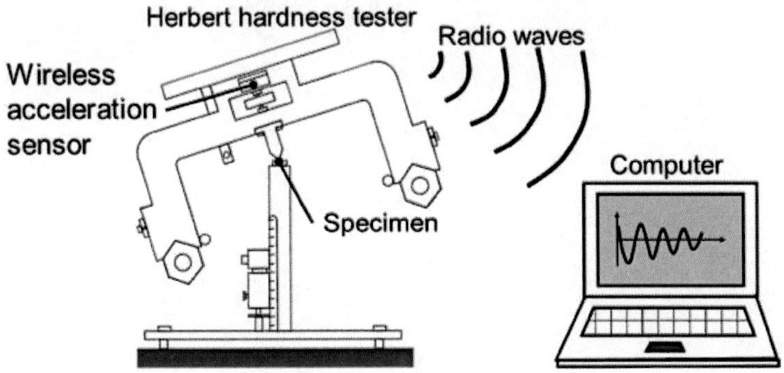

Fig. 10 Herbert hardness testing system using a wireless acceleration sensor

4 Further Improvement to the Herbert Hardness Tester

As an attempt for further improvement, a wireless acceleration sensor is incorporated in the Herbert hardness tester (see Fig. 10). That provides the device with excellent portability. Just bringing the tester and a computer would enable on-site measurement.

5 Conclusions

In conclusion, the present study has demonstrated the following results for popular use of the Herbert hardness tester.

(1) A proposal for a new measure of hardness defined as the "Matsubara hardness" given by the natural angular frequency/damping factor; and
(2) Development of a new Herbert hardness tester incorporating a wireless acceleration sensor.

References

1. BS EN ISO 1522:2006 (2007) Paints and varnishes—Pendulum damping test
2. Anonymous (1923) The "pendulum" hardness tester. The Engineer, April 13, 390–391
3. Herbert EG (1923) Some recent developments in hardness testing. The Engineer, June 29, 686–687
4. Benedicks C, Christiansen V (1924) Investigations on the Herbert pendulum hardness tester. J Iron Steel Inst 110:219–248
5. Matsubara M, Sakamoto K (2012) Improved Herbert hardness tester. Exp Tech 36(3):73–76

Microstructure and Abrasive Wear of Particle-Filled Composites

Yusuf Şahin and H. Şahin

Abstract This paper describes the influence of SiC and Al_2O_3 particles on the wear behaviour of epoxy including PTFE particles. Particle-reinforced epoxy composites (PRCs) covering 5 wt%SiC, 10 wt%SiC and 10 wt%Al_2O_3 particles plus 5 wt%PTFE were fabricated successfully by mixing technique. The tribological property of PRCs at dry sliding was studied against SiC abrasives on a conventional pin-on-disc machine using Box-Behnken method at different factors. Microstructures were examined through Scanning Electron Microscope (SEM). In addition, second order regression model of weight loss was derived from the main parameters and their interactions. SiC particles indicated a better performance than that of Al_2O_3 particles due to enough bonding between the fibers and the resin. Furthermore, analysis of variance showed that linear contributions were about 47.60% (load (28.36%) had a considerable effect on the mass loss, followed by speed (11.09%) and material types (8.15%) for PRCs, respectively. However, square effect was about 42.0% while two-way interactions had no effect on the mass loss.

Keywords Epoxy · SiC/al_2o_3/PTFE composite · Microstructure · Load · Speed · Wear · Response surface methodology

1 Introduction

Composite materials consist of matrix as a base material and some types of reinforcement as a second phase. A large variety of fibers are available as a reinforcement for the composites because most fibers have a higher strength/stiffness and lower density. The reinforcement can be shaped in the form of fibre, particle, whiskers or

Y. Şahin (✉)
Faculty of Engineering and Architecture, Nişantaşı University, Nişantaşı, İstanbul,, Turkey
e-mail: yusuf.sahin@nisantasi.edu.tr

H. Şahin
Devlet Malzeme Ofisi Genel Müdürlüğü İsmet İnönü Bulvarı, No. 18 Yücetepe Çankaya, Ankara, Turkey
e-mail: mechenghakan@gmail.com

M. Abdel Wahab (ed.), *Proceedings of the 8th International Conference on Fracture, Fatigue and Wear*, Lecture Notes in Mechanical Engineering,
https://doi.org/10.1007/978-981-15-9893-7_46

woven type. Epoxy is the most commonly preferable polymer for the composite due high resistance to degradation over and lower shrinkage. However, there are some problems in practical applications as bearing materials for pure forms of polymers because of its lower mechanical properties [1]. Therefore, polymer based composite materials have found an increasing usage in many engineering fields ranging from airspace structures to automotive, and marine applications owing to their strength, modulus, lightness, low density, easy of processing, availability of wider choice of systems and stabile behaviour at high temperature [2, 3]. Frictional properties of polymer composites including different particles were investigated at dry wear. The fillers improved the tribological resistance of polyimide [4] and epoxy [2, 5–8], PA6/PEEK [9, 10], polyetherimide/phenolic/vinyl ester [1, 11, 12]. Subbaya et al. [13] studied the wear behaviour of carbon filled epoxy (C-E) including particle types of graphite at abrasives. The lowest wear was observed for the composite when introduced 10 wt% graphite. Rajashekaraiah et al. [14] investigated various types of fillers like short glass, carbon fibers and PTFE, SiC, and Al_2O_3 particulates for thermoplastic matrix. The results of abrasive wear showed that thermoplastic-filled PTFE composite exhibited a better abrasion resistance.

RSM is powerful tool, commonly used to provide improvements for product quality. Mishra [15] studied the RSM model for predicting the wear behaviour of bagasses-fiber reinforced polymers at abrasive sliding. Effects of volume fraction, load and velocity on the wear property under abrasive were analyzed, and the accuracy of RSM was verified with some data. Şahin et al. [16] investigated the tribology of carbon-reinforced composites including nano scale powders. RSM was applied to determine effects of various factors on the mass loss under abrasive sliding. The load was significant than the others [17]. Shipway and Ngao [18] reported that the types of polymer is critically important for the abrasive wear behaviour of polymers. Liu et al. [19] adopted the statistical approach to study the influence of some variables like distance, load/speed on the wear behaviour of polyamide and UHMWPE matrix. Chang et al. [20] indicated that load, followed by distance and speed were most effective factors on the wear behaviour of UHMWPE composites, respectively. A tribological model for Titanium alloy was successfully developed under two different (dry and cryogenic) environments by RSM approach in terms of load, speed and temperature [21, 22]. There has been very limited works conducted at the tribological property of polymer composites using RSM [7, 20–22] and no study reported on tribology of composites including nano PTFE particle using RSM. Therefore, the effects of main factors and material hardness on the wear of PRCs covering various weight fraction of SiC and Al_2O_3 particles were investigated under abrasive conditions.

Table 1 The box-Behnken design for the weight loss of composites

Main variables	Code	Levels of factors		
		−1	0	+1
Applied load (N)	L	5	10	15
Rota.speed (rpm)	S	80	140	200
Mater. types/hardness (SiC/Al_2O_3)	M	10wt%SiC + 5wt% PTFE + PE	5wt%SiC + 5wt% PTFE + PE	+10wt%Al_2O_3 + 5wt% PTFE + PE
Rockwell B hardness (HRB)	HB	97	88	91

2 Experimental Procedure

2.1 *Materials and Sample Preparation*

In manufacturing of composites, SiC/Al_2O_3/PTFE particles were used as reinforcements while epoxy resin (bisphenol-A epoxy) was used as a matrix in this work, and the resulting epoxy resin was coded as EP. Its tensile modulus and tensile strength were about 3.6 GPa and 48 MPa, respectively. The filler sizes of SiC, Al_2O_3 and PTFE were about 6 μm, 6 μm and 40–60 nm, respectively. The weight fraction of PTFE was about 5wt%. It was stirred with a mixer at a time of 45 min to distribute the particles of ceramics and PTFE particles evenly in the epoxy. The catalyzed resin mixtures were molded to form composite plate (Ø60 ×6.0 mm). Details of fabrication of composites and formulations were shown in Table 1. SiC, SiC, Al_2O_3 particles were selected as fillers in range of 5, 10, and 10 wt%, respectively.

2.2 *Microstructure*

Microstructures of the epoxy composites were studied through SEM. Cross-sections of the specimens were mounted in the resin and polished to 6 μm with diamond paste. In order to achieve a good electrical conductivity, all the samples were sputtered by gold using ion sputtering device before subjecting them to SEM.

2.3 *Plan of Experiments*

Experimental testing was performed to determine the influence of working factors for the tribological property of PRCs. The parameters selected were load (L), rotational speed (S) and hardness (H). The selected working factors were shown in Table 2. It

Table 2 The hardness of PRCs

Sample Code	Matrix content (wt%)	Reinforcement content (wt%)	PTFE content (wt%)	Rockwell B hardness (HRB)
EP	95	–	5	41
5SiC	90	SiC (5)	5	88
10SiC	85	SiC (10)	5	97
$10Al_2O_3$	85	Al_2O_3 (10)	5	91

is exhibited three types of the materials including PTFE/SiC/Al_2O_3 reinforcements. RSM was preferred to develop an optimum level of the independent variables using Box-Behnken Design (BBD). In this study, 15 experiments were made to construct the model for the tribological property of the epoxy composites. The design and independent variables were shown in Table 2. The quadratic mathematical equation of the weight loss of the composites with independent control factors and their interactions on response was formed through RSM. A quadratic equation can be modelled with the least square method in the following.

$$Y = b_o + \sum_{i=1}^{3} b_i X_i + \sum_{i=1}^{3} b_{ii} X_i^2 + \sum_{i=1}^{2} \sum_{j>1}^{3} b_{ij} X_i X_j + \varepsilon \tag{1}$$

where Y = the corresponding response variable, Xi and Xj = the dimensionless coded control parameters, b = a constant or coefficient of regression, bi = the linear coefficient, bii = the quadratic coefficient, bij = the interaction coefficient and the residual (ε) measures the experimental errors of observations.

2.4 *Wear Test*

For evaluating the tribological property of epoxy polymeric composites, a pin/on/disc type wear machine was employed for these experiments. Specimen's size was a square shape (5 mm × 5 mm), made from its epoxy composites. The samples were forced with SiC emery paper on the hardened disc by means of a cantilever mechanism. Tests were lasted for total time of 1 min. Specimens were loaded under 5, 10 and 15 N with three different spindle speeds of 80, 140 and 200 rpm. After finishing test, worn pin was cleaned in acetone, then dried. Finally, samples were weighed using a sensitive microbalance. Each test was repeated at least three.

3 Results and Discussion

3.1 *Microstructures*

Figure 1 a indicates the microstructure of a 5wt% SiC particles in the epoxy resin in addition to 5wt% PTFE at lower magnification. It is observed a distribution of some particles in the resin while numbers of SiC/PTFE particles were agglomerated under higher magnification (Fig. 1b), leading to lower hardness. Aggregations were obvious for this microstructure because the surface free energy increased with decreasing the nano-particle sizes.

Figure 1 c shows the microstructure of a 10 wt% SiC particle-reinforced epoxy composite at lower magnification, but containing 5wt% PTFE as well, which was about 40–100 nm particle size, but SiC particles was about 5–8 μm. It did not include any porosities in the epoxy matrix. Higher magnification of SEM image in Fig. 1 d indicates a formation of well bonding interface between the particles and the epoxy resin. In other words, there was no debonding between the particles and the matrix at this higher magnification due to the mixing ratio of epoxy resin/hardener and their compatibility. Figure 1 e indicates the microstructure of a 10wt% Al_2O_3 particle-reinforced epoxy composites. This lower magnification image exhibited the particle's distributions in resin matrix. The distributions of them seemed to be normal with some observations of more porosities in the epoxy, which were more obvious in Fig. 1 f under a higher magnification that led to lower hardness. Its particle sizes were observed to be about 5–8 μm at this micrograph.

3.2 *Hardness*

The hardness of the epoxy/composites was exhibited in Table 2. The epoxy's hardness itself was measured as 41 HRB. The hardness of the resin increased considerably with the introduction of three different particles. The adding 5 and 10 wt% SiC into the epoxy resulted in 114 and 137% increases in the hardness values, respectively. This might be due to the achievement of enough bonding between the resin matrix and the SiC particles. For 10wt% Al_2O_3 particles, however, increase in the hardness was about 122% in compared to the epoxy due to some porosities in the structure. In other words, the hardness's ranged from 88 HRB, 91 HRB to 97 HRB for 5 wt% SiC, 10wt% Al_2O_3 and 10 wt% SiC particle-reinforced polymer composites, respectively.

3.3 *Wear*

The experimental plan and abrasive wear results of the composites under various variables and surface plots were indicated in Fig. 2. The tests were performed at

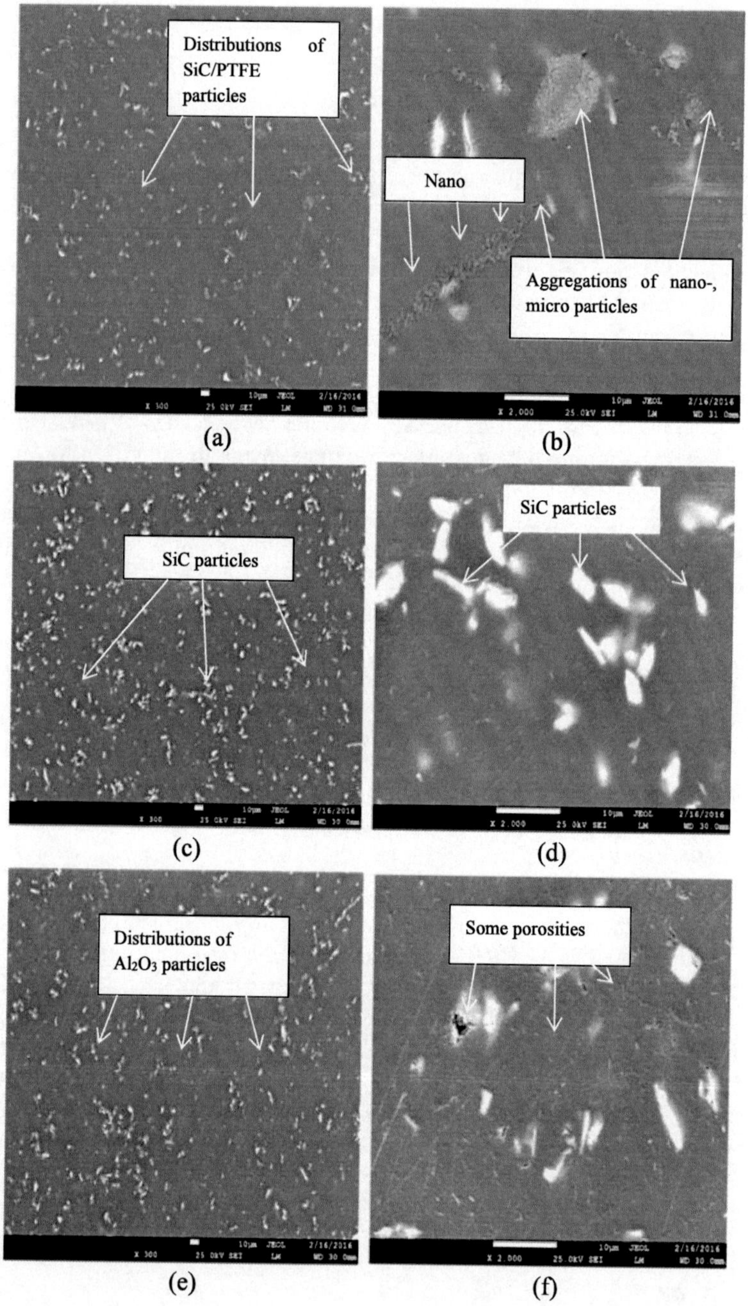

Fig. 1 Typical microstructures of particle-filled epoxy composites (cross-section). **a** 5wt% SiC/PTFE particles, **b** Higher magnification of Fig. 1a, indicating agglomerations of nano PTFE particles and a few SiC particles, **c** 10 wt% SiC/PTFE particles, **d** Higher magnification of Fig. 1c, showing enough interface bonding, **e** 10wt% Al_2O_3/PTFE particle, **f** Higher magnification of Fig. 1e, exhibiting some porosities in the epoxy resin

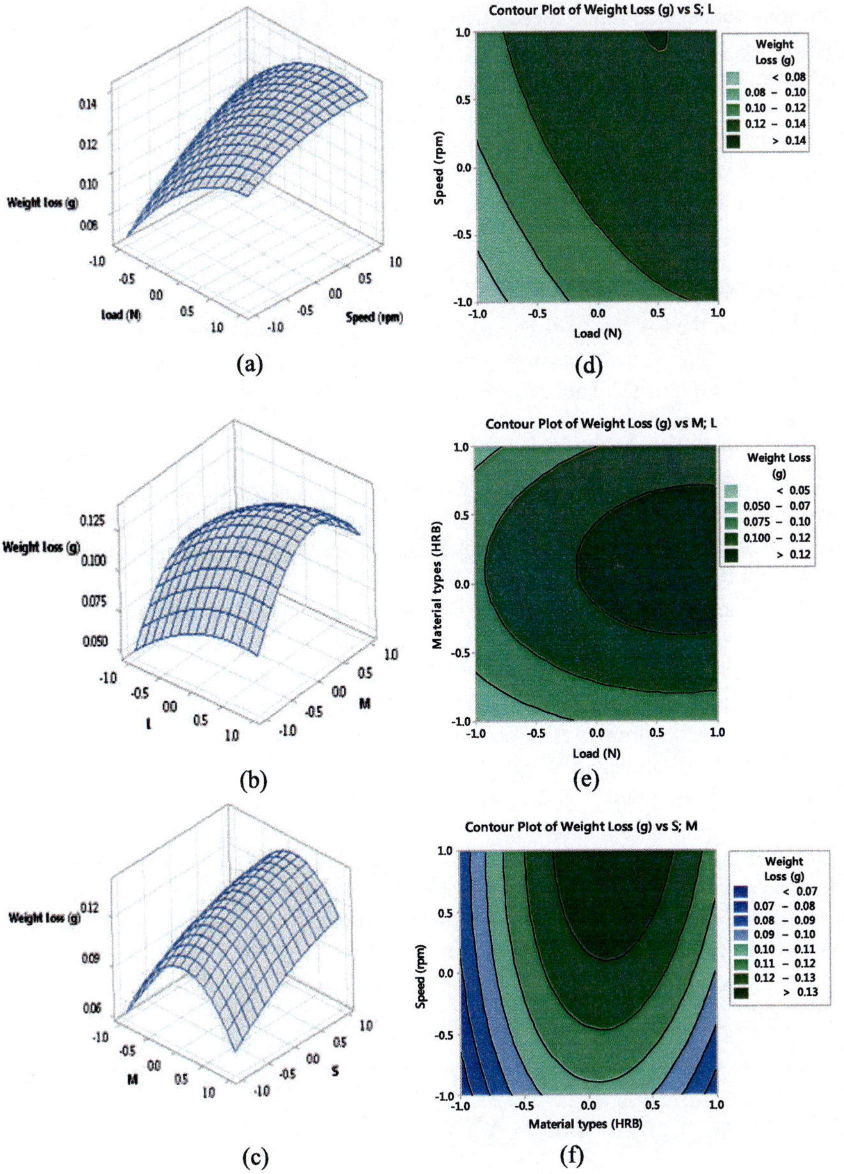

Fig. 2 3D surface plots of the mass loss of PRCs. **a** Weight loss versus Load and Speed, **b** Weight loss versus Load&Material hardness, **c** Weight loss versus Hardness and Speed. Contour plots of the mass loss of composites. **d** Weight loss versus L&S, **e** Weight loss versus M&L, **f** Weight loss versus S&M

different loads, rotational speeds using three types of polymeric composites. The 10wt%SiCp-reinforced epoxy composite revealed the lowest wear than those of others due to getting the achievement of good interface bonding between the particles/matrix using these particles (Fig. 1 d). In other words, the wear resistance was better for experimental run 15 owing to having the first-level load (5 N), the medium speed (140 rpm), and hardness (97 HRB). Figure 2 a showed WL versus L&S while Fig. 2 b indicated WL versus L&M, and Fig. 2 c exhibited WL versus M&S. The weight loss increased with load/speed for tested materials, but an increase in the mass loss was more affected by the load than by the speed due to depth of deformation ability on the tested samples (Fig. 2 a).

However, the weight loss increased for EP/Al_2O_3 reinforced composites because of including some porosities in the resin while decreased with changing to EP/SiCp reinforced composite. The deforming and fracturing the asperities of the softer surface increased with speed and load since the penetration ability of abrasives (SiC) fixed on disc increased when pin surface was softer. For main control variables, factor L indicated the largest influence on the wear resistance (Fig. 2a). It is followed with factor S and factor M (type of material/hardness), respectively (Fig. 2 b and c). Similar results were concluded on the tribological behaviour of polymeric matrix composites [16, 17, 19, 20, 22].

It is contradictory to other people studies conducting at composites. The sliding distance was more effective than those of other factors [2, 6, 20]. However, weight fractions/grit sizes were found to be the most significant factor on the wear behaviour of epoxy under abrasives [7, 9, 13]. The effects of some parameters on the abrasive wear of composite samples as contour plots were exhibited in Figs. 2 d, e and f, respectively. Figure 2 d indicated Weight Loss versus L&S while Fig. 2 e showed Weight Loss versus M&L, and Fig. 2 f exhibited Weight Loss versus S&M, respectively. This graph indicated that material's behaviour acted as a concave type because of using PTFE plus EP matrix when combined with SiC particles. Therefore, the epoxy reinforced with higher weight fraction of SiC showed the less weight loss than that of the others. The improvements in the tribological behaviour might be because of the microstructure, mechanical strength of composites. The lowest weight loss of epoxy composite was obtained from 5 N load, 140 rpm speed and 60 m sliding distance test conditions.

3.4 Analysis of Variance (ANOVA)

Results of ANOVA for tribological behaviour of composites at different conditions were tabulated in Table 3. Confidence level is 95% for this analysis. This table presented P values and indicated a considerable effect of each parameters for this model. It means that the smaller value of P signified that the regression coefficient is highly significant. The probability value was 0.011, 0.056 and 0.086 for L, S and M, respectively. Among these independent variables, factor L showed a considerable

Table 3 ANOVA for the mass loss of PRCs

Source	DF	Adj SS	Adj MS	F-Value	P^c (Contr.%)	P-Value
Model	9	0.011411	0.001268	5.64	91.288	0.036
Linear	3	0.005950	0.001983	8.83	47.6	0.019
L	1	0.003545	0.003545	15.78	28.36	0.011
S	1	0.001386	0.001386	6.17	11.088	0.056
M	1	0.001019	0.001019	4.54	8.152	0.086
Square	3	0.005250	0.001750	7.79	42	0.025
L*L	1	0.000423	0.000423	1.88	3.37	0.228
S*S	1	0.000097	0.000097	0.43	0.77	0.540
M*M	1	0.005008	0.005008	22.30	39.90	0.005
2-Way Interac.	3	0.000211	0.000070	0.31	1.675	0.816
L*S	1	0.000064	0.000064	0.28	0.512	0.616
L*M	1	0.000056	0.000056	0.25	0.448	0.638
S*M	1	0.000091	0.000091	0.41	0.728	0.552
Error	5	0.001123	0.000225		8.984	
Lack-of-Fit	3	0.001010	0.000337	5.98		0.147
Pure Error	2	0.000113	0.000056			
Total	14	0.012534				

DF: degrees of freedom, SS: sequential sum of squares, Adj SS: adjusted sum of squares, Adj MS: adjusted mean squares

influence on wear because P-value was lower than 0.05, while other two factors had an insignificant influence on the response.

Moreover, the quadratic M revealed the most significant effect (0.005), followed by the load (0.011) and finally the speed (0.056) while two-way interactions, the quadratic L and S were the less significant, respectively. Furthermore, P-values of the lack of fit for weight loss was about 0.147, which was not significant.

3.5 *Multiple Regression Model*

Multiple regression model was obtained through statistical software "Minitab16" to find out the correlation between material behaviour and tribological process parameters. The regression equation for the weight loss (WL) was shown as follows:

$$
\begin{aligned}
Weight\ Loss\ (WL) &= 0.12433 + 0.02105L + 0.01316S + 0.01129M \\
&- 0.01070L * L - 0.00513S * S - 0.03683M * M \\
&- 0.00400L * S + 0.00375L * M + 0.00478S * M \qquad (2)
\end{aligned}
$$

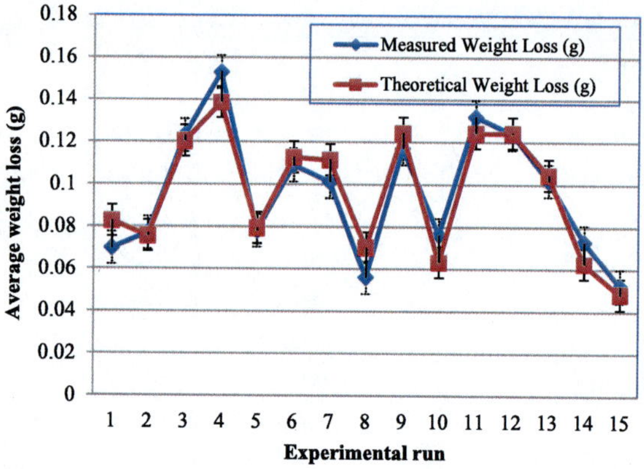

Fig. 3 Comparison of measured and theoretical prediction for mass loss of PRCs

where WL = weight loss (g/cm^3), L = load (N), S = speed (rpm), and M = material hardness (HRB). This WL was obtained in terms of un-coded units. The WL of the composites could be calculated with substituting the recorded values of the variables for Eq. (2). This equation shows that the coefficients of load, speed and material's hardness were positive. Load was the main variable on the mass loss, followed by speed while hardness was not effective. Predicted weight losses were listed in Fig. 3 for measured and predicted values of variables. The predicted values were found to be close to experimentally measured value. The lowest weight loss of 0.052 g was obtained for the trial 15 because of using L1, S1 and H1. The average error was about 8.3%, which were included with error bars because more standard deviations were observed for trials 8 and 10. It was considered that less than the 10% error was reasonable. According to the prediction of Eq. (2), its WL was about 0.0482, which was the lower value in the obtained experimental result of 0.052 and 0.050.

This table showed the differences between measured and predicted were found in appropriate value. It was ranged from 3.66 to 7.81%, which confirmed the previous results.

4 Conclusions

Application of response surface methodology on the wear behaviour of PRCs was presented.

1- Epoxy composites containing 5 wt% SiC, 10 wt% SiC and 10wt% Al_2O_3 particles that all associated with 5 wt% PTFE nano-particles were manufactured successfully by polymer mixing method.
2- The microstructures exhibited that the particle's distributions in the epoxy resin were uniform. Hardness of the SiC particle-reinforced composite was slightly larger than that of Al_2O_3 particle-reinforced composite due to not having some porosities.

3- A better wear resistance was found for the SiC particles in the epoxy than that of the others because enough interface bonding between the particles and resin was achieved.
4- It was indicated that the tribological property was highly influenced with the load, followed with the speed among other control factors, but the hardness had no effect on the abrasive wear. Their linear contributions were 47.60% (load at 28.36%, speed at 11.09% and hardness at 8.15%, respectively) while square was 42.0%.
5- Furthermore, second order model was developed using the testing parameters and interaction effects and the maximum error was about 7.81%.

References

1. Harsha AP, Tewari US (2003) Two-body and three-body abrasive wear behaviour of polyaryletherketone composites. Polym Test 22:403–418
2. Suresha B, Ramesh BN, Subbaya KM, Kumar BNR, Chandramohan G (2010) Influence of graphite filler on two body abrasive wear behaviour of carbon fabric reinforced epoxy composites. Mater Des 31:1833–1843
3. Zhang X, Pei X, Wang Q (2009) Friction and wear properties of polyamide matrix composites reinforced with short basalt fibers. J Appl Polym Sci 111:2980–2985
4. Zhang X, Pei X, Wang Q (2009) Friction and wear properties of basalt fibre reinforced/solid lubricants filled polyimide composites under different sliding conditions. J Appl Polym Sci 114:1746–1752
5. Basawarajappa S, Joshi AGK, Arun V, Kumar AP, Kumar PM (2010) Three-body abrasive wear behaviour of polymer matrix composites filled with SiC particles. Polymer-Plast Technol Eng 49:8–12
6. Chairman CA, Kumaresh SP (2012) Three-body abrasive wear behavior of basalt and glass fabric reinforced epoxy composites. Appl Mech Maters 121–126:534–538
7. Şahin Y (2015) Analysis of abrasive wear behaviour of PTFE composite using Taguchi's technique. Cogent Eng 2:1000510
8. Patnaik A, Satapathy A, Biswas S (2010) Investigations on three-body abrasive wear and mechanical properties of particulate filled glass epoxy composites. Malays Polym J 5:37–48
9. Zsidai L, Kátai L (2016) Abrasive wear and abrasion testing of PA 6 and PEEK composites in small-scale model system. Acta Polytech Hung 13(6)
10. Friedrich K, Schlarb AK, Briscoe BJ (ed) (2008) Tribology of polymeric nanocomposites. Tribol Inter Eng Ser 55
11. Bijwe J, Indumathi J, Ghosh AK (2002) On the abrasive wear behaviour of fabric-reinforced polyetherimide composites. Wear 253:768–777
12. Mishra P, Acharya SK (2010) Anisotropy abrasive wear behaviour of bagasse fiber reinforced polymer composite. Int J Eng Sci Technol 2:104–112
13. Subbaya KM, Rajendra N, Varadarajan YS, Suresha B (2012) Multiple response optimization of three-body abrasive wear behaviour of graphite filled carbon-epoxy composites using Grey-based Taguchi approach. J Miner Mater Character Eng 11:876–884
14. Rajashekaraiah H, Mohan S, Kumari Pallathadka P, Suresha B (2014) Dynamic mechanical analysis and three-body abrasive wear behaviour of thermoplastic copolyester elastomer composites. Adv Tribol 2014, Article ID 210187
15. Mishra P (2012) Statistical analysis for the abrasive wear behaviour of bagasse fiber reinforced polymer composite. Int J Appl Res Mech Eng (IJARME) ISSN, 2:2231–5950

16. Şahin Y, Erdil E, Şahin H (2017) The tribological wear behavior of carbon fabric-reinforced epoxy composites. In: Proceedings of the international symposium of mechanism and machine science, 2017 AzCIFToMM-Azerbaijan Technical University 11–14 September 2017, Baku, Azerbaijan
17. Chang BP, Nasir AHM, Khan RBA (2015) Optimization on wear performance of UHMWPE composites using response surface methodology. Tribol Int 88:252–262
18. Shipway PH, Ngao NK (2003) Microscale abrasive wear of polymeric materials. Wear 255:742–750
19. Liu C, Ren L, Arnell RD, Tong J (1999) Abrasive wear behaviour of particle reinforced ultrahigh molecular weight polyethylene composites. Wear 225:199–204
20. Chang BP, Akil HM, Affendy MG, Khan A (2014) Comparative study of wear performance of particulate and fiber reinforced nano-ZnO/Ultra-high molecular weight polyethylene hybrid composites using response surface methodology. Mater Des 63:805–819
21. El-Tayeb NSM, Yap TC, Brevern PV (2013) Wear characteristics of titanium alloy Ti54 for cryogenic sliding applications. Tribol Int 43:2345–2354
22. El-Tayeb NSM, Yap TC, Venkatesh VC, Brevern PV (2009) Modelling of crygenic frictional behaviour of titanium alloys using response surface methodology approach. Mater Des 30:4032–4034

Experimental Characterization of Vibration on Gearing Mechanism Using Taguchi Approach

H. I. Mirzayev, Yusuf Şahin, A. Z. Khalilov, and R. S. Valili

Abstract Experimental characterization of the vibration analysis were investigated in a spur gearbox using Taguchi method (TM). The influences of different parameters such as rotational speed, torque, and center distance deviation on the frequency and acceleration of the tested samples were investigated. It is shown that frequency for normal gear changed from approximately 592–1184 Hz, corresponding values of accelerations were about 0.093–0.143 m/s^2, respectively. In the case of failure gear, however, frequency changed from 592 to 740 Hz, corresponding values were about 0.287–0.53 m/s^2, respectively. The experimental design results exhibited that center distance deviation was the effective factor for both frequency and acceleration, followed by speed and torque, respectively. Optimized process condition was first level of speed (600 rpm), first level of torque (1000 N.mm), and second level of center distance (0.0 mm).

Keywords Gearbox · Vibration · Gear failure · Design of experiment · Taguchi model · Frequency · Acceleration

---

H. I. Mirzayev · A. Z. Khalilov
Baku Engineering University, Khirdalan, Baku, Azerbaijan
e-mail: humirzeyev@beu.edu.az

A. Z. Khalilov
e-mail: anarxelilov95@gmail.com

Y. Şahin (✉)
Faculty of Engineering and Architecture, Nisantasi University, Nisantası, İstanbul, Turkey
e-mail: yusuf.sahin@nisantasi.edu.tr

R. S. Valili
Polytechnic University of Milan, Milan, Italy
e-mail: rashadvalili@hotmail.com

M. Abdel Wahab (ed.), *Proceedings of the 8th International Conference on Fracture, Fatigue and Wear*, Lecture Notes in Mechanical Engineering,
https://doi.org/10.1007/978-981-15-9893-7_47

1 Introduction

Spur gearboxes widely used in the transmission mechanisms such as automobiles, cranes, conveyors and other machines [1–3]. Recent years a lot of constructions of spur gearbox pocket reducers have been designed for lifting and drilling machines and elevators. For the elevator's hoisting mechanism designed and optimized the spur-bevel pocket gearbox [4]. Decreasing of the vibration in the gear boxes are very important for the elevators, machine cutting tools, automobiles and many types of other machines. A systematic consideration has been taken of the influence of many factors on the vibration signals generated by a system in which a planetary gearbox is included [5]. These considerations give the basis for vibration signal interpretation, development of the means of condition monitoring and for the scenario of the degradation of the planetary gearbox. Real measured vibration signals obtained in the industrial environment are processed. The signals are recorded during normal operation of the diagnosed objects, namely planetary gearboxes, which are a part of the driving system used in a bucket wheel excavator, used in lignite mines. It is found that a planetary gearbox in bad condition is more susceptible to load than a gearbox in good condition.

Each gear damages monitoring technique has its merits and limitations. The other work aim [6] is to investigate experimentally the sensitivity and robustness of the currently well-accepted techniques: phase and amplitude demodulation, beta kurtosis and wavelet transformation. Four gear test cases are used: healthy gears, cracked, filed and chipped gears. The vibration signal is measured on the gearbox housing and processed, online under three filtering conditions: general signal average, overall residual and dominant meshing frequency residual. The test results show that beta kurtosis is a very reliable time-domain diagnostic technique. Phase modulation is very sensitive to gear imperfections, but other information should be used to confirm its diagnostic results. Continuous wavelet transform provides a good visual inspection, especially when residual signals are used. The diagnosis based only on dominant meshing frequency residual, however, should not be used independently for gear health condition monitoring, it may give false alarms.

Vibration occurs due to several reasons such as pure manufacturing, wrong manufacturing technique, a rough surface finish and during mounting/fixing process, not applying proper assembly, clearance and looseness of parts and so on. The vibration always causes to noise, pure power transmission, and also leading to earlier wear. Wear always exists from commence of usage. As the vibration, occurrence of the wear is linked to manufacturing and mounting issues. Both wear and vibration are getting intensive and decrease the machine service duration whilst using machines actively [7]. During the design stage, the main focus point is the stress calculation on the transmission parts and optimization of dimensions. However, while manufacturing, dimension errors occur on mechanical parts. Small tolerances are kept for inaccuracy of parts. These tolerances start from 1 μm through CNC machining to a few micrometers. Inaccuracies might be resulted in unsuccessful fittings of the machine parts. In best case, vibrations will occur. This project is related to vibrations in the machine parts, especially in gearboxes. Gear failures might be seen in various circumstances and environments. This can be arisen during several stages such as design, manufacturing and operational process. There are various modes of failures that could be seen in gears such as wear, pitting, scoring, plastic flow and tooth fracture. Taguchi provides a best combination of input parameters with using less experiment. This method investigates the different parameters affecting on the mean and variance of the process performance characteristics. In this method, 9 experiments are enough to get the qualitative information [8].

According to the Taguchi method, the experimental results were transformed into SNR with un-coded or coded values. SNR was used to determine the deviation of the quality characteristics from the desired values. The gearbox that used in the experiments has several features that adjustments of inaccuracies are possible. These are the changing the center distances and exchanging gear with failure one. The experiments are carried out with normal gears and failed/failure gears for the spur gear sets. The purpose of this study is, thus, to investigate the possible consequences of geometrical failures including speed, torque and center distance in gears on the whole gearbox mechanism. Taguchi approach was adopted for analysis of the results.

2 Experimental Details

2.1 Materials/Equipments

Main purpose of doing this experiment was to inspecting how tooth profile may affect the work performance of the gearbox. In the ideal state, none of devices should not be generated a vibration. In reality, vibrations exist due to cyclic forces. These forces may be generated because of accurate dimensions of transmission components. As a result of vibrations, power loss occurs. The experiment was performed in Mechanical Engineering Laboratory of Baku Engineering University. Experiment is about measuring vibrations in parallel axis gears (spur gears). PT500 laboratory unit, from GUNT Hamburg, is shown in Fig. 1, which is used for measuring and investigating the vibrations in the gear boxes. Also PT500 laboratory units provides loading to output power and as a result, simulation of real action of transmissions can be done easily.

For the experiments, 4 type of units are used:

1. PT500e–V1.1. Machinery diagnostic system, base unit.
2. PT500.04e–V1.1. Computerized vibration analyzer.
3. PT500.05e–V1.0. Brake and Load unit.
4. PT500.15–V1.0. Damage to gears kit.

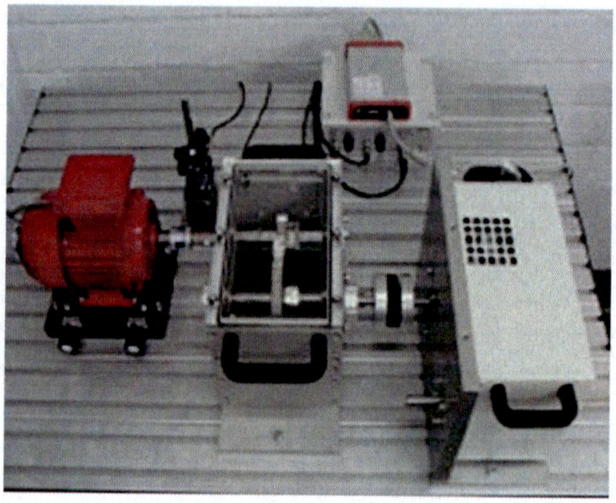

Fig. 1 General view of the experiment device

2.2 Experimental Design

Taguchi methodology (TM) uses a statistical measure of performance to evaluate the optimal parameter settings. In the present work, three input parameters such as motor speed (S), loading torque (L), and center distance between gears (D) on the mean of the frequency and acceleration as control parameters are studied using the TM. The TM requires minimum three levels of each factor for the calculation of the regression coefficients. The total 9 experimental trials are carried out according to the Taguchi orthogonal array design matrix. In this method, input parameters are controllable and multi-factor analysis is made for optimization of vibration control parameters for given responses. In this study, a lower-was-better characteristic was requested because the smallest wear rate of the tested materials was of interest. Experimental design, their codes and levels of control parameters are shown in Table 1.

This table indicates that the experimental plan has three levels. Each combination of experiments is repeated three to acquire more accurate results in the process. Motor speed, load torque and center distance are input parameters. 600, 1200, and 1800 rpm used for motor speed while 1, 2, and 3 N.m used for loading, but center distance is changed using an eccentric bearing flange among 0, −0.45, and 0.45 mm (Table 2). Experiment is done with 2 gear sets: spur gear set with normal gear tooth, spur gear set with gear damaged tooth. Thus, a number of experiments can be increased from 9 to 18. PT500.04e software provides acceleration vs. frequency graphs. Also, acceleration amplitude vs. frequency data can be obtained from this graph.

Table 1 Control factors and their levels (un-coded)

Symbol	Control parameters	Unit	Level 1	Level 2	Level 3
L	Torque	N.mm	1000	2000	3000
S	Motor speed	rpm	600	1200	1800
D	Center distance	mm	−0.45	0	0.45

Table 2 Taguchi experimental variables, their levels and experimental design

Level	Rotational speed (S, rpm)	Torque- (T, N.mm)	Center distance deviation (D, mm)
L1	600	1000	−0.45
L2	600	2000	0
L3	600	3000	0.45
L4	1200	1000	0
L5	1200	2000	0.45
L6	1200	3000	−0.45
L7	1800	1000	0.45
L8	1800	2000	−0.45
L9	1800	3000	0

From every graph, for every acceleration channel (there are 2 acceleration channels), 50 data were taken from more than 1300 data. Generally, 3600 data were taken during the experiment. After finishing the experiment, average data were obtained in addition to Taguchi regression equations. The data has been analyzed with this method in order to calculate an average amplitude and frequency for each process condition. The frequency and amplitude rates were taken in 5 samples consisting of 10 values from 2 channels and then calculated.

3 Results and Discussion

3.1 *Variation of Frequency and Amplitude*

Figure 2a, b shows the variations of frequency and amplitude for ideal spur or failure spur gear in terms of experimental condition, respectively. According to this figure, frequency of the vibration for the failure gears approximately two times greater than the ideal one. Further, more fluctuations were observed for the failure gear due to having got more damage of the gear surface. This is more obvious, especially for L2, L3 and L8 trial. A comparison of amplitude rate of the vibrations for gear sets with ideal design and damaged ones was given in Fig. 2b. Some fluctuations were also appeared in this figure, when tested at a trial of L6 and L8 run. Both spur gears indicated similar trends, but amplitude for the failure gear was bigger than that of ideal gear.

Graphical representation of the experimental results was exhibited in Fig. 3. This can be categorized for two tooth profiles. Figure 3a, b indicates the variation of acceleration versus time for the ideal spur gears, which was tested at the trial number L7 while Fig. 3c, d shows the variation of acceleration versus time for the failure of spur gears, but again trial number was L7 samples. The acceleration in the first

Fig. 2 Variations of frequency and amplitude for ideal spur gear or failure spur gear in terms of experimental condition

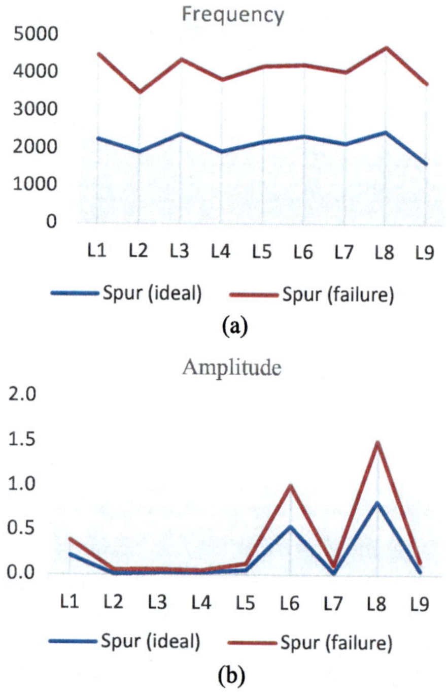

figure was ranged from -2 to 2 m/s^2, but time was variable. The second figure indicates that the acceleration was ranged from -10 to 10 m/s^2, respectively. In other words, frequencies change from approximately 592, 740, 838 and 1184 Hz, which are corresponded values of accelerations were about 0.093, 0.086, 0.134 and 0.143 m/s^2, respectively. Its limiting value is about 0.08. However, in the case of failure gear, frequency changes from 592 Hz to 740 Hz, its corresponded values were about 0.287–0.53 m/s^2, respectively.

Its limiting value is found to be approximately 0.3, which is a quite higher than that of previous one. The results of average values of frequency and amplitude that obtained from vibration analyzer under different speed, torque and center distance deviation on L9 orthogonal design are shown in Table 3.

3.2 *Main Effect Plots for Frequency*

The effects of main control factors are shown in Fig. 4 for means of frequency. Frequency of the vibration also increases with increasing of the shaft speed. Changing the speed from 600 to 1800 rpm seems to be smooth, but when increasing the speed from 1200 to 1800 rpm, the frequency of vibration sharply increases. This is normal because in reality with increasing the rotation speed of the shaft may be increased

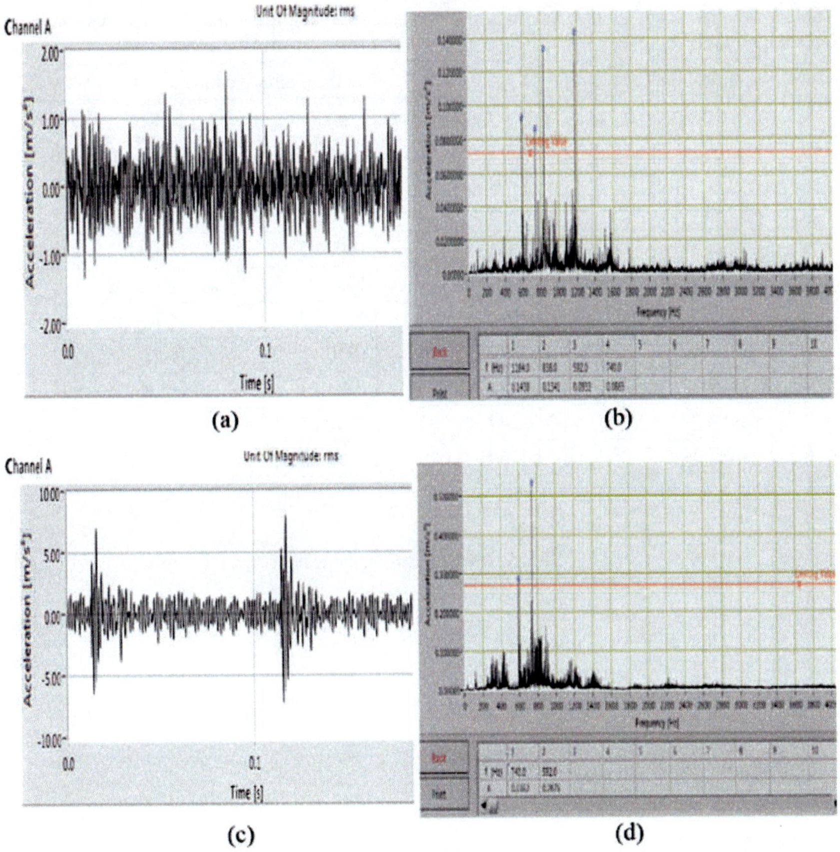

Fig. 3 Graphical representation of the experimental results. **a, b** Variations of acceleration for the ideal spur gears (trial number L7), **c, d** variations of acceleration for the failure spur gears (trial number L7)

the vibration frequency. Changing of the torque and center distance deviation also influenced to the frequency. For the small values of the torque, frequency is high. Due to low pressure between the contacting surfaces, a real surface contact area is so small that relative motion of the elements observed with high vibration and noise because this is attributed to contact ratios and roughness of the tooth face. For the middle values of the torque, however, increasing loads decreased the clearances between the friction surfaces. This action can be leading to the reduction of the frequency, but for the high values of the torque, forces acting on the rotating elements increases slightly. Therefore, the vibration frequency can also be increased. Similar results are found with previous work on the shaved gears and ground gears [6].

When the center distance deviation is zero frequency of vibration must be low because the shaft is well balanced. But if there is positive or negative center distance deviation, shaft balance may be destroyed since the concentration of the loads acting

Table 3 The results of average values of the frequency and amplitude output, which was obtained from the vibration analyzer

Level	Width ideal gears		Width failure gears	
	Frequency—F, Hz	Acceleration—A, m/s^2	Frequency—F, Hz	Acceleration—A, m/s^2
L1	2237.87	0.213481	2257.09	0.176340
L2	1900.46	0.008478	1587.22	0.048891
L3	2375.30	0.022194	1997.85	0.039935
L4	1916.10	0.023794	1931.97	0.029035
L5	2177.24	0.050147	2023.77	0.077361
L6	2329.12	0.543187	1910.41	0.467115
L7	2140.53	0.024417	1924.23	0.084963
L8	2456.20	0.822474	2261.25	0.680414
L9	1627.54	0.044641	2130.98	0.105025

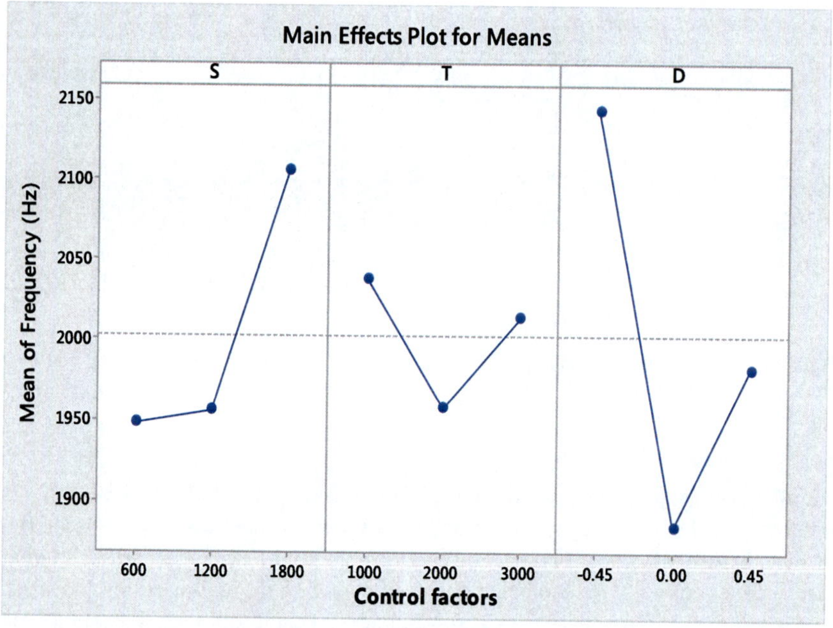

Fig. 4 Main effect plots for frequency of mean values

on the contacted zone/teeth increases, and as a result, this influenced to the vibration. It should be noted that the sensitivity of the center distance deviation was slightly higher than that of the rotational speed. The mean response referred the average values of frequency and acceleration performance under various testing conditions and their levels. Among the control factors, the deviation of center distance showed the highest

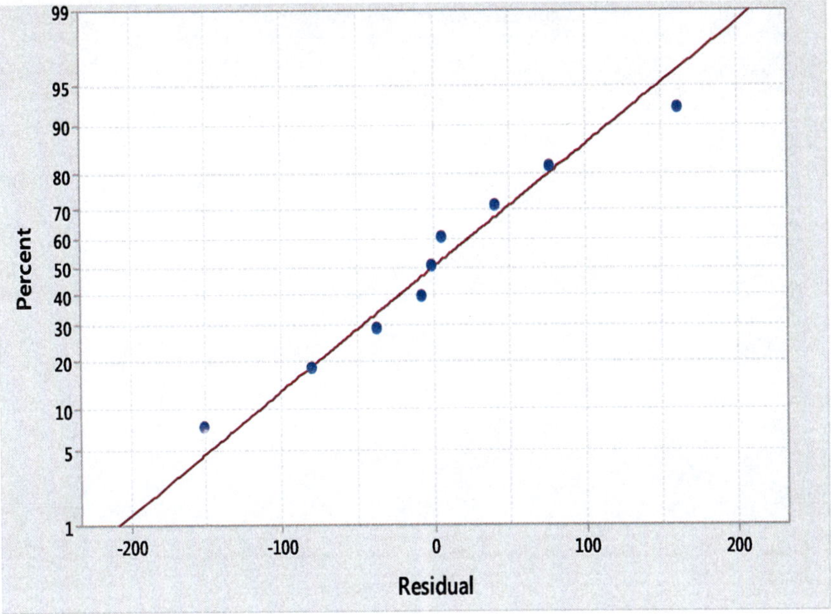

Fig. 5 Normal probability plot for tested both gears

influence on the frequency and acceleration because of destroying the balance of the mechanical system, but the factor S and factor T were followed, respectively. The optimum values are obtained when tested at S1 (600 rpm), T2 (2000 N.mm) and D2 (0.0 mm) conditions.

Figure 5 indicates the normal probability plot for tested both gears for cumulative distributions of residuals. These residuals are calculated from theoretical (prediction) and experimental (measured) results. This difference is defined as a residual. The residualis calculated for each run. The distributions of errors seem to be normal because of distributing evenly for both positive and negative along the experimental run. It indicates that the model proposed assumes an adequate.

3.3 *Main Effect Plots for Acceleration*

Main effect plots for the accelerations were plotted in Fig. 6 as a function of the control parameters. It increased linearly with the rotation speed of the shaft. Unlike to changing the frequency, acceleration of vibration takes its maximum value in average rates of torque at 2000 N.mm.

Large amount of changes are obtained from variation of center distance deviation. When the center distance deviation is zero, the acceleration reached its minimum value in terms of mean, but frequency up its maximum value. Its speed increased

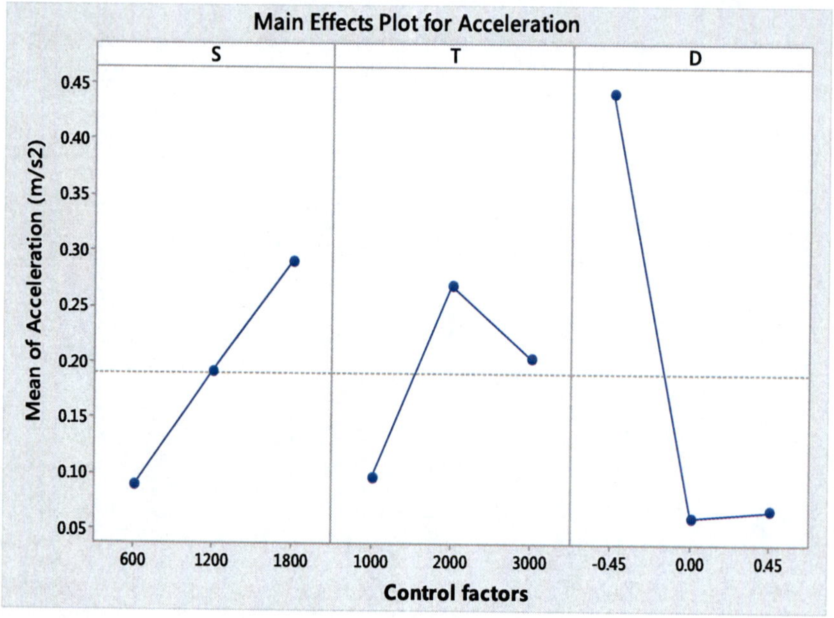

Fig. 6 Main effect plots for acceleration of mean values

linearly but torque increased up to certain point sharply and then decreased slightly. Optimum values are achieved from the lower mean values. For acceleration, again the optimum values are obtained at 600 rpm, 1000 N.mm torque and 0.0 mm center distance conditions. Further, it is observed that the decrease of center distance between the gears, and medium load lead to additional high vibration and noise. In addition, higher speed causes to higher vibration, which leading to increases the severity of vibration in the gear set.

Figure 7 shows the normal probability plot for both gear samples. These residuals are calculated from theoretical and experimental results. The residuals are calculated for each run and ranked in ascending order. The error distributions seems to be normal because they are distributed evenly in both positive and negative along the experimental trials. The residual is lower than $\pm$ 0.2. It indicates that the model assumes the adequate.

4 Conclusions

The following conclusions were drawn from the vibration analysis and geometrical failure in gearboxes.

The experimental results indicated that failure of tooth profile affected the environment and the performance considerably in terms of noise, vibration and failure

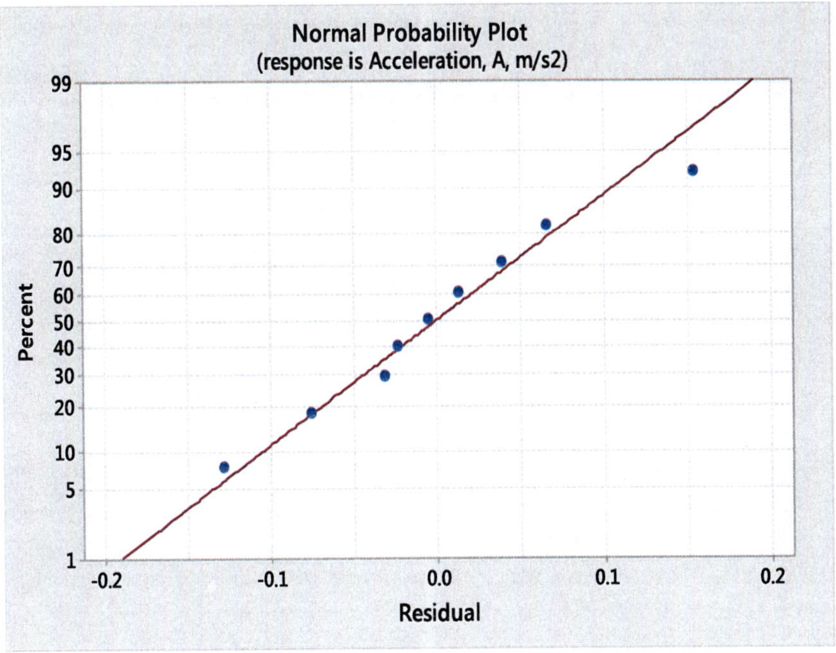

Fig. 7 Normal probability plots

of the parts. Even low speed and moderate torque might cause to vibration and wear. The longer distances between the gear centers maybe appeared early wear. It is shown that frequency for the ideal gear change from 592 to 1184 Hz, which are corresponding values of accelerations were about 0.093–0.143 m/s^2, respectively. For the case of failure gear, however, frequency changes from 592 to 740 Hz, its corresponded values were about 0.287–0.53 m/s^2, respectively. This is a quite higher than that of previous one. Moreover, it is found that center distance deviation was the effective factor for both mean frequency and acceleration, followed by speed and torque, respectively. Optimized processing condition of the frequency or acceleration was first level of speed (600 rpm), first level of torque (1000 N.mm), and second level of center distance (0.0 mm).

References

1. Khurmi RS, Gupta JKA (2005) In: Textbook of machine design, Eurasia Publishing House (PVT.) ltd. Ram Nagar New Delhi, 110–055, pp 1230
2. Childs P (2004) Mechanical Design, 2nd edn. Elsevier l.t.d., pp 359
3. Mirzayev H, Charkasov S, Mirza A (2016) Optimization of the bevel-spur gearbox. J Qafqaz Univ Mech Indus Eng Baku 4(1):42–88

4. Bartelmus W, Zimroz R (2009) Vibration condition monitoring of planetary gearbox under varying external load. Mech Syst Signal Process 23(1):246–257 Elsevier
5. Wang WQ, Ismail F, Golnarghi MF (2001) Assessment of gear damage monitoring techniques using vibration measurements. Mech Syst Signal Process 15(5):905–922 Elsevier
6. Chen Y, Ishibashi A (2006) Investigation of the noise and vibration of planetary gear drive. www.geartechnology.com. Jan/February, pp 48–55
7. Şahin Y, Patrick DB (2020) Tribological behaviour of basalt fabric reinforced epoxy composites. J. Balkan Tribol Assoc 26:135–157
8. Şahin Y, Şahin F (2018) Dry wear behavior of basalt/carbon fabric-reinforced epoxy composite by Taguchi method. In: Lecture notes in mechanical engineering: proceedings of the 7th international conference on fracture fatigue and wear, pp 581–590

Fretting Wear Effect on Fretting Fatigue by Findley Parameter in Mixed Slip Regime

S. Wang and Magd Abdel Wahab

Abstract When there is a small amplitude relative slip between two contacting parts, fretting happens. Wear, fatigue and corrosion are the three main damages caused by fretting. In reality, these three damages interact with each other, which is not emphasized in current research. Findley parameter (FP) is a commonly used parameter that can be utilized to analyze fretting wear. In this paper, the wear profile updating effect on fatigue is analyzed by FP in mixed slip regime of fretting. A finite element (FE) model of a cylinder-on-flat configuration is built to investigate the FP difference between models with and without wear profile updating. The results showed that FP in the model with wear profile updating was lower compared with the model without wear profile updating near the contact edge. It can be concluded that fretting wear shows a positive effect on fretting fatigue. Moreover, the predicted life shows a good agreement with experiments.

Keywords Findley parameter · Fretting wear · Fretting fatigue · Partial slip

1 Introduction and Background

Fretting is a phenomenon that happens between two contacting parts when there is a relative slip [1]. Fretting damages like fretting wear and fretting fatigue are sometimes detrimental to the parts of the machine [2, 3]. With the increment of relative slip, we can classify fretting as three regimes: stick regime, mixed slip regime and gross sliding regime [4, 5]. In gross sliding regime, wear is the leading damage, while in stick and mix regime fatigue plays a key role [6–11]. In partial slip regime, there a competence between wear and fatigue. The initiation of micro crack can be moved away when the wear is significant, which can prolong the parts lifetime. Mostly, the wear profile updating effect on fatigue in fretting process is ignored. In most research work, it is assumed that there is no wear, which is not the real case.

S. Wang (✉) · M. Abdel Wahab
Department of Electromechanical, Systems and Metal Engineering, Ghent University, Ghent, Belgium
e-mail: Shengjie.wang@ugent.be

M. Abdel Wahab (ed.), *Proceedings of the 8th International Conference on Fracture, Fatigue and Wear*, Lecture Notes in Mechanical Engineering,
https://doi.org/10.1007/978-981-15-9893-7_48

Two wear quantification models are commonly used for fretting wear analysis: Archard's model and dissipated energy model [8, 10, 12]. Dissipated energy model shows its advantage by considering the coefficient of friction (COF). The equation of dissipated energy is as follows:

$$\Delta h_i = K_E \int_0^T q_i(x) ds_i(x) \tag{1}$$

where K_E is coefficient of wear, Δh_i is the local wear depth, T is one cycle period and $s_i(x)$ and $q_i(x)$ are incremental relative slip and shear stress on the surface position x.

Many multiaxial fatigue criteria have been used for fretting fatigue analysis. Critical plane approaches and continuum damage mechanics (CDM) are mostly used due to their applicability and generality [13]. FP criterion is firstly proposed by Findley in Ref. [14], which is a typical critical plane method to calculate direction and position of crack nucleation in early stage [15]. FP is given by [16]:

$$FP = \frac{\Delta \tau_{max}}{2} + k_1 \sigma_n^{max} \tag{2}$$

where $\Delta \tau_{max}$ and σ_n^{max} are the maximum range of shear stress and the maximum normal stress on the critical plane in one fretting cycle respectively, and k_1 is determined by the fatigue limit in tension σ_{f-1} and torsion τ_{f-1} when the stress ratio R is equal to -1. k_1 is determined from:

$$\frac{\sigma_{f-1}}{\tau_{f-1}} = \frac{2}{1 + \frac{k_1}{\sqrt{1+k_1^2}}} \tag{3}$$

To determine the angle of the critical plane, θ varying from $-90°$ to $90°$ is used to get the maximum shear stress range on one point as shown in Fig. 1.

FP is used to predict the crack initiation time and the equation for lifetime prediction is given by [17, 18]:

$$FP = \tau_f'(2N_i)^{b'} \tag{4}$$

where τ_f' is the coefficient of shear fatigue strength, N_i is the crack initiation time and b' is exponent of torsion fatigue strength.

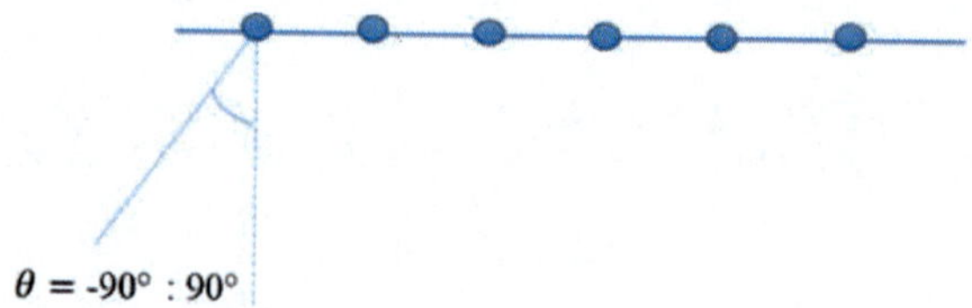

Fig. 1 Sketch of the angle of the critical plane on one point

When considering wear, the contact is always changing during fretting process. FP parameter and predicted life vary with cycles. To consider this, Miner's rule is employed, as proposed in Ref. [19].

$$V_D = \sum_{j=1}^{\frac{N_t}{\Delta N}} \frac{\Delta N}{N_{i,j}}$$

where V_D is the accumulated value of damage, ΔN is the cycles jumped to optimise the simulation time, N_t is the gross cycles number and $N_{i,j}$ is predicted nucleation lifetime for jth jump cycle.

This paper is presented as follows. After the introduction and background, experimental setup and finite element model are presented. Secondly, FP, which is employed to analyse wear effect on fretting fatigue is described. Then, the results are discussed and at last, conclusions are drawn.

2 Experimental Setup and Finite Element Model

The experimental setup is shown in Fig. 2. A configuration of cylinder-on-flat is applied in the reported experiments. The specific dimensions of the cylinder and specimen can be found in Ref. [20]. The material is heat treated Ti–6Al–4 V. Normal load, $P = 1334$ N is exerted on two pads' top. Oscillatory displacement, $D = 36$ μm is exerted by dual actuator system, which was first used in Ref. [21] and the relative slip between contact parts is measured by an extensometer. Bulking force is used at end of the specimen. The relation between applied displacement and local slip is given by [22]:

$$\delta_{applied} = \delta_{local} \frac{C_{COF}}{C_{rig}} \tag{5}$$

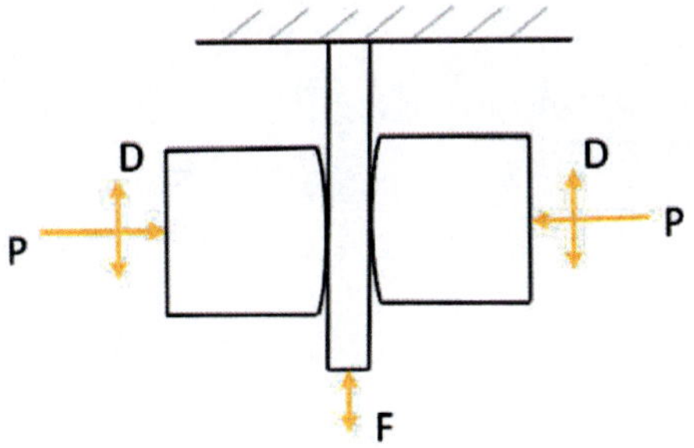

Fig. 2 Experimental setup [20]

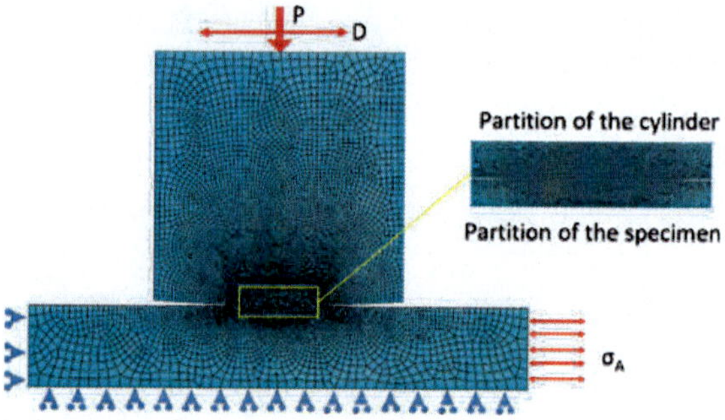

Fig. 3 Mesh and boundary conditions of the FE model

where C_{COF} is the coefficient affected by the COF and C_{rig} is a coefficient determined by experimental setup. $\frac{C_{COF}}{C_{rig}} = 16$ is employed based on the experiments.

To simplify the model, half of the experimental setup is used for simulation. The mesh and boundary conditions in commercial FE modelling software ABAQUS are presented in Fig. 3.

CPE4 elements are assigned for the whole model for the feasibility of the ALE domain and high accuracy. Element size on contact surface is 10 μm × 10 μm. To make the simulation effective, partitions with finer mesh are used in contact area. To simulate wear profile updating in fretting process, ALE adaptive mesh domain is applied in the partition of the specimen. The change in nodal y coordinate is implemented by subroutine UMESHMOTION. The axial stress, σ_A is applied on the right surface of the flat. The ratio between maximum and minimum axial stress is 0.03. The maximum axial stress is 550 MPa. The loading history is described in Fig. 4. To optimize the simulation time, $\Delta N = 1000$ cycles is employed as the jump cycle.

Ti-6Al-4 V was employed in the referenced experimental setup. In FE model, the Young's modulus, E and the Poisson ratio, υ are is 126 GPa and 0.32, respectively. The coefficient of friction between cylinder and flat is 0.8 [20]. k_1 in Eq. (2) ranges from 0.2 to 0.3; herein 0.3 is selected in this paper [23]. The energy wear coefficient K_E in Eq. (1) is 2.75×10^{-8} MPa^{-1} [22]. Yield stress σ_y is 965 MPa and the tension stress curve is take from experimental data [24]. Isotropic hardening is applied in this

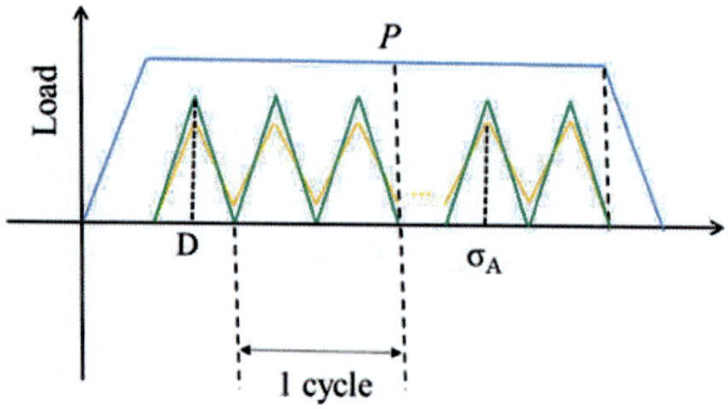

Fig. 4 Loading history of the FE model

Fig. 5 Flow chart of life prediction by FP

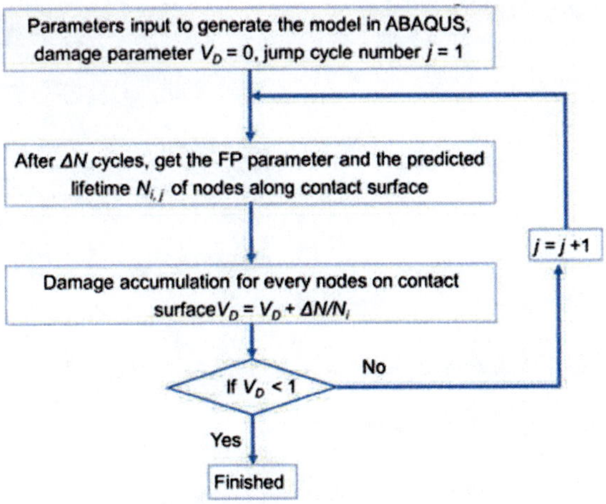

FE model. Fatigue strength coefficient in shear, $\tau_f^{'}$ can be derived from an estimation equation as [25]:

$$\tau_f^{'} = \frac{\sigma_f^{'}}{\sqrt{3}} \tag{6}$$

where $\sigma_f^{'}$ is the fatigue strength coefficient in tension. $\sigma_f^{'} = 2030$ MPa is used from Ref. [26]. In Ref. [25], exponent of torsion fatigue strength, b' is approximately equal to exponent of torsion fatigue strength, b for titanium alloys. Herein $b' \approx b = -0.104$ is employed as in Ref. [26]. Good correlations were found between analytical solution and the simulation results of the shear stress and contact pressure on contacting surface.

Flowchart for life prediction is shown in Fig. 5. In simulation process, the mesh is updated after each increment based on Eq. (1). After each jump cycle, the job will be restarted at the end of last increment, when V_D is not greater than 1.

3 Results and Discussion

In our previous work [27], wear profile characteristics in partial slip regime were analyzed. In this paper, the effect of fretting wear on fretting fatigue is discussed. In Fig. 6, the schematic of the cylinder and specimen configuration after wear is shown. With the wear in partial slip regime, stick zone gradually support normal load [28]. Shrinkage of the stick zone is also discovered in Ref. [28]. In Ref. [29], the localized plasticity is analyzed in fretting process without considering wear. The variation of maximum principle plasticity strain with cycles along contact surfaces when considering wear is shown in Fig. 7. As shown in the figure, we can conclude

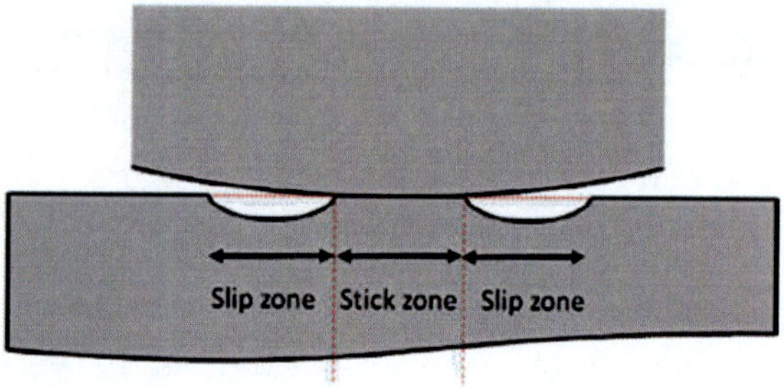

Fig. 6 Schematic of the configuration after wear [30]

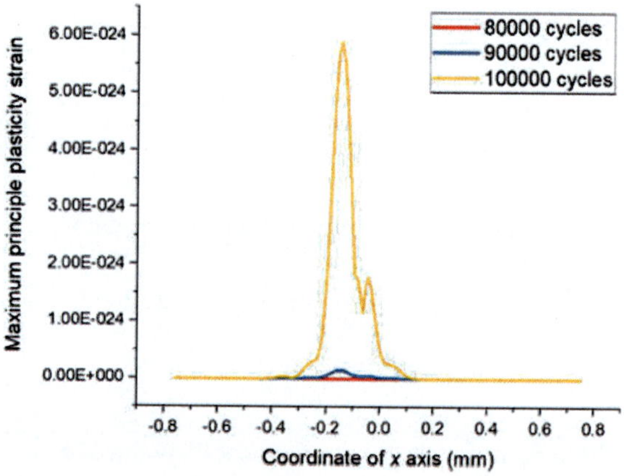

Fig. 7 Variation of maximum principle plasticity strain with cycles, when $D = 36\ \mu\text{m}$

that with the wear aggravation, principle plastic strain becomes greater. A good agreement with the result in Ref. [28] is found, i.e. the stick zone supports more and more normal load when considering wear. Moreover, the maximum plastic strain is not at the contact center. This is caused by the different wear profile on both sides. The stick zone center is not in the contact center in first cycle anymore.

To determine the maximum FP location variation, half contact width a is used. The half contact width for cylinder on flat configuration can be derived from [21]:

$$a = \sqrt{\frac{4PR}{\pi E^{*} L}} \tag{7}$$

where R is the radius of the cylinder, P is the load applied on the top of the cylinder and L is the thickness of the specimen in out of plane direction. E^{*} can be obtained from:

$$\frac{1}{E^{*}} = \frac{1 - \nu_1^2}{E_1} + \frac{1 - \nu_2^2}{E_2}$$

where E_1, E_2, ν_1 and ν_2 are Young's modulus and Poisson's ratio of the cylinder and specimen, respectively.

The variation of FP with number of cycles is shown in Fig. 8. From the figure, it can be concluded that maximum FP moves from contact edge to the wear edge. Because of the wear, the center of the wear profile, which is also close to the contact edge at the beginning bear less load from cylinder. This causes the decrease of FP near the contact edge.

Accumulated damage value with number of cycles is described in Fig. 9. In 1000 cycle, the maximum V_D is located near the contact edge. With evolution of wear profile, the maximum damage value moves to the wear scar edge. And after 32,000

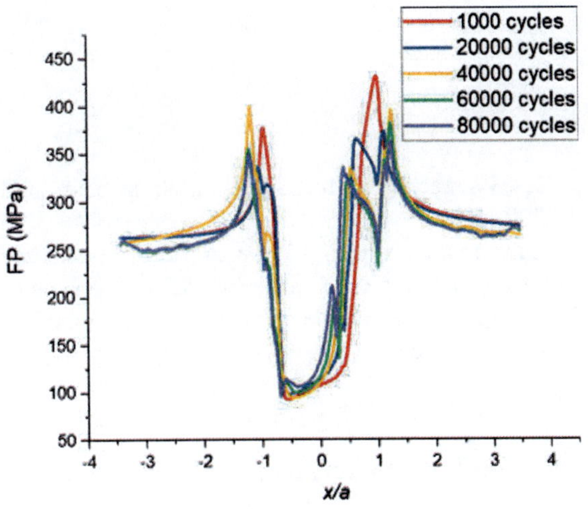

Fig. 8 Variation of FP with cycles

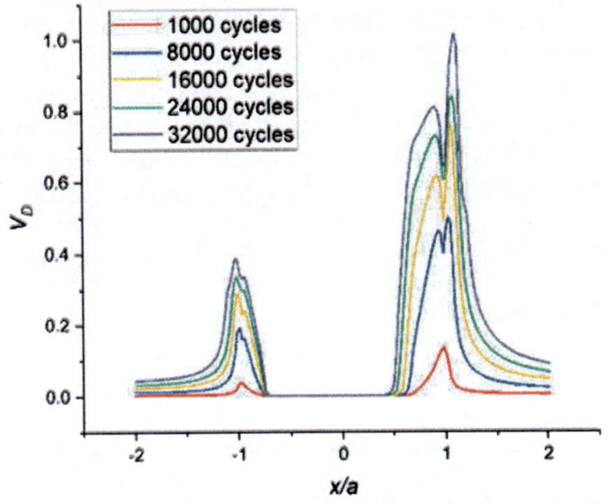

Fig. 9 Accumulated damage value V_D with cycles

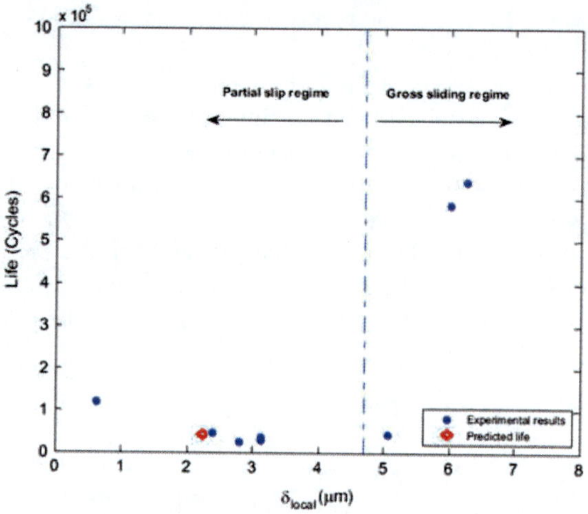

Fig. 10 Comparison between experimental and predicted lifetime

cycles, V_D reaches 1. The fatigue initiation occurs at the maximum V_D position after 32,000 cycles.

Local relative slip, $\delta_{local} = 2.25\ \mu\text{m}$ can be derived from Eq. (5). The predicted life is compared with the experimental life as shown in Fig. 10 [20, 31]. From the figure, we can see that the predicted life of FE model shows a good agreement with experimental data.

4 Conclusions and Future Work

In this paper, the FP is employed to investigate the wear effect on fretting fatigue in mixed slip regime. The results shows that when wear profile updating is considered, the maximum FP and damage value both moves from the edge of contact to wear scar edge. The fretting wear effect on fretting fatigue is positive. And because of the material removal in slip zone, stick zone support more normal load and plastic strain appears in stick zone with the aggravation of wear. In the next step, more prediction cases should be considered and analyzed to characterize this phenomenon.

References

1. Vingsbo O, Söderberg S (1988) On fretting maps. Wear 126:131–147
2. Ovcharenko A, Etsion I (2009) Junction growth and energy dissipation at the very early stage of elastic-plastic spherical contact fretting. J Tribol 131:031602
3. Fouvry S, Kapsa P, Vincent L (1996) Quantification of fretting damage. Wear 200:186–205
4. Yan S, Zhang D, Shirong G (2010) Effect of fretting amplitudes on fretting wear behavior of steel wires in coal mines. Mining Sci Technol (China) 20:803–808

5. Li J, Lu Y (2013) Effects of displacement amplitude on fretting wear behaviors and mechanism of Inconel 600 alloy. Wear 304:223–230
6. Leonard BD, Ghosh A, Sadeghi F, Shinde S, Mittelbach M (2014) Third body modeling in fretting using the combined finite-discrete element method. Int J Solids Struct 51:1375–1389
7. Leonard BD, Patil P, Slack TS, Sadeghi F, Shinde S, Mittelbach M (2011) Fretting wear modeling of coated and uncoated surfaces using the combined finite-discrete element method. J Tribol 133:021601
8. McColl I, Ding J, Leen S (2004) Finite element simulation and experimental validation of fretting wear. Wear 256:1114–1127
9. Pereira K, Yue T, Wahab MA (2017) Multiscale analysis of the effect of roughness on fretting wear. Tribol Int 110:222–231
10. Yue T, Abdel Wahab M (2016) A numerical study on the effect of debris layer on fretting wear. Materials 9:597
11. Yue T, Wahab MAJTI (2017) Finite element analysis of fretting wear under variable coefficient of friction and different contact regimes. 107:274–282
12. Fouvry S, Duo P, Perruchaut P (2004) A quantitative approach of Ti–6Al–4 V fretting damage: friction, wear and crack nucleation. Wear 257:916–929
13. Bhatti NA, Wahab MAJTI (2018) Fretting fatigue crack nucleation: a review. 121:121–138
14. Findley WN (1959) A theory for the effect of mean stress on fatigue of metals under combined torsion and axial load or bending. J Eng Indus 81:301–305
15. Szolwinski MP, Farris TN (1998) Observation, analysis and prediction of fretting fatigue in 2024-T351 aluminum alloy. Wear 221:24–36
16. Bhatti NA, Wahab MA (2017) A numerical investigation on critical plane orientation and initiation lifetimes in fretting fatigue under out of phase loading conditions. Tribol Int 115:307–318
17. Park J, Nelson D (2000) Evaluation of an energy-based approach and a critical plane approach for predicting constant amplitude multiaxial fatigue life. Int J Fatigue 22:23–39
18. Bhatti NA (2018) Numerical modelling of fretting fatigue crack nucleation under out-of-phase loading. Ghent University
19. Madge J, Leen S, Shipway P (2008) A combined wear and crack nucleation–propagation methodology for fretting fatigue prediction. Int J Fatigue 30:1509–1528
20. Jin O, Mall S (2004) Effects of slip on fretting behavior: experiments and analyses. Wear 256:671–684
21. Hills DA (1994) Mechanics of fretting fatigue. Wear 175:107–113
22. Madge J, Leen S, McColl I, Shipway P (2007) Contact-evolution based prediction of fretting fatigue life: effect of slip amplitude. Wear 262:1159–1170
23. Cardoso R, Doca T, Néron D, Pommier S, Araújo J (2019) Wear numerical assessment for partial slip fretting fatigue conditions. Tribol Int 136:508–523
24. Handbook M (1998) Metallic materials and elements for aerospace vehicle structures. Military Handbook No. MIL-HDBK-5H, Section 5
25. Sharifimehr S, Fatemi A (2018) Evaluation of estimation methods for shear fatigue properties and correlations with uniaxial fatigue properties for steels and titanium alloys. In: MATEC web of conferences, EDP Sciences, pp 16012
26. Llavori I, Zabala A, Urchegui M, Tato W, Gómez X (2019) A coupled crack initiation and propagation numerical procedure for combined fretting wear and fretting fatigue lifetime assessment. Theor Appl Fract Mech 101:294–305
27. Wang S, Wahab MA (2020) Effect of loading conditions in fretting fatigue on wear characteristics. In: Proceedings of the 13th international conference on damage assessment of structures, Springer, pp 659–665
28. Ghosh A, Leonard B, Sadeghi F (2013) A stress based damage mechanics model to simulate fretting wear of Hertzian line contact in partial slip. Wear 307:87–99
29. Talemi RH, Wahab MA (2012) Finite element analysis of localized plasticity in Al 2024-T3 subjected to fretting fatigue. Tribol Trans 55:805–814

30. Ghosh A, Wang W, Sadeghi F (2016) An elastic–plastic investigation of third body effects on fretting contact in partial slip. Int J Solids Struct 81:95–109
31. Shen F, Hu W, Meng Q (2015) A damage mechanics approach to fretting fatigue life prediction with consideration of elastic–plastic damage model and wear. Tribol Int 82:176–190

Effects of Ti-6Al-4 V Surface Condition on the Performance of DFL Lifetimes

E. Laolu-Balogun, S. P. Owen, S. Read, G. Pattinson, P. H. Shipway, and K. T. Voisey

Abstract In this study, the impact of different substrate surface conditions on the performance of an epoxy-bonded dry film lubricant (DFL) and their fretting lifetimes are evaluated. Ti-6Al-4 V fretting samples with various surface preparation methods; ground (as received), grit blasted (at a range of severity levels) and shot peened samples were coated with a MoS_2 loaded epoxy based DFL and implemented in fretting tests against like on like cylindrical specimen with a radii of 15 mm which were also coated. Distinct damage behaviours and coating lifetimes are observed for each of the surface preparation methods tested. Results indicate the various damage mechanisms are associated with initial substrate surface conditions which influence the stability and longevity of the system. Having established a correlation between DFL lifetime and substrate surface conditions, the next step in this work is to identify the specific surface features which promote the longevity of DFL. It must be noted that the substrate surface considered in the study is the DFL/substrate interface, i.e. the surface of the specimen prior to DFL coating, as opposed to the outer surface of the coated specimen.

Keywords Dry film lubricant · Fretting · Surface

1 Introduction

Fretting is the small amplitude oscillatory motion that occurs between contacting surfaces which leads to material degradation and damage. This is particularly prevalent in the dovetail joint of aero-engine fan blades [1] due to the geometrical interlocking structure, operating conditions and material composition. Dry film lubricants (DFLs) are applied on the contacting surfaces as a sacrificial layer to supress the

E. Laolu-Balogun (✉) · P. H. Shipway · K. T. Voisey
Faculty of Engineering, University of Nottingham, Nottingham NG7 2RD, UK
e-mail: emxel2@exmail.nottingham.ac.uk

S. P. Owen · S. Read · G. Pattinson
Rolls-Royce Plc, Derby, UK

M. Abdel Wahab (ed.), *Proceedings of the 8th International Conference on Fracture, Fatigue and Wear*, Lecture Notes in Mechanical Engineering,
https://doi.org/10.1007/978-981-15-9893-7_49

effects of fretting. DFLs however, have a finite lifetime and require intermediary applications before reaching the point of failure and the exposure of metal–metal contact [2]. The advantage of a polymer bonded DFL is that the DFL is better retained within the contact area where it is required compared to an un-bonded DFL where the DFL can more easily be redistributed outside of the contact area.

Mechanical properties of the substrate, surface morphology, distribution of contact stress and the adhesion strength of MoS_2 polymer bonded DFL coatings are all thought to affect fretting wear [3]. The performance and lifetimes of such DFLs have been linked to various substrate conditions, with roughness being a key factor [4]. This has led to various substrate pre-treatments being employed. The roughness of the substrate surface, and the way in which the DFL crystallites are arranged on it, are thought to have a considerable impact on DFL lifetimes [5]. The average roughness, Ra, is the most frequently quoted surface roughness measure [6]. However, Ra does not capture full details of the surface hence other measures such as Sa and 3D surface scans are also relevant.

Initial surface roughness and texture direction have been found to have an influence on the fretting wear; affecting the friction torque and accumulated dissipated energy which at first increase and then decreases with increasing surface roughness [7]. Xu et al. [8] investigated the fretting wear life of bonded MoS_2 lubricant coatings on steel substrates with Ra values of about 0.5 and 2.0 μm. With an applied normal load of 600 N and a displacement of 10 μm, an increase in the number of cycles required to reach a CoF (coefficient of friction) of 0.25 was observed for the rougher substrate surfaces. The author also noted the rougher surface enhanced the bond strength between coating and substrate which also had a beneficial effect in wear at later fretting stage. This was explained as due to residual MoS_2 coating being retained in the valleys of the rough surface.

Surface roughening through sandblasting has successfully improved bond strength between MoS_2 polymeric coatings and substrates [9, 10]. Zhou et al. [11] reported that enhancing substrate hardness, increasing coating thickness, sandblasting substrate surface, increasing temperature to solidify coating and reducing relative humidity all improved fretting behaviour of polymer bonded MoS_2 coatings.

In the work by Barman et al. [12], the damage mechanisms throughout the DFL system lifetime were categorized into three stages. The initial Stage I is characterised by a high ECoF (energy coefficient of friction) which gradually dissipates into a durable low ECoF phase termed Stage II where a thin layer of mixed DFL-metallic content is proposed to exist during this low friction regime. This eventually gives way to Stage III where metal–metal contact with limited influence of DFL dominates, in this unstable regime, there is a rapid increase of ECoF until the failure point is reached.

It has been shown that initial interface substrate conditions have a considerable impact on fretting systems with DFL coatings. Systems with higher Ra values have shown significant effectiveness against fretting, however, there is no understanding of what the optimal surface roughness is. This work addresses that issue. Surface roughness and parameters concerning the topographical interface and their influences on the fretting wear behaviours of polymer bonded MoS_2 DFL, particularly the Stage II low friction regime [12], are explored.

2 Experimental Methodology

Fretting wear tests were conducted with Ti-6Al-4 V to represent the material comprising the fan blade and disk in an aero-engine dovetail joint.

Samples were machined from a Ti-6Al-4 V plate to make up the cylinder-on-flat testing configuration which resulted in a contact length (w) of 10 mm (determined by the width of the flat sample) with the cylindrical sample having a radius of 15 mm. The experimental programme implements various surface treatment methods which are induced on the flat and round specimens for self-mated tests, i.e. identical surface conditions for matched flat and round specimens:

1. Ti-PL237—Plain (as received) testing specimens where no surface treatment has been imposed.
2. GB-PL237—Grit blasted specimens with aluminium oxide grit (Airblast, UK; grade NK36) with a grit size of 0.125– 0.149 mm at a range of severity pressures; 1 bar, 3 bar and 9 bar
3. SP-PL237—Shot peened specimens where the as received flat and round Ti-6Al-4 V specimens underwent a steel shot peened surface treatment with an intensity of 6–10A, 230 R and 200% coverage.

Following surface treatments, deposition of the DFL proceeded on both flat and cylindrical specimen using a spray gun. The paint referred to as DFL is the commercially available epoxy-based DFL paint (procured from Indestructible Paints Ltd, Birmingham, UK under the trade name PL237-R2), in which the primary solid lubricant phase in the DFL is MoS_2. Once sprayed and cured, DFL coating thickness (t) measurement was done using a PosiTector 6000 eddy current thickness measurement. All samples had a DFL coating thickness of 50 ± 5 μm. MoS_2 particle size has been measured at 2–4 μm through examination of high magnification SEM images.

Fretting wear tests were carried out on a servo-hydraulic fretting testing machine, Fig. 1. A linear variable displacement transducer (LVDT) mounted on the base of the hydraulic actuator continuously measures the applied displacement while an instrumented strain gauge load cell measures the total shear force applied to both contacts.

Fretting test conditions (amplitude, frequency and contact pressure) are given in Table 1. These were selected to be representative of those experienced within aero-engine dovetail joints.

During the fretting tests, the tangential force and applied displacement were recorded at a rate of 200 samples per fretting cycle. The tangential force-displacement data were plotted to form fretting loops of the type presented in Fig. 2. The coefficient of friction is given by:

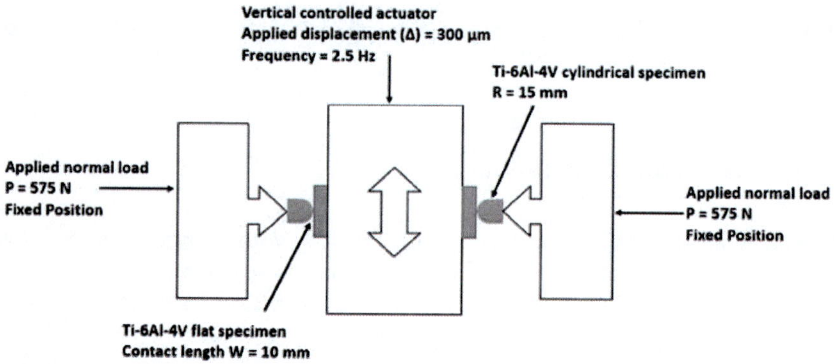

Fig. 1 Electro servo hydraulic fretting rig operating schematics

Table 1 Fretting test conditions

Normal Load (P) / N	575
Displacement amplitude Δ* / μm	300
Frequency (f) / Hz	2.5
Temperature (T) / °C	20 ± 3 (ambient)
Relative humidity (%)	30 ± 5 (ambient)

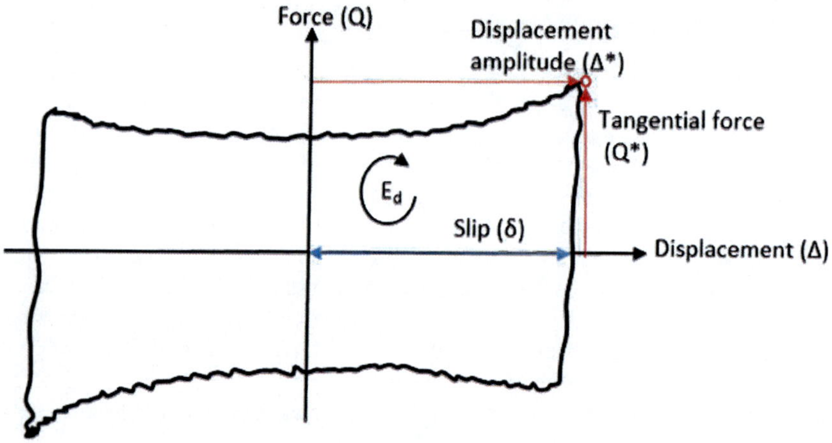

Fig. 2 Schematic diagram of fretting loop analysis

$$CoF = \frac{Q^*}{P} \quad (1)$$

where Q^* is the maximum tangential force recorded in each cycle and P is the constant normal load applied. For each condition examined, tests to completion

Table 2 Surface parameters

Surface conditions	Sa–Surface roughness (μm)	Sz–Maximum height (μm)	Sp–Maximum peak height (μm)	Sv–Maximum pit depth (μm)
Ti (as received)	0.321	5.31	1.96	3.18
SP	0.613	10.3	4.24	6.08
1 bar GB	1.23	20.7	8.06	12.7
3 bar GB	1.6	22.4	9.34	13
9 bar GB	2.77	35.9	19.8	16.2

were terminated when the CoF reached a value of 0.7. Interrupted tests were also conducted on selected samples to facilitate a better understanding of the wear process in the different stages of fretting lifetimes.

Fouvry and Paulin [13] proposed the concept of energy coefficient of friction (ECoF) as a better representation of fretting contact through the energy dissipated over a fretting loop (E_d).

$$ECoF = \frac{E_d}{4\delta^* P}$$

where δ* is the slip amplitude (distinct form the applied displacement amplitude, Δ^*). This work is focussed on fretting damage, energy dissipation is highly relevant to this, and hence the results are presented in terms of ECoF.

Surface characterisation and profilometry analysis was carried out using an Alicona G4/G5 Infinite Focus system. A defined area of 800 × 800 μm was acquired for surface parameter measurements (as seen in Table 2) which were calculated in accordance with ISO 25178 using a cut-off wavelength of 160 μm for S-L filters. Further 3D measurements (also with the Alicona system) generated the surface profiles of the wear scars from the flat specimen after fretting tests. The average wear scar depth profiles reported are the average of 50 profile lines.

A FEI Quanta600 MLA scanning electron microscope with energy dispersive x-ray (EDX) analysis capability was used to obtain plan and cold mounted cross-sectional views of the fretted titanium and carry out EDX analysis.

It must be noted that the surface considered in the study is the DFL/substrate interface, i.e. the surface of the specimen prior to DFL coating, as opposed to the outer surface of the coated specimen.

3 Results

The surface topography of the surface interface conditions tested are presented in Fig. 3. The plain (as received) Ti-6Al-4 V is characterised as having low surface roughness and unidirectional lays. No significant peaks or depths appear on this

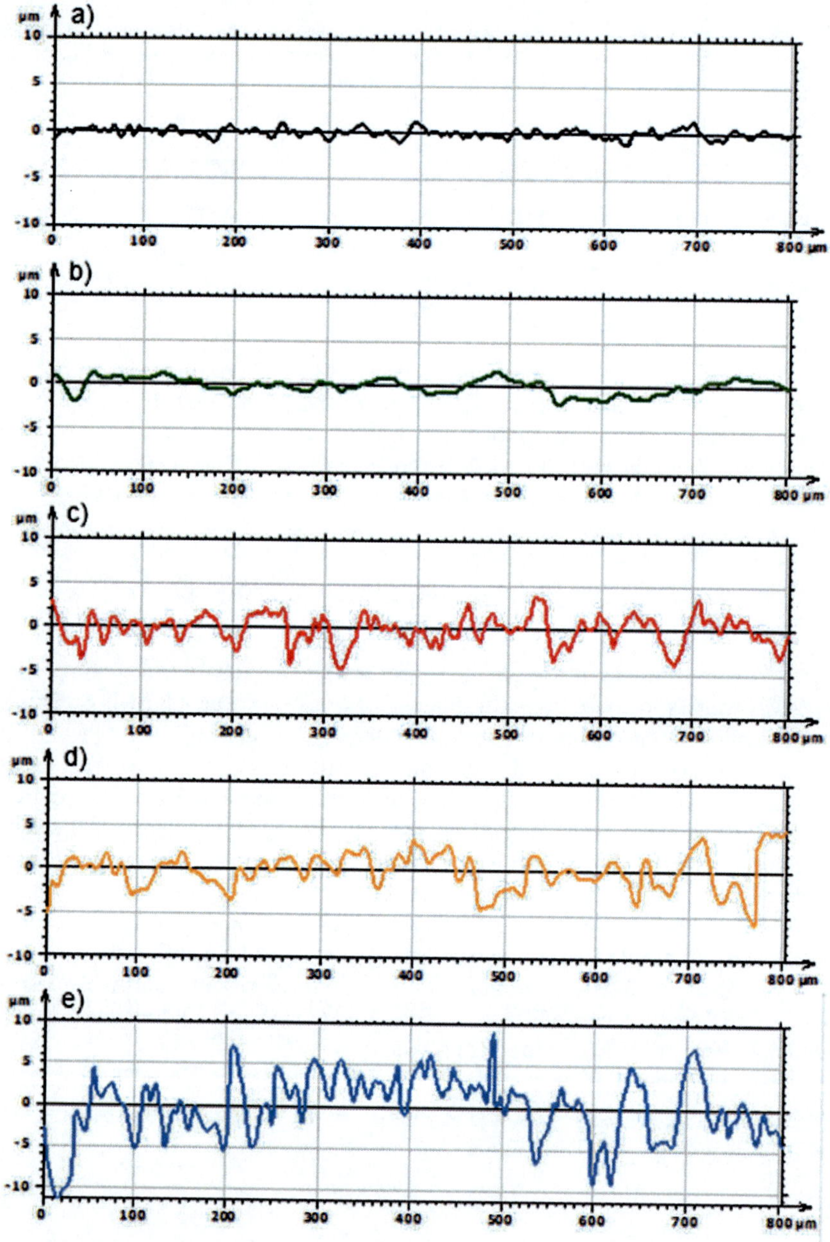

Fig. 3 Line profiles of initial surface conditions of **a** plain as received Ti-6Al-4 V surface **b** shot peened surface **c** 1 bar grit blasted surface **d** 3 bar grit blasted surface and **e** 9 bar grit blasted surface

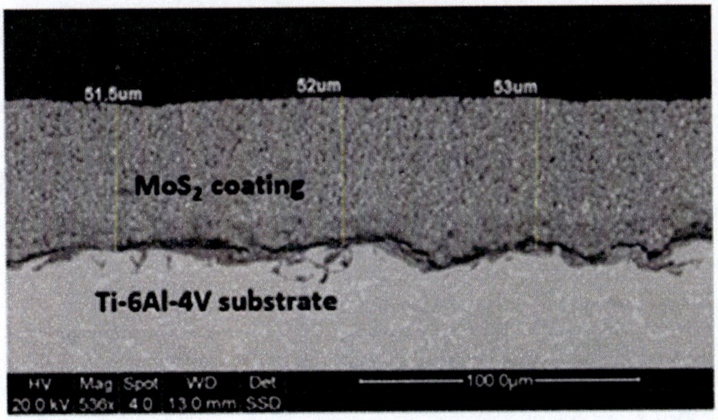

Fig. 4 Cross sectional BSE image of the DFL thickness measurement on grit blasted flat specimen

surface. The shot peened surface topography is characterised by smooth groves and pitted holes with smaller levelled peaks. Of the three surface conditions, grit blasted surfaces produce the highest roughness (seen in Table 2) as well as having the most pronounced topographical features with sharper peaks, low depths, and deeper groves. Grit blasted surface also provide the most irregularities with a spectrum of different heights and depths across the surface. In Fig. 4, a cross sectional SEM image of the grit blasted surface with a target DFL coating of 50 of ± 5 μm prior to fretting test can be observed.

Figure 5. depicts ECoF against number of cycles, documenting the DFL lifetimes of each surface condition. The results suggest DFL lifetime is associated with the interface conditions as decreased fretting lifetimes correlate with decreasing interface roughness while longer lifetimes are achieved by rougher interfaces.

Evolution of ECoF for the self-mated grit blasted DFL systems, referred to as GB-PL237 from here on, reaches failure at 80–140 k cycles whilst exhibiting a distinct behaviour of low ECoF, below 0.3, before the incremental increase of ECoF which eventually leads to failure. For shot peened systems, denoted as SP-PL237, DFL lifetime lies in the range between 30 and 40 k cycles. The damage behaviour associated with this system begins with an initial rise in ECoF at the start of the test which quickly dissipates and drops to a minimum of 0.35, followed by a rapid and unstable rise to reach failure. The system with the lowest surface roughness, Ti-PL237, fails under 25 k cycles.

Two representative surface conditions; 3 bar grit blasted system (3 GB-PL237) and a shot peened system (SP-PL237) are analysed in more detail, including interrupted tests that were stopped before the failure point. In the 3 GB-PL237 system, the interruption point was at 70 k cycles, within the low ECoF, Stage II, phase (Fig. 6a). Completed fretting tests on the 3 GB-PL237 system had worn to a maximum depth of 100 μm (Fig. 6b), well below the original DFL thickness of 50 μm; suggesting the coating had been completely worn through to failure with metal to metal contact now occurring on the fretting contacting surfaces. The wear depth profile of the

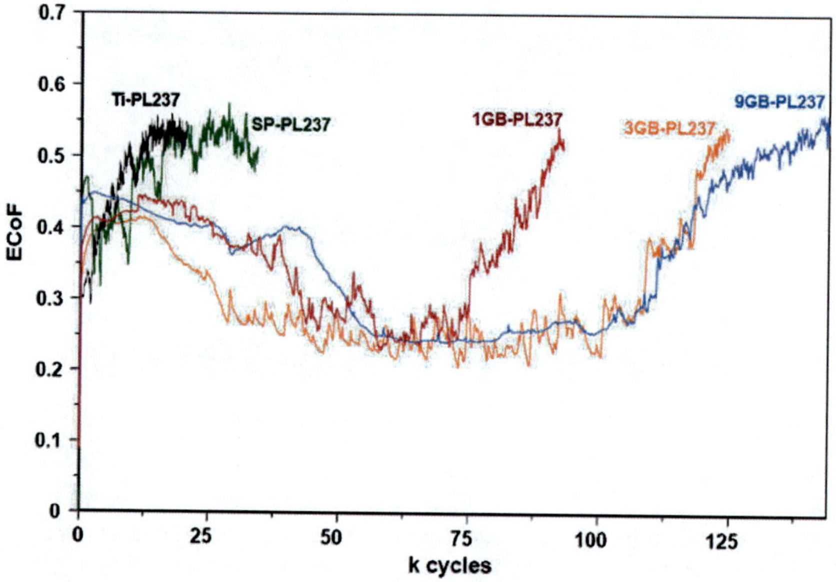

Fig. 5 Evolution of ECoF with number of cycles for Plain Ti-6Al-4 V (Ti-PL237), shot peened (SP-PL237) and grit blasted surfaces (1 GB-PL237, 3 GB-PL237, 9 GB-PL237) tested until coating failure

interrupted 3 GB-PL237 test reaches a smaller maximum depth of 60 μm, meaning that the wear scar depth approximately coincides with the original DFL/substrate interface; implying some DFL material could remain on the surface in the low ECoF phase.

Figure 7 displays low and high magnification plan view BSE micrographs images of the interrupted 3 GB-PL237 test wear scar surface corresponding to low friction stage. EDX analysis indicates the presence of both metal and DFL at the wear scar surface. Figure 8 presents a high magnification cross sectional BSE image with EDX point analysis identifying the chemical composition of the low friction wear surface.

Interrupted and completed fretting tests were also conducted for the SP-PL237 system. Figure 9a displays the evolution of ECoF for completed and interrupted tests on the SP-PL237 DFL system. The wear scar profiles (Fig. 9b) show a maximum depth of 125 μm for the completed test and 76 μm for the interrupted test. Both these depths exceed the initial DFL coating thickness of 50 μm. This suggests the removal of bulk DFL material had already occurred early in the fretting lifetime at 8 k cycles in the initial period of the ECoF evolution.

The BSE plan view images from the interrupted SP-PL237 tests in Fig. 10a show a low magnification area of the worn scar: two distinct phases can be observed with a bright area on the edges of the wear scar and a darker area dominating the central contacting surface.

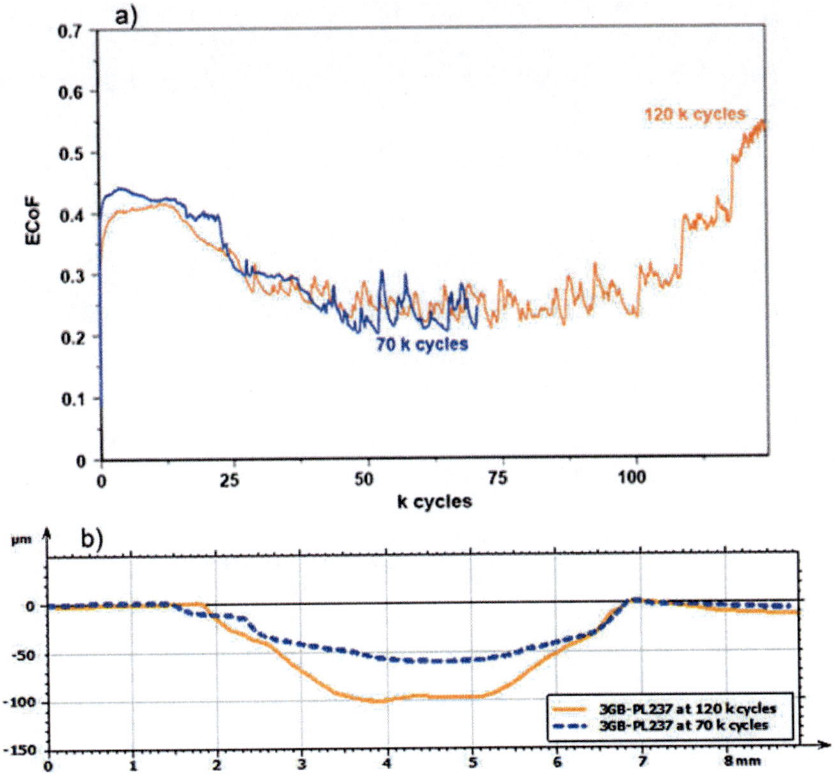

Fig. 6 **a** Evolution of ECoF against number of cycles of 3 GB-PL237 system for both completed and interrupted test. **b** Average wear profiles of corresponding completed and interrupted tests of 3 GB-PL237 system

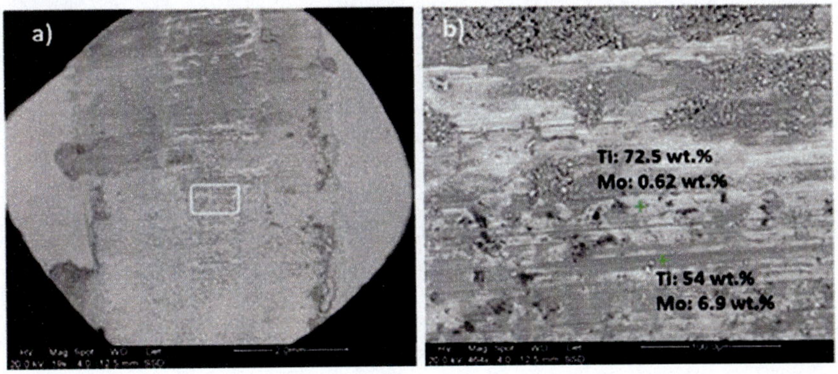

Fig. 7 BSE-SEM images of interrupted 3 GB-PL237 tests showing, **a** plan view low magnification image of interrupted wear scar **b** high magnification of indicated area with EDX point analysis

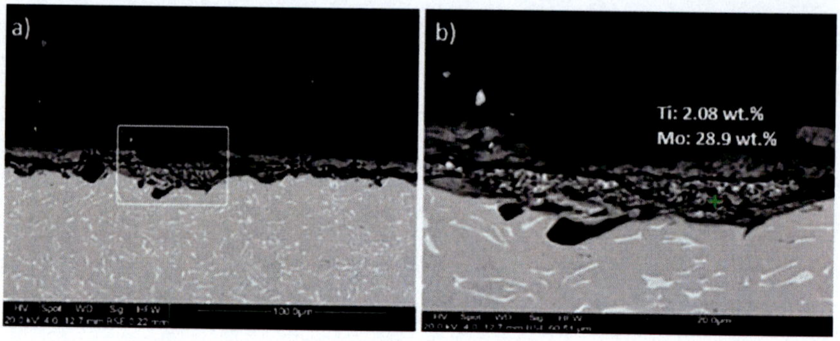

Fig. 8 BSE micrographs of interrupted GB-PL237 wear scar cross section. **a** Low magnification of worn area, **b** high magnification of selected area along with EDX data

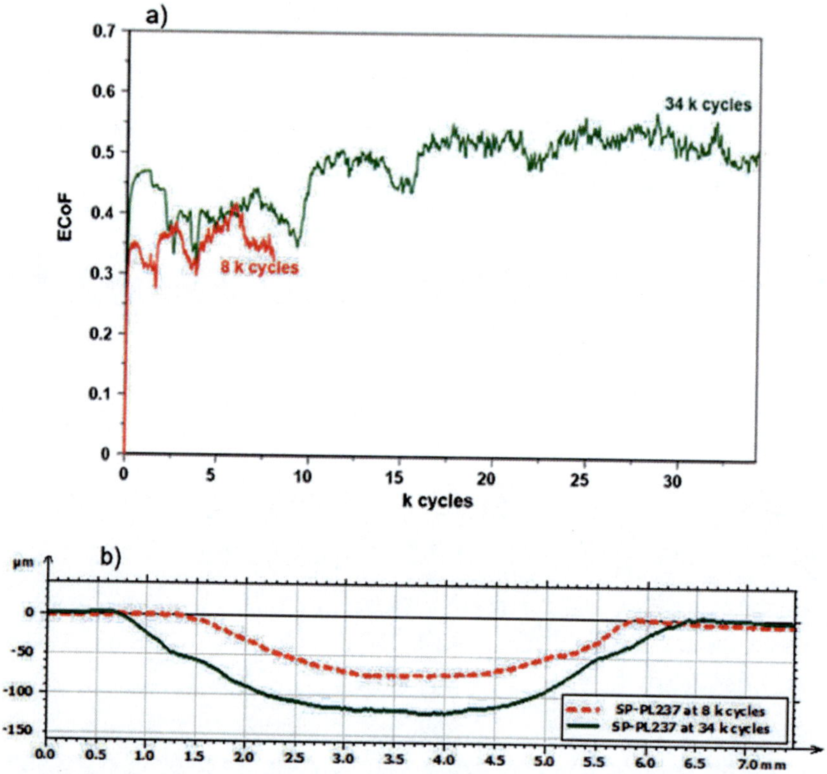

Fig. 9 **a** Evolution of ECoF against number of cycles on SP-PL237 system for both completed and interrupted test. **b** Average wear profiles of corresponding completed and interrupted tests

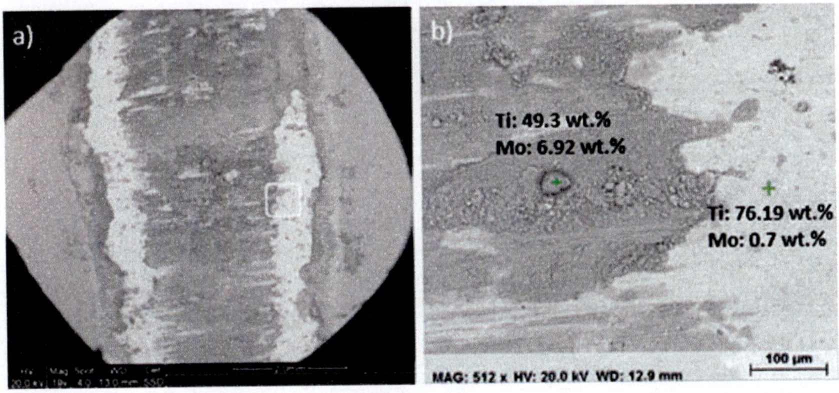

Fig. 10 BSE-SEM plan view images of SP-PL237 interrupted at 8 k cycles. **a** Low magnification image of interrupted fretting tests and **b** high magnification of indicated area along with EDX point analysis

EDX point analysis in Fig. 10b shows that the darker areas in the central region of the wear scar show the presence of both titanium and molybdenum, indicating the DFL material is still present on the surface of the wear scar shown. EDX results from the bright area on the outer region of the worn scar reveals an area completely dominated by titanium rich material with minimal traces of any molybdenum. Corresponding cross-sectional images were acquired for the SP-PL237 interrupted tests in Fig. 11 which show the wear scar is dominated by oxidised titanium material.

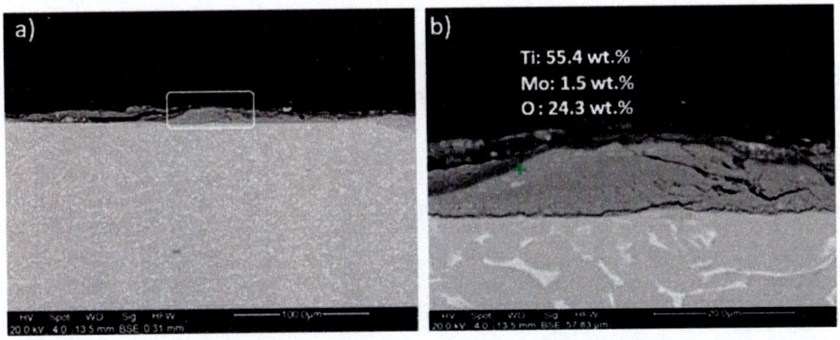

Fig. 11 BSE micrographs of interrupted SP-PL237 wear scar cross section. **a** Low magnification of worn area, **b** high magnification of selected area along with EDX data

4 Discussion

This work investigates how individual DFL/substrate interface characteristics, as shown in Fig. 3, affect damage behaviour of DFL fretting wear system. It must be noted that all roughness results quoted refer to this interface, as opposed to the outer surface of the coated system.

This work adopts the DFL fretting lifetime stages established in previous work [12] (i.e. Stage I, II, III) and illustrated in Fig. 12.

Fu et al. [14] confirmed that rougher surfaces have a higher plasticity index than smooth surfaces, and so, plastic deformation will occur at the tips of the asperities. This idea relates to the wear mechanism taking place at the DFL/substrate interface where metallic asperities in the interface are thought to undergo plastic deformation (while under fretting loads) and form an adhesive wear mixture consisting of both metal and DFL, from the reservoirs residing within the pits of the surface. This mixed material is considered to be a thin smooth layer between the contacting surfaces which delays metal–metal contact while maintaining a low ECoF (labelled as Stage II in Fig. 12). The proposed mechanism is stable elastic shearing on the surface once the metal-DFL mixed material has been fully established.

Cross sectional imaging of the interrupted (70 k cycles) 3 GB-PL237 test in Fig. 8 shows how DFL reservoirs are generated by the roughness of the interface. EDX elemental analysis confirms substantial DFL remaining within the pits in the interface. This material combines with the metal from the asperities to produce the mixed metal-DFL material described above. The flattening of the initially rough metal at the interface achieved by the load combined with lubricating effects due to

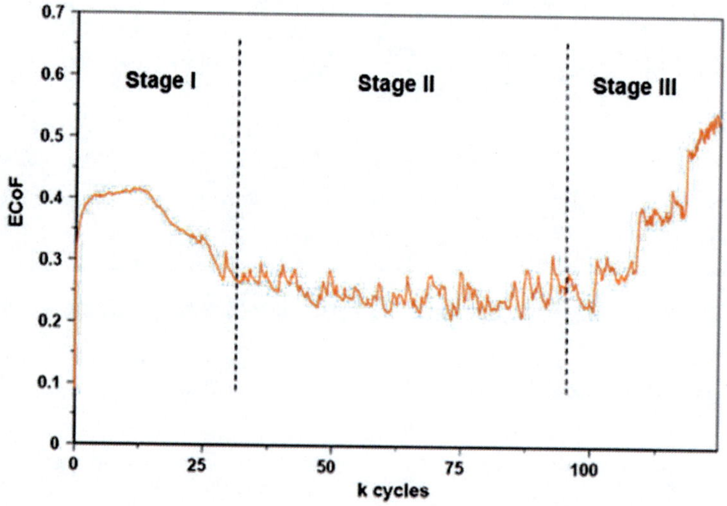

Fig. 12 Categorization of behavior stages within the overall lifetime of 3 GB-PL237 test

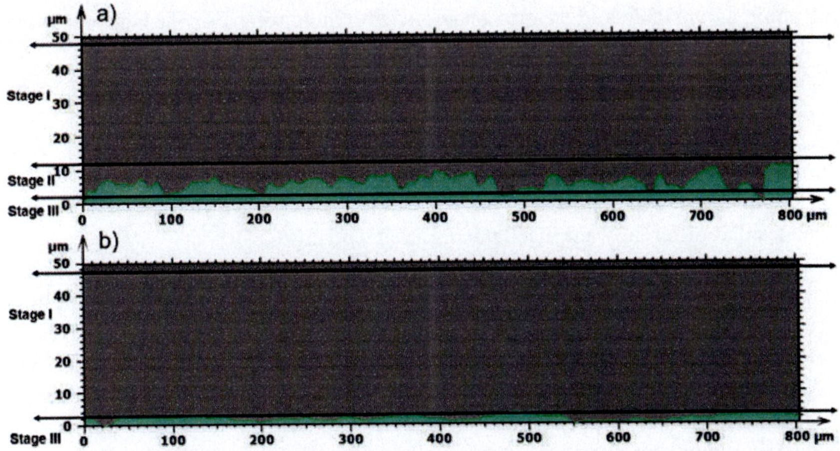

Fig. 13 Schematic diagram of relative DFL thickness in reference to substrate topographical features on the interface surface outlining the dominating stages in the overall lifetime of the system and the point when they begin. **a** is 3 GB-PL237 system and **b** is the SP-PL237 system

the DFL component of the mixed layer leads to progressive decrease of both wear rate and coefficient of friction.

Topographical parameters including the maximum pit depth (Sv) and maximum peak heights (Sp) are therefore proposed to play an influential role on the damage mechanism occurring in the 3 GB-PL237 system.

This leads to the proposition that an initially rougher substrate interface, such as that of the 3 GB-PL237 system in Fig. 13a, leads to a better and more stable DFL fretting system.

The fully established smooth metal-DFL layered interface produces the low friction, stable phase of Stage II (Fig. 12) where elastic shearing is thought to dominate the wear mechanism. Signs of this shearing mechanism amid the low friction phase can be observed in Fig. 7b where plan view SEM images of the interrupted (at 70 k cycles) 3 GB-PL237 test show streaks of DFL material overlaying the underlying titanium substrate. This low ECoF phase (Stage II) and wear mechanism is only evident from fretting test samples with the rough (grit blasted) interfaces; consistent with the requirement for the DFL/substrate interface to feature both peak metallic asperities and pits that act as DFL reservoirs.

Minimal resistance to the fretting damage can be observed from cross sectional images seen in Fig. 10 for the SP-PL237 system. EDX analysis confirms depletion of DFL material on the surface with oxide debris and titanium asperities being dominant after just 8 k cycles. This is consistent with the average depth profiles in Fig. 9b which show the interrupted SP-PL237 test to have a wear depth of 76 µm (below the initial DFL thickness of around 50 µm), indicating removal of DFL material had already occurred in the early stages of the test. Evolution of ECoF in Fig. 9a, shows that after this point, an unstable and erratic behaviour with increasing ECoF ensues resulting in ultimate failure, a characteristic behaviour of Stage III.

Average wear depth profile of the higher interface roughness from the interrupted 3 GB-PL237 test (Fig. 6b) shows that at 70 k cycles, the wear depth is 60 μm deep. This a deeper than the initial DFL thickness, however despite a significantly large number of cycles run this system has a markedly lower wear depth than the 76 μm reached in the SP-PL237 test after only 8 k cycles. This large difference in wear rate indicates that different wear mechanisms are occurring.

The shot peened substrate interface is relatively smooth, featuring a maximum peak height (Sp) of 4.24 μm and maximum pit depth (Sv) of 6.08 μm. The low surface roughness (Sa) of 0.6 μm is mainly dominated by smooth low grooves as seen in Fig. 3b. The lack of both asperities to undergo plastic shearing and break-off, and DFL retaining pits prevents the creation of the mixed metal-DFL layer associated with the stable low friction Stage II. The lack of formation of the low CoF mixed metal-DFL layer accelerates the wear rate in the initial stages, resulting in relatively rapid penetration of the DFL to produce metal to metal contact. This explains why results from 3 GB-PL237 tests features the low CoF Stage II whereas the SP-PL237 tests do not. It is also clear that the features of the interface directly affect the wear mechanisms, and hence the overall system lifetimes. Figure 13 highlights which parts of the DFL-metal system are involved in the different stages of the fretting lifetime.

The evolution of ECoF against number of cycles (Fig. 5) displays the variety of GB-PL237 systems with the various initial surface conditions tested. These set of tests shows the various GB-PL237 systems both outperform as received and shot peened surfaces by having longer lifetimes and have distinctly different ECoF evolutions. All grit blasted surfaces, with varying degrees of roughness, show similar behaviours with a high initial ECoF which decreases to a steady low friction phase for a number of cycles (Stage II) before reaching the failure stage where the behaviour increasingly become erratic and friction coefficient gets higher until failure point is met. The requirement of a rough interface for Stage II to occur has already been explained. The link between lifetime and actual roughness of the grit blasted interface has not.

All GB-PL237 systems follow the same pattern, with fretting lifetimes and the rate of damage varying depending on the interfacial roughness generated by different severities of grit blasting: 1 bar (1 GB-PL237) produced a Sa of 1.23 μm; 3 bar (3 GB-PL237) to give a Sa of 1.6 μm; 9 bar (9 GB-PL237) resulted in a Sa of 2.77 μm (Table 2). 1 GB-PL237, with the lowest interfacial roughness of the grit blasted samples of 1.23 μm, had the shortest fretting lifetime of 100 k cycles. Increasing the interfacial roughness to 1.6 μm increased the fretting lifetime to 120 k cycles. However, further increase Sa to 2.77 μm resulted in only incremental further increase in the longevity of the system to 140 k cycles. These results show that there is a minimum roughness threshold, lying between that of the shot peened and 1 bar grit blasted conditions, that is required for the Stage II formation mechanism to work. The results also suggest that there is an upper roughness threshold with grit blasting, beyond which, there is no further significant lifetime extension.

5 Conclusion

The effects of initial surface conditions on the performance of an epoxy bonded DFL fretting systems have been explored. The observed fretting wear damage mechanisms are discussed. This study has established a correlation between DFL lifetimes and interfacial roughness. Different fretting wear mechanisms have been identified and associated with the different stages of damage evolution and ECoF behaviours. The wear mechanisms in each surface conditions occur at various paces through each system to which rougher surface experience a phase of elastic shearing and low friction behaviours while smoother surfaces are characterised by having an abrasive wear mechanism and instability throughout the fretting lifetime. It can be concluded that:

(1) DFL fretting lifetimes and stability are dependent on the roughness and topographical features of substrate interface.
(2) A rough DFL/substrate interface is required to form the durable low friction Stage II regime which increases the overall lifetime of the system. Metallic asperities and DFL retaining pits are both required.
(3) Initial wear of bulk DFL material occurs with SP and GB systems, however, different mechanisms takeover once the counter-body reaches the metallic peaks to lead to plastic deformation of said peaks to be mixed and adhered with DFL reservoirs and form a durable interface layer between contacting surfaces.
(4) GB-PL237 system with higher roughness only leads to a slight increase in DFL lifetime.

This work confirms that DFL fretting lifetimes is closely linked to the DFL/substrate interfacial roughness. Further studies to isolate effects of material and geometry are needed to fully understand how to optimise such systems.

Acknowledgements This work is funded by Rolls-Royce plc and the EPSRC through an industrial CASE award.

References

1. Cressman D, Tury B, Doll GL (2016) Effects of surface treatments and coatings on tribological performance of Ti-6Al-4V in the mixed fretting and gross slip regimes. Surf Coat Technol 290:110–115
2. Fouvry S, Paulin C (2014) An effective friction energy density approach to predict solid lubricant friction endurance: Application to fretting wear. Wear 319(1):211–226
3. Xu GH, Zhu M, Liu J, Zhou Z, Liang H (2003) The effect of pre-treatment of substrate on fretting tribological behavior of MoS2 coatings. Wear 255(1–6):246–252
4. Matthews A, Leyland A, Holmberg K, Ronkainen H (1998) Design aspects for advanced tribological surface coatings. Surf Coat Technol 100–101:1–6
5. Busch C, Dresel W, Heckler R-P (2012) Lubricants, lubricating greases. Ullmann's Encycl Ind Chem 21:547–571

6. Hutchings I, Shipway P, Hutchings I, Shipway P (2017) Surface topography and surfaces in contact. Tribology 7–35
7. Lu W et al (2017) Influence of surface topography on torsional fretting wear under flat-on-flat contact. Tribol Int 109:367–372
8. Xu J, Zhu MH, Zhou ZR, Kapsa P, Vincent L (2003) An investigation on fretting wear life of bonded MoS2solid lubircant coatings in complex conditions. Wear 255(1–6):253–258
9. Xu J, Zhou ZR, Zhang CH, Zhu MH, Luo JB (2007) An investigation of fretting wear behaviors of bonded solid lubricant coatings. J Mater Process Technol 182(1–3):146–151
10. Ye Y, Chen J, Zhou H (2009) An investigation of friction and wear performances of bonded molybdenum disulfide solid film lubricants in fretting conditions. Wear 266(7–8):859–864
11. Zhu MH, Zhou ZR (2001) An investigation of molybdenum disulfide bonded solid lubricant coatings in fretting conditions. Surf Coat Technol 141(2–3):240–245
12. Barman K, Shipway PH, Voisey KT, Pattinson G (2017) Evolution of damage in MoS2-based dry film lubricants (DFLs) in fretting wear—the effect of DFL thickness and contact geometry. Prog Org Coat 105:67–80
13. Fouvry S, Duó P, Perruchaut P (2004) A quantitative approach of Ti-6Al-4V fretting damage: friction, wear and crack nucleation. Wear 257(9–10):916–929
14. Fu Y, Wei J, Batchelor AW (2000) Some considerations on the mitigation of fretting damage by the application of surface-modification technologies. J Mater Process Technol 99(1–3):231–245

Predicting Rebound of Ellipsoidal Granules Using SPH

Dhairya R. Vyas, Sharen J. Cummins, Murray Rudman, Gary W. Delaney, Paul W. Cleary, and Devang V. Khakhar

Abstract The accuracy of SPH in predicting rebound kinematics in granular flow applications involving non-spherical granules is investigated. For this, the rebound behavior of an ellipsoidal granule impacting an elastic substrate is analyzed for different impact angles and for different initial orientations of the granule. The friction modeling capabilities of SPH are investigated by comparing the rebound spin predicted by the SPH simulations (a) without any special friction model and (b) using Coulomb's friction model, with DEM results. This study presents the potential of using SPH in analysing granular flow applications and brings to light the limitations associated with the existing friction modeling capabilities in SPH.

Keywords Ellipsoids · Granular flow · Friction · DEM · SPH

1 Introduction

Granular flow of particles is encountered in a large number of industrial, agricultural and mining applications. These interacting granules often have different shapes, sizes, and materials, and can undergo a wide range of processes [1]. These can range from compaction and storage of small capsule or tablet-shaped medicines in pharmaceutical industries [2, 3] to crushing and transport of much larger randomly shaped rocks in mining applications [4, 5]. The design of the equipment used in these

D. R. Vyas (✉)
IITB-Monash Research Academy, Indian Institute of Technology Bombay, Mumbai 400076, India
e-mail: dhairya.vyas@monash.edu

S. J. Cummins · G. W. Delaney · P. W. Cleary
CSIRO Data61, Private Bag 10, Clayton, VIC 3168, Australia

M. Rudman
Department of Mechanical and Aerospace Engineering, Monash University, Vic 3800, Australia

D. V. Khakhar
Department of Chemical Engineering, Indian Institute of Technology Bombay, Mumbai 400076, India

M. Abdel Wahab (ed.), *Proceedings of the 8th International Conference on Fracture, Fatigue and Wear*, Lecture Notes in Mechanical Engineering,
https://doi.org/10.1007/978-981-15-9893-7_50

applications requires an understanding of collisions, deformation and breakage of the interacting granules. Experimental methods used to analyse these applications are often expensive and, in some cases, can be intrusive which makes it difficult to obtain the information inside the flow domain [6]. Numerical simulations provide a cost-efficient and non-intrusive alternative to analyze these phenomena in granular systems.

A numerical method widely used to analyze such applications is the Discrete Element Method (DEM) [7]. The collisions and flow of both spherical and non-spherical granules in several applications, including hoppers [8], mills [4] and crushers [5] has been modelled accurately with DEM. To model these collisions, DEM assumes a soft-sphere approach where the local deformations are assumed to be small and are expressed in terms of overlaps between the granules. These overlaps are then used to calculate the contact forces and subsequently update the velocity, spin and positions of the interacting granules. To calculate these forces, idealized contact force models are assumed. These can be simple force models like the linear spring model [9] or much more comprehensive models like the Hertz-Mindlin and Deresiewicz's (HMD) model [10]. However, since DEM is a single point method, these models do not account for the spatially varying force distributions and/or continually changing shape, which can be important in applications like metal compaction and chemical mechanical planarization. While attempts have been made to model local force variation using distributed contacts in DCDEM [11] and FEM-DEM [12], more work is required to model deformation. Therefore, other methods which spatially discretize the interacting granules and can model deformation need to be explored.

Smooth Particle Hydrodynamics (SPH) [13] is one such meshfree method. The spatial discretization of the interacting objects provides SPH the ability to model the local force distribution and deformation of the interacting objects [14–16]. The accuracy of predicting deformation in high velocity impacts using SPH has been investigated in several studies [16–18]. However, studies of the rebound behavior of non-spherical granules in low velocity impacts using SPH, are scarce. The aim of this study is therefore to investigate the accuracy of the existing SPH approaches to model the rebound kinematics of an ellipsoidal granule impacting an elastic substrate. For ellipsoidal granules, since DEM is able to reproduce rebound kinematics accurately, we will first simulate the problem using the CSIRO's DEM code [4, 8, 19, 20] and then use those results as a benchmark for investigating the SPH results.

In SPH implementations, two methods are commonly used to model the contact and frictional forces between the two impacting bodies [21, 22]. In the first method, the conventional SPH approach is used and no additional modifications are made to incorporate friction. Here, the only frictional dissipation is due to the artificial viscosity. This was used by Li et al. [21] to analyse oblique impacts in cold spray simulations. We will call this SPH-NF (No Friction) model. In the second method implemented by Gutfraind and Savage [22], the frictional force is assumed to be proportional to the normal force. We will denote this method as SPH-CF (Coulomb's Friction) model [15, 23, 24]. The accuracy of both these methods will be investigated.

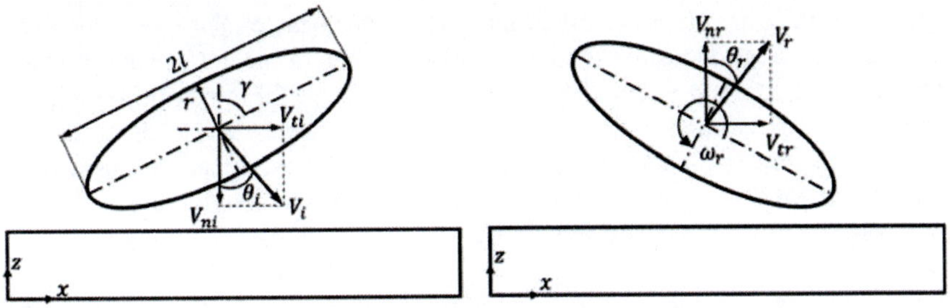

Fig. 1 Schematic representing the pre and post-impact configuration of a granule impacting an elastic substrate

2 Problem Statement

The accuracy of the two SPH friction modeling approaches is investigated by analysing the rebound kinematics of an ellipsoidal granule obliquely impacting an elastic substrate. The pre- and post-impact configurations of the system are shown in Fig. 1a, b respectively. The impacting granule shown in Fig. 1 is a 3D prolate spheroid with minor axis $r = 25$ mm and major axis $l = 50$ mm. The granule impacts a flat elastic substrate with a normal velocity $V_{ni} = 5$ m/s and the tangential velocity V_{ti} changes with the impact angle θ_i. It is assumed that the ellipsoid has zero spin before the impact. The orientation of the granule with respect to the impacted surface is expressed in terms of γ, which is the angle made by the major axis with respect to the vertical. So, for $\gamma = \pm 90$, the major axis will be horizontal and for $\gamma = 0$, it will be vertical. The granule rebounds at an angle θ_r with velocity V_r and spin ω_r.

The properties of the granule and the substrate are chosen to be identical. Their density $\rho = 2650\,\text{kg/m}^3$, Young's modulus $E = 70\,\text{GPa}$, and Poisson's ratio $\upsilon = 0.3$. The coefficient of friction is chosen as $\mu = 0.1$.

3 DEM Modeling

Since DEM can predict rebound kinematics for this problem accurately, it will be used as a benchmark with which the SPH results will be compared. We, therefore, begin by analyzing the above problem using DEM. The specific implementation of the DEM method used here has been developed by CSIRO and uses the super-quadric approach used in [8, 20, 25] to model the particle shape. Here, the equation of the particle in the local frame is given by

$$\left(\frac{x}{a}\right)^m + \left(\frac{y}{b}\right)^m + \left(\frac{z}{c}\right)^m = 1, \tag{1}$$

where, a, b and c are the aspect ratios and m is the shape factor. For $m = 2$, the shape is ellipsoidal, as m increases the particles become more blocky. For the problem described in Sect. 2, $a = b = r = 25$ mm and $c = l = 50$ mm with $m = 2$. The identification of the contacts and calculation of the overlaps is done using the method described in [26]. To model the normal and tangential forces, a linear spring model is used. The normal component of force (F_n) is determined by multiplying the overlap (δ) with the normal spring stiffness (k_n)

$$F_n = k_n \delta. \tag{2}$$

Since the collision is elastic, no dashpot is required. The tangential components of force ($\boldsymbol{F}_t$) are given by

$$\boldsymbol{F}_t^n = \boldsymbol{F}_t^{n-1} + k_t \boldsymbol{v}_t \Delta t, \tag{3}$$

and if the magnitude $F_t^n > \mu F_n$ then as per Coulomb's criteria $F_t^n = \mu F_n$. Here, the vector force $\boldsymbol{F}_t$ and velocity $\boldsymbol{v}_t$ are defined in the tangent plane at the contact point. The tangential spring stiffness is k_t, the coefficient of friction is μ and the incremental time step is Δt. The summation term represents the force increments corresponding to tangential elastic deformation of the contacting surfaces and the μF_n term represents Coulomb's friction limit beyond which pure slip occurs. The k_t/k_n ratio is chosen as

$$\frac{k_t}{k_n} = \frac{(1 - \upsilon_1)/G_1 + (1 - \upsilon_2)/G_2}{(1 - \upsilon_1/2)/G_1 + (1 - \upsilon_2/2)/G_2}, \tag{4}$$

which is a common approach in DEM simulations [27] and has been demonstrated to match the experimental results [28]. While k_t is dependent on k_n using the above correlation, the appropriate value of k_n still needs to be determined. To do this we examine how the rebound spin changes with k_n. Two cases are considered; in the first case, the major axis is parallel to the substrate ($\gamma = 90°$) and in the second case it is inclined ($\gamma = 30°$). The variation of rebound spin ω_r with k_n for impact angle $\theta_i = 5°$ is shown in Table 1. Since there is no appreciable change after $k_n = $ 1e10, this stiffness is used for the further DEM simulations.

Table 1 Effect of k_n on rebound spin (ω_r) for $\gamma = 30°$ and $\gamma = 90°$ with $\theta_i = 5°$

k_n	1e6	1e7	1e8	1e9	**1e10**	1e11	1e12
ω_r ($\gamma = 30°$)	−157.6	−163.3	−162.7	−162.3	**−162.01**	−162.03	−162.06
ω_r ($\gamma = 90°$)	−19.96	−16.25	−15.08	−15.34	**−15.43**	−15.47	−15.49

4 DEM Results

The effect of the impact angle and the initial orientation of the granule on the granule's rebound spin is analysed. The impact angles chosen are $\theta_i = 5°, 15°, 25°, 35°$ and $45°$ (tangential component of impact velocity in $+x$ direction) and the initial orientations chosen are $\gamma = -90°, -60°, -30°, 0°, 30°, 60°$, and $90°$. We are motivated to study inclinations in both positive and negative directions. The reason for this is best explained by referring to Fig. 2. Here two configurations of the granule are shown (a) with $\gamma > 0°$ and (b) with $\gamma < 0°$. In the initial stage of the contact, when the granule has a positive tangential velocity in the contact zone, for both the configurations, the tangential frictional force (F_t) will induce spin in the clockwise direction. However, while the normal force (F_n) in the first case ($\gamma > 0°$) will induce spin in the clockwise direction, in the second case ($\gamma < 0°$), it will spin the granule in the anticlockwise direction. Therefore, for $\gamma > 0°$, both the moments will be initially in the same direction. In contrast, for $\gamma < 0°$, initially they will be in the opposite directions and the relative magnitudes of these will govern the direction of spin.

For the setup described in Sect. 2, the variation of the final rebound spin with the granule's orientation for different impact angles is shown in Fig. 3. For $\gamma = 0°$ and $\gamma = \pm 90°$, the normal force acts on the center of gravity, and therefore the only spin induced is because of the frictional forces. For inclined granules ($\gamma \neq 0°$ or $\pm 90°$), the orientation effects come into play and a significantly higher spin is observed. For $\gamma > 0°$ that is $\gamma = 30°$ and $60°$, a clockwise (negative) spin is observed. This is expected, since for $\gamma > 0°$ both normal and the initial tangential forces have moments in the clockwise direction. For these angles, the spin is also found to vary with the impact angle. This is because while the normal forces are not strongly dependent on the impact angle (and depend mainly on the orientation), the tangential frictional forces vary with the impact angle, because of their dependence on the tangential velocity. As the tangential velocity increases with impact angle, the tangential forces also increase resulting in an increase in spin magnitude. Note that as the collision progresses for $\gamma > 0°$, and the granule begins to spin, the tangential velocity of the granule at the contact zone will reverse (because of spin) and the direction of the

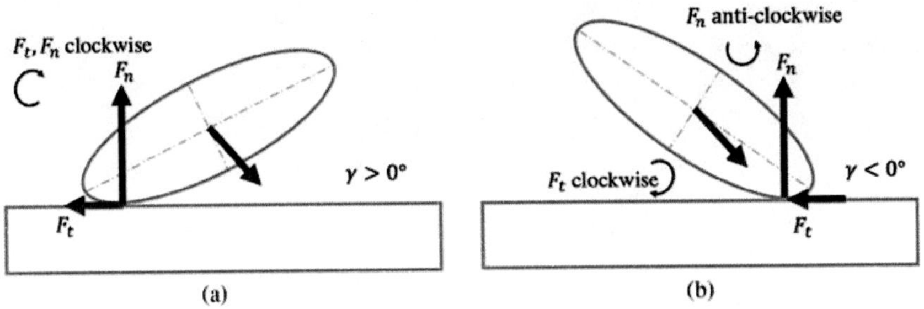

Fig. 2 Schematic representing the direction of the normal force (F_n) and tangential frictional force (F_t) at the beginning of the contact for the two orientations **a** $\gamma > 0°$ and **b** $\gamma < 0°$ of the granule

Fig. 3 Variation of rebound spin of the granule with the initial orientation of the granule for different impact angles predicted using DEM

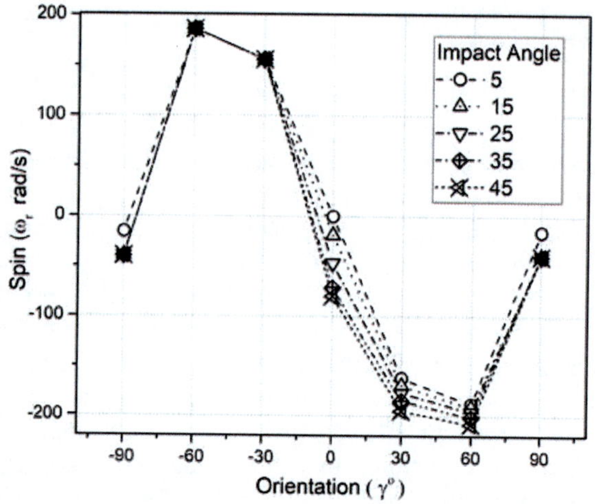

frictional force will also reverse. This is shown in Fig. 4 and Fig. 5 for $\theta_i = 5°$ and 45° and $\gamma = 30°$ and 60°. As seen in these cases, for $\gamma > 0°$ while initially, the tangential force was acting in -*x*-direction, as the collision progresses, it will become zero and then act in the +*x*-direction. This is seen for both $\theta_i = 5°$ and 45°. However, since the tangential velocity for $\theta_i = 45°$ is higher than that for $\theta_i = 5°$, the force reversal for $\theta_i = 45°$ takes longer and therefore for $\theta_i = 45°$ the tangential frictional forces continue to increase the spin magnitude (in clockwise direction) for a longer duration.

For $\gamma < 0°$, that is $\gamma = -30°$ and $\gamma = -60°$, the moment of the normal force is in the anti-clockwise direction but the moment for the initial tangential force is in the clockwise direction. Unlike the $\gamma > 0°$ case, here, the moments are acting in the opposite directions. Since the spin observed is positive (in Fig. 3), it is evident that the

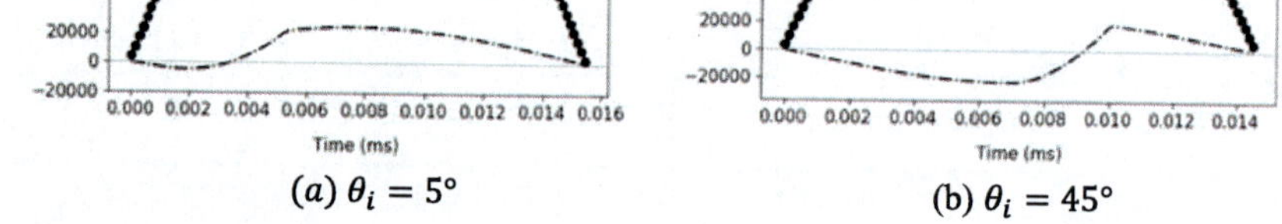

Fig. 4 Variation of normal (F_n) and tangential (F_t) forces for a granule having orientation $\gamma = 30°$ and impacting at angle **a** $\theta_i = 5°$ and **b** $\theta_i = 45°$

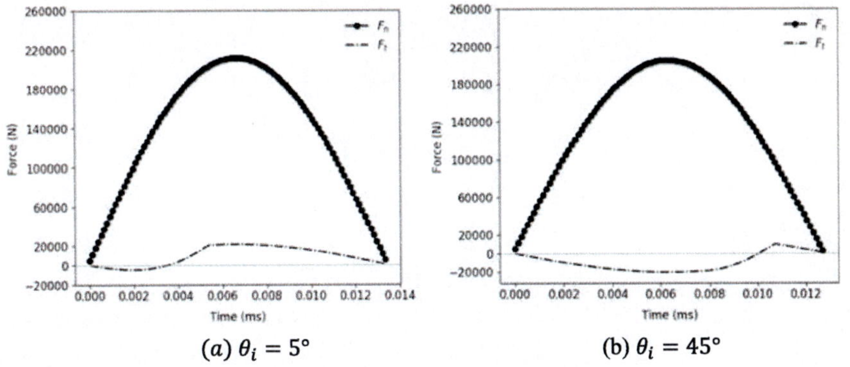

Fig. 5 Variation of normal (F_n) and tangential (F_t) forces for a granule having orientation $\gamma =$ 60° and impacting at angle **a** $\theta_i = 5°$ and **b** $\theta_i = 45°$

normal forces (acting in the anti-clockwise direction) are more dominant. The impact angle has minimal effect on the spin for these cases. This is because in this case the tangential velocity of the granule at the contact zone will always be in the $+x$ direction (as spin is anti-clockwise) and therefore the tangential friction force (acting in the -x-direction) will reach Coulomb's limit sooner (as there is no force reversal) resulting in slip for all the chosen impact angles. This is seen in Figs. 6 and 7 where for $\theta_i = 5°$ and 45° and $\gamma = -30°$ and $-60°$ the tangential force is seen to be in the slip regime throughout the entire collision duration. Also, note in Fig. 3, the magnitudes of spin $|\omega_r(\gamma = +30)| > |\omega_r(\gamma = -30)|$ and $|\omega_r(\gamma = +60)| > |\omega_r(\gamma = -60)|$. This occurs because for $\gamma > 0°$ while the moments of both the forces are initially in the same direction and assisting each other, for $\gamma < 0°$ they are in opposite direction and opposing each other throughout the entire contact duration. These DEM results (in Fig. 3) will be used as a benchmark to assess the SPH results.

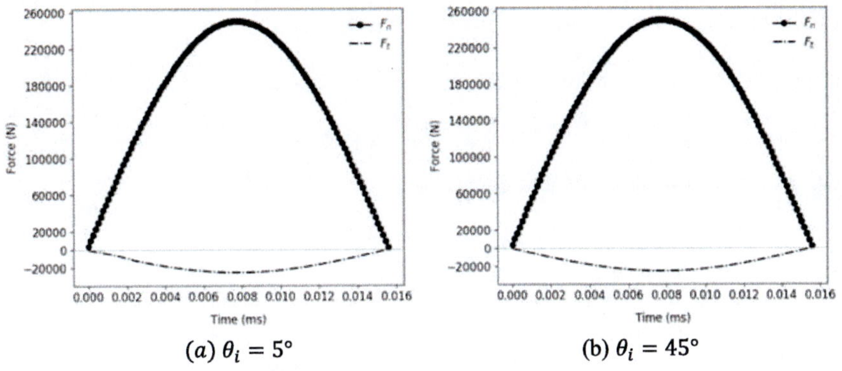

Fig. 6 Variation of normal (F_n) and tangential (F_t) forces for a granule having orientation $\gamma = -$ 30° and impacting at angle **a** $\theta_i = 5°$ and **b** $\theta_i = 45°$

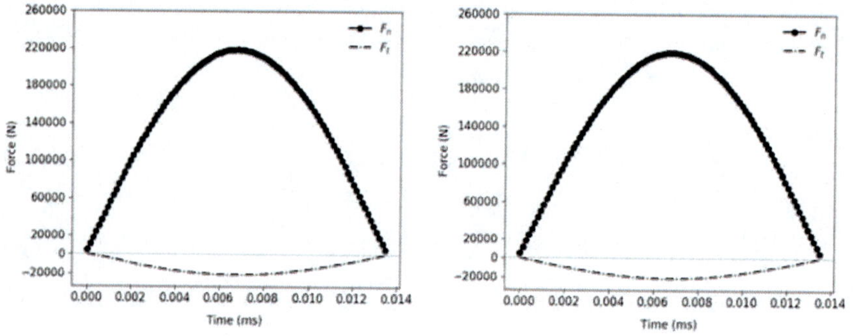

(a) $\theta_i = 5°$ (b) $\theta_i = 45°$

Fig. 7 Variation of normal (F_n) and tangential (F_t) forces for a granule having orientation $\gamma = -60°$ and impacting at angle **a** $\theta_i = 5°$ and **b** $\theta_i = 45°$

5 SPH-Modeling

5.1 SPH Formulation

Smoothed Particle Hydrodynamics (SPH) is a meshfree Lagrangian technique used for solving partial differential equations using a set of moving interpolation points, often termed as particles. It was developed initially for gravitational gas dynamic problems in astrophysics [29] but over the last four decades has been extended to analyze the problems of incompressible fluid dynamics [30] and solid mechanics [31]. A good overview of the technique can be found in the review paper of Monaghan [13].

A brief description of the SPH variant implemented in the code developed by CSIRO and employed in this paper is provided here; the details of the implementation can be found in [13, 31–34].

The equations solved here are the mass conservation equation:

$$\frac{d\rho}{dt} = -\rho \nabla \cdot \boldsymbol{v} \tag{5}$$

and the momentum conservation equation,

$$\frac{d\boldsymbol{v}}{dt} = \frac{1}{\rho} \nabla \cdot \boldsymbol{\sigma} + \boldsymbol{g}. \tag{6}$$

Here, ρ is the material density, $\boldsymbol{v}$ the velocity field, $\boldsymbol{\sigma}$ the stress tensor and $\boldsymbol{g}$ the gravity vector. In the momentum equation Eq. (6) the stress tensor is split into a pressure P and deviatoric stress tensor $\boldsymbol{S}$ via

$$\boldsymbol{\sigma} = -P\boldsymbol{I} + \boldsymbol{S}. \tag{7}$$

The pressure is calculated using the following equation of state

$$P_a = c_a^2(\rho_a - \rho_0) \tag{8}$$

where ρ_a is the particle a's current density, ρ_0 is a reference density (problem dependent) and the speed of sound c_a is given by

$$c_a = \sqrt{\frac{K}{\rho_0}} \tag{9}$$

where K is the bulk modulus. We assume that Hooke's law is valid for the elastic impacts analysed here with a material's shear modulus given by G. Evolution for the components of the deviatoric stress $\boldsymbol{S}$ is calculated using the Jaumann rate [33] equation,

$$\frac{dS^{ij}}{dt} = 2G\left(\dot{\epsilon}^{ij} - \frac{1}{3}\delta^{ij}\dot{\epsilon}^{ij}\right) + S^{ik}\Omega^{kj} + \Omega^{ik}S^{kj}, \tag{10}$$

where the rate of strain tensor is

$$\dot{\epsilon}^{ij} = \frac{1}{2}\left(\frac{\partial v^i}{\partial x^j} + \frac{\partial v^j}{\partial x^i}\right), \tag{11}$$

and the rate of rotation tensor is

$$\Omega^{ij} = \frac{1}{2}\left(\frac{\partial v^i}{\partial x^j} - \frac{\partial v^j}{\partial x^i}\right). \tag{12}$$

For interactions between particles of two different bodies the deviatoric stress term in Eq. (7) is set to zero, since there is no shear stress between them. The SPH discretized form of the continuity equation Eq. (5) for particle a is

$$\frac{d\rho_a}{dt} = \sum_b m_b(\boldsymbol{v}_a - \boldsymbol{v}_b) \cdot \nabla_a W_{ab}. \tag{13}$$

Here the function $W_{ab} = W(\boldsymbol{r}_{ab}, h)$, with $\boldsymbol{r}_{ab} = \boldsymbol{r}_a - \boldsymbol{r}_b$, is a smoothing kernel that weights nearby particles more strongly than distant particles. In practice the smoothing kernel has compact support, i.e., $W(|\boldsymbol{r}|) = 0$ for $|\boldsymbol{r}| > 2h$ where h is a "smoothing length" that is specified a priori. Here, the cubic spline kernel [13] is chosen for W as it has been shown to be suitable for solid mechanics applications [35].

The SPH discretization of Eq. (6) for particle a is

$$\frac{d\boldsymbol{v}_a}{dt} = \sum_b m_b \left(\frac{\boldsymbol{\sigma}_a}{\rho_a^2} + \frac{\boldsymbol{\sigma}_b}{\rho_b^2} + \Pi_{ab}\boldsymbol{I} \right) \cdot \nabla_a W_{ab} + \boldsymbol{g} \tag{14}$$

where, Π_{ab} is an artificial viscosity term as calculated in [31]. In addition, we used a tensile instability correction term of the type proposed by [31] with a coefficient of 0.5. The gradients of the velocity needed to calculate $\dot{\epsilon}^{ij}$ and Ω^{ij} are determined using

$$\left(\frac{\partial v^i}{\partial x^j} \right)_a = \sum_b \frac{m_b}{\rho_b} (v_b^i - v_a^i) \frac{\partial W_{ab}}{\partial x_a^j} . \tag{15}$$

Finally, the Jaumann rate Eq. (10) is written by substituting in the values of $\dot{\epsilon}^{ij}$ and Ω^{ij} found via the interpolations presented in Eq. (15). Individual values of $\dot{\epsilon}^{ij}$ and Ω^{ij} at each particle b are found and then multiplied.

The XSPH smoothing [13] is adopted to update the position ($\boldsymbol{r}$) of the particle

$$\frac{d\boldsymbol{r}_a}{dt} = \boldsymbol{v}_a + \epsilon \sum_b \frac{m_b}{\rho_{ab}} (\boldsymbol{v}_b - \boldsymbol{v}_a) W_{ab} \tag{16}$$

where $\rho_{ab} = (\rho_a + \rho_b)/2$ and $\epsilon = 0.5$. This smoothing results in a particle advection velocity closer to the average velocity in the neighbourhood and reduces the disorder [13]. The integration is performed using the improved Euler explicit integration scheme [13] with timestep Δt calculated using the Courant condition,

$$\Delta t = \min_a \left(\frac{0.5h}{c_a} \right). \tag{17}$$

For the problem described in Sect. 2, the only boundary condition required is an immovable row of particles at the base of the domain.

In SPH implementations used for modelling collisions, two methods have been commonly used to model the contact and frictional forces at the interface of the two impacting bodies, we denote these as SPH-NF and SPH-CF. In SPH-NF [21], the conventional SPH approach is used and no special modifications are made to incorporate friction. Here, the only frictional dissipation is due to the artificial viscosity. In the SPH-CF model implemented by Gutfraind and Savage [22], an imaginary radius is assigned to the SPH particles on the granule's and the substrate's surface and overlap between these is calculated. The normal force is calculated by multiplying the overlap with a proportionality parameter and the tangential force is obtained by multiplying the normal force with the coefficient of friction. This is similar to the slip regime of DEM, however, while in DEM the net force between two impacting bodies was obtained at a single contact point, here the granule and the substrate are discretized into many SPH particles and the force acts individually between the interacting particle pairs of the two bodies. For this study, the imaginary radius for the surface particles is chosen as the SPH smoothing length $h = 1.2p$, where p is the initial particle separation. The normal overlap between the particles is then calculated using

$$\delta_n = (h - dx), \tag{18}$$

where dx is the distance between the particle centers. The normal force is then calculated by multiplying the normal overlap with a proportionality parameter (β), which needs to be adjusted to avoid large overlaps between the particles [22],

$$F_n = \begin{cases} 0, & \delta_n < 0 \\ \beta\delta_n, & \delta_n \geqslant 0. \end{cases} \tag{19}$$

The direction of the normal force is in the direction of normal to the tangent plane between the two particles. The tangential force is then calculated by multiplying the normal force with the coefficient of friction μ, and is acting in the direction opposite to the tangential velocity of the interacting particles

$$F_t = \mu F_n. \tag{20}$$

Which is then distributed to the two tangential directions based on the velocity components, for example when the tangential velocity is zero in one of the two tangential directions, the tangential force is also set to zero in that direction. These normal and tangential forces calculated between the particle pairs are then applied to the individual particles.

To compare the rebound spin predicted by SPH with DEM (Fig. 3), the total angular momentum $\boldsymbol{L}$ of the granule, the net moment of inertia $\boldsymbol{I}$ for the granule and the granule's rebound spin are calculated as follows

$$\boldsymbol{L} = \sum_l \boldsymbol{r}^l \times \left(m^l \boldsymbol{v}^l\right), \tag{21}$$

$$\boldsymbol{I} = \sum_l \boldsymbol{r}^l \otimes \left(m^l \boldsymbol{r}^l\right), \tag{22}$$

$$\boldsymbol{\omega}_r = \boldsymbol{I}^{-1}\boldsymbol{L}, \tag{23}$$

where l represents a summation over the SPH particles in the granule and $\boldsymbol{r}^l$ is the radius vector of the lth particle from the granule's center. The spin around the y-axis, ω_r is the y-component of $\boldsymbol{\omega}_r$.

5.2 *SPH Model*

The SPH setup of the problem described in Sect. 2 is shown in Fig. 8. The flat elastic substrate is constructed from uniformly spaced particles and has a thickness equal to half of the granule's minor radius (12.5 mm). Since for the chosen velocity ($V_{ni} = 5$ m/s) the elastic deformation is expected to be in microns, this thickness is expected to be sufficient for the results to be independent of the boundary effects from the bottom surface. To ensure an almost uniform particle separation and an accurate representation of the surface curvature, the granule is constructed using a polar coordinate system by discretizing the radial direction and the zenith and azimuthal angles such that the distance between two neighbouring particles is close to p. The dimensions and material properties of the granule are the same as those outlined in Sect. 2.

To ensure that the initial particle separation (p) is sufficiently small, a resolution study is performed. The change in rebound spin for an impact angle $\theta_i = 5°$ and two orientations ($\gamma = 30°$ and $\gamma = 90°$) is monitored as the resolution (initial particle separation) is changed. The results of this for SPH-NF and SPH-CF are presented in Fig. 9. Here, the initial particle separation p is expressed as a fraction of the length of the minor axis r. Since no significant changes are observed after $p = r/50$, this resolution is chosen to analyse the results in Sect. 6.

6 Results

To evaluate the accuracy of the SPH-NF and SPH-CF models in modeling rebound kinematics for an ellipsoidal granule impacting an elastic substrate, the rebound spin predicted by these models is compared with that predicted using DEM. This is

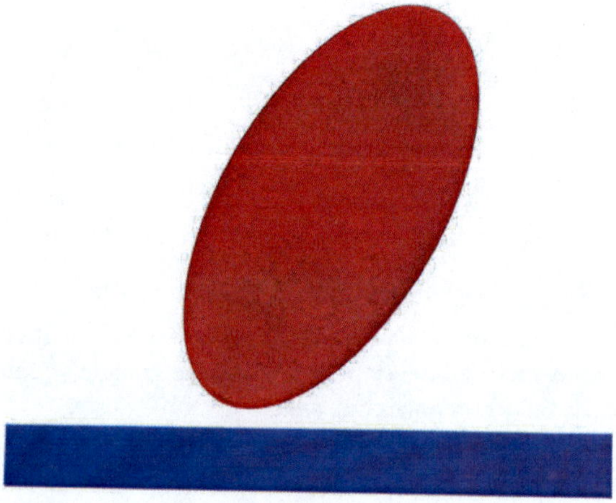

Fig. 8 Setup of an ellipsoidal granule impacting an elastic substrate, with the small red and blue spheres representing SPH particles in the granule and the substrate respectively

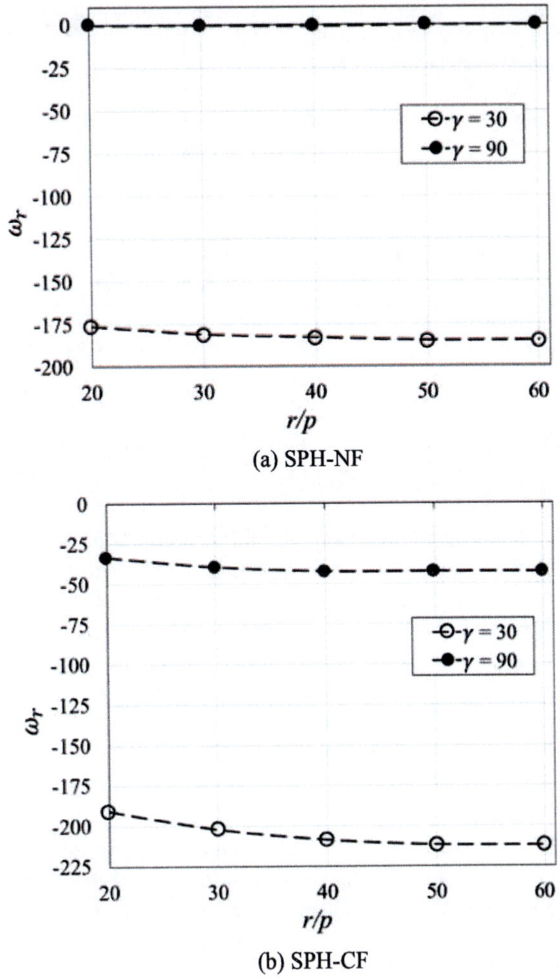

Fig. 9 Resolution study: Effect of initial particle separation on the rebound spin of the granule ($\theta_i = 5°$) for **a** SPH-NF and **b** SPH-CF

done for different initial orientations of the granule (γ), impacting the substrate at different impact angles (θ_i). These inclination angles and impact angles have been chosen identical to those chosen in Sect. 4. The results for these models are presented in the next two subsections.

6.1 SPH-NF Results

The rebound spin predicted using SPH-NF for different combinations of initial orientation and impact angles is shown in Fig. 10. In SPH-NF, there is no explicit friction

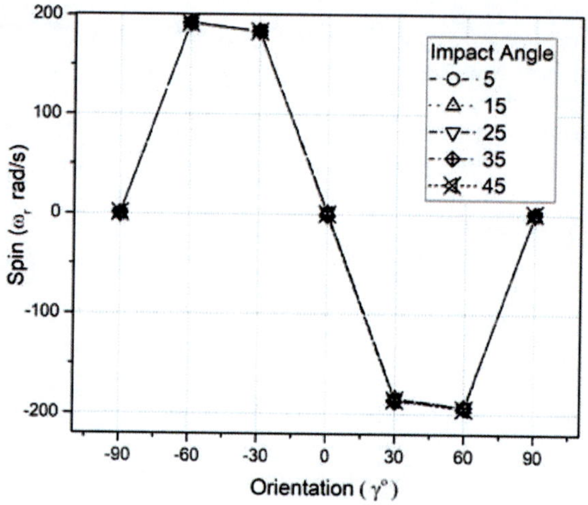

Fig. 10 Variation of the rebound spin of the granule with the initial orientation for different impact angles predicted using SPH-NF

model implemented and the only dissipation is due to the artificial viscosity. For $\gamma = 0°$ and $\pm 90°$ for all impact angles an almost zero spin is observed in Fig. 10; this suggests that the artificial viscosity hardly has any contribution in friction modeling. A similar observation was made by Maveyraud et al. [36].

For these orientations ($\gamma = 0°$ and $\pm 90°$), the DEM results seen in Fig. 3 exhibit a dependence on the impact angle. For DEM, with an increase in impact angle, the spin gradually increases until a specific angle and then becomes constant. Before this angle, there is a combination of stick and slip and beyond this only slip occurs. SPH-NF results, however, show no dependence on the impact angle. It is therefore evident that SPH-NF can neither model stick nor slip. The spin predicted for inclined cases ($\gamma = \pm 30°$ and $\pm 60°$) is similar to that predicted using DEM, though again, the change in frictional forces due to changing impact angle is not observed. Also note that unlike DEM, here $|\omega_r(\gamma = +30°)| = |\omega_r(\gamma = -30°)|$ and $|\omega_r(\gamma = +60°)| = |\omega_r(\gamma = -60°)|$. This is again a result of SPH-NF's inability to model friction. These frictional effects, however, can be of importance while modeling rebound kinematics in granular applications. We next investigate the accuracy of the SPH-CF model.

6.2 *SPH-CF Results*

In SPH-CF, the normal force is calculated by multiplying the overlaps with a proportionality parameter (β) as shown in Eq. (19). This proportionality parameter is chosen such that the overlaps are "sufficiently small" [22] and therefore its choice is not unique. Before evaluating the accuracy of SPH-CF, we investigate the effect of the choice of this proportionality parameter on the results. For an impact angle $\theta_i = 5°$, and two orientations of the granule $\gamma = 30°$ and $\gamma = 90°$, the variation of the

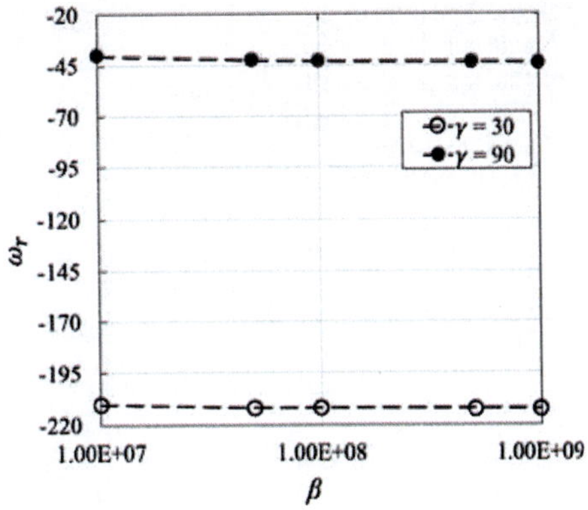

Fig. 11 Effect of the selection of the proportionality parameter β on the rebound spin for SPH-CF

rebound spin with β is presented in Fig. 11. For the range of values chosen $\beta = 1e7$, 5e7, 1e8, 5e8 and 1e9, there is no significant effect of the choice of β on the rebound spin.

To further investigate the effect of the choice of β, the variation of substrate deformation with β was investigated. Note that the primary focus of the work is to assess the rebound kinematics of the impacting granule, and here the substrate deformation is only being analysed to select β. For the problem described in Sect. 2 for $\gamma = 90°$, the maximum substrate deformation predicted using Hertz's theory is 120.9 μm [27]. The predictions made by SPH-CF for different values of β are shown in Table 2. Since for $\beta = 5e7$, the predictions are closest to that of Hertz's theory, this value of β is chosen for further simulations. Note that the deformation predicted by SPH-CF depends strongly on the choice of β (as seen in Table 2), therefore if β is chosen only to "avoid large overlaps" as suggested by Gutfraind and Savage [22], the predicted deformation may not be accurate, since for all these cases shown in Table 2 the overlaps are less than 1% of granules minor axis length.

Having selected β, the effect of the impact angle and the initial orientation of the granule on the rebound spin is investigated. The results of this are shown in Fig. 12.

In Fig. 12, for $\gamma = 0°$ and $\pm 90°$, for all the impact angles, a non-zero spin is predicted. This is identical to the spin corresponding to pure slip cases observed at higher impact angles in the DEM results. For the inclined cases, the spin values are again similar to the DEM values, however, for $\gamma > 0°$, the spin magnitude is higher than that predicted using DEM, as the frictional forces are over-predicted here (as it does not model sticking and force reversals). The spin magnitudes $|\omega_r(\gamma = +30°)| > |\omega_r(\gamma = -30°)|$ and $|\omega_r(\gamma = +60°)| > |\omega_r(\gamma = -60°)|$. This is because for $\gamma > 0°$, the moments of normal and tangential forces are in the

Table 2 Effect of β on maximum substrate deformation for SPH-CF

β	1e7	**5e7**	1e8	5e8	1e9
$u_z(\mu m)$	78.4	**148**	174	181	195

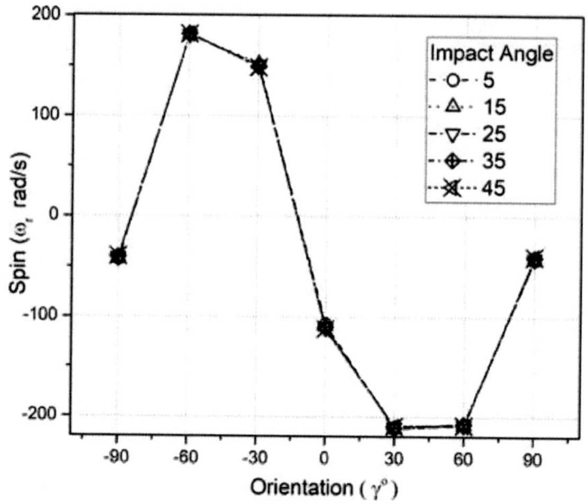

Fig. 12 Variation of the rebound spin of the granule with the initial orientation for different impact angles predicted using SPH-CF

same directions and therefore these two assist each other, while for $\gamma < 0$ they act in opposite directions and resist each other. This was also seen in DEM results (Fig. 3). This shows that SPH-CF is more accurate than the SPH-NF model. However, there are still some differences observed between the DEM and SPH-CF results. While the DEM model captures the complete stick-slip phenomenon, SPH-CF assumes pure slip for all impact angles and therefore the effect that the impact angle has on the results is not observed. This leads to unphysical results especially at lower impact angles where this model predicts significantly higher spin than expected. In summation, the SPH-CF does need some improvement but is still significantly better than SPH-NF in predicting rebound kinematics for the oblique impact of ellipsoidal granules on an elastically deformable substrate.

7 Conclusion

For a low velocity oblique impact of an inclined ellipsoidal granule on an elastic substrate, SPH's prediction of the rebound spin of the granule is compared with the DEM results. Since, for modeling the contact friction in SPH, two commonly used methods are SPH-NF (No special treatment for friction) and SPH-CF (Coulomb's friction), the accuracy of these two methods is investigated. It is observed that both the normal and tangential forces contribute to the spin of an inclined granule. Both the SPH approaches accurately model the contribution of the normal force. The tangential frictional forces are underpredicted by SPH-NF since it neither models stick nor slip but are more accurate for SPH-CF, especially for slip dominated collisions, because of its assumptions of pure slip. However, for stick dominated collisions SPH-CF overpredicts the spin because of its inability to model the stick-slip phenomenon. For positive orientations, none of the two methods capture the effect of the impact angle. The predictions compare better for negative orientations for SPH-CF as it is a slip dominated collision. However, for more general cases (where the stick component is more critical) more accurate models need to be developed. This work investigates the utility of existing friction modeling approaches in SPH and brings to light the limitations associated with these, which when overcome will result in increased accuracy and applicability of the method for analysing granular collisions and the resulting deformation in arbitrarily shaped particles.

References

1. Duran J (2000) Sands, powders, and grains: an introduction to the physics of granular materials. Springer, New York
2. Wu SJ, (Calvin) Sun C (2007) Insensitivity of compaction properties of brittle granules to size enlargement by roller compaction. J Pharmaceutical Sci 96(5):1445–1450. https://doi.org/10.1002/jps.20929
3. Yusof YA, Ng SK, Chin NL, Talib RA (2010) Compaction pressure, wall friction and surface roughness upon compaction strength of Andrographis paniculata tablets. Tribol Int 43(5):1168–1174. https://doi.org/10.1016/j.triboint.2009.12.020
4. Delaney GW, Cleary PW, Morrison RD, Cummins S, Loveday B (2013) Predicting breakage and the evolution of rock size and shape distributions in Ag and SAG mills using DEM. Miner Eng 50–51:132–139. https://doi.org/10.1016/j.mineng.2013.01.007
5. Delaney GW, Morrison RD, Sinnott MD, Cummins S, Cleary PW (2015) DEM modelling of non-spherical particle breakage and flow in an industrial scale cone crusher. Miner Eng 74:112–122. https://doi.org/10.1016/j.mineng.2015.01.013
6. Vu-Quoc L, Zhang X, Walton OR (2000) A 3-D discrete-element method for dry granular flows of ellipsoidal particles. Comput Methods Appl Mech Eng 187(3):483–528. https://doi.org/10.1016/S0045-7825(99)00337-0
7. Cleary PW (2010) DEM prediction of industrial and geophysical particle flows. Particuology 8(2):106–118. https://doi.org/10.1016/j.partic.2009.05.006
8. Cleary PW, Sawley ML (2002) DEM modelling of industrial granular flows: 3D case studies and the effect of particle shape on hopper discharge. Appl Math Model 26(2):89–111. https://doi.org/10.1016/S0307-904X(01)00050-6
9. Walton OR (1984) Application of molecular dynamics to macroscopic particles. Int J Eng Sci 22(8):1097–1107. https://doi.org/10.1016/0020-7225(84)90110-1
10. Thornton C, Cummins SJ, Cleary PW (2011) An investigation of the comparative behaviour of alternative contact force models during elastic collisions. Powder Technol 210(3):189–197. https://doi.org/10.1016/j.powtec.2011.01.013
11. Cummins SJ, Cleary PW (2011) Using distributed contacts in DEM. Appl Math Model 35(4):1904–1914. https://doi.org/10.1016/j.apm.2010.10.019
12. Latham JP, Munjiza A, Garcia X, Xiang J, Guises R (2008) Three-dimensional particle shape acquisition and use of shape library for DEM and FEM/DEM simulation. Miner Eng 21(11):797–805. https://doi.org/10.1016/j.mineng.2008.05.015
13. Monaghan JJ (2005) Smoothed particle hydrodynamics. Rep Prog Phys 68(8):1703–1759. https://doi.org/10.1088/0034-4885/68/8/R01
14. Hosono N, Saitoh TR, Makino J, Genda H, Ida S (2016) The giant impact simulations with density independent smoothed particle hydrodynamics. Icarus 271:131–157. https://doi.org/10.1016/j.icarus.2016.01.036
15. Takaffoli M, Papini M (2012) Material deformation and removal due to single particle impacts on ductile materials using smoothed particle hydrodynamics. Wear 274–275:50–59. https://doi.org/10.1016/j.wear.2011.08.012

16. Hayhurst CJ, Clegg RA (1997) Cylindrically symmetric SPH simulations of hypervelocity impacts on thin plates. Int J Impact Eng 20(1):337–348. https://doi.org/10.1016/S0734-743X(97)87505-7
17. Rabczuk T, Eibl J (2003) Simulation of high velocity concrete fragmentation using SPH/MLSPH. Int J Numer Meth Eng 56(10):1421–1444. https://doi.org/10.1002/nme.617
18. Johnson GR, Petersen EH, Stryk RA (1993) Incorporation of an SPH option into the EPIC code for a wide range of high velocity impact computations. Int J Impact Eng 14(1):385–394. https://doi.org/10.1016/0734-743X(93)90036-7
19. Cleary PW (2008) The effect of particle shape on simple shear flows. Powder Technol 179(3):144–163. https://doi.org/10.1016/j.powtec.2007.06.018
20. Cleary PW, Morrison RD (2016) Comminution mechanisms, particle shape evolution and collision energy partitioning in tumbling mills. Miner Eng 86:75–95. https://doi.org/10.1016/j.mineng.2015.12.006
21. Li WY, Yin S, Wang X-F (2010) Numerical investigations of the effect of oblique impact on particle deformation in cold spraying by the SPH method. Appl Surf Sci 256(12):3725–3734. https://doi.org/10.1016/j.apsusc.2010.01.014
22. Gutfraind R, Savage SB (1997) Smoothed particle hydrodynamics for the simulation of broken-ice fields: Mohr–Coulomb-type rheology and frictional boundary conditions. J Comput Phys 134(2):203–215. https://doi.org/10.1006/jcph.1997.5681
23. Takaffoli M, Papini M (2012) Numerical simulation of solid particle impacts on Al6061-T6 part I: three-dimensional representation of angular particles. Wear 292–293:100–110. https://doi.org/10.1016/j.wear.2012.05.028
24. Takaffoli M, Papini M (2012) Numerical simulation of solid particle impacts on Al6061-T6 Part II: materials removal mechanisms for impact of multiple angular particles. Wear 296(1–2):648–655. https://doi.org/10.1016/j.wear.2012.07.022
25. Cleary PW (2004) Large scale industrial DEM modelling. Eng Comput 21(2/3/4), 169–204. https://doi.org/10.1108/02644400410519730
26. Cleary PW, Stokes N, Hurley J (1997) Efficient collision detection for three dimensional super-ellipsoidal particles. In: Proceedings of 8th international conference on field programmable logic and applications, pp 1–7
27. Johnson KL (1985) Contact mechanics. Cambridge University Press, Cambridge
28. Johnson KL (1995) Surface interaction between elastically loaded bodies under tangential forces. In: Proceedings of the royal society of London. Series A. Mathematical and physical sciences
29. Gingold RA, Monaghan JJ (1977) Smoothed particle hydrodynamics: theory and application to non-spherical stars. Mon Not R Astron Soc 181(3):375–389. https://doi.org/10.1093/mnras/181.3.375
30. Cummins SJ, Rudman M (1999) An SPH projection method. J Comput Phys 152(2):584–607. https://doi.org/10.1006/jcph.1999.6246
31. Gray JP, Monaghan JJ, Swift RP (2001) SPH elastic dynamics. Comput Methods Appl Mech Eng 190(49–50):6641–6662. https://doi.org/10.1016/S0045-7825(01)00254-7
32. Randles PW, Libersky LD (1996) Smoothed particle hydrodynamics: some recent improvements and applications. Comput Methods Appl Mech Eng 139(1–4):375–408. https://doi.org/10.1016/S0045-7825(96)01090-0
33. Lemiale V, King PC, Rudman M, Prakash M, Cleary PW, Jahedi MZ, Gulizia S (2014) Temperature and strain rate effects in cold spray investigated by smoothed particle hydrodynamics. Surf Coat Technol 254:121–130. https://doi.org/10.1016/j.surfcoat.2014.05.071
34. Das R, Cleary PW (2009) Simulating brittle fracture of rocks using smoothed particle hydrodynamics (Chapter 1). In: AM Korsunsky (ed) Current themes in engineering science 2008, Edited volume. America Institute of Physics, pp 1–12
35. Libersky LD, Petschek AG, Carney TC, Hipp JR, Allahdadi FA (1993) High strain lagrangian hydrodynamics: a three-dimensional SPH code for dynamic material response. J Comput Phys 109(1):67–75. https://doi.org/10.1006/jcph.1993.1199

36. Maveyraud C, Benz W, Sornette A, Sornette D (1999) Solid friction at high sliding velocities: an explicit three-dimensional dynamical smoothed particle hydrodynamics approach. J Geophys Res Solid Earth 104(B12):28769–28788. https://doi.org/10.1029/1999JB900217

Micro-Abrasive Wear Behavior Study of an Intermetallic Material—Fe–30Al–6Cr (at.%) Under Conditions of Room and Moderate Temperatures: A Comparison

Eduardo K. T. M. Silva, Jorge H. Luna-Domínguez, Vikas Verma, and Ronaldo Câmara Cozza

Abstract Present work focus on research of tribological resistance of an intermetallic material (Fe–30Al–6Cr—at.%), seeking correlations between wear volume, friction coefficient and temperature. Abrasive experiments were performed with specimens of an iron aluminide alloy against AISI 52100 steel ball and abrasive particles of silicon carbide in glycerin. An individual study was done with respect to their characteristics in terms of SEM-EDS analysis. Different test conditions were defined and the abrasive slurry was, continuously, supplied between the specimen and the ball. Values of tangential force and normal force were acquired simultaneously, for "*ball – abrasive particles – specimen*" tribological system. Systematic studies of the occurrences of the micro-abrasive wear modes, friction and wear generated during tests were done. Moderate temperature favored a larger degree of plastic deformation than removal of material, reducing the wear rate and decreasing glycerin viscosity, which facilitated the movement of the abrasive particles and, consequently, reduced the friction coefficient. Wear volume presented a rising behavior with increase in sliding distance at room and moderate temperatures. Present research explored the potential of an intermetallic material as structural material subjected to moderate temperatures.

Keywords Iron aluminide alloy · Ball-cratering wear testing · Rolling abrasion

E. K. T. M. Silva · R. C. Cozza (✉)
University Center FEI—Educational Foundation of Ignatius "Priest Sabóia de Medeiros", São Bernardo do Campo, SP, Brazil
e-mail: rcamara@fei.edu.br

J. H. Luna-Domínguez
Universidad Autónoma de Tamaulipas, Tamaulipas, Mexico

V. Verma
National University of Science and Technology (NUST) MISiS, Moscow, Russia

R. C. Cozza
CEETEPS-State Center of Technological Education "Paula Souza", Mauá, SP, Brazil

M. Abdel Wahab (ed.), *Proceedings of the 8th International Conference on Fracture, Fatigue and Wear*, Lecture Notes in Mechanical Engineering,
https://doi.org/10.1007/978-981-15-9893-7_51

1 Introduction

In recent decades, advantages of use of intermetallic materials in mechanical and metallurgical applications have been observed, mainly aluminide alloys of iron have been researched, primarily due to the possibility of use as structural materials subjected to elevated temperatures [1].

These intermetallic materials exhibit a high concentration of aluminum [1], which form an adherent layer of alumina, protecting materials against corrosion and oxidation under elevated temperatures [1–3]; it also exhibits greater thermodynamic stability than other oxides, as chromium oxide (Cr_2O_3). The intermetallic materials have smaller densities and exhibit excellent mechanical and metallurgical properties [3–5].

The mechanical properties of intermetallic materials can be controlled by varying aluminum percentage, type of heat treatment and the grain size [6, 7], and as a function of these parameters, they can be applied to specific mechanical and metallurgical needs.

Generally, in carbon-steels, decreases in the yield strength stress and the ultimate stress were observed as a function of the increasing temperature; this phenomenon is reported with the maintenance of the value of the Longitudinal Elastic Modulus. However, with the insertion of iron aluminides (Fe_3Al), it is possible to reach a limit of, approximately, 500 °C, while the magnitudes of yield strength stress and ultimate stress either remain constant or increase [8–11].

Regarding wear tests, the tribological test by rotative ball is widely adopted in researches focusing on micro-abrasive wear of various classifications of materials. Figure 1 shows the principle of the tribological test by rotative ball, where a rotative ball is forced against the tested material and an abrasive slurry is fed between the sphere and the specimen during the tests.

"*Wear craters*" are generated on the tested material and Fig. 2 shows an image of a wear crater, together with a technical indication of the crater diameter (*b*), which is commonly measured using optical microscopy. The wear volume (*V*) can

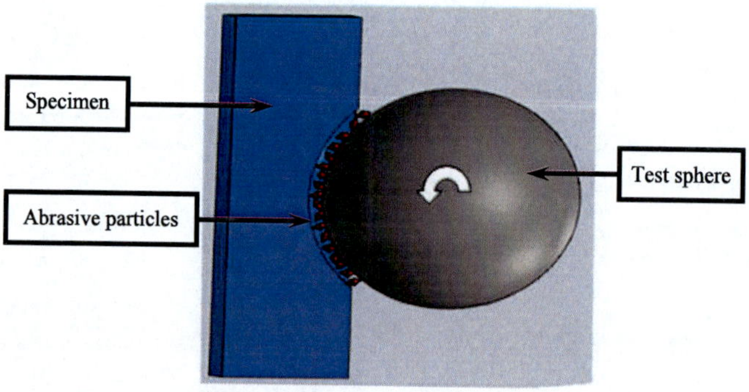

Fig. 1 Ball-cratering tribological test: Its operating principle

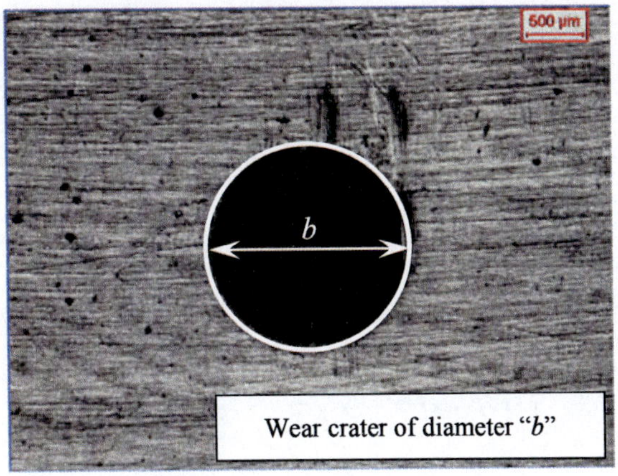

Fig. 2 Wear crater from a micro-abrasive wear test by rotative ball: "*b*" is the diameter of the wear crater

be determined as a function of "*b*", by Eq. (1) [12], being *R* the test sphere radius.

$$V \cong \frac{\pi b^4}{64R} \quad \text{for}\, b \ll R \tag{1}$$

Generally, two micro-abrasive wear modes are reported on the surface of the wear crater: "*grooving abrasion*" is observed when the abrasive particles slide on the specimen surface (Fig. 3a [13]) and "*rolling abrasion*" is related when the abrasive particles roll along the specimen surface (Fig. 3b).

As published in Reference [14], occurrence of abrasive wear deteriorates the properties and lifespan of engineering components, which are influenced by operational conditions and material properties. Then, the prospective applications of intermetallic materials require characterization and quantification of their tribological properties.

Analyzing the scientific literature regarding to wear of intermetallic materials, is possible to find many works [15–23]. However, researches on the micro-abrasive

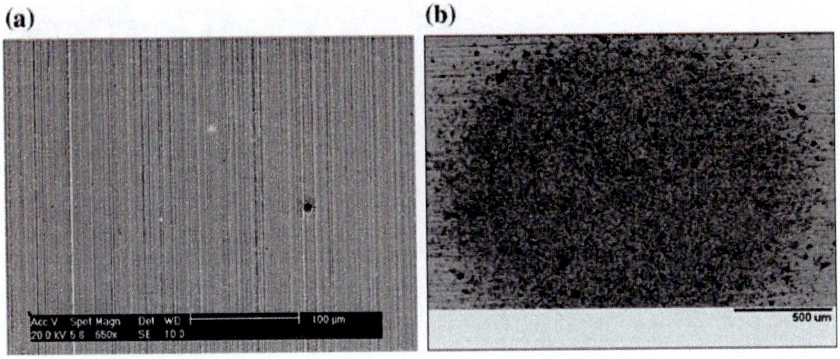

Fig. 3 Micro-abrasive wear modes: **a** "*grooving abrasion*" [13] and **b** "*rolling abrasion*"

wear behavior of intermetallic materials under room and moderate temperatures are scarce. Then, considering the importance of intermetallic materials in many industrial applications [24–31], the purpose of this work is to characterize the micro-abrasive wear performance of an intermetallic material under room and moderate temperatures, seeking correlations between wear volume, friction coefficient and temperature.

2 Materials and Methods

2.1 Tribometer

Analyzing possible (micro-)abrasive wear test methods to conduct this research, "*ball-cratering*" procedure provided a favorable technical condition to conduct micro-abrasive wear study under room and moderate temperatures, due to its mechanical set.

Thus, a ball-cratering wear test equipment of "*free-ball*" mechanical configuration (Fig. 4) was used to conduct tribological tests at room and moderate temperatures. To heat the intermetallic material and conduct experiments above room temperature, a device consisting of an aluminum specimen-support block containing a 1.2 kW electrical heating element was coupled to equipment. Additionally, two load cells were used: one lead cell to control the "*normal force—N*" and one load cell to measure the "*tangential force—T*".

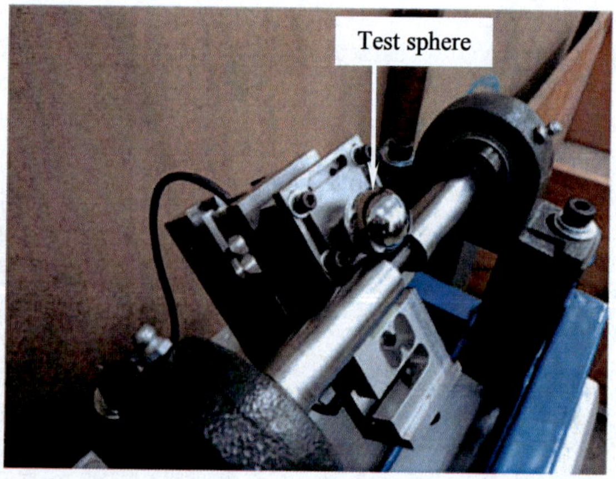

Fig. 4 Ball-cratering equipment of "*free-ball*" configuration

Chemical Element	% (in weight)
C	1.04
Mn	0.35
Si	0.25
Cr	1.45
Fe	Balance

Fig. 5 AISI 52100 bearing steel sphere quenched in oil at 860 °C and stress-relieved at 200 °C for 1 h: its microstructure and chemical composition [32–34]

2.2 *Tested Intermetallic Material*

The tested material was an intermetallic material—Fe–30Al–6Cr (at.%), researched in the "*as cast*" state.

The tested iron aluminide alloy was characterized by SEM and EDS.

2.3 *Counter-Body*

An AISI 52100 bearing steel ball was used as counter-body; its diameter was $D =$ 1" ($D = 25.4$ mm—*standard size*).

Figure 5 [32–34] shows an image of its microstructure with the respective chemical composition.

2.4 *Abrasive Slurries*

The abrasive slurry was composed by abrasive particles of silicon carbide and glycerin.

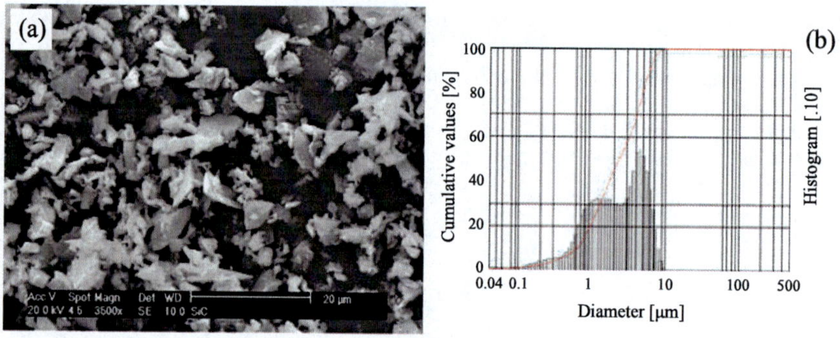

Fig. 6 Abrasive particles of SiC [33, 34]: **a** image obtained by scanning electron microscopy (SEM) and **b** its particle size distribution

Table 1 Hardness of the materials used in this research [7, 8, 32–34]

	Material	Hardness—*H* [HV]	
Tested material	Fe–30Al–6Cr (at.%) intermetallic material	381.6	[7, 8]
Test sphere	AISI 52100 bearing steel	856	[32–34]
Abrasive material	Black silicon carbide—SiC	1886–1937	[33, 34]

SiC material presented an average particle size of 3 μm and an angular shape. Figure 6 [33, 34] shows a scanning electron micrograph of the abrasive material (Fig. 6a) and its particle size distribution (Fig. 6b).

2.5 Hardness of the Materials

Table 1 presents the hardness (H) of the materials used in this research (intermetallic material [7, 8], test ball [32–34] and abrasive particles [33, 34]).

2.6 Experiments

Table 2 presents the tribological parameters defined for the performed tribological tests.

One value for normal force (N) was selected, $N = 0.40$ N, together with one value of abrasive slurry concentration (C), $C = 35\%$ SiC + 65% glycerin (volumetric values) and two values for the test temperatures (Te), $Te_1 = 29$ °C (room temperature) and $Te_2 = 250$ °C (moderate temperature).

Table 2 Test parameters established for the tribological tests

Normal force	[N]	N	0.4	
Abrasive slurry concentration	(% vol)	C	35% SiC + 65% glycerin	
Temperature of test	[°C]	Te_1	29	(room)
		Te_2	250	(moderate)
Test sphere rotational speed	[rpm]	n	75	
Tangential sliding velocity of the test sphere	[m/s]	v	0.1	
Sliding distance	[m]	S_1	6	
		S_2	12	
		S_3	24	
Test time	[min]	t_1	1	
		t_2	2	
		t_3	4	

The test ball rotational speed was $n = 75$ rpm, previously selected by other Researchers [35–37]. For $n = 75$ rpm and a ball of diameter $D = 1$", its tangential sliding velocity is $v = 0.1$ m/s.

Three values of sliding distance (S) were established: $S_1 = 6$ m, $S_2 = 12$ m and $S_3 = 24$ m, with the following corresponding test times (t): $t_1 = 1$ min, $t_2 = 2$ min and $t_3 = 4$ min. Three tribological tests were conducted for each value of sliding distance, according to the following test sequence: 6–6–24–12–12–24–24–12 and 6 m. In total, 18 tribological tests were conducted: nine experiments at $Te_1 = 29$ °C (room temperature) and nine experiments at $Te_2 = 250$ °C (moderate temperature).

All tribological experiments were conducted without interruption. Additionally, the test temperature of $Te_2 = 250$ °C was monitored using an infrared thermometer.

2.7 *Results Analysis*

With the aim of measuring the dimension "*b*" of each wear crater and reporting the occurrences of the abrasive wear modes, subsequently at the end of wear tests, the wear craters were analyzed by optical microscopy and scanning electron microscopy, respectively.

Then, with the values of "*b*", "*T*" and "*N*", volume of wear (V) and friction coefficient (μ) were determined using Eqs. (1) and (2), respectively.

$$\mu = \frac{T}{N} \tag{2}$$

Finally, graphs of $V = f(S, Te)$ and $\mu = f(t, Te)$ were plotted to shows the results obtained.

3 Results and Discussion

3.1 Intermetallic Material Characterization

Figures 7a [7, 8] and 7b [8] are the micrographs (SEM) of the iron aluminide alloy studied; its surface was etched using Villela's reagent (composition of 95 ml of ethanol, 5 ml of HCl and 1 g of picric acid). It consists of an aluminide matrix (D03 ordered) reinforced by a continuous network of eutectic chromium carbides at the interdendritic regions, tentatively identified as M_7C_3 based on the chromium/iron ratio measured by EDS, where Figs. 7c [8] and 7d [8] show the results obtained, for the "*as cast*" state and region of carbides.

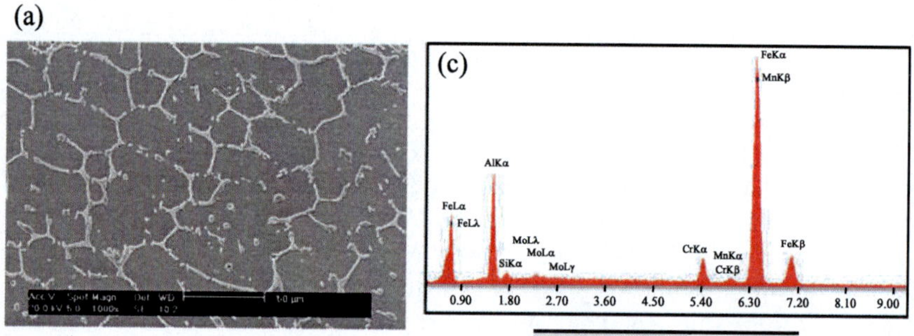

Chemical Element	% (in weight)
Al	14.14
Cr	4.95
Mo	0.75
C	0.66
Fe	78.57

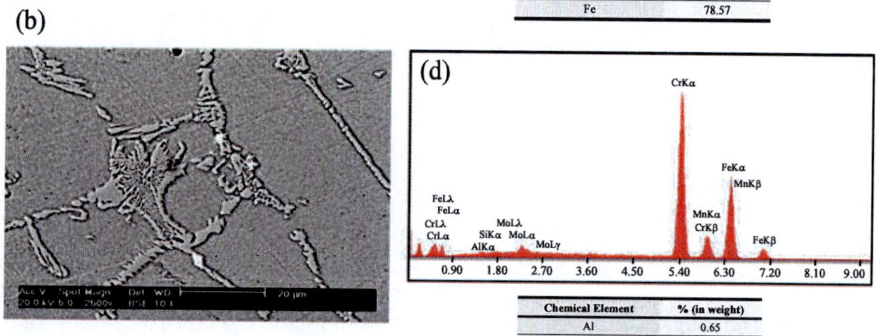

Chemical Element	% (in weight)
Al	0.65
Cr	54.22
Mo	2.18
Si	0.37
Fe	41.92

Fig. 7 Fe-30Al-6Cr (at.%) intermetallic material: **a–b** Its microstructure [7, 8] and EDS for the **c** "*as cast*" state [8] and **d** region of carbides [8]

3.2 Micro-Abrasive Wear Modes Analysis

Figure 8a shows a wear crater produced during the ball-cratering tribological experiments by rotative ball. Figure 8b, c show images obtained by SEM, where the occurrence of rolling abrasion is observed for the conditions of $Te_1 = 29$ °C and $Te_2 = 250$ °C, respectively.

Independent of the sliding distance and test temperature, in all experimental conditions, the wear craters exhibited the occurrence of rolling abrasion. This tribological behavior is in qualitative agreement with the results reported in the literature [35–40], including reports that low values of normal forces and high values of abrasive slurries concentrations favor the occurrence of rolling abrasion.

The greater propensity of rolling abrasion action in the tests conducted under elevated temperature can be explained by the fact that upon increasing the temperature, the specimen material became more ductile, facilitating its indentation by the SiC abrasive particles.

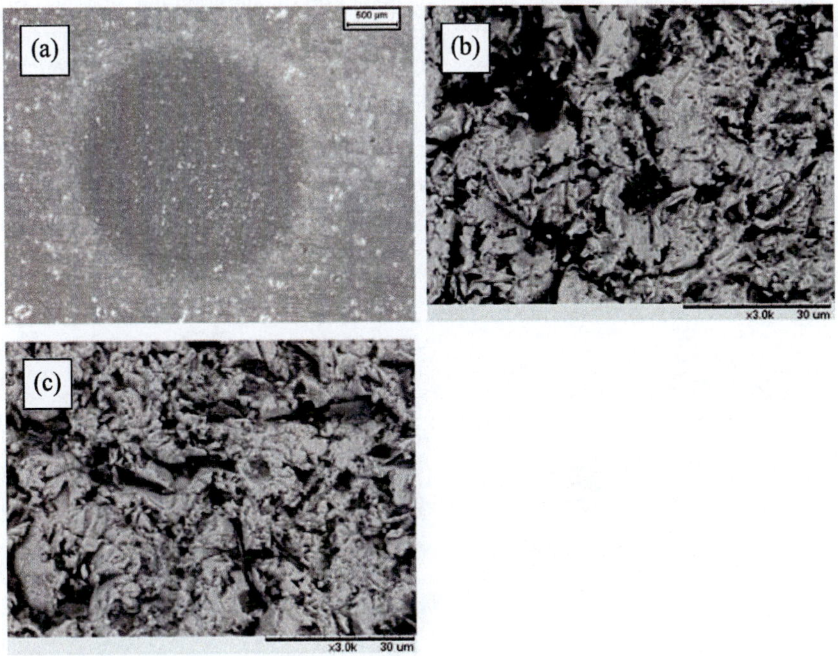

Fig. 8 **a** Wear crater produced during the ball-cratering tribological experiments ($Te_1 = 29$ °C) and the action of rolling abrasion for the conditions of **b** $Te_1 = 29$ °C and **c** $Te_2 = 250$ °C

3.3 Micro-Abrasive Wear Modes Analysis

The behaviors of the wear volume (*V*) as a function of the sliding distance (*S*) and temperature (*Te*) – $V = f(S, Te)$ are presented in Fig. 9.

In both cases, the wear volume presented a rising behavior with the increase of the sliding distance. In addition, the reproducibility obtained in the experiments of this work was directly influenced by the action of rolling abrasion in the wear craters, which, as reported by Bose and Wood [41], tends to produce results with relatively reliable reproducibility.

By analyzing Fig. 9, it is possible to note that the wear volume decreased with the increase of the test temperature—*Te*.

The rolling abrasion, which is the abrasive wear observed in all wear craters of this work, can be related to the "*cutting*" and "*micro-fatigue*" [42] that occurs due to repetitive rolling of the abrasive particles on the surface of the wear crater, e.g., the wear that occurs due to plastic deformation [43].

With the increase in the temperature, the material of the specimen exhibits larger ductility, consequently favoring a larger degree of plastic deformation than removal of material. This high capacity of plastic deformation that the specimen material acquires with the increase of the test temperature favors the occurrence of "*micro-fatigue*" (due to the actions of the abrasive particles) which, in addition to being the wear mechanism of smaller severity when compared with the "*cutting*", is consequently associated with smaller wear rates. To remove a certain quantity of material under the occurrence of micro-fatigue (plastic deformation), a larger number of abrasive particles must slide or to roll in the same region of wear.

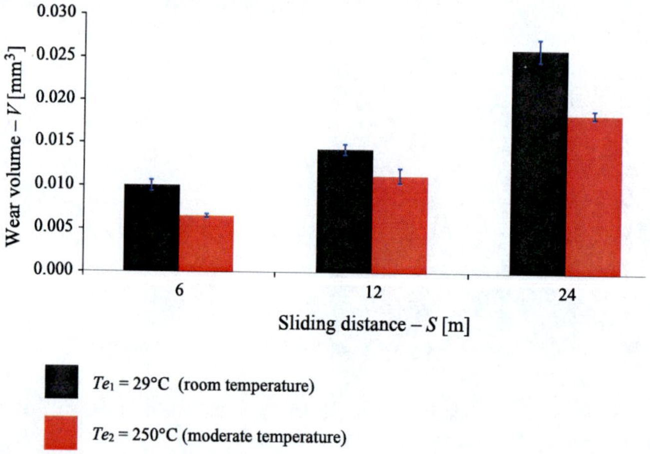

Fig. 9 Wear volume (*V*) as a function of the sliding distance (*S*) and the temperature of test (*Te*) – $V = f(S, Te)$

3.4 Friction Coefficient Analysis

Figure 10 shows the graphics of the friction coefficient (μ) as a function of the test time (t) and test temperature (Te) – $\mu = f(t, Te)$.

Independent of the test condition, the friction coefficient tended to a constant behavior as a function of the test time and temperature.

The temperature significantly influenced the values of the friction coefficient, and in a general overview, the coefficient of friction obtained under $Te_1 = 29$ °C was found to be greater than the values of μ obtained under $Te_2 = 250$ °C. With the increase in the temperature, the specimen material becomes more ductile; thus, a greater amount of abrasive particles is required to rub on a wear region to remove the same quantity of material. In addition, an abrasive particle is more likely to cut at room temperature, while an abrasive particle just deforms plastically under an elevated temperature. Finally, with the increase in the temperature, the viscosity of the glycerin decreased, facilitating the movement of the abrasive particles and, consequently, decreasing the friction coefficient.

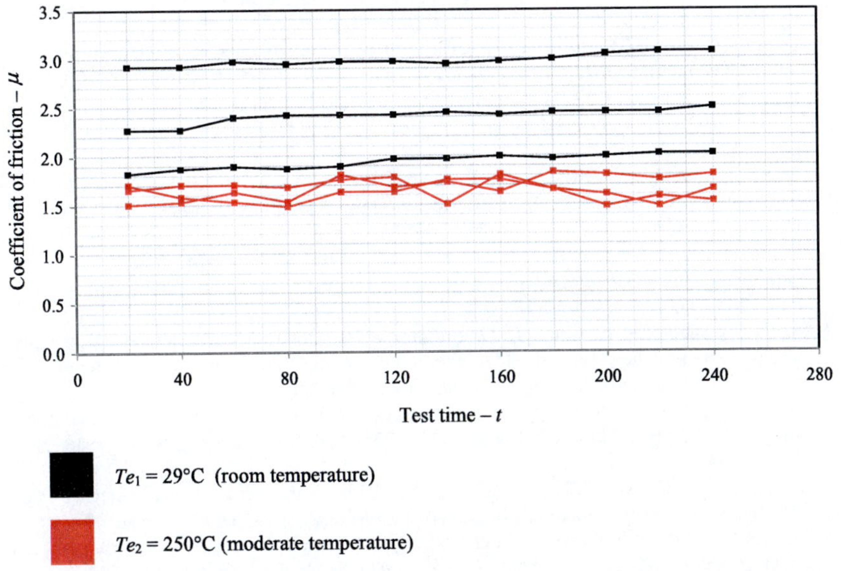

Fig. 10 Graphics of the friction coefficient (μ) as a function of the test time (t) and test temperature (Te) – $\mu = f(t, Te)$

4 Conclusions

Present research explored the potential of an iron aluminide alloy as structural materials subjected to moderate temperatures.

The following points can be highlighted in this research:

- The action of the "*rolling abrasion*" micro-abrasive wear mode was independent of the temperature of test;
- The temperature significantly influences the tribological behavior of the Fe-30Al-6Cr (at.%) intermetallic material studied: as the temperature increased, the wear resistance of the alloy increased;
- Wear volume and friction coefficient: as the temperature increased, the wear volume and the friction coefficient decreased due to the higher ductility acquired by the material specimen and lower viscosity of the glycerin.

References

1. Morris DG, Morris-Muñoz MA (1999) The influence of microstructure on the ductility of iron aluminides. Intermetallics 7(10):1121–1129. https://doi.org/10.1016/S0966-9795(99)00038-2
2. Risanti DD, Sauthoff G (2011) Microstructures and mechanical properties of Fe-Al-Ta alloys with strengthening Laves phase. Intermetallics 19(11):1727–1736. https://doi.org/10.1016/j.intermet.2011.07.008
3. Hanus P, Bartsch E, Palm M, Krein R, Bauer-Partenheimer K, Janschek P (2010) Mechanical properties of a forged Fe-25Al-2Ta steam turbine blade. Intermetallics 18(7):1379–1384. https://doi.org/10.1016/j.intermet.2009.12.035
4. Wu D, Baker I, Munroe PR, George EP (2007) The yield strength anomaly of single-slip-oriented Fe-Al single crystals. Intermetallics 15(2):103–107. https://doi.org/10.1016/j.intermet.2006.03.007
5. Krein R, Schneider A, Sauthoff G, Frommeyer G (2007) Microstructure and mechanical properties of Fe_3Al-based alloys with strengthening boride precipitates. Intermetallics 15(9):1172–1182. https://doi.org/10.1016/j.intermet.2007.02.005
6. Herrmann J, Inden G, Sauthoff G (2003) Deformation behavior of iron-rich iron-aluminium alloys at high temperatures. Acta Mater 51(11):3233–3242. https://doi.org/10.1016/S1359-6454(03)00144-7
7. Borges DFL, Espinosa DCR, Schön CG (2014) Making iron aluminides out of scrap. J Mater Res Technol 3(2):101–106. https://doi.org/10.1016/j.jmrt.2013.12.002
8. Borges DFL (2010) Processing and characterization of iron aluminides made from recycled raw materials (master's thesis, Polytechnic School of the University of São Paulo, São Paulo—SP, Brazil, 2010). Available online: http://www.teses.usp.br/teses/disponiveis/3/3133/tde-25102011-122243/pt-br.php
9. Guan X, Iwasaki K, Kishi K, Yamamoto M, Tanaka R (2004) Dry sliding wear behavior of Fe-28Al and Fe-28Al-10Ti alloys. Mater Sci Eng A 366(1):127–134. https://doi.org/10.1016/j.msea.2003.09.049
10. Zhang X, Ma J, Fu L, Zhu S, Li F, Yang J, Liu W (2013) High temperature wear resistance of Fe-28Al-5Cr alloy and its composites reinforced by TiC. Tribol Int 61:48–55. https://doi.org/10.1016/j.triboint.2012.12.005

11. Sharma G, Limaye PK, Ramanujan RV, Sundararaman M, Prabhu N (2004) Dry-sliding wear studies of Fe3Al-ordered intermetallic alloy. Mater Sci Eng, A 386(1–2):408–414. https://doi.org/10.1016/j.msea.2004.07.053
12. Rutherford KL, Hutchings IM (1997) Theory and application of a micro-scale abrasive wear test. J Test Eval JTEVA 25(2):250–260. doi.org/doi.org/10.1520/JTE11487J
13. Cozza RC, Tanaka DK, Souza RM (2009) Friction coefficient and abrasive wear modes in ball-cratering tests conducted at constant normal force and constant pressure—preliminary results. Wear 267(1–4):61–70. https://doi.org/10.1016/j.wear.2009.01.055
14. Sapate SG, Selokar A, Garg N (2010) Experimental investigation of hardfaced martensitic steel under slurry abrasion conditions. Mater Des 31(8):4001–4006. https://doi.org/10.1016/j.matdes.2010.03.009
15. Itoi T, Mineta S, Kimura H, Yoshimi K, Hirohashi M (2010) Fabrication and wear properties of Fe_3Al-based composites. Intermetallics 18(11):2169–2177. https://doi.org/10.1016/j.intermet.2010.07.014
16. Jóźwiak S, Karczewski K (2009) Influence of aluminum oxides on abrasive wear resistance of Fe-50 at.% Al intermetallics sinters. J Alloy Compd 482(1–2):405–411. https://doi.org/10.1016/j.jallcom.2009.04.034
17. Malafaia AMS, Milan MT, Omar M, Muñoz Riofano RM, de Oliveira MF (2010) Oxidation and abrasive wear of Fe-Si and Fe-Al intermetallic alloys. J Mater Sci 45(19):5393–5397. https://doi.org/10.1007/s10853-010-4591-4
18. Johnson M, Mikkola DE, March PA, Wright RN (1990) The resistance of nickel and iron aluminides to cavitation erosion and abrasive wear. Wear 140(2):279–289. https://doi.org/10.1016/0043-1648(90)90090-W
19. Ahmadian M, Wexler D, Chandra T, Calka A (2005) Abrasive wear of WC-FeAl-B and WC-Ni_3Al-B composites. Int J Refract Metal Hard Mater 23(3):155–159. https://doi.org/10.1016/j.ijrmhm.2004.12.002
20. Mosbah AY, Wexler D, Calka A (2005) Abrasive wear of WC-FeAl composites. Wear 258(9):1337–1341. https://doi.org/10.1016/j.wear.2004.09.061
21. Alman DE, Hawk JA, Tylczak JH, Doğan CP, Wilson RD (2001) Wear of iron-aluminide intermetallic-based alloys and composites by hard particles. Wear 251(1–12):875–884. https://doi.org/10.1016/S0043-1648(01)00745-1
22. Hawk JA, Alman DE (1997) Abrasive wear of intermetallic-based alloys and composites. Mater Sci Eng, A 239–240:899–906. https://doi.org/10.1016/S0921-5093(97)00681-3
23. Maupin HE, Wilson RD, Hawk JA (1992) An abrasive wear study of ordered Fe_3Al. Wear 159(2):241–247. https://doi.org/10.1016/0043-1648(92)90307-T
24. Aoki K (1990) Ductilization of $L1_2$ intermetallic compound Ni_3Al by microalloying with boron: materials transactions. JIM 31(6):443–448. https://doi.org/10.2320/matertrans1989.31.443
25. Deevi SC, Sikka VK (1996) Nickel and iron aluminides: an overview on properties, processing, and applications. Intermetallics 4(5):357–375. https://doi.org/10.1016/0966-9795(95)00056-9
26. Morris DG (1998) Possibilities for high-temperature strengthening in iron aluminides. Intermetallics 6(7–8):753–758. https://doi.org/10.1016/S0966-9795(98)00028-4
27. Schneibel JH, George EP, Anderson IM (1997) Tensile ductility, slow crack growth, and fracture mode of ternary B2 iron aluminides at room temperature. Intermetallics 5(3):185–193. https://doi.org/10.1016/S0966-9795(96)00087-8
28. Dobeš F, Milička K (2010) Estimation of ductility of Fe-Al alloys by means of small punch test. Intermetallics 18(7):1357–1359. https://doi.org/10.1016/j.intermet.2009.11.002
29. Stein F, Schneider A, Frommeyer G (2003) Flow stress anomaly and order-disorder transitions in Fe_3Al-based Fe-Al-Ti-X alloys with X = V, Cr, Nb, or Mo. Intermetallics 11(1):71–82. https://doi.org/10.1016/S0966-9795(02)00187-5
30. Morris DG, Chao J, Garcia Oca C, Muñoz-Morris MA (2003) Obtaining good ductility in an FeAl intermetallic. Mater Sci Eng A 339(1–2):232–240. https://doi.org/10.1016/S0921-5093(02)00108-9

31. Bystrzycki J, Fraczkiewicz A, Łyszkowski R, Mondon M, Pakiela Z (2010) Microstructure and tensile behavior of Fe-16Al-based alloy after severe plastic deformation. Intermetallics 18(7):1338–1343. https://doi.org/10.1016/j.intermet.2010.01.014
32. Cozza RC (2013) A study on friction coefficient and wear coefficient of coated systems submitted to micro-scale abrasion tests. Surf Coat Technol 215:224–233. https://doi.org/10.1016/j.surfcoat.2012.06.088
33. Cozza RC, Tanaka DK, Souza RM (2011) Friction coefficient and wear mode transition in micro-scale abrasion tests. Tribol Int 44(12):1878–1889. https://doi.org/10.1016/j.triboint.2011.08.006
34. Cozza RC (2014) Influence of the normal force, abrasive slurry concentration and abrasive wear modes on the coefficient of friction in ball-cratering wear tests. Tribol Int 70:52–62. https://doi.org/10.1016/j.triboint.2013.09.010
35. Adachi K, Hutchings IM (2003) Wear-mode mapping for the micro-scale abrasion test. Wear 255(1–6):23–29. https://doi.org/10.1016/S0043-1648(03)00073-5
36. Adachi K, Hutchings IM (2005) Sensitivity of wear rates in the micro-scale abrasion test to test conditions and material hardness. Wear 258(1–4):318–321. https://doi.org/10.1016/j.wear.2004.02.016
37. Trezona RI, Allsopp DN, Hutchings IM (1999) Transitions between two-body and three-body abrasive wear: influence of test conditions in the microscale abrasive wear test. Wear 225–229:205–214. https://doi.org/10.1016/S0043-1648(98)00358-5
38. Mergler YJ, in't Veld AH (2003) Micro-abrasive wear of semi-crystalline polymers. Tribol Res Des Eng Syst 41:165–173. https://doi.org/10.1016/S0167-8922(03)80129-3
39. Umemura MT, Varela LB, Pinedo CE, Cozza RC, Tschiptschin AP (2019) Assessment of tribological properties of plasma nitrided 410S ferritic-martensitic stainless steels. Wear 426–427:49–58. https://doi.org/10.1016/j.wear.2018.12.092
40. Allsopp DN, Hutchings IM (2001) Micro-scale abrasion and scratch response of PVD coatings at elevated temperatures. Wear 251(1–12):1308–1314. https://doi.org/10.1016/S0043-1648(01)00755-4
41. Bose K, Wood RJK (2005) Optimum tests conditions for attaining uniform rolling abrasion in ball cratering tests on hard coatings. Wear 258(1–4):322–332. https://doi.org/10.1016/j.wear.2004.09.018
42. Axén N, Jacobson S, Hogmark S (1994) Influence of hardness of the counterbody in three-body abrasive wear—an overlooked hardness effect. Tribol Int 27(4):233–241. https://doi.org/10.1016/0301-679X(94)90003-5
43. Fang L, Liu W, Du D, Zhang X, Xue Q (2004) Predicting three-body abrasive wear using Monte Carlo methods. Wear 256(7–8):685–694. https://doi.org/10.1016/S0043-1648(03)00464-2

Acoustic Wear Monitoring During the Milling of Tool Steel for Machining 4.0

Oluwole Olufayo and Khaled Abou-El-Hossein

Abstract For some years, the concept of Industry 4.0 has been a focal point of interest. Initial deliberations have often been the subject of careful academic research and visionary ideas. However, more recent developments have shown that machining 4.0 is the technology of the future in solving complex autonomous manufacturing and production issues. This technology will permeate the manufacturing sphere with the industrial concept of "Internet of Things", powered by data and nested with machining. To efficiently digitize and automate the machining industry, it is essential to obtain data that can be precisely correlated to machining process condition. As the cutting tool is central to any machining operation, tool condition monitoring is one of the possible approaches used for data generation, production control and to optimize production economics. The research work presented herewith utilizes acoustic emission (AE) signals as a sensing method to provide adequate information on the cutting condition and wear during the end-milling operation of tool steel. A signal processing sequence consisting of feature selection, extraction and correlation process of the AE wave was performed to determine features highly correlated to the tool wear. These features were then fed to a neural network for generating prediction models for the cutting tool condition. The results identified a higher correlation between time-frequency AE features and tool wear as well as distinct time-domain features such as RMS and standard deviation.

Keywords Machining 4.0 · Tool condition monitoring · Acoustic emission

O. Olufayo
Department of Mechanical Engineering, École de Technologie Supérieure, H3C 1K3 Quebec, Canada

O. Olufayo · K. Abou-El-Hossein (✉)
Department of Mechatronics, Nelson Mandela University/Institution, Port Elizabeth, South Africa
e-mail: khaled.abou-el-hossein@mandela.ac.za

M. Abdel Wahab (ed.), *Proceedings of the 8th International Conference on Fracture, Fatigue and Wear*, Lecture Notes in Mechanical Engineering,
https://doi.org/10.1007/978-981-15-9893-7_52

1 Introduction

Due to the pressing need for process optimization and sustainability in manufacturing, the implementation of the concept of industry 4.0 from a machining perspective is essential. Commonly manufacturing comprises of processes that traditionally are not interconnected. However, under industry 4.0, individual steps of the value chain are linked and intercommunicate. The part meant for production is designed for manufacturing using leading-edge analytics and established on data acquired from diverse sources such as sensors. Sensors are devices or portion of the system used to detect machining conditions, observe the process evolution, in an aim to establish performance, as well as control process parameters [1]. Incorporated sensor-based tooling solutions permit the extraction of data from machining for optimization. These are aimed at improving productivity with increased metal removal rates, optimized machine utilization, traceability, economic efficiency, product quality as well as create sustainability in production.

1.1 Automation of the Shop Floor Manufacturing Process

The cybernation of shop floor manufacturing is important (Fig. 1). Yet, today's implementation often falls short in suitably monitoring the peculiarity of its operations. Adequate machining experience is often essential in a conventional workshop for useful data retrieval from machining. However, though an adept workforce is vital to a company's performance, the isolation of this skillset from machining analytics highlights underlying challenges in the implementation of machining 4.0. The dependence of this novel concept on an acute understanding of the science of machining is vital in identifying data sources based on selected output condition. One of such technical condition is wear prediction, which generates defects on the workpiece and threatens automation of the process [2]. To estimate tool condition, various

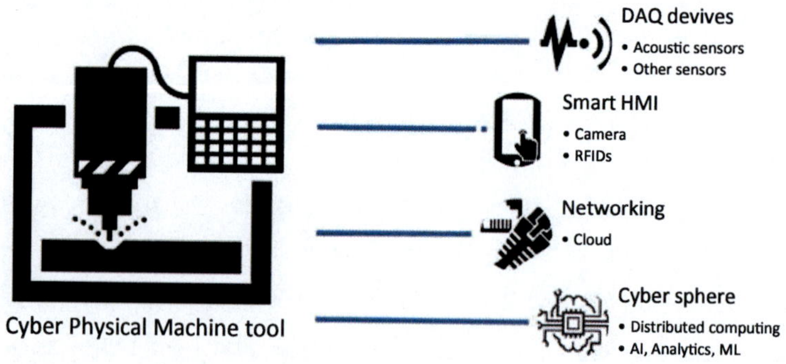

Fig. 1 Cybernation of the manufacturing shop floor

approaches and methodologies based on various choices of sensors (e.g. acoustic, force, vibration and temperature) have been employed by different researchers [3]. The next section presents a tool condition monitoring approach based on sensor fusion.

1.2 Tool Monitoring

As the name implies, tool condition monitoring (TCM) observes the state of the tool during cutting to pre-empt adverse conditions. In machining, key performance indicators such as the rate system availability and performance as well as the quality of the production line need to be an integral part of any condition monitoring system [4]. The performance in this context addresses the on-time and downtime ratios. Though the excessive wear generated on tools during machining are sadly inevitable challenges associated with on-line downtime monitoring, the wear formation on the tool can be observed and mitigated.

The TCM framework. The design of a tool monitoring process as an inclusive precautionary scheme in automated shop floor manufacturing can be perceived as a categorization model to determine tool states. The framework in Fig. 2 illustrates the multi-stage process for the machining features exploration.

Direct and indirect sensing. Toll monitoring techniques can be categorized based on the parameter monitored, state of the machine as well as the type of sensor. Based on sensor types, techniques employed have been categorized two classes i.e. direct and indirect sensing techniques [3]. Despite its lower accuracy, indirect sensing has been utilized in a lot of applications in the industry. These come as a result of its simplicity and profitability. In contrast to direct sensing, this technique observes the parameters of the process associated with the state of the tool. The indirect methodology also uses both statistical and analytical modelling to draw its conclusions on tool wear. Some notable sensing methods associated with it are cutting force, cutting vibration,

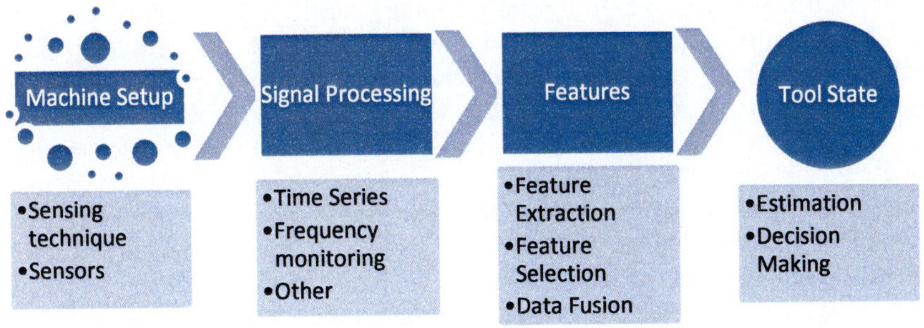

Fig. 2 Tool monitoring (TCM) Framework

Table 1 Sensing methods used in different operations

	Drilling	Forging	Gear	Grinding	Milling	Reaming	Turning
Acoustic emission	[5]	[6, 7]	[8]	[9, 10]	[11–13]	[14]	[15]
Current	[16]		[17]	[17]	[18]		[19]
Cutting force	[16, 20, 21]	[6]	[22]	[23]	[24]		[25–27]
Gap (sensor)	[28]				[29]		[30]
Image					[31]		
Sound	[20]		[32]		[33]		[34]
Spindle power				[35]	[36, 37]		[38]
Vibration	[21]		[32]	[39]	[40]		[19]

acoustic emissions, motor current, cutting temperature. Table 1 shows a summary of sensing methods used in different operations.

Real-time and offline monitoring. Acoustic emission (AE) is a powerful method in monitoring the cutting tool state [41]. AE is particularly effective with the indirect method owing to their non-intrusiveness and fast dynamic response [42]. Monitoring using AE could be implemented real-time or off-line. Real-time continued monitoring enables prompt and instant wear identification. However owing to the influence emerging from setting conditions, this technique has faced numerous challenges in industrial implementation.

Literature highlights the gains of acoustic monitoring owing to its high-frequency range, ease of sensors placement, reduced cost and improved sampling speed [43]. Furthermore, based its properties, AE does not hamper cutting and signals can easily be filtered from environmental noise. It has been classified as an efficient sensing technique suitable for machining [44].

Acoustic emission (AE) sensing. A critical review of the use of AE in machining was undertaken by Kishawy [45]. In their study, they substantiate the premise pointing to AE as a viable TCM technique in machining. Numerous other research studies in recent decades further validate the present-day enthusiasm for this technique [5, 27, 45, 46]. AE in machining emanates from the strain energy released during tool-workpiece action, which in turn results from the plastic deformation occurring in metal cutting. In essence, the AE signals are transient elastic waves from the localized energy release within the material. This can be measured using piezoelectric instruments which transform mechanical stress waves to electrical signals [47].

AE signals of continuous form are often generated from the plastic deformation in ductile materials [48] _ENREF_1 from the progressive wear formed on the tool. In contrast, AE signals of burst form are mostly identified during sudden engagement impact and cracking in metalworking [49]. As AE signals are often as a result of plastic deformation, burst AE form are identified as more efficient in identifying fractures while continued signal forms provide a steady information stream on the machining state. Additional studies have also demonstrated the relationship between elevated AE signals peaks to tool failure [50]. Other works have employed multi-sensing fusion approaches which comprise of AE signals and other sensing techniques in tool state predictions [18, 51, 52].

Sensor fusion. Multisensory fusion is often used with tool monitoring systems [53]. Dimla [54] describes this process as the use of several signals from diverse origins to assess the same parameter. During the fusion process, an overwhelming amount of features can be generated. Furthermore, noise signals from the process could infiltrate and influence its correlation efficiencies. Consequently, utilizing an appropriate sensing and extraction techniques as well as identifying highly correlated signals with reduced noise ratio is important. Cho et al. [51] in their study used a multi-sensor approach consisting of vibration, cutting forces, acoustic emission and spindle power in the determination of the tool state used a total of 68 time-domain and 67 frequency-domain features. These were obtained and integrated into a neural network (NN). In their study, different fusion methods were evaluated namely; decision level fusion (DLF) and feature level fusion (FLF). From their results, it was shown that FLF greatly improved the accuracy of the tool condition classification with the highest accuracy recorded using a subset of 25 correlated features [51]. Uekita, Takaya [18] used AE and spindle current to investigate the tool deterioration process. In their study, it was found that the AE signals did generate pertinent information on the tool crack propagation as well as wear progression. Segreto et al. [55] in their approach, implemented a multi-sensor pattern recognition tool for robot-assisted polishing. The sensing approach used was the spindle motor current, AE signals and strain sensing between the tool and the robot arm. Their study applies different modes of feature extraction namely; conventional and advanced wavelet packet transform (WT). Their results also indicate a higher success rate in prediction using WT. A neural network was trained based on the selected features to determine response value.

Multisensory integration of correlated features provides more sensitivity and reliability in prediction. Sick [56] in his study posits that this system is essential to establish an efficient monitoring system. However, some of the applications of this process with neural networks have yielded worse outcomes. This comes from the over-generalization of the output by an excessive pattern learning [56].

In this study, seeks to establish an efficient manufacturing process for machining 4.0 using acoustic emission as the sensing technology of choice in the estimation of tool state. It presents the system dynamic flow a smart machine shop floor, set to the milling production of smart products made of tool steel. As far as the author's know, the study of the cyber-physical machine tool (CPMT) powered by highly correlated AE features is a new area of research in manufacturing production.

2 Experimental Setup and Result Discussion

The modified cyber-physical machine tool in the shop floor consisted of a Deckel Maho DMU 40 CNC machine (5 axis, 12,000 RPM spindle speed) mounted with three Kistler AE sensors with a bandpass frequency ranging from 50 kHz to 1 MHz.

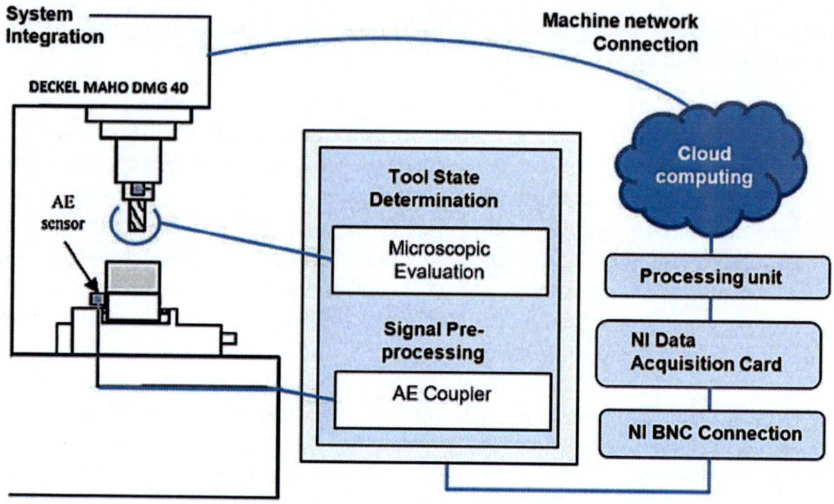

Fig. 3 The framework of modified DMU 40 CNC machine tool

The AE sensors were linked to the BNC 2110 block of the National Instrument (NI) via the piezotron coupler for data acquisition. Figure 3 shows the production framework of the CPMT setup.

The integrated acoustic sensors were affixed on the workpiece to supply a continuous elastic energy signal. These sensors are sensitive to the material properties (e.g. gauge thickness) [57]. Also, literature shows that the signal quality differs based on its location within the machine [49].

2.1 *Experimental Plan*

The workpiece is frequently used for hot and cold applications. A rectangular block of dimensions 200 × 100 × 40 of the workpiece was cut using a combination of cutting parameters shown in Table 2. A 25 mm end-mill indexable tool was used with TiAlN-PVD-coated carbide inserts. Repeated longitudinal cuts along the length of the workpiece were performed (Fig. 4).

The output response. The response parameter observed during the test is the tool life. This response is estimated based on the wear progression of the tool. Axinte [58] described tool life as the length of time to machine a product before tool failure. This parameter is indicative of the useful life of the tool which is affected by the wear rate on the cutting tool edge. This factor bears importance to the establishment of machining 4.0 as a key performance indicator integral to a smart production system. Therefore, tool life can be estimated through the rate of wear formed on the tool, which gradually deteriorates until intolerable levels. To estimate this factor, the use

Table 2 Experimental cutting parameters

Description		Depth of cut (mm)	Speed (m/min)	Feed Rate (mm/min)	Workpiece	Passes	Total time (min)	Total length (m)
Experiments	1	2	200	200	Tool steel	119	23.8	23.8
	2			250		134	26.8	26.8
	3			300		125	25	25
	4		230	300		123	24.6	24.6
	5			250		106	21.2	21.2
	6			200		72	14.4	14.4
	7		170	300		132	26.4	26.4
	8			250		135	27	27
	9			200		113	22.6	22.6

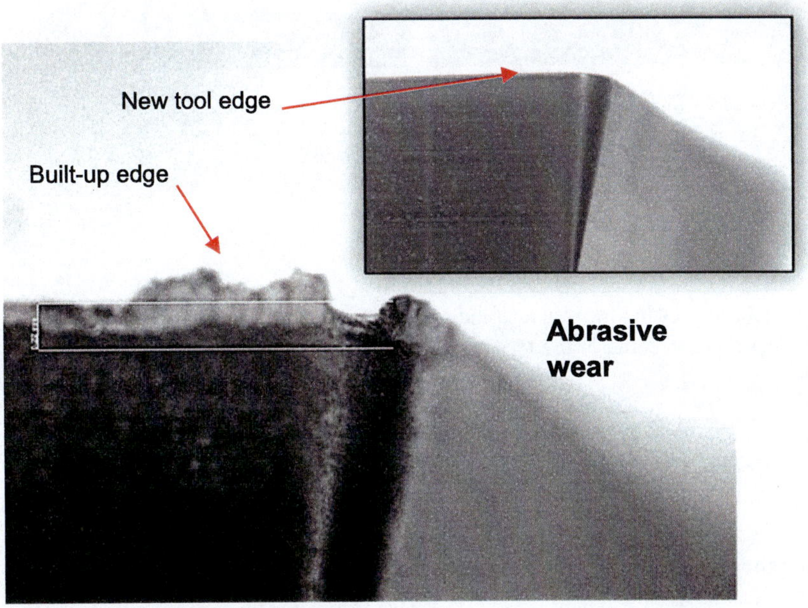

Fig. 4 New and worn cutting tool edge

of empirical equations using cutting parameters, have been commonly studied by researchers [59, 60], however, these are not employed in this study. To determine the state of production for the CPMT the wear classifications in Table 3 shows the different tool wear classification considered in this study. The industrial wear value

Table 3 Tool wear classification

Class	Flank wear (Vb)	Tool state
0	Vb < 0.1	New tool
1	0.1 < Vb < 0.2	Moderately Worn
2	0.3 > Vb > 0.2	Worn tool

Fig. 5 Sample wear seen on the tool inserts

of 300 μm for cutting tools is adapted to determine worn tooling. Figure 5 displays wear observed on some of the tool inserts.

2.2 *Signal Processing*

AE signal rectification can be performed with the addition of signal treatment blocks such as filtering, amplification and conversion. Figure 6 shows the AE process chain used in the study.

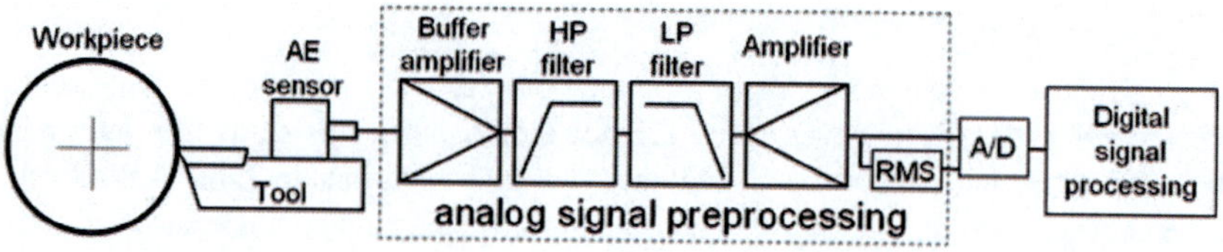

Fig. 6 The AE process chain [61]

Extensive processing of AE waveforms is essential to mitigate the effects of its non-stationary and erratic transient nature. Three basic stages in processing were performed on the AE signals; namely; feature extraction, feature selection, decision making. The signal is then decomposed into features which generate correlated information vital to machining. These generated features act as inputs to decision-making models as well as a parameter for visual analyses of trends. Caesarendra and Tjahjowidodo [62] categorized feature extraction methods into time-domain, frequency-domain, time-frequency domain, phase-space dissimilarity measurement and other feature extraction methods. Table 4 shows the pre-processing extraction steps performed on the experimental AE signal for the various stages of the tools.

Time and frequency domain features extraction. A frequently used tool in the analysis of AE data is the time series model [62]. Time series presents a collection of numerical observations organised in a natural sequence often linked to a specific time interval. Statistical analysis is usually used to extract relevant information from time-series data. Some examples of such statistical methods are the Autoregressive (AR) and Autoregressive Moving-Average (ARMA) model. Literature [63] indicates correlations between AR residual signal in AE monitoring wear conditions. Pontuale et al. [41] employed the use of histograms of new and worn tools with time series amplitudes. From their study with these log-log histograms, they highlight some power-law physiognomies present in the data. The study by Cho et al. [51] highlights some commonly used features such as root mean square, arithmetic mean, kurtosis, variance, signal power, burst rate, and crest factor.

Frequency and time-frequency features which vary in spectral energy distribution also bare importance to machining conditions. Spectral energy which is determined by calculating the power spectral density (PSD) provides information on the state of the machining process and surges or declines based on wear level. Several frequency features can be extracted from the energy power band and may provide correlations to output response. Table 5 shows some frequency domain features extracted.

Additional features. Additional features can be obtained from the AE wave using wavelet transform. Wavelet analysis as a signal processing technique has drawn a lot of interest [39, 64], It has been used in numerous systems and processes namely communications, image systems, transient signal, as well as other signal processing systems [55]. Wavelet analysis works with the expansion of frequency band signals. The wavelets are localized in time and frequency permitting close-fitting between the function represented and their coefficients [65]. Figure 7 shows the wavelet decomposition levels frequency band used in the study.

Feature Selection. Owing to the high amount of features which can be obtained in the divers' domains during extraction, a process of selection is directed to optimally reduce the feature set. The coefficient correlation approach was utilised to assess the dependence of the signal with a specific value. The Pearson correlation equation for signal features is seen in Eq. (1).

Table 4 Processing stages of AE signal for different tool states

	Initial wear	Moderate wear	Worn tool
Raw AE signal	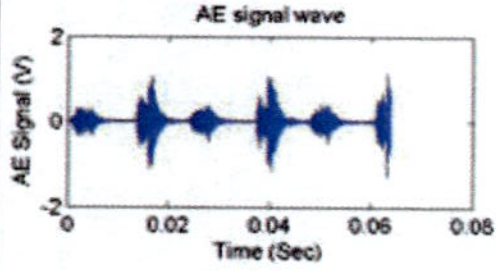	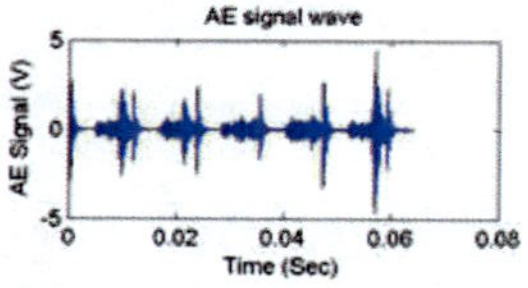	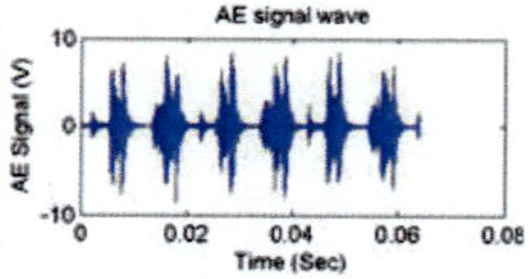
Filtered AE signal	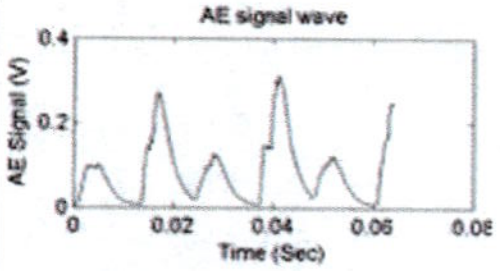	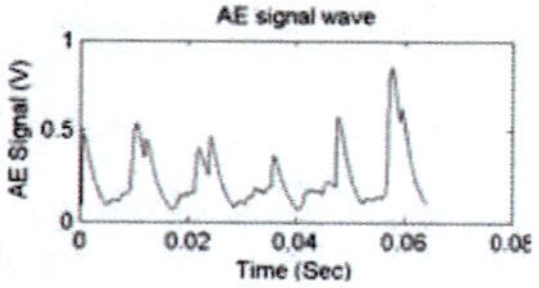	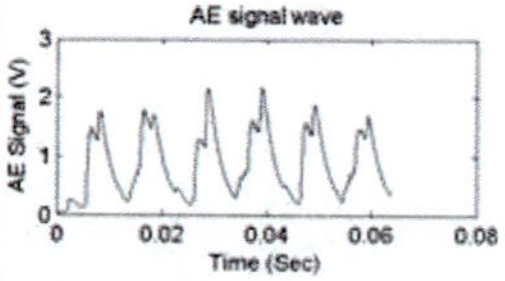
Power spect. Distr	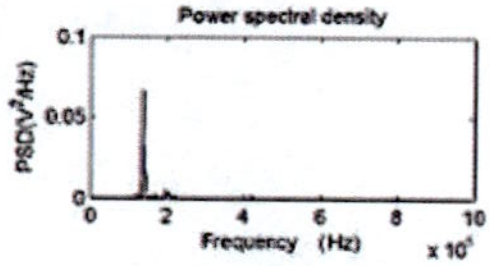	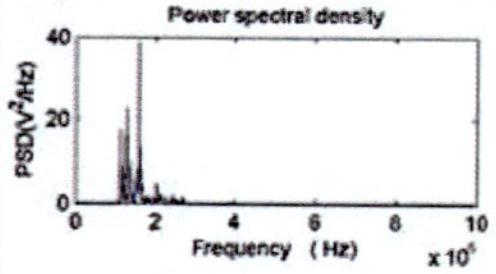	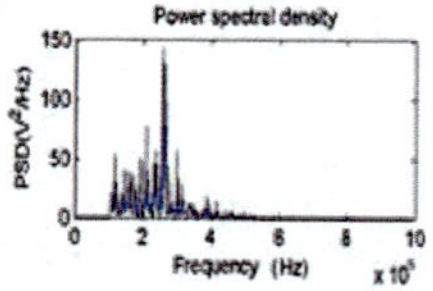
Energy distribution	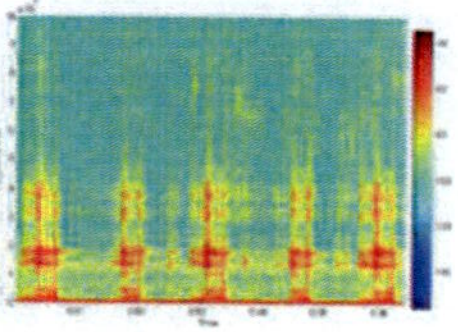	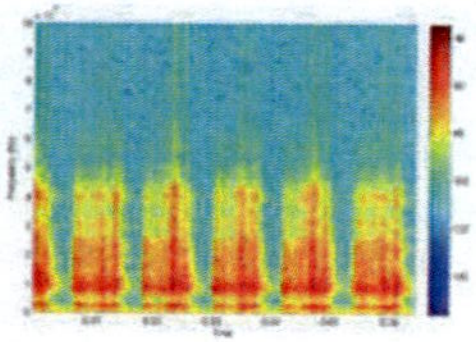	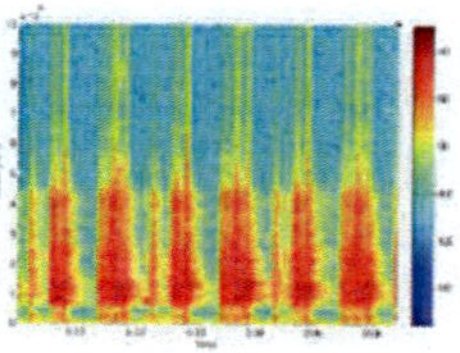

Table 5 Notable band power spectrum frequency domain features [51]

Frequency domain features	Description
Mean band power (MBP)	$MBP = \frac{1}{n}\sum_{i=1}^{n} S(f)_i$
Variance band power (VBP)	$VBP = \frac{\sum_{i=1}^{n}(S(f)_i - MBP)^2}{n-1}$
Skewness band power (SkBP)	$SkBP = \frac{1}{n}\frac{\sum_{i=1}^{n}(S(f)_i - MBP)^3}{VBP^{3/2}}$
Kurtosis band power (KuBP)	$KuBP = \frac{1}{n}\frac{\sum_{i=1}^{n}(x_i - \mu)^4}{\sigma^3}$
Relative spectral peak[a] (RSPBP)	$RSPBP = \frac{1}{n}\frac{\sum_{i=1}^{n}(x_i - \mu)^3}{\sigma^3}$

[a]Band power spectrum

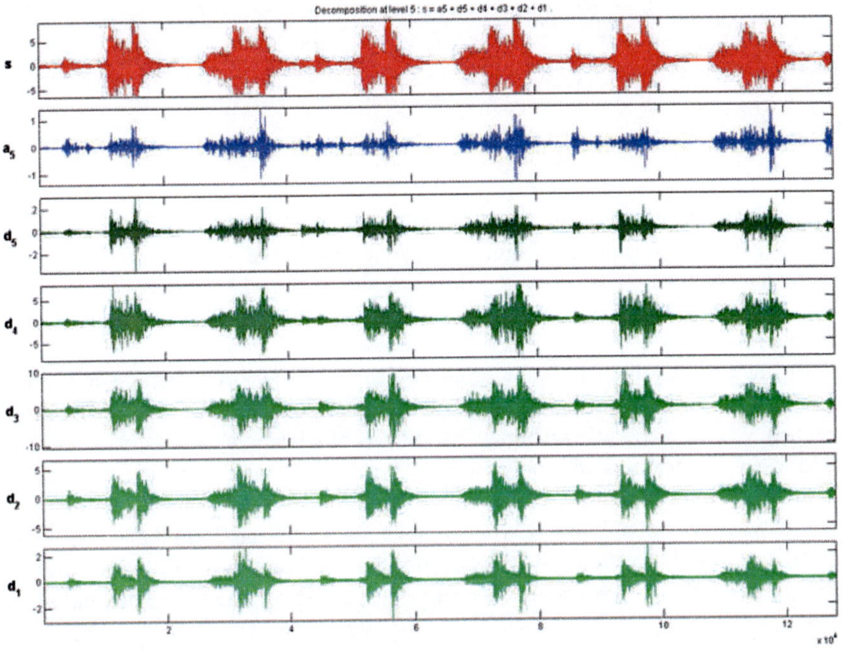

Fig. 7 AE signal wavelet analysis decomposition levels

$$\rho = \frac{X_i \max \sum_i (X_i - \bar{X})(Y_i - \bar{Y})}{X_i \min \sqrt{\sum_i (X_i - \bar{X})^2} \sqrt{\sum_i (Y_i - \bar{Y})^2}}. \quad (1)$$

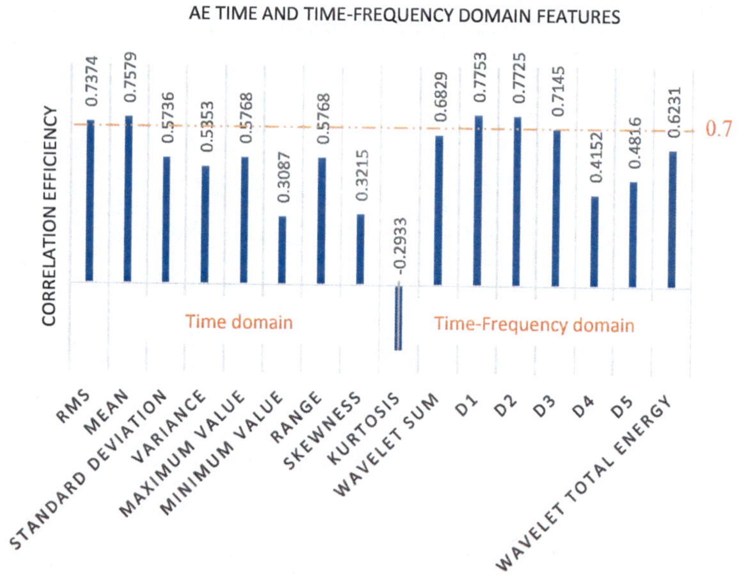

Fig. 8 Correlation efficiencies of AE features

where $\bar{X}$ and $\bar{Y}$ represent the mean of features X and output wear Y, respectively; X_i are features extracted from signals; Y_i are tool wear values, ρ is the correlation coefficient. When$|\rho|$ draws near value one, the relation between X and Y draws near-linear, and when ρ is close to zero Y is independent of X. The results from the correlations of each feature are displayed in Fig. 8.

Figure 8 shows the correlation efficiency across all features. A threshold of 0.7 was used for the selection of features used for prediction. From this analysis, time-frequency features revealed a higher correlation efficiency of 64% when compared to time-domain features with only 48%.

From the spectrogram of the energy distribution shown in Table 4, an increase energy signature is observed in lower bands between 62 and 250 kHz which correspond to D1–D3 wavelets. In these sections, all experiments exhibited a higher prevalence of energy density indicating the significance of these bands to tool state.

2.3 Intelligent Decision Making

Smart shop floors are powered by advanced software platforms able to incorporate and link the various virtual twin components of the manufacturing system in the cybersphere for process management. These platforms which could be localized or cloud-based integrate new advanced analysis techniques such as neural

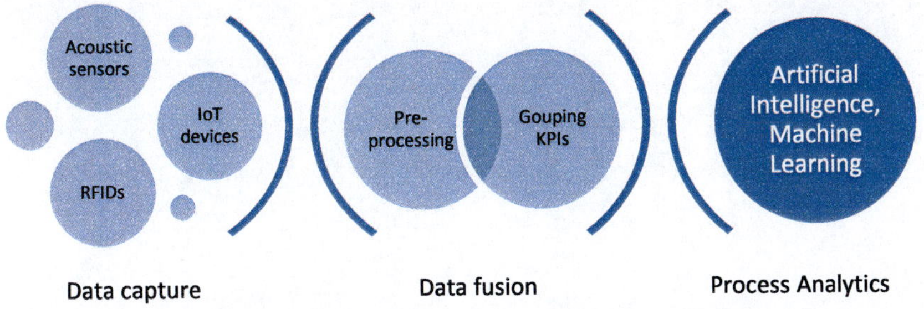

Fig. 9 Phased process flow for smart manufacturing decision

networks, fuzzy logic, swarm optimization, machine learning algorithms in deciphering information for predictions (Fig. 9). Each of these evolutionary tools varies in efficiency based on certain factors such as data size, inputs, design and parameters. Neural networks are often used as predictive tools in research studies due to their accruing performance with bulky data size [13, 51]. The next sub-sections present the prediction of the tool state using NN and cloud integration.

Neural network framework. There is a need for easy interconnection solutions for the implementation of machining 4.0. These solutions could form new technologies which form part of the new CPMT design of the future. In this study, the NN framework was designed on the data acquisition platform software LabView® for computing and prediction. Selected key features with higher correlation efficiencies were used to classify wear output states. This analysis was performed by training a data set of 160 input through a feed-forward Back propagation (BPNN) neural network with a sigmoid activation function. An adaptive learning rate of 0.3 was adopted and training cycles were run for a thousand epoch. The classification output of the function classifies the selected input signals in a value within the range of 0–1. Figure 10 shows the front-end of the processing interface for network training. From the gradient descent curve, a very low error value was observed for classification of tool state.

Simulation tests. Simulations to confirm the classification efficiency of the network was performed. Parametric tests of varying cutting conditions indicated with a binary output of 0 0 1 the tool's wear state. Neural network approximations may also reflect some noise of the process through minor indication in other output states during classification (Table 6).

Results from Table 6, indicate an adequate classification of new tools (0 0 1), moderately worn tools (0 0 1) and worn state (1 0 0) with respective binary outputs. Minimal noise was observed as well in alternate digits during classification. Misclassification errors in NN predictions can occur due to a generalisation error, however, they are mitigated with continued real-time data input. Following the successful estimation of tool state AE signals, The integration of the decision state of the network is fed back to the CPMT processing unit for control of the tool replacement state.

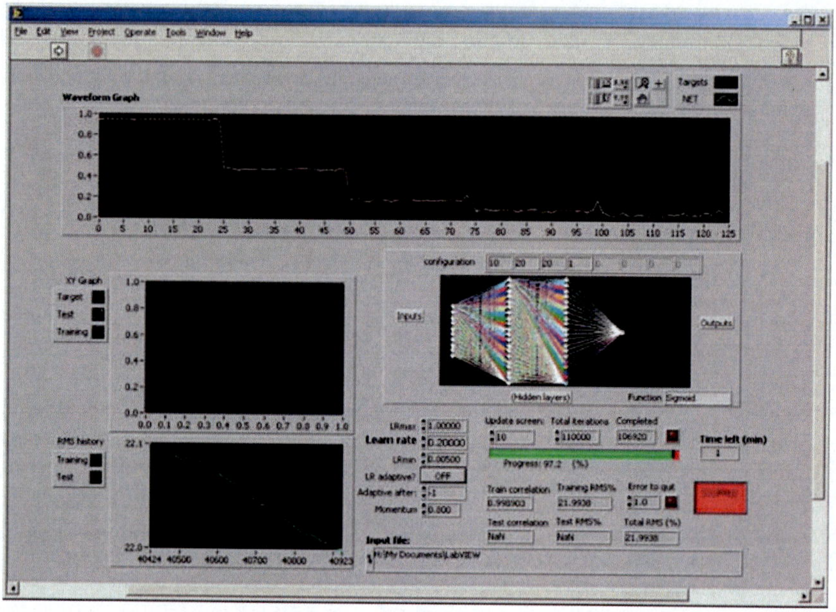

Fig. 10 Front-end of the neural network training tool

Table 6 Validation of the neural network with the various stages of wear

Initial wear	Moderate wear	Worn tool
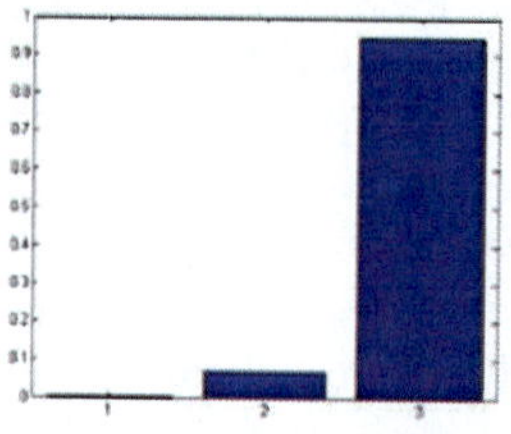	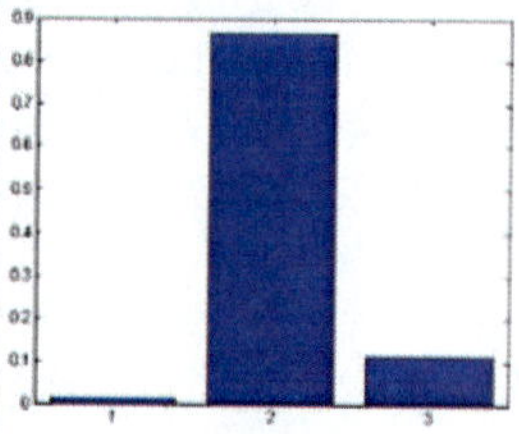	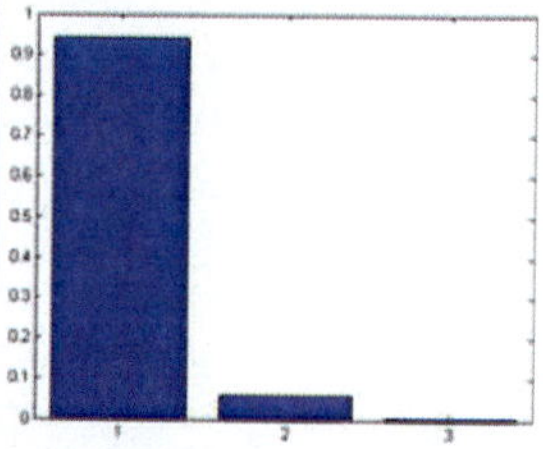
$v = 200$ m/min and $f =$ 200 mm/min	$v = 170$ m/min and $f =$ 250 mm/min	$v = 230$ m/min and $f =$ 300 mm/min
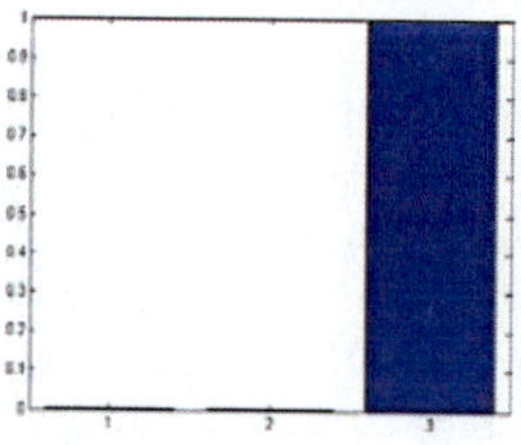	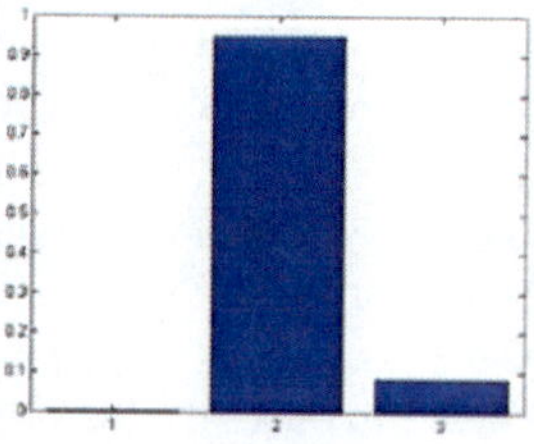	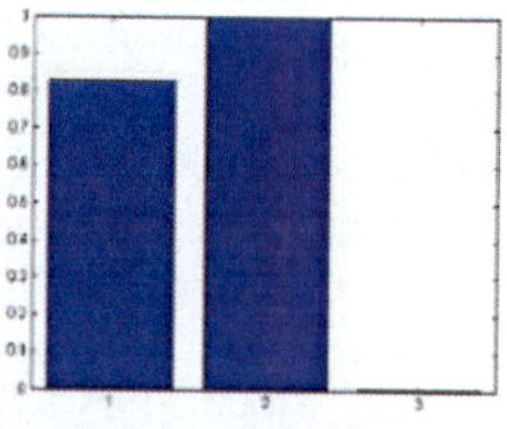
$v = 200$ m/min and $f =$ 250 mm/min	$v = 200$ m/min and $f =$ 300 mm/min	$v = 170$ m/min and $f =$ 200 mm/min

Cloud-based Integration. A cloud-based adaptation of this system permits the creation of a shared ecosphere of information for improved production. This cyber-physical data cloud performs the archiving, warehousing, searching and manipulation of sensing data to provide actionable information. As such, it permits not only the optimization of the machining process but also the autonomous scheduling of cutting tools and process intelligence which adheres to quality product requirements. Cloud-based implementations of evolutionary solutions greatly optimize the processing of data originating from tools, factories and machines. Various types of analytics are thus possible and easily applicable using the cloud integration platform namely; predictive analytics which is used to quantify cause and effects features relevant to distinctive outputs and prescriptive analytics for optimized decisions for machine control (e.g. cutting parameters control as well as tool change schedules).

In already optimized machining processes it may be more challenging to identify room for further improvement. However, the advantage cloud computing offers could be found in the availability of advanced tools and computational capacity to provide an expansive optimization of the whole manufacturing process, all the way from planning and design to in-machining and machining analysis. Such systems could also evolve to include a cognitive computing ability for self-diagnosis utilizing reinforcement learning algorithms.

3 Conclusion

This study highlights acoustic emission as a viable tool in the creation of a smart shop floor. A multi-sensing AE monitoring approach using various acquisition points are fusion and employed to ascertain the state of milling production. Though, recent studies indicate a rise in the use of the sensing technique, a comprehensive interpretation of the vast information an AE signal wave contains is yet to be determined.

Despite modern applications of heterogeneous multi-sensing approaches, these processes are characterized by an exponentially increasing "big data" which incorporates added difficulty in processing and prediction. Furthermore, these necessitate the development of adequate algorithms to meet the computational needs of the exponentially growing data set. As automation in productions is powered by information from sensors, efficiency and optimization of this process are required to mitigate costs and computational demands associated. The present study conducts a comprehensive signal processing analysis on the fused AE signal to identify correlated, time, time-frequency and other features to tool state.

References

1. Olufayo OA, Abou-El-Hossein K, van Niekerk T (2011) Online tool wear monitoring using acoustic emission. Paper presented at the 4th robotics and mechatronics conference of South Africa (ROBMECH 2011), South Africa, 23–25 Nov 2011
2. Wang J, Yan J, Li C, Gao RX, Zhao R (2019) Deep heterogeneous GRU model for predictive analytics in smart manufacturing: application to tool wear prediction. Comput Ind 111:1–14
3. Mohanraj T, Shankar S, Rajasekar R, Sakthivel NR, Pramanik A (2020) Tool condition monitoring techniques in milling process—a review. J Mater Res Technol 9(1):1032–1042. https://doi.org/10.1016/j.jmrt.2019.10.031
4. Wohlers B, Dziwok S, Pasic F, Lipsmeier A, Becker M (2020) Monitoring and control of production processes based on key performance indicators for mechatronic systems. Int J Prod Econ 220:107452. https://doi.org/10.1016/j.ijpe.2019.07.025
5. Klocke F, Döbbeler B, Pullen T, Bergs T (2019) Acoustic emission signal source separation for a flank wear estimation of drilling tools. Procedia CIRP 79:57–62. https://doi.org/10.1016/j.procir.2019.02.011
6. Kong LX, Nahavandi S (2002) On-line tool condition monitoring and control system in forging processes. J Mater Process Technol 125–125:464–470
7. Behrens B-A, Bouguecha A, Buse C, Wölki K, Santangelo A (2016) Potentials of in situ monitoring of aluminum alloy forging by acoustic emission. Arch Civil Mech Eng 16(4):724–733. https://doi.org/10.1016/j.acme.2016.04.012
8. Caso E, Fernandez-del-Rincon A, Garcia P, Iglesias M, Viadero F (2020) Monitoring of misalignment in low speed geared shafts with acoustic emission sensors. Appl Acoust 159:107092. https://doi.org/10.1016/j.apacoust.2019.107092
9. Liu C-S, Li Y-A (2018) Evaluation of grinding wheel loading phenomena by using acoustic emission signals. Int J Adv Manuf Technol 99(5):1109–1117. https://doi.org/10.1007/s00170-018-2513-9
10. Mirifar S, Kadivar M, Azarhoushang B (2020) First steps through intelligent grinding using machine learning via integrated acoustic emission sensors. J Manuf Mater Process 4(2):35
11. Gaja H, Liou F (2016) Automatic detection of depth of cut during end milling operation using acoustic emission sensor. Int J Adv Manuf Technol 86(9):2913–2925. https://doi.org/10.1007/s00170-016-8395-9
12. Sachin Krishnan P, Rameshkumar K, Krishnakumar P (2020) Hidden Markov modelling of high-speed milling (HSM) process using acoustic emission (AE) signature for predicting tool conditions. In: Li L, Pratihar DK, Chakrabarty S, Mishra PC (eds) Advances in materials and manufacturing engineering, Springer, Singapore, pp 573–580
13. Olufayo O, Abou-El-Hossein K (2015) Tool life estimation based on acoustic emission monitoring in end-milling of H13 mould-steel. Int J Adv Manuf Technol 81(1):39–51. https://doi.org/10.1007/s00170-015-7091-5
14. Mathews PG, Shunmugam MS (1999) Condition monitoring in reaming through acoustic emission signals. J Mater Process Technol 86:81–86
15. Chethan YD, Ravindra HV, Krishnegowda YT (2019) Optimization of machining parameters in turning Nimonic-75 using machine vision and acoustic emission signals by Taguchi technique. Measurement 144:144–154. https://doi.org/10.1016/j.measurement.2019.05.035
16. Al-Sulaiman FA, Baseer MA, Sheikh AK (2005) Use of electrical power for online monitoring of tool condition. J Mater Process Technol 166:364–371
17. Liu Y, Wang X, Lin J, Zhao W (2016) Early chatter detection in gear grinding process using servo feed motor current. Int J Adv Manuf Technol 83(9):1801–1810. https://doi.org/10.1007/s00170-015-7687-9
18. Uekita M, Takaya Y (2017) Tool condition monitoring for form milling of large parts by combining spindle motor current and acoustic emission signals. Int J Adv Manuf Technol 89(1):65–75. https://doi.org/10.1007/s00170-016-9082-6

19. Morgan J, O'Donnell GE (2017) Multi-sensor process analysis and performance characterisation in CNC turning—a cyber physical system approach. Int J Adv Manuf Technol 92(1–4):855–868
20. Aliustaoglu C, Ertunc HM, Ocak H (2009) Tool wear condition monitoring using a sensor fusion model based on fuzzy inference system. Mech Syst Signal Pr 23:539–546
21. Harun MHS, Ghazali MF, Yusoff AR (2017) Analysis of tri-axial force and vibration sensors for detection of failure criterion in deep twist drilling process. Int J Adv Manuf Technol 89(9):3535–3545. https://doi.org/10.1007/s00170-016-9344-3
22. Svahn M, Andersson C, Vedmar L (2016) Prediction and experimental verification of the cutting forces in gear form milling. Int J Adv Manuf Technol 82(1):111–121. https://doi.org/10.1007/s00170-015-7309-6
23. Batako ADL, Tsiakoumis V (2015) An experimental investigation into resonance dry grinding of hardened steel and nickel alloys with element of MQL. Int J Adv Manuf Technol 77(1):27–41. https://doi.org/10.1007/s00170-014-6380-8
24. Bhattacharyya P, Sengupta D, Mukhopadhyay S, Chattopadhyay AB (2006) Current signal based continuous on-line tool condition estimation in face milling. IEEE (1-4244-0726-5/06)
25. Sun Jie WYS, Soon HG, Rahman M, Zhigang W (2008) Identification of feature set for effective tool condition monitoring—a case study in titanium machining. In: 4th IEEE conference on automation science and engineering, Key Bridge Marriott, Washington DC, USA, 23–26 Aug 2008
26. Gao H, Xu M, Shi X, Huang H (2009) Tool wear monitoring based on localized fuzzy neural networks for turning operation. In: 2009 6th international conference on fuzzy systems and knowledge discovery, pp 417–420
27. Ahmed YS, Alam MS, Arif AFM, Veldhuis SC (2019) Use of acoustic emission and cutting force signals to monitor built-up edge formation in stainless steel turning. Int J Adv Manuf Technol 103(5):2257–2276. https://doi.org/10.1007/s00170-019-03607-3
28. Paktinat H, Amini S (2018) Numerical and experimental studies of longitudinal and longitudinal-torsional vibrations in drilling of AISI 1045. Int J Adv Manuf Technol 94(5):2577–2592. https://doi.org/10.1007/s00170-017-0893-x
29. Bang Y-b, Lee K-m, Oh S (2005) 5-axis micro milling machine for machining micro parts. Int J Adv Manuf Technol 25(9):888–894. https://doi.org/10.1007/s00170-003-1950-1
30. Khajehzadeh M, Akhlaghi M, Razfar MR (2014) Finite element simulation and experimental investigation of tool temperature during ultrasonically assisted turning of aerospace aluminum using multicoated carbide inserts. Int J Adv Manuf Technol 75(5):1163–1175. https://doi.org/10.1007/s00170-014-6163-2
31. Abolghasemi Fakhri M, Bordatchev EV, Tutunea-Fatan OR (2012) An image-based methodology to establish correlations between porosity and cutting force in micromilling of porous titanium foams. Int J Adv Manuf Technol 60(9):841–851. https://doi.org/10.1007/s00170-011-3647-1
32. Younes R, Ouelaa N, Hamzaoui N, Djebala A (2015) Experimental study of real gear transmission defects using sound perception. Int J Adv Manuf Technol 76(5):927–940. https://doi.org/10.1007/s00170-014-6313-6
33. Yan R, Tang X, Peng FY, Wang Y, Qiu F (2016) The effect of variable cutting depth and thickness on milling stability for orthogonal turn-milling. Int J Adv Manuf Technol 82(1):765–777. https://doi.org/10.1007/s00170-015-7418-2
34. Seemuang N, McLeay T, Slatter T (2016) Using spindle noise to monitor tool wear in a turning process. Int J Adv Manuf Technol 86(9–12):2781–2790
35. Forbrigger C, Warkentin A, Bauer R (2018) Improving the performance of profile grinding wheels with helical grooves. Int J Adv Manuf Technol 97(5):2331–2340. https://doi.org/10.1007/s00170-018-2098-3
36. Binsaeid S, Asfour S, Cho S, Onar A (2009) Machine ensemble approach for simultaneous detection of transient and gradual abnormalities in end milling using multisensor fusion. J Mater Process Technol 209:4728–4738

37. Xu G, Chen J, Zhou H, Yang J, Hu P, Dai W (2019) Multi-objective feedrate optimization method of end milling using the internal data of the CNC system. Int J Adv Manuf Technol 101(1):715–731. https://doi.org/10.1007/s00170-018-2923-8
38. Qiu J, Ge R (2019) A theoretical model and experimental investigation to predict and evaluate cutting capability of turning center based on material removal rate. Int J Adv Manuf Technol 104(9):3287–3302. https://doi.org/10.1007/s00170-018-2006-x
39. Thomazella R, Lopes WN, Aguiar PR, Alexandre FA, Fiocchi AA, Bianchi EC (2019) Digital signal processing for self-vibration monitoring in grinding: a new approach based on the time-frequency analysis of vibration signals. Measurement 145:71–83
40. Zhou C, Guo K, Yang B, Wang H, Sun J, Lu L (2019) Singularity analysis of cutting force and vibration for tool condition monitoring in milling. IEEE Access 7:134113–134124. https://doi.org/10.1109/ACCESS.2019.2941287
41. Pontuale G, Farrelly FA, Petri A, Pitolli L (Published Online 2002) A statistical analysis of acoustic emission signals for tool condition monitoring (TCM). Acoust Res Lett Online. https://doi.org/10.1121/1.1532370
42. Zhou JH, Pang CK, Zhong ZW, Lewis FL (2011) Tool wear monitoring using acoustic emissions by dominant-feature identification. Ieee T Instrum Meas 60(2):547–559. https://doi.org/10.1109/Tim.2010.2050974
43. Sun J, Hong GS, Rahman M, Wong YS (2004) Identification of feature set for effective tool condition monitoring by acoustic emission sensing. Int J Prod Res 42(5):901–918
44. xiqing M, Chuangwen X (2009) Tool wear monitoring of acoustic emission signals from milling processes. In: 2009 1st international workshop on education technology and computer science, IEEE. https://doi.org/10.1109/etcs.2009.105
45. Kishawy HA, Hegab H, Umer U, Mohany A (2018) Application of acoustic emissions in machining processes: analysis and critical review. Int J Adv Manuf Technol 98(5):1391–1407. https://doi.org/10.1007/s00170-018-2341-y
46. Ribeiro Filho SLM, de Oliveira JA, Lauro CH, Brandão LC (2019) Monitoring of microturning process using acoustic emission signals. J Braz Soc Mech Sci Eng 41(6):271
47. Mix P (2005) Introduction to nondestructive testing: a training guide, 2nd edn. Wiley, Hoboken
48. Lee S, Kim J (2012) Acoustic emission monitoring of AFM nano scratching for ductile and brittle materials. Information 15:3477–3484
49. Olufayo O, Abou-El-Hossein K (2015) Tool life estimation based on acoustic emission monitoring in end-milling of H13 mould-steel. Int J Adv Manuf Tech 81(1–4):39–51. https://doi.org/10.1007/s00170-015-7091-5
50. Balsamo V, Caggiano A, Jemielniak K, Kossakowska J, Nejman M, Teti R (2016) Multi sensor signal processing for catastrophic tool failure detection in turning. Procedia CIRP 41:939–944. https://doi.org/10.1016/j.procir.2016.01.010
51. Cho S, Binsaeid S, Asfour S (2010) Design of multisensor fusion-based tool condition monitoring system in end milling. Int J Adv Manuf Technol 46(5):681–694. https://doi.org/10.1007/s00170-009-2110-z
52. Segreto T, Simeone A, Teti R (2012) Sensor fusion for tool state classification in nickel superalloy high performance cutting. Procedia CIRP 1:593–598. https://doi.org/10.1016/j.procir.2012.05.005
53. Kene AP, Choudhury SK (2019) Analytical modeling of tool health monitoring system using multiple sensor data fusion approach in hard machining. Measurement 145:118–129
54. Dimla DE (2000) Sensor signals for tool-wear monitoring in metal cutting operations—a review of methods. Int J Mach Tool Manu 40:1073–1098
55. Segreto T, Karam S, Teti R (2017) Signal processing and pattern recognition for surface roughness assessment in multiple sensor monitoring of robot-assisted polishing. Int J Adv Manuf Technol 90(1):1023–1033. https://doi.org/10.1007/s00170-016-9463-x
56. Sick B (2002) On-line and indirect tool wear monitoring in turning with artificial neural networks: A review of more than a decade of research. Mech Syst Signal Pr 16(4):487–546. https://doi.org/10.1006/mssp.2001.1460

57. Jayakumar T, Mukhopadhyay CK, Venugopal S, Mannan SL, Raj B (2005) A review of the application of acoustic emission techniques for monitoring forming and grinding processes. J Mater Process Technol 159(1):48–61. https://doi.org/10.1016/j.jmatprotec.2004.01.034
58. Axinte DA, Belluco W, De Chiffre L (2001) Reliable tool life measurements in turning - an application to cutting fluid efficiency evaluation. Int J Mach Tool Manu 41(7):1003–1014
59. Olufayo O, Songmene V, Kenné J-P, Ayomoh M (2020) Modelling for cost and productivity optimisation in sustainable manufacturing: a case of dry versus wet machining of mould steels. Int J Prod Res:1–20. https://doi.org/10.1080/00207543.2020.1778207
60. Ahmed ZJ, Prickett PW, Grosvenor RI (2016) The difficulties of the assessment of tool life in CNC milling. In: 2016 international conference for students on applied engineering (ICSAE), 20–21 Oct 2016. pp 452–457. https://doi.org/10.1109/icsae.2016.7810234
61. Jemielniak K (2013) Advanced signal processing (unabridged)
62. Caesarendra W, Tjahjowidodo T (2017) A review of feature extraction methods in vibration-based condition monitoring and its application for degradation trend estimation of low-speed slew bearing. Machines 5(4):21
63. Li X (2002) A brief review: acoustic emission method for tool wear monitoring during turning. Int J Mach Tools Manuf 42:157–165
64. Benkedjouh T, Zerhouni N, Rechak S (2018) Tool wear condition monitoring based on continuous wavelet transform and blind source separation. Int J Adv Manuf Technol 97(9–12):3311–3323
65. Goryunova MV, Kuleshova LS, Khakimova AI (2017) Application of signal analysis for diagnostics. In: 2017 international conference on industrial engineering, applications and manufacturing (ICIEAM), 16–19 May 2017. pp 1–5. https://doi.org/10.1109/icieam.2017.8076487

Study of Wear on AISI E52100 Steel Using a Lithium Complex Grease and a Calcium Sulfonate Grease

J. F. Márquez-Santiago, M. Vite-Torres, and E. A. Gallardo-Hernandez

Abstract The purpose of this work was to study the antiwear and extreme pressure properties of two synthetic greases, one of them consists of lithium complex as a thickener, it is classified in the standard ISO 6743-9 as LX and the other has calcium sulfonate as a thickener and it is classified as CS. Both greases are widely used to lubricate bearings under high loads and high speeds operating conditions. Experimental wear results in show that the two greases have similar wear protection. However, extreme pressure tests show that calcium sulfonate grease offers better performance at high contact pressures, because this can work in short times at a maximum contact pressure of 8741 MPa without the presence of the seizure wear mechanism, while the lithium complex supports lower contact pressure around 7.08 GPa with features of seizure.

Keywords Seizure · Complex of lithium · Sulfonate of calcium

1 Introduction

An important function of lubricating greases is to avoid that the elements of the machinery or mechanical systems have not whatever kind of interaction among their surfaces only during of start and stop of equipment. Besides, the greases need to keep adhered to surfaces towards preventing leaks due to gravity or centrifugal conditions, they must also tolerate contact pressures when high load are applied without suffering deformations or squeezing. The flow capacity of grease is determined by the viscosity of the base oil and the type and percentage of soap used in the formulation. Also, temperature, pressure, shear stress as well as the frequency it is exposed [1].

Gears, bearings, cams, and chains generally operate on elastohydrodynamic lubrication causing high contact pressures due to the crushing of the lubricant film between the surfaces. Specifically, the exponential increase in loads applied in the trials for

J. F. Márquez-Santiago · M. Vite-Torres (✉) · E. A. Gallardo-Hernandez
Instituto Politécnico Nacional, Grupo de Tribología, ESIME UZ, Av. IPN s/n, Col. Lindavista, Ciudad de México, México
e-mail: drmanuelvite9@hotmail.com

M. Abdel Wahab (ed.), *Proceedings of the 8th International Conference on Fracture, Fatigue and Wear*, Lecture Notes in Mechanical Engineering,
https://doi.org/10.1007/978-981-15-9893-7_53

Extreme Pressure (EP) is because the normal loads or the viscosity of the lubricant film increased linearly, and an exponential effect occurs in the increase in the thickness of the lubricating film in the elastohydrodynamic region, thus guaranteeing separation of the surfaces. Also, some additives dependent on temperature and not linearly activated in an EP grease.

There are greases with soap thickeners and greases with non-soap thickeners. In both groups, the base oil can consist of mineral oil or synthetic oil [2, 3]. LX is the highest quality lithium soap grease manufactured with vegetable greasy acids, oils highly refined minerals. The additives have extraordinary resistance characteristics under normal operating circumstances, extreme situations, and high temperatures. Some physical properties can be then mentioned as dropping point ~240 °C [2, 3] according to standard ASTM D-2265 [4] and the viscosity of base oil to 40–100 °C is ~14 cSt [2, 3] by ASTM D-445 [5]. Besides, it possesses good texture and excellent pumpability, recommended by the systems centralized, due to its high resistance to washing water, it is proposed to applications where the equipment is affected by contamination with water. Also, it can be used in gear boxes including [2, 3]. The lubricant grease produced from calcium sulfonate is constituted by refined mineral oil with thickener calcium sulfonate and extreme pressure additives offering better performance lubrication compare to grease commonly used at high mechanical and thermal stability [6]. However, both greases working at low or high temperatures above 300 °C. It offers optimum protection for materials and equipment in service which will give a lower frequency of machine stops for maintenance and lubrication, its main application is for bearing lubrication equipment at high temperatures including applications such as automotive, agricultural, construction, food, industry, mining, papermaking and steel mills. The Applications CS in the automotive industry includes chassis specifically, ball joints, universal joints, and wheel bearings. Besides CS is recommended, for lubricating bearings, sliding surfaces. Also, this grease has a higher dripping point, >300 °C [6], according to ASTM D-2265 standard, and the viscosity of the base oil at 40–100 °C, >1700 cSt [7], according to ASTM D-445, this allows it to operate the CS at high temperatures because it has a higher density in its microstructure compared to the microstructure of the LX, which is less dense. The microstructure of a grease soap is created in the process of crystallization when the thickening agent, the particles, and the substances together formed the grease. If the relation (length/diameter) of the fibers formed is higher, the performance at high temperatures is better because resistance to the flow of the grease increases [8] also, the small number of empty spaces significantly reduce water penetration thus improving protecting against oxide and corrosion [9] The microstructure of the fibers of CS and LX are shown in Fig. 1a, b.

On the other hand, the world production of lubricating greases decreased from 1.20 M Tonnes in 2017 to 1.17 M Tonnes in 2018. The Caribbean, together with Central and South America, produced about 3% of the world production of lubricating greases in 2018 [11, 12], see Fig. 2. The Greases with lithium thickeners and lithium complex predominate the producing countries with 72% (850,000 Tonnes) of the total lubricating grease produced worldwide, during which greases with anhydrous calcium thickeners and sulfonate of calcium covered 9% in 2018 [11, 12], see Fig. 3.

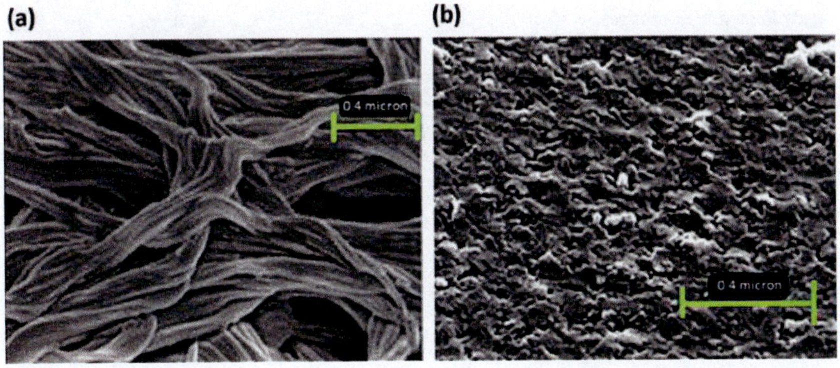

Fig. 1 Microstructure of greases. a Lithium complex. b Calcium sulfonate. [10]

Fig. 2 Percentages of production of lubricating greases around the world in 2018

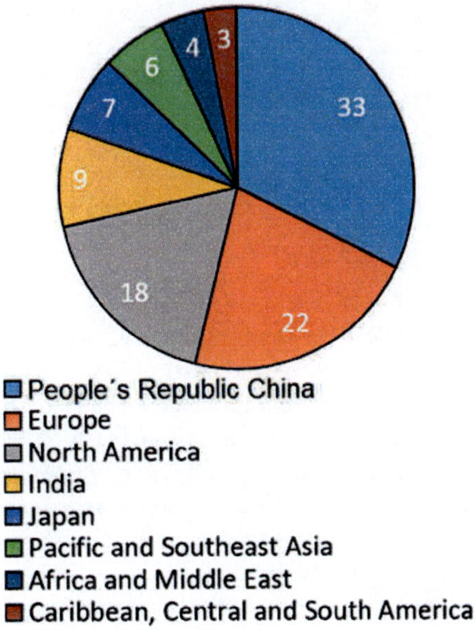

Analysis of production from 2012 to 2018 determined that the grease production with a lithium thickener decreased, while the grease with a lithium complex thickener increased.

On the other hand, the grease production with hydrated calcium thickener decreased while the grease production with anhydrous calcium thickener and calcium sulfonate increased. Besides, new thickener technologies such as polyurea and aluminum complex developed and grew in this period [11, 12].

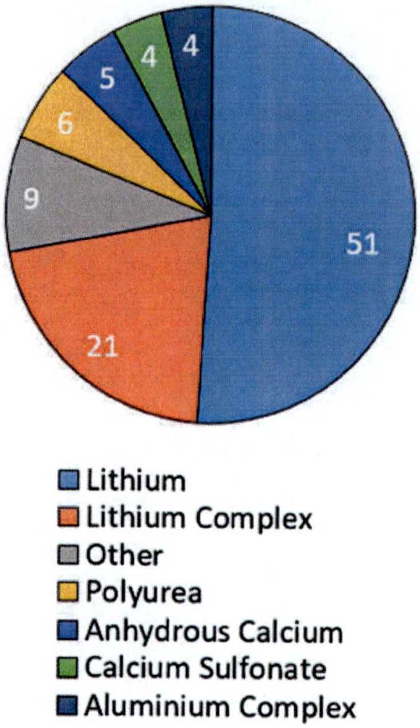

Fig. 3 Percentages of types of lubricating greases produced around the world in 2018

1.1 Four Ball Test

Four-ball tribometer is used to evaluate the mechanical properties of oils and greases lubricating under conditions of extreme and not extreme pressure. This tribometer allows to carry out lubricated tests. [13]. The geometric configuration consists of a tetrahedral arrangement of four AISI E52100 steel spheres. Three of them, are put inside bottom of a pot and strongly tight, while the fourth upper sphere applies the load and speed constantly, so the fourth sphere on the top is sliding over the three other spheres being the contact points submerged into a lubricating substance, as shown in Fig. 4a, b.

1.2 Normal Contact Force

Figure 5 shows the diagram of the components of normal force at a point of contact.

Normal contact force F_n, is calculated using a component F_y:

$$F_y = \frac{1}{3}\left(F + F_g\right) \tag{1}$$

where

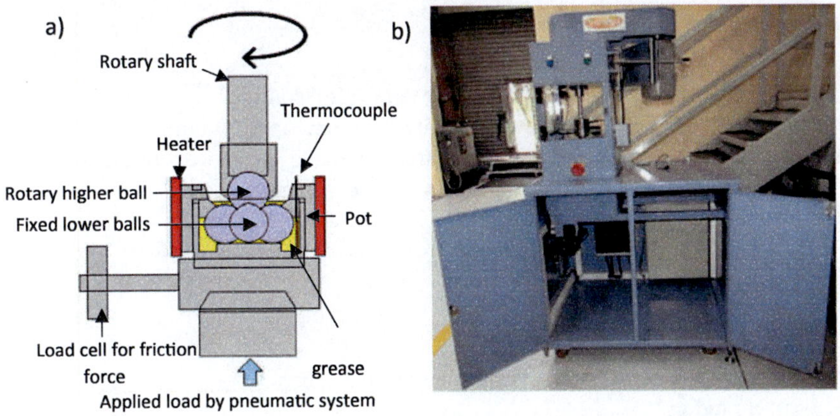

Fig. 4 Four ball tribometer. **a** Diagram of the test machine, **b** Tribological machine with automatic control was designed, manufactured by IPN, Tribology Group, SEPI- ESIME UZ [14, 15]

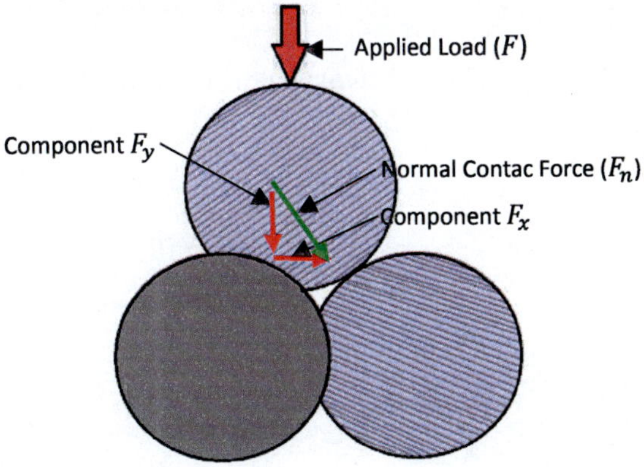

Fig. 5 Diagram of the components of normal force at a point of contact

F_y Vertical component of normal force.
F_g The weight of one ball.
F Applied force.

The normal contact force is determinate by:

$$F_n = \frac{F_y}{\cos(\beta)} = \frac{1}{3}\frac{\left(F + F_g\right)}{\cos(\beta)} \tag{2}$$

The four balls are made of AISI E52100 Steel and the same size, so the angle between F_y and F_n is $\beta = 35.26°$.

Finally, the normal contact force is:

$$F_n = 0.4082\left(F + F_g\right) \tag{3}$$

1.3 Contact Area

Towards determine the contact area of the tetrahedral configuration is necessary that the fourth ball do not experience severe adhesive wear. Figure 6, shows of the diagram forces analysis to determine the contact area generated by the wear trace.

The contact area is calculated using the next expression:

$$D = \sqrt{R^2 - r^2} \tag{4}$$

where

D Height of the deformation to the center of the sphere.
R Sphere radius.
r Wear scar radius

The area of a sphere of radius R, therefore

$$A = 4\pi R^2 \tag{5}$$

Consequently, the approximate contact area A_C between the contact of two spheres is:

$$\begin{aligned} A_C &= 2\pi R^2 - 2\pi RD \\ A_C &= 2\pi R(R - D) \end{aligned} \tag{6}$$

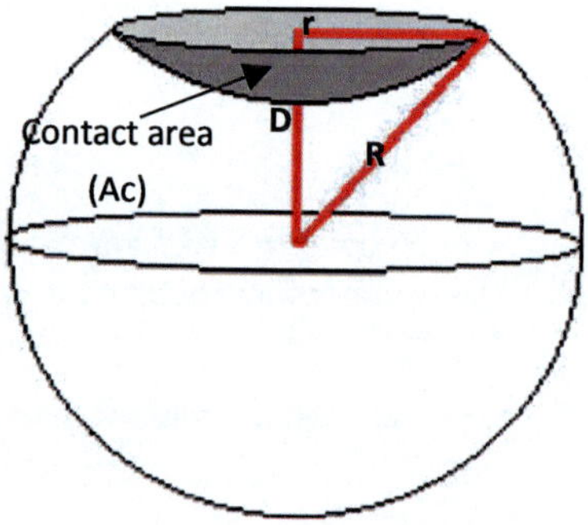

Fig. 6 Contact area

Table 1 Parameters of wear tests

Load applied	40 kg
Temperature	75 °C
Angular speed	1200 rpm
Time	1 h.

Table 2 Parameters of extreme pressure test

Load applied	Exponencial increase of load
Temperature	27 °C
Angular speed	1770 rpm
Time	10 s

2 Experimental Tests

2.1 Friction and Wear Tests

The tests are presented according to the method described in the ASTM D 2266 standard [16]. Five tests were carried out for both greases used, to obtain an average value of the generated wear on the steel spheres, the test parameters are showed in Table 1.

2.2 Extreme Pressure Tests

The grease is subjected to extreme operating conditions, the behavior mainly depends on the characteristics of lubricating oil, and the type grease thickener employed, welding produced at the end, which causes the broken of the fibers and the release of the oil, it burns due to the high temperatures generated by friction. [8]. The extreme pressure tests were performed according to the ASTM D 2596 standard [17]. Test parameters are showed in Table 2.

3 Results

3.1 Results of Wear Tests

Figures 7 and 8 show the friction coefficients behavior for the series of tests carried out with CS and LX.

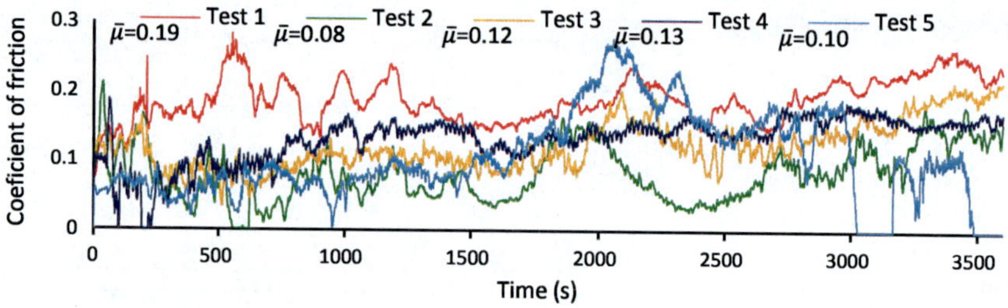

Fig. 7 Coefficients of friction of the tests with LX

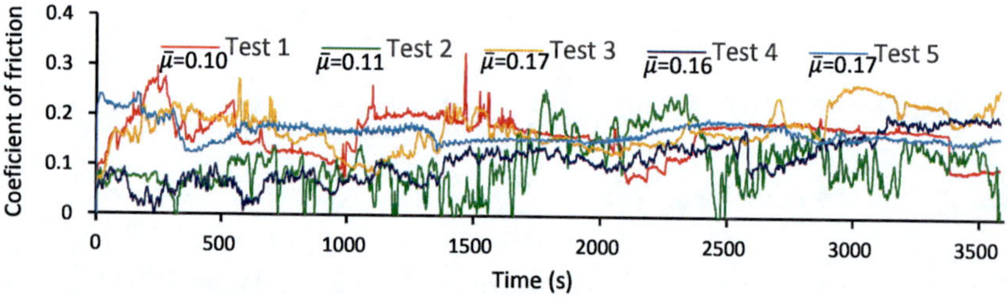

Fig. 8 Coefficients of friction of the tests with CS

The values of the average friction coefficients of trials carried out with LX, see Fig. 7, had a mean $\bar{\mu} = 0.12$ and while the trials with CS had a mean $\bar{\mu} = 0.14$, see Fig. 8.

By another hand, the averages of the wear scar diameters of the five wear tests performed with LX that shown in Table 3, had an average value of $\bar{d} = 0.509$ mm., while the tests performed with CS resulted with $\bar{d} = 0.523$ mm.

The close approximation of the average of friction coefficients and the diameters of the wear scars in both greases reveal a similar behavior after the wear tests.

Table 3 Wear tests results

Test number	CS (wear scar diameter) (mm)	Contact Pressure at the end of the test of CS (MPa)	LX (wear scar diameter) (mm)	Contact Pressure at the end of the test of LX (MPa)
1	0.4150	945	0.5539	708
2	0.4572	857	0.5223	751
3	0.5662	692	0.5229	750
4	0.5270	744	0.5385	728
5	0.5836	672	0.4800	817

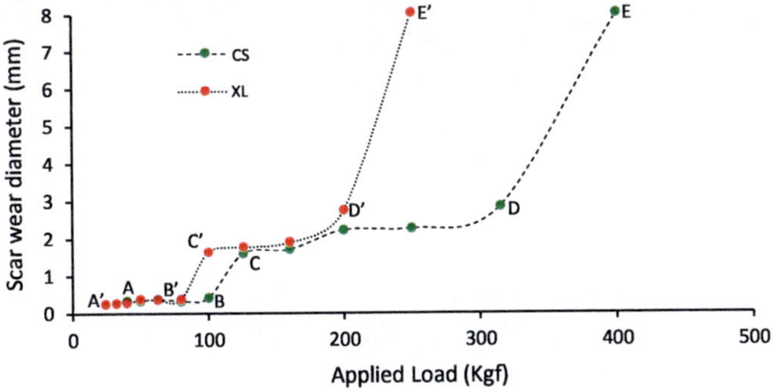

Fig. 9 Phases of surface damage CS and LX while the load was increased

3.2 *Results of Extreme Pressure Tests*

The seizure mechanism in ASTM D 4176 standard is defined as the fusion of two surfaces that are in relative motion, which effect the immobility of the surfaces. On the other hand, Lim and Ashby redefined this term in a more precise way, not considering the temperature effects (melt wear), adhesive wear delamination wear (wear dominated by plasticity), and the regime where loads and sliding velocities are so low (ultra-soft wear).

Figure 9 shows areas without seizure phenomenon conformed by the trajectories from A to B for CS meanwhile A´ to B´ by LX the areas with incipient seizure are located from B to C for CS and from B´ to C´ for LX, immediate seizure regions were located from C to D for CS and from C´ to D´ for LX. The melting points were at E for CS and E´ for LX.

On the other hand, It is necessary to calculate the mean contact pressure to the end extreme pressure tests, was used the Eq. (3), and the normal contact force obtained for each of the wear scars finally, the areas contact found using Eqs. (4) and (6). The graphic method developed is presented in Fig. 10, then load increases implicate a contact pressure increases reaching a maximum point, after both falls sharply due to the contact area rise precisely in the regions near the beginning of the seizure wear phenomenon.

4 Discussion

The term seizure is defined as a process where high contact pressure caused by the increasing of applied load induce to the union of small surfaces areas by the process of metallic flow, where the surfaces sliding causes fracture at union points, rejoining continuously during the interaction of the surfaces. The damaged surface generates a nominal area where a seizure point occurs [18].

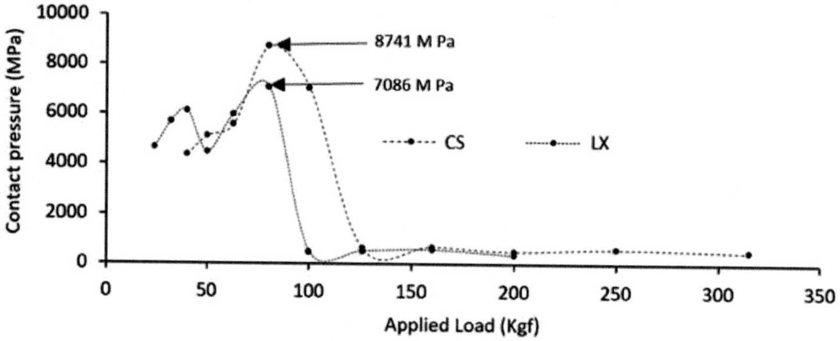

Fig. 10 Contact pressure in the extreme pressure tests. 4 results and discussion

The images showed in Table 4 present in detail the wear mechanisms in the scars generated. When CS grease is subjected to loads from 40 to 80 kg, mainly micro-cutting can be observed on the steel balls surfaces. In contrast, when LX grease is subjected to loads from 24 to 50 kg, mechanisms micro-cutting is presented in the same way, and there is also micro-plowing for all tests in this load range. It applied a load of 126 kg for CS grease another of 100 kg in LX grease, appearing the wear mechanisms such as pitting on the surfaces for both cases. When reaching the critical loads of 126 and 100 kg with CS and LX greases respectively by another hand seizure phenomenon starting to create stresses and fractures and plowing and regions with material plastically deformed these wear mechanisms severely grown up when load increased. Using 400 kg as loading in CS another LX grease of 315 kg, the soaps fibers in both cases broken up to release the oil, it evaporated quickly by the substantial increase in temperature owing to sliding contact, simultaneously is strongly welding and interrupting sliding between these surfaces.

The conditions of extreme pressure trials present data relevant showed that both greases reached their maximum contact pressure level, and working on interacting surfaces without seizure damage when using a load of 80 kg. In case that employs a load 126 kg in CS grease and a load 100 kg used to LX grease, then contact pressure starting to severe damage on the fibers of the structure of the lubricating soap, releasing the oil, this diminishes the performance of lubrication over the surface, and is presented seizure phenomenon, for this reason, increases the generated damage. But if the load to reach 400 kg for CS grease and 250 kg for LX grease, this produces a very high contact pressure causes that friction to increment, and the temperature grows much, so the released oil evaporated instantaneously, and the surfaces were weld strongly.

Table 4 Mechanisms of damage on wear scars in extreme pressure tests

<table>
<tr><th>Wears scars with CS</th><th>Wears scars with XL</th></tr>
<tr><td>40 Kg (Non-seizure)
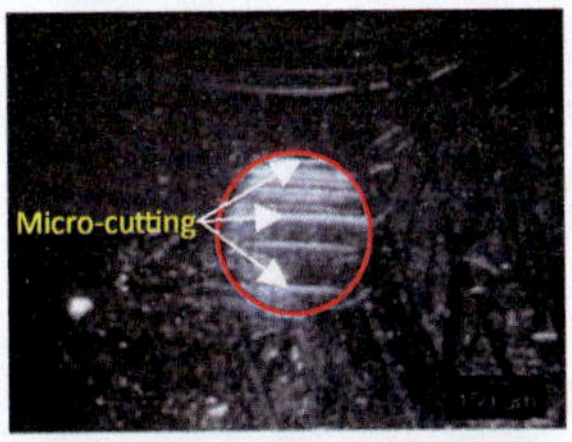
</td><td>24 Kg (Non-seizure)
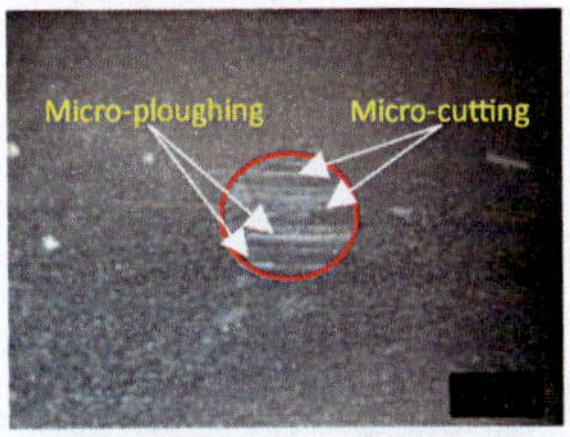
</td></tr>
<tr><td>50 Kg (Non-seizure)
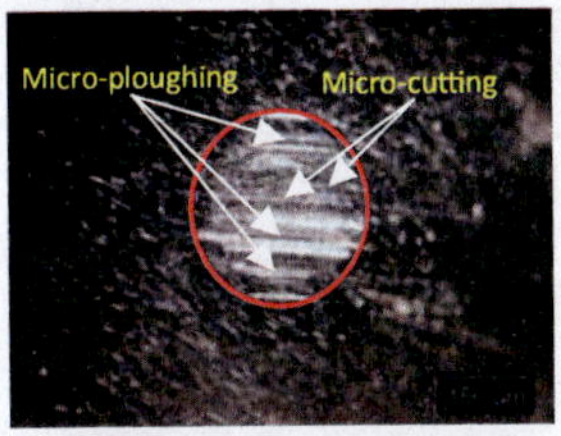
</td><td>32 Kg (Non-seizure)
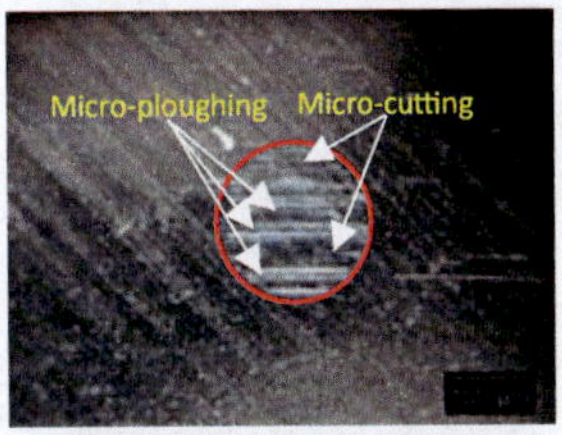
</td></tr>
<tr><td>63 Kg (Non-seizure)
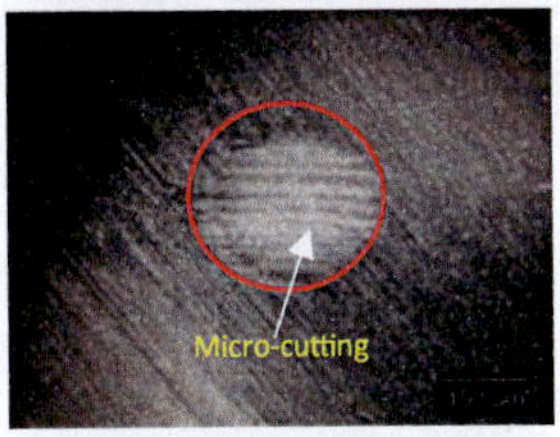
</td><td>40 Kg (Non-seizure)
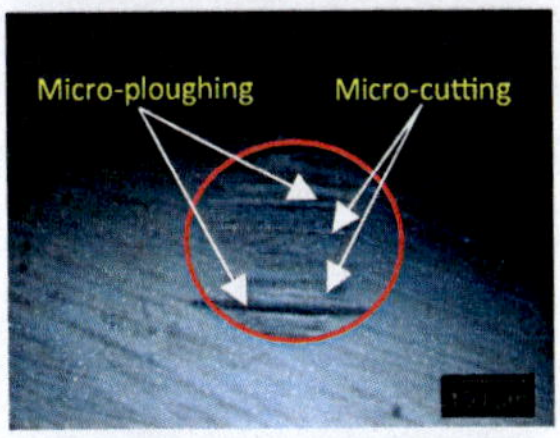
</td></tr>
<tr><td>80 Kg (Non-seizure)
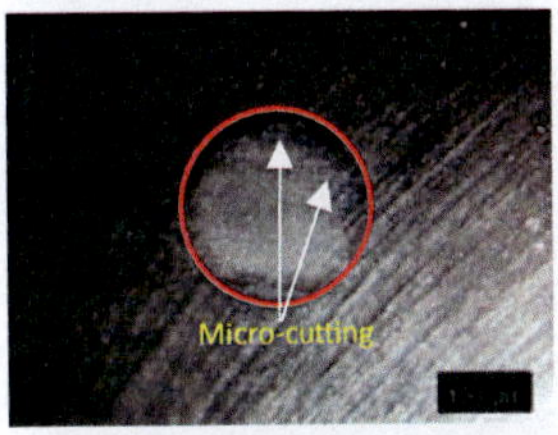
</td><td>50 Kg (Non-seizure)
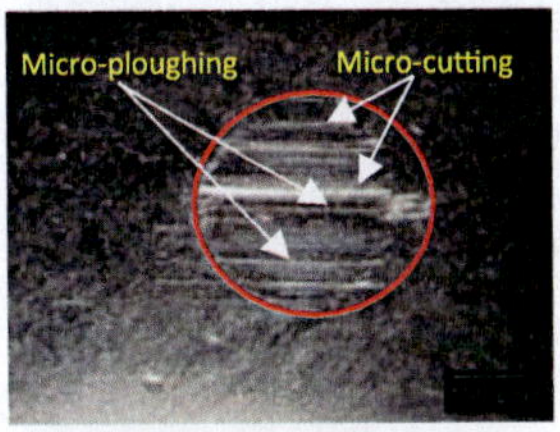
</td></tr>
<tr><td>100 Kg (Non-seizure)
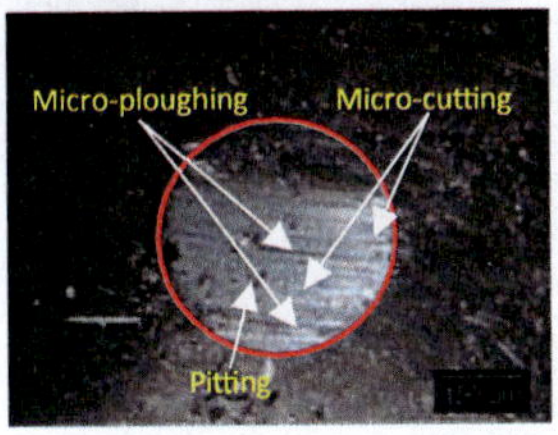
</td><td>63 Kg (Non-seizure)
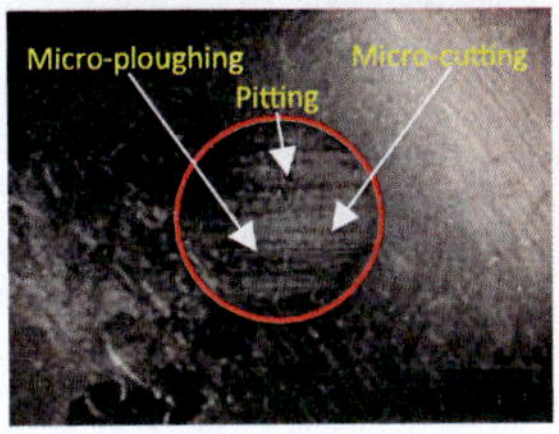
</td></tr>
</table>

(continued)

Table 4 (continued)

<table>
<tr><th colspan="2">Wears scars with CS</th><th colspan="2">Wears scars with XL</th></tr>
<tr><td>126 Kg (Seizure)</td><td>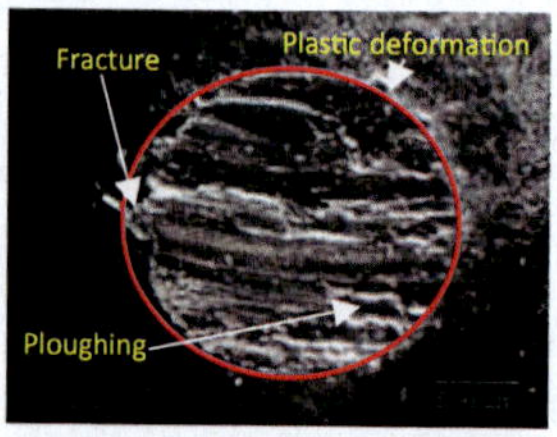</td><td>80 Kg (Non-seizure)</td><td>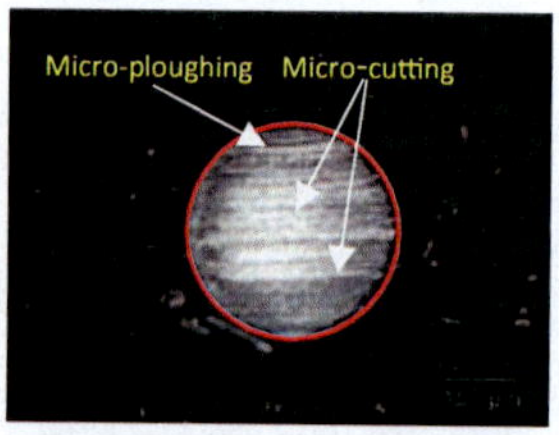</td></tr>
<tr><td>160 Kg (Seizure)</td><td>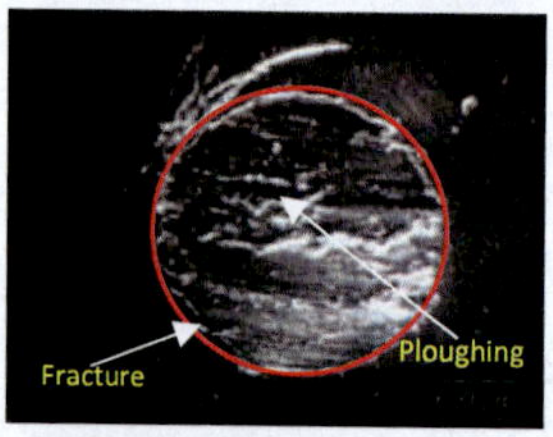</td><td>100 Kg (Seizure)</td><td>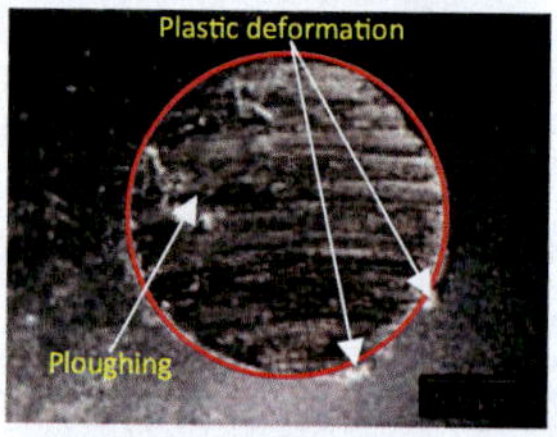</td></tr>
<tr><td>200 Kg (Seizure)</td><td>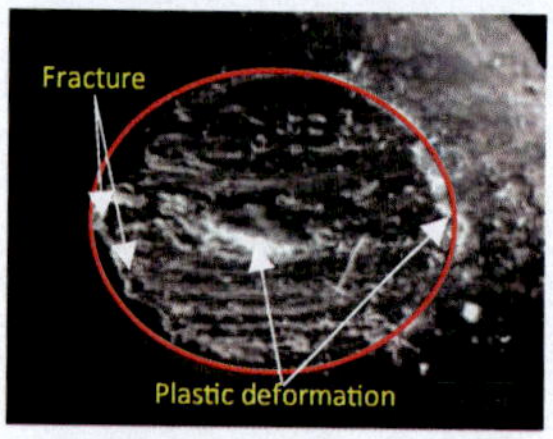</td><td>126 Kg (Seizure)</td><td>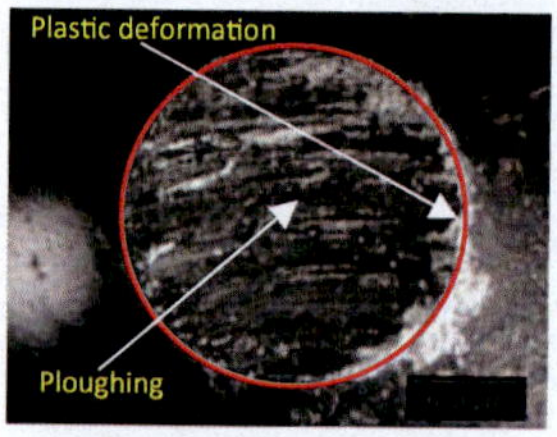</td></tr>
<tr><td>250 Kg (Seizure)</td><td>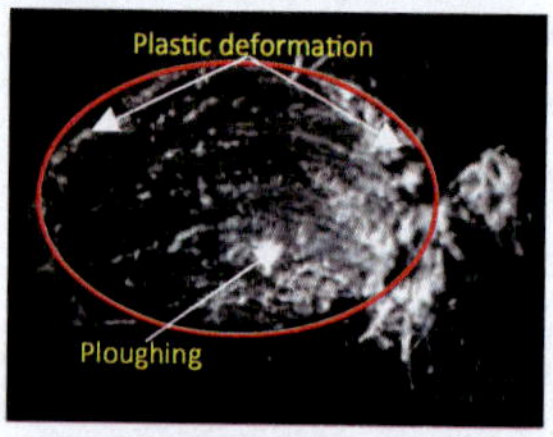</td><td>160 Kg (Seizure)</td><td>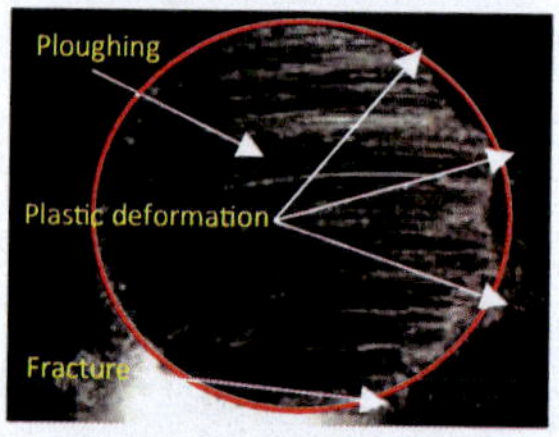</td></tr>
<tr><td>315 Kg (Seizure)</td><td>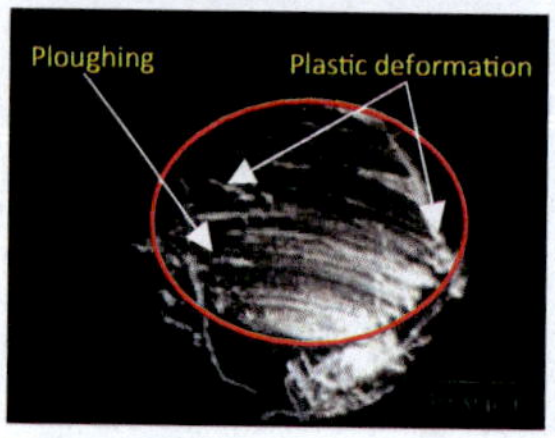</td><td>200 Kg (Seizure)</td><td>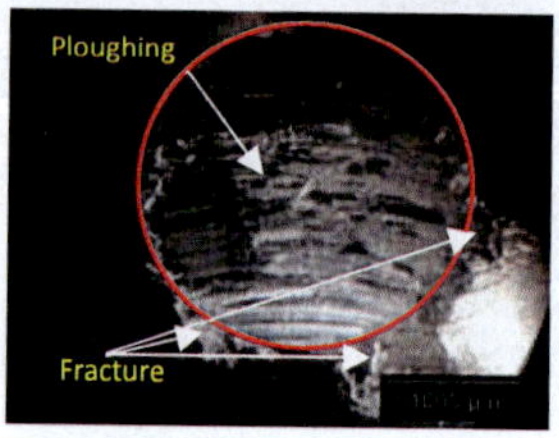</td></tr>
</table>

5 Conclusions

The thickeners derived from Lithium and Calcium are the most used to produce a better quantity of lubricating greases worldwide, using it with vegetable oil could contribute significantly to the environment.

In the wear tests of greases with the CS and XL thickeners, none significant difference was seen in the behavior of the coefficient of friction. On the other hand, the wear scar diameters were of acceptable size but without significant differences.

The structure of thickener contributes to improving the physicochemical, and tribological properties of lubricating grease.

For those that present a higher density in its fibers have better tribological properties, especially in high contact pressures, so grease CS thickener have better performed better in high contact pressure.

References

1. Totten GE (2017) ASM handbook, volume 18: friction, lubrication, and wear technology. ASM international, USA
2. Technical Manual Grease Company Luboks www.luboks.com.ar
3. Technical Manual Oil and Grease Company Lubral https://www.lubral.com/productos_lubral/lubra-grease-litio-ep/
4. ASTM D 2265-19 Standard test method for dropping point of lubricating grease over wide temperature range, https://www.astm.org/Standards/D2265.htm. Last Accessed 27 Jan 2020
5. ASTM D 445-19a Standard test method for kinematic viscosity of transparent and opaque standard test method for dropping point of lubricating grease over wide temperature. https://www.astm.org/Standards/D445. Last Accessed 2 Feb 2020
6. Technical Manual of Sulfonate Calcium Grease (EP 2) Company Vistony
7. Technical Manual Grease, Company Ceran and Plus, Spain
8. Martín Alonso JE (2010) Uso de polímeros reciclados y convencionales como modificadores de las propiedades reológicas de grasas lubricantes (Ph.D. Thesis), Universidad de Huelva (2010), ESPAÑA
9. Del Mundo DMN (2017) Manote Sutheerawattananonda: Influence of fat and oil type on the yield, physico-chemical properties, and microstructure of fat, oil, and grease (FOG) deposits. Water Res 124:308–319
10. Grasas de litio. https://www.lubral.com/cuales-son-las-opciones-para-las-grasas-de-litio/. Last Accessed 10 Dec 2019
11. Assess technical qualities of grease to keep up with market demand, https://eu.eventscloud.com/file_uploads/7b36504eeded58067b2df0f853e88df2_1650-TechnicalQualitiesofGrease_ABlack20191108.pdf. Last Accessed 10 Mar 2020
12. NLGI. https://www.nlgi.org/. Last Accessed 10 Jan 2020
13. Suresha B, Hemanth G, Rakesh A, Adarsh KM (2020) Tribological behaviour of neem oil with and without graphene nanoplatelets using four-ball tester. Adv Tribol. (Article ID 1984931)
14. Figueroa Guadarrama MA (2010) Diseño mecánico integral y manufactura de los componentes mecánicos de la máquina tribológica de cuatro bolas para ensayos lubricados (Master's Thesis). Instituto Politécnico Nacional, SEPI, ESIME UA, Ciudad de México (2010), MÉXICO
15. Márquez Santiago JF (2019) Validación de un sistema de control novedoso para una máquina tribológica de cuatro bolas (Master's Thesis). Instituto Politécnico Nacional, SEPI, ESIME UZ, Ciudad de México (2019), MÉXICO

16. ASTM D 2266-01 Standard test method for wear preventive characteristics of lubricating grease (four-ball method). https://www.astm.org/Standards/D2266.htm. Last Accessed 23 June 2019
17. ASTM D 2596-15 Standard test method for measurement of extreme-pressure properties of lubricating grease (four-ball method). https://www.astm.org/DATABASE.CART/HISTORICAL/D2596-14.htm. Last Accessed 23 June 2019
18. Westbrook SR, Shah RJ (2003) Fuels and lubricants handbook: technology, properties, performance and testing, 1st edn. ASTM International, USA

Surface Roughness and Normal Force Effects on the Sliding and Rolling Behavior of POM-H Rolls

Leonhard Kilian Doppelbauer, Philipp Siegfried Stelzer, and Zoltán Major

Abstract Polymer bearings are increasingly used due to their simple mass production and good tribological properties. The current contribution presents a macroscopic study of the effects of surface roughness and normal force on the sliding and rolling behavior of POM-H rolls. The counter-face roughness Ra was varied from 0.01 to 0.5 μm and the normal force from 150 up to 350 N. A polynomial regression model was used to analyze the effects of the two parameters and their interaction. The experiments showed that the coefficient of sliding friction depends strongly on the roughness with a distinct minimum at around 0.1 μm and only slightly on the normal force. In contrast, the coefficient of rolling resistance increases strongly with higher normal forces. From 150 to 350 N the coefficient of rolling resistance doubles in value. The roughness has a minor influence in rolling. The difference between sliding and rolling expresses also in the opposite interaction effects of the roughness and normal forces.

Keywords Polymer · Surface roughness · Sliding · Rolling

1 Introduction

Rolling elements made from various thermoplastic polymers play a major role in modern bearing designs. Polymer rolls can not only be manufactured cost-effectively with large batch production methods like injection molding, but they also offer a variety of other advantages such as corrosion resistance, low weight and self-lubrication [1–4]. For these reasons, plastic bearings are widely used in many practical engineering applications from energy conversion systems to aerospace applications [5–7]. The materials used for polymer bearings are either high-performance polymers such

L. K. Doppelbauer (✉) · P. S. Stelzer · Z. Major
Institute of Polymer Product Engineering, Johannes Kepler University Linz, Altenbergerstraße 69, 4040 Linz, Austria
e-mail: leonhard.doppelbauer@jku.at
URL: https://www.jku.at/institut-fuer-polymer-product-engineering/

M. Abdel Wahab (ed.), *Proceedings of the 8th International Conference on Fracture, Fatigue and Wear*, Lecture Notes in Mechanical Engineering,
https://doi.org/10.1007/978-981-15-9893-7_54

as polytetrafluoroethylene (PTFE) and polyetheretherketone (PEEK) or engineering polymers such as polyamide (PA) and polyoxymethylene (POM). High-performance plastics are expensive and thus not suitable for mass production. However, POM offers good tribological properties at a moderate price. Therefore, it is widely used for mass production of injection molded components in many tribological applications.

While POM-H homopolymers initially show better short-term mechanical properties (higher stiffness) than copolymers (POM-C), they tend to degrade faster over time. For some applications this could lead to performance reduction fairly quickly. High stiffness and dimensional stability, favorable tribological properties (low friction and wear rate) and moderate creep make POM to an attractive bearing material [6, 8, 9]. Furthermore, it is assumed that POM reveals a pronounced self-lubricating effect which allows low friction coefficients at high loads over a long sliding or rolling time [10].

Naturally, a deeper understanding of the factors that influence the friction process is of prime theoretical and practical importance. Two specific factors that have been discussed intensively in the literature for polymers are surface roughness of the counterpart and normal force. Particularly interesting experimental investigations of these two factors for sliding were carried out by [11–16]. In terms of rolling Harrass et al. [17] provided a comprehensive experimental study of the tribological properties under rolling contact of various thermoplastic polymers. In further consequence, the contact as well as the global and local deformation characteristics of polymer rolls need to be concerned [18, 19].

The objective of this research was a macroscopic study of the effects of surface roughness R_a and normal force F_n on the sliding and rolling behaviour of POM-H rolls. A statistical design of experiments was carried out to analyze the coefficients of sliding friction and rolling resistance by means of polynomial regression models. The regression models help to establish a functional relationship and the interactions between R_a and F_n. The results of this contribution could be helpful in the further development and optimization of rolling bearings made of thermoplastic polymers.

The above experiments were designed distinctly as short-term experiments and we intentionally neglected the long-term effects such as transfer film formation on the counterpart and time dependent viscoelastic deformation of the rolls. However, due to the inherent viscoelasticity of POM-H the rolls reveal creep which results in a macroscopic flattening of the cylindrical surface [18, 19]. This flattening may significantly affect the rolling resistance. For the sake of simplicity this flattening was neglected in the recent study in our short-term experiments. The experimental results made the development of corresponding simulation models possible. We are going to analyze above additional effects in a forthcoming simulation study of the sliding and rolling behavior of the thermoplastic rolls.

2 Experimental

2.1 Material and Specimens

Injection molded cylindrical rolls made of the thermoplastic homopolymer polyoxymethylene POM-H were used in the current study. The elastoplastic material properties were measured according to DIN EN ISO 527-2 [20] with a Young's modulus of 2439 ± 6.4 MPa, a Poisson's ratio of 0.37 ± 0.01 and a yield strength of 83 ± 3.5 MPa. The process induced shrinkage and warpage resulted in a concave shape of the rolls (nominal diameter Ø= 5.90 mm, length = 6.00 mm) with a maximum difference in diameter of 0.04 mm in the middle of the specimen, shown in Fig. 1. The surface roughness value of the rolls was R_a 0.5 µm. The roughness was determined both tactically (MarTalk, Mahr GmbH, Göttingen) and optically (Alicona InfiniteFocus, Alicona Imaging GmbH, Graz) because of measurement difficulties due to the curved white surface. However, the two different methods gave the same results.

The counterparts of the rolls were flat friction tracks with different R_a values ranging from 0.01 to 0.5 µm. Again both tactical and optical methods were used in order to ensure an exact roughness measurement of the counterparts. The roughness values and the used materials for the counterparts are summarized in Table 1. The friction tracks made of tool steel (1.2343) were wire eroded to provide a surface with isotropic friction properties. To realize the glass and steel sheet friction tracks small pieces of these materials were glued onto racks made from tool steel.

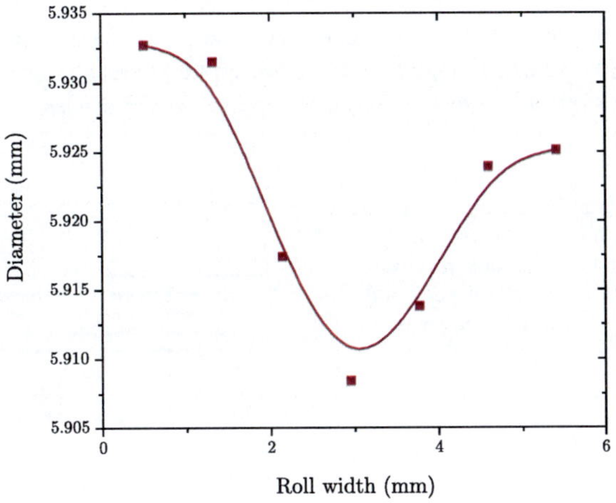

Fig. 1 The profile of the roll diameter shows the volumetric shrinkage caused by the injection molding process. The points correspond to the actual measured values with an approximated spline for better visualization

Table 1 Materials and surface roughness values of the used friction tracks

R_a (μm)	$R_{a,roll}$ / R_a	Materials
0.01	50	Soda-lime glass
0.1	5	Steel (1.2343)
0.15	$3.\dot{3}$	Zinc galvanized sheet steel
0.25	2	Steel (1.2343)
0.5	1	Steel (1.2343)

2.2 *Test Setup*

All tribological investigations were performed with a UMT Tribolab test rig (Bruker Corporation, Billerica) in a temperature-controlled room at RT 22 ± 1 °C and 50 % rel. humidity. The relative movement between the two friction partners was realized by a linear drive. The normal and tangential forces were measured using a 2D load cell of type DFH-100G with a resolution of 50 mN (Bruker Corporation, Billerica). A system with interchangeable inserts for the friction tracks was developed to enable a quick variation of the surface roughness. These can be mounted on the spring assembly as well as on the carriage by means of a bracket. The spring assembly serves as overload protection for the load cell which is sensitive to impact loads. The schematic structure is shown in Fig. 2 and was extended and modified depending on the test requirements.

A customized measurement script with three main steps was used for the tests. In the first step the test setup moves fast in z direction and stops at 1 mm above the specimen. In the second step the measuring head slowly contacts the roll and builds up the desired radial load. With the parameters shown in Table 2 (left), a targeted overshoot of the contact force is applied to reach the desired load level more quickly as shown in Fig. 3. Since the application of the load is force-controlled, a break-off

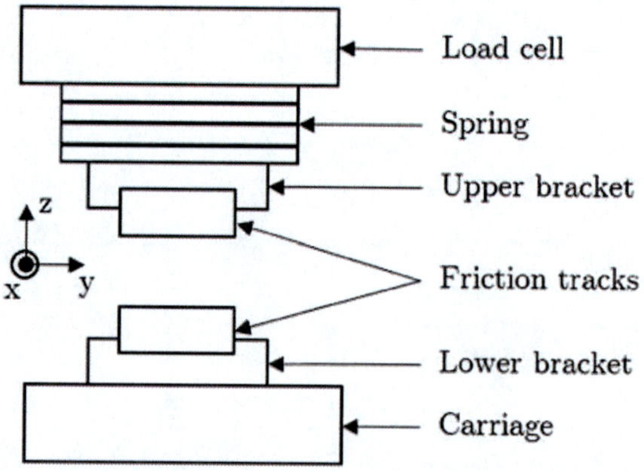

Fig. 2 Schematic structure of the test bench with interchangeable inserts to allow quick variation of the track surface roughness

Table 2 Parameters of the measurement script

Parameter (step 2)	Value	Parameter (step 3)	Value
Touch force	5 N	Load Fz	150/200/250/300/350 N
Pre touch speed	5 mm/s	Speed	10 mm/s
Touch speed	5 mm/s	Displacement in x direction	40 mm
Tracking speed	0.02 mm/s	Data rate	100 Hz
		Cycles	3
		Tracking speed	0.02 mm/s

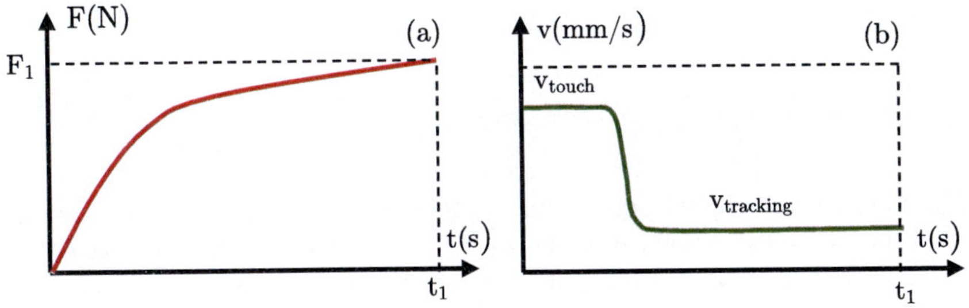

Fig. 3 Load (**a**) and velocity profile (**b**) for the targeted overshoot of the radial load. The time stamp t_1 marks the end of step 2 and the beginning of step 3

condition is introduced in case the test specimen is lost and to prevent a collision of the friction tracks. In the third step the carriage begins to move in x direction and the measurement starts. To exclude transient effects from the measurement the recording of data starts after the third cycle. The parameters of step 3 are shown in Table 2 (right). After finishing the measuring process the test rig moves into home position to enable the withdrawal of the specimen.

2.3 Sliding Test

In the sliding tests, the roll was clamped on one side to prevent the roll from rotating. Therefore, the lower friction track was replaced by a mounting for the roll which essentially consists of an undersized pocket. At the beginning of the sliding test the roll was pressed into this pocket as shown in Fig. 4a. The travel paths were adjusted so that the forward and backward movements do not overlap each other, see Fig. 4b. This was achieved by a lateral offset after the respective forward or backward movement, whereby the measuring head was lifted.

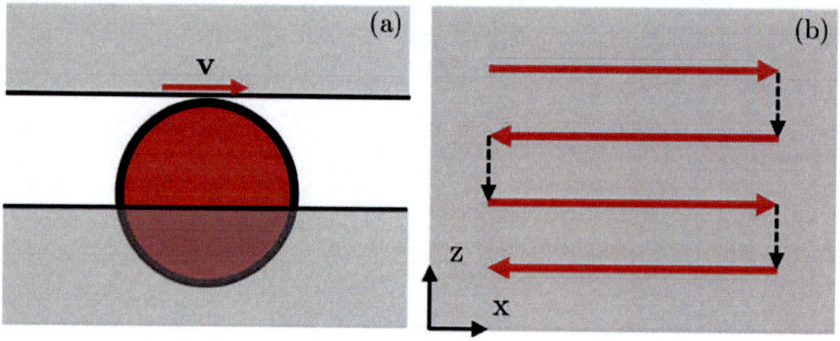

Fig. 4 Roll pressed into an undersized pocket to prevent rolling (**a**), travel path of the measuring head with lateral offset (**b**)

2.4 Rolling Test

For the rolling tests, a single roll was radially loaded between two equal friction plates while the lower plate was moved relatively to the upper one. The translational motion of the lower plate produced the desired rolling motion. To realize a full rotation of the roll the lower plate must move twice the period of the roll curve cycloid which equals to $4\pi R$. Since the actual radius R varies depending on the test specimen, a slightly larger travel distance of the theoretical value was selected for the tests. This ensured that at least one full rotation of the roll was made before the direction of movement was changed in the test.

The roll was aligned with a prism before the start of each experiment to guarantee reproducible results. In that way the roll was always aligned parallel to the rolling direction and the starting position remained the same for each experiment (Fig. 5). The radial load was force-controlled throughout the entire test. The translational movement of the friction tracks are displacement-controlled.

2.5 Test Plan and Data Analysis

In order to be able to detect possible interactions between the different test parameters, a statistical design of experiment as shown in Table 3 was created. Although different materials were used as counter-faces for the friction tracks, they were not evaluated as a separate factor. For each setting (R_a: 0.01–0.5 μm, F_n: 150–350 N), six replicate tests were carried out in sliding. Each sliding test consisted of three partial measurements. For rolling 10 replicate tests were conducted for each testing configuration.

For the data analysis the slight offset of the force signal had to be corrected. In addition, the sections around the reversion of the testing direction were removed to guarantee a constant testing speed. Since the data evaluation works with points and

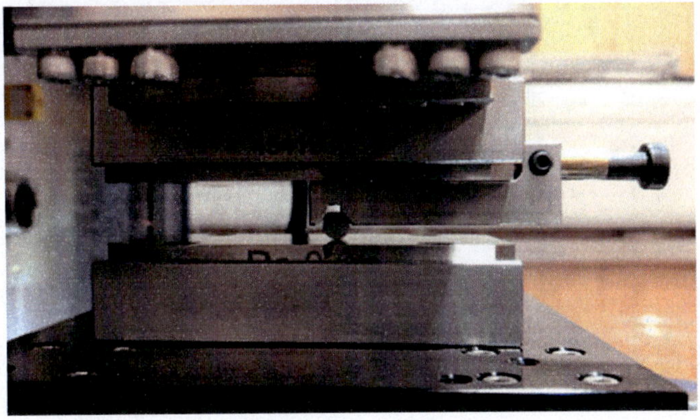

Fig. 5 Roll on a friction track aligned by a prism

Table 3 Experimental design plan for sliding (left) and rolling (right)

R_a (μm)	Sliding (sample size 6)			Rolling (sample size 10)				
	Load F_n (N)			Load F_n (N)				
0.01	150	250	350	150	200	250	300	350
0.1	150	250	350	150	200	250	300	350
0.15	150	250	350	150	200	250	300	350
0.25	150	250	350	150	200	250	300	350
0.5	150	250	350	150	200	250	300	350

their position, the number of points to be removed around the reversal point had to be calculated. Therefore Eq. 1 was used, where v is the testing speed, f the data rate and s the length of the reversal section.

$$Np_{removed}(f, s, v) = \frac{fs}{v} \tag{1}$$

The distance required to overcome the inertia of the test setup after reversing the testing direction was determined experimentally and evaluated for every load level. The tangential force was than calculated as the arithmetic mean of its absolute values. In Fig. 6 a typical signal with its offset and the filtered sections can be seen.

With the values for the tangential and normal force the coefficient of friction μ and the rolling resistance Π were evaluated. With the corresponding values for R_a and F_n as factors a polynomial regression model was fitted (confidence interval 95 %). Polynomial regression functions were chosen due to the good fit in the examined measurement range and to analyze the data with the possible interactions of the parameters. An extrapolation beyond the investigated range is not possible due to the selected approach functions. In order to develop a sufficiently good

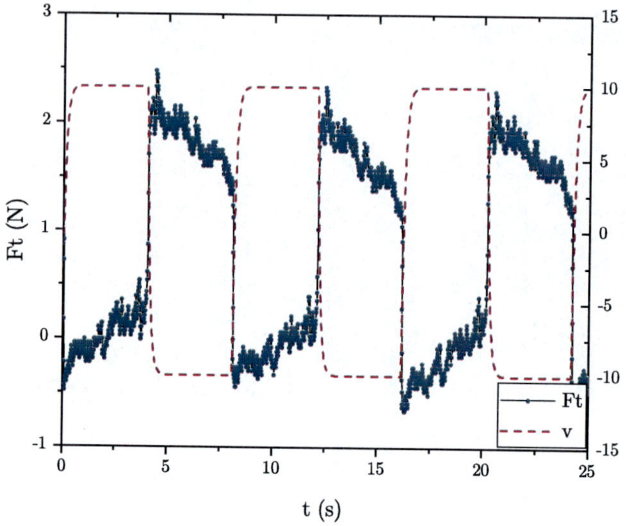

Fig. 6 Typical force signal of a rolling test with the corresponding velocity

model in the test range, polynomial functions up to degree three were used. In contrast, the sum of the powers of the crossed model terms was limited to three (e.g. $f(x, y) = c_j \cdot x^n \cdot y^m | (n + m) \leq 3$).

3 Results and Discussion

3.1 *Sliding*

The data of the friction coefficient μ depending on the normal force F_n with the roughness R_a as a group parameter are presented in Fig. 7. While the measurements for R_a 0.01 and 0.1 μm show a decreasing trend for μ, the data for v indicates the opposite. In contrast, Fig. 8 shows the dependence on R_a with F_n as group parameter. The values of μ decrease at first but increase again with a distinct minimum at R_a 0.1 μm for all load levels of F_n. This trend is amplified at higher normal forces, possibily due to the deviation from the ideal cylindrical shape of the rolls. The concave shape introduces a concentration of contact at the roll ends. Less contact happens on the side surfaces with only local contacts at roughness peaks. Higher normal forces would lead to more deformation and thus a more even contact distribution along the roll. The same dependency of μ on R_a demonstrated in this work for POM-H was found by the authors of [11] for the polymers PTFE and PE.

The polynomial regression functions (red curve) were used to fit the data and to establish a functional relationship between μ and the parameters R_a and F_n due

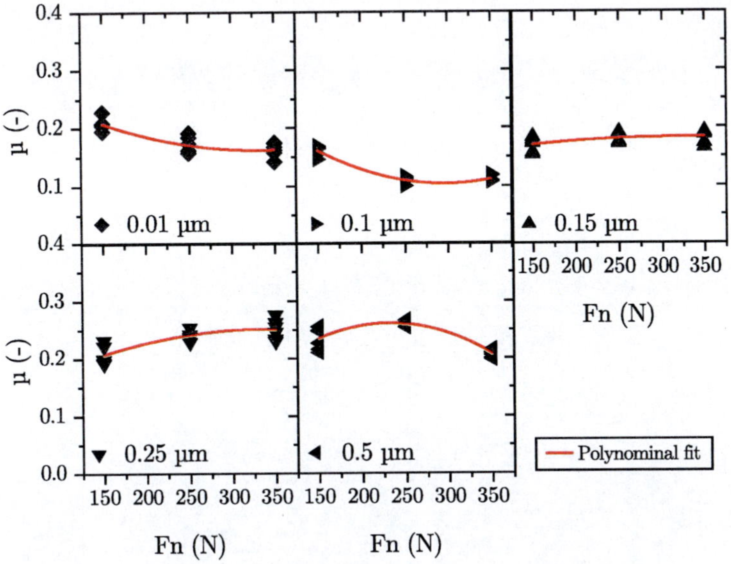

Fig. 7 The friction coefficient μ versus normal force F_n with the surface roughness R_a as group parameter

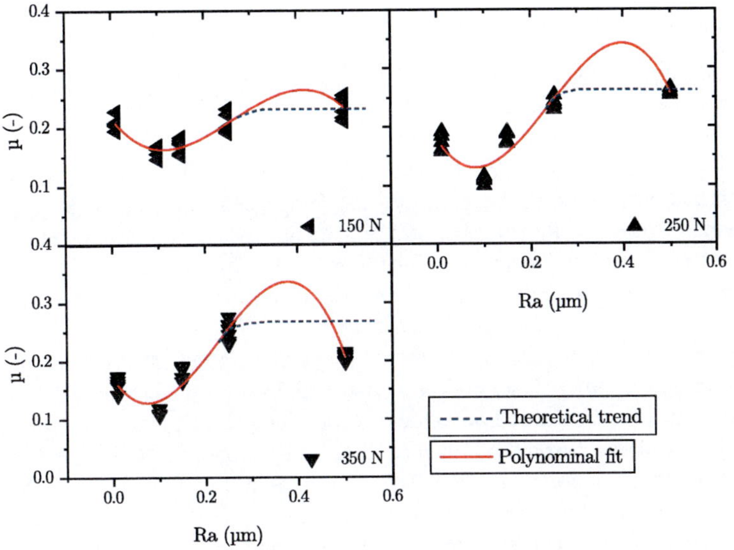

Fig. 8 The friction coefficient μ versus the surface roughness R_a with the normal force F_n as group parameter

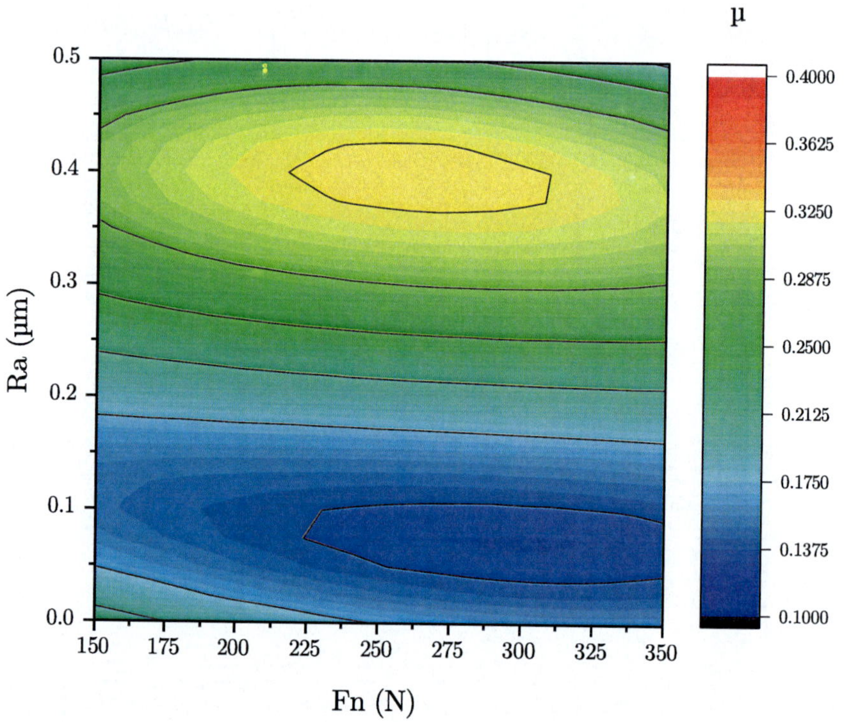

Fig. 9 Contour plot of the regression model for the friction coefficient μ as a function of the normal force F_n and the surface roughness R_a

to the good correlation in the measured test range. However, the local maximum between the measured R_a values of 0.25 and 0.5 μm does not necessarily exist and is rather an artifact of the polynomial functions. Alternatively, the blue line indicates a theoretical trend. We believe that the true trend converges towards a plateau value of μ.

Figure 9 presents the regression model for μ in a contour plot. The polynomial regression model provides a good fit of R-sq. 83.55% for the coefficient of sliding friction. The low gradient in horizontal direction indicates an almost constant influence of F_n. In comparison the influence of R_a is far more pronounced and shows a distinct minimum at around 0.1 μm. The maximum in the contour plot might be an artefact of the used third order functions as discussed above. The gradient of μ in vertical direction changes with the applied load level and clearly indicates an interaction of the two parameters R_a and F_n. These findings offer compelling evidence that the roughness of the counter-face is the dominant factor in the sliding friction of POM-H rolls and that the normal force plays a subordinate role.

3.2 Rolling

The data of the rolling resistance Π depending on F_n with R_a as a group parameter is presented in Fig. 10. It can be easily seen that Π increases strongly for higher values of F_n. The strong influence of the normal force can be traced back to the definition of the rolling resistance. It is strongly related to the contact radius r_c, which is a function of the normal force F_n. The asymmetry of the contact forces during rolling is primarily a cause of the non-linearly increasing contact stresses on the contact surfaces in an ideal cylindrical body. The process-induced concave shape of the roll further complicates the contact problem with non-continuous contact surface and localized stress concentrations at the roll edges.

Figure 11 shows a comparable behavior of the roughness dependency as in the sliding test but with a less distinctive influence of R_a. The curvature slowly decreases and the entire curve is shifted to higher values of Π as the load increases. This indicates an interaction between R_a and F_n as in sliding but in a reciprocal way (compare Figs. 8 and 11). With increasing deformation of the roll this interactional effect decreases and is superposed by the dominating effect of F_n.

Figure 12 displays the polynomial regression model for Π as a contour plot. This regression provides a good fit of R-sq. 95.61%. F_n is the dominant factor with a big horizontal gradient for the rolling resistance. In terms of R_a, as with sliding, a minimum can be found at 0.1 μm which becomes weaker with rising F_n. The flattening of the curve at higher loads is due to the interaction terms of the regression

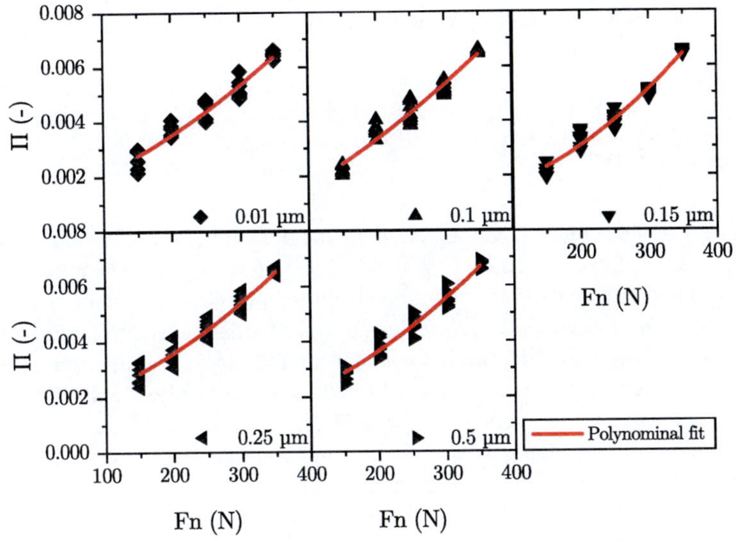

Fig. 10 The coefficient of rolling resistance Π versus normal force F_n with the surface roughness R_a as group parameter

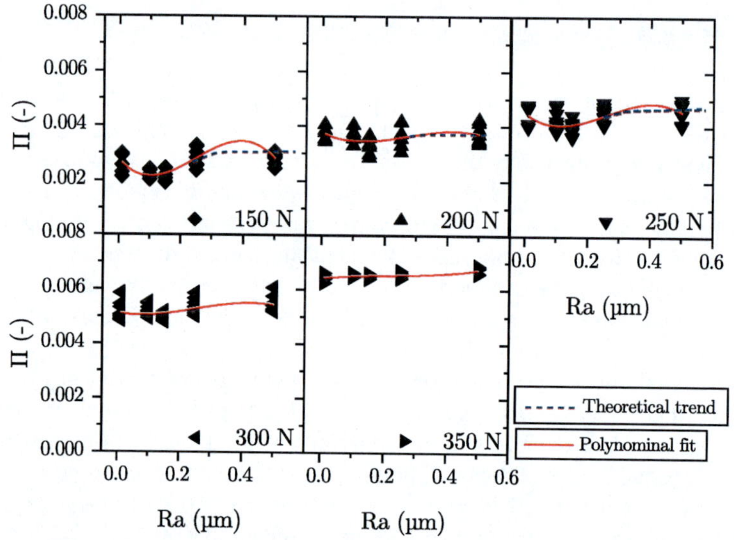

Fig. 11 The friction coefficient Π versus the surface roughness R_a with the normal force F_n as group parameter

model. This effect can be observed particularly well at the two extreme points for the surface roughness (R_a 0.1 µm and 0.4 µm). Contrary to the coefficient of sliding friction, the effect of the normal force dominates over the roughness in rolling. It is likely that the adhesive forces play a minor role than when sliding and thus the lower effect of the surface roughness.

4 Conclusions

The effects of the counter-face roughness R_a and the normal force F_n on the sliding friction μ and rolling resistance Π of POM-H were investigated. The experimental data was analyzed by means of a polynomial regression model to account for possible interactions. The experiments suggest that there is a fundamental difference between sliding and rolling regarding the influence of the roughness and the normal force. While μ was revealed to be strongly affected by R_a but only slightly by F_n, the opposite holds true for the rolling resistance Π. However, the functional relationship of the roughness dependency with a distinct minimum at a R_a value of around 0.1 µm was found to be the same for sliding and rolling.

Furthermore, interactions between the two parameters exist in both cases but they express differently. Based on the curvature of the fits for μ and Π as a function of R_a the interaction with F_n can be illustrated well. On the one hand, the curvature of μ increases with increasing load level; while on the other hand, the curvature of

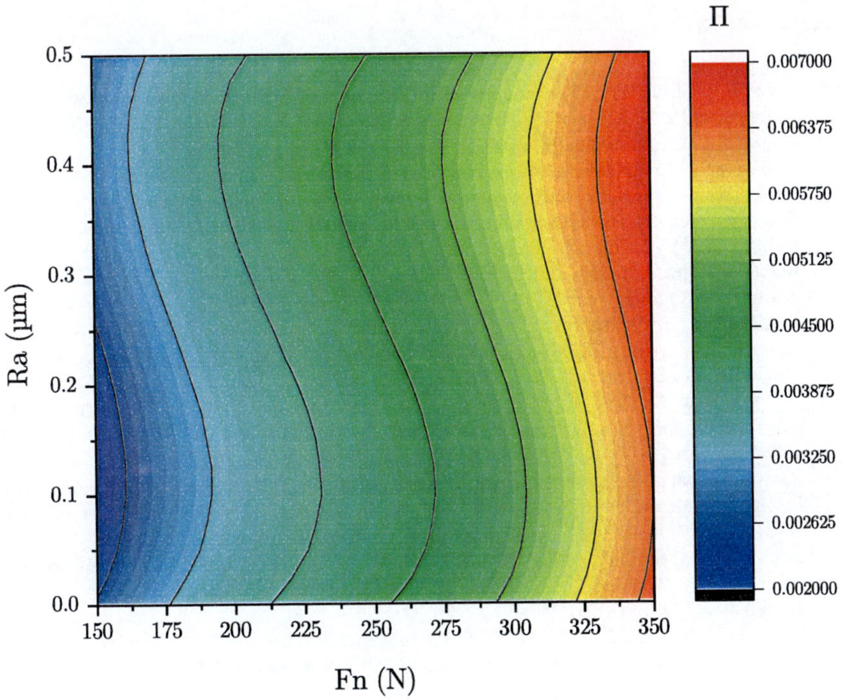

Fig. 12 Contour plot of the regression model for the coefficient of rolling resistance Π as a function of the normal force F_n and the surface roughness R_a

Π decreases. In addition, the entire curve for Π is shifted to higher values when increasing F_n.

Lacking a model based on the underlying physics, it must be emphasized that the polynomial regression was used due to the good correlation in the observed range with a relatively high number of replicate tests (sliding R-sq. 85%, rolling R-sq. 95%). However, the polynomial functions introduce an artificial local maximum between the measured R_a values of 0.25 and 0.5 µm for both sliding and rolling. We believe that the true trend of the friction coefficient and the rolling resistance would converge to a plateau value for higher R_a. Further experiments or a model describing the physics for the influence of the roughness and normal force on the sliding and rolling behavior of polymers are necessary to establish the true functional relationship. Microstructural effects like the transfer film formation and kinetics might explain the interaction between the roughness and normal force on a physical basis.

References

1. Koike H et al (2012) Self-lubrication of PEEK polymer bearings in rolling contact fatigue under radial loads. In: Tribology international, vol 49, pp 30–38. ISSN: 0301-679X. https://doi.org/10.1016/j.triboint.2011.12.005
2. Koike H et al (2012) Observation of crack propagation in PEEK polymer bearings under water-lubricated conditions. In: Machine design and manufacturing engineering, vol 566. Advanced materials research. Trans Tech Publications Ltd, pp 109–114. https://doi.org/10.4028/www.scientific.net/AMR.566.109
3. Berer M, Major Z, Pinter G (2013) Elevated pitting wear of injection molded polyetheretherketone (PEEK) rolls. Wear 297:1052–1063. https://doi.org/10.1016/j.wear.2012.11.062
4. Koike H et al (2013) Observation of wear on PEEK-PTFE hybrid radial bearings. In: Advanced materials and engineering materials II, vol 683. Advanced materials research. Trans Tech Publications Ltd, pp 385–390. https://doi.org/10.4028/www.scientific.net/AMR.683.385
5. Dong Y et al (2015) Design of special plastic bearings and their application in renewable energy conversion system. Open Mater Sci J 9:203–209. https://doi.org/10.2174/1874088X01509010203
6. Samyn P et al (2007) Wear transitions and stability of polyoxymethylene homopolymer in highly loaded applications compared to small-scale testing. In: Tribology international 40(5):819–833. ISSN: 0301-679X. https://doi.org/10.1016/j.triboint.2006.08.003
7. Buck V (1986) Self-lubricating polymer cages for space-proofed bearings: performance and roundness. In: Tribology international 19(1):25–28. ISSN: 0301-679X. https://doi.org/10.1016/0301-679X(86)90091-5
8. Sadık Ünlü B, Atik E, Köksal E (2009) Tribological properties of polymer-based journal bearings. Mater Design 30(7):2618–2622. ISSN: 0261-3069. https://doi.org/10.1016/j.matdes.2008.11.018
9. Benabdallah HS (1997) Reciprocating sliding friction and contact stress of some thermoplastics against steel. J Mater Sci 32(19):5069–5083. https://doi.org/10.1023/A:1018609215248
10. Samyn P, Baets PD (2005) Friction and wear of acetal: a matter of scale. In: Wear 259(1). In: 15th international conference on wear of materials, pp 697–702. ISSN: 0043-1648. https://doi.org/10.1016/j.wear.2005.02.055
11. Tanaka K, Nagai T (1985) Effect of counterface roughness on the friction and wear of polytetrafluoroethylene and polyethylene. In: Ludema KC (ed) Wear of materials, pp 397–404
12. Menezes PL et al (2011) Friction and transfer layer formation in polymer-steel tribo-system: role of surface texture and roughness parameters. In: Wear 271(9). 18th international conference on wear of materials, pp 2213–2221. ISSN: 0043-1648. https://doi.org/10.1016/j.wear.2010.12.047
13. Rojsatean J et al (2017) Friction characteristics of self-lubricating ABS under different surface roughnesses and temperatures. In: Tribol Int 109:229–237. ISSN: 0301-679X. https://doi.org/10.1016/j.triboint.2016.12.055
14. Panda S, Sarangi M, Roy Chowdhury SK (2019) Examinations on PEEK wear debris accumulation over counter surfaces in room and vacuum sliding environments. Polym Test 77:105880. ISSN: 0142-9418. https://doi.org/10.1016/j.polymertesting.2019.04.027
15. Quaglini V et al (2009) Influence of counterface roughness on friction properties of engineering plastics for bearing applications. Mater Design 30(5):1650–1658. ISSN: 0261-3069. https://doi.org/10.1016/j.matdes.2008.07.025
16. Chen W et al (2020) Effect of load on the friction and wear characteristics of Si3N4-hBN ceramic composites sliding against PEEK in artificial seawater. Tribol Int 141:105902. ISSN: 0301-679X. https://doi.org/10.1016/j.triboint.2019.105902
17. Harrass M, Friedrich K, Almajid AA Tribological behavior of selected engineering polymers under rolling contact. Tribol Int 43(3):635–646. ISSN: 0301-679X. https://doi.org/10.1016/j.triboint.2009.10.003. http://www.sciencedirect.com/science/article/pii/S0301679X0900303X

18. Berer M, Major Z (2010) Characterization of the global defor- mation behavior of engineering plastics rolls. Int J Mech Mater Design 6:1–9. https://doi.org/10.1007/s10999-010-9111-9
19. Berer M, Major Zoltan (2012) Characterisation of the local deformation behaviour of engineering plastics rolls. Strain 48:225–234. https://doi.org/10.1111/j.1475-1305.2011.00816.x
20. EN ISO 527-2:2012 Plastics—determination of tensile properties—part 2: test conditions for moulding and extrusion plastics (ISO 527-2:2012); German version EN ISO 527-2:2012. 6-2012. https://doi.org/10.31030/1860304

Numerical Analysis of the Wind Turbine Pitch Bearing Raceway Tribo-contact due to Cyclic Loading Under Constant Pitch Angle

David Cubillas, Mireia Olave, Iñigo Llavori, Ibai Ulacia, Jon Larrañaga, Aitor Zurutuza, and Arkaitz Lopez

Abstract Pitch control movements in wind turbines are commonly short and slow. Moreover, some operating conditions may result in no pitch movement for large time periods. Variable loads coming from wind and the own weight of the blades will turn into micro-oscillations at ball to raceway contact as consequence of ring and contact deformations. Control oscillating movements have been extensively studied in literature. However, no research has been found relating motionless operating point with wear and false brinelling damage. In the present work, a comprehensive analysis at ball-raceway contact taking a 4-point contact ball bearing of 5 MW NREL offshore reference wind turbine as case of study. A simplified finite element model allows to simulate different bearing parameters and loading conditions in order to determinate their effects on the ball and raceway contact behavior. The numerical results highlight the limitations of classic analytical contact mechanics equations which do not predict any tangential stress for non-conformal spherical bodies of same material under normal loading. Furthermore, results reveal the occurrence of micro-slip once a pressure threshold is reached and the effects of some bearing parameters such as raceway conformity or contact friction coefficient. As consequence of this analysis, it is numerically demonstrated that a cyclic normal load without any rolling can lead to local frictional wear, false brinelling or fretting corrosion damage at the bearing raceway.

Keywords Wind power · False brinelling · Pitch bearing · Wear damage · Fretting wear

D. Cubillas (✉) · M. Olave
Ikerlan Technology Research Centre, Basque Research and Technology Alliance (BRTA), Pº J. M. Arizmendiarrieta, 2, 20500 Arrasate/Mondragón, Spain
e-mail: dcubillas@ikerlan.es

I. Llavori · I. Ulacia · J. Larrañaga
Departmento de Mecánica Y Producción Industrial, Escuela Politécnica Superior de Mondragon Unibertsitatea, Arrasate, Spain

A. Zurutuza · A. Lopez
Laulagun Bearings, Harizti Industrialdea, 201-20212 Olaberria, Spain

M. Abdel Wahab (ed.), *Proceedings of the 8th International Conference on Fracture, Fatigue and Wear*, Lecture Notes in Mechanical Engineering,
https://doi.org/10.1007/978-981-15-9893-7_55

1 Introduction

Recent years wind power has experienced a high growth and future expectations are really promising. Higher quality and knowledge are demanded in all turbine components to satisfy current lifetime and reliability requirements. Pitch bearings link hub and blade in wind power turbines and allow control to adjust pitch position in order to achieve the best efficiency, reducing structural loads and helping to starting up at low wind speeds.

Supported loads are large and movements between the rings are low and slow. This operating mode is the cause for false brinelling and fretting corrosion raceway bearing damage developing [1]. Low oscillating movements promote the lubricant to be evacuated from ball-raceway contact and fretting is developed forming black magnetite debris [1, 2]. Godfrey [3] and Errichello [4] focused their efforts establishing the main differences between false brinelling, brinelling and fretting corrosion. First research approaching this issue is attributed to Almen [5] in 1937 when he reported bearing car damage after shipping. Grebe [6] studied standstill marks and established three different damage zones that were produced by different tribological agents. Barthou [7] and Grebe [8] test small oscillating movements in bearings under different operating parameters and lubricants. Varied parameters like amplitude oscillation angle, frequency, temperature, lubricant and applied load have been studied and its damaged effects correlated. Yue et al. [9] have studied the effect of loads on fretting wear and reviewed recent numerical methods on fretting wear simulation [10]. Fabian [11], Korsho [12, 13] have made an approach to damage problems with the finite element method (FEM) model wear using Archard or frictional energy as wear damage indicators. Stammler et al. [14] considers the influence of tribolayers formation on oscillating applications and confirms the influence of oscillating sequence on raceway wear damage.

Even pitch control commonly leads to oscillating movements, some operating conditions are responsible of large periods without any movement. Pitch control defines blade position as function of wind speed, rotor angular speed and blade position among others and every control will works in a different way. However, pitch control commonly remains static under low wind speed.

In this work an analysis on the effect of the cyclic loading process is achieved and it is demonstrated that led to tangential stress at contact than finally result in microslip regions. These tangential stresses were already been studied in detail in previous studies. Carter [15] studied wheel-railway contact and predicted tangential stresses on rolling. Mildlin [16] and Catetaneo [17] improved Hertz formulation considering friction effects on similar spheres in contact. They solved the problem of two similar spheres in contact subjected to a tangential force acting on one of them without any rolling. Goodman [18] extended the formulations with the so-called Goodman approximation. This formulation considers the effect of dissimilar materials of the spheres in contact. Bentall and Johnson [19] improved Goodman formulation by adding the coupling effect and Kalker [20] gave solution to the transient problem.

However, no formulation contemplates any tangential stresses when two spherical contact surfaces of the same material are bringing into contact under normal loading.

This paper is organised on three sections. Firstly, a numerical methodology for the contact analysis of a ball bearing and the raceway through finite element model is presented. Secondly the results are presented for different conformity aspect ratio and bearing sizes and a discussion about the limitations of classic contact mechanics formulations for complex geometrics and finally, conclusions are presented in which it is concluded that initial hypothesis of normal variable loading on bearings develop fretting damaged is reaffirmed.

2 Methodology

2-rows and 4-point contact ball slewing bearing and NREL 5 MW case of study were used as reference in this work. Inner ring remains bolted to the blade while the other ring is to the hub. This bearing works through two diagonals (see Fig. 1a) that depending on the direction of the force or moment, load is transmitted by one of the two contact diagonals as shown Fig. 1b. Initial contact angle is theoretically 45°, however, contact angle will change as load function due to the elastic contact deformation and ring flexibility [21, 22].

When pitch angle is constant under low speed wind series, variable loading still occurs due to wind series. At contact interface, this variable loading is responsible of two effects. Firstly, it leads to contact deformation and therefore contact ellipse size modification (Fig. 1). Due to this movement, the lubricant is pushed out form the contact area. Additionally, contact surfaces deformation would produce micro-slip regions. Secondly (see Fig. 2), as previously described, rings and contact deformation lead to contact angle variations, so rolling contact effects are expected [23].

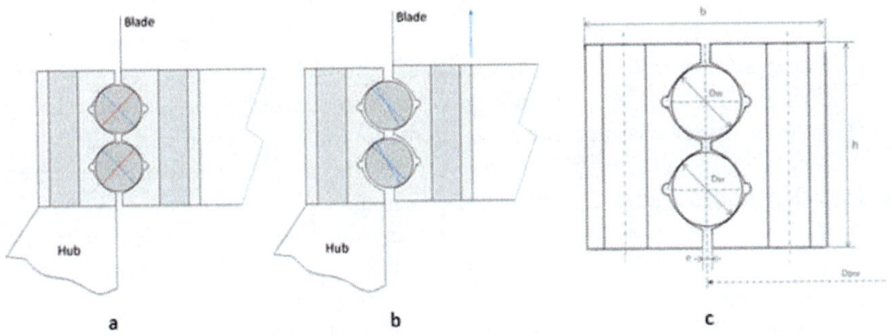

Fig. 1. 2-rows and 4-point contact ball slewing bearing. **a** Initial 2 diagonals contact. **b** 1 diagonal works under axial force. **c** Bearing main dimensions

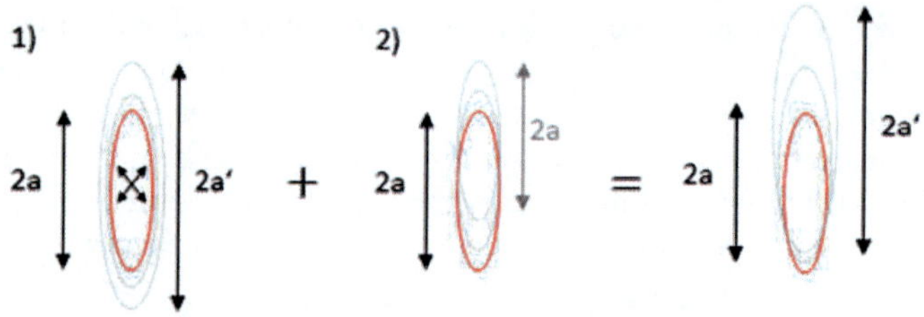

Fig. 2 Variable loading contact effect

Table 1 Bearing reference dimensions

Parameter	Dimension (mm)
D_{pw}	3610.00
D_w	65.00
b	200.00
h	200.00
e	10.00

For a first analysis, variable loading behavior was studied individually considering only contact deformation effect due to change in dimension of the contact ellipse. Rolling effects were neglected in this first approach.

Main geometrical data is represented in Fig. 1c and listed in Table.1, where D_{pw} is the main bearing diameter, D_w is the ball diameter, b is the width of one sector, h is the height of one sector, e is the space between rings and c is the conformity ratio as per Eq. (1).

$$c = \frac{D_p}{2 \cdot D_w} \tag{1}$$

where, D_p, is the raceway diameter.

The analysis is based on a FEM model. The model simulates a 1/4 ball being pressed against a static ring so no rolling is considered. Similar models can be found in [13, 14] for roller bearings and wear prediction having a good correlation with laboratory tests. In this case, simplifications from the complete geometry were taken and it was assumed planes 1, 2 and 3 to be symmetry planes (see Fig. 3). Figure 3c shows the FE-model mesh. Special attention was paid to the contact with a symmetric behavior and augmented Lagrange formulation. Second order hexahedron elements were used for the entire model (628,501 nodes) and finer mesh was placed at contact (14,400 nodes). The contact detection is based on nodal point—normal from contact surface (projection-based method) and an isotropic friction model based on Coulomb coefficient of friction. The simulation was divided in 200 increments. 42CrMo4 is a commonly used [24] material for pitch bearings, elastic module $E = 210$ GPa and poison's ratio $\nu = 0.3$.

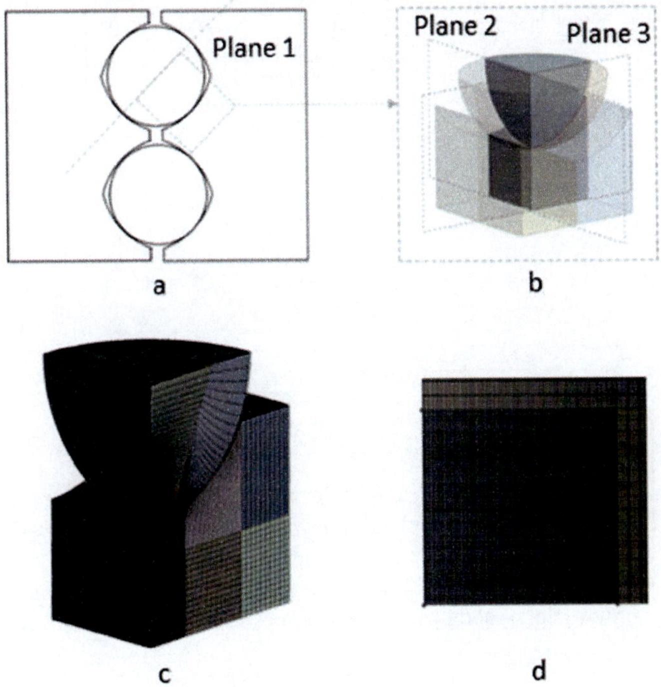

Fig. 3 Simplified model for ball-raceway contact analysis

A Full parametric model and mesh allowed to obtain results for different bearing parameters automatically. Different conformity ratios for typical values recommended by NREL for pitch bearings and steel-steel frictional coefficients [24] were also simulated to see the effects on the contact as listed in Table 2.

Normal variable loading was applied through a vertical displacement on plane 1 and frictionless supports boundary condition are set in symmetry planes. NREL

Table 2 Bearing simulation parameters

D_w [mm]	c	Dp [mm]	Coefficient of friction
65.00	0.51	66.30	0.10
	0.52	67.60	0.10
	0.53	68.90	0.10
	0.54	70.20	0.10
	0.52	67.60	0.10
	0.52	67.60	0.05
	0.52	67.60	0.15
	0.52	67.60	0.20

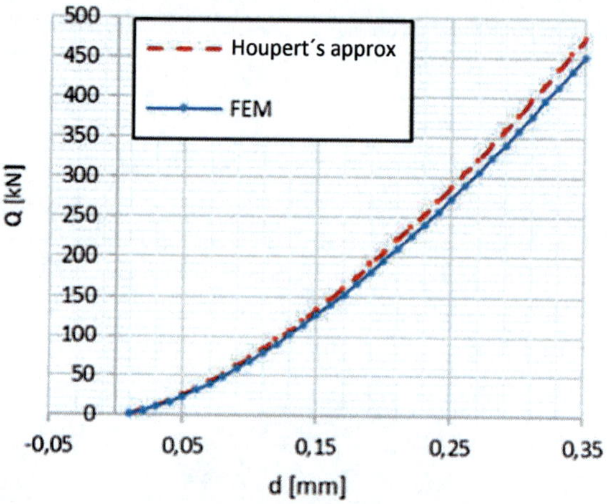

Fig. 4 Contact stiffness evolution under normal load process

guidelines [24] consider 4200 MPa as limit for maximum contact pressure on raceway bearings. Considering this, a complete cycle was applied (from 0–4200–0 MPa) for full range cycle to analyze contact effects (Fig. 4).

Houpert's approximation [25, 26] can be used to calculate load (Eq. 2), major (Eq. 3) and minor (Eq. 4) semi-axes dimension as function of pressure and bearing parameters.

$$Q = \left(\frac{P_{\max} \cdot D_w^{\frac{3}{2}}}{1805 \cdot (2c - 1)^{0.2117}}\right)^3 \tag{2}$$

$$a = 1.3085(2c - 1)^{-0.4091} \cdot \frac{1 - v^2}{E} \cdot \left(\frac{D_w}{2}\right)^{\frac{1}{3}} \cdot Q^{\frac{1}{3}} \tag{3}$$

$$b = 1.1687(2c - 1)^{-0.1974} \cdot \frac{1 - v^2}{E} \cdot \left(\frac{D_w}{2}\right)^{\frac{1}{3}} \cdot Q^{\frac{1}{3}} \tag{4}$$

Finally, the necessary displacement as function of mayor semi axes can be calculated as per equation (Eq. 5).

$$d = \sqrt{\left(\frac{d_p}{2}\right)^2 - a^2} - \sqrt{\left(\frac{d_w}{2}\right)^2 - a^2} - \frac{d_p - d_w}{2} \tag{5}$$

3 Results and Discussion

As expected, initial contact deforms and evolves from a simple point to an ellipse contact area (Fig. 5). Contact stiffness, ellipse dimensions, and pressure distribution

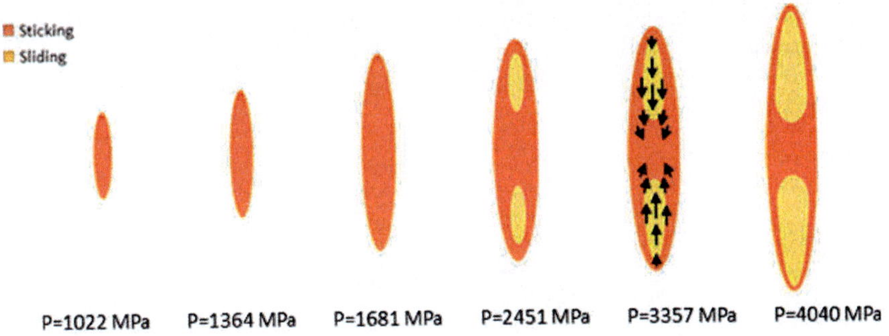

Fig. 5 Contact status evolution. $D_{pw} = 65.00$ mm, $c = 0.52$ and $\mu = 0.1$ and different contact pressure

were verified with Harris [24, 27] and Hertz [28] formulation. Maximum error was found at maximum load moment when pressure value differs in 4.3% from analytical formulations. Figure 6 shows FEM load prediction (blue line) and Hertz prediction (dashed red line). Differences are related with boundary conditions applied in FEM while Hertz prediction is based on an elastic half-space assumption.

As shown in Fig. 5, contact surfaces were in stick at initial stages of loading process. However, once a threshold pressure was reached, slip regions arise because of high tangential stresses on contact zone, as shown in Fig. 6. At this point, two slip regions were developed at a point near both ellipse focus and slip area grew as load was increased. Slip direction points to the center of the ellipse contact as consequence of the contact deformation.

These numerical results highlight the limitations of classic contact mechanics formulations for complex geometric as no formulation contemplates any tangential stresses when two spherical contact surfaces of the same material are bringing into contact under normal loading. Figure 7 shows the absolute tangential stress distribution and stresses direction at ring surface for different contact pressure values. A

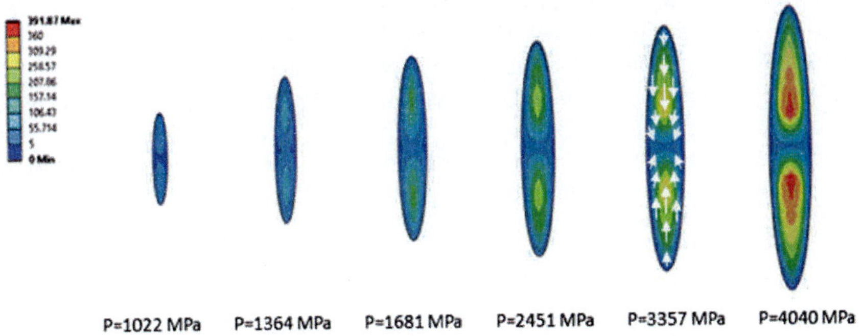

Fig. 6 Contact tangential stresses at ring surface. $D_{pw} = 65.00$ mm, $c = 0.52$ and $\mu = 0.1$ and different contact pressures

Fig. 7 Tangential stress profile on ellipse semi-axes *a*. $D_{pw} = 65.00$ mm, $c = 0.52$ and $\mu = 0.1$ and $p = 20$

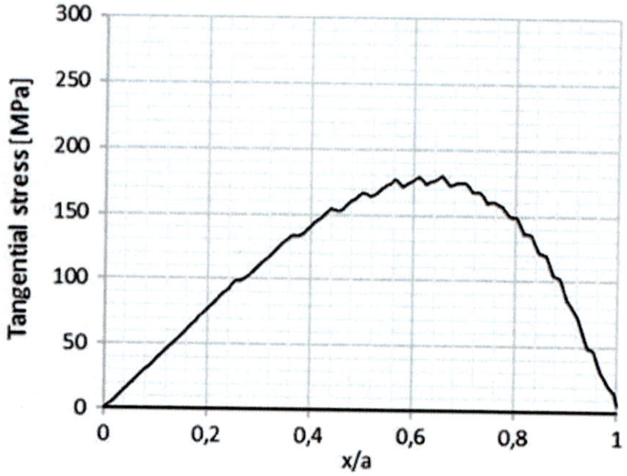

higher load leads to higher tangential stresses. Stresses grows from ellipse center to some a point close to the ellipse focus and vanishes at outline. Figure 7 shows tangential stresses profile along mayor semi-axes a and tangential stress direction pointing to the center of the contact ellipse where a maximum value is reached at $0.65a$.

As previously described, during loading process, pressure and tangential stress grew until Coulomb's limit was reached (see Fig. 8, dashed red line). Once the threshold pressure was reached (i.e., Coulomb's limit), any load increase led to a micro-slip and therefore wear could occur.

Figure 9 shows tangential stress evolution with pressure for different conformity ratios. Conformity has a considerable impact on contact tangential stresses. A lower conformity, which means more similar ratio, leads to higher tangential stresses for the same contact pressure.

Conformity also affects slip pressure threshold. Figure 10 shows the relation between different conformity ratios. Closer values to 0.5 led to lower pressure threshold, thus micro-slip tends to happen more easily for lower conformity values.

As mentioned before, sliding occurs when Coulomb's limit is reached and therefore there is a relation between friction coefficient and the threshold pressure. A lower value of friction coefficient lead to a lower pressure threshold. From simulation results, threshold pressure is represented for different frictional coefficients in Fig. 11.

The unloading process will stop the micro-slip at the very beginning moment and the tangential stresses start decreasing. However, during this process, tangential stresses are developed in the opposite direction and finally micro-slip occurred at the outer side of the contact as shown in Fig. 12 and wear damage would appear and the unloading process does not help relubrication at contact zone.

Results show a reciprocating effect on the tangential stresses that are responsible sliding zones. This slip area (Fig. 13) is in a concrete area of the contact ellipse and it remains stationary over the cycles what may accelerate the fretting damage

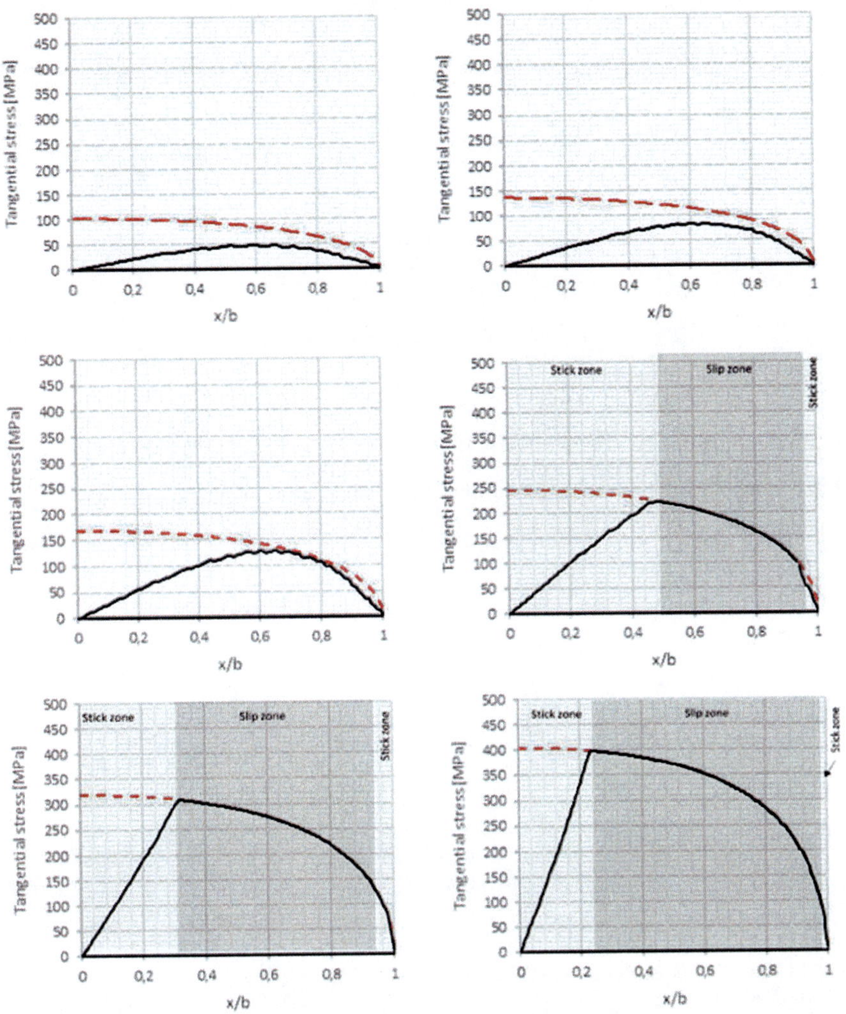

Fig. 8 Tangential stress evolution. Conformity = 0.52. Friction coefficient = 0.1 and different pressure values

at this zone. Moreover, the reciprocating movement is responsible of evacuating the lubricant out of the contact that aggravate the damage.

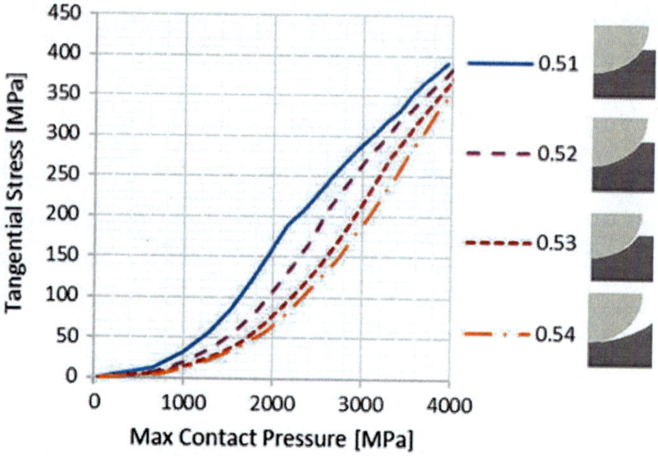

Fig. 9 Tangential stress behavior for different conformity ratios

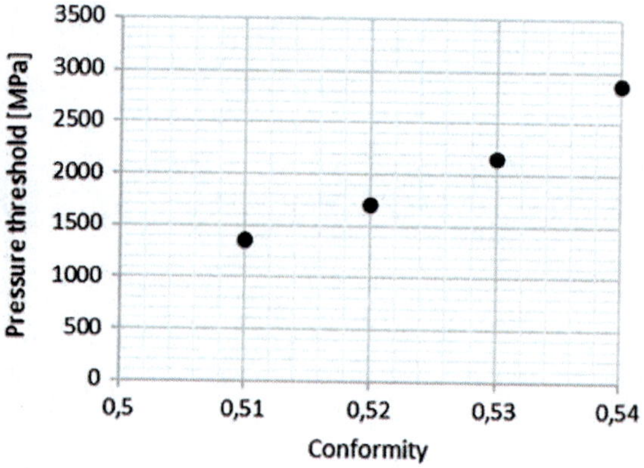

Fig. 10 Pressure threshold relation with conformity ratio

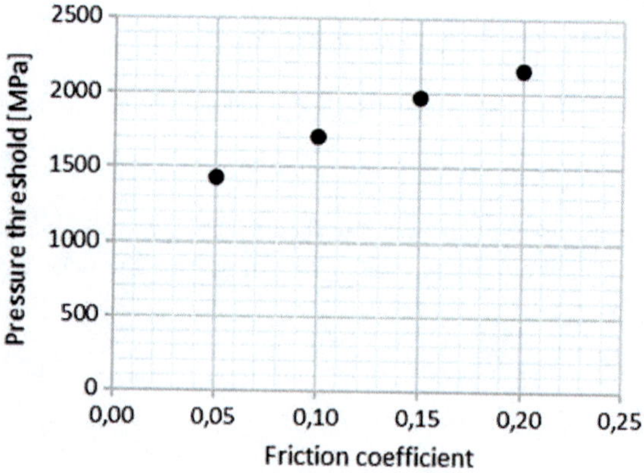

Fig. 11 Pressure threshold relation with friction coefficient

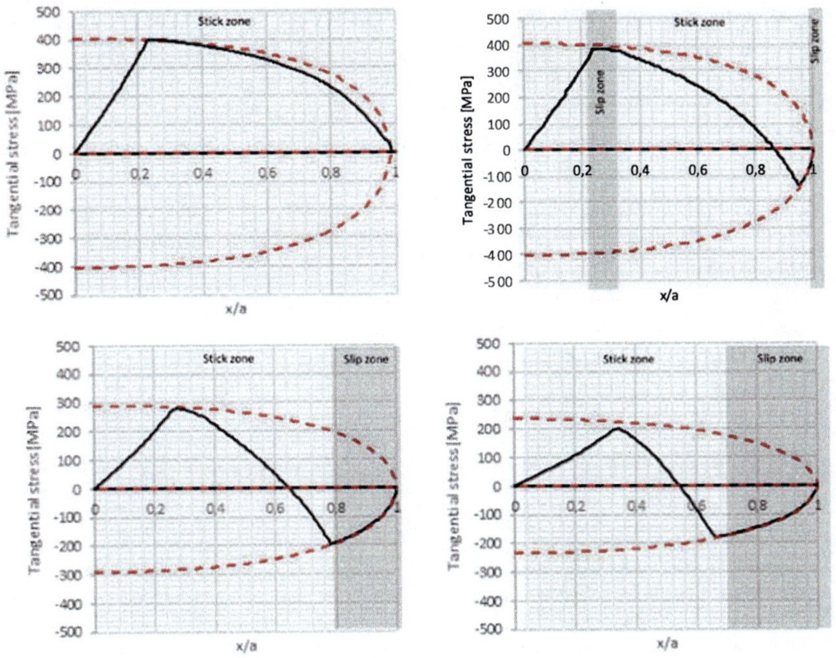

Fig. 12 Tangential stress evolution on unloading process. Conformity = 0.52. Friction coefficient = 0.1 and different pressure values

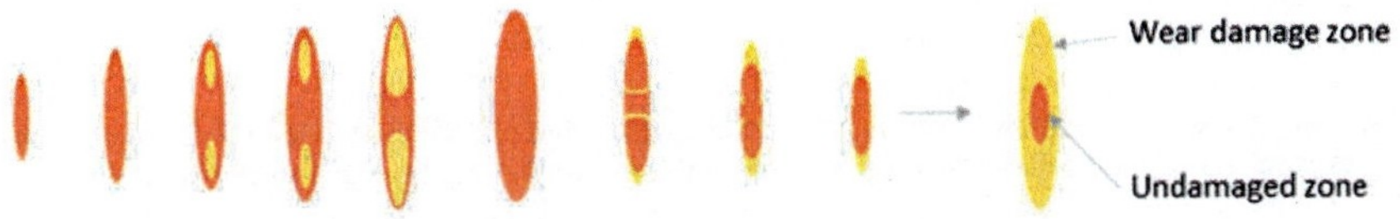

Fig. 13 Complete cycle stick/slip zone

4 Conclusions and Future Work

The aim of this work was to study the effects of ball to raceway contact under pure normal loading. To this end, a parametric finite element model was developed to carry out a numerical analysis and the following conclusions were drawn:

- The main hypothesis that a variable contact loading would lead to false brinelling wear is reaffirmed. Contact starts at the central point and grows to an elliptical area. This effect ejects the lubricant out of the contact zone as previous mentioned and accelerate the fretting damage.
- Simulation results show the limitations of classical mechanical contact theories where tangential stresses are not predicted when two dissimilar and non-conformal

bodies of same material are in contact under a pure normal loading. Tangential stresses grow with the applied load and they lead to micro-slips areas.

- According to the numerical results, micro-slips occur over a pressure threshold that is related with the conformity and the friction coefficient.
- Once the pressure threshold is reached and micro-slip occurs in the loading process, the unloading process will also develop micro-slip.
- It was found the existence of micro-slip under 4-point contact ball bearing realistic operating conditions and therefore the existence of fretting damage like false brinelling.

References

1. ISO (2004) ISO 15243: Rolling bearings—damage and failures—terms, characteristics and causes
2. Godfrey D (1999) Iron oxides and rust (hydrated iron oxides) in tribology. Tribol Lubric Technol 55(2):33–37
3. Godfrey D (2003) Fretting corrosion or false brinelling? Tribol Lubric Technol 12:28–31
4. Errichello R (2004) Another perspective: false brinelling and fretting corrosion. Tribol Lubric Technol 60(4):34–36
5. Almen JO, Lubricants and false brinelling of ball and roller bearings. Mech World Eng Rec
6. Grebe M, Feinle P, Hunsicker W (2011) Various influence factors on the development of standstill marks (false-brinelling effect). Frict Wear Wear Prot Int Symp Frict Wear Wear Prot 2008, Aachen, Germany, pp. 290–300
7. Barthou C, Vannes B, Girodin D, Pierantoni M, Sauger E (1998) Methodology of characterisation of the raceway/lubricant/ball contacts submitted to vibrations and degraded by false brinelling. In: Tribology for energy conservation of tribology series, proceedings of the 24th Leeds-Lyon symposium on tribology, vol 34, Elsevier. pp. 389–398
8. Berthier Y, Play D (1982) Wear mechanisms in oscillating bearings. Wear 75(2):369–387
9. Wang, S, Abdel Wahab M, Effect of loading conditions in fretting fatigue on wear characteristics
10. Tongyan Yue, Magd Abdel Wahab, A review on fretting wear mechanisms, models and numerical analyses. In: State Grid State Grid Xinyuan Maintenance Branch, Beijing, China. Xinyuan Maintenance Branch, Beijing, China.
11. Schwack F, Prigge F, Poll G (2018) Finite element simulation and experimental analysis of false brinelling and fretting corrosion. Tribol Int 126(April):352–362
12. Fallahnezhad K, Liu S, Brinji O, Marker M, Meehan PA (2019) Monitoring and modelling of false brinelling for railway bearings. Wear 424–425(February):151–164
13. Fallahnezhad K, Brinji O, Desai A, Meehan PA (2019) The influence of different types of loading on false brinelling. Wear 440–441(October):203097
14. Stammler M, Poll G, Reuter A (no date) The influence of oscillation sequences on rolling bearing wear, pp. 19–25
15. Carter FW (1926) On the action of a locomotive driving wheel. Proc R Soc Lond Ser A 112:151–157. https://doi.org/10.1098/rspa.1926.0100
16. Mindlin Compilace of elastic bodies in contact. J Appl Mech 16:259
17. Cattaneo C (1938) Sul contatto di due corpi elastici: distribuzione locale degli sforzi Accademia Nazionale Lincei Rendiconti Ser. 6. L'Accademia, Roma
18. Goodman LE, Contact stress analysis of normally loaded rough spheres, Department of Aeronautics and Engineering Mechanics, University of Minnesota, Minneapolis, Minn Mem ASME

19. Bentall RH, Johnson KL, Slip in the rolling contact of two dissimilar elastic rollers. University Engineering Department, Trumpington Street, Cambridge, England
20. Kalker, JJ Transient phenomena in two elastic cylinders rolling over each other with dry friction. Mathematical Institute, University of Technology, Delft, The Netherlands
21. Olave M, Sagartzazu X, Damian J, Serna A (2010) Design of four contact-point slewing bearing with a new load distribution procedure to account for structural stiffness. ASME J Mech Des 132(2):021006
22. Amasorrain JI, Sagartzazu X, Damián J (2003) Load distribution in a four contact-point slewing bearing. Mech Mach Theory 38:479–496. https://doi.org/10.1016/S0094-114X(03)00003-X
23. Peiyu He1, Rongjing Hong1, Pitch bearing/raceway fretting: influence of contact angle. In: Proc IMechE C J Mech Eng Sci, School of Mechanical and Power Engineering, Nanjing Technical University, Nanjing, China
24. Harris TA, Rumbarger JH, Butterfield CP (2009) Wind turbine design guideline DG03: yaw and pitch rolling bearing life. NREL/TP-500–42362
25. Houpert L (2001) An engineering approach to hertzian contact elasticity—Part I. Trans ASME 123, https://doi.org/10.1115/1.1308043
26. Houpert L (2001) An engineering approach to non-hertzian contact elasticity—Part II. Trans ASME 123:589–594. https://doi.org/10.1115/1.1308042
27. Harris TA, Kotzalas MN (2006) Advanced concepts of bearing technology. CRC, Rolling Bearing Analysis
28. Hertz H (1896) On the contact of elastic solids. Misc Pap 146–162